THE LOGIC OF CHEMICAL SYNTHESIS

THE LOGIC OF CHEMICAL SYNTHESIS

E. J. COREY AND XUE-MIN CHENG

Department of Chemistry
Harvard University
Cambridge, Massachusetts

WILEY

JOHN WILEY & SONS

New York • Chichester • Brisbane • Toronto • Singapore

Library of Congress Cataloging in Publication Data:

Corey, E. J.
The logic of chemical synthesis.

Bibliography: p.
1. Chemistry, Organic—Synthesis. I. Cheng, Xue-Min. II. Title.

QD262.C577 1989 547'.2 89-5335
ISBN 0-471-50979-5

Printed in the United States of America

10 9 8 7 6 5 4 3

35.00

Dedicated to past and present members of the Corey Research Family

PREFACE

The title of this three-part volume derives from a key theme of the book—the logic underlying the rational analysis of complex synthetic problems. Although the book deals almost exclusively with molecules of biological origin, which are ideal for developing the fundamental ideas of multistep synthetic design because of their architectural complexity and variety, the approach taken is fully applicable to other types of carbon-based structures.

Part One outlines the basic concepts of retrosynthetic analysis and the general strategies for generating possible synthetic pathways by systematic reduction of molecular complexity. Systematic retrosynthetic analysis and the concurrent use of multiple independent strategies to guide problem solving greatly simplify the task of devising a new synthesis. This way of thinking has been used for more than two decades by one of the authors to teach the analysis of difficult synthetic problems to many hundreds of chemists. A substantial fraction of the intricate syntheses which have appeared in the literature in recent years have been produced by these individuals and their students. An effort has been made to present in Part One the essentials of multistrategic retrosynthetic analysis in a concise, generalized form with emphasis on major concepts rather than on specific facts of synthetic chemistry. Most of the key ideas are illustrated by specific chemical examples. A mastery of the principles which are developed in Part One is a prerequisite to expertise in synthetic design. Equally important is a command of the reactions, structural-mechanistic theory, and reagents of carbon chemistry. The approach in Part One is complementary to that in courses on synthetic reactions, theoretical carbon chemistry, and general organic chemistry.

Part Two, a collection of multistep syntheses accomplished over a period of more than three decades by the Corey group, provides much integrated information on synthetic methods and pathways for the construction of interesting target molecules. These syntheses are the result of synthetic planning which was based on the general principles summarized in Part One. Thus, Part Two serves to supplement Part One with emphasis on the methods and reactions of synthesis and also on specific examples of retrosynthetically planned syntheses.

The reaction flow-charts of Part Two, and indeed all chemical formulae which appear in this book, were generated by computer. The program used for these drawings was *ChemDraw*™ adapted for the *Macintosh*® personal computer by Mr. Stewart Rubenstein of these Laboratories from the molecular graphics computer program developed by our group at Harvard in the 1960's (E. J. Corey and W. T. Wipke, *Science*, **1969**, *166*, 178-192) and subsequently refined.

Part Three is intended to balance the coverage of Parts One and Two and to serve as a convenient guide to the now enormous literature of multistep synthesis. Information on more than five hundred interesting multistep syntheses of biologically derived molecules is included. It is hoped that the structural range and variety of target molecules presented in Part Three will appeal to many chemists.

Synthesis remains a dynamic and central area of chemistry. There are many new principles, strategies and methods of synthesis waiting to be discovered. If this volume is helpful to our many colleagues in the chemical world in their pursuit of discovery and new knowledge, a major objective of this book will have been met.

CONTENTS OF PART ONE

GENERAL APPROACHES TO THE ANALYSIS OF COMPLEX SYNTHETIC PROBLEMS

CHAPTER ONE
The Basis for Retrosynthetic Analysis

CHAPTER TWO
Transform-Based Strategies

CHAPTER THREE
Structure-Based and Topological Strategies

CHAPTER FOUR
Stereochemical Strategies

CHAPTER FIVE
Functional Group-Based and Other Strategies

CHAPTER SIX
Concurrent Use of Several Strategies

CONTENTS OF PART TWO

SPECIFIC PATHWAYS FOR THE SYNTHESIS OF COMPLEX MOLECULES

CHAPTER SEVEN
Macrocyclic Structures

CHAPTER EIGHT
Heterocyclic Structures

CHAPTER NINE
Sesquiterpenoids

CHAPTER TEN
Polycyclic Isoprenoids

CHAPTER ELEVEN

Prostanoids

CHAPTER TWELVE

Leukotrienes and Other Bioactive Polyenes

CONTENTS OF PART THREE

GUIDE TO THE ORIGINAL LITERATURE OF MULTISTEP SYNTHESIS

CHAPTER THIRTEEN

THE LOGIC OF CHEMICAL SYNTHESIS

PART ONE

GENERAL APPROACHES TO THE ANALYSIS OF COMPLEX SYNTHETIC PROBLEMS

CHAPTER ONE

The Basis for Retrosynthetic Analysis

1.1 Multistep Chemical Synthesis

The chemical synthesis of carbon-containing molecules, which are called *carbogens* in this book (from the Greek word *genus* for family), has been a major field of scientific endeavor for over a century.* Nonetheless, the subject is still far from fully developed. For example, of the almost infinite number and variety of carbogenic structures which are capable of discrete existence, only a minute fraction have actually been prepared and studied. In addition, for the last century there has been a continuing and dramatic growth in the power of the science of constructing complex molecules which shows no signs of decreasing. The ability of chemists to synthesize compounds which were beyond reach in a preceding 10-20 year period is dramatically documented by the chemical literature of the last century.

As is intuitively obvious from the possible existence of an astronomical number of discrete carbogens, differing in number and types of constituent atoms, in size, in topology and in three dimensional (stereo-) arrangement, the construction of specific molecules by a single chemical step from constituent atoms or fragments is almost never possible even for simple structures. Efficient synthesis, therefore, requires *multistep construction* processes which utilize at each stage chemical reactions that lead specifically to a single structure. The development of carbogenic chemistry has been strongly influenced by the need to effect such multistep syntheses successfully and, at the same time, it has been stimulated and sustained by advances in the field of synthesis. Carbon chemistry is an information-rich field because of the *multitude of known types of reactions* as well as the number and diversity of possible compounds. This richness provides the chemical methodology which makes possible the broad access to synthetic

---

References are located on pages 92-95. A glossary of terms appears on pages 96-98.

* The words *carbogen* and *carbogenic* can be regarded as synonymous with the traditional terms *organic compound* and *organic*. Despite habit and history, the authors are not comfortable with the logic of several common chemical usages of *organic*, for example *organic synthesis*.

carbogens which characterizes today's chemistry. As our knowledge of chemical sciences (both fact and theory) has grown so has the power of synthesis. The synthesis of carbogens now includes the use of reactions and reagents involving more than sixty of the chemical elements, even though only a dozen or so elements are commonly contained in commercially or biologically significant molecules.

1.2 Molecular Complexity

From the viewpoint of chemical synthesis the factors which conspire to make a synthesis difficult to plan and to execute are those which give rise to structural complexity, a point which is important, even if obvious. Less apparent, but of major significance in the development of new syntheses, is the value of understanding the roots of complexity in synthetic problem solving and the specific forms which that complexity takes. *Molecular size, element* and *functional-group content, cyclic connectivity, stereocenter content,* chemical *reactivity,* and structural *instability* all contribute to *molecular complexity* in the synthetic sense. In addition, other factors may be involved in determining the difficulty of a problem. For instance, the density of that complexity and the novelty of the complicating elements relative to previous synthetic experience or practice are important. The connection between specific elements of complexity and strategies for finding syntheses is made in Section 1.8.

The successful synthesis of a complex molecule depends upon the analysis of the problem to develop a feasible scheme of synthesis, generally consisting of a pathway of synthetic intermediates connected by possible reactions for the required interconversions. Although both *inductive/associative* and *logic-guided* thought processes are involved in such analyses, the latter becomes more critical as the difficulty of a synthetic problem increases.[1] Logic can be seen to play a larger role in the more sophisticated modern syntheses than in earlier (and generally simpler) preparative sequences. As molecular complexity increases, it is necessary to examine many more possible synthetic sequences in order to find a potentially workable process, and not surprisingly, the resulting sequences are generally longer. Caught up in the excitement of finding a novel or elegant synthetic plan, it is only natural that a chemist will be strongly tempted to start the process of reducing the scheme to practice. However, prudence dictates that many alternative schemes be examined for relative merit, and persistence and patience in further analysis are essential. After a synthetic plan is selected the chemist must choose the chemical reagents and reactions for the individual steps and then execute, analyze and optimize the appropriate experiments. Another aspect of molecular complexity becomes apparent during the execution phase of synthetic research. For complex molecules even much-used standard reactions and reagents may fail, and new processes or options may have to be found. Also, it generally takes much time and effort to find appropriate reaction conditions. The time, effort, and expense required to reduce a synthetic plan to practice are generally greater than are needed for the conception of the plan. Although rigorous analysis of a complex synthetic problem is extremely demanding in terms of time and effort as well as chemical sophistication, it has become increasingly clear that such analysis produces superlative returns.[1]

Molecular complexity can be used as an indicator of the frontiers of synthesis, since it often causes failures which expose gaps in existing methodology. The realization of such limitations can stimulate the discovery of new chemistry and new ways of thinking about synthesis.

1.3 Thinking About Synthesis

How does a chemist find a pathway for the synthesis of a structurally complex carbogen? The answer depends on the chemist and the problem. It has also changed over time. Thought must begin with perception—the process of extracting information which aids in logical analysis of the problem. Cycles of perception and logical analysis applied reiteratively to a target structure and to the "data field" of chemistry lead to the development of concepts and ideas for solving a synthetic problem. As the reiterative process is continued, questions are raised and answered, and propositions are formed and evaluated with the result that ever more penetrating insights and more helpful perspectives on the problem emerge. The ideas which are generated can vary from very general "working notions or hypotheses" to quite sharp or specific concepts.

During the last quarter of the 19th century many noteworthy syntheses were developed, almost all of which involved benzenoid compounds. The carbochemical industry was launched on the basis of these advances and the availability of many aromatic compounds from industrial coal tar. Very little planning was needed in these relatively simple syntheses. Useful synthetic compounds often emerged from exploratory studies of the chemistry of aromatic compounds. Deliberate syntheses could be developed using associative mental processes. The starting point for a synthesis was generally the most closely related aromatic hydrocarbon and the synthesis could be formulated by selecting the reactions required for attachment or modification of substituent groups. Associative thinking or thinking by analogy was sufficient. The same can be said about most syntheses in the first quarter of the 20th century with the exception of a minor proportion which clearly depended on a more subtle way of thinking about and planning a synthesis. Among the best examples of such syntheses (see next page) are those of α-terpineol (W. H. Perkin, 1904), camphor (G. Komppa, 1903; W. H. Perkin, 1904), and tropinone (R. Robinson, 1917).[2] During the next quarter century this trend continued with the achievement of such landmark syntheses as the estrogenic steroid equilenin (W. Bachmann, 1939),[3] protoporphrin IX (hemin) (H. Fischer, 1929),[2,4] pyridoxine (K. Folkers, 1939),[5] and quinine (R. B. Woodward, W. von E. Doering, 1944) (page 4).[6] In contrast to the 19th century syntheses, which were based on the availability of starting materials that contained a major portion of the final atomic framework, these 20th century syntheses depended on the knowledge of reactions suitable for forming polycyclic molecules and on detailed planning to find a way to apply these methods.

In the post-World War II years, synthesis attained a different level of sophistication partly as a result of the confluence of five stimuli: (1) the formulation of detailed electronic mechanisms for the fundamental organic reactions, (2) the introduction of conformational analysis of organic structures and transition states based on stereochemical principles, (3) the development of spectroscopic and other physical methods for structural analysis, (4) the use of chromatographic methods of analysis and separation, and (5) the discovery and application of new selective chemical reagents. As a result, the period 1945 to 1960 encompassed the synthesis of such complex molecules as vitamin A (O. Isler, 1949), cortisone (R. Woodward, R. Robinson, 1951), strychnine (R. Woodward, 1954), cedrol (G. Stork, 1955), morphine (M. Gates, 1956), reserpine (R. Woodward, 1956), penicillin V (J. Sheehan, 1957), colchicine (A. Eschenmoser, 1959), and chlorophyll (R. Woodward, 1960) (page 5).[7,8]

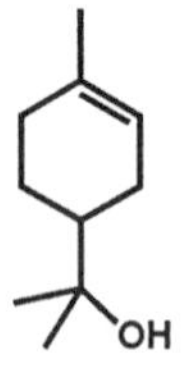

α-Terpineol
(Perkin, 1904)

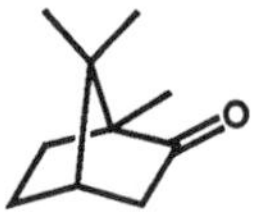

Camphor
(Komppa, 1903; Perkin, 1904)

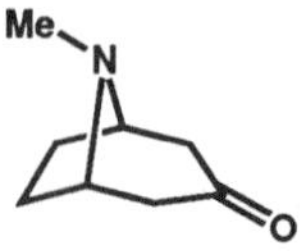

Tropinone
(Robinson, 1917)

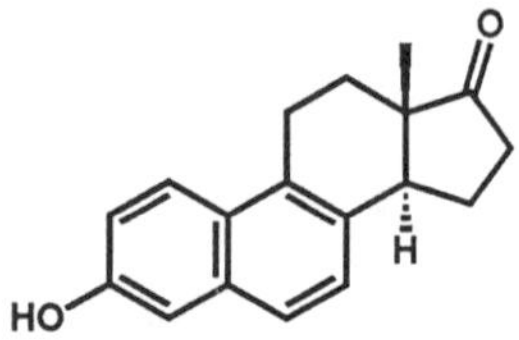

Equilenin
(Bachmann, 1939)

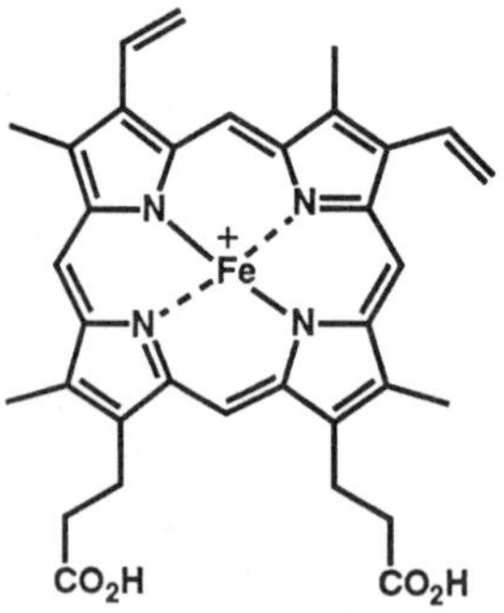

Hemin
(Fischer, 1929)

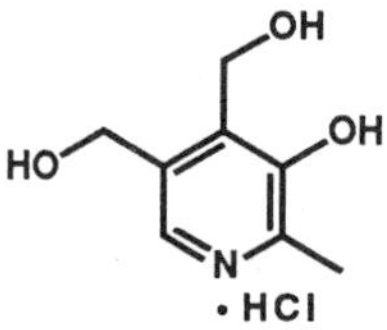

Pyridoxine Hydrochloride
(Folkers, 1939)

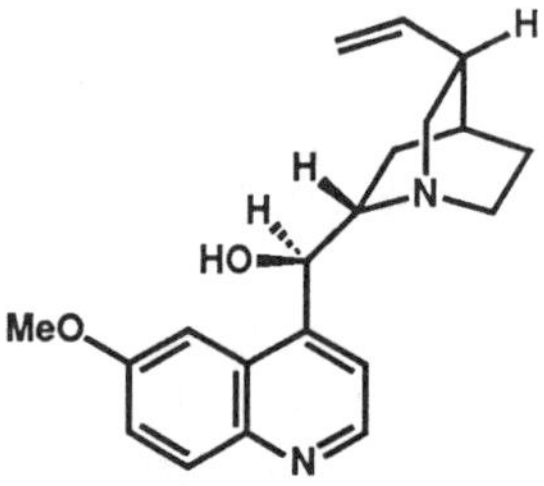

Quinine
(Woodward, Doering, 1944)

The 1950's was an exhilarating period for chemical synthesis—so much so that for the first time the idea could be entertained that no stable carbogen was beyond the possibility of synthesis at some time in the not far distant future. Woodward's account of the state of "organic" synthesis in a volume dedicated to Robert Robinson on the occasion of his 70th birthday indicates the spirit of the times.[9] Long multistep syntheses of 20 or more steps could be undertaken with confidence despite the Damocles sword of synthesis—only one step need fail for the entire project to meet sudden death. It was easier to think about and to evaluate each step in a projected synthesis, since so much had been learned with regard to reactive intermediates, reaction mechanisms, steric and electronic effects on reactivity, and stereoelectronic and conformational effects in determining products. It was possible to experiment on a milligram scale and to separate and identify reaction products. It was simpler to ascertain the cause of difficulty in a failed experiment and to implement corrections. It was easier to find appropriate selective reagents or reaction conditions. Each triumph of synthesis encouraged more ambitious undertakings and, in turn, more elaborate planning of syntheses.

However, throughout this period each synthetic problem was approached as a special case with an individualized analysis. The chemist's thinking was dominated by the problem under consideration. Much of the thought was either unguided or subconsciously directed. Through the 1950's and in most schools even into the 1970's synthesis was taught by the presentation of a series of illustrative (and generally unrelated) cases of actual syntheses. Chemists who learned synthesis by this "case" method approached each problem in an ad hoc way. The intuitive search for clues to the solution of the problem at hand was not guided by effective and consciously applied general problem-solving techniques.[8]

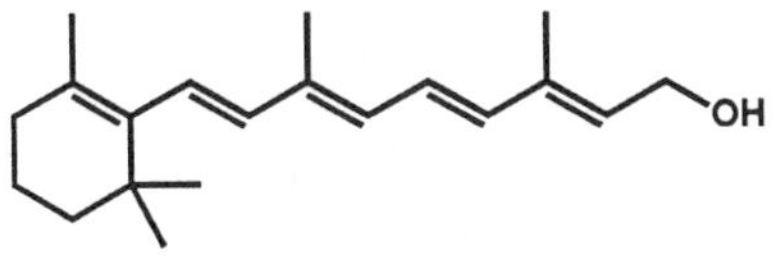
OH

Vitamin A
(Isler, 1949)

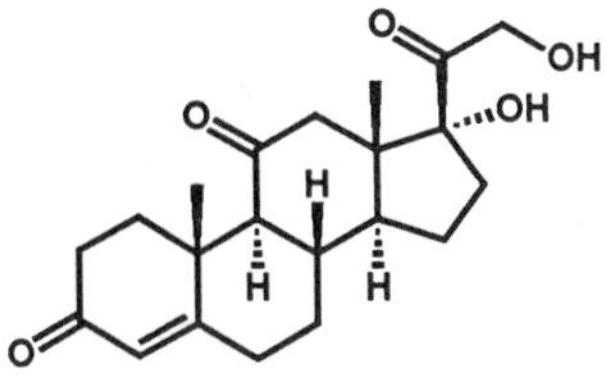
O
OH
OH
O
H
H
H
O

Cortisone
(Woodward, Robinson, 1951)

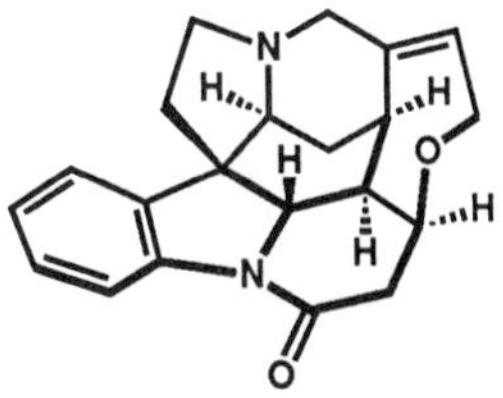
N
H
H
O
H
H
H
N
O

Strychnine
(Woodward, 1954)

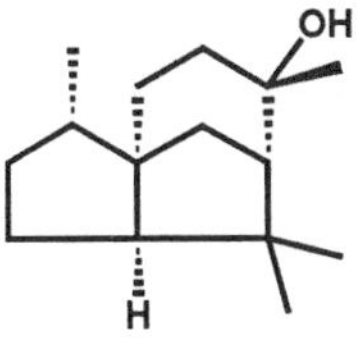
OH
H

Cedrol
(Stork, 1955)

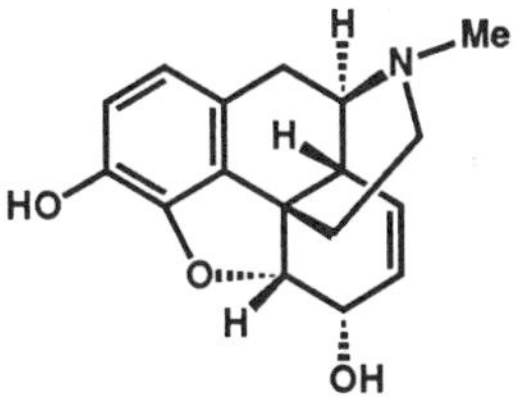
H
N
Me
H
HO
O
H
OH

Morphine
(Gates, 1956)

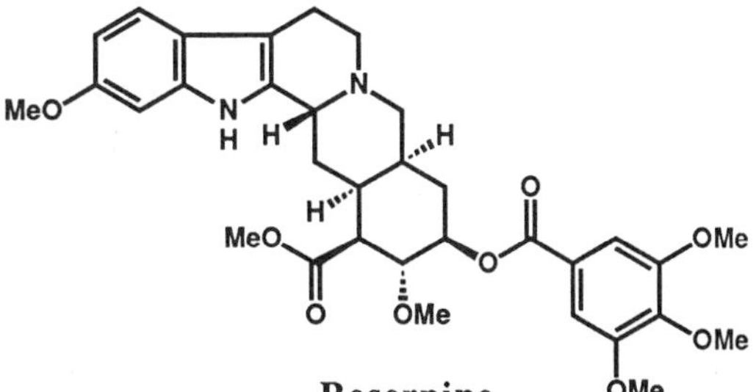
MeO
N
N
H
H
H
H
MeO
O
O
OMe
O
OMe
OMe
OMe

Reserpine
(Woodward, 1956)

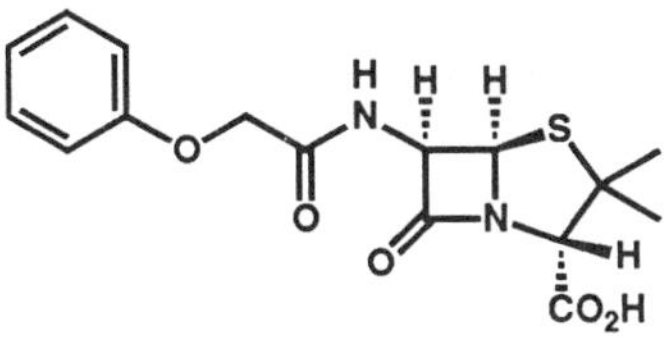
O
H
H
H
N
S
O
O
N
H
CO_2H

Penicillin V
(Sheehan, 1957)

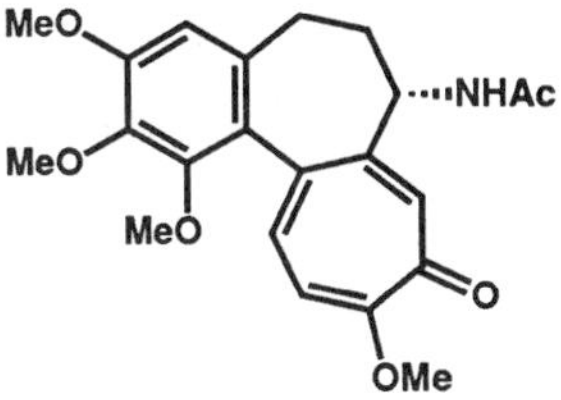
MeO
MeO
MeO
NHAc
O
OMe

Colchicine
(Eschenmoser, 1959)

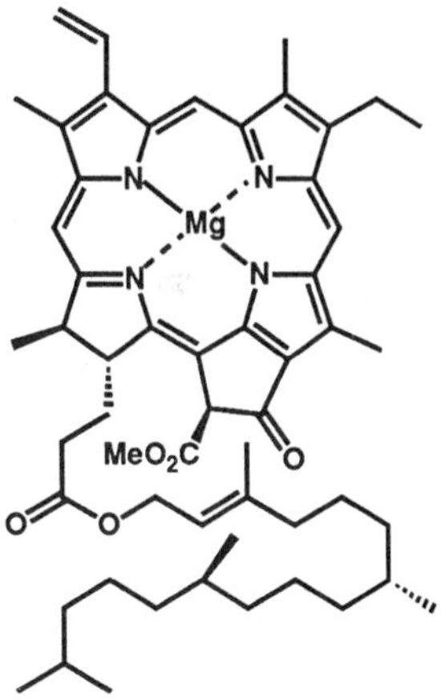
N
N
Mg
N
N
MeO_2C
O
O
O

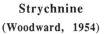

Chlorophyll
(Woodward, 1960)

1.4 Retrosynthetic Analysis

In the first century of "organic" chemistry much attention was given to the structures of carbogens and their transformations. Reactions were classified according to the types of substrates that underwent the chemical change (for example "aromatic substitution," "carbonyl addition," "halide displacement," "ester condensation"). Chemistry was taught and learned as transformations characteristic of a structural class (e.g. phenol, aldehyde) or structural subunit type (e.g. nitro, hydroxyl, α,β-enone). The natural focus was on chemical change in the direction of chemical reactions, i.e. reactants → products. Most syntheses were developed, as mentioned in the preceding section, by selecting a suitable starting material (often by trial and error) and searching for a set of reactions which in the end transformed that material to the desired product (synthetic target or simply TGT). By the mid 1960's a different and more systematic approach was developed which depends on the perception of structural features in *reaction products* (as contrasted with starting materials) and the manipulation of structures in the reverse-synthetic sense. This method is now known as *retrosynthetic* or *antithetic* analysis. Its merits and power were clearly evident from three types of experience. First, the systematic use of the general problem-solving procedures of retrosynthetic analysis both simplified and accelerated the derivation of synthetic pathways for any new synthetic target. Second, the teaching of synthetic planning could be made much more logical and effective by its use. Finally, the ideas of retrosynthetic analysis were adapted to an interactive program for computer-assisted synthetic analysis which demonstrated objectively the validity of the underlying logic.[1,8,10] Indeed, it was by the use of retrosynthetic analysis in each of these ways that the approach was further refined and developed to the present level.

Retrosynthetic (or *antithetic*) analysis is a problem-solving technique for transforming the structure of a *synthetic target* (TGT) molecule to a sequence of progressively simpler structures along a pathway which ultimately leads to simple or commercially available starting materials for a chemical synthesis. The transformation of a molecule to a synthetic precursor is accomplished by the application of a *transform*, the exact reverse of a *synthetic reaction*, to a target structure. Each structure derived antithetically from a TGT then itself becomes a TGT for further analysis. Repetition of this process eventually produces a tree of intermediates having chemical structures as nodes and pathways from bottom to top corresponding to possible synthetic routes to the TGT. Such trees, called EXTGT trees since they grow out from the TGT, can be quite complex since a high degree of branching is possible at each node and since the vertical pathways can include many steps. This central fact implies the necessity for control or guidance in the generation of EXTGT trees so as to avoid explosive branching and the proliferation of useless pathways. Strategies for control and guidance in retrosynthetic analysis are of the utmost importance, a point which will be elaborated in the discussion to follow.

1.5 Transforms and Retrons

In order for a transform to operate on a target structure to generate a synthetic predecessor, the enabling structural subunit or *retron*[8] for that transform must be present in the target. The basic retron for the Diels-Alder transform, for instance, is a six-membered ring containing a π-bond, and it is this substructural unit which represents the minimal *keying* element for transform function in any molecule. It is customary to use a double arrow (⇒) for the retrosynthetic direction in drawing transforms and to use the same name for the transform as is appropriate to the reaction. Thus the carbo-Diels-Alder transform (tf.) is written as follows:

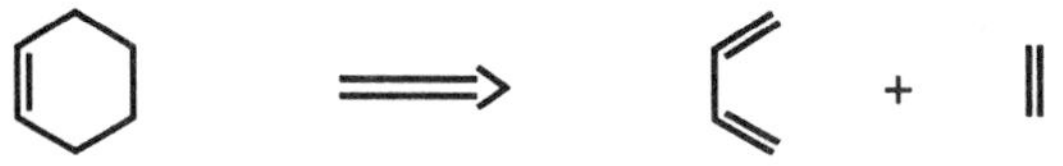

Carbo-Diels-Alder Transform

The Diels-Alder reaction is one of the most powerful and useful processes for the synthesis of carbogens not only because it results in the formation of a *pair* of bonds and a six-membered ring, but also since it is capable of generating selectively one or more stereocenters, and additional substituents and functionality. The corresponding transform commands a lofty position in the hierarchy of all transforms arranged according to simplifying power. The Diels-Alder reaction is also noteworthy because of its broad scope and the existence of several important and quite distinct variants. The retrons for these variants are more elaborate versions, i.e. *supra retrons,* of the basic retron (6-membered ring containing a π-bond), as illustrated by the examples shown in Chart 1, with exceptions such as (c) which is a composite of addition and elimination processes.

Given structure **1** as a target and the recognition that it contains the retron for the Diels-Alder transform, the application of that transform to **1** to generate synthetic precursor **2** is straightforward. The problem of synthesis of **1** is then reduced retrosynthetically to the simpler

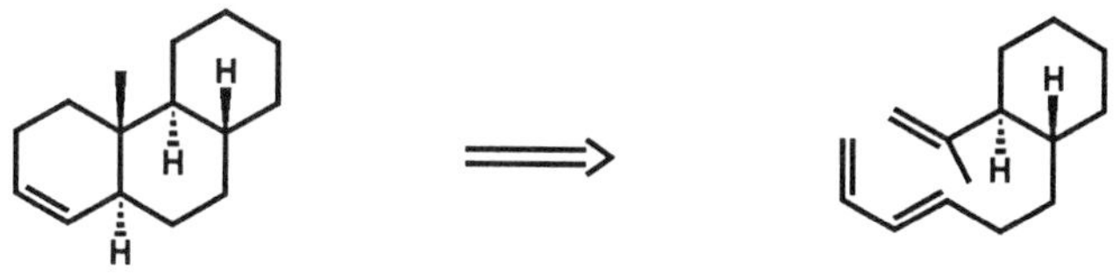

1 **2**

task of constructing **2**, assuming the transform **1** ⇒ **2** can be validated by critical analysis of the feasibility of the synthetic reaction. It is possible, but not quite as easy, to find such retrosynthetic pathways when only an incomplete or *partial retron* is present. For instance, although structures such as **3** and **4** contain a 6-membered A ring lacking a π-bond, the basic Diels-Alder retron is easily established by using well-known transforms to form **1**. A 6-membered ring lacking a π-bond, such as the A ring of **3** or **4**, can be regarded as a *partial*

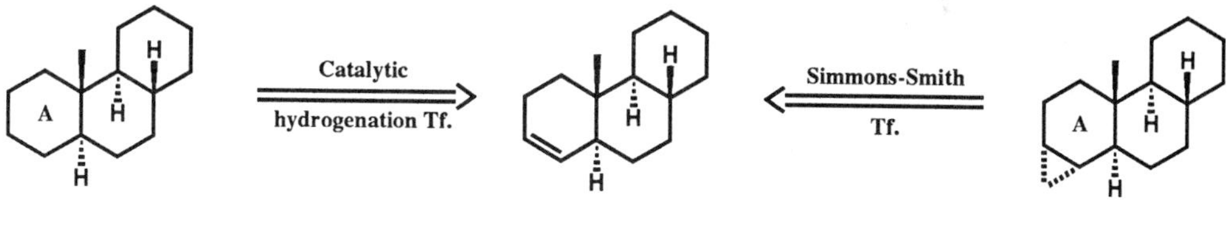
Catalytic
hydrogenation Tf.
Simmons-Smith
Tf.
A
H

3 **1** **4**

retron for the Diels-Alder transform. In general, partial retrons can serve as useful keying elements for simplifying transforms such as the Diels-Alder.

Additional keying information can come from certain other structural features which are present in a retron- or partial-retron-containing substructure. These *ancillary keying* elements can consist of functional groups, stereocenters, rings or appendages. Consider target structure **5**

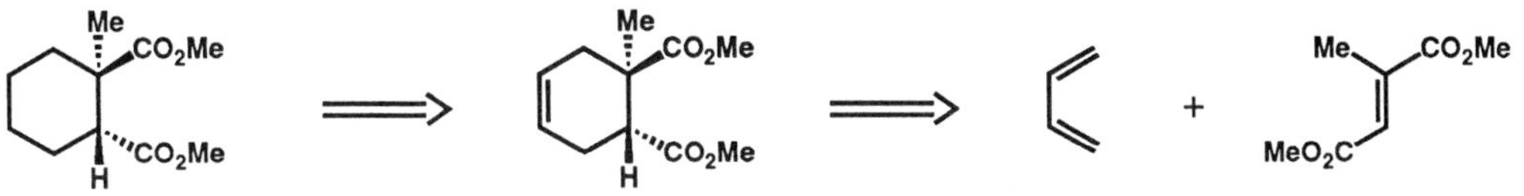
Me
CO2Me
CO2Me
H
MeO2C

5

which contains, in addition to the cyclic partial retron for the Diels-Alder transform, two adjacent stereocenters with electron-withdrawing methoxycarbonyl substituents on each. These extra

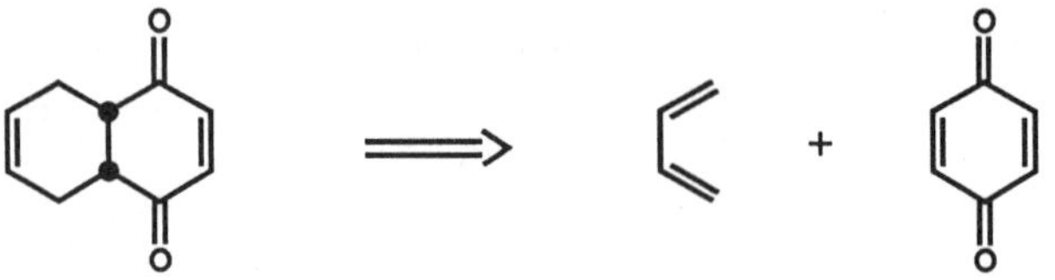

(a)

Quinone-Diels-Alder Tf.

(b)

o-Quinonemethide-Diels-Alder Tf.

(c)

Diels-Alder-1,4-Cycloelimination Composite Tf.

(d)

Benzyne-Diels-Alder Tf.

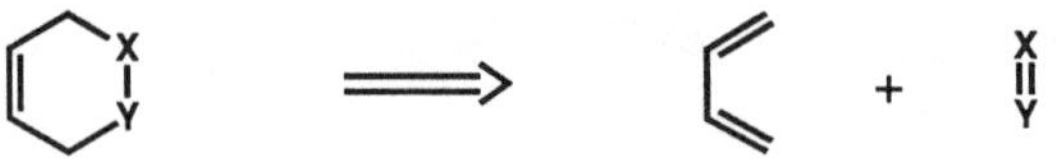

(e)

Heterodienophile-Diels-Alder Tf.
(X and/or Y = heteroatom)

Chart 1. Types of Diels-Alder Transforms

keying elements strongly signal the application of the Diels-Alder transform with the stereocenters coming from the dienophile component and the remaining four ring atoms in the partial retron coming from butadiene as shown. Ancillary keying in this case originates from the fact that the Diels-Alder reaction proceeds by stereospecific suprafacial addition of diene to dienophile and that it is favored by electron deficiency in the participating dienophilic π-bond.

In the above discussion of the Diels-Alder transform reference has been made to the minimal retron for the transform, extended or *supra* retrons for variants on the basic transform, *partial* retrons and *ancillary keying* groups as important structural signals for transform application. There are many other features of this transform which remain for discussion (Chapter 2), for example techniques for *exhaustive* or *long-range* retrosynthetic search[11] to apply the transform in a subtle way to a complicated target. It is obvious that because of the considerable structural simplification that can result from successful application of the Diels-Alder transform, such extensive analysis is justifiable. Earlier experience with computer-assisted synthetic analysis to apply systematically the Diels-Alder transform provided impressive results. For example, the program OCSS demonstrated the great potential of systematically generated intramolecular Diels-Alder disconnections in organic synthesis well before the value of this approach was generally appreciated.[1,11]

On the basis of the preceding discussion the reader should be able to derive retrosynthetic schemes for the construction of targets **6**, **7**, and **8** based on the Diels-Alder transform.

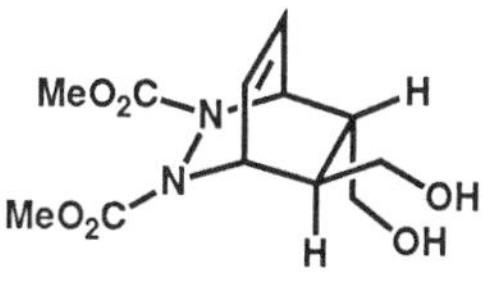

MeO2C
N
N
MeO2C
H
OH
OH
H

6

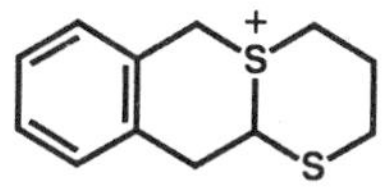

S+
S

7

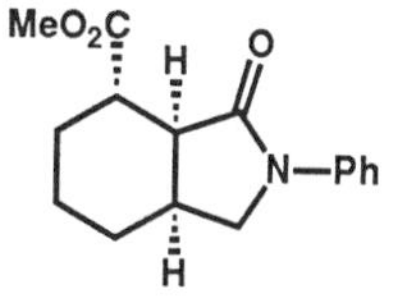

MeO2C
H
O
N—Ph
H

8

1.6 Types of Transforms

There are many thousands of transforms which are potentially useful in retrosynthetic analysis just as there are very many known and useful chemical reactions. It is important to characterize this universe of transforms in ways which will facilitate their use in synthetic problem solving. One feature of major significance is the overall effect of transform application on molecular complexity. The most crucial transforms in this respect are those which belong to the class of *structurally simplifying* transforms. They effect molecular simplification (in the retrosynthetic direction) by *disconnecting* molecular skeleton (chains (CH) or rings (RG)), and/or by removing or disconnecting functional groups (FG), and/or by removing (R) or disconnecting (D) stereocenters (ST). The effect of applying such transforms can be symbolized as CH-D, RG-D, FG-R, FG-D, ST-R or ST-D, used alone or in combination. Some examples of carbon-disconnective simplifying transforms are shown in Chart 2. These are but a minute sampling from the galaxy of known transforms for skeletal disconnection which includes the full range of transforms for the disconnection of acyclic C-C and C-heteroatom bonds and also cyclic C-C and C-heteroatom or heteroatom-heteroatom bonds. In general, for complex structures containing many stereorelationships, the transforms which are *both* stereocontrolled and disconnective will be more significant. Stereocontrol is meant to include both diastereo-control and enantio-control.

TGT STRUCTURE	RETRON	TRANSFORM	PRECURSOR(S)
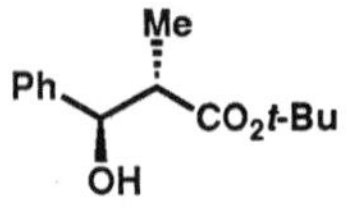	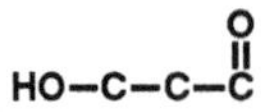	(*E*)-Enolate Aldol	
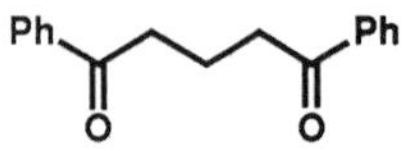	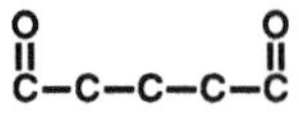	Michael	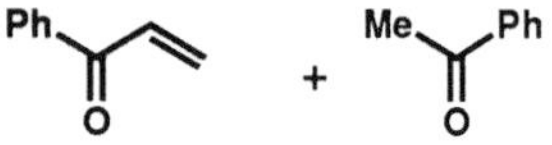
Et_3COH	EtCOH	Orgmet. Addn. to Ketone	Et_2CO + EtMet
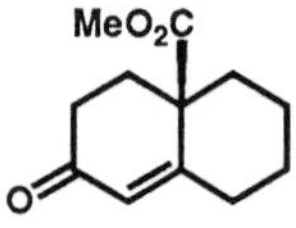	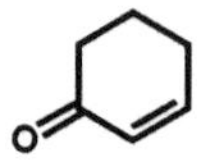	Robinson Annulation (Aldol + Michael)	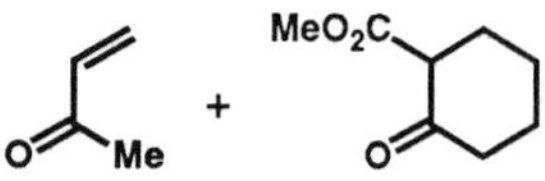
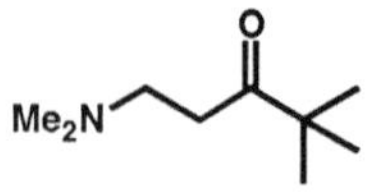	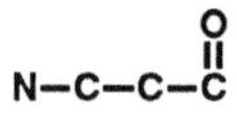	Mannich (Azaaldol)	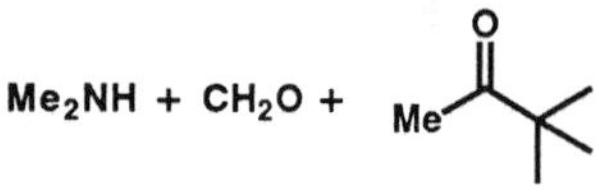
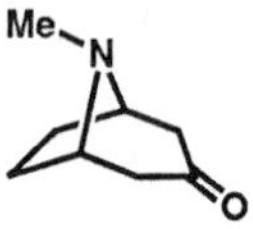	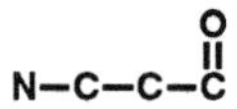	Double Mannich	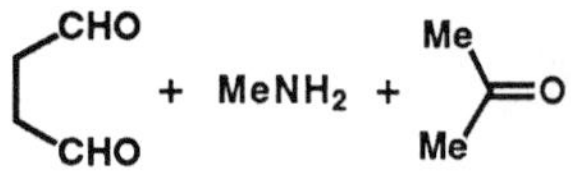
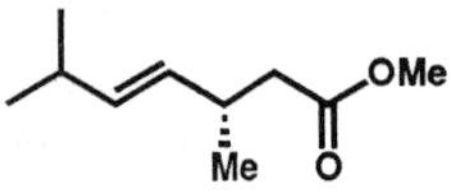	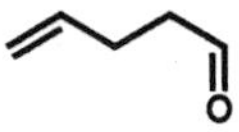	Claisen Rearrangement	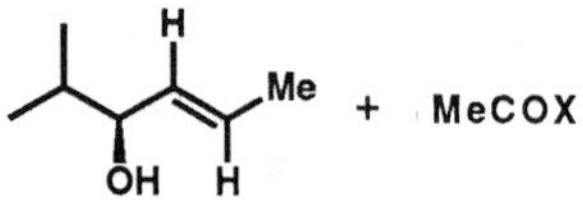
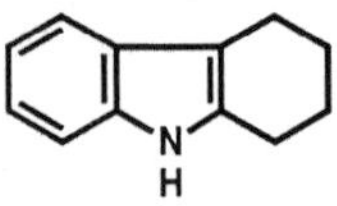	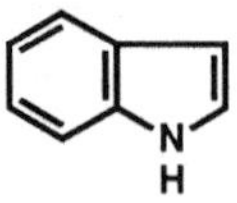	Fischer Indole	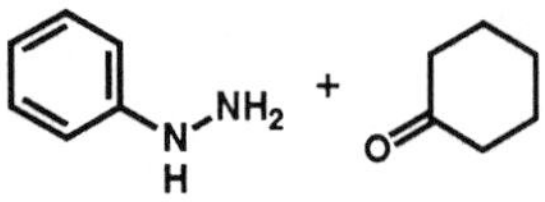
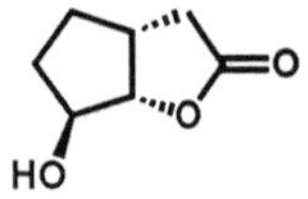	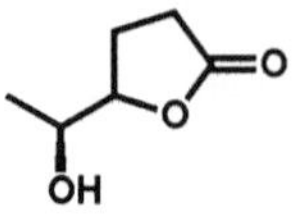	Oxy-lactonization of Olefin	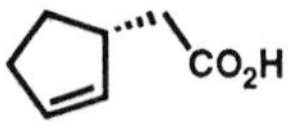

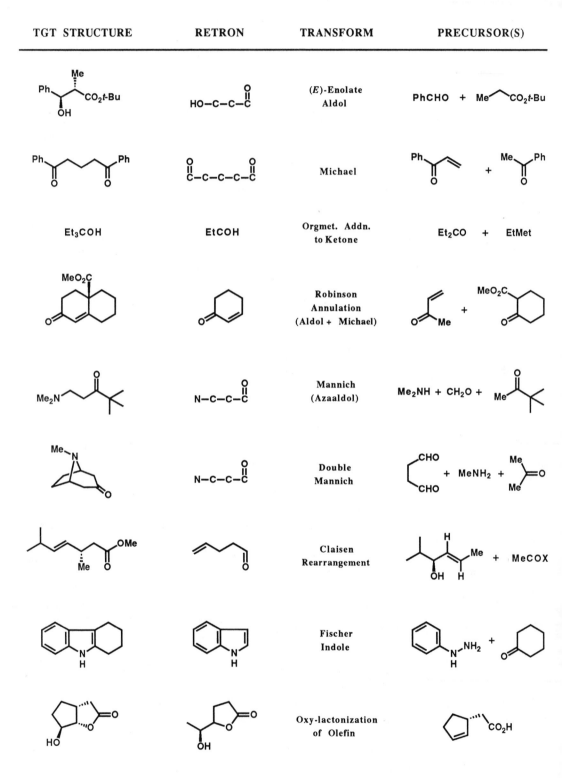

Chart 2. Disconnective Transforms

Transforms may also be distinguished according to retron type, i.e. according to the critical structural features which signal or *actuate* their application. In general, retrons are composed of the following types of structural elements, singly or in combination (usually pairs or triplets): hydrogen, functional group, chain, appendage, ring, stereocenter. A specific interconnecting path or ring size will be involved for transforms requiring a unique positional relationship between retron elements. For other transforms the retron may contain a variable path length or ring of variable size. The classification of transforms according to retron type serves to organize them in a way which facilitates their application. For instance, when confronted with a TGT structure containing one or more 6-membered carbocyclic units, it is clearly helpful to have available the set of all 6-ring-disconnective transforms including the Diels-Alder, Robinson annulation, aldol, Dieckmann, cation-π cyclization, and internal S_N2 transforms.

The reduction of stereochemical complexity can frequently be effected by stereoselective transforms which are not disconnective of skeletal bonds. Whenever such transforms also result in the replacement of functional groups by hydrogen they are even more simplifying. Transforms which remove FG's in the retrosynthetic direction without removal of stereocenters constitute another structurally simplifying group. Chart 3 presents a sampling of FG- and/or stereocenter-removing transforms most of which are not disconnective of skeleton.

There are many transforms which bring about essentially no change in molecular complexity, but which can be useful because they modify a TGT to allow the subsequent application of simplifying transforms. A frequent application of such transforms is to generate the retron for some other transform which can then operate to simplify structure. There are a wide variety of such non-simplifying transforms which can be summarized in terms of the structural change which they effect as follows:

1. *molecular skeleton:* connect or rearrange
2. *functional groups:* interchange or transpose
3. *stereocenters:* invert or transfer

Functional group interchange transforms (FGI) frequently are employed to allow simplifying skeletal disconnections. The examples **9** ⇒ **10** and **11** ⇒ **12** + **13**, in which the initial FGI transform plays a critical role, typify such processes.

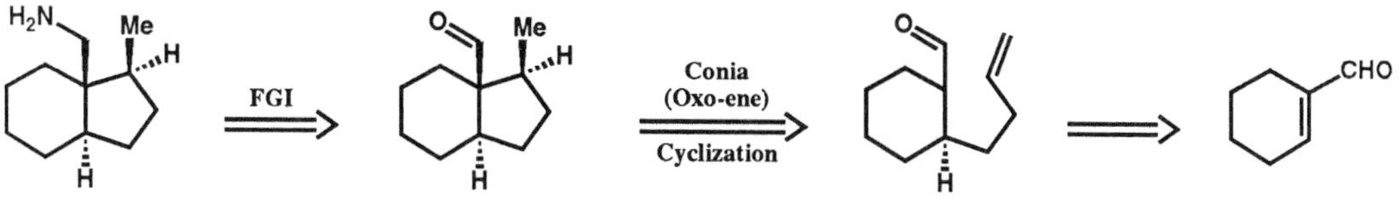

H2N
Me
H
FGI
Me
H
Conia
(Oxo-ene)
Cyclization
H
CHO

9 **10**

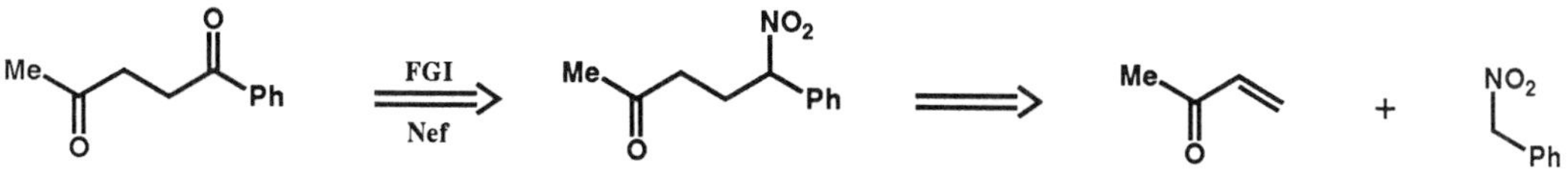

Me
Ph
FGI
Nef
NO2
Me
Ph
Me
NO2
Ph

11 **12** **13**

STRUCTURE	RETRON	TRANSFORM	PRECURSOR
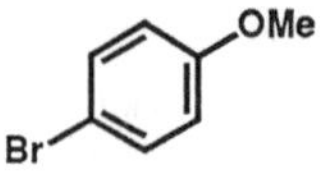	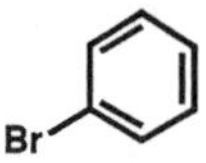	Aromatic Bromination	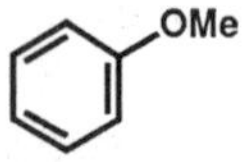
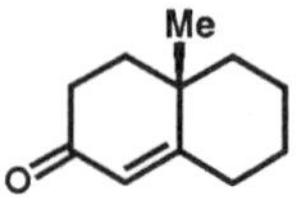	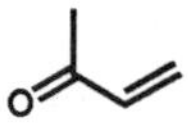	Allylic Oxidation of CH_2 to C=O	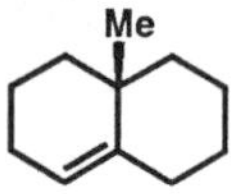
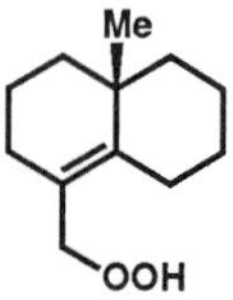	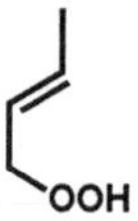	Allylic Oxidation by ${}^1\Delta_g O_2$, with C=C Transposition	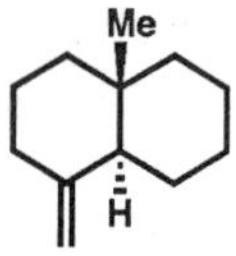
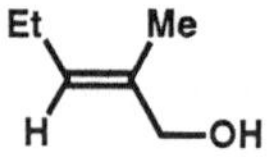		Allylic Oxidation by SeO_2	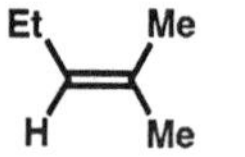
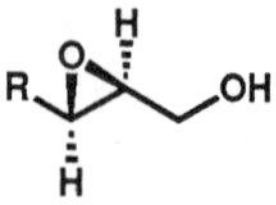	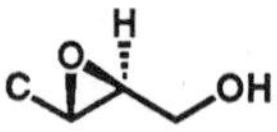	Sharpless Epoxidation with (*R*,*R*)-(+)-DET	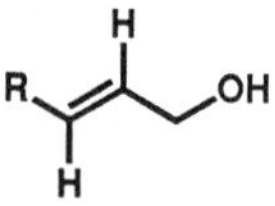
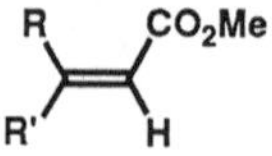	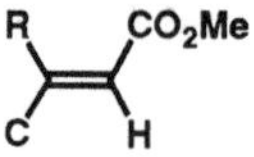	*cis*-Addition of R'_2CuMet to C≡C	
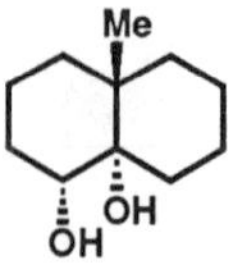	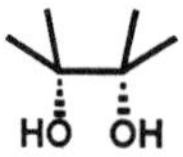	*cis*-Hydroxylation of C=C	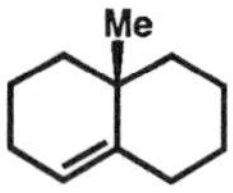
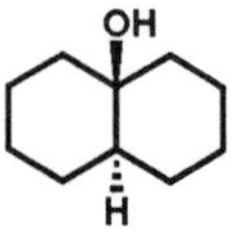		"O" Insertion into C-H (O_3 or RuO_4)	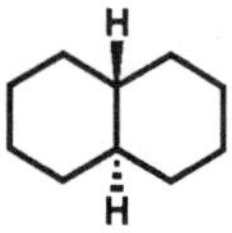
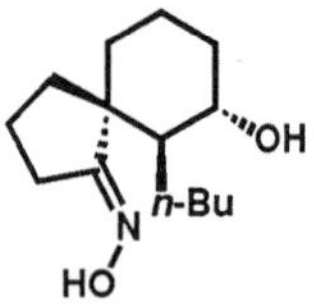	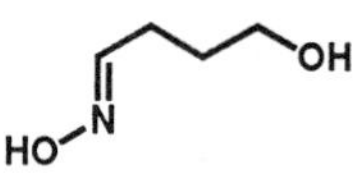	Barton Functionalization	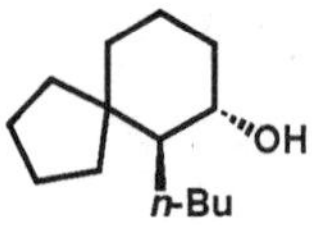
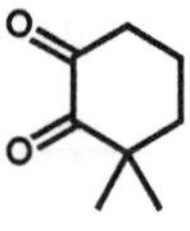	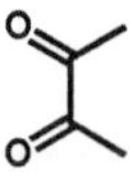	Oxidation of Ketones by SeO_2	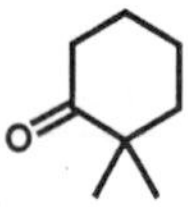
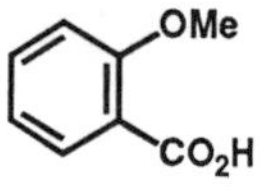	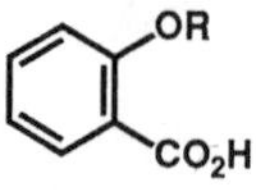	*o*-Metallation (RLi) and Carboxylation	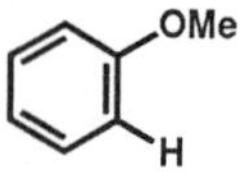

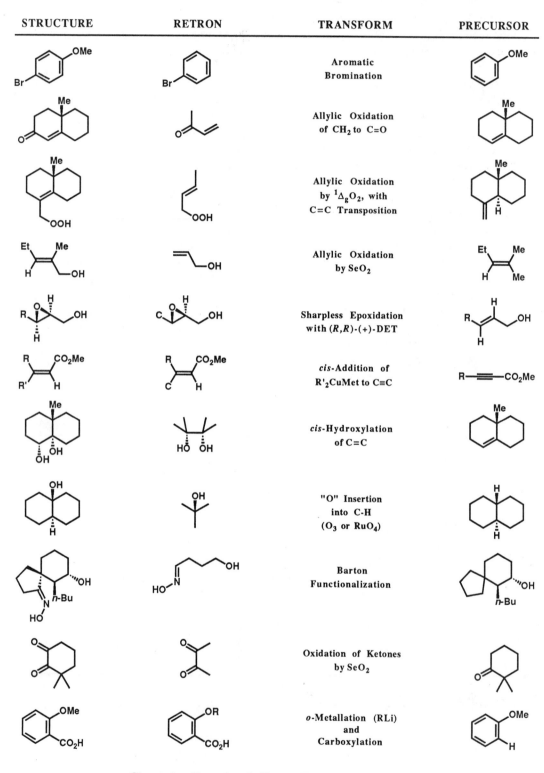

Chart 3. Functional Group Removing Transforms

The transposition of a functional group, for example carbonyl, C=C or C≡C, similarly may set the stage for a highly effective simplification, as the retrosynthetic conversion of **14** to **15** + **16** shows.

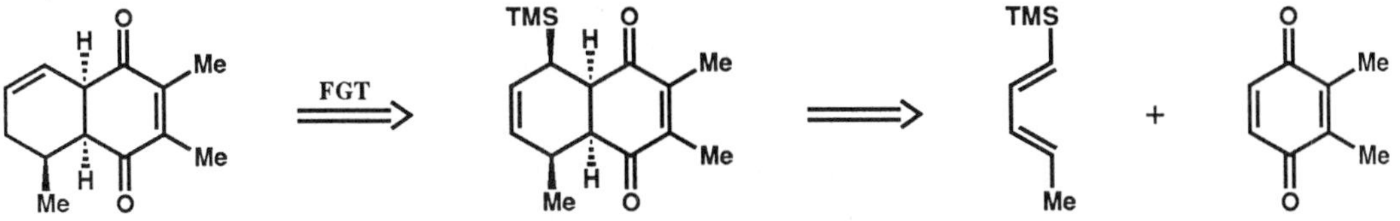
FGT

14 **15** **16**

Rearrangement of skeleton, which normally does not simplify structure, can also facilitate molecular disconnection, as is illustrated by examples **17** ⇒ **18** + **19** and **20** ⇒ **21**.

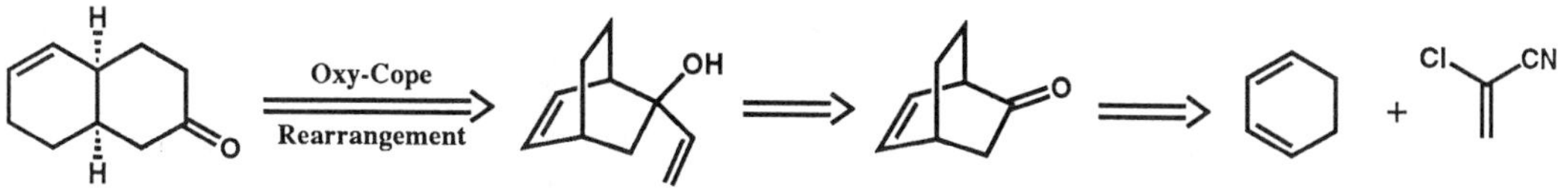
Oxy-Cope
Rearrangement

17 **18** **19**

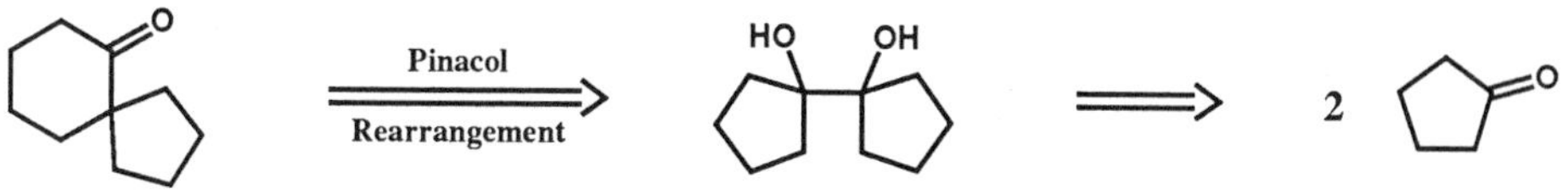
Pinacol
Rearrangement
2

20 **21**

The last category of transforms in the hierarchy of retrosynthetic simplifying power are those which increase complexity, whether by the addition of rings, functional groups (FGA) or stereocenters. There are many such transforms which find a place in synthesis. The corresponding synthetic reactions generally involve the removal of groups which no longer are needed for the synthesis such as groups used to provide stereocontrol or positional (regio-) control, groups used to provide activation, deactivation or protection, and groups used as temporary bridges. The retrosynthetic addition of functional groups may also serve to generate the retron for the operation of a simplifying transform. An example is the application of hydrolysis and decarboxylation transforms to **22** to set up the Dieckmann retron in **23**.

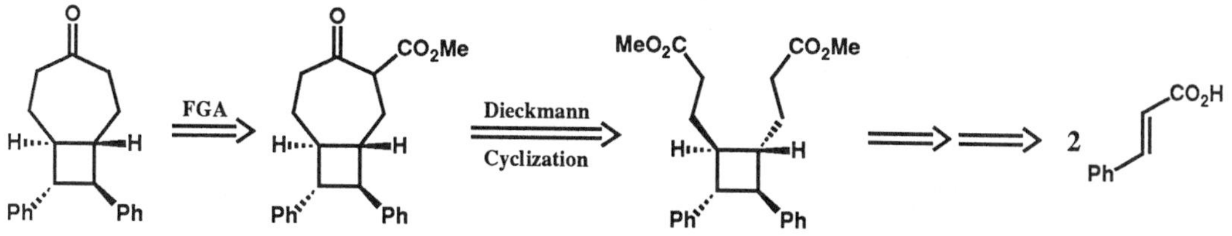
FGA
Dieckmann
Cyclization
2

22 **23**

Dechlorination transforms are also commonly applied, e.g. **24** ⇒ **25** ⇒ **26** + **27**.

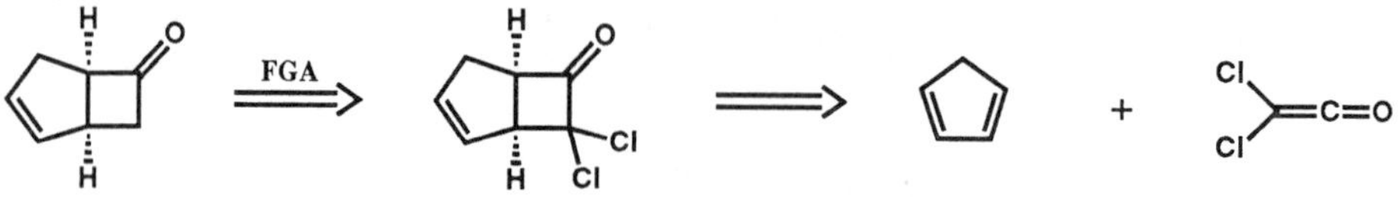

24 **25** **26** **27**

The following deamination transform, **28** ⇒ **29**, illustrates how FGA can be used for positional control for a subsequent aromatic FG removal (FGR) transform, **29** ⇒ **30**.

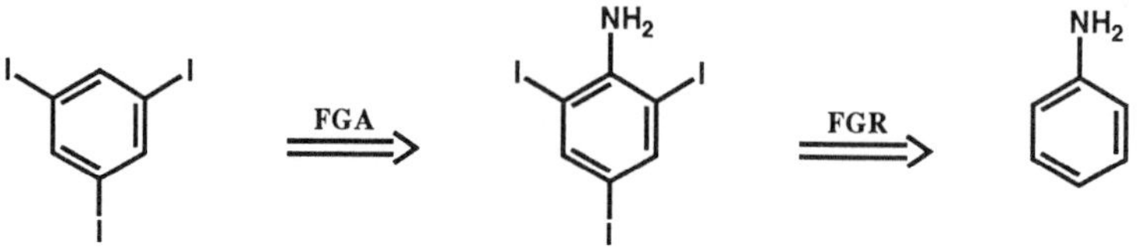

28 **29** **30**

Desulfurization is an important transform for the addition of a temporary bridge (**31** ⇒ **32**).

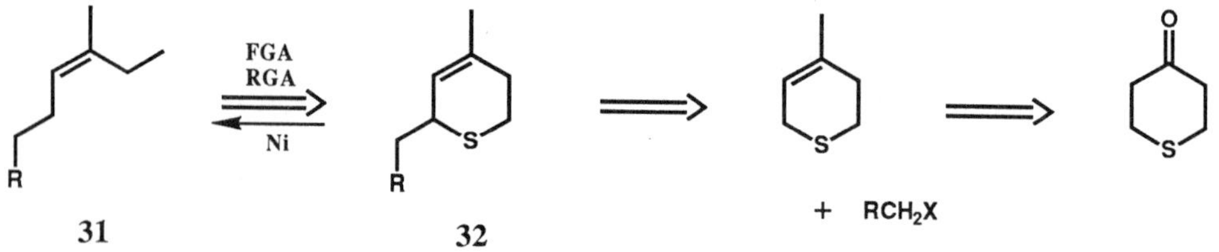

31 **32**

Retrosynthetic addition of elements such as sulfur, selenium, phosphorous or boron may be required as part of a disconnective sequence, as in the Julia-Lythgoe *E* olefin transform as applied to **33**.

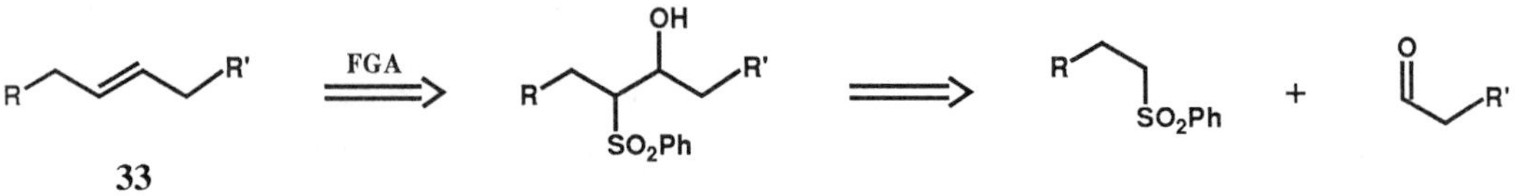

33

The frequent use of chiral controller or auxiliary groups in enantioselective synthesis (or diastereoselective processes) obviously requires the addition of such units retrosynthetically, as illustrated by the antithetic conversion **34** ⇒ **35**.

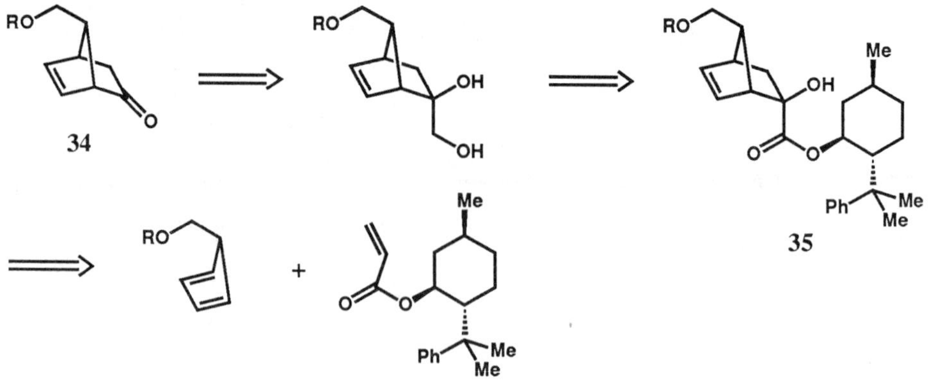

34 **35**

1.7 Selecting Transforms

For many reasons synthetic problems cannot be analyzed in a useful way by the indiscriminate application of all transforms corresponding to the retrons contained in a target structure. The sheer number of such transforms is so great that their undisciplined application would lead to a high degree of branching of an EXTGT tree, and the results would be unwieldy and largely irrelevant. In the extreme, branching of the tree would become explosive if all possible transforms corresponding to *partial retrons* were to be applied. Given the complexity and diversity of carbogenic structures and the vast chemistry which supports synthetic planning, it is not surprising that the intelligent selection of transforms (as opposed to opportunistic or haphazard selection) is of utmost importance. *Fundamental to the wise choice of transforms is the awareness of the position of each transform on the hierarchical scale of importance with regard to simplifying power and the emphasis on applying those transforms which produce the greatest molecular simplification.* The use of non-simplifying transforms is only appropriate when they pave the way for application of an effectively simplifying transform. The unguided use of moderately simplifying transforms may also be unproductive. It is frequently more effective to apply a powerfully simplifying transform for which only a partial retron is present than to use moderately simplifying transforms for which full retrons are already present. On this and many other points, analogies exist between retrosynthetic analysis and planning aspects of games such as chess. The sacrifice of a minor piece in chess can be a very good move if it leads to the capture of a major piece or the establishment of dominating position. In retrosynthetic analysis, as in most kinds of scientific problem solving and most types of logic games, the recognition of strategies which can direct and guide further analysis is paramount. A crucial development in the evolution of retrosynthetic thinking has been the formulation of general retrosynthetic strategies and a logic for using them.

1.8 Types of Strategies for Retrosynthetic Analyses

The technique of systematic and rigorous modification of structure in the retrosynthetic direction provides a foundation for deriving a number of different types of strategies to guide the selection of transforms and the discovery of *hidden* or *subtle* synthetic pathways. Such strategies must be formulated in general terms and be applicable to a broad range of TGT structures. Further, even when not applicable, their use should lead to some simplification of the problem or to some other line of analysis. Since the primary goal of retrosynthetic analysis is the reduction of structural complexity, it is logical to start with the elements which give rise to that complexity as it relates to synthesis. As mentioned in section 1.2 on *molecular complexity*, these elements are the following: (1) molecular size, (2) cyclic connectivity or topology, (3) element or functional group content, (4) *stereocenter* content/density, (5) centers of high chemical reactivity, and (6) kinetic (thermal) instability. It is possible to formulate independent strategies for dealing with each of these complicating factors. In addition, there are two types of useful general strategies which do not depend on molecular complexity. One type is the *transform-based* or *transform-goal* strategy, which is essentially the methodology for searching out and invoking effective, powerfully simplifying transforms. The other variety, the *structure-goal* strategy, depends on the guidance which can be obtained from the recognition of possible starting materials or key intermediates for a synthesis.

An overarching principle of great importance in retrosynthetic analysis is the *concurrent use of as many of these independent strategies as possible.* Such parallel application of several strategies not only speeds and simplifies the analysis of a problem, but provides superior solutions.

The actual role played by the different types of strategies in the simplification of a synthetic problem will, of course, depend on the nature of the problem. For instance, in the case of a TGT molecule with no rings or with only single-chain-connected rings (i.e. neither bridged, nor fused, nor spiro) but with an array of several stereocenters and many functional groups, the role played by topological strategies in retrosynthetic analysis will be less than for a more topologically complex polycyclic target (and the role of stereochemical strategies may be larger). For a TGT of large size, for instance molecular weight of 4000, but with only isolated rings, the disconnections which produce several fragments of approximately the same complexity will be important.

The logical application of retrosynthetic analysis depends on the use of higher level strategies to guide the selection of effective transforms. Chapters 2-5 which follow describe the general strategies which speed the discovery of fruitful retrosynthetic pathways. In brief these strategies may be summarized as follows.

1. Transform-based strategies—long range search or look-ahead to apply a powerfully simplifying transform (or a tactical combination of simplifying transforms) to a TGT with certain appropriate keying features. The retron required for application of a powerful transform may not be present in a complex TGT and a number of antithetic steps (subgoals) may be needed to establish it.

2. Structure-goal strategies—directed at the structure of a potential intermediate or potential starting material. Such a goal greatly narrows a retrosynthetic search and allows the application of bidirectional search techniques.

3. Topological strategies—the identification of one or more individual bond disconnections or correlated bond-pair disconnections as strategic. Topological strategies may also lead to the recognition of a key substructure for disassembly or to the use of rearrangement transforms.

4. Stereochemical strategies—general strategies which remove stereocenters and stereorelationships under stereocontrol. Such stereocontrol can arise from transform-mechanism control or substrate-structure control. In the case of the former the retron for a particular transform contains critical stereochemical information (absolute or relative) on one or more stereocenters. Stereochemical strategies may also dictate the retention of certain stereocenter(s) during retrosynthetic processing or the joining of atoms in three-dimensional proximity.

5. Functional group-based strategies. The retrosynthetic reduction of molecular complexity involving functional groups (FG's) as keying structural subunits takes various forms. Single FG's or pairs of FG's (and the interconnecting atom path) can (as retrons) key directly the disconnection of a TGT skeleton to form simpler molecules or signal the application of transforms which replace functional groups by hydrogen. Functional group interchange (FGI) is a commonly used tactic for generating from a TGT retrons which allow the application of simplifying transforms. FG's may key transforms which stereoselectively remove stereocenters, break topologically strategic bonds or join proximate atoms to form rings.

CHAPTER TWO

Transform-Based Strategies

2.1 Transform-Guided Retrosynthetic Search

The wise choice of appropriate simplifying transforms is the key to retrosynthetic analysis. Fortunately methods are available for selecting from the broad category of powerful transforms a limited number which are especially suited to a target structure. This selection can be made in a logical way starting with the characterization of a molecule in terms of complexity elements and then identifying those transforms which are best suited for reducing the dominant type of complexity. For instance, if a TGT possesses a complex cyclic network with embedded stereocenters, the category of ring-disconnective, stereoselective transforms is most relevant. With such a target structure, the particular location(s) within the cyclic network of strategic disconnection possibilities, as revealed by the use of topological strategies, can further narrow the list of candidate transforms. At the very least, topological considerations generally produce a rough ordering of constituent rings with regard to disconnection priority. For each ring or ring-pair to be examined, a number of disconnective transforms can then be selected by comparison of the retrons (or supra-retrons) and ancillary keying elements for each eligible transform with the region of the target being examined. Since the full retron corresponding to a particular candidate transform is usually not present in the TGT, this analysis amounts to a comparison of retron and TGT for partial correspondence by examining at least one way, and preferably all possible ways, of mapping the retron onto the appropriate part of the TGT. From the comparison of the various mappings with one another, a preliminary assignment of relative merit can be made. With the priorities set for the group of eligible transforms and for the best mappings of each onto a TGT molecule, a third stage of decision making then becomes possible which involves a multistep

---

References are located on pages 92-95. A glossary of terms appears on pages 96-98.

retrosynthetic search for each transform to determine specific steps for establishing the required retron and to evaluate the required disconnection. That multistep search is driven by the goal of applying a particular simplifying transform (T-goal) to the TGT structure. The most effective T-goals in retrosynthetic analysis generally correspond to the most powerful synthetic constructions.

2.2 Diels-Alder Cycloaddition as a T-Goal

There are effective techniques for rigorous and exhaustive long-range search to apply each key simplifying transform. These procedures generally lead to removal of obstacles to transform application and to establishment of the necessary retron or supra-retron. They can be illustrated by taking one of the most common and powerful transforms, the Diels-Alder cycloaddition. The Diels-Alder process is frequently used at an early stage of a synthesis to establish a structural core which can be elaborated to the more complex target structure. This fact implies that retrosynthetic application of the Diels-Alder T-goal can require a deep search through many levels of the EXTGT tree to find such pathways, another reason why the Diels-Alder transform is appropriate in this introduction to T-goal guided analysis.

Once a particular 6-membered ring is selected as a site for applying the Diels-Alder transform, six possible [4 + 2] disconnections can be examined, i.e. there are six possible locations of the π-bond of the basic Diels-Alder retron. With ring numbering as shown in **36**, and

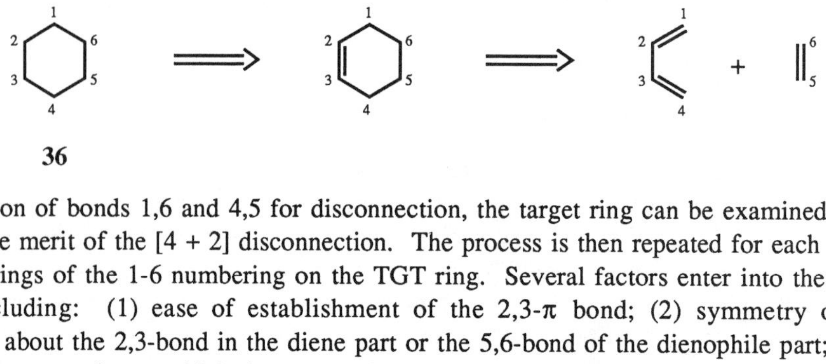

1
2
6
3
5
4
1
2
6
3
5
4
1
2
3
4
+
6
5

36

specification of bonds 1,6 and 4,5 for disconnection, the target ring can be examined to estimate the relative merit of the [4 + 2] disconnection. The process is then repeated for each of the other five mappings of the 1-6 numbering on the TGT ring. Several factors enter into the estimate of merit, including: (1) ease of establishment of the 2,3-π bond; (2) symmetry or potential symmetry about the 2,3-bond in the diene part or the 5,6-bond of the dienophile part; (3) type of Diels-Alder transform which is appropriate (e.g. quinone-Diels-Alder); (4) positive (i.e favorable) or negative (i.e. unfavorable) substitution pattern if both diene and dienophile parts are unsymmetrical; (5) positive or negative electronic activation in dienophile and diene parts; (6) positive or negative steric effect of substituents; (7) positive or negative stereorelationships, e.g. 1,4, 1,6, 4,5, 5,6; (8) positive or negative ring attachments or bridging elements; and (9) negative unsaturation content (e.g. 1,2-, 3,4-, 4,5- or 1,6-bond aromatic) or heteroatom content (e.g Si or P). For instance, an *o*-phenylene unit bridging ring atoms 1 and 3, or 2 and 6, would be a strongly negative element. Alternatively a preliminary estimate is possible, once the 2,3-π bond is established for a particular ring orientation, by applying the transform and evaluating its validity in the synthetic direction. Again, positive and negative structural factors can be identified and evaluated.

The information obtained by this preliminary analysis can be used not only to set priorities for the various possible Diels-Alder disconnections, but also to pinpoint obstacles to transform application. Recognition of such obstacles can also serve to guide the search for specific retrosynthetic sequences or for the highest priority disconnections. At this point it is likely that all but 1 or 2 modes of Diels-Alder disconnection will have been eliminated, and the retrosynthetic search becomes highly focused. Having selected both the transform and the mapping onto the TGT, it is possible to sharpen the analysis in terms of potentially available dienophile or diene components, variants on the structure of the intermediate for Diels-Alder disconnection, tactics for ensuring stereocontrol and/or position control in the Diels-Alder addition, possible chiral control elements for enantioselective Diels-Alder reaction, etc.

2.3 Retrosynthetic Analysis of Fumagillol (37)

The application of this transform-based strategy to a specific TGT structure, fumagillol (**37**),[12] will now be described (Chart 4). The Diels-Alder transform is a strong candidate as T-goal, not only because of the 6-membered ring of **37**, but also because of the 4 stereocenters in that ring, and the clear possibility of completing the retron by introducing a π-bond retrosynthetically in various locations. Of these locations π-bond formation between ring members *d* and *e* of **37**, which can be effected by (1) retrosynthetic conversion of methyl ether to hydroxyl, and (2) application of the OsO_4 *cis*-hydroxylation transform to give **39**, is clearly of high merit. Not only is the Diels-Alder retron established in this way, but structural simplification is concurrently effected by removal of 2 hydroxyl groups and 2 stereocenters. It is important to note that for the retrosynthetic conversion of **37** to **39** to be valid, site selectivity is required for the synthetic steps **39** $\rightarrow$ **38** and **38** $\rightarrow$ **37**. Selective methylation of the equatorial hydroxyl at carbon *e* in **38** is a tractable problem which can be dealt with by taking advantage of reactivity differences between axial and equatorial hydroxyls. In practice, selective methylation of a close analog of **38** was effected by the reaction of the mono alkoxide with methyl iodide.[12] Use of the cyclic di-*n*-butylstannylene derivative of diol **38** is another reasonable possibility.[13] Selective *cis*-hydroxylation of the *d-e* double bond in **39** in the presence of the trisubstituted olefinic bond in the 8-carbon appendage at *f* is a more complex issue, but one which can be dealt with separately. Here, two points must be made. First, whenever the application of a transform generates a functional group which also is present at one or more other sites in the molecule, the feasibility of the required selectivity in the corresponding synthetic reaction must be evaluated. It may be advantageous simply to note the problem (one appropriate way is to box those groups in the offspring which are duplicated by transform operation) and to continue with the T-goal search, leaving the resolution of the selectivity problem to the next stage of analysis. Second, goal-directed retrosynthetic search invariably requires a judicious balance between the *complete* (immediate) and the *partial* (deferred) resolution of issues arising from synthetic obstacles such as interfering functionality.

Assuming that the synthetic conversion of **39** to **37** is a reasonable proposition, the Diels-Alder disconnection of **39** can now be examined. Clearly, the direct disconnection is unworkable since allene oxide **40** is not a suitable dienophile, for several reasons. But, if **39** can be modified retrosynthetically to give a structure which can be disconnected to an available and suitably reactive equivalent of allene oxide **40**, the Diels-Alder disconnection might be viable. Such a possibility is exemplified by the retrosynthetic sequence **39** $\Rightarrow$ **43** + **44**, in which R* is a *chiral control element (chiral controller or chiral auxiliary).*[14,15] This sequence is especially interesting since the requisite diene (**44**) can in principle be generated from **45** by enantioselective epoxidation (see section 2.8). Having derived the possible pathway **37** $\Rightarrow$ $\Rightarrow$ **45** the next stage of refinement is reached for this line of analysis. All of the problems which had been noted, but deferred, (e.g. interference of the double bond of the ring appendage) have to be resolved, the

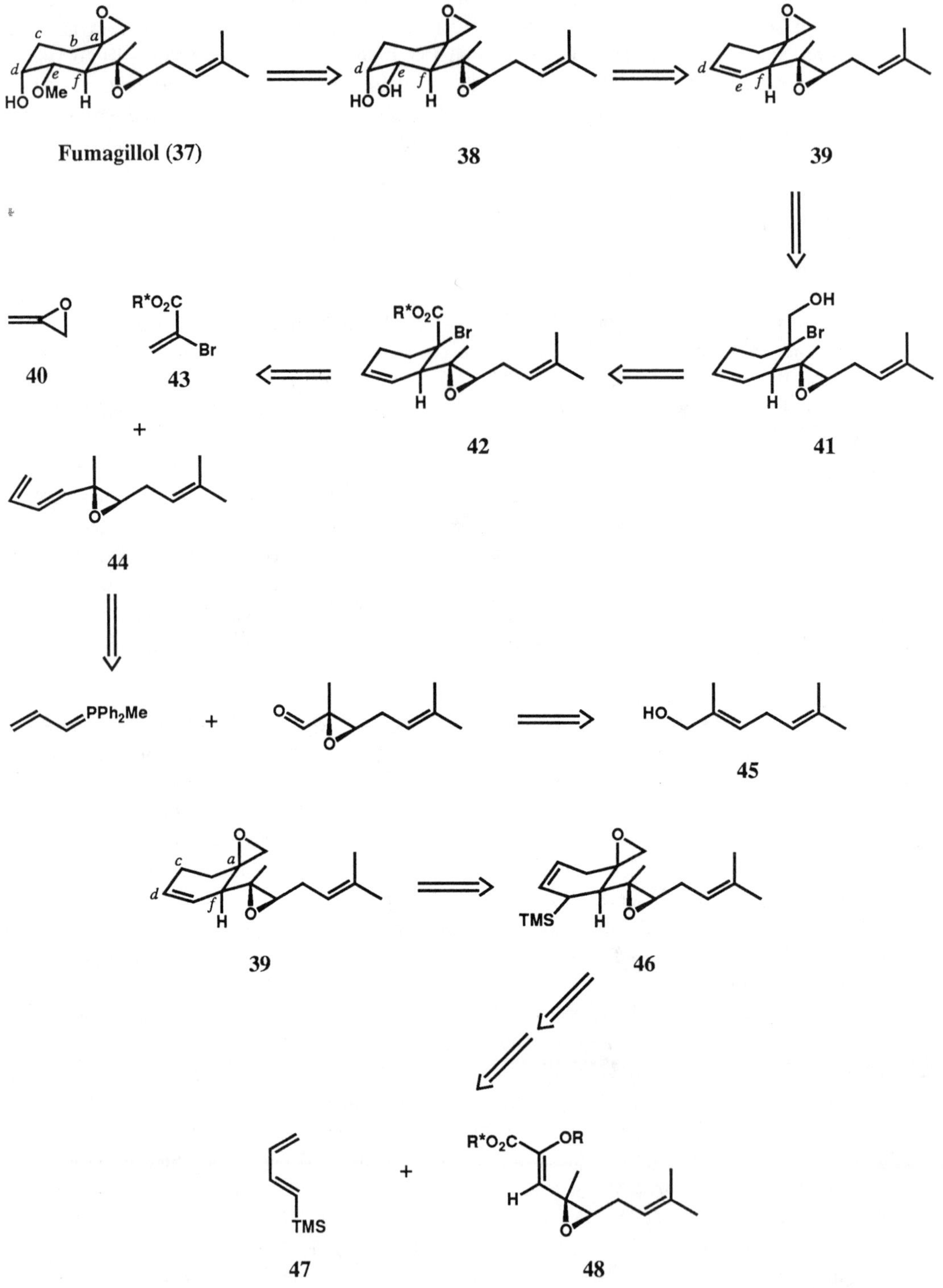
Fumagillol (37)
38
39
40
43
42
41
44
45
46
47
48
R*O2C
Br
OH
PPh2Me
TMS
OR
HO
OMe
H

Chart 4

feasibility of each synthetic step must be scrutinized, and the sequence optimized with regard to specific intermediates and the ordering of steps. Assuming that a reasonable retrosynthetic pathway has been generated, attention now must be turned to other Diels-Alder disconnection possibilities.

The retrosynthetic establishment of the minimal Diels-Alder retron in **39** by the removal of two oxygen functions and two stereocenters is outstanding because retron generation is accomplished concurrently with structural simplification. It is this fact which lent priority to examining the disconnection pathway via **39** over the other 5 alternatives. Of those remaining alternatives the disconnection of *a-b* and *e-f* bonds of **37** is signalled by the fact that centers *a* and *f* are carbon-bearing stereocenters which potentially can be set in place with complete predictability because of the strict suprafacial *(cis)* addition course of the Diels-Alder process with regard to the dienophile component. This disconnection requires the introduction of a π-bond between the carbons corresponding to *c* and *d* in **37**. Among the various ways in which this might be arranged, one of the most interesting is from intermediate **39** by the transposition of the double bond as indicated by **39** ⇒ **46**. From **46** the retrosynthetic steps leading to disconnection to form **47** and **48** are clear. Although Diels-Alder components **47** and **48** are not symmetrical, there are good mechanistic grounds for a favorable assessment of the cycloaddition to give **46**.

In the case of target **37** two different synthetic approaches have been discovered using a transform-based strategy with the Diels-Alder transform as T-goal. Although it is possible in principle that one or more of the other 4 possible modes of Diels-Alder disconnection might lead to equally good plans, retrosynthetic examination of **37** reveals that these alternatives do not produce outstanding solutions. The two synthetic routes to **37** derived herein should be compared with the published synthetic route.[14]

The analysis of the fumagillol structure which has just been outlined illustrates certain general aspects of T-goal driven search and certain points which are specific for the Diels-Alder search procedure. In the former category are the following: (1) establishing *priority* among the various modes of transform application which are possible in principle; (2) recognizing ancillary keying elements; (3) dealing with obstacles to transform application such as the presence of interfering FG's in the TGT or the creation of duplicate FG's in the offspring structure; and (4) the replacement of structural subunits which impede transform application by equivalents (e.g., using FGI transforms) which are favorable. In the latter category it is important to use as much general information as possible with regard to the Diels-Alder reaction in order to search out optimal pathways including: (1) the generation of Diels-Alder components which are suitable in terms of availability and reactivity; (2) analysis of the pattern of substitution on the TGT ring to ascertain consistency with the orientational selectivity predicted for the Diels-Alder process; (3) analysis of consistency of TGT stereochemistry with Diels-Alder stereoselectivities; (4) use of stereochemical control elements; and (5) use of synthetic equivalents of invalid diene or dienophile components. Additional examples of the latter include $H_2C{=}CH{-}COOR$ or $H_2C{=}CHNO_2$ as ketene equivalents or $O{=}C(COOEt)_2$ as a CO_2 equivalent.

Further analysis of the fumagillol problem under the T-goal driven search strategy can be carried out in a similar way using the other ring disconnective transforms for 6-membered rings. Among those which might be considered in at least a preliminary way are the following: (1) internal S_N2 alkylation; (2) internal acylation (Dieckmann); (3) internal aldol; (4) Robinson annulation; (5) cation-π-cyclization; (6) radical-π-cyclization; and (7) internal pinacol or acyloin closure. It is also possible to utilize T-goals for the disconnection of the 8-carbon appendage attached to carbon *f* of **37**, *prior* to ring disconnection, since this is a reasonable alternative for topological simplification. Disconnection of that appendage-ring bond was a key step in the synthesis of ovalicin, a close structural relative to fumagillol.[16]

2.4 Retrosynthetic Analysis of Ibogamine (49)

Mention was made earlier of the fact that many successful syntheses of polycyclic target structures have utilized the Diels-Alder process in an early stage. One such TGT, ibogamine (**49**, Chart 5), is an interesting subject for T-goal guided retrosynthetic analysis. The Diels-Alder transform is an obvious candidate for the disconnection of the sole cyclohexane subunit in **49** which contains carbons *a-f*. However, direct application of this transform is obstructed by various negative factors, including the indole-containing bridge. Whenever a TGT for Diels-Alder disconnection contains such obstacles, it is advisable to invoke other ring-disconnective transforms to remove the offending rings. As indicated in Chart 5 the Fischer-indole transform can be applied directly to **49** to form tricyclic ketone **50** which is more favorable for Diels-Alder transform application. Examination of the various possible modes of transform application reveals an interesting possibility for the disconnection in which carbons *a, b, c,* and *d* originate in the diene partner. That mode requires disconnection of the *c-f* bridge to form **51**. From **51** the retron for the quinone-Diels-Alder transform can be established by the sequence shown in Chart 5 which utilizes the Beckmann rearrangement transform to generate the required *cis*-decalin system. Intermediate **52** then can be disconnected to *p*-benzoquinone and diene **53**. It is even easier to find the retrosynthetic route from **49** to **53** if other types of strategies are used concurrently with the Diels-Alder T-goal search. This point will be dealt with in a later section. A synthesis related to the pathway shown in Chart 5 has been demonstrated experimentally.[17]

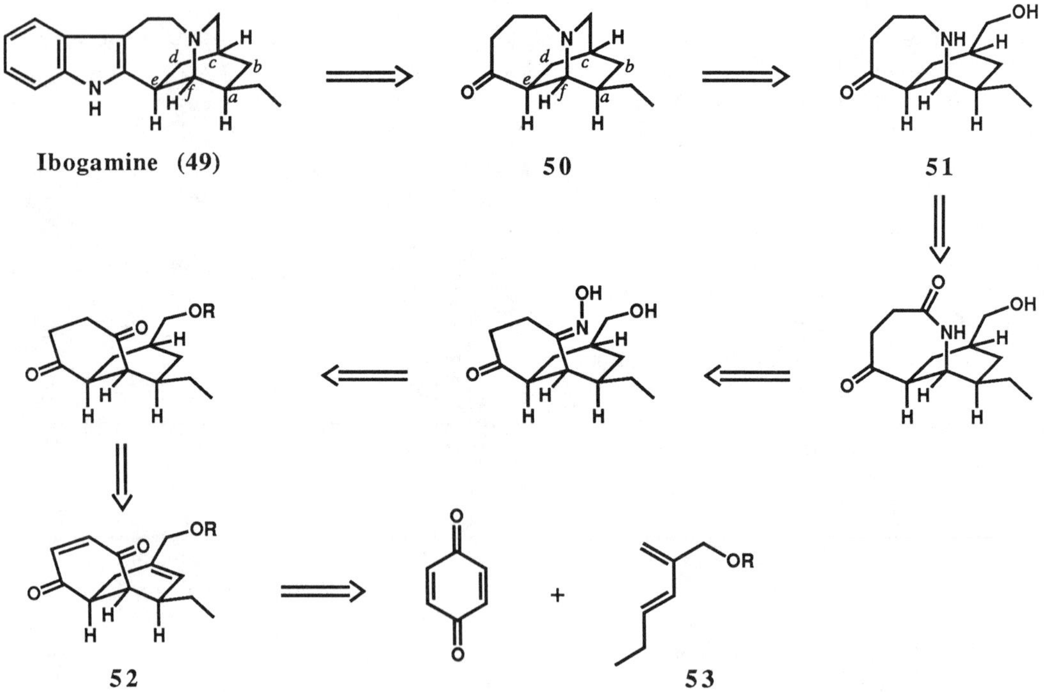

Ibogamine (49)
50
51
52
53

Chart 5

2.5 Retrosynthetic Analysis of Estrone (54)

Estrone (**54**, Chart 6) contains a full retron for the *o*-quinonemethide-Diels-Alder transform which can be directly applied to give **55**. This situation, in which the Diels-Alder transform is used early in the retrosynthetic analysis, contrasts with the case of ibogamine (above), or, for example, gibberellic acid[18] (section 6.4), and a Diels-Alder pathway is relatively easy to find and to evaluate. As indicated in Chart 6, retrosynthetic conversion of estrone to **55** produces an intermediate which is subject to further rapid simplification. This general synthetic approach has successfully been applied to estrone and various analogs.[19]

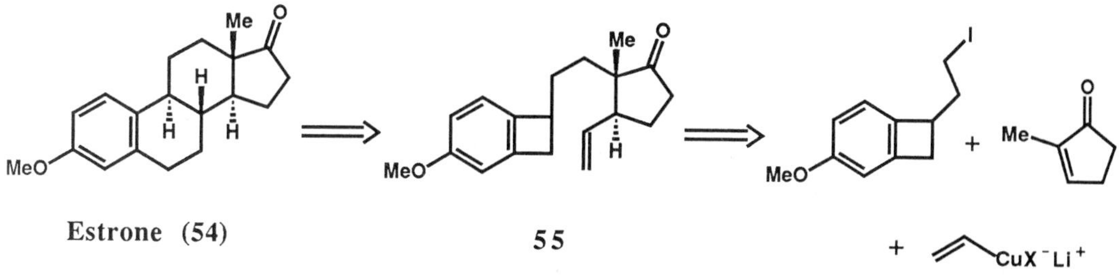

Me
O
H
H
H
MeO
Estrone (54)
55
I
Me
+
CuX⁻Li⁺

Chart 6

2.6 Retrosynthetic Analysis by Computer Under T-Goal Guidance

The derivation of synthetic pathways by means of computers, which was first demonstrated in the 1960's,[1,8,10] became possible as a result of the confluence of several developments, including (1) the conception of rigorous retrosynthetic analysis using general procedures, (2) the use of computer graphics for the communication of chemical structures to and from machine, and tabular machine representations of such structures, (3) the invention of algorithms for machine *perception* and comparison of structural information, (4) the establishment of techniques for storage and retrieval of information on chemical transforms (including retron recognition and keying), and (5) the employment of general problem-solving strategies to guide machine search. Although there are enormous differences between the problem-solving methods of an uncreative and inflexible, serial computer and those of a chemist, T-goal-driven retrosynthetic search works for machines as well as for humans. In the machine program a particular powerfully simplifying transform can be taken as a T-goal, and the appropriate substructure of a TGT molecule can be modified retrosynthetically in a systematic way to search for the most effective way(s) to establish the required retron and to apply the simplifying transform. Chart 7 outlines the retrosynthetic pathways generated by the Harvard program LHASA during a retrosynthetic search to apply the Robinson annulation transform to the TGT valeranone (**56**).[20] Three different retrosynthetic sequences were found by the machine to have a sufficiently high merit rating to be displayed to the chemist.[20] The program also detected interfering functionality (boxed groups). Functional group addition (FGA) and interchange (FGI) transforms function as subgoals which lead to the generation of the Robinson-annulation goal retron. The synthetic pathways shown in Chart 7 are both interesting and different from published syntheses of **56**.[21] The machine analysis is facilitated by the use of subservient T-subgoal strategies which include the use of chemical subroutines which are effectively

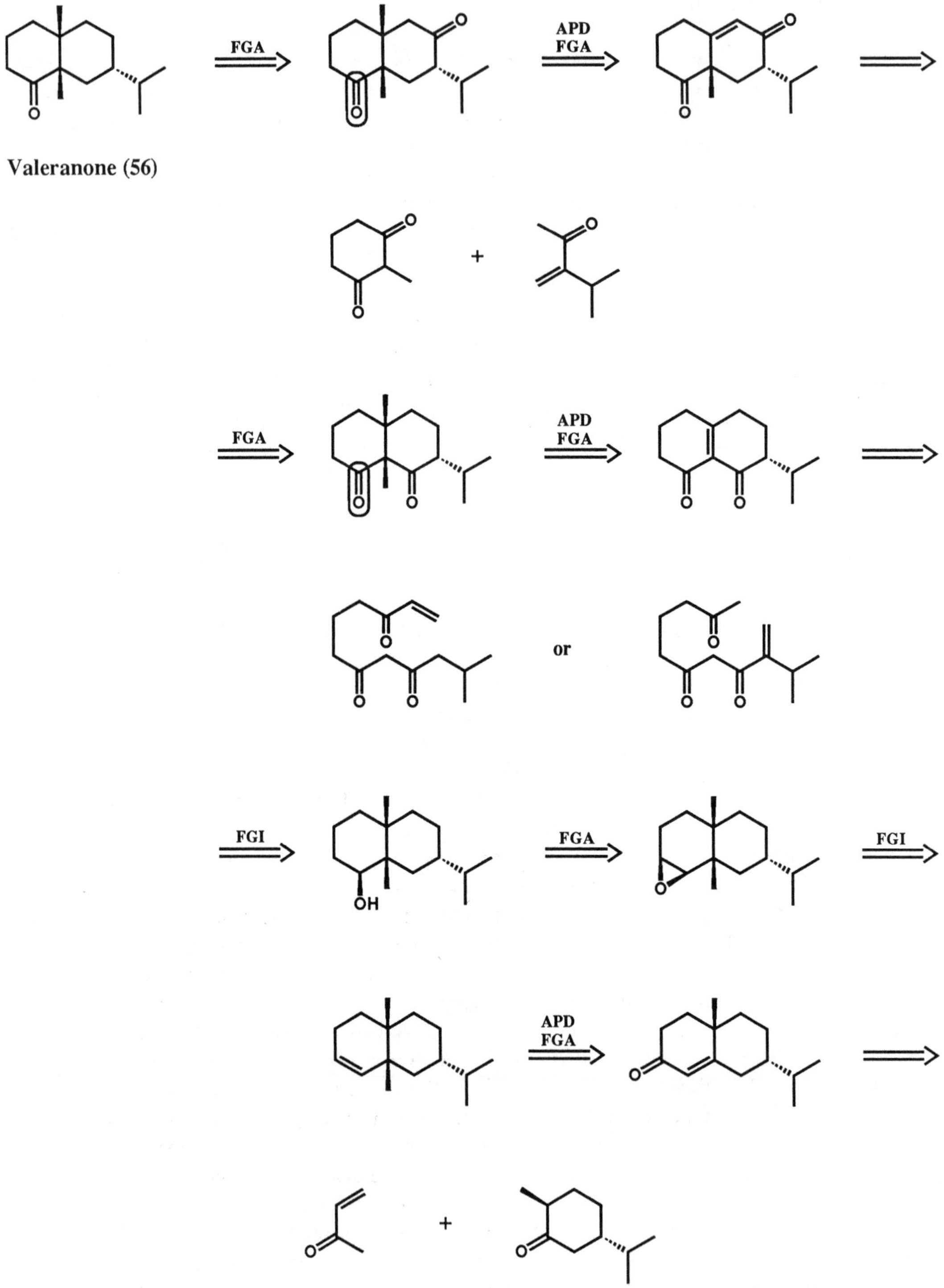
Valeranone (56)
FGA
APD
FGA
+
FGA
APD
FGA
or
FGI
OH
FGA
FGI
APD
FGA
+

Chart 7

standard combinations of transforms for removing obstacles to retron generation or establishing the α,β-enone subunit of the Robinson annulation retron. The program systematically searches out every possible mapping of the enone retron onto each 6-membered ring with the help of a general algorithm for assigning in advance relative priorities. Such machine analyses could in principle be made very powerful given the following attributes: (1) sufficiently powerful machines and substructure matching algorithms, (2) completely automatic subgoal generation from the whole universe of subgoal transforms, (3) parallel analysis by simultaneous search of two or more possible retron mappings, and (4) accurate assessment of relative merit for each retrosynthetic step. Altogether these represent a major challenge in the field of machine intelligence, but one which may someday be met.

2.7 Retrosynthetic Analysis of Squalene (57)

Squalene (**57**) (Chart 8) is important as the biogenetic precursor of steroids and triterpenoids. Its structure contains as complicating elements six trisubstituted olefinic linkages, four of which are *E*-stereocenters. Retrosynthetic analysis of **57** can be carried out under T-goal guidance by selecting transforms which are both C-C disconnective and stereocontrolled. The appropriate disconnective T-goals must contain in the retron the *E*-trisubstituted olefinic linkage. One such transform is the Claisen rearrangement, which in the synthetic direction takes various forms, for example the following:

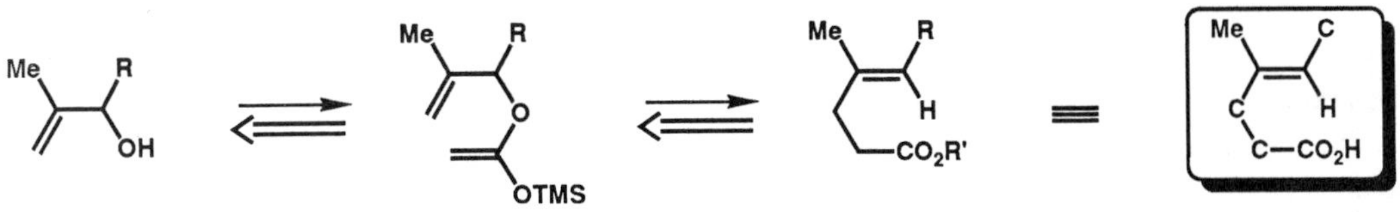

Me
R
OH
Me
R
O
OTMS
Me
R
H
CO_2R'
Me
C
C
H
C—CO_2H

Claisen Retron

The retron for the Claisen rearrangement transform (see above) is easily established by the application of a Wittig disconnection at each of the equivalent terminal double bonds of **57**

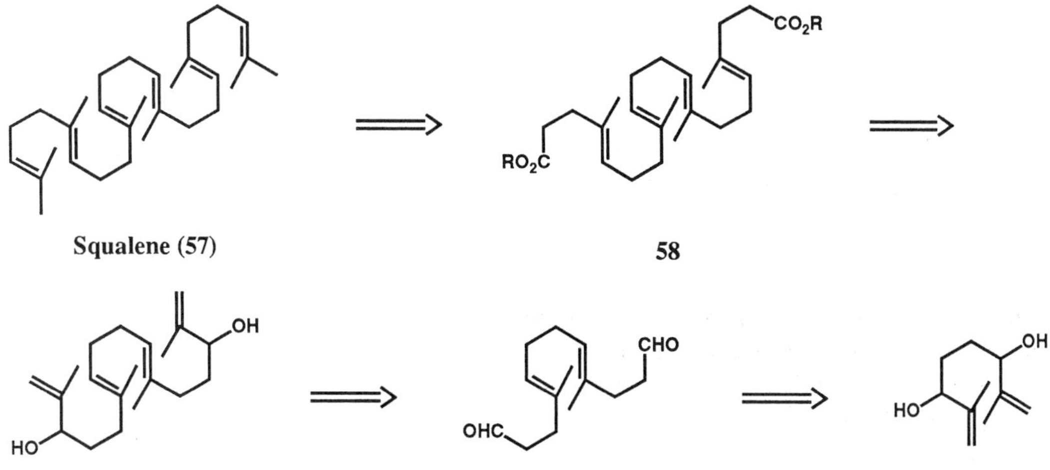

CO_2R
RO_2C
OH
HO
CHO
OHC
OH
HO

Squalene (57) **58** **59** **60** **61**

Chart 8

followed by functional group interchange, CHO ⇒ COOR, to form **58**. Application of the Claisen rearrangement transform to **58** generates **59** which can be disconnected by the organometallic-carbonyl addition transform to give **60**. A second application of the combination of CHO ⇒ COOR FGI and Claisen transform produces **61**, an easily available starting material. This type of Claisen rearrangement pathway, which can also be derived by computer analysis,[22] has been demonstrated experimentally.[23]

2.8 Enantioselective Transforms as T-Goals

In recent years a number of methods have been developed for the enantioselective generation of stereocenters by means of reactions which utilize chiral reagents, catalysts, or controller groups that are incorporated into a reactant.[24] Such processes are especially important for the synthesis of chiral starting intermediates and for the establishment of stereocenters at non-ring, hetero-ring, or remote locations. Many of the corresponding transforms can serve effectively as simplifying T-goals to guide multistep retrosynthetic search. A good example is the Sharpless oxidation process, which can be used for the synthesis of a chiral α,β-epoxycarbinol either from an achiral allylic alcohol or from certain chiral allylic alcohols with kinetic resolution. The asymmetric epoxidations (AE) without and with kinetic resolution (KR) are illustrated by the conversions, **62** → **63** and **64** → **65** using a mixture of *(R,R)*-(+)-diisopropyl tartrate [(+)-DIPT] and $Ti(OiPr)_4$ (**66**) as catalyst.[25] In the case of **63** two vicinal stereocenters are established, whereas three contiguous stereocenters are developed in **65**. The retrosynthetic search procedure to apply the Sharpless oxidation transform is directed at the generation of either the two- or three-stereocenter retron from a TGT which may have 1, 2 or 3 stereocenters on a 3-carbon path. In general the α,β-epoxycarbinol retron can be mapped onto a TGT 3-carbon

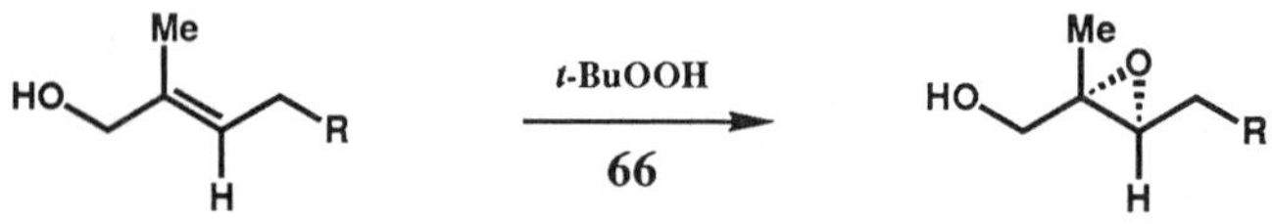

Me
HO
R
H
t-BuOOH
66
Me
O
HO
R
H

62 **63** (AE)

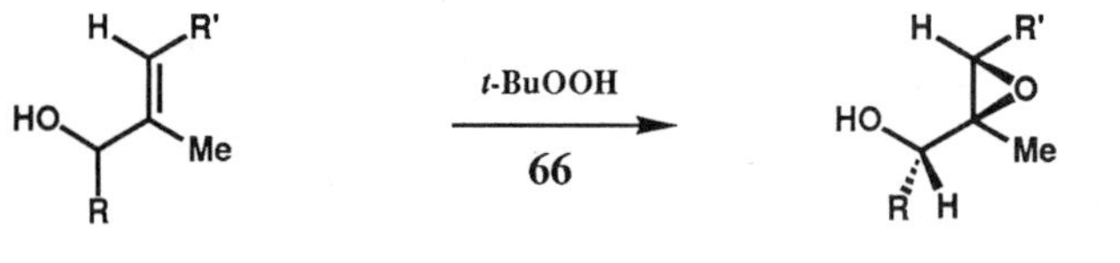

H
R'
HO
Me
R
t-BuOOH
66
H
R'
O
HO
Me
R
H

64 **65** (AE/KR)

subunit in two possible ways, both of which need to be evaluated by systematic search to effect the appropriate change using subgoal transforms. The systematic T-goal guided search method can be illustrated by TGT structure **67** (Chart 9), an intermediate for the synthesis of the polyether antibiotic X-206.[26]

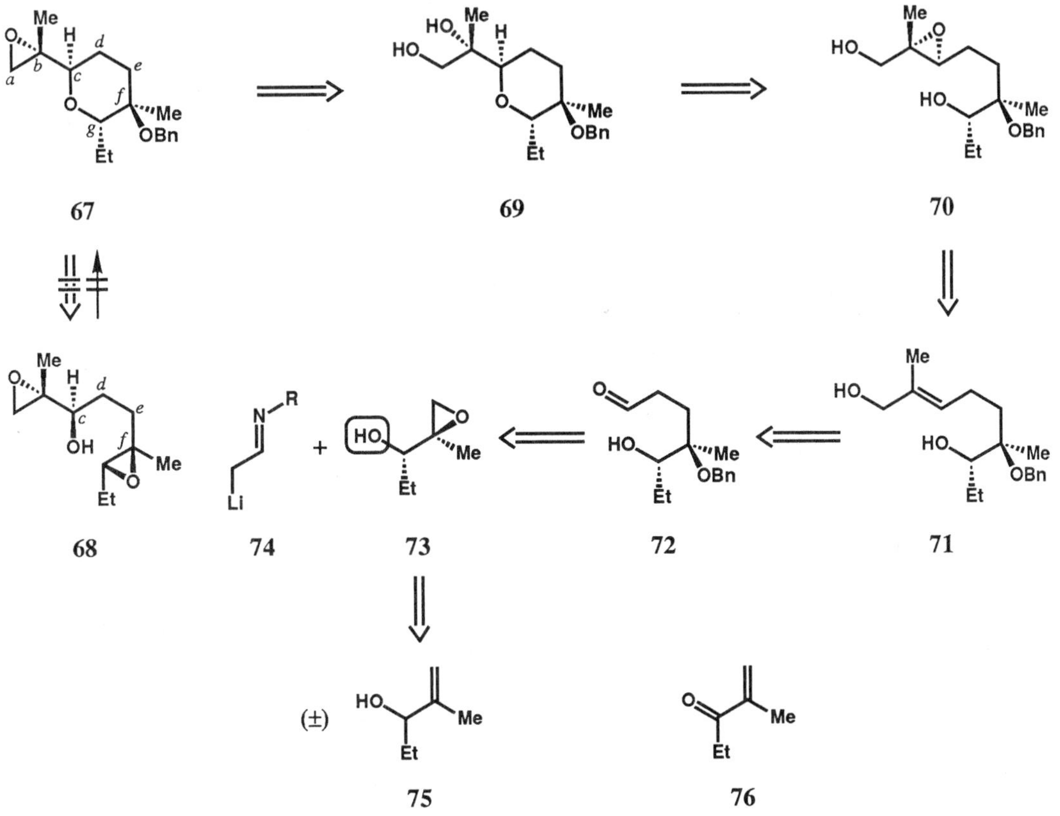

67
69
70
68
74
73
72
71
(±)
75
76

Chart 9

There are two vicinal pairs of stereocenters in **67**, one at skeletal atoms *b* and *c* and the other at atoms *f* and *g*. The former dictates mapping the 3-carbon, oxiranylcarbinol unit either on atoms *a, b* and *c* or on atoms *b, c* and *d,* whereas the latter calls for mapping onto atoms *e, f* and *g*. Implied in these possibilities is the suggestion that an eventual disconnection of bond *d-e* might be strategic. The simplest way to generate the oxiranylcarbinol retron might appear to be by the application of the epoxide-S_N2 (hydroxyl) transform which converts **67** to **68**. However, this is an invalid transform since the corresponding reaction would be disfavored relative to the alternative closure to form a tetrahydrofuran ring. There is, however, a valid 2-step process for mapping the retron on atoms *a, b* and *c* the other way around, as is shown by the sequence **67** ⇒ **69** ⇒ **70**. The synthetic conversion of **70** to **69** is clearly a favored pathway, which makes **70** a valid intermediate. The Sharpless oxidation transform converts **70** to **71**. Intermediate **71** can be converted retrosynthetically in a few steps via **72** to **73**, which contains the Sharpless oxidation retron, and a 2-carbon nucleophile such as **74** (protection/deprotection required). Application of the AE (KR) transform to **73** produces the readily available (±) alcohol **75**. Alternatively the chiral form of **75** might be obtained by enantioselective reduction of **76** and then converted by an AE process to the required **73**. The retrosynthetic T-goal guided generation of the synthetic pathway from **75** to **67** is illustrative of a general procedure which can be applied to a large number of stereoselective transforms. Algorithms suitable for use by computer have been developed for such transforms, including aldol, Sharpless oxidation, halolactonization of unsaturated acids, and others.[8,27] The synthesis of **67** from precursor **75** has been accomplished by a route which is essentially equivalent to that shown in Chart 9.

2.9 Mechanistic Transform Application

The mechanistic application of transforms constitutes another type of transform-based strategy, which is especially important when coupled with retrosynthetic goals such as the realization of certain strategic skeletal disconnections. The transforms which are suitable for mechanistic application, and which might be described as *mechanistic transforms*, generally correspond to reactions which proceed in several steps via reactive intermediates such as carbocations, anions, or free radicals. With strategic guidance, such as the breaking of a certain bond or bond set, or the removal of an obstacle to T-goal application, a specific subunit in the TGT is converted to a reactive intermediate from which the TGT would result synthetically. Then other reactive intermediates are generated mechanistically (by the exact mechanistic reverse of the *reaction* pathway) until the required structural change is effected, at which point a suitable precursor of the last reactive intermediate, i.e. an initiator for the reaction, is devised. An example of this mechanistic approach to molecular simplification is shown in Chart 10 for the

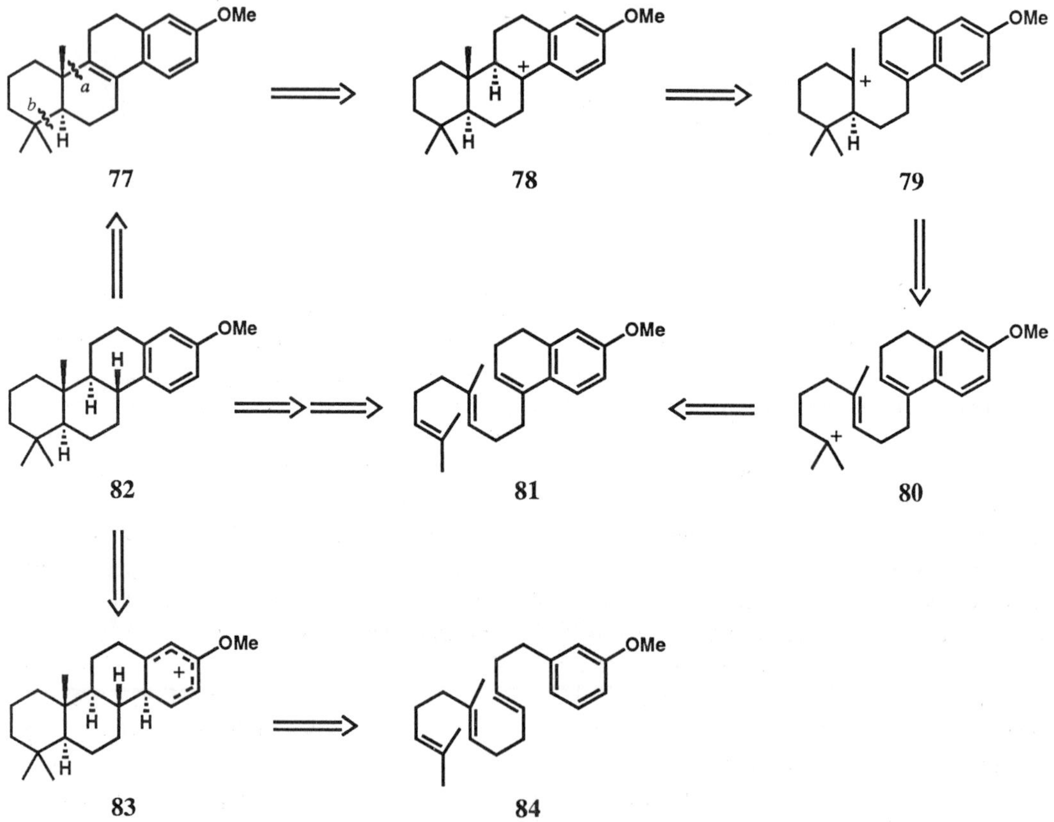
OMe
a
b
H
77
78
79
80
81
82
83
84
+

Chart 10

cation-π-cyclization transform as applied to target **77**. The retron for the cation π-cyclization transform can be defined as a carbocation with charge beta to a ring bond which is to be cleaved. Given the guidance of a topological strategy (Chapter 3) which defines as strategic the disconnection of bonds *a* and *b* in **77**, generation of cation **78** then follows. Disconnection of **78** affords **79** which can be simplified further to cation **80**. Having achieved the goal-directed topological change, it only remains to devise a suitable precursor of **80** such as **81**. It is only slightly more complicated to derive such a retrosynthetic pathway for TGT molecule **82** since this structure can be converted to **77** by the alkali metal-ammonia π-reduction transform or converted to cation **83** and sequentially disconnected to **84**.

An example of an analogous retrosynthetic process for ring disconnection via radical intermediates is outlined for target structure **85** in Chart 11. In the case of **85** the disconnection of two of the 5-membered rings and the removal of stereocenters are central to molecular simplification. One of the appropriate T-goals for structures such as **85** is the radical-π-cyclization transform, the mechanistic use of which will now be outlined. There are several versions of this transform with regard to the keying retron, one of the most common being that which follows:

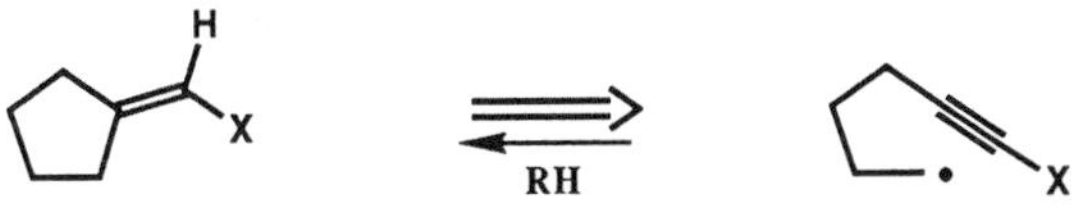
H
X
RH
X

When this type of transform is applied mechanistically to **85**, retron generation is simple, for example by the change **85** ⇒ **86**, and the sequence **86** ⇒ **90** disconnects two rings and provides an interesting synthetic pathway. Radical intermediate **88**, which is disconnected at β-CC bond *a* to produce **89**, may alternatively be disconnected at the β-CC bond *b* which leads to a different, but no less interesting, pathway via **91** to the *acyclic* precursor **92**. The analysis in Chart 11 is intended to illustrate the mechanistic transform method and its utility; it is not meant to be exhaustive or complete.

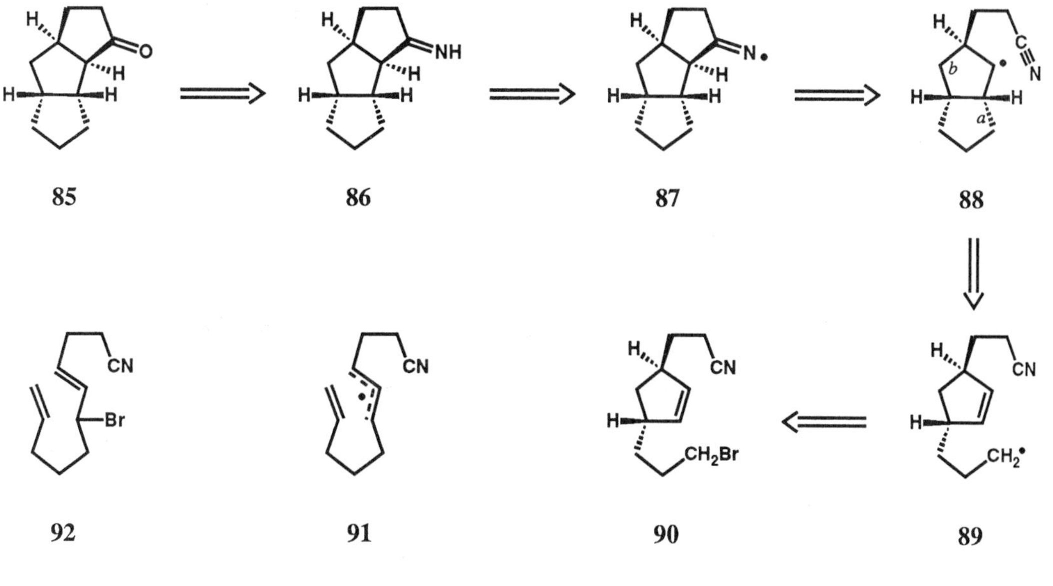
85
86
87
88
89
90
91
92
NH
N•
CN
Br
CH_2Br
$CH_2^{\bullet}$

Chart 11

There are a number of other types of uses of mechanistic transforms which can be of importance in retrosynthetic simplification. For instance if direct application of mechanistic transforms fails to produce a desired molecular change, the replacement of substructural units (usually functional groups) by synthetic equivalents may be helpful since it can allow an entirely new set of transforms to function. To take a specific example, retrosynthetic replacement of carbonyl by $HC\text{-}NO_2$ or $HC\text{-}SO_2Ph$ often provides new anionic pathways and disconnections. The use of synthetic equivalents together with the mechanistic mode of transform application can lead to novel synthetic pathways and even to the suggestion of possible new methods and processes for synthesis. For illustration, the synthesis of intermediate **90** (from Chart 11) will be considered. Two interesting and obvious synthetic equivalents of **90** are **93** and **94** (Chart 12). Intermediate **93** can be transformed mechanistically via **95** to **96** and lithium diallylcopper. Similarly **94** can be converted to potential precursors **97** and **98**. Both pathways are interesting for consideration. It is worth mentioning an important, but elementary aspect of mechanistic transform application. The retrosynthetic mechanistic changes occur in the direction of *higher energy* structures (endergonic change). It is not unusual that small or strained ring systems will be generated in retrosynthetic precursors by mechanistic transform application. Thus, the rich chemistry of strained systems can be accessed by this, and related, straightforward retrosynthetic approaches.

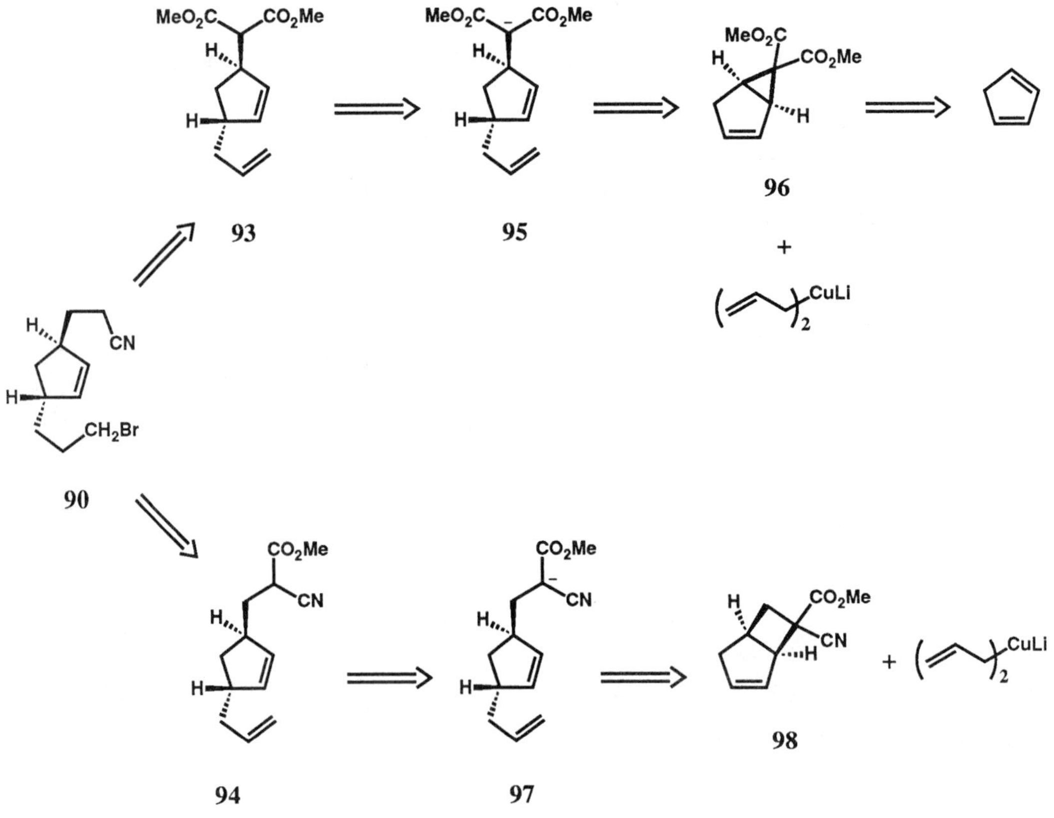

MeO2C
CO2Me
H
93
95
96
+
CuLi
2
CN
CH2Br
90
94
97
98

Chart 12

2.10 T-Goal Search Using Tactical Combinations of Transforms

It is useful to think about synthetic processes which can be used together in a specific sequence as multistep packages. Such standard reaction combinations are typified by the common synthetic sequences shown in Chart 13. In retrosynthetic analysis the corresponding transform groupings can be applied as *tactical combinations.*

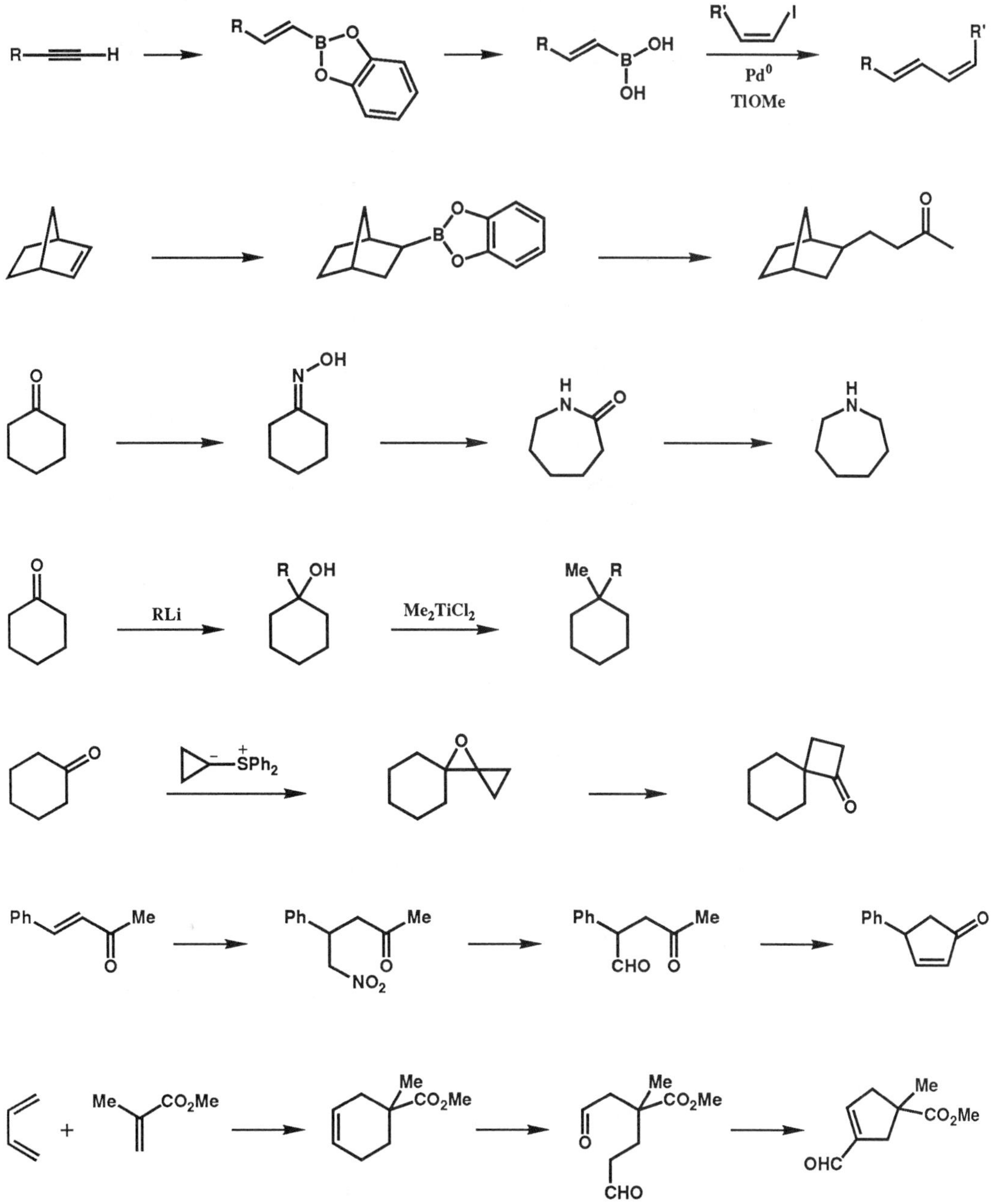

R'
I
Pd⁰
TlOMe
RLi
Me₂TiCl₂

Chart 13

The most interesting tactical combinations of transforms are those in which two or more simplifying transforms have the property of working directly in sequence. Three such simplifying transforms, for instance Q, R and S, will form a powerful tactical combination when Q acts on a TGT to produce the full retron for R which in turn functions to produce the full retron for S. There are all sorts of useful tactical combinations of transforms which differ with regard to simplifying power of the individual transforms. Not uncommonly one or more non-simplifying transforms may be used in tactical combination with one or more simplifying transforms. The recognition and use of simplifying tactical combinations, including and beyond the repertoire of standard combinations, facilitates retrosynthetic analysis. Some tactical combinations of transforms which have been used very effectively are listed in Chart 14. A dictionary of combinations such as those appearing in Charts 13 and 14 is a useful aid in retrosynthetic planning.

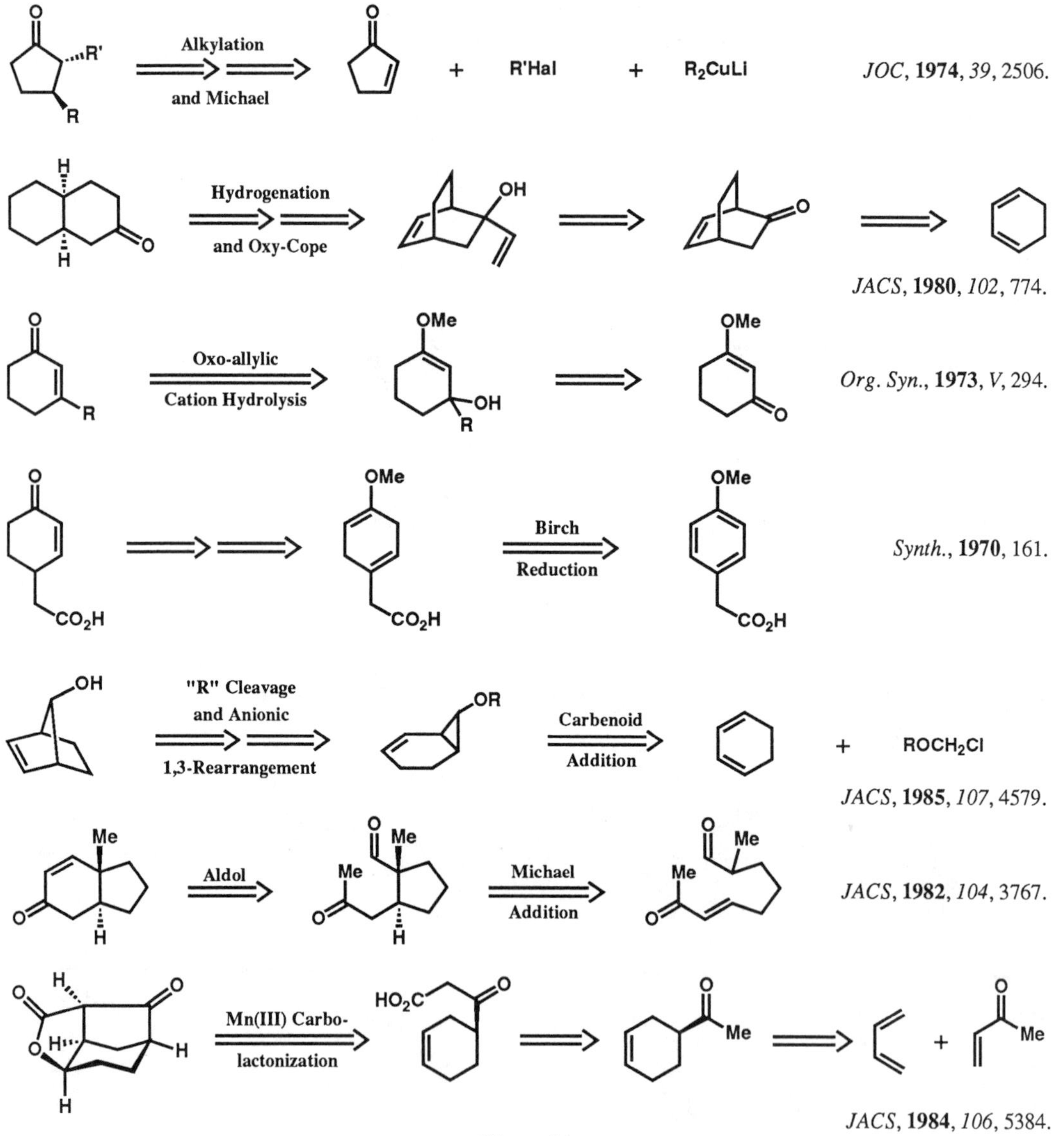

Alkylation
and Michael
R'Hal
R2CuLi
JOC, 1974, 39, 2506.
Hydrogenation
and Oxy-Cope
JACS, 1980, 102, 774.
Oxo-allylic
Cation Hydrolysis
Org. Syn., 1973, V, 294.
Birch
Reduction
Synth., 1970, 161.
"R" Cleavage
and Anionic
1,3-Rearrangement
Carbenoid
Addition
ROCH2Cl
JACS, 1985, 107, 4579.
Aldol
Michael
Addition
JACS, 1982, 104, 3767.
Mn(III) Carbo-
lactonization
JACS, 1984, 106, 5384.

Chart 14

CHAPTER THREE

Structure-Based and Topological Strategies

3.1 Structure-goal (S-goal) Strategies

The identification of a potential *starting material* (SM), *building block, retron-containing subunit,* or *initiating chiral element* can lead to a major simplification of any synthetic problem. The structure so derived, in the most general sense a structure-goal or S-goal, can then be used in conjunction with that of the target to guide a *bidirectional search*[10] (combined retrosynthetic/synthetic search), which is at once more restricted, focused and directed than a purely antithetic search. In many synthetic problems the presence of a certain type of subunit in the target molecule coupled with information on the commercial availability of compounds containing that unit suggests, more or less directly, a potential starting material for the synthesis. The structure of the TGT *p*-nitrobenzoic acid can be mapped onto simple benzenoid hydrocarbons to suggest toluene or benzene as a starting material, or *SM-goal*. One can readily derive the SM-goals shown below for the anxiolytic agent buspirone (**99**) which is simply a linear

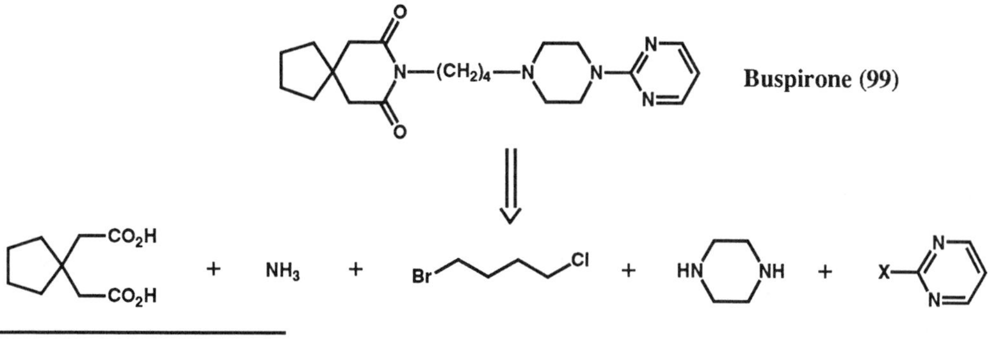

O
N—(CH2)4—N
N
N
N
O
CO2H
CO2H
+
NH3
+
Br
Cl
+
HN
NH
+
X
N
N

Buspirone (99)

References are located on pages 92-95. A glossary of terms appears on pages 96-98.

collection of readily disconnected building blocks. Most of the early syntheses of "organic" chemistry were worked out by this mapping-disconnection approach. The first commercial synthesis of cortisol (**100**) started with the available and inexpensive deoxycholic acid (**101**). Unfortunately, because of the number of structural mismatches, more than thirty steps were

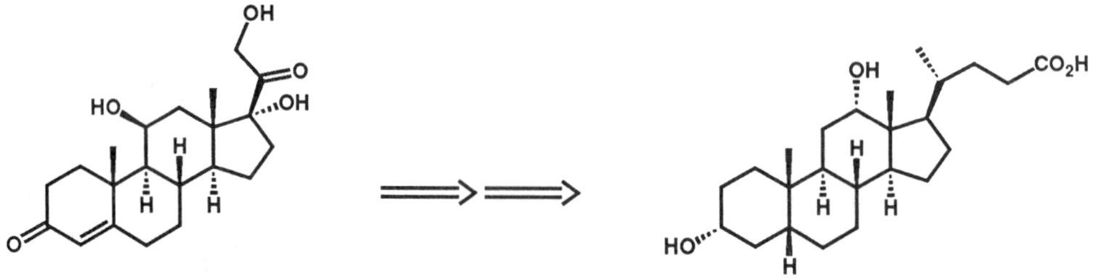

Cortisol (100) **101**

required for this synthesis.[28] Although SM-goals are the result of matching structures of various possible starting materials with the TGT structure, it can be useful to consider also *partial or approximate matches*. For example, heptalene (**102**) shows an approximate match to naphthalene and a somewhat better one with naphthalene-1,5-dicarboxylic acid which contains

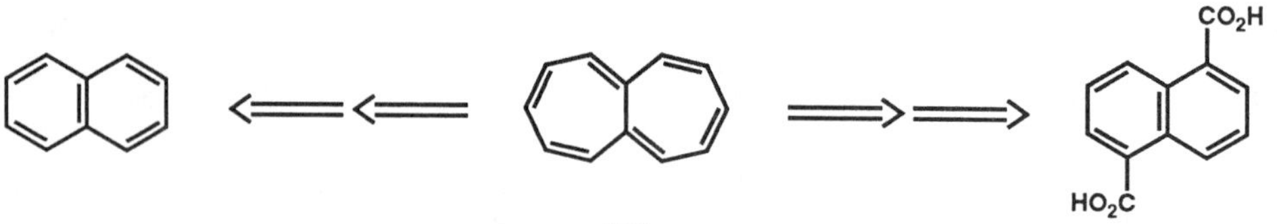

102

all of the necessary carbon atoms. Heptalene has been synthesized successfully from the latter.[29] Stork's synthesis of (±)-cedrol (**103**) from the previously known 2,2-dimethyl-3,5-di-(ethoxycarbonyl) cyclopentanone (**104**) is another example of a not-so-obvious SM-goal which proved useful.[30]

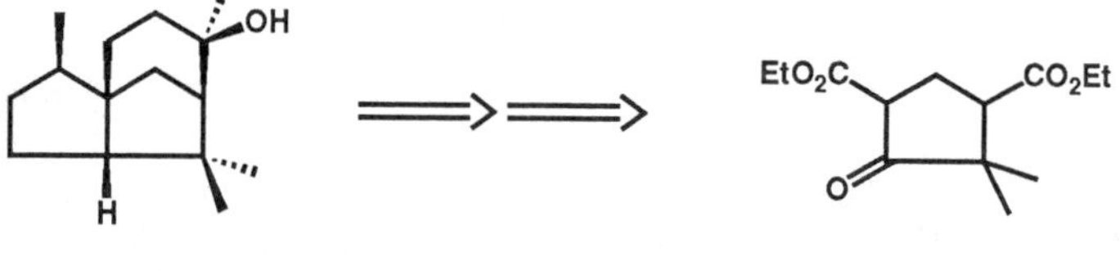

(±)-Cedrol (103) **104**

The example given above of the selection of deoxycholic acid as a SM for the synthesis of cortisol also illustrates the use of a chiral natural substance as synthetic precursor of a chiral TGT. Here the matching process involves a mapping of individual stereocenters as well as rings, functional groups, etc. The synthesis of helminthosporal (**105**) from (+)-carvone (**106**)[31] and the synthesis of picrotoxinin (**107**) from (-)-carvone (**108**)[32] amply demonstrate this approach employing terpenes as chiral SM's.

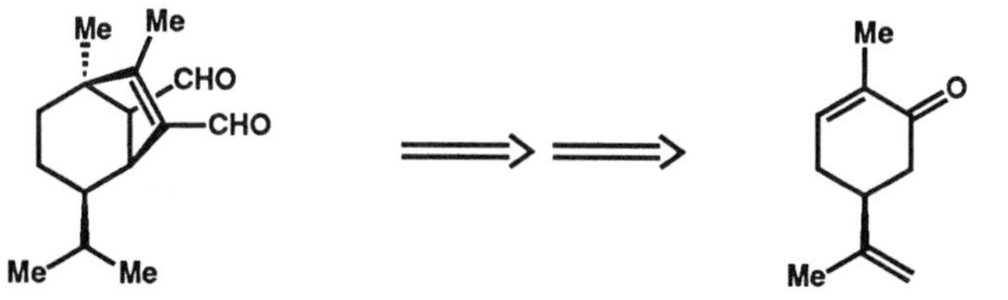
Me
Me
CHO
CHO
Me
Me
Me
O
Me

Helminthosporal (105) **(+)-Carvone (106)**

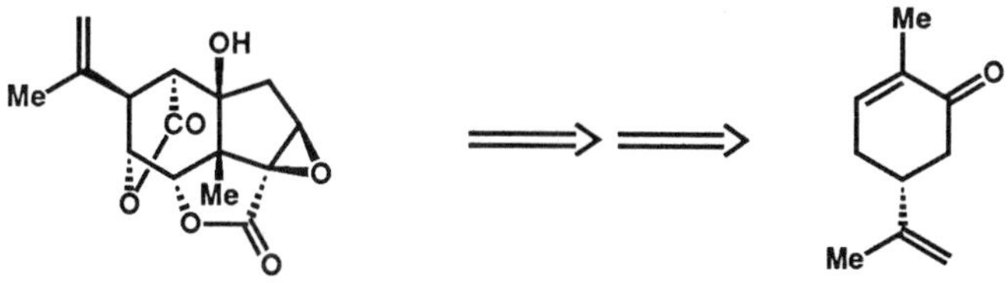
OH
Me
CO
O
Me
O
O
O
Me
O
Me

Picrotoxinin (107) **(-)-Carvone (108)**

The use of carbohydrates as SM's has greatly expanded in recent years, and many cases have been summarized in a text by Hanessian.[33] Several examples of such syntheses are indicated in Chart 15. Other commercially available chiral molecules such as α-amino acids or α-hydroxy acids have also been applied widely to the synthesis of chiral targets as illustrated by the last two cases in Chart 15.

Methodology for the enantioselective synthesis of a broad range of chiral starting materials, by both chiral catalytic and controller-directed processes, is rapidly becoming an important factor in synthesis. The varied collection of molecules which are accessible by this technology provides another type of chiral S-goal for retrosynthetic analysis.

The identification by a chemist of potentially useful S-goals entails the comparison of a target structure (or substructure) with potential SM's to ascertain not only matches and mismatches but also similarities and near matches between subunits. The process requires extensive information concerning available starting materials or building blocks and compounds that can be made from them either by literature procedures or by standard reactions. Fortunately, the organized literature of chemistry and the enormous capacity of humans for visual comparison of structures combine to render this a manageable activity. Also, it often becomes much easier to generate useful S-goals for a particular complex TGT after a phase of retrosynthetic disconnection and/or stereocenter removal. Such molecular simplification may concurrently be guided by a hypothetical S-goal which is only incompletely or roughly formulated (e.g. "any monosaccharide" or "any cyclopentane derivative").

Structural subgoals may be useful in the application of transform-based strategies. This is especially so with structurally complex retrons which can be mapped onto a target in only one or two ways. It is often possible in such cases quickly to derive the structure of a possible intermediate in a trial retrosynthetic sequence. For instance, with **109** as TGT the quinone-Diels-Alder transform is an obvious T-goal. The retron for that transform can readily be mapped

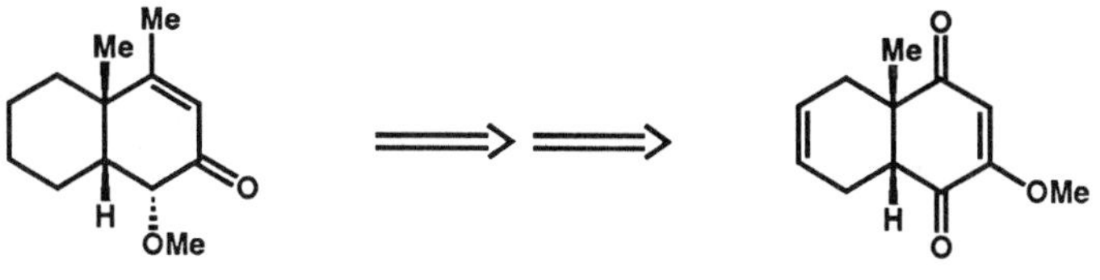
Me
Me
O
H
OMe
O
Me
OMe
H
O

109 **110**

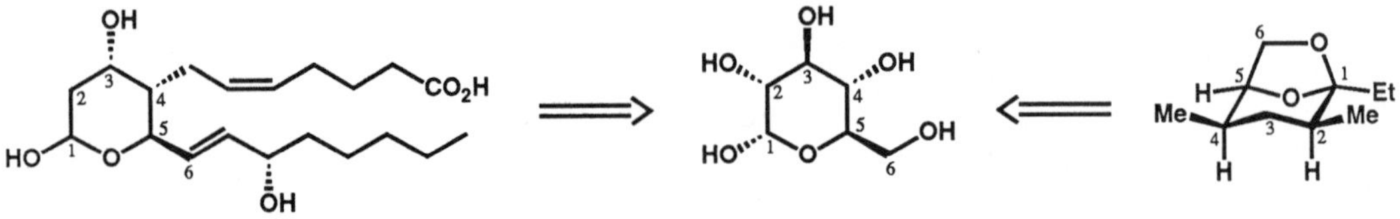

Thromboxane B_2
CJC, **1977**, *55*, 562.
TL, **1977**, 1625.

D-Glucose

(-)-α-Multistriatin
CJC, **1982**, *60*, 327.
JOC, **1982**, *47*, 941.

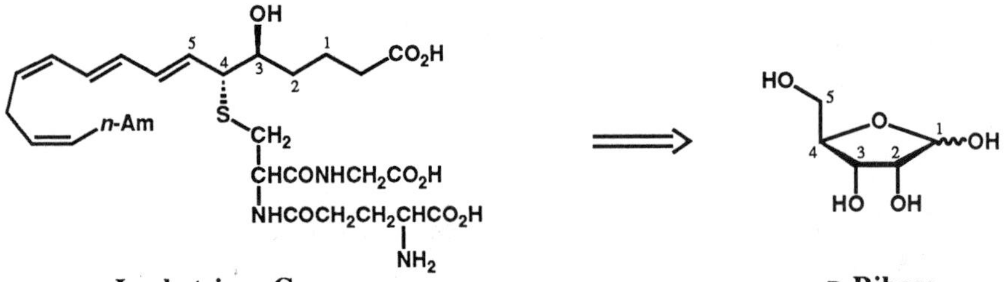

Leukotriene C_4

D-Ribose

JACS, **1980**, *102*, 1436.

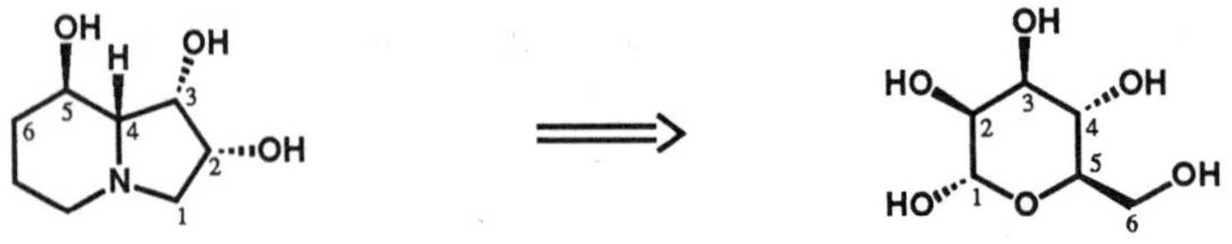

Swainsonine

D-Mannose

TL, **1984**, *25*, 1853.

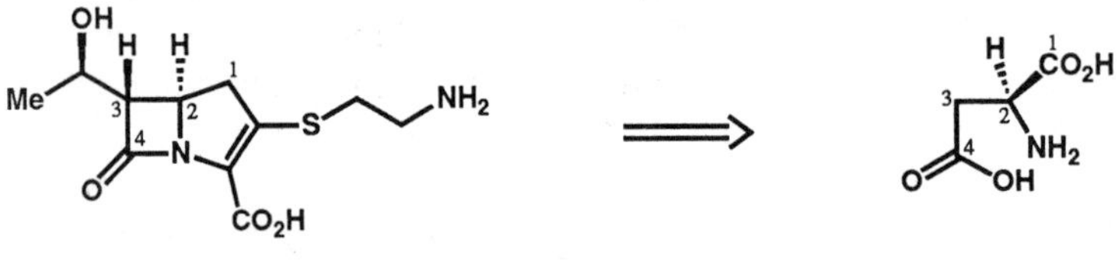

Thienamycin

L-Aspartic Acid

JACS, **1980**, *102*, 6161.

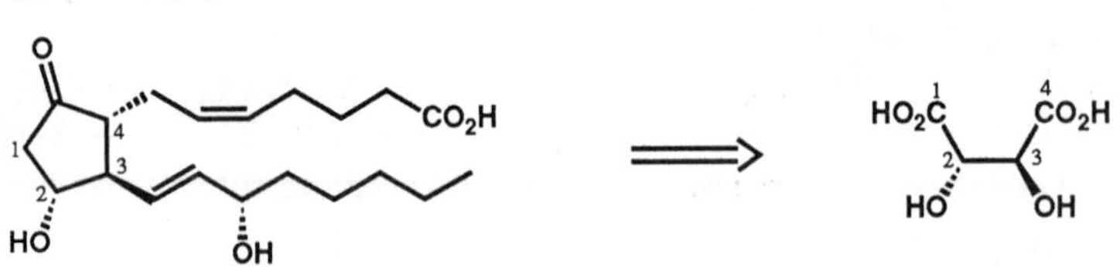

Prostaglandin E_2

(*S*,*S*)-(-)-Tartaric Acid

TL, **1986**, *27*, 2199.

Chart 15

onto the nucleus of **109** to produce **110**. Since **110** was generated by modifying TGT **109** simply to introduce a substructural retron, it can be described as a substructure goal (SS-goal) for retrosynthetic analysis. The connection between **109** and **110** can be sought by either retrosynthetic search or bidirectional search.

3.2 Topological Strategies

The existence of alternative bond paths through a molecular skeleton as a consequence of the presence of cyclic subunits gives rise to a topological complexity which is proportional to the degree of internal connectivity. Topological strategies are those aimed at the retrosynthetic reduction of connectivity.

Topological strategies guide the selection of certain bonds in a molecule as strategic for disconnection and play a major role in retrosynthetic analysis when used concurrently with other types of strategies. Strategic disconnections, those which lead most effectively to retrosynthetic simplification, may involve *non-ring* bonds, or *ring* bonds in *spiro, fused* or *bridged* ring systems. Possible strategic disconnections can be derived with the help of general criteria for each topological type. The search for strategic disconnections is conducted not only for the primary target molecule but also for precursors derived from it at lower levels of the EXTGT tree. Generally several different strategic disconnections can be identified for each target, and all have to be examined even when it is possible to assign rough priority values based on topology. This is so because the best retrosynthetic disconnections usually are those which are independently indicated by several strategies rather than just one. As topological simplification is achieved retrosynthetically, new sets of strategic disconnections will develop. Certain of these disconnections may be non-executed carryovers from preceding retrosynthetic steps which remain as strategic. The ordering of strategic disconnections is largely dictated by the mix of strategies used to guide retrosynthetic analysis.

It is also possible to identify certain bonds or certain rings in a structure as strong candidates for retrosynthetic *preservation*, i.e. not to be disconnected retrosynthetically. This bond category, the opposite of the class of bonds which are strategic for disconnection, generally includes bonds within *building block* substructures such as an *n*-alkyl group or a benzene or naphthalene ring. Many of the bonds in a molecule will be in neither the *strategic* nor the *preserved* category.

When topological strategies are used concurrently with other types of strategic guidance several benefits may result including (1) reduction of the time required to find excellent solutions; (2) discovery of especially short or convergent synthetic routes; (3) effective control of stereochemistry; (4) orientational (regiochemical) selectivity; (5) minimization of reactivity problems; and (6) facilitation of crucial chemical steps.

3.3 Acyclic Strategic Disconnections

In the case of TGT structures which are acyclic or which contain isolated rings, the disconnection of non-ring bonds must be examined to identify those disconnections which may be most effective on topological grounds. However, for such acyclic disconnections the topological factors may be overshadowed by other structural considerations. For instance, if a powerful stereosimplifying disconnective transform, such as stereospecific organometallic addition to carbonyl or aldol, can be applied directly, such a disconnection may be as good as or better than those which are suggested on a purely topological basis. In this discussion bonds in the strategic and preserved categories will be considered together.

The most useful general criteria for the assessment of acyclic strategic disconnections are summarized below. Most of these are based on the retrosynthetic preservation of building blocks and expeditious reduction of molecular size and complexity.

1. Alkyl, arylalkyl, aryl, and other building-block type groups should not be internally disconnected (preserved bonds).
2. A disconnection which produces two identical structures or two structures of approximately the same size and structural complexity is of high merit. Such disconnections may involve single or multiple bonds.
3. Bonds between carbon and various heteroatoms (e.g. O, N, S, P) which are easily generated synthetically are strategic for disconnection. Specific bonds in this category are ester, amide, imine, thioether, and acetal.
4. For aryl, heteroaryl, cycloalkyl and other building-block type rings which are pendant to the major skeleton, the most useful disconnections are generally those which produce the largest available building block, e.g. $C_6H_5CH_2CH_2CH_2CH_2$ rather than simply C_6H_5 (this is essentially a special case of rule 1, above).
5. The dissection of skeletally embedded cyclic systems (i.e. rings within chains) into molecular segments is frequently best accomplished by acyclic bond disconnection, especially when such rings are separated by one or more chain members. Such acyclic bonds may be attached directly (i.e. *exo)* to a ring, or 1, 2, or 3 bonds removed from it, depending on the type of ring which is involved.
6. Skeletal bonds directly to remote stereocenters or to stereocenters removed from functional groups by several atoms are preserved. Those between non-stereocenters or double bonds which lie on a path between stereocenters are strategic for disconnection, especially if that path has more than two members.
7. Bonds along a path of 1, 2, or 3 C atoms between a pair of functional groups can be disconnected.
8. Bonds attached to a functional group origin or 1, 2, or 3 removed can be disconnected.
9. Internal *E*- or *Z*-double bonds or double bond equivalents can be disconnected.

3.4 Ring-Bond Disconnections—Isolated Rings

In general the advantage of disconnecting isolated rings (i.e. rings which are not spiro, fused or bridged) varies greatly depending on structure. For example, "building block" rings such as cycloalkyl, aryl or heteroaryl which are singly connected to the major skeletal structure, i.e. which are essentially cyclic appendages on the major skeleton, clearly should not be disconnected. This is also the case for aryl or heteroaryl rings which are internal to the main skeleton (i.e. with two or more connections to a ring and the major skeleton). At the other extreme, however, is the disconnection of easily formed heterocyclic rings such as lactone, cyclic acetal or ketal, or lactam which may be very useful if such a ring is within the major skeleton and especially if it is centrally located. Thus, it follows that the value of disconnecting a monocyclic structural subunit depends on the nature of that ring, the number of connections to the major skeleton and the location of the ring within the molecule. Even when there is an advantage in disconnecting a ring, the precise nature of such a disconnection may be better determined by the use of ring transforms as T-goals.

Listed below are some types of disconnections which have strategic value.

1. Disconnection of non-building-block rings which are embedded in a skeleton and also centrally located, either by breaking one bond or a pair of bonds. The one-bond disconnections which are of value are: (a) bonds between C and N, O or S; and (b) bonds leading to a totally symmetrical, locally symmetrical, or linear skeleton. The bond-pair disconnections which are most effective in simplification are those which generate two structures of roughly equal complexity.

2. Disconnection of easily formed rings such as lactone, hemiketal or hemiacetal embedded in the skeleton but in a non-central location.

In general the less centrally a non-building-block carbocyclic ring is located within the skeleton, the less value will attach to its disconnection. In a structure with several isolated rings embedded within the main skeleton, the most strategic ring for disconnection topologically will be the most centrally located, especially if it is a size which allows two-bond disconnection (usually 3-, 4- or 6-membered rings).

3.5 Disconnection of Fused-Ring Systems

Strategic considerations based on topological analysis of cyclic structures become more significant as the numbers of rings and interconnections between such rings increase. Polycyclic structures in which two or more rings are fused together have long occupied an important place in synthesis, since they are common and widely distributed in nature, especially for 5- and 6-membered rings. A set of helpful topological guidelines can be formulated for the simplification of such fused cyclic networks. The degree to which such purely topological strategies contribute to the search for effective synthetic pathways for fused-ring TGT's will vary from one TGT to another since there is a major dependence on structural parameters other than connectedness, for instance ring sizes, stereorelationships between rings, functionality, and the synthetic accessibility of the precursors which are generated.

The formal procedures for analysis of alternative modes of disconnection of fused-ring systems are facilitated by the use of a *standard nomenclature* for various types of *key bonds* in such structures. A number of useful terms are illustrated in formulas **111-114**, which have been constructed arbitrarily using rings of the most common sizes, 5 and 6. Structures are shown for

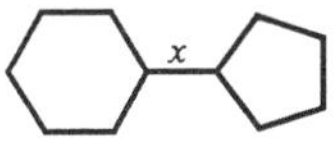
x

111

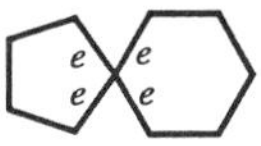
e
e
e
e

112

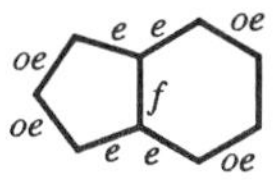
e
e
oe
oe
f
oe
e
e
oe

113

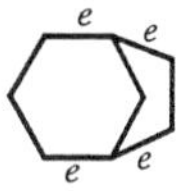
e
e
e
e

114

the following types of ring pairs: *directly joined* (**111**, no common atoms, but directly linked), *spiro* (**112**, one and only one common atom), *fused* (**113**, one and only one common bond, *f*, a *fusion* bond), and *bridged* (**114**, more than one common bond). The 5- and 6-membered rings in structures **113** and **114** are termed *primary* rings, whereas the *peripheral* rings which correspond to deletion of the fusion bond *f* in **113** (9-membered) and the bridged atom in **114** (7-membered) are termed *secondary* rings. Other bonds which are defined and indicated in the fused bicycle **113** in addition to fusion (*f*) are: *exendo* (*e*, exo to one ring and endo to another); *offexendo* (*oe*, off, or from, an exendo bond). Exendo bonds are also indicated for the *spiro* bicycle **112** and the *bridged* bicycle **114** (*exendo* bonds for primary rings). Bicycle **111** contains a bond (*x*) which is *exo* to each of the two rings. Fusion bonds may be of several types as illustrated by **115** (not directly linked), **116** (directly linked), **117** (contiguous), and **118** (cyclocontiguous).

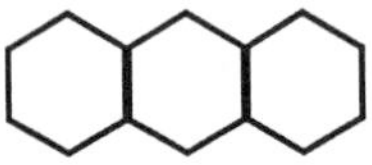

115

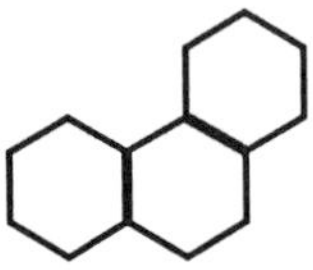

116

117

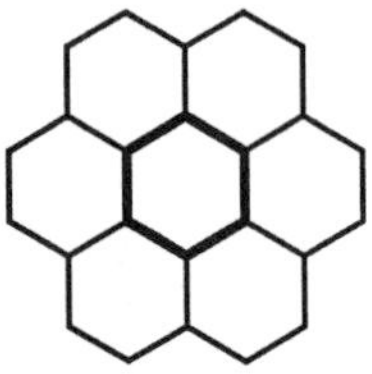

118

The most generally useful topological criteria for the effective disconnection of a network of fused rings fall into several categories. In the examples which follow most rings are arbitrarily chosen as 5- or 6-membered, and the term *ring* refers to a *primary* ring.

1. The first type of guide to the disconnection of fused rings derives from the general principle that the cleavage of a target structure into two precursors of nearly the same complexity and size is a desirable goal. Such disconnections involve the most centrally located ring(s) and the cleavage of two *cocyclic* bonds (i.e., in the same primary ring) which are exendo to a fusion bond (non-contiguous type), especially bonds involving the heteroatoms O, N and S.

2. Building-block rings (e.g. benzenoid) which are terminal are *not* disconnected; central benzenoid rings in a polycyclic system may be eligible for disconnection especially if adjacent rings are benzenoid or not readily disconnectible.

3. Disconnection of a *cocyclic pair* of bonds, especially in a central ring, may be strategic if there is a cycloaddition transform which is potentially applicable to breaking that bond pair. Such bond-pair disconnections generally involve a fusion bond (preferably non-cyclocontiguous) and a cocyclic offexendo bond (one bond away) and they result in the cleavage of *two* rings. Examples of such disconnections include *a,a'* in **119-122**. In each case *a* is the fusion bond and *a'* the offexendo bond (note, the latter may

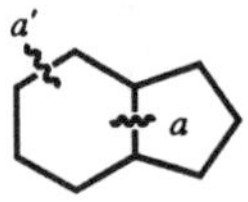

a'
a

119

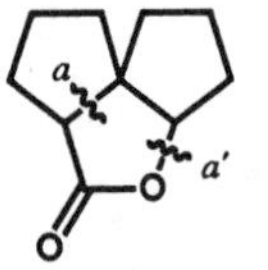

a
a'
O
O

120

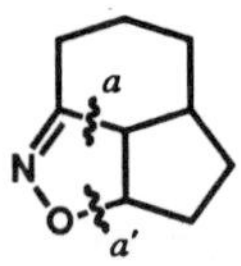

a
N
O
a'

121

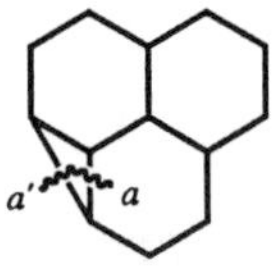

a'
a

122

concurrently be an exendo type). The ring containing the offexendo bond must be of a size (3, 4, 5, or 6-membered) to accommodate the retron for a particular cycloaddition transform. Bonds *a* and *a'* should be *cis* to one another if the bond which joins them is in a ring of size 3-7 (as in **120 - 122**). Otherwise suprafacial disconnection is not possible without prior *trans* ⇒ *cis* stereomutation. There are many syntheses which have been designed around such retrosynthetic bond-pair disconnections. One interesting example is that of carpanone (**123**) which utilized the disconnection shown.[34]

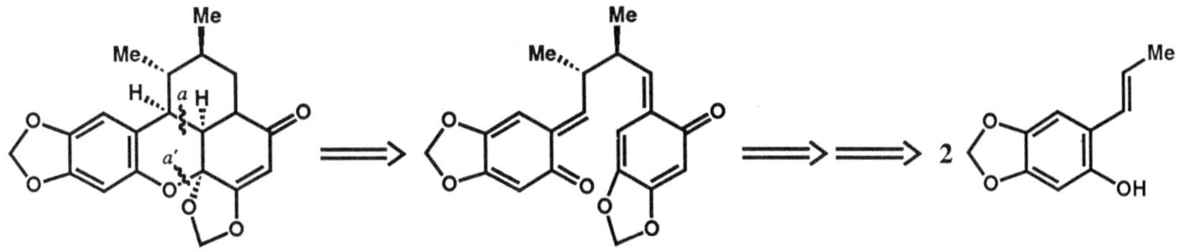
Me
Me
H
a
H
O
O
O
a'
O
O
Me
Me
O
O
O
O
O
Me
2
O
O
OH

Carpanone (123)

4. All possible [2+1] disconnections of fused 3-membered rings and [2+2] disconnections of fused 4-membered rings are strategic.
5. Fusion bonds are not candidates for strategic one-bond disconnection if such disconnection generates a ring of greater than 7 members.
6. Cocyclic vicinal exendo bonds, especially in centrally-located rings may be selected for topologically strategic disconnection. Structures **124** and **125** are provided for illustration. One reason for the effectiveness of such disconnections is that it can signal the application of various annulation transforms. The broken bonds may involve heteroatoms such as N, O and S.

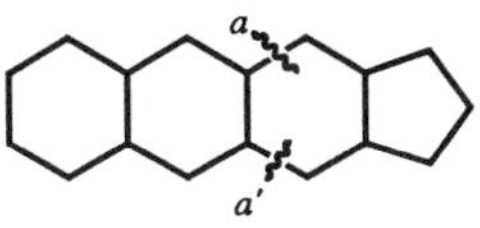
a
a'

124

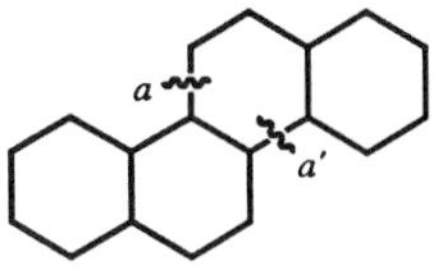
a
a'

125

7. Fused ring structures with sequences of contiguous exendo and fusion bonds in alternation may be strategic for disconnection. Such structures may be converted to linear or nearly linear precursors by cleavage of the successive exendo bonds, as shown in **126** ⇒ **127**. Disconnections such as this can serve to guide the application of polyannulation transforms (e.g. cation-π-cyclization to fused target structures).

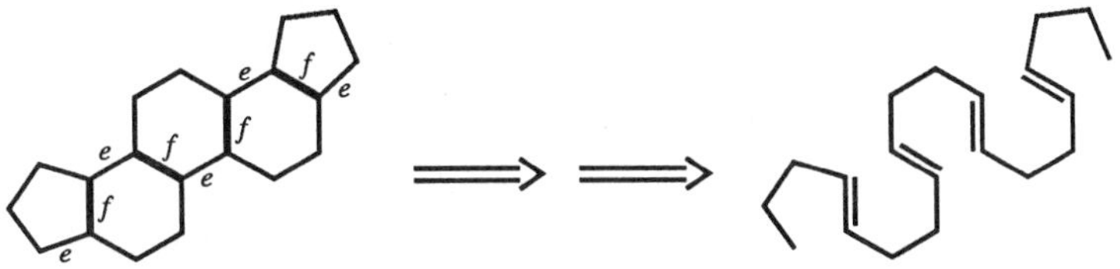
e
f
e
f
e
f
e
f
e

126 **127**

Other procedures for generating chains from polycyclic fused ring systems and for disconnecting fused rings which use simple graph theoretical approaches have been described.[35] They make use of the *dual* of the molecular graph, i.e. the figure generated by drawing a line between the centers of each fused ring pair through the corresponding fusion bond.[35]

8. As with isolated rings, individual heterorings in fused systems which are synthetic equivalents of acyclic subunits, e.g. lactone, ketal, lactam, and hemiketal, can be disconnected.

9. Disconnections which leave stereocenters on newly created appendages are not strategic unless the stereocenters can be removed with stereocontrol prior to the disconnection (see section 4.3).

3.6 Disconnection of Bridged-Ring Systems

Networks composed of bridged rings are the most topologically complex carbogenic structures. In such systems there is a great difference in the degree of retrosynthetic simplification which results from disconnection of the various ring bonds. For these reasons effective general procedures for the identification of strategic bond disconnections are more crucial than for other skeletal types. An algorithm has been developed for the perception by computer of the most strategic disconnections for bridged networks.[35] The method is also adaptable for human use; a simple version of the procedure for the selection of individual bonds (as opposed to bond-pair disconnections) is outlined here.

The individual bonds of a bridged ring system which are eligible for inclusion in the set of strategic bond disconnections are those which meet the following criteria.

1. A strategic bond must be an exendo bond within a primary (i.e. non-peripheral, or non-perimeter) ring of 4-7 members and exo to a primary ring larger than 3-membered.

2. A disconnection is not strategic if it involves a bond common to two bridged primary rings and generates a new ring having more than 7 members. Thus the disconnections shown for **128** and **129** are allowed by rules 1 and 2, whereas those shown for **130** and **131** are not.

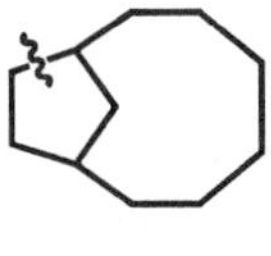

128

129

130

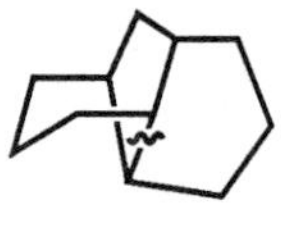

131

3. A strategic bond must be endo to (within) a ring of *maximum bridging*. Within a bridged network the ring of maximum bridging is usually that synthetically significant ring containing the greatest number of bridgehead atoms. For example the 5-membered ring in **131** is the ring of maximum bridging. Synthetically significant rings for this purpose is the set of all primary rings plus all secondary rings less than 7-membered. Bridgehead atoms are those at the end of the common path for two bridged primary rings.

4. If the disconnection of a bond found to be strategic by criteria 1-3 produces a new ring appendage bearing stereocenters, those centers should be removed if possible (by stereocontrolled transforms) before the disconnection is made.

5. Bonds within aromatic and heteroaromatic rings are not strategic.

6. Heterobonds involving O, N and S which span across or otherwise join fused, spiro or bridged rings are strategic for disconnection, whether or not in a ring of maximum bridging. This category includes bonds in cyclic functional groups such as ketal, lactone, etc.

Chart 16 shows some specific examples of strategic one-bond disconnections (bold lines) in bridged ring systems.[35] It will be seen that these favored disconnections tend to convert bridged systems to simple fused ring structures, to avoid the generation of large rings, to minimize the retrosynthetic formation of appendages, and to remove stereocenters.

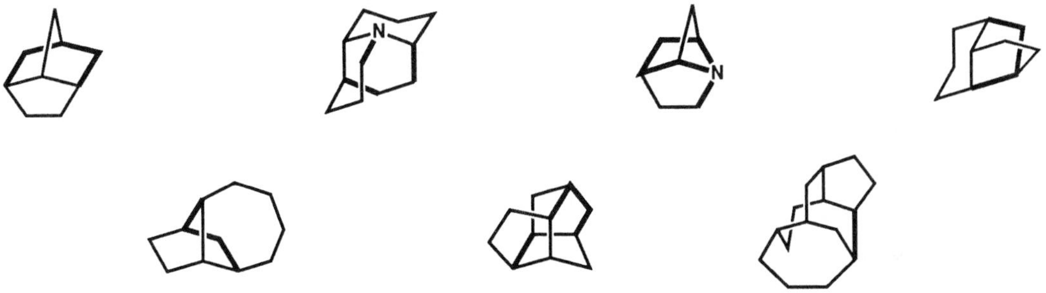

N
N

Chart 16

Strategic bond-pair disconnections in bridged ring systems owe their existence to the operability of certain intramolecular cycloadditions such as [2+1] carbenoid or nitrenoid addition to C=C, [2+2] π-π cycloaddition, [3+2] dipolar-π cycloaddition, and [4+2] Diels-Alder cycloaddition. Bond-pair disconnections at 3- and 4-membered rings containing adjacent bridgeheads are strategic since they correspond to [2+1] and [2+2] product structures. Similarly 5-membered heterocyclic rings containing at least two bridgehead atoms (1,3-relationship) can be doubly disconnected at 1,3-bonds to bridgeheads. Finally, 6-membered rings containing 1,4-related bridgehead atoms are strategic for double disconnection at one of the two-atom bridges between these bridgeheads. The disconnections shown in **132** - **134** illustrate such strategic bond-pair cleavage.

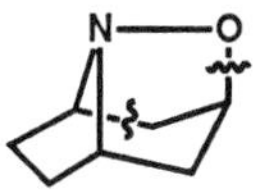

N
O

132

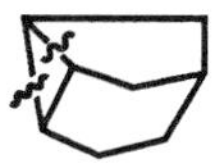

133

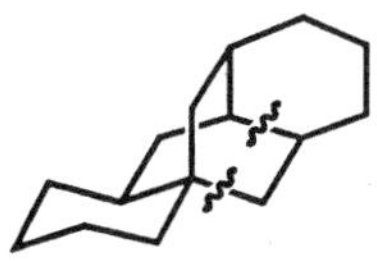

134

3.7 Disconnection of Spiro Systems

Carbocyclic spiro ring pairs generally are disconnected in two ways: (1) one-bond disconnection at an exendo bond (**135**, *a* or *a*'), or (2) bond-pair disconnection at an exendo bond and a cocyclic bond *beta* to it (**136**, *aa*' or *aa*"), (**137**, *aa*' or *aa*"). More complex networks consisting of spiro rings and also fused or bridged rings can be treated by a combination of the above guidelines and the procedures described in the foregoing sections for fused and bridged ring systems. For instance, the preferred disconnections of structure **138** are the one-bond cleavages of bond *f* or *h* and the bond-pair cleavage of bonds *a* and *g* or *d* and *e*.

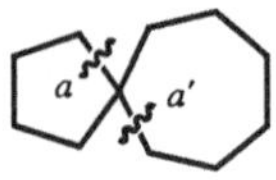
a
a'

135

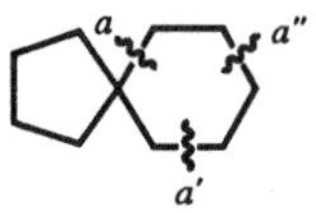
a
a''
a'

136

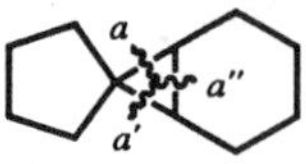
a
a''
a'

137

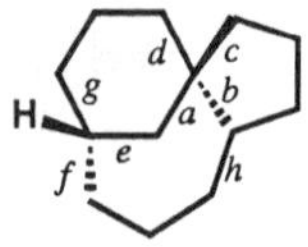
H
d
c
b
g
a
e
f
h

138

3.8 Application of Rearrangement Transforms as a Topological Strategy

Another useful topological strategy is the modification of the topology of a TGT by rearrangement to achieve any of several goals. Examples of such goals are cases (1) and (2) which follow.

(1) Antithetic conversion of a TGT by molecular rearrangement into a symmetrical precursor with the possibility for disconnection into two identical molecules. This case can be illustrated by the application of the Wittig rearrangement transform which converts **139** to **140** or the pinacol rearrangement transform which changes spiro ketone **141** into diol **142**.

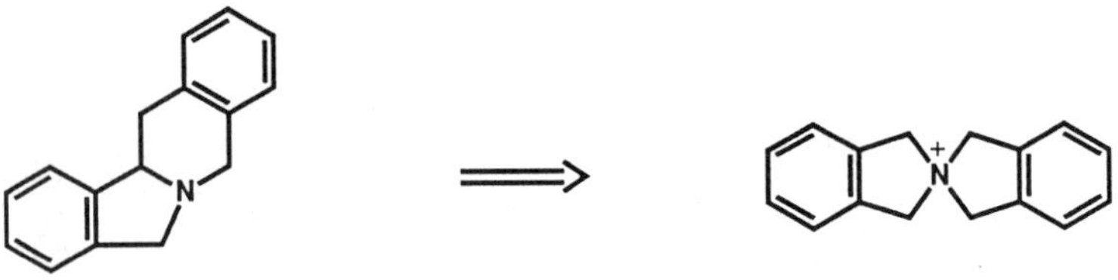
N
N
+

139 **140**

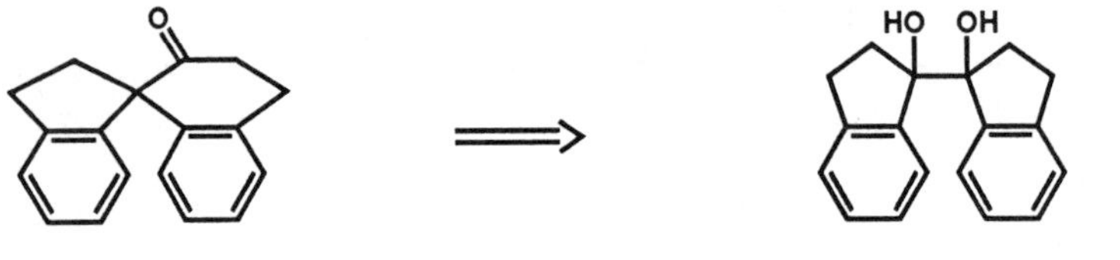
O
HO OH

141 **142**

(2) A more common and general version of case (1) is the modification of a TGT by application of a rearrangement transform to give a precursor which is readily disconnected into two or more major, but not identical, fragments. Often such rearrangements modify ring size of a TGT ring for which few disconnective transforms are available to form a new ring of a size which can be disconnected relatively easily. Some examples are presented in Chart 17.

3.9 Symmetry and Strategic Disconnections

Structural symmetry, either in a target molecule or in a subunit derived from it by antithetic dissection, can usually be exploited to reduce the length or complexity of a synthesis.

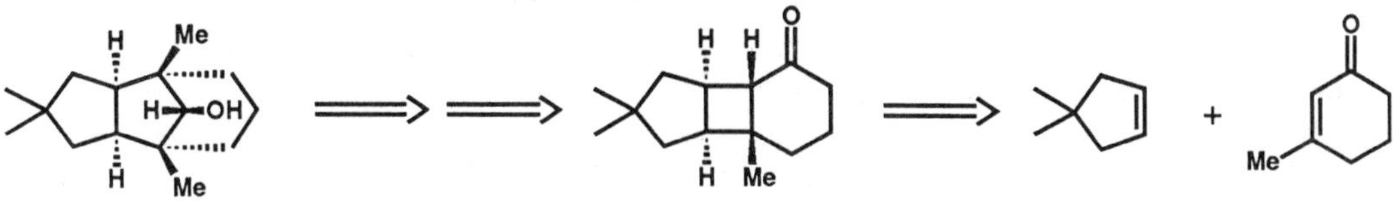

JACS, **1964**, *86*, 1652.

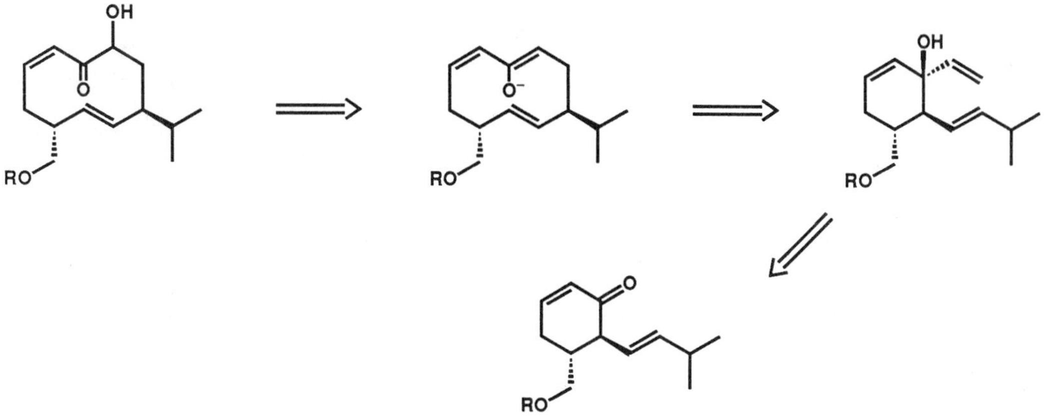

JACS, **1979**, *101*, 2493.

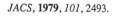

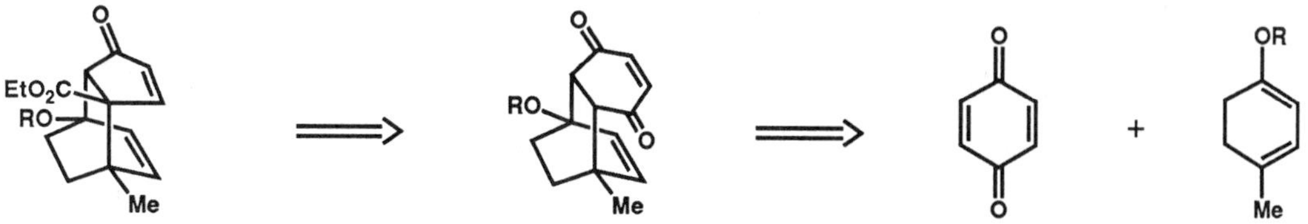

JACS, **1980**, *102*, 3654.

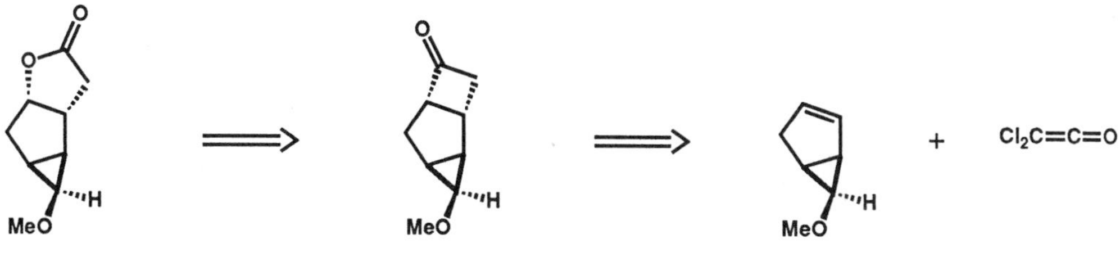

TL, **1970**, 307.

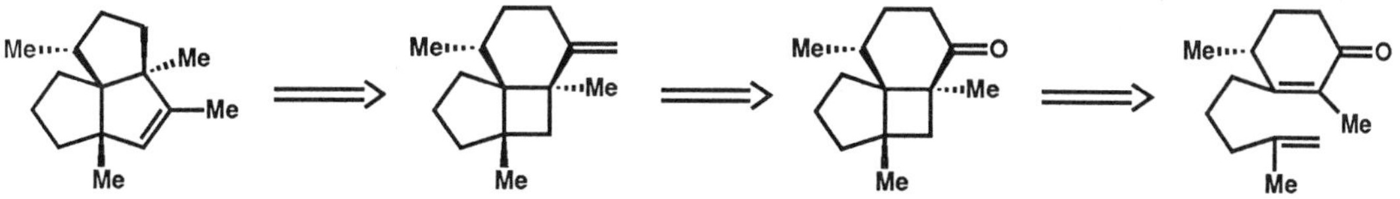

JACS, **1981**, *103*, 82.

Chart 17

For example, the symmetry of squalene (**57**), a synthesis of which was discussed in Section 2.7, can be used to advantage in two different ways. As indicated above in Section 3.3 on *acyclic strategic disconnections* (criterion 2), disconnection *a* which generates two identical fragments is clearly strategic since it leads to maximum convergence. However, double disconnections such as *b* and *b*' are also strategic since they produce three fragments which can be joined in a single

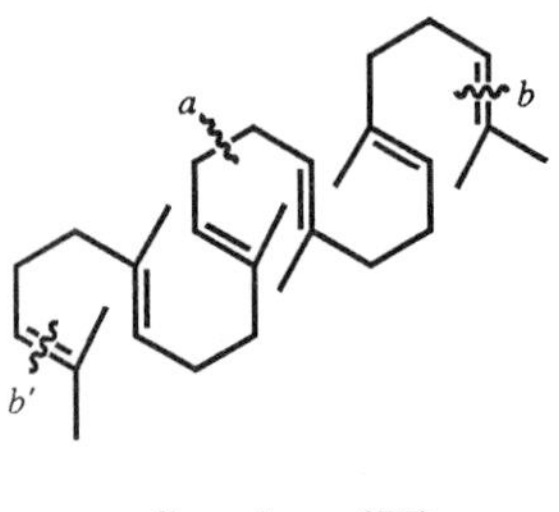

Squalene (57)

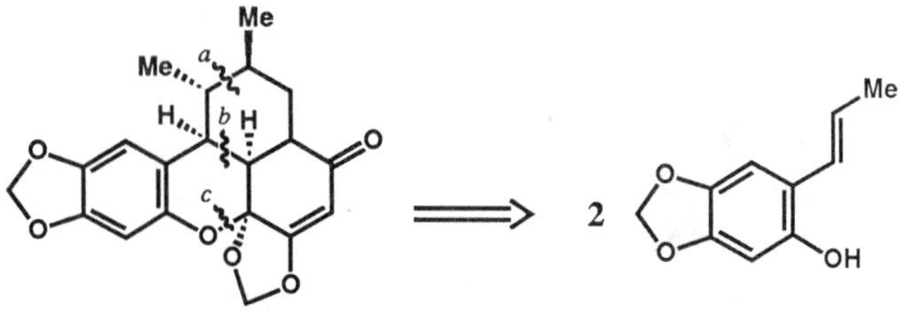

Carpanone (123)

synthetic operation and take advantage of the identical character of the terminal subunits of squalene. The fact that Wittig disconnection of two isopropylidine groups produces a symmetrical C_{24}-dialdehyde adds further to the merit of retrosynthetic cleavage at *b* and *b*'.

A molecule which is not itself symmetrical may still be cleavable to identical precursors. Examples of such cases are carpanone (**123**)[34a] and usnic acid.[34b]

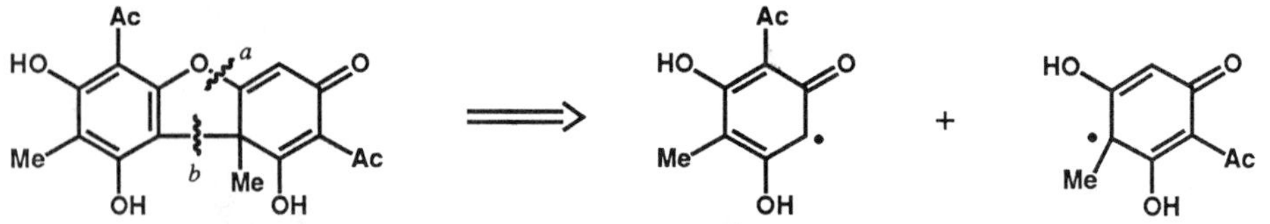

Usnic Acid

The antibiotic rifamycin provides an example of a different and more common situation in which a target structure which has no overall symmetry has imbedded within it a C_2-symmetrical or nearly symmetrical substructure that, in turn, can be converted retrosynthetically to either a C_2-symmetrical precursor or a pair of precursors available from a common intermediate.[36]

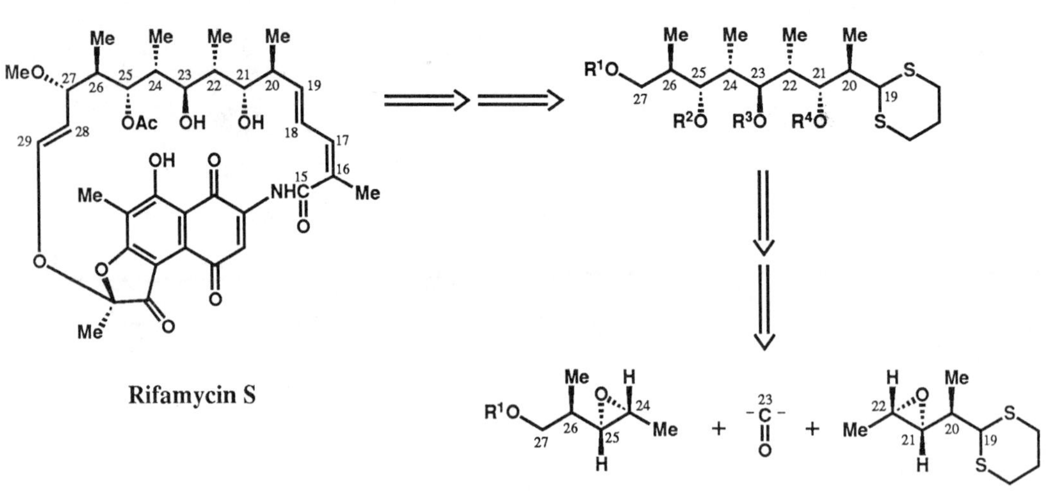

Rifamycin S

CHAPTER FOUR

Stereochemical Strategies

4.1 Stereochemical Simplification—Transform Stereoselectivity

The direct goal of stereochemical strategies is the reduction of stereochemical complexity by the retrosynthetic elimination of the stereocenters in a target molecule. The greater the number and density of stereocenters in a TGT, the more influential such strategies will be. The selective removal of stereocenters depends on the availability of stereosimplifying transforms, the establishment of the required retrons (complete with defined stereocenter relationships), and the presence of a favorable spatial environment in the precursor generated by application of such a transform. The last factor, which is of crucial importance to stereoselectivity, mandates a bidirectional approach to stereosimplification which takes into account not only the TGT but also the retrosynthetic precursor, or reaction *substrate*. Thus both retrosynthetic and synthetic analyses are considered in the discussion which follows.

The creation of a stereocenter in the synthetic direction may be controlled sterically by one or more stereocenters in a *substrate*. Such control is especially common in structures which are conformationally rigid, foremost among which are cyclohexane derivatives. Because the preexisting stereocenters are linked to the reaction site by a rigid and relatively short path in such a case, a sufficiently strong steric bias can result so that one particular diastereomeric product is favored, even for a reaction which has no intrinsic stereospecificity. Generally reactions in which a minor reactant or reagent carries no stereochemical information (e.g. $NaBH_4$ or *n*-BuLi) will be stereoselective only if the substrate (major reactant) displays a strong spatial bias. In retrosynthetic terms the corresponding transforms operate stereoselectively only under spatial control by the substrate; their retrons need not contain stereochemical information. Two examples are provided in the retrosynthetic sequence **143** ⇒ **145**. The carbonyl reduction

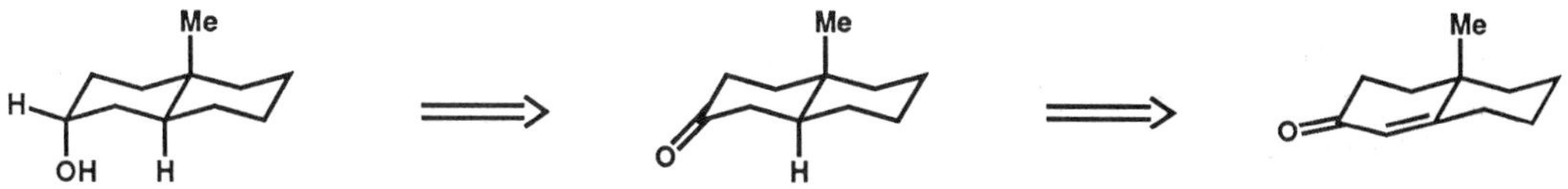
Me
H
OH
H
Me
O
H
Me
O

143 **144** **145**

References are located on pages 92-95. A glossary of terms appears on pages 96-98.

transform **143** ⇒ **144** can be applied provided that a reagent is available (e.g. $LiBHEt_3$) which renders sufficient the spatial bias within **144** favoring approach to only one π-face of the keto group. Similarly, the α,β-enone reduction transform **144** ⇒ **145** is valid for the reagent $Li\text{-}NH_3$ because of substrate spatial bias. Although the stereoselectivity of such processes depends fundamentally on mechanism and relative energetics of competing diastereogenic pathways, reaction mechanism *per se* does not induce stereoselectivity.

There are also reactions which show stereoselectivity primarily because of mechanism rather than spatial bias of substrate. For instance, the conversion of an olefin to a 1,2-diol by osmium tetroxide mechanistically is a cycloaddition process which is strictly suprafacial. The hydroxylation transform has elements of both substrate and mechanism control, as illustrated by the retrosynthetic conversion of **146** to **147**. The validity of the retrosynthetic removal of both

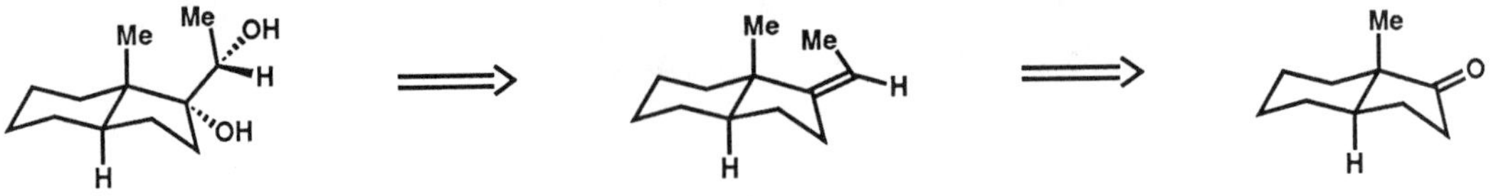

Me
Me
OH
H
OH
H
Me
Me
H
H
Me
O
H

146 **147** **148**

stereocenters depends on the intrinsic stereoselectivity of the transform (for the stereorelationship between the two centers) and also on the Z-olefinic structure and spatial bias of substrate **147**. The Z-olefinic stereocenter of **147** can be removed by application of the Wittig olefination transform to generate **148** under substrate spatial control.

There is also a category of intramolecular reactions/transforms which involves total mechanistic stereocontrol with conformationally restricted structures, for example the halolactonization transform **149** ⇒ **150** and the internal cycloaddition **151** ⇒ **152**. These transforms are relatively insensitive to steric effects and are powerfully stereosimplifying. In

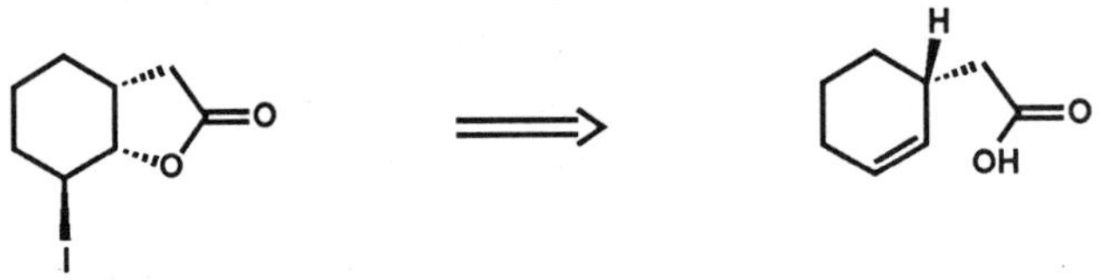

O
O
I
H
O
OH

149 **150**

H
O
H
O
H
H
O
O
CHN₂

151 **152**

general the more rigid a target structure is, the more feasible the selective removal of stereocenters will be if mechanism-controlled transforms are available.

The retrosynthetic elimination of olefinic stereocenters (*E* or *Z*) was illustrated above by the conversion **147** ⇒ **148** under substrate spatial control. It is also possible to remove olefinic stereocenters under transform mechanism control. Examples of such processes are the retrosynthetic generation of acetylenes from olefins by transforms such as *trans*-hydroalumination ($LiAlH_4$), *cis*-hydroboration (R_2BH), or *cis*-carbometallation

(Me_3Al-Cp_2ZrCl_2 or R_2CuLi). In such cases of *cis* addition, stereoselectivity originates from a dominant cycloaddition mechanism.

Stereoelectronic control also plays a role in mechanistic stereoselectivity. One such case is the very fundamental S_N2 process which proceeds rigorously with inversion of configuration at carbon. Because of that intrinsic and predictable stereoselectivity, the C-C disconnective S_N2 displacement transform is very important even though it does not directly reduce the number of stereocenters, e.g. **153** ⇒ **154**.

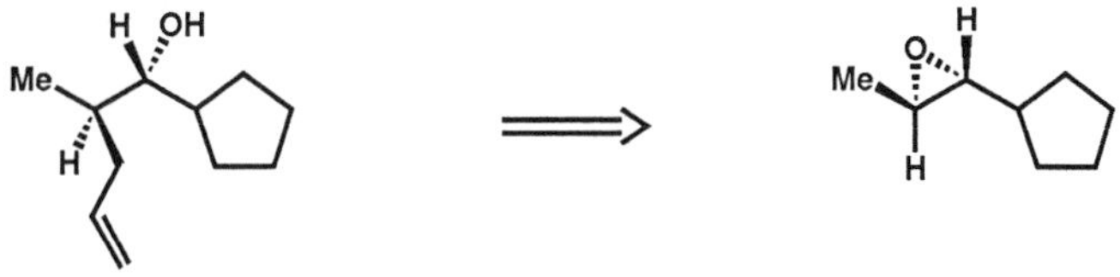
H OH
Me
H
H
Me
O
H

153 **154**

Examples were given above of stereocontrol due to substrate bias of a steric nature. Substrate bias can also result from coordinative or chelate effects. Some instances of coordinative (or chelate) substrate bias are shown retrosynthetically in Chart 18.

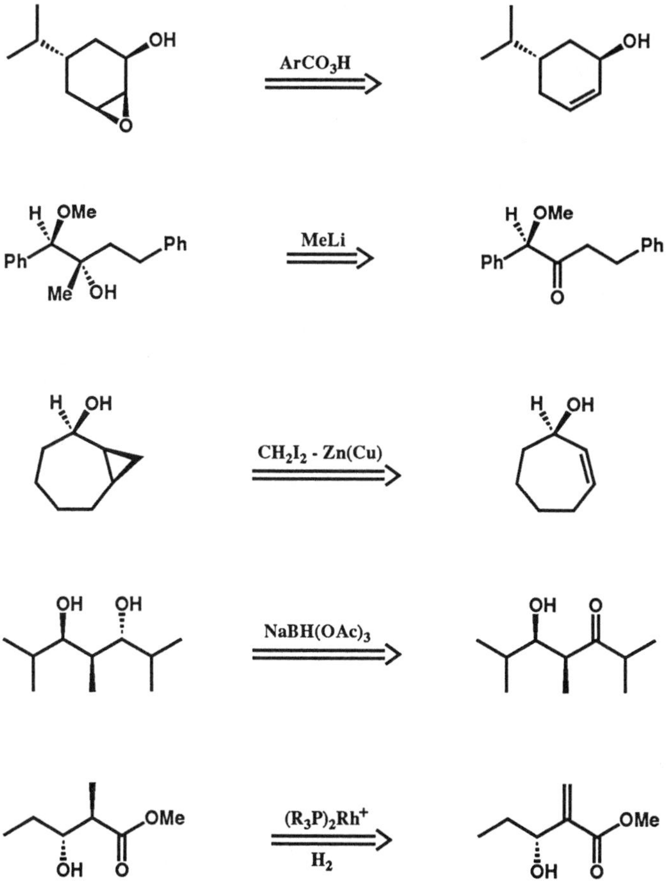
OH
O
ArCO3H
OH
H OMe
Ph
Ph
Me OH
MeLi
H OMe
Ph
Ph
O
H OH
CH2I2 - Zn(Cu)
H OH
OH OH
NaBH(OAc)3
OH O
OMe
OH O
(R3P)2Rh+
H2
OMe
OH O

Chart 18

Spatial and/or coordinative bias can be introduced into a reaction substrate by coupling it to an auxiliary or controller group, which may be achiral or chiral. The use of chiral controller groups is often used to generate enantioselectively the initial stereocenters in a multistep synthetic sequence leading to a stereochemically complex molecule. Some examples of the application of controller groups to achieve stereoselectivity are shown retrosynthetically in Chart 19.

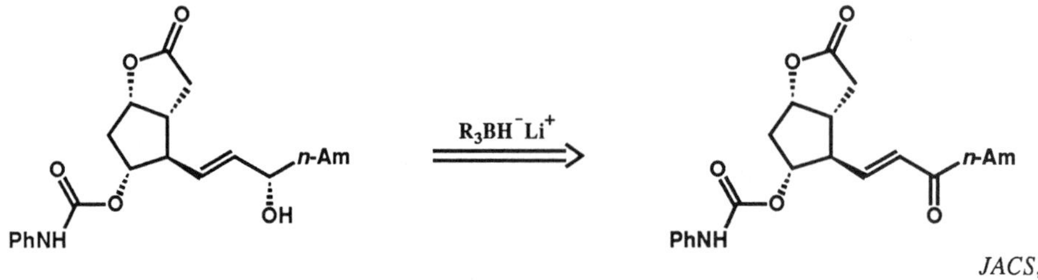

PhNH
n-Am
OH
$R_3BH^-Li^+$
PhNH
n-Am

JACS, **1972**, *94*, 8616.

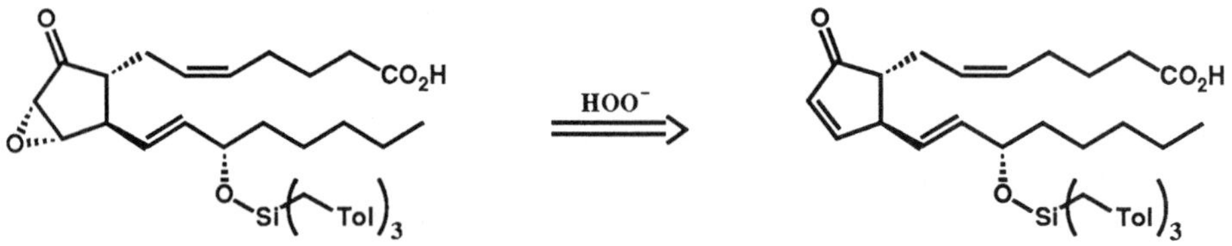

CO2H
Si
Tol
HOO⁻
CO2H
Si
Tol

JOC, **1973**, *38*, 3187.

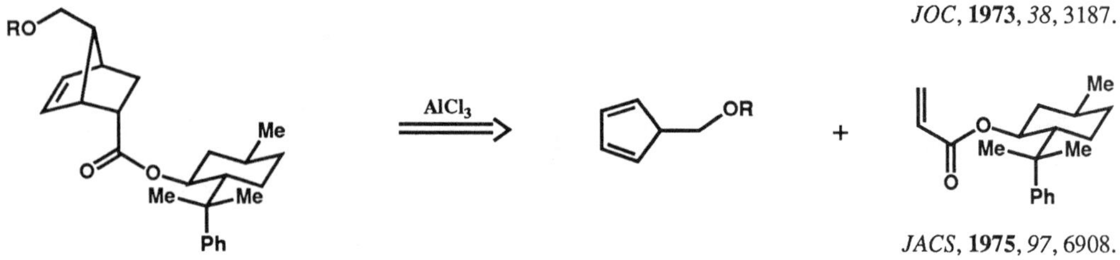

RO
Me
Me
Me
Ph
AlCl3
OR
+
Me
Me
Me
Ph

JACS, **1975**, *97*, 6908.

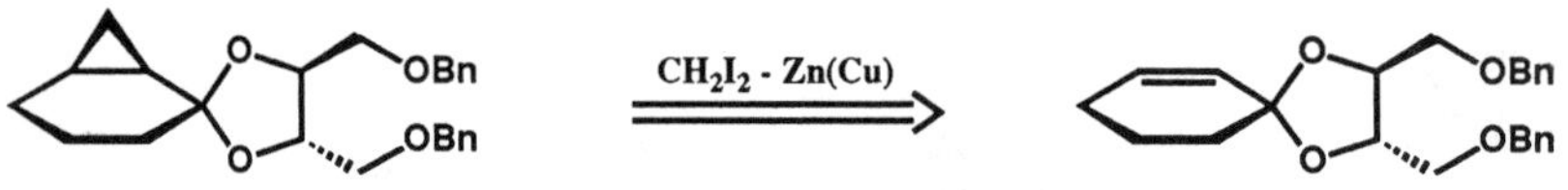

OBn
OBn
CH2I2 - Zn(Cu)
OBn
OBn

JACS, **1985**, *107*, 8254, 8256.

Me
Me
Ph
Me
Me
H
Ph
PhCu • BF3
Me
Me
Ph
Me
Me

HCA, **1981**, *64*, 2808.

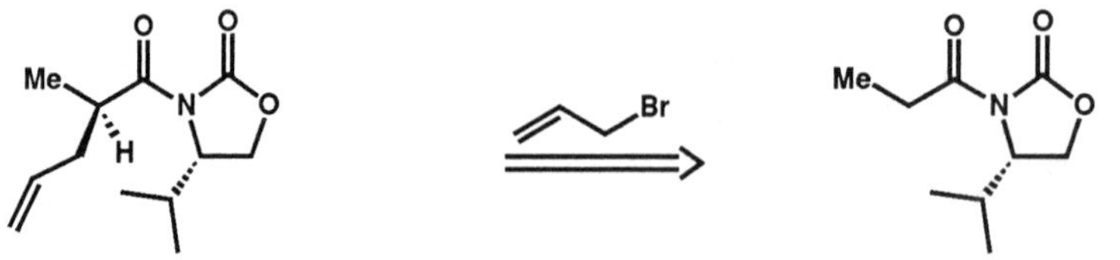

Me
H
N
O
Br
Me
N
O

JACS, **1982**, *104*, 1737.

Chart 19

There are a number of powerful synthetic reactions which join two trigonal carbons to form a CC single bond in a stereocontrolled way under proper reaction conditions. Included in this group are the aldol, Michael, Claisen rearrangement, ene and metalloallyl-carbonyl addition reactions. The corresponding transforms are powerfully stereosimplifying, especially when rendered enantioselective as well as diastereoselective by the use of chiral controller groups. Some examples are listed in Chart 20.

Enantioselective processes involving chiral catalysts or reagents can provide sufficient spatial bias and transition state organization to obviate the need for control by substrate stereochemistry. Since such reactions do not require substrate spatial control, the corresponding transforms are easier to apply antithetically. The stereochemical information in the retron is used to determine which of the enantiomeric catalysts or reagents are appropriate and the transform is finally evaluated for chemical feasibility. Of course, such transforms are powerful because of their predictability and effectiveness in removing stereocenters from a target.

In summary, modern synthetic methodology allows the stereoselective generation of one, two, or even more stereocenters in a single reaction with or without spatial control by the substrate. The application of transforms to retrosynthetic simplification of stereochemistry requires the selection of transforms on the basis of both structural and stereochemical information in the target and also validation of the corresponding synthetic processes by analysis for both chemical feasibility and stereoselectivity.

4.2 Stereochemical Complexity—Clearable Stereocenters

A practically infinite range of stereochemical situations can be found in carbogenic structures. This fact complicates both the definition of stereochemical complexity and the formulation of useful and general strategies for retrosynthetic simplification. In a rough way, stereocomplexity depends on the number of stereocenters present in a molecule and their spatial and topological locations relative to one another. For purposes of formulating retrosynthetic strategies, however, a number of other factors have to be taken into account. In stereosimplification, the elimination of stereocenters or stereorelationships between centers can be achieved by skeletal disconnection or simply by non-dissective conversion of one or more stereocenters to non-stereocenters, for example $C(sp^3) \Rightarrow C(sp^2)$ or HC=CH *(Z)* ⇒ C≡C. Since stereocontrol is of paramount importance for an effective synthetic plan, this element is crucial for selection of strategies or individual transforms, or for the evaluation of retrosynthetic steps. The power of synthetic planning depends on the ability to estimate levels of stereocontrol as well as the ability to predict whether an efficient chemical change can be realized in practice. *If a stereocenter or stereorelationship can be selectively eliminated by a transform with stereocontrol, that center can be regarded or described as clearable*, CL for short. Otherwise, that stereocenter or stereorelationship must be recognized as *non-clearable* (nCL). During retrosynthetic analysis, this distinction must be made constantly for the chiral subunits in a target structure. Whenever a stereocenter is removed retrosynthetically, its clearability must be confirmed. The use of the CL and nCL designations facilitates the discussion of retrosynthetic stereosimplifying strategies. The idea of clearability also sharpens the evaluation of stereocomplexity for purposes of retrosynthetic planning. Obviously, it is not just the absolute number of stereocenters in a structure which correlates with synthetic stereocomplexity, but the number which are not *directly or obviously clearable.*

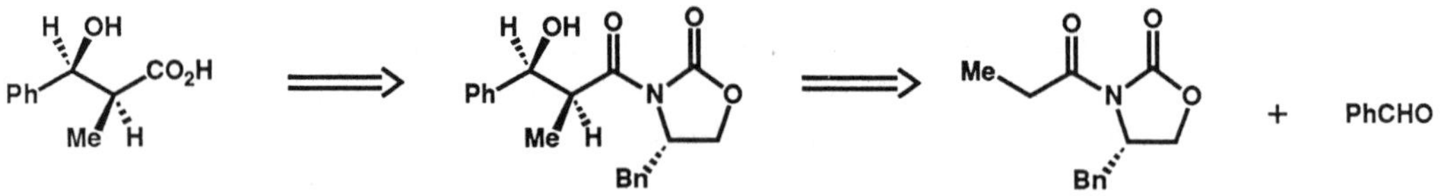
H OH
Ph
CO2H
Me H
Bn
Me
PhCHO

TL, **1982**, *23*, 807.

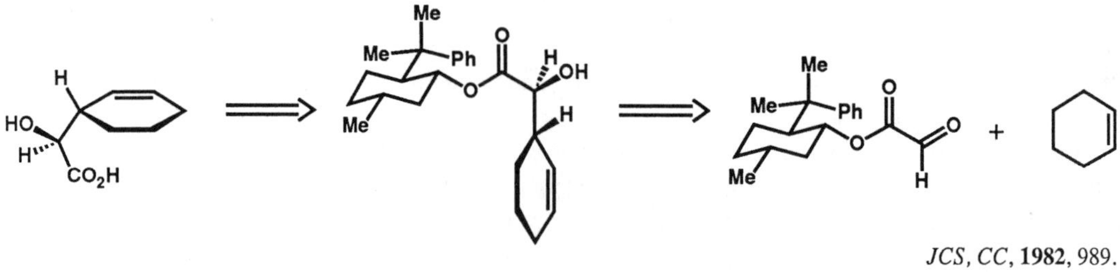
HO
H
CO2H
Me
Ph
OH
Me

JCS, CC, **1982**, 989.

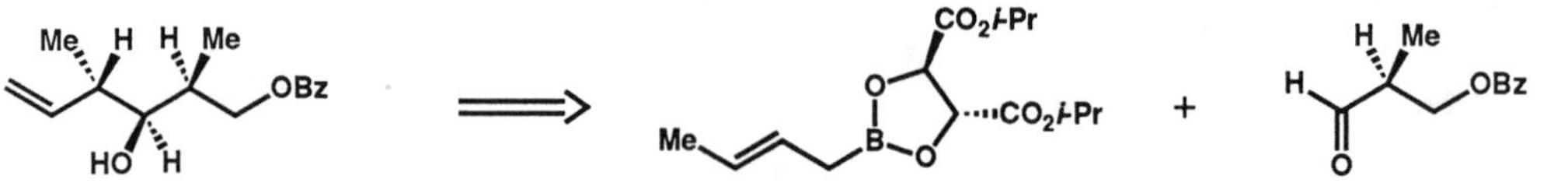
Me H H Me
OBz
HO H
CO2i-Pr
B
Me
H
O

JOC, **1987**, *52*, 316.

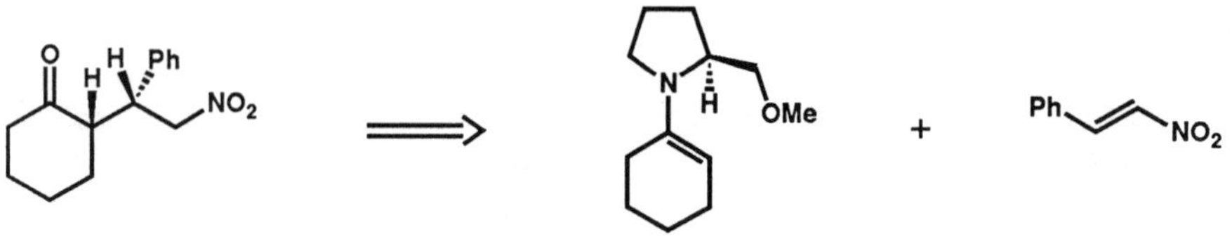
O
H H Ph
NO2
N
H
OMe
Ph

HCA, **1982**, *65*, 1637.

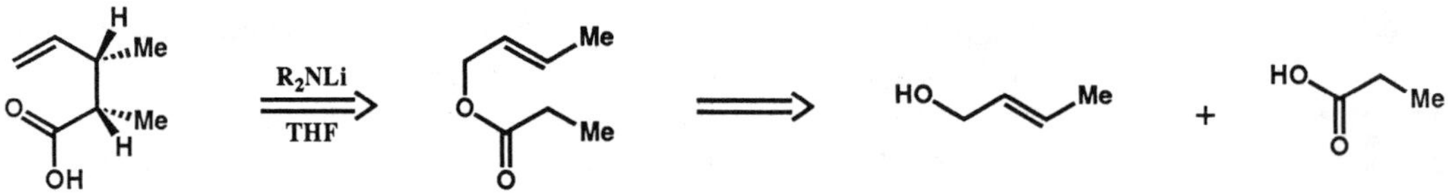
H
Me
OH
R2NLi
THF
HO
Me

JACS, **1976**, *98*, 2868.

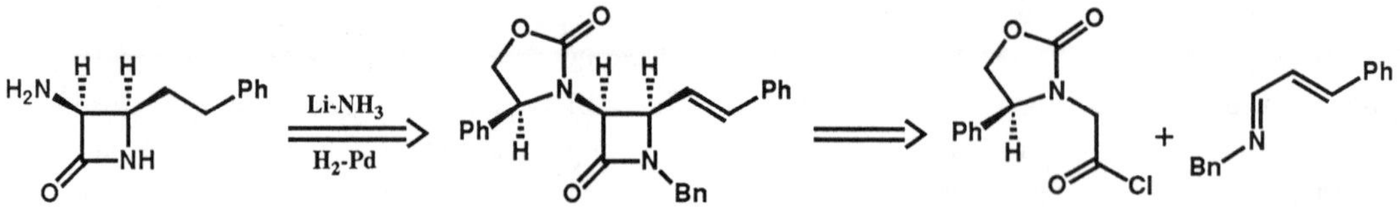
H2N
H H
Ph
NH
Li-NH3
H2-Pd
Bn
Cl
N

TL, **1985**, *26*, 3783, 3787.

Chart 20

Whether a stereocenter is clearable or non-clearable ultimately is a function of both the range of transforms available in the arsenal of synthesis (which fortunately has been increasing at a significant rate) and the structural factors in a target and its synthetic predecessor. The spatial and skeletal location of stereocenters relative to one another and the conformational flexibility of the intervening paths are two such factors. The detailed nature of the individual stereocenters is also closely connected to clearability. Individual stereocenters can be distinguished both with respect to atom and chiral type and with respect to attached substituents. For example, for tetrahedral carbon, the following variations of stereocenter type strongly affect clearability: (1) number of attached hydrogens (0 or 1); (2) type and number of functional group attachments; (3) topological status, i.e. cyclic or non-cyclic, and for the former the number of primary rings in which the stereocenter is held; (4) number of attached stereocenters (0, 1, 2, 3, or 4); (5) chemical invertibility; (6) retrosynthetic convertibility to an sp^2 carbon of a core functional group such as C=C, C=O, or C=N; and (7) dissymmetry of the immediate spatial environment. The last factor relates to the question of whether substrate spatial bias is likely to provide stereocontrol for that particular center.

Some examples of directly clearable (CL) and non-clearable (nCL) stereocenters with respect to a particular transform follow.

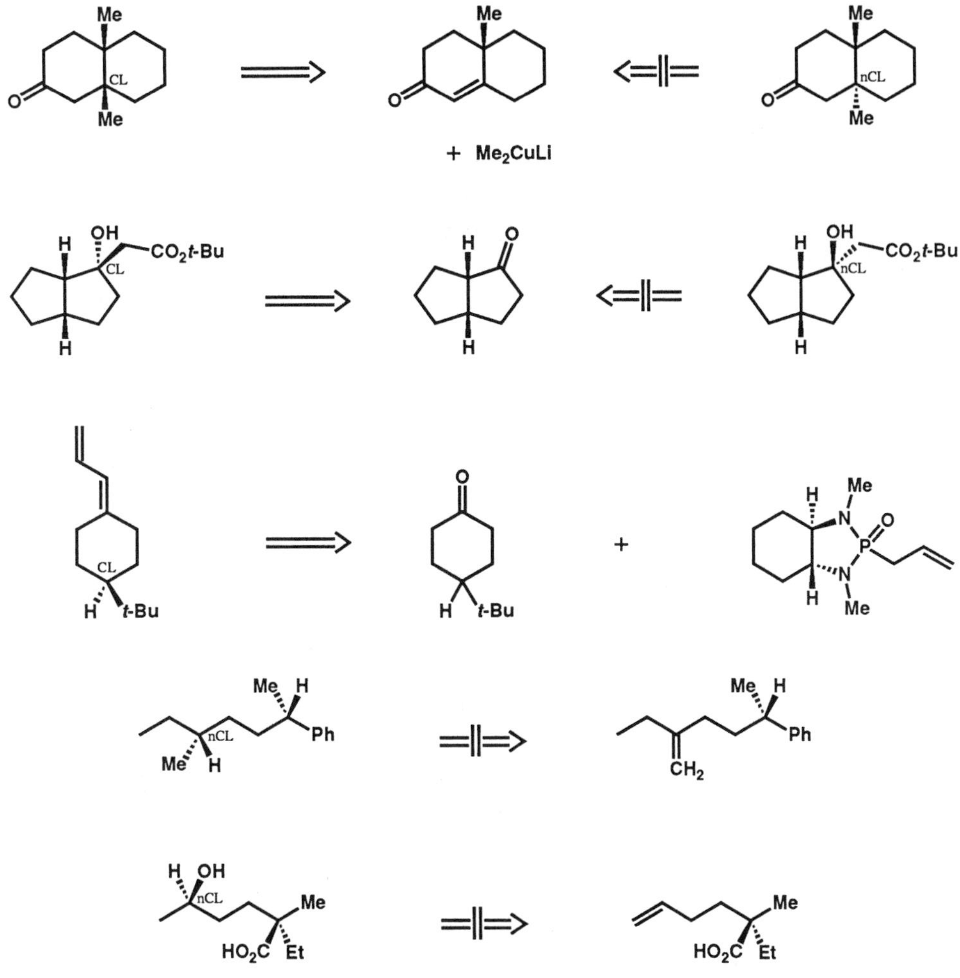
Me
CL
Me
Me
+ Me2CuLi
Me
nCL
Me
OH
CO2t-Bu
CL
nCL
t-Bu
Ph
CH2
HO2C
Et

Relationships between stereocenters vary between two extremes. On the one hand, stereocenters may interact strongly in a spatial sense if they are directly joined, proximate to one another, or part of a compact rigid-ring structure. On the other hand, two stereocenters which are remote from one another and/or flexibly connected may be so independent that one cannot be used to provide substrate spatial control for the other. Nonetheless, this latter type of stereorelationship may still be clearable if the target molecule can be disconnected to divide the two stereocenters between two precursors or if an appropriate enantioselective transform is available.

Another strategic variable enters if a target contains a chiral substructural unit which can be associated with an available chiral building block, of either natural or enantioselective synthetic origin. Obviously the presence of such subunits has a bearing on which stereocenters are strategic for clearing and which should be retained retrosynthetically. This factor can influence the selection of strategic bond disconnections or T-goals as well as the optimum timing of retrosynthetic stereocenter elimination.

When stereochemical complexity is embedded in topological complexity, such as in complex polycyclic structures, the stereochemical strategies which are most effective are those which are linked to both complexities. A decidedly different strategic approach is appropriate for topologically simpler systems.

4.3 Stereochemical Strategies—Polycyclic Systems

Stereosimplifying strategies which are appropriate to polycyclic systems deal with carbon stereocenters in different ways depending on both stereocenter type and the environment of the stereocenter in the target molecule. It is advantageous to start the discussion with stereocenters which are embedded in the ring system. In general these are either tertiary (3 °) or quaternary (4 °) and they may be in one ring or shared by two or three primary rings. They may carry functional groups, appendages, or hydrogen, and they may be linked to other stereocenters.

1. *Stereocenters in a single ring*

In polycyclic systems, one or more rings usually can be identified for retrosynthetic preservation if they qualify, for various reasons, as possible monocyclic precursors of the target. In general these *preserve rings* are not highly functionalized, and most commonly they are 5- or 6-membered. One or two stereocenters in such a ring generally qualify as "preserve" stereocenters and should not be removed retrosynthetically. Other reasons for eligibility in the preserve category for stereocenters are the following: (1) the ring or subunit within the ring matches with an available chiral starting material; (2) the ring contains one or two stereocenters which may be established by enantio-controlled processes; and (3) the ring contains one or two stereocenters bearing appendages which can cause spatial bias to control diastereoselectivity in subsequent synthetic steps leading to new stereocenters. Some examples of stereocenters which are eligible for preserve status are the starred atoms of targets **107** and **155**.

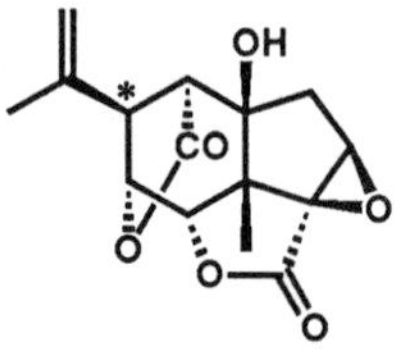
OH
CO
O
O
O
O

107

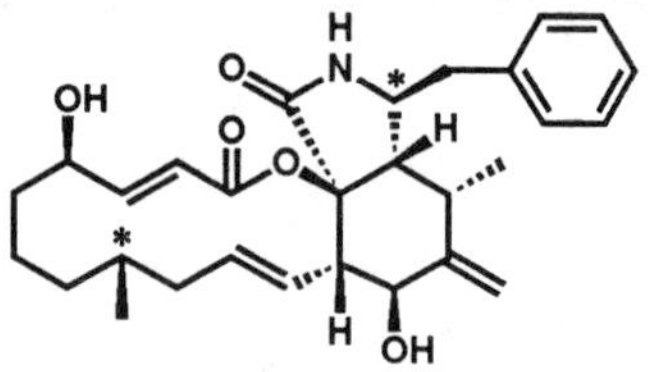
H
N
O
OH
O
O
H
H
OH

155

At the opposite end of the topological spectrum are stereocenters in terminal rings that are eligible for disconnection. Stereocenters in such rings which are elements of a retron or partial retron for a ring disconnective transform should be cleared preferentially by application of that transform. Examples **156, 157** and **158** illustrate such stereocenters (starred).

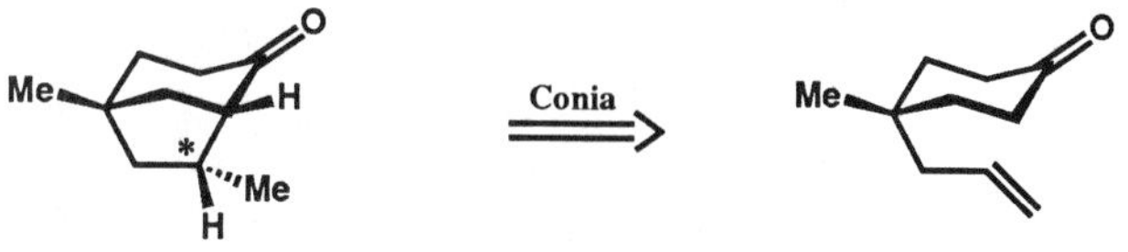
Conia

156

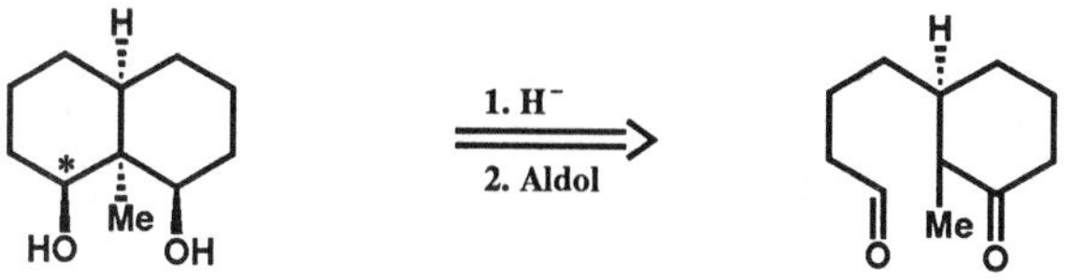
1. H⁻
2. Aldol

157

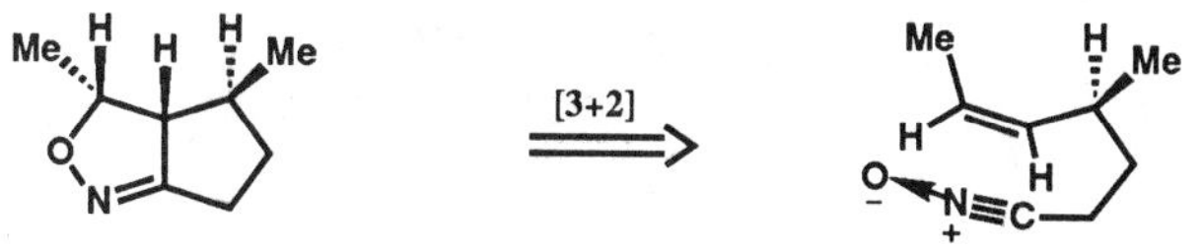
[3+2]

158

Stereocenters in a ring which can be severed by a disconnective transform, but which are not part of the retron, can be eliminated prior to disconnection if they are clearable. Removal of such stereocenters may convert a non-strategic bond into a strategic one. Stereocenters should also be cleared if that sets up the retron for a disconnective transform. Such strategic stereocenter eliminations commonly involve transforms which remove a 3° or 4° stereocenter to generate C=C (endo- or exocyclic), C=O or C=N. Elimination of two clearable vicinal stereocenters to generate C=C retrosynthetically is strategically indicated whether or not that leads to a disconnectable retron.

The following types of stereocenters are also strategic for removal, if clearable: (1) in only one ring and at an exendo strategic bond; (2) bearing a complex or functionalized appendage; (3) bearing a pair of FG's; and (4) bearing a FG or appendage in a theromodynamically less stable orientation (e.g. axial on a 6-membered ring with a *cis*-axial substituent). This list is not all inclusive but can be expanded in a general way by *extension to any clearable stereocenter which obstructs the application of a simplifying transform.*

2. *Stereocenters common to two or more rings*

Such stereocenters are commonly eliminated coincidentally with ring disconnection. In the case of fused rings with one or two hetero exendo bonds, clearance of a fusion stereocenter is strategic. Retrosynthetic clearance of two adjacent fusion stereocenters with concomitant formation of an olefinic linkage is also strategic and is often possible with heteroatom substituents. Elimination of stereocenters within two rings is also strategic if it results in a retron for a simplifying transform.

4.4 Stereochemical Strategies—Acyclic Systems

Acyclic chains, or their heterocyclic equivalents, which contain numerous stereocenters present a challenge both to synthetic methodology and to synthetic planning. Commonly, substrate steric bias is insufficient to control stereochemistry of many of the available methods for skeletal assembly. As a result, it may not be possible to assess with confidence the clearable or non-clearable status of individual stereocenters. Reliance on intrinsically diastereo- and enantioselective transforms must be heavy. As described in Section 4.1 the stereoselectivity of such transforms rests on the synthetic use of chiral controller groups and reagents as well as on mechanism- or coordination-based stereoselectivity. Alternatively, the identification of a set of nCL stereocenters with a potential chiral starting material may provide another approach to stereosimplification. Some of the most useful general guides for the reduction of stereochemical complexity in acyclic systems are listed below.

1. Reduce stereochemical complexity and molecular size concurrently by applying diastereoselective or enantioselective transforms which are also disconnective.
2. Apply stereoselective transforms to clear stereocenters by removal or interchange of functional groups with the establishment of the retron for a disconnective transform, especially with retrosynthetic generation of the core groups C=C, C=O or C=N.
3. Apply stereoselective, FG-transposing transforms which convert a 1,4- or other 1,n-stereorelationship into a 1,2-stereorelationship and allow the operation of disconnective transforms.
4. Apply disconnective transforms which do not alter stereocenters but which separate stereocenters by molecular cleavage.
5. Identify segments of nCL stereocenters which in some degree map onto available chiral predecessors in order to generate a chiral S-goal.
6. Identify segments containing nCL stereocenters which can be dissected to give precursors of C_2-symmetry.[36]
7. Apply stereoselective transforms to reduce the number of reactive functional groups, especially those of sufficient reactivity to cause interference with stereocontrolled C-C disconnective transforms.
8. Use connective transforms to convert an achiral chain segment of nCL centers into a conformationally fixed ring containing CL stereocenters (Section 5.7).

The following sequences of transforms can be derived with the help of strategic guides 1-8 listed above.

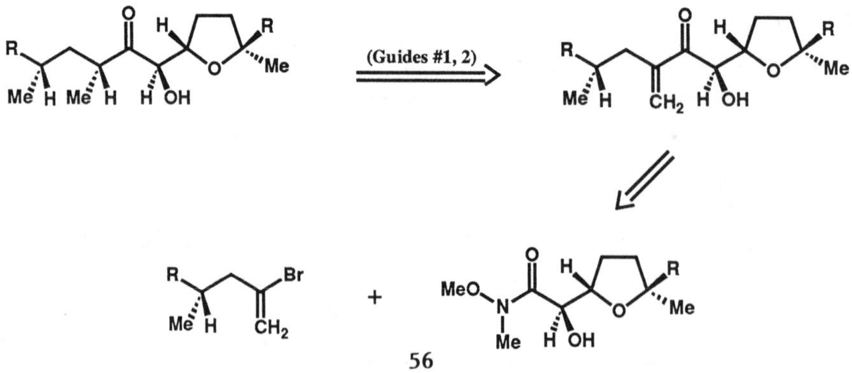

(Guides #1, 2)

Ph
Ph
H
Me
HO
H
H
Me
(Guide #4)
Ph
Br
H
Me
+
O
H
Ph
H
Me

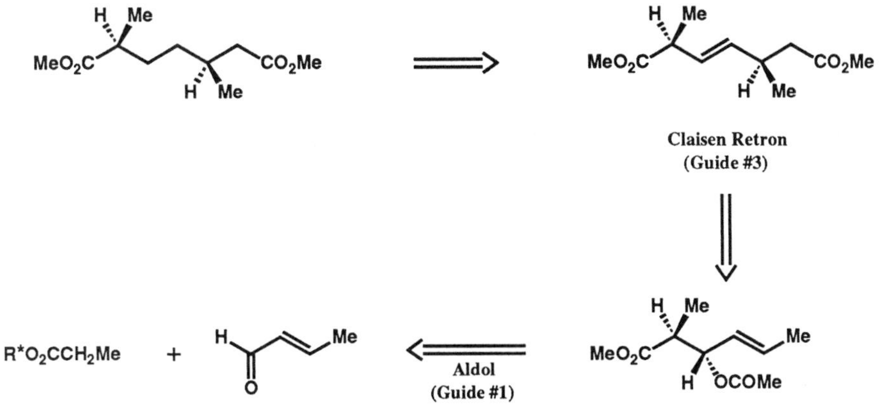
MeO2C
CO2Me
H
Me
H
Me
Claisen Retron
(Guide #3)
R*O2CCH2Me
+
Me
O
Aldol
(Guide #1)
OCOMe

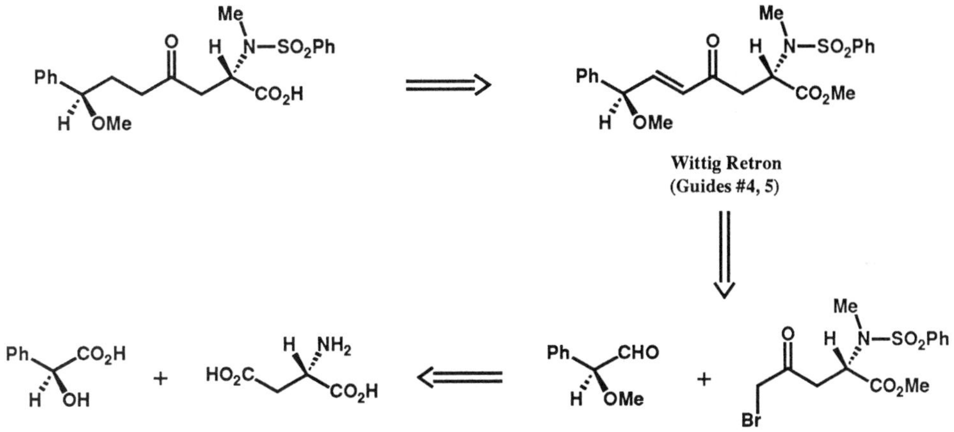
Ph
H
OMe
Me
N—SO2Ph
CO2H
CO2Me
Wittig Retron
(Guides #4, 5)
Ph
CO2H
H
OH
+
HO2C
NH2
CHO
Br

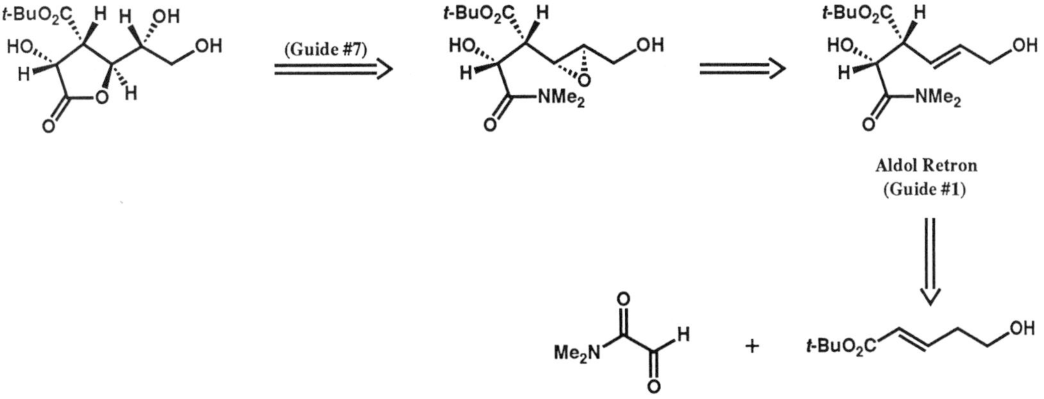
t-BuO2C
HO
H
OH
OH
(Guide #7)
NMe2
Aldol Retron
(Guide #1)
Me2N
+

CHAPTER FIVE

Functional Group-Based and Other Strategies

5.1 Functional Groups as Elements of Complexity and Strategy

The concept of functional groups allows the organization of vast amounts of information on chemical reactions, reactivity, and mechanisms, and provides a valuable framework for understanding in the chemical sciences.[37] Functional groups, those frequently-occurring collections of atoms that give rise to characteristic chemical behavior, are also very important to retrosynthetic analysis. They are the source of one form of molecular complexity and a basis for strategies dealing with that complexity. Since functional groups are essential elements in many retrons,[38] they are crucial to transform as well as strategy selection.

A small number of the most common functional groups occupy a dominant position with regard to antithetic analysis and to synthesis itself. For instance, C=C (olefinic), C=O (ketone or aldehyde), C≡C, C-OH, carboxyl, amino, nitro, and cyano groups are absolutely central in synthesis because of the number and importance of the interconversions which they allow. There are a large number of less versatile functional groups, for example, N=N (azo), S-S (disulfide), and phosphine which are crucial to synthesis mainly when they occur as components of a target structure. Other of the less central or "peripheral" groups are common along synthetic paths to target structures which lack them because they are associated with useful reagents or provide activation or control in chemical processes. Examples of such groups are halide, selenoxide, phosphonium, sulfone, trimethylsilyl, and borane. Many peripheral functional groups are concatenations of the more fundamental groups, for instance enamine, 1,2-diol, N-nitrosourea, β-hydroxy-α,β-enone, and guanidine. Another aspect of the system of functional

References are located on pages 92-95. A glossary of terms appears on pages 96-98.

groups is the existence of variants or subclasses for many individual functional group types.[39] For instance, in the case of the hydroxyl group, several subclasses can be distinguished, including the following: primary, secondary, tertiary OH, or concatenated OH (phenolic, enolic, carboxylic, hydroperoxidic), or H-bonded OH.

A useful extension of the functional group system is the inclusion of compositionally disparate groups into super-families which display qualitatively similar electronic behavior, as exemplified by the superset of all electron withdrawing groups, which includes carbonyl, cyano, sulfonyl, and nitro, or the analogous set of all electron donating groups with unshared electron pairs (alkoxy, amino, etc.). Further broadening of the functional group framework produces other useful supersets, for instance, the set of all FG's which can lead to carbocations by (1) heterolysis, (2) protonation or Lewis acid coordination, and (3) oxidative electron transfer. Similar supersets can be constructed for the generation of other types of reactive intermediates, including carbon radicals or carbon anions. This last type of superset can be expanded logically to include non-traditional FG components, such as C-H groups which can be deprotonated either because of anion stabilization or coordinative assistance to deprotonation (as for *o*–methoxyphenyl). Finally, even 6-π aromatic ring systems can be classed as polycentric FG's with appropriate subclasses such as pyridine, furan, etc.

Many retron subunits contain only a single functional group (1-Gp), while others consist of a pair of functional groups separated by a specific path (2-Gp), or a collection of several functional groups (n-Gp).[38] Functional groups, therefore, are important in the process of transform selection. The systematic application of FG-keyed disconnections is one of the most elementary strategies in retrosynthetic analysis, and is especially useful when employed concurrently with topological or stereochemical strategies. Transforms may be classed according to the number of FG's in their retrons, for example as 1-Gp or 2-Gp transforms.

Functional groups contribute to synthetic complexity for a number of reasons. Clearly, as the total number of FG's in a target increases so does structural complexity. Synthetic complexity also mounts rapidly because problems arising from reaction selectivity or functional-group interference can become formidable. The selective modification of only one of several chemically similar functional groups in a molecule can be especially difficult. Functional group-based strategies, therefore, have to deal not only with methods for the systematic reduction of the number of functional groups but also with the removal of interfering functionality and the avoidance of selectivity problems during the retrosynthetic generation of precursors. These and similar considerations underscore the strategic importance of functional groups. Several useful functional group-based strategies follow.

5.2 Functional Group-Keyed Skeletal Disconnections

Two of the most frequently used transform types for skeletal simplification are the 1-Gp and 2-Gp disconnective transforms. Disconnections keyed by 1-Gp transforms are highly strategic if they effect any of the following changes: (1) break a strategic bond in a ring, appendage or chain; (2) remove a stereocenter under mechanistic or substrate stereocontrol; (3) establish a retron for a disconnective or simplifying transform; (4) generate a new strategic bond disconnection mode.

The category of 2-Gp-keyed transforms which disconnect C-C bonds is among the most important of all transform types. These transforms, especially in their stereoselective versions, are workhorses of retrosynthetic planning as their names alone attest: aldol, Michael, Claisen,

Dieckmann, Claisen or oxy-Cope rearrangement, Mannich, Friedel-Crafts acylation, etc. They are sufficiently useful to justify a strategy in which each pair of FG's which are connected by a path containing strategic bond(s) or stereocenter(s) is examined to find or generate the retron for a 2-Gp disconnective transform.[35,38] (For a molecule containing n FG's there are $n(n-1)/2$ possible FG pairs.) If the target subunit being examined contains a partial retron for the 2-Gp disconnection, other FG-keyed transforms such as functional group interchange (FGI) or functional group addition (FGA) may be necessary to generate the required retron.

A simple example of this use of 2-Gp disconnections is shown for **159**. In this illustration

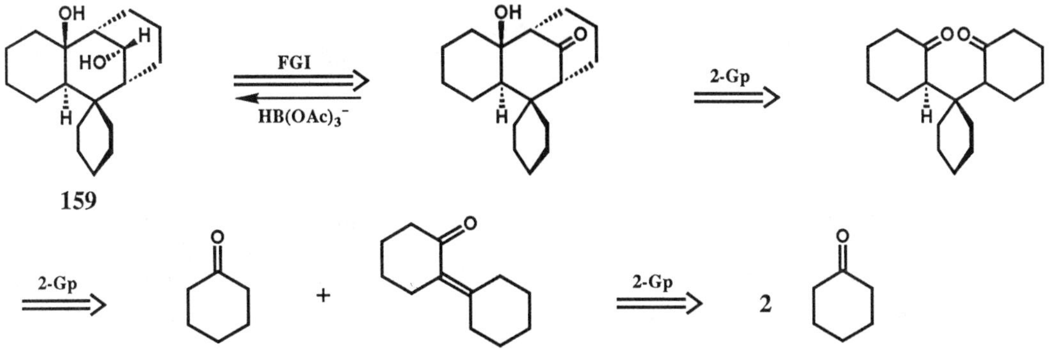

OH
HO
H
FGI
HB(OAc)3⁻
2-Gp
O
2-Gp
+
2-Gp
2

159

the initial disconnection is guided by a topological (strategic disconnection) strategy and the application of the aldol transform is readily found by a part match to the aldol retron, which can be converted to a full match by the carbonyl reduction transform (stereoselective version with $NaBH(OAc)_3$ as reagent[40,41]).

With strategic bond guidance, it is easy to find 2-Gp transform disconnections even if neither FG of an effective retron is present. In the case of the bridged aldehyde **160**, recognition of the strategic bond shown (in bold face) keys FGI processes in both directions from the bond, which successfully establish the aldol retron leading to molecular disconnection by a sequence of aldol and Michael transforms, to generate a simple chiral precursor.[31]

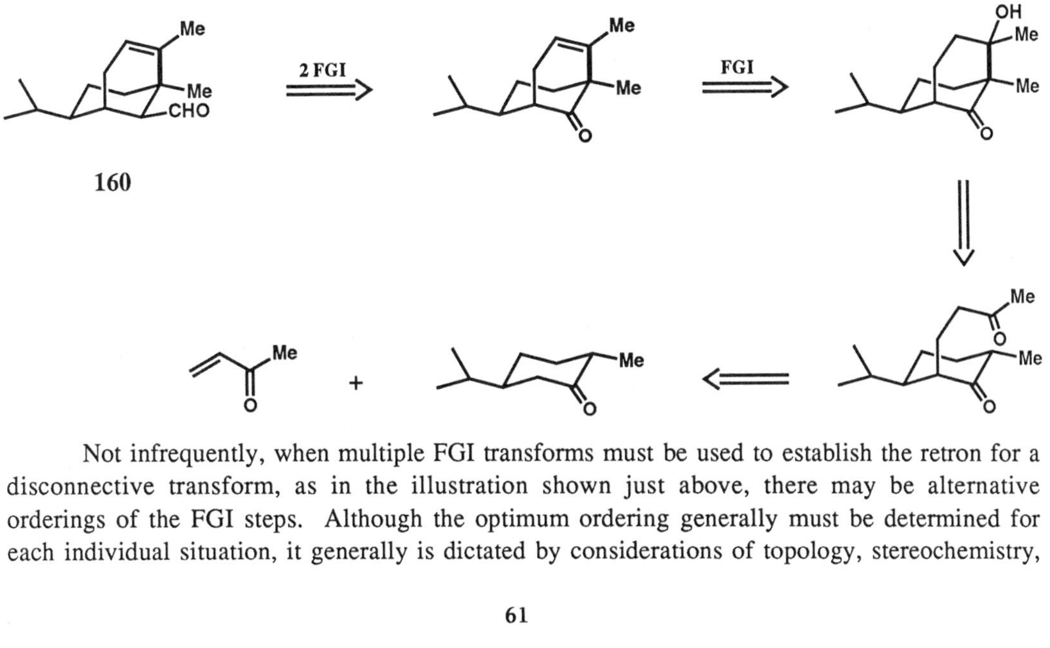

Me
Me
CHO
2 FGI
FGI
OH
O
+

160

Not infrequently, when multiple FGI transforms must be used to establish the retron for a disconnective transform, as in the illustration shown just above, there may be alternative orderings of the FGI steps. Although the optimum ordering generally must be determined for each individual situation, it generally is dictated by considerations of topology, stereochemistry, and possible interfering functionality (see Section 5.10). Even when two FG-keyed disconnections can be applied directly to a TGT, both possible orderings may require consideration. For instance the tetracyclic lactam **161** can be disconnected to a tricyclic precursor by application of Michael and amide condensation transforms, in either order. In this instance stereochemical and positional factors favor application of the Michael transform before amide formation transform.[42]

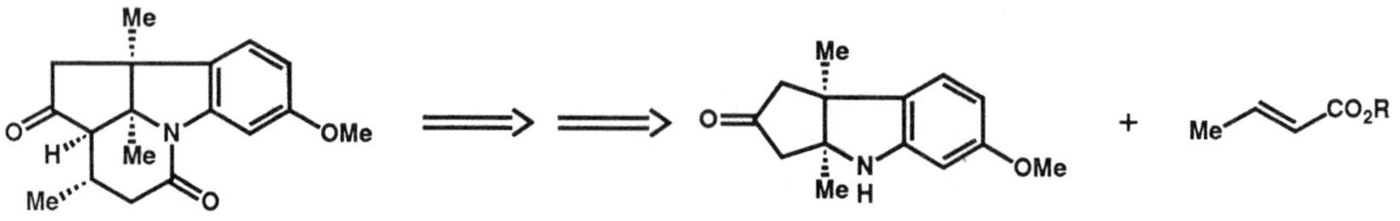

Me
O
H
Me
N
Me
O
OMe
Me
O
N
Me H
OMe
+
Me
CO₂R

161

5.3 Disconnection Using Tactical Sets of Functional Group-Keyed Transforms

Retrosynthetic analysis can often be enhanced by strategies built around tactical combinations of FG-keyed transforms which together produce molecular simplification in a coordinated (but subtle) way. The concept of tactical combinations of transforms and a few examples of such combinations have been described in Section 2.10. The use of FG-keyed tactical combinations may be illustrated by a selection of specific applications.

Target **162** cannot be directly simplified. However, it contains the full retron for the FG-keyed tactical combination which converts it to precursor **163** and thence to simple molecules.[43] Obviously, this approach requires that information on such tactical combinations be available in a form which facilitates its use.

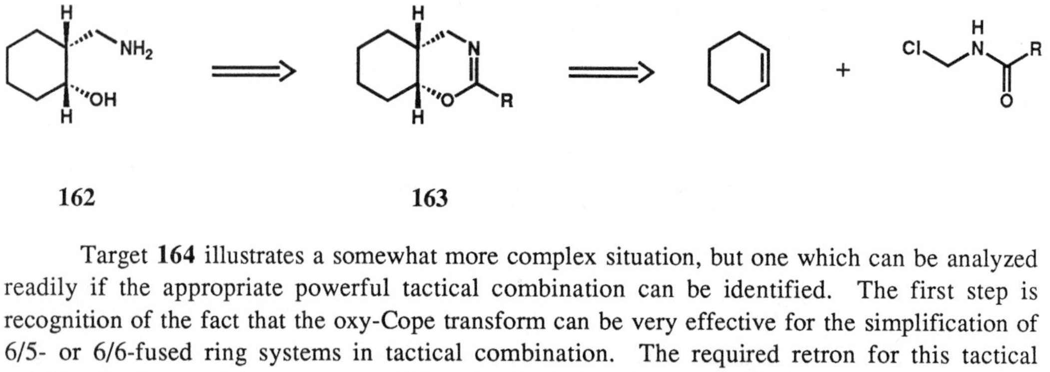

H
NH₂
OH
H
H
N
O
R
H
+
H
Cl
N
R
O

162 **163**

Target **164** illustrates a somewhat more complex situation, but one which can be analyzed readily if the appropriate powerful tactical combination can be identified. The first step is recognition of the fact that the oxy-Cope transform can be very effective for the simplification of 6/5- or 6/6-fused ring systems in tactical combination. The required retron for this tactical combination is a cyclohexanone to which a ring is *cis*-fused at carbons 4 and 5, with an offexendo double bond attached to carbon 4 of the cyclohexanone unit, as shown in **166**. It is clear that **164** is a good candidate for the application of the oxy-Cope transform since it contains all the elements of the retron except for the C=C subunit. This is easily established (**164** ⇒ **165**). Subsequent application of the oxy-Cope transform and the organometallic carbonyl addition transform generate simple synthetic precursors as shown.[44] This example shows the importance of knowing how to use simultaneously FG, ring-pair and stereochemical keys to identify tactical combinations of simplifying transforms.

The retrosynthetic simplification of **164** which has just been outlined was initiated by the application of a [3,3] sigmatropic rearrangement transform. Other tactical combinations involve the use of such rearrangement transforms to link a pair of disconnective transforms.

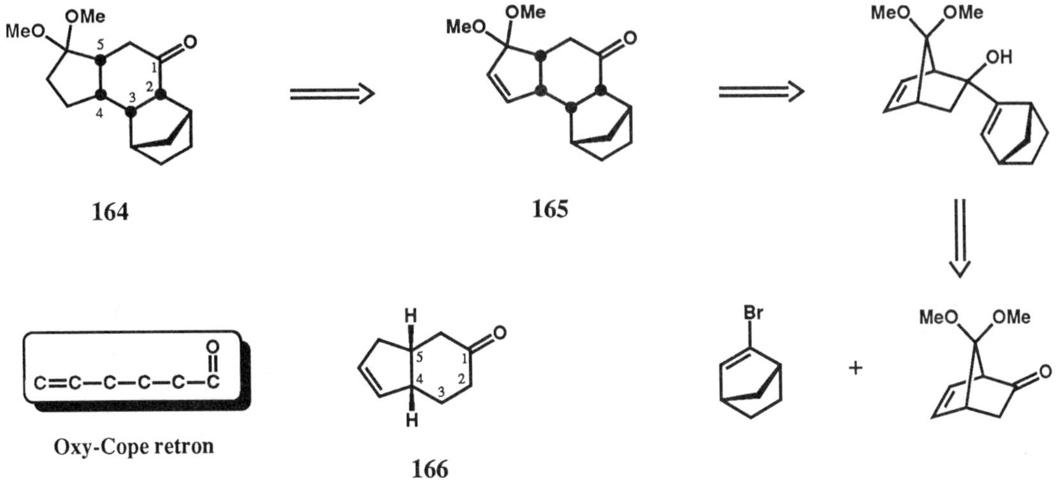
MeO
OMe
5
1
2
3
4
O
Br
+
C=C—C—C—C—C
H

164 **165** **166**

Oxy-Cope retron

Disconnections which generate a biallyl subunit (**167**) or heteroatom equivalent allow the [3,3] sigmatropic rearrangement transform to be applied. If that rearrangement produces the retron for a simplifying transform, a successful strategic approach may result. Target **168** provides just such a case.[45] The Mannich transform can be directly applied in a mechanistic way to generate intermediate **169** and precursor molecule **170** plus formaldehyde. However, it will be noted that intermediate **169** contains a full retron for the aza-Cope rearrangement transform, application of which leads to **171**. The further disconnection of **171** is easily achieved, for example by either of the pathways shown. It is important to note that the keying for this particular tactical

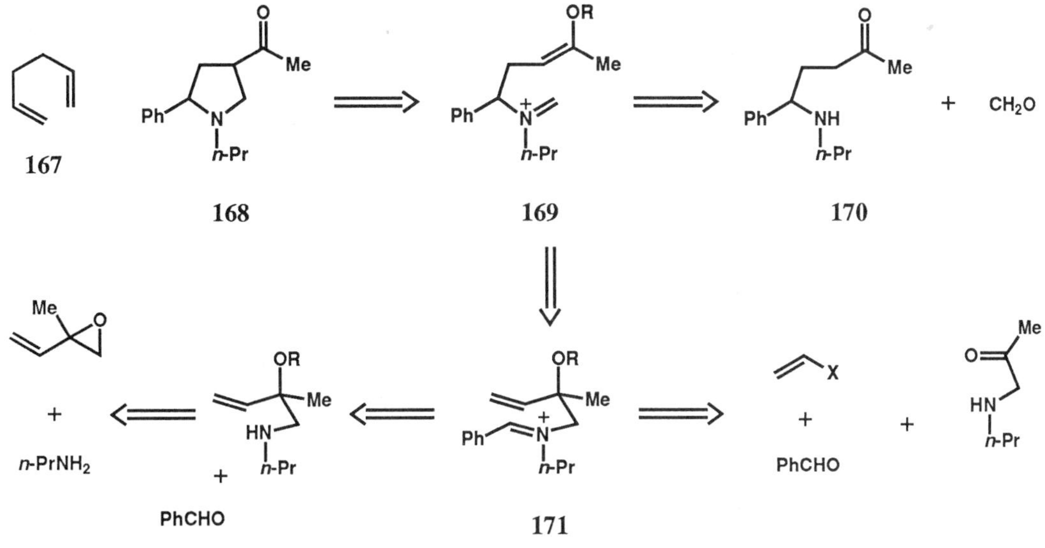
O
OR
Me
Ph
N
NH
n-Pr
+
CH2O
HN
X
n-PrNH2
PhCHO

167 **168** **169** **170** **171**

combination can stem from mechanistic transform application involving reaction intermediate (**169**) which contains a crucial retron link not found in the stable Mannich precursor **170**. This example is instructive since it demonstrates that important benefits can be derived by the application of transforms in a mechanistic way and the concurrent examination of intermediates for the presence of retrons for key transforms. On the other hand, the retrosynthetic pathways shown for **168** via **169** and **171** may be also generated simply by recognizing that **168** contains the retron **172** which keys the tactical combination of the Mannich, [3,3] sigmatropic rearrangement,

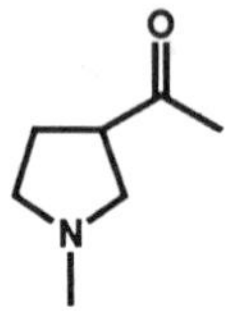

172

and C-C or C-N disconnective transforms. There is also an overlap between the use of FG-keyed tactical combinations and T-goal strategies based on initial keying by ring or ring-pair units which are partial retrons for transforms that disconnect rings. The retrosynthetic simplification of the desoxy analogs of **164** and **172**, for instance, could be guided by a T-goal search based only on keying by the rings as partial retrons without recourse to the functional groups contained in the full retrons. The same is true for many important simplifying tactical combinations of transforms which are also keyed by retrons containing a ring or ring-pair as well as FG(s). The use of tactical combinations of transforms as T-goals for strategic guidance has been discussed in Section 2.10 on transform-based strategies.

5.4 Strategic Use of Functional Group Equivalents

The possibilities for assembling carbogenic structures have sharply increased in recent years with the advent of various synthetic strategies for generating the equivalents of carbon electrophiles and nucleophiles which are unavailable in structures containing simple core functional groups such as carbonyl, amino, or hydroxyl. For example, acyl anions (RCO^-), which for various reasons are not readily available, can be utilized in synthesis in modified or equivalent forms. Several acyl anion synthetic equivalents have become standard reagents, for instance the reagents **173 a-e**,[46] which react with carbon electrophiles to afford products that can be converted

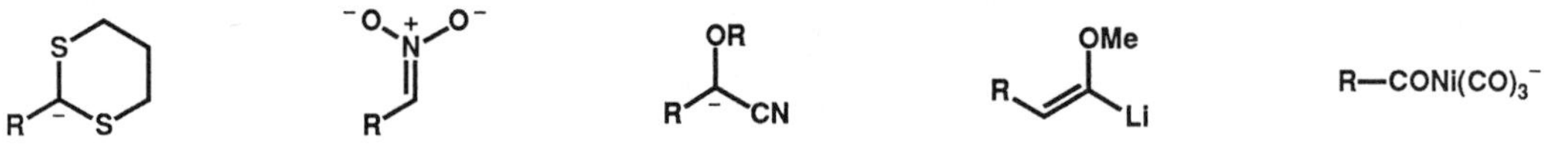

S
R
S
-O
+
N
O-
R
OR
R
CN
OMe
R
Li
R—CONi(CO)$_3^-$

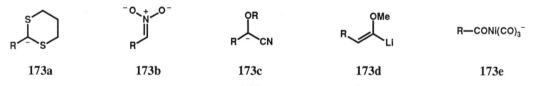

173a **173b** **173c** **173d** **173e**

to carbonyl compounds. Other useful reagents function as equivalents of β-acylcarbanions ($RCOCH_2CH_2^-$), α-acylcarbocations ($RCOCH_2^+$), α-aminocarbanions ($H_2NCH_2^-$), and α-alkoxycarbanions ($ROCH_2^-$). The impact of such methodology on retrosynthetic analysis is to expand the number of valid disconnections through the replacement of a core functional group such as carbonyl by an equivalent which allows a bond disconnection not otherwise attainable. In a retrosynthetic sense, if a disconnection is identified as strategic but is not permitted by the particular core functional group present, the replacement of that group by an equivalent which allows or *actuates* (Section 1.6) bond cleavage becomes a subgoal objective. (The term *actuate* is a convenient antithetic equivalent of *activate* in the synthetic direction). The following are useful FG equivalents of the keto group: $R_2CO \equiv R_2CHNO_2$, $R_2CH(S{=}O)R'$, $R_2C(SR')_2$, R_2CHNH_2, $R_2C{=}NOH$, $R_2C(OH)COOH$.

As a simple example consider target **174** and the specific objective of disconnecting the bond joining the ring and the acetylenic appendage. Heterolytic antithetic dissection of that bond produces the fragments (synthons) pictured, of which **175** and **176** are unavailable as synthetic reagents. To achieve disconnection *a* the 1-butynyl unit must be converted into a retrosynthetic precursor that can be cleaved to an available electrophile which serves as an equivalent of **175**. One possibility is shown by the sequence **174** ⇒ **177** ⇒ **178**.[47] Bond cleavage according to alternative *b* necessitates the use of a practical equivalent of the keto carbocation **176**. Such a pathway via **179** and **180** is outlined for the case of α-bromo oxime **181** as the carbocation equivalent.[48] It is usually possible to devise valid sequences which utilize functional group

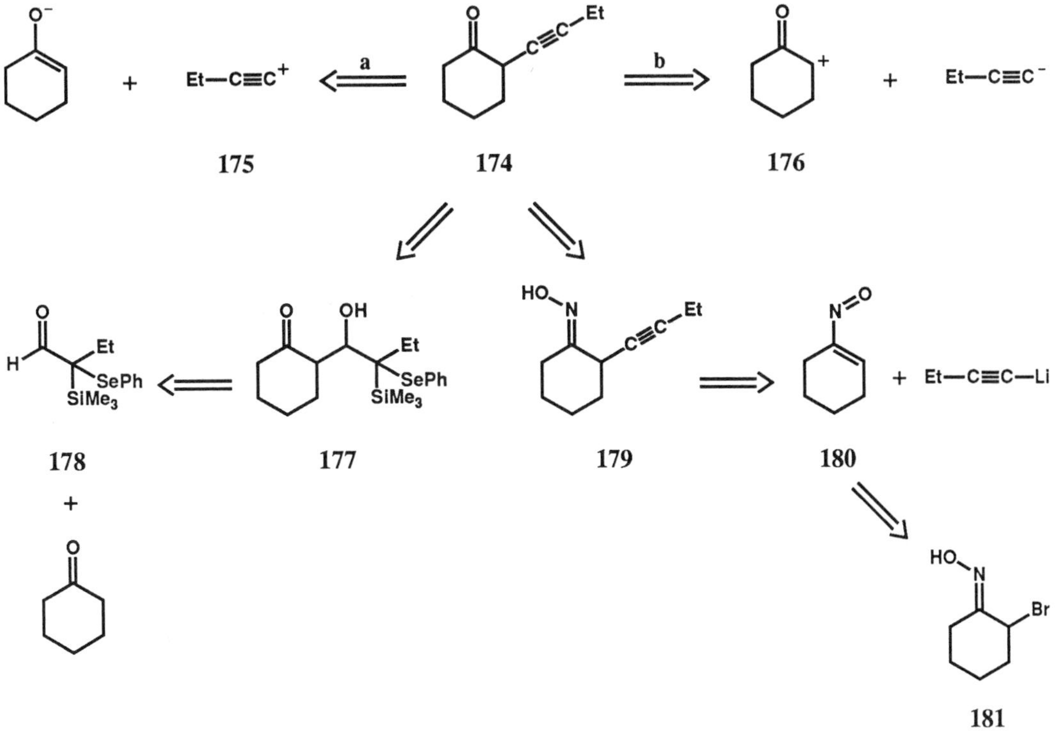

O⁻
+
Et—C≡C⁺
a
O
C≡C
Et
b
O
+
+
Et—C≡C⁻
175
174
176
O
H
Et
SePh
SiMe₃
O
OH
Et
SePh
SiMe₃
HO
N
C≡C
Et
N=O
+ Et—C≡C—Li
178
177
179
180
+
O
HO
N
Br
181

equivalents by a bidirectional procedure that includes the following elements, once the specific bond dissection has been defined: (1) replacement of the *obstructing* FG by an equivalent from which it can easily be made (usually in one or two steps); (2) analysis of the precursors so generated to determine whether they can be disconnected to known or preparable reagents; and (3) use of the knowledge of known or potential equivalents of the unavailable synthon that results from heterolytic disconnection of the chosen strategic bond to guide the selection of an equivalent FG, or to test the validity of the bond-forming process in the synthetic direction. The recognition of the need for valid synthetic equivalents stems from the identification of an obstacle to a particular synthetic disconnection, the use of mechanistic reasoning, and the perception or gathering of relevant chemical information. A sampling of some synthetic equivalents which are useful for the elimination of obstacles to crucial bond disconnection is presented in Chart 21.

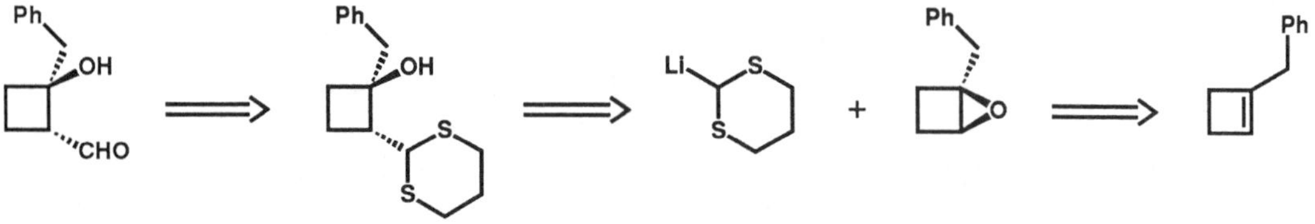
Ph
OH
CHO
Ph
OH
S
S
Li
S
S
+
Ph
O
Ph

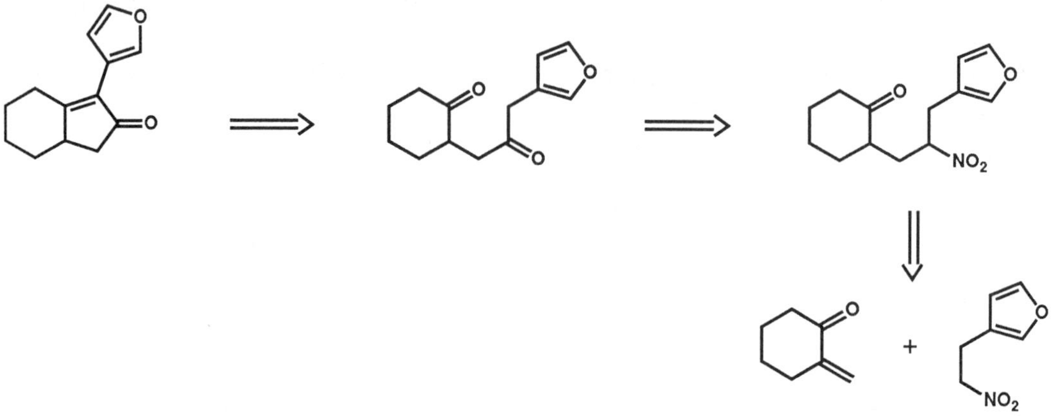
O
O
O
O
O
O
O
NO2
O
+
O
NO2

JACS, **1987**, *109*, 918.

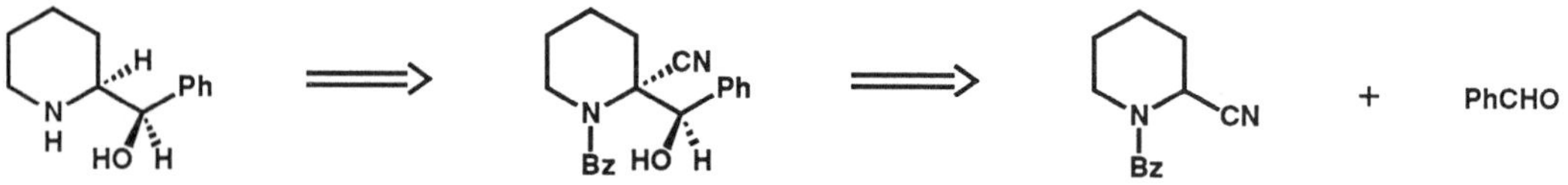
H
Ph
N
H
HO
H
CN
Ph
N
Bz
HO
H
N
CN
Bz
+
PhCHO

TL, **1979**, 771.

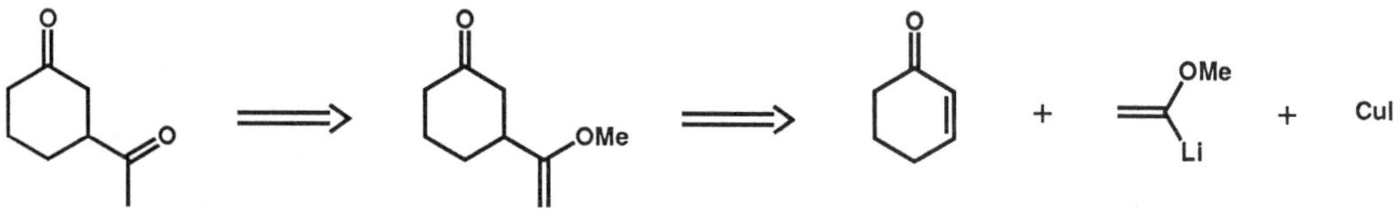
O
O
O
OMe
O
+
OMe
Li
+
CuI

JACS, **1974**, *96*, 7125.

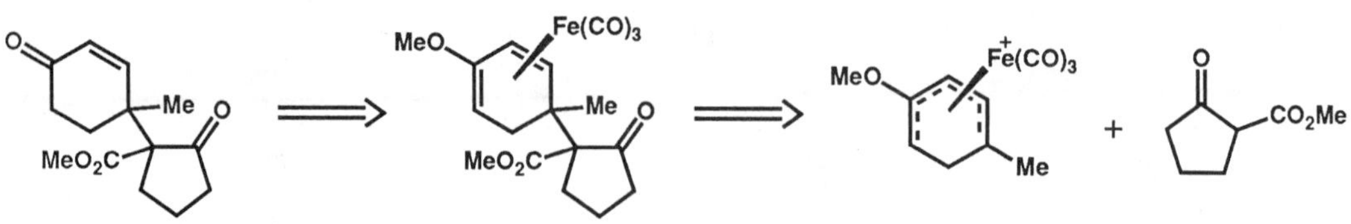
O
Me
O
MeO2C
MeO
Fe(CO)3
Me
O
MeO2C
MeO
Fe(CO)3
Me
+
O
CO2Me

JACS, **1981**, *103*, 6686.

Chart 21

5.5 Acyclic Core Group Equivalents of Cyclic Functional Groups

In preceding sections on the use of topological strategies for the simplification of polycyclic networks, it was mentioned that various cyclic functional groups are so readily formed from acyclic precursors that they are essentially equivalent to the acyclic structures. The retrosynthetic cleavage of such easily formed cyclic functional groups as ketal, hemiketal, acetal, hemiacetal, lactone, lactam, imine, ether, and thioether can lead to highly effective simplification whenever it generates core functionality (especially carbonyl, C=C, or hydroxyl) which keys disconnective transforms or powerful tactical combinations. A simple but typical example of the use of this FG-based strategy is provided by target **182**, a dilactone whose acyclic equivalent **183**, is readily transformed into simple precursors as shown.[42,49]

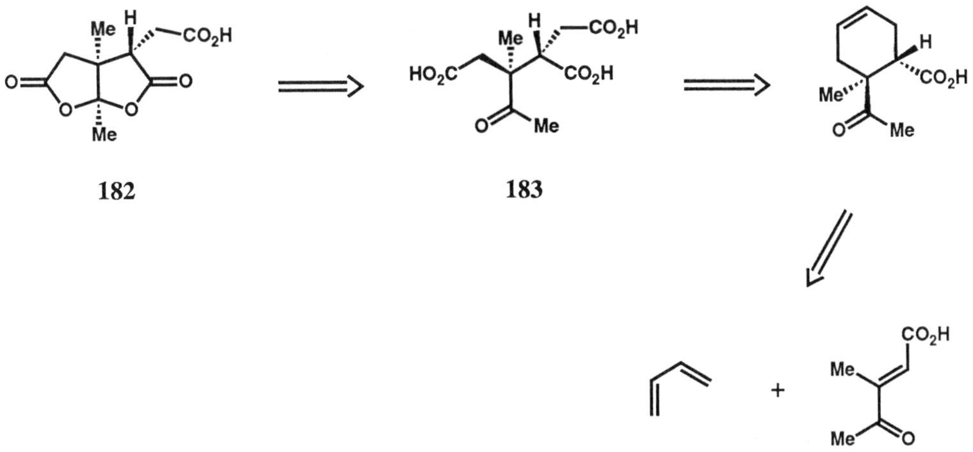

182 **183**

A second example of this strategy is the retrosynthetic simplification of **184** via the equivalent **185** by subsequent application of the tactical combination of retroaldol and [2 + 2] photocycloaddition transforms.[50] Various targets which are structurally related to **184** have also

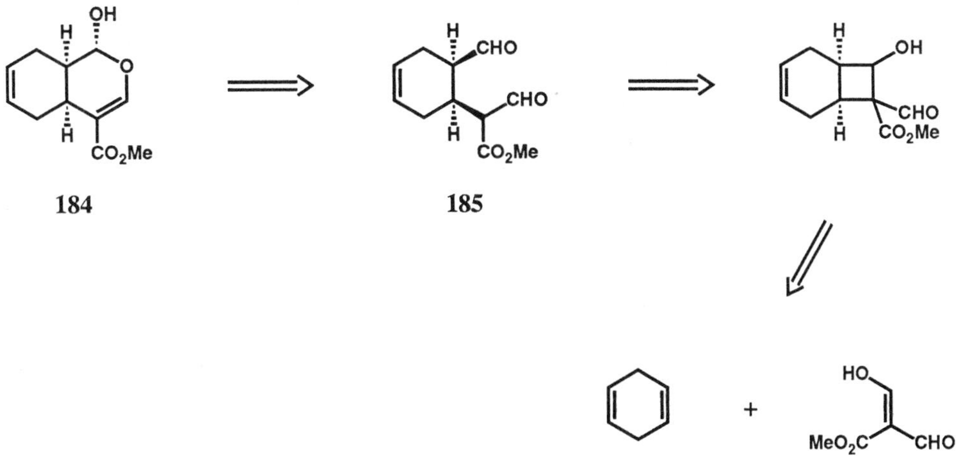

184 **185**

been synthesized by interesting alternative routes based on concurrent heteroring and C-C disconnection. For instance nepatalactone (**186**) can be simplified via intermediates **187** and **188** to **189** and **190** using, for example, the enamine-α,β-enal cycloaddition transform as a strategic bond-pair disconnection.[51-54]

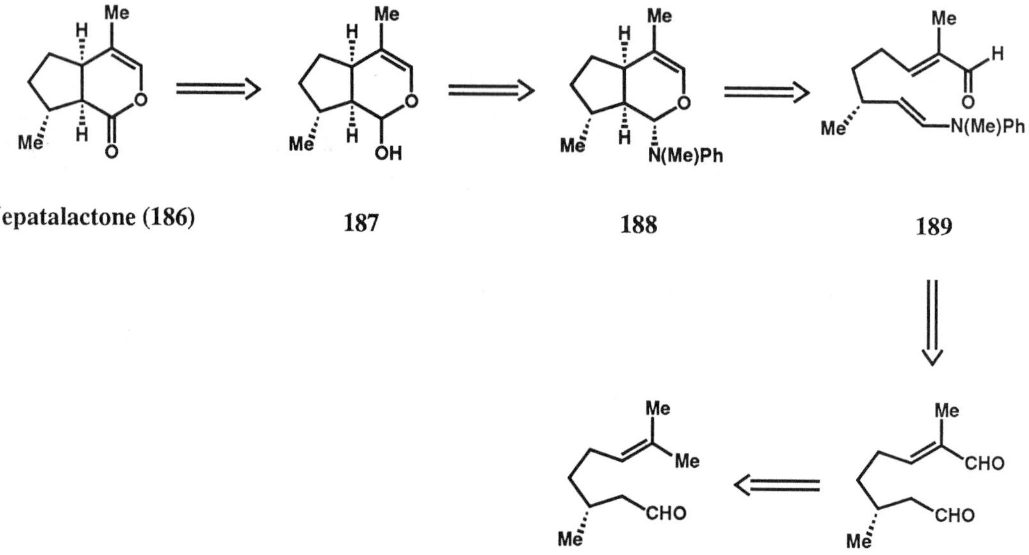
Me
H
O
N(Me)Ph
OH
CHO

Nepatalactone (186) **187** **188** **189**

(+)-Citronellal **190**

5.6 Functional Group-Keyed Removal of Functionality and Stereocenters

The simplification of stereochemical and functional group-based complexity is commonly possible using transforms which are keyed by functional groups. Antithetic removal of functional groups often provides the additional benefit of diminishing problems of functional group interference or selectivity and, for this reason, can be crucial to the valid application of simplifying transforms. Such functional group removal is often coincident with the removal of stereocenters. The most common type of functional group removal involves adjacent functional groups and is keyed by a FG pair. The 2-Gp transforms shown in Chart 22 are typical of many in this category, which may also involve stereocenter (at C*) removal.

Retrosynthetic functional group removal (often with concomitant stereocenter removal) in structures containing two non-adjacent but spatially proximate FG's may be valid using intramolecular functionalization transforms. Whenever such a retrosynthetic step is possible, it may lead to the derivation of simple synthetic routes. This area of synthetic methodology, which is still underdeveloped, is exemplified by the transforms outlined in Chart 23. Even more underdeveloped is the methodology of intermolecular selective functionalization of carbogens at sites remote from or not activated by other functionality. It is in this area where enzymic reactions such as cytochrome P-450-mediated hydroxylation at specific unactivated sites give enzymic biosynthesis a substantial advantage over chemical synthesis, at least at present.

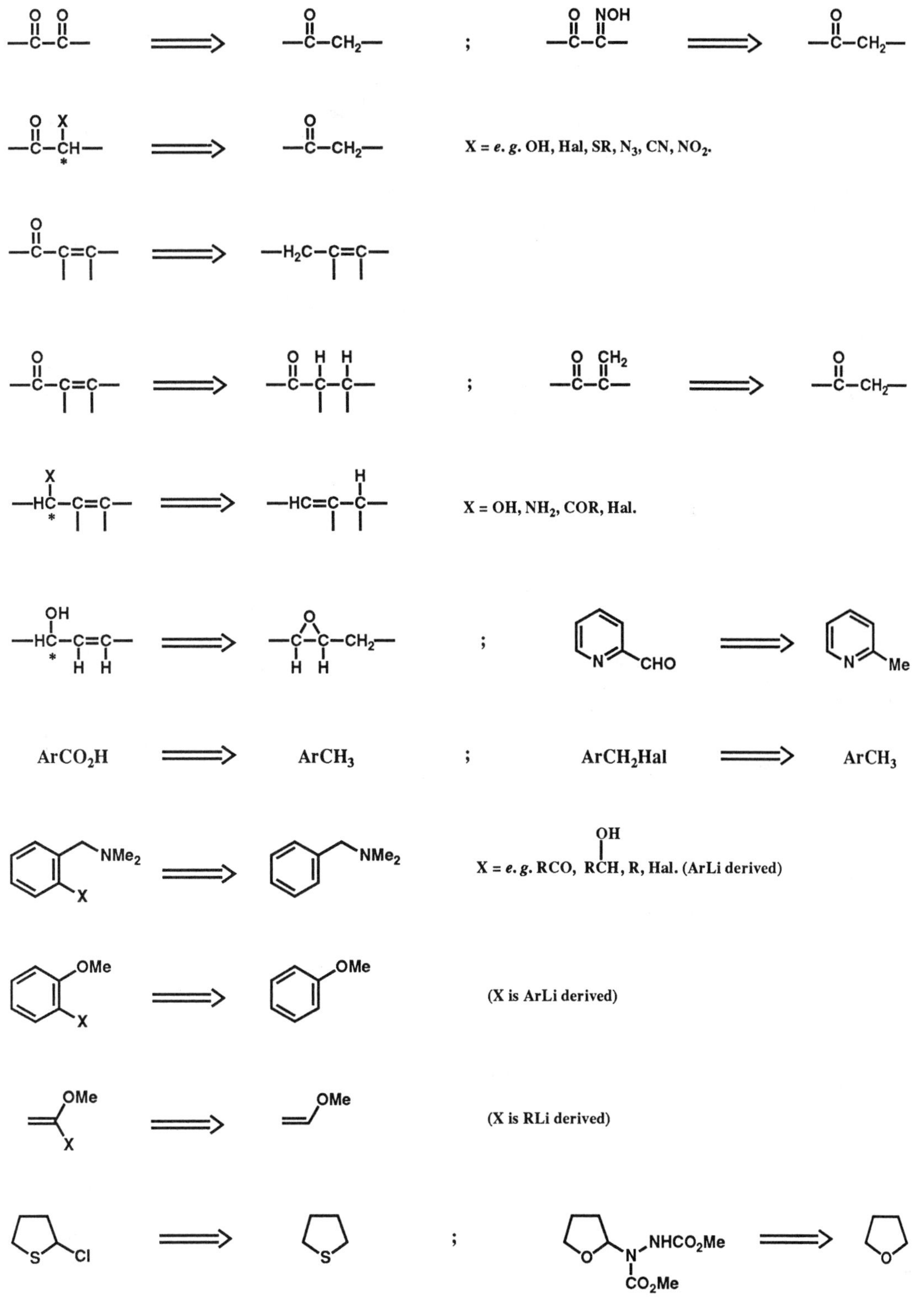
X = e. g. OH, Hal, SR, N3, CN, NO2.
X = OH, NH2, COR, Hal.
ArCO2H
ArCH3
ArCH2Hal
ArCH3
X = e. g. RCO, RCH(OH), R, Hal. (ArLi derived)
(X is ArLi derived)
(X is RLi derived)

Chart 22

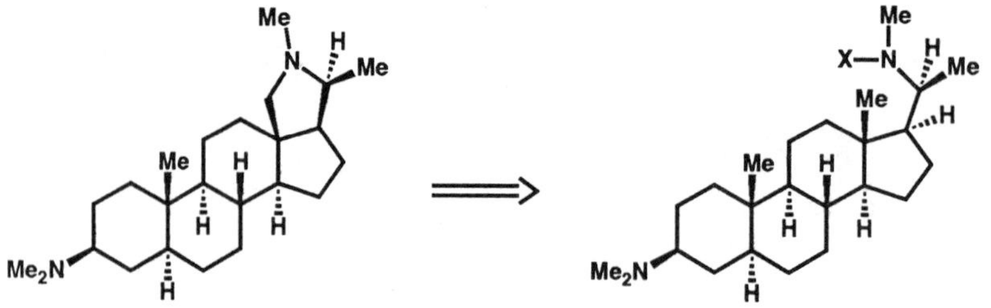

X = Cl ⇒ X = H

JACS, **1959**, *81*, 5209.
Experientia, **1960**, *16*, 169.

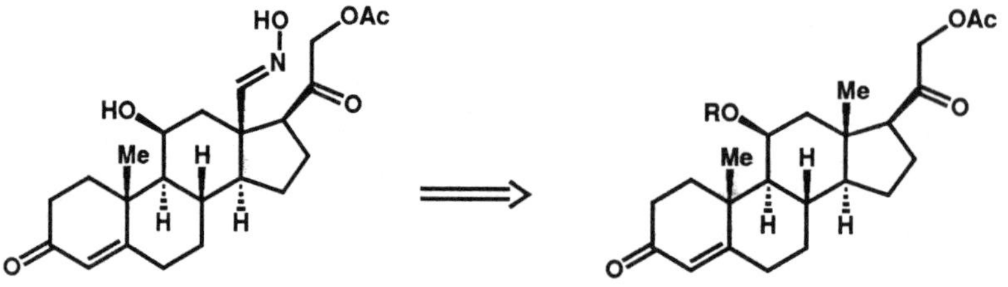

R = NO ⇒ R = H

JACS, **1960**, *82*, 2641.
JACS, **1961**, *83*, 4083.
Adv. Photochem., **1964**, *2*, 263.

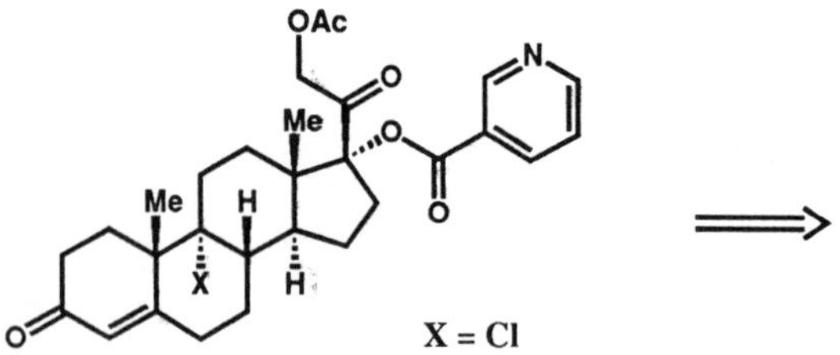

X = Cl

X = H

JACS, **1987**, *109*, 3799.

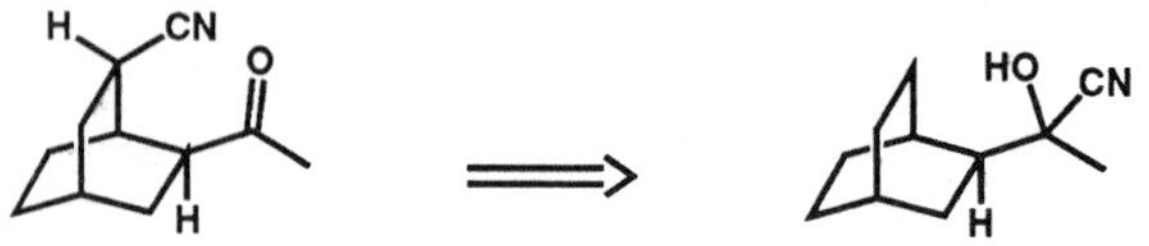

Synth. **1971**, 501.

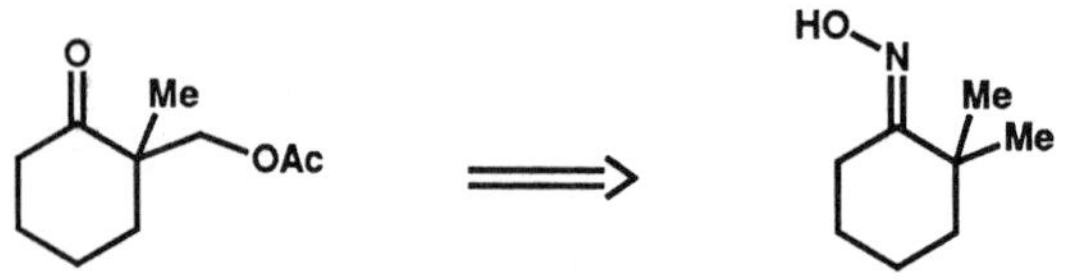

JCS, CC, **1985**, 126.

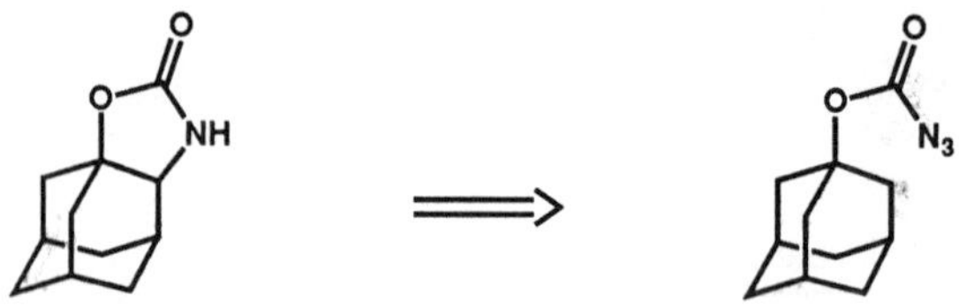

JCS, CC, **1967**, 563.
JCS, CC, **1976**, 668.

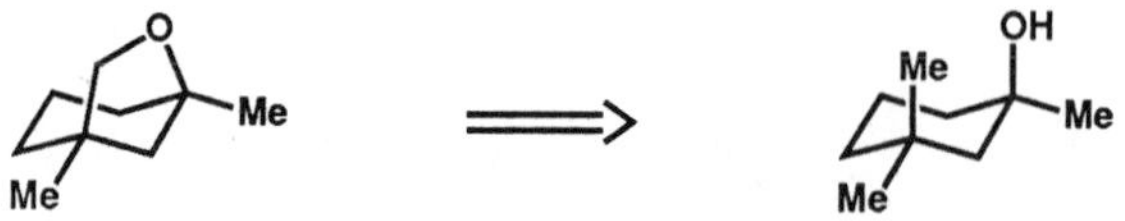

JACS, **1958**, *80*, 6686.
JOC, **1965**, *30*, 3216.

Chart 23

5.7 Functional Groups and Appendages as Keys for Connective Transforms

The formation of carbon-carbon bonds is an important operation in synthesis, and so it may appear paradoxical at first that the cleavage of carbon-carbon bonds is also an indispensable element in synthetic planning. There are many different reactions for the fission of carbon-carbon bonds, the most useful being those which cleave endocyclic single or double bonds. A small sampling of such processes is collected in Chart 24 (for C-C) and Chart 25 (for C=C). Inspection of these reactions reveals that the synthetically interesting consequences of ring cleavage include the following: (1) generation of a pair of functional groups from one FG; (2) elevation of FG reactivity levels; (3) conversion of a ring to one or two appendages on another ring in a stereochemically unambiguous way; (4) generation of a larger ring from a pair of smaller ones; (5) creation of olefinic stereocenters stereoselectively; and (6) conversion of stereocenters on rigid rings to stereocenters on flexible rings or chains.

In the retrosynthetic direction, the transforms corresponding to these reactions are *connective*. Although the application of such connective transforms creates an additional ring, this structural complication may be more than balanced by the reduction in functional group-based or stereochemical complexity. As applied to targets possessing relatively inaccessible medium or large rings, these connective transforms can result in fused or bridged rings of readily available size or type.

The application of simplifying connective transforms is often suggested by a pair of structural features, such as two ring appendages or a ring with its appendage, suitably functionalized for connection to form a *synthetically accessible* ring. Other keying units include medium or large rings, or olefinic stereocenters. The retrosynthetic application of connective transforms is illustrated by three examples which follow.

The γ–appendage of 2-cyclohexenone **191** cannot be directly disconnected by an alkylation transform. (γ–Extended enolates derived from 2-cyclohexenones undergo alkylation α- rather than γ- to the carbonyl group). However, **191** can be converted to **192** by application of the retro-Michael transform. The synthesis of **192** from methoxybenzene by way of the Birch reduction product **193** is straightforward.[55] Another synthesis of **191** (free acid) is outlined in

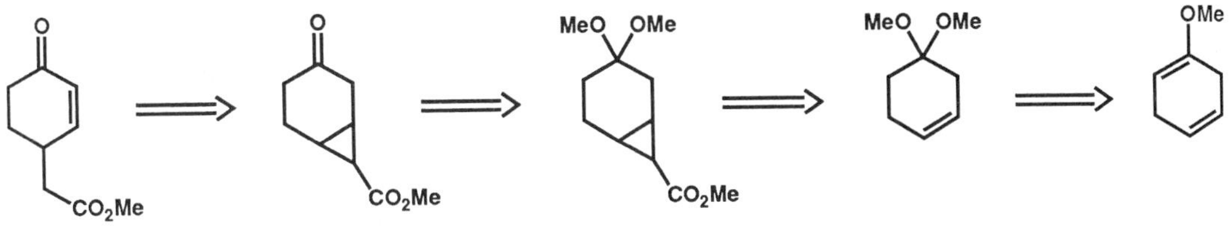
O
CO2Me
O
CO2Me
MeO OMe
CO2Me
MeO OMe
OMe

191 **192** **193**

Chart 14 which presents a number of powerfully simplifying tactical combinations of transforms as T-goals for strategic guidance. Connective transforms which form 3- and 4-membered rings are frequent components of such tactical combinations. There are many connective transforms for the generation of 3- and 4-membered rings, a consequence of the rich collection of chemical reactions now known for the cleavage of strained 3- and 4-membered rings via radical, anionic or cationic intermediates. Such retrosynthetic connections can also be deduced by use of the mechanistic mode of transform application. However they are arrived at, connective transforms provide a powerful and effective approach to FG-keyed appendage removal, often with stereocontrol.

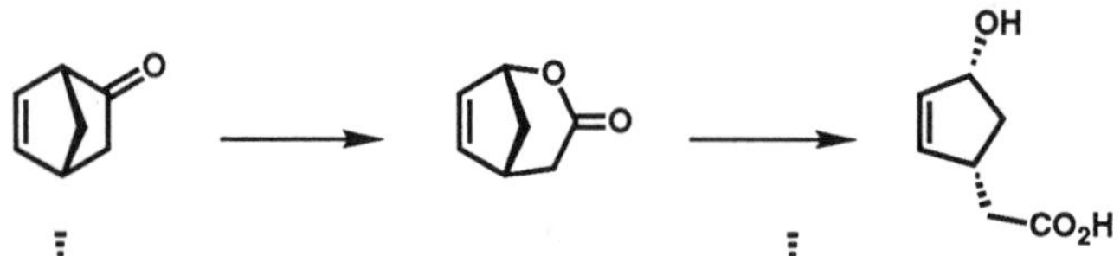
OH
O
CO2H

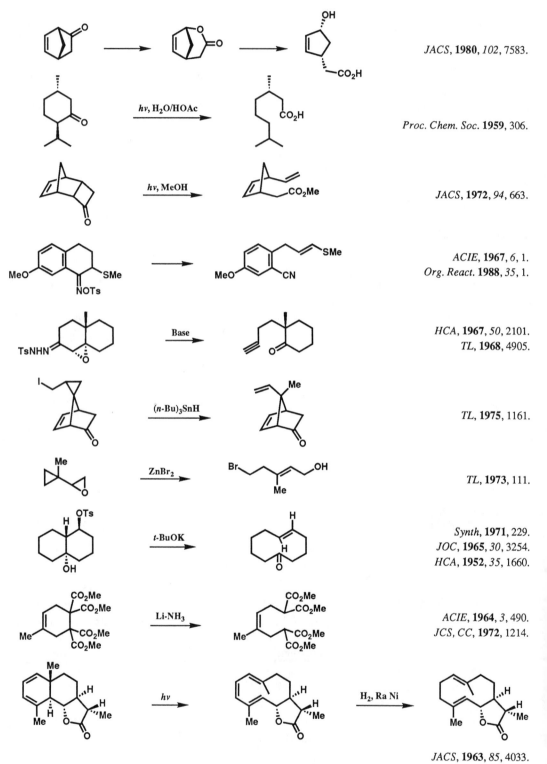

JACS, **1980**, *102*, 7583.

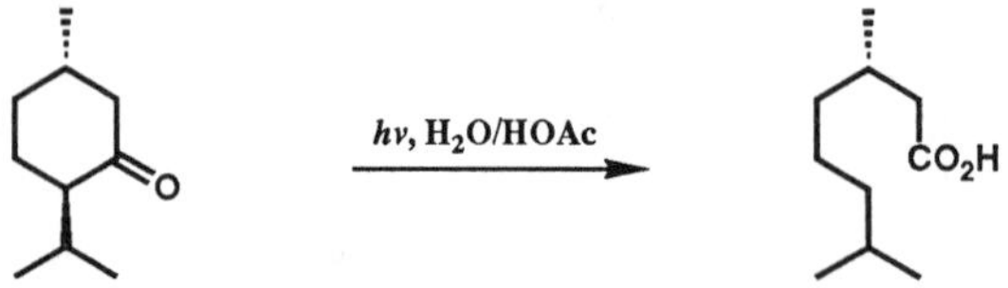
hv, H2O/HOAc
CO2H

Proc. Chem. Soc. **1959**, 306.

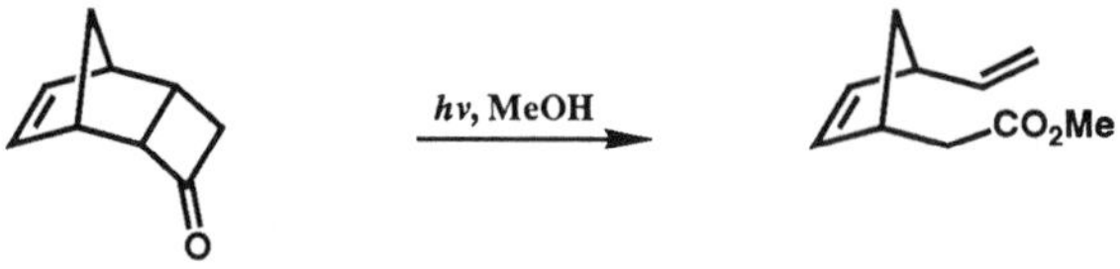
hv, MeOH
CO2Me

JACS, **1972**, *94*, 663.

MeO
SMe
NOTs
CN

ACIE, **1967**, *6*, 1.
Org. React. **1988**, *35*, 1.

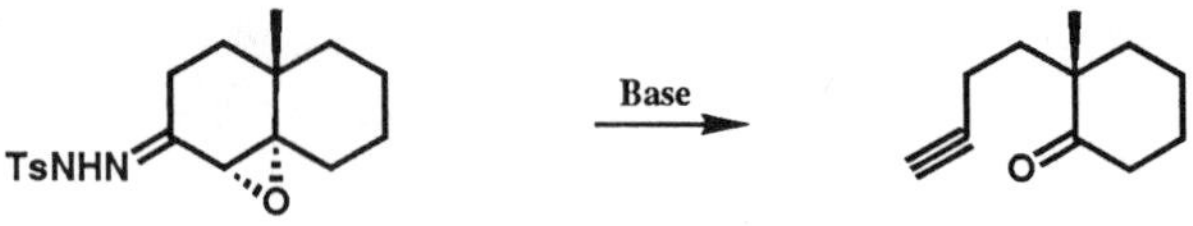
TsNHN
Base

HCA, **1967**, *50*, 2101.
TL, **1968**, 4905.

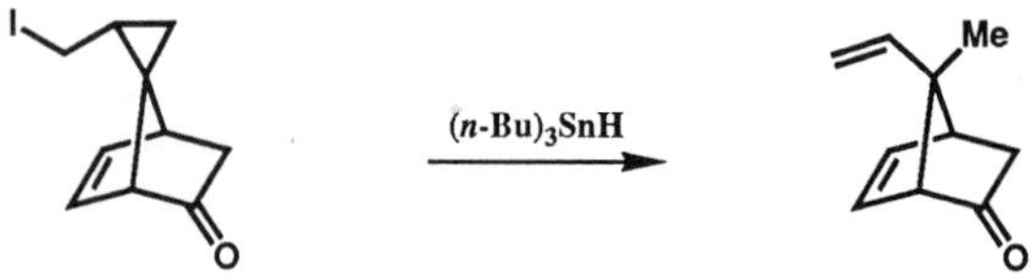
(n-Bu)3SnH
Me

TL, **1975**, 1161.

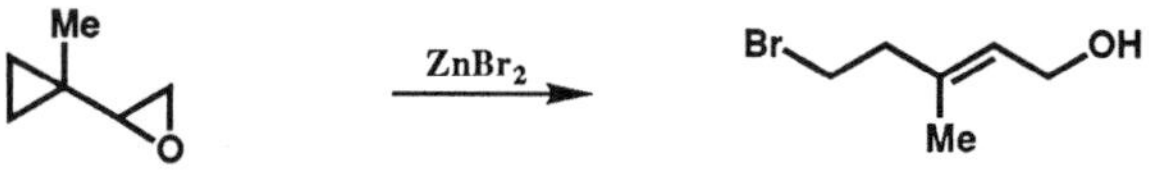
Me
ZnBr2
Br
OH

TL, **1973**, 111.

OTs
H
OH
t-BuOK

Synth, **1971**, 229.
JOC, **1965**, *30*, 3254.
HCA, **1952**, *35*, 1660.

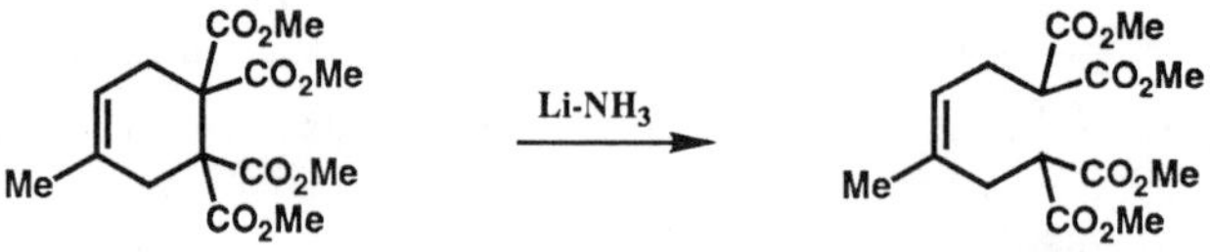
CO2Me
Me
Li-NH3

ACIE, **1964**, *3*, 490.
JCS, CC, **1972**, 1214.

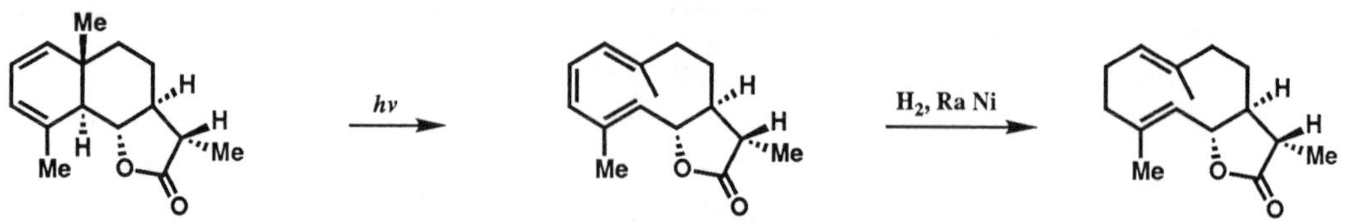
Me
H
hv
H2, Ra Ni

JACS, **1963**, *85*, 4033.

Chart 24

The most important connective transforms in retrosynthetic analysis are the family of C=C cleavage transforms, including one-step (e.g. ozonolytic) or two-step (e.g. OsO_4 followed by $Pb(OAc)_4$) (Chart 25). There are many elegant syntheses of challenging molecules which depend on such processes. Two examples will provide an idea of the underlying retrosynthetic approach.

The *cis*-fused lactone **194** is equivalent to hydroxy acid **195** which can be cyclized by a C=C oxidative cleavage transform to afford enol ether **196** after functional group interchange ($CH_2OH \Rightarrow CHO$). Direct disconnection of the remaining ring appendage under stereocontrol is then possible by use of the organocopper—α,β-enone conjugate addition transform.[56,57] Although it is also possible in principle to arrive at the same starting materials by initial application of the Baeyer-Villiger transform to **194** (forming **197**), the position selectivity of the Baeyer-Villiger process is problematic.

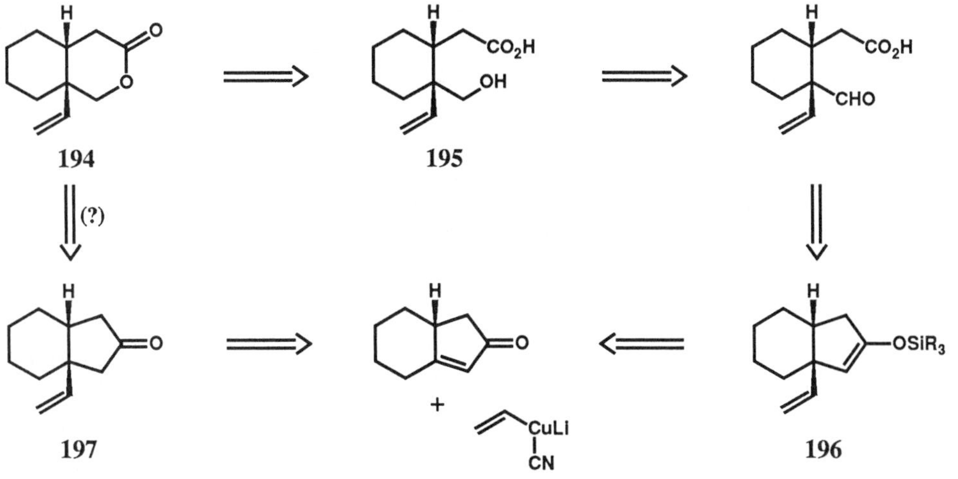
H
O
O
194
H
CO2H
OH
195
H
CO2H
CHO
(?)
H
O
197
H
O
+
CuLi
CN
H
OSiR3
196

Aldehyde **198** served as a key intermediate in a synthesis of the alkaloid ajmaline. The Mannich aminomethylation transform triggers disconnection of two bonds in **198** to form dialdehyde **199**, which by connective transform application can be converted to cyclopentene **200**.[58,59] The reduction in functional group reactivity and in structural complexity are both apparent by comparison of **198** and **200**.

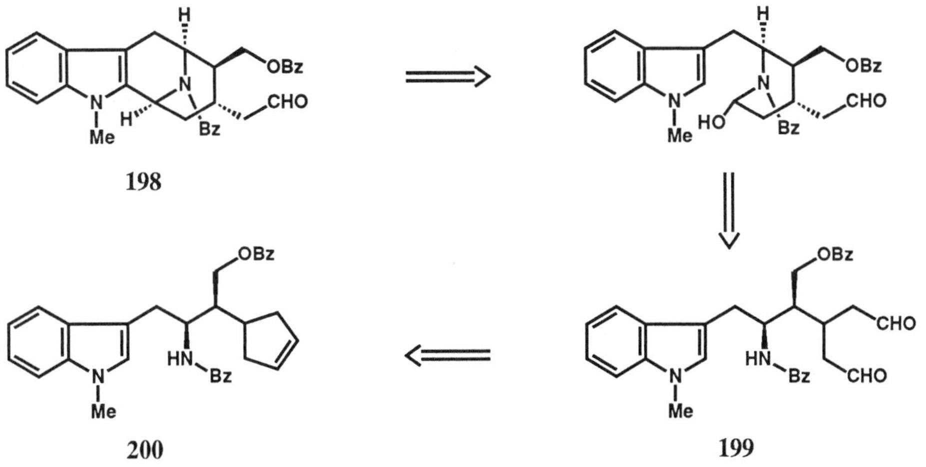
H
OBz
N
CHO
N
H
Bz
Me
198
H
OBz
N
CHO
N
HO
Bz
Me
OBz
HN
Bz
N
Me
200
OBz
CHO
HN
Bz
CHO
N
Me
199

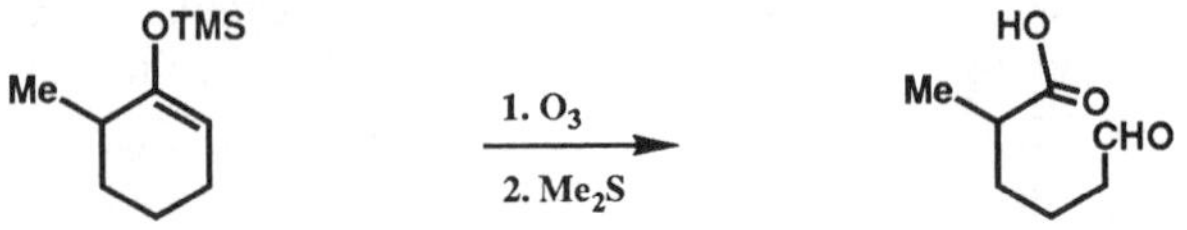
OTMS
Me
1. O3
2. Me2S
HO
Me
O
CHO

TL, **1974**, 2027.

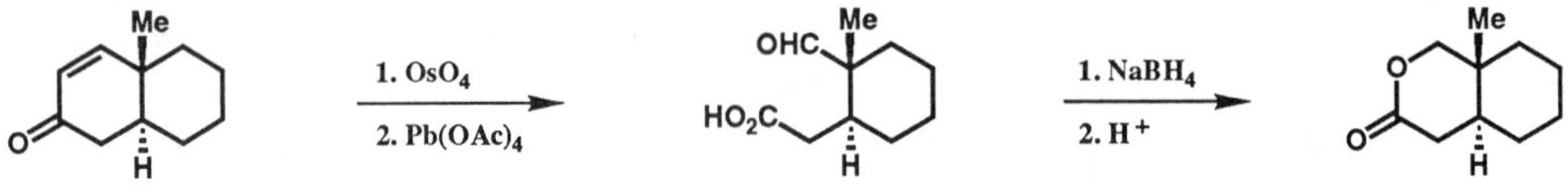
Me
O
H
1. OsO4
2. Pb(OAc)4
OHC
HO2C
1. NaBH4
2. H+

JACS, **1977**, *99*, 6066.

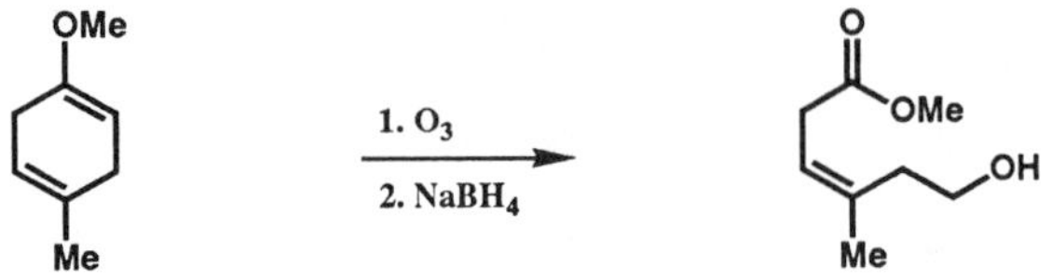
OMe
Me
1. O3
2. NaBH4
O
OMe
OH

JACS, **1968**, *90*, 5618.

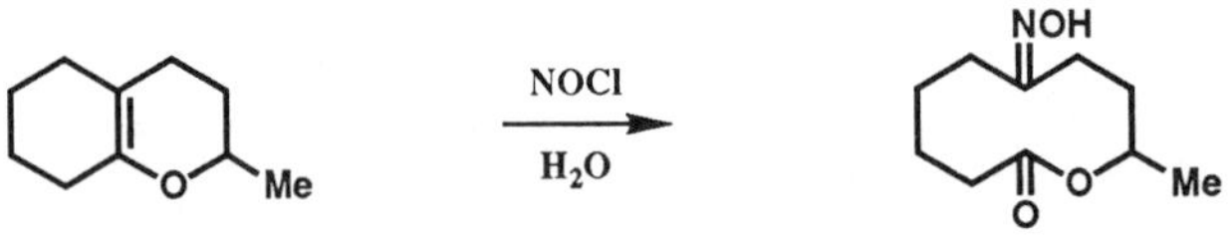
O
Me
NOCl
H2O
NOH

Synth, **1981**, 49.

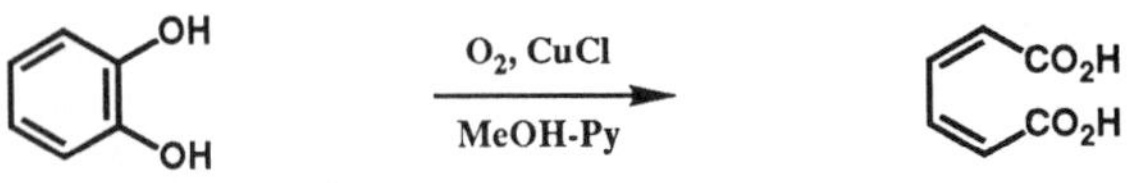
OH
OH
O2, CuCl
MeOH-Py
CO2H
CO2H

JOC, **1976**, *41*, 3992.

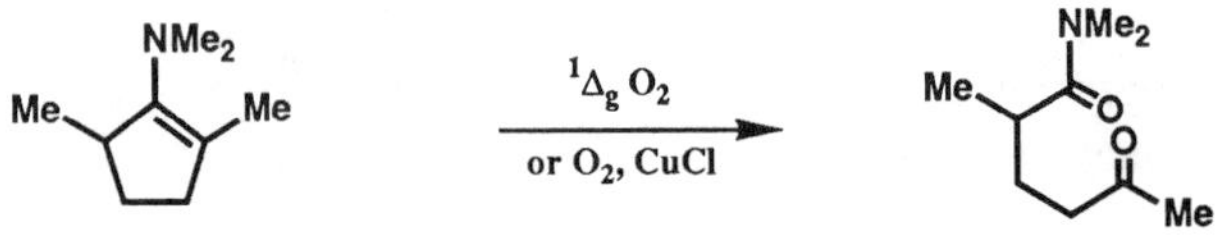
NMe2
Me
Me
1Δg O2
or O2, CuCl
O

JCS, CC, **1969**,314.

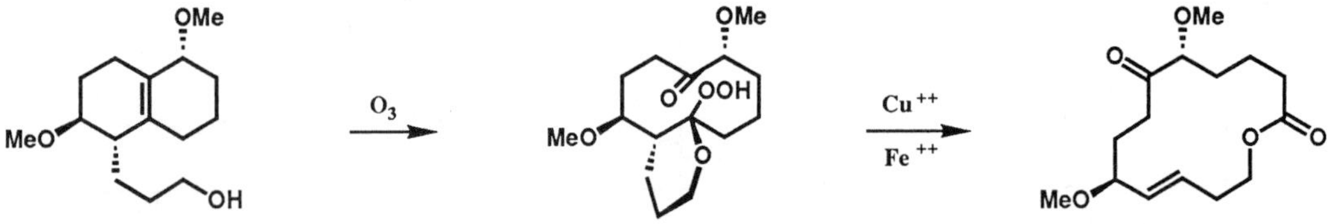
OMe
MeO
OH
O3
OOH
O
Cu++
Fe++

JACS, **1985**, *107*, 2980.

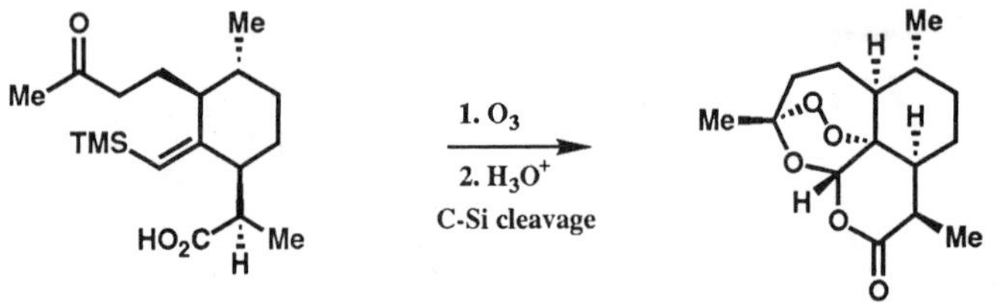
O
Me
TMS
HO2C
H
1. O3
2. H3O+
C-Si cleavage

TL, **1987**, *28*, 4629.

Chart 25

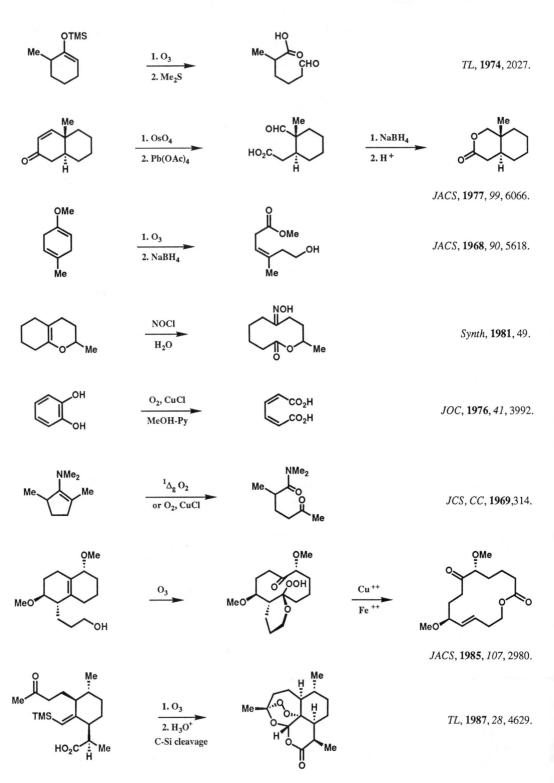

A pair of appendages which are in spatial proximity suggests the application of connective transforms, even if such appendages are not functionalized. Typical of retrosynthetic connections of this type are those shown for targets **201**[60] and **202**.[61]

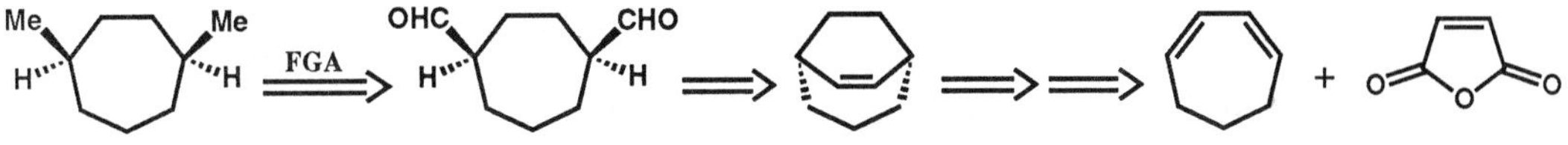

Me
Me
H
H
FGA
OHC
CHO
H
H
+
O
O
O

201

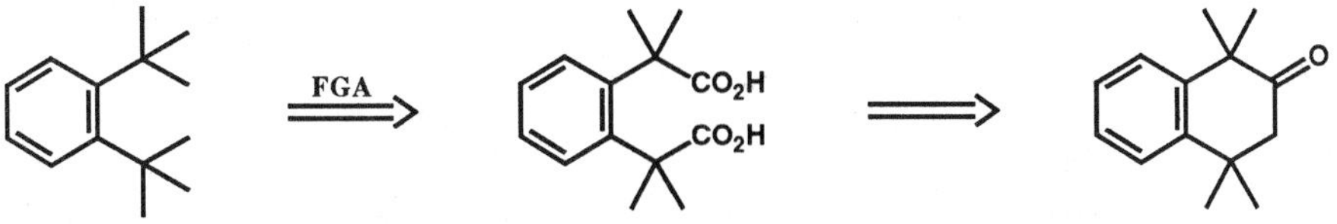

FGA
CO₂H
CO₂H
O

202

Algorithms for machine-generated syntheses by application of connective transforms to target structures containing appendages, medium-sized rings, etc. have been described.[62]

5.8 Functional Group-Keyed Appendage Disconnection

Appendages attached to rings are a source of structural complexity in a large number of interesting target structures. Direct disconnection of the bond which attaches an appendage to a ring is frequently the most effective way of removing that appendage, and often the transform for such cleavage is keyed by nearby functionality. The combination of appendage and functional group defines the retrons for such appendage disconnective transforms. Examples of such appendage disconnections are given below for targets **203**, **204**, and **205**.[63]

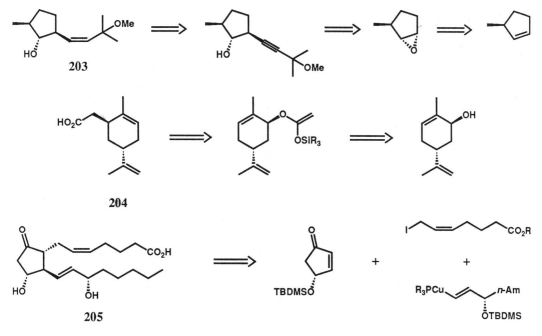

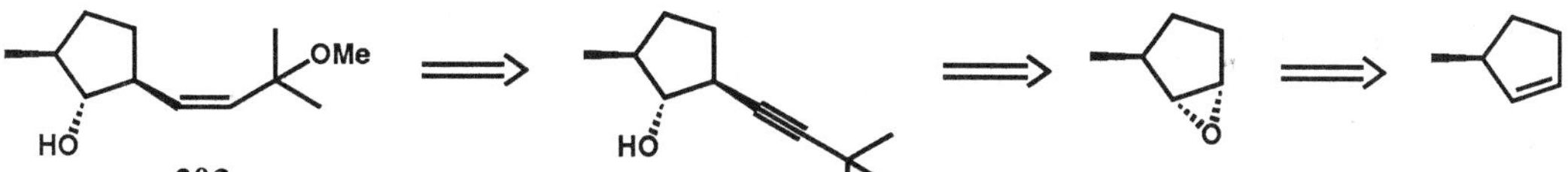

OMe
HO
HO
OMe
O

203

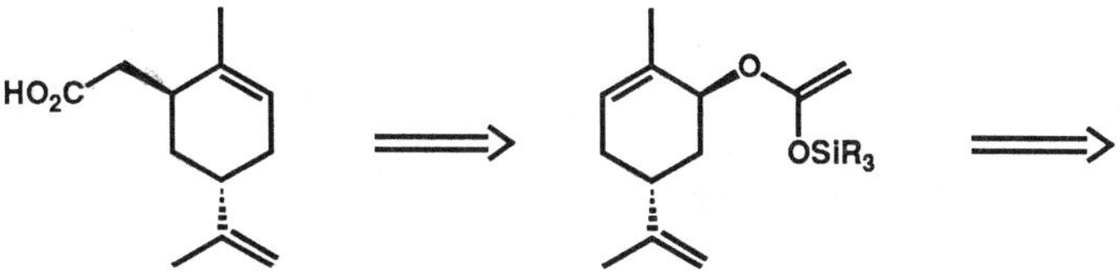

HO₂C
O
OSiR₃
OH

204

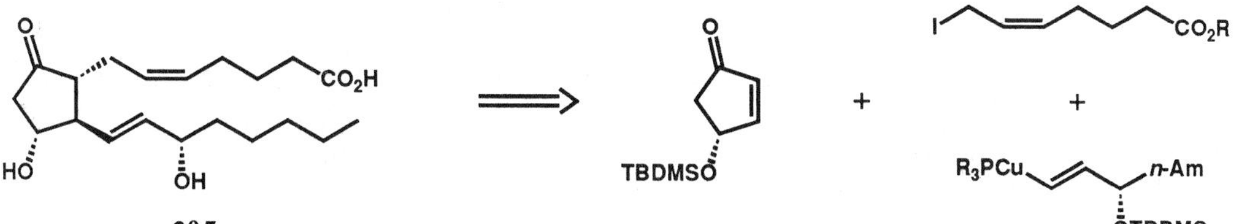

O
CO₂H
HO
OH
O
TBDMSO
+
I
CO₂R
+
R₃PCu
n-Am
OTBDMS

205

Disconnections within an appendage are generally advantageous only when they are directly applicable and also bring other benefits, for example paving the way for the application of simplifying transforms or establishing useful functionality near a ring.

5.9 Strategies External to the Target Structure

Useful strategies for the synthesis of a particular target molecule may result from idiosyncratic or external circumstances which do not apply generally to all synthetic problems. For example, the synthesis of an unstable or sensitive molecule requires planning precautions to ensure that the synthetic steps are not inconsistent with the sensitivity of the reaction products. The impact on synthetic design of such requirements can be considerable especially if the sensitivity or reactivity of the target is incompletely or only roughly known. For instance, the first synthesis of prostaglandin E, (**206**), which was carried out without quantitative knowledge of the sensitivity of the β-hydroxycyclopentanone subunit to acidic or basic conditions, was designed to generate that subunit in the mildest possible way (pH 5 at 0° in water) from imine **207** and amine **208**. That constraint, in turn, sharply restricted the type of synthetic plan which could be used.[64] Subsequently, with a sufficient supply of synthetic prostaglandins, the

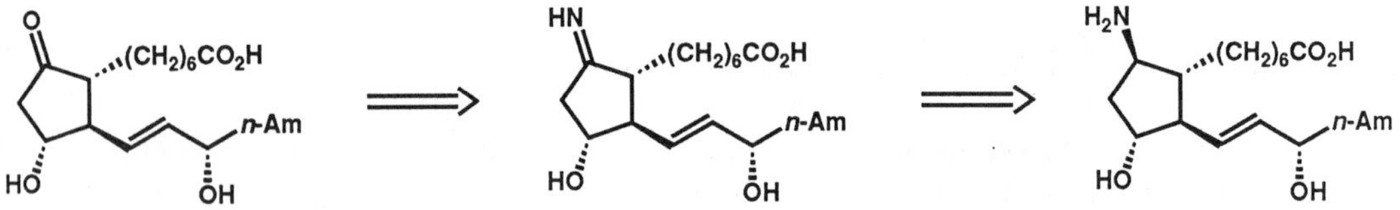

O
(CH2)6CO2H
n-Am
HO
OH
HN
H2N

206 **207** **208**

chemistry of these substances was explored in more detail with the result that other and more effective synthetic routes could be devised.[64e] The strategy for one of these drew further on the special requirements of the problem, specifically the desirability of synthesizing all members of the large family of prostaglandins (e.g. **209** and **210**) from a common synthetic intermediate, lactone **211**.[64e]

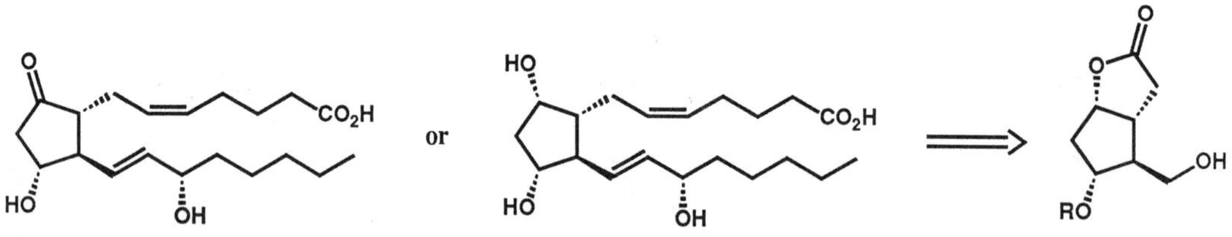

O
CO2H
HO
OH
or
RO

209 **210** **211**

The last-mentioned property of a synthetic process, i.e. versatility, is frequently an important consideration in research on optimally effective therapeutic agents in which the synthesis of a large series of structural analogs from a single intermediate is desirable. This aspect of the problem-solving environment can play a decisive role in synthetic design.

Another external consideration which may strongly influence retrosynthetic analysis is connected with the available body of knowledge on chemical transformations of a target structure. In general the more of such information which is available, the easier the task of synthesis will be, especially if the last stages of a synthesis can be tested with derivatives of the target structure. This option exists for readily available naturally occurring substances, but clearly not for rare or non-existent species. The total synthesis of ryanodine (**212**) was facilitated by the availability of a transformation product, anhydroryanodol (**213**, produced by treatment of ryanodol with acid) and the demonstration that epoxidation of anhydroryanodol (**213**) produces **214** which upon treatment with lithium in liquid ammonia forms ryanodol.[65] In summary, knowledge of the chemistry of a target molecule generally simplifies the discovery of a workable synthetic route.

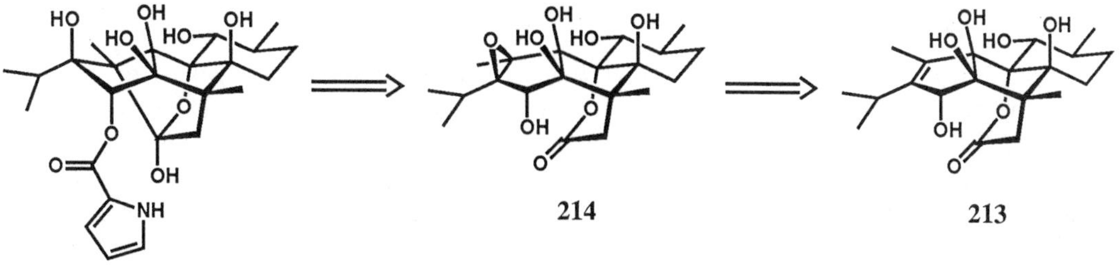

214

213

Ryanodine (212)

The recognition of substructural units within a TGT which represent obstacles to synthesis often provides major strategic input. Certain other strategies result from the requirements of a particular problem with regard to economic, logistical, or legal restrictions on intermediates or starting materials. A TGT which resists retrosynthetic simplification may require that new methodology be developed for a synthesis. Such a program of research can lead to the invention of new chemical processes.

One important human problem-solving strategy is the application of 'imagination' or 'intelligent use of a chain of hypotheses' to guide the search for an effective line of retrosynthetic analysis. This inductive problem-solving dimension has been discussed previously.[66,67]

"The synthetic chemist is more than a logician and strategist; he is an explorer strongly influenced to speculate, imagine, and even to create. These added elements provide the touch of artistry which can hardly be included in a cataloguing of the basic principles of synthesis, but they are very real and extremely important."

"The proposition can be advanced that many of the most distinguished synthetic studies have entailed a balance between two different research philosophies, one embodying the ideal of a deductive analysis based on known methodology and current theory, and the other emphasizing innovation and even speculation. The appeal of a problem in synthesis and its attractiveness can be expected to reach a level out of all proportion to practical considerations whenever it presents a clear challenge to the creativity, originality, and imagination of the expert in synthesis." [66,67]

It is not surprising that multistep synthesis of challenging and complex target molecules is an engine for the discovery of new synthetic principles and novel methodology which may have very broad application. Just as each component of structural complexity can signal a strategy for synthesis, each obstacle to the realization of a chemical synthesis presents an opportunity for scientific discovery.

5.10 Optimization of a Synthetic Sequence

The principles and strategies of retrosynthetic analysis facilitate the discovery of possible synthetic pathways to a complex target structure. Exhaustive and systematic analysis over a period of time generally reveals a surprisingly large number and diversity of such candidate routes. Such extensive analysis, which is a prerequisite in modern synthetic planning, leads to an important phase of synthesis which may be described as the evaluation of alternative synthetic pathways to assess relative merit. The first step in this evaluation is the critical analysis of each possible synthetic pathway in the synthetic direction in detail in order to estimate the potential success and efficiency of each step. It is then necessary to derive for each pathway an optimum ordering of the individual synthetic steps since the best ordering cannot be ascertained (and is certainly not guaranteed) during initial generation of a retrosynthetic sequence, and is only possible after the derivation of a complete pathway. Alternative orderings of the same steps usually differ in merit because the efficiency of individual reactions, the need for protective groups or activative groups, stereoselectivity, and interference by reactive functionality all vary with the precise structure of a reaction substrate. In general an ordering which promises better overall yields, fewer control steps (protective or activative), more practical reagent requirements, and higher certainty of successful execution, is much to be preferred.

The situation with regard to protective groups has been discussed previously.[39,68]

"The deactivation or masking of functional groups is a very important part of synthetic methodology and practice. It may involve the use of an externally derived protective group or take advantage of a connection to another unit within the molecule. The use of these 'control operations' and others (e.g., activation of functional groups or the introduction of directing groups) often contributes crucially to the experimental realization of a synthetic plan. Generally, the larger the number of functional groups in the molecule to be synthesized, the more likely will be the need for and the greater the importance of functional group protection. This area of organic synthesis is characterized less by the availability of a small number of ideal protective groups than by the alternative of a large arsenal of protective groups each with a definitely restricted range of applicability. A protective group must convert some functional group to a form which will not cause interference with reactions aimed at modifying other units in the molecule. Further, there is a reciprocal requirement that the various units in the molecule not interfere with the attachment or removal of a protective group." [39]

In an ideal chemical synthesis, functionality created in a particular step never interferes with succeeding steps. In practice, it is often impossible to carry out one or more of the succeeding steps directly because of incompatibility between existing functional groups and the reagents required for these succeeding transformations. Occasionally, reordering the steps in the synthesis and/or changing the reagents will remove such problems of incompatibility, but often the chemist must resort to protection and subsequent deprotection of the interfering functionality.

Some of the most elegant examples of functional group protection are those in which a group can be masked by reaction with another functional group in the molecule. However, this so-called 'internal protection' process is less common than 'external protection,' in which the protective group is derived from the protecting reagent rather than from the reactant itself. The methodology of external protection has advanced considerably in recent years, and very useful compilations of protective groups for a variety of functional group types exist.[69]

"The process of selecting a protective group involves a number of discrete steps. First, the proposed scheme is summarized, with reactants, reaction conditions, and products delineated for each synthetic step. Next, the relative reactivities of the functional groups in each reactant and product are evaluated, and potentially interfering groups are identified. For each such interfering group, a number of possible protective groups are considered. Each candidate protective group is attached hypothetically, and its reactivity toward the reaction conditions for successive steps is evaluated. If the proposed protective group is not stable toward all these reaction conditions, it is rejected.

"A number of other factors must also be taken into consideration. Candidate protective groups must stand up not only to reaction conditions for the succeeding steps in the synthesis but also to the conditions for addition and removal of other protective groups. In addition, the reactivity of each candidate protective group as a reagent itself must be considered, not only toward unprotected functionality but also toward other protective groups present at each step. The optimum stage for addition and removal of each protective group must be chosen, with the possibility that a single protective group may, if carried through several steps, serve to protect a functional group in two or more steps in which that functional group is expected to be interfering. Often the initially chosen set of protective groups for a synthesis will be changed to allow for simultaneous protection of more than one functional group or for simultaneous removal of more than one protective group. Finally, it may be possible to minimize the number of protective groups or to permit the use of certain desirable protective groups by reordering the steps in the original synthetic scheme." [68]

It is a common experience in synthetic chemistry that a truly optimal ordering of a synthetic route may not be possible in the planning stage, but may have to determined experimentally. The precise information necessary for the complete and unambiguous evaluation of each step in a possible synthesis is hardly ever available. Nonetheless it is clearly wise to try to optimize a synthetic plan on the basis of available information before the experimental approach begins. Such an effort may suggest certain preliminary or "model" experiments that can be helpful in the choice or refinement of a synthetic plan. It is also obviously desirable to devise and consider alternate or bypass paths for each problematic step of a synthetic sequence.

CHAPTER SIX

Concurrent Use of Several Strategies

6.1 Multistrategic Retrosynthetic Analysis of Longifolene (215)

A foregoing discussion of *types of strategies for retrosynthetic analysis* (Section 1.8) stressed the importance of the concurrent use of as many independent strategies as possible in devising retrosynthetic pathways. *In general the greater the number of strategies which are used in parallel to develop a line of analysis, the easier the analysis and the simpler the emerging plan is likely to be.*[67] This *multistrategic approach* to the solution of problems in chemical synthesis is illustrated in this Chapter by a number of specific examples, starting with longifolene (**215**),[70] a molecule of some historical interest both from the point of view of structure determination and synthesis. The dominating complexity of structure **215**, with its bridge across two bridges and quaternary appendages, is topological. Of the bonds which are strategic for disconnection, *a*, *b*, and *c* (see Section 3.6 and Chart 16), *a* stands out since its cleavage reduces the cyclic network to a simple 6-7 *cis*-fused ring system with no increase in appendage count. This disconnection is also stereochemically strategic since it results in the removal of two stereocenters, both of which are unambiguously clearable. There are a number of C-C disconnective transforms which can be studied in turn as T-goals to effect this retrosynthetic cleavage, including internal Michael, internal S_N2 enolate alkylation, and cation-olefin addition. Examination of each of these as T-goals suggests bicyclic precursor structures **216**, **217** and **218**. Of these **216** was especially attractive since the required Michael cyclization found precedent[71a] and since **216** maps closely onto the readily prepared Wieland-Miescher ketone (**219**),[71b] the mismatch being correctable by addition of two carbons and ring expansion of **219**. Ketone **219** therefore serves as a logical S-goal. The design of the first synthesis of **215** depended on this reasoning which essentially incorporates simultaneously topological, stereochemical, functional group, and transform-based strategies (see Section 9.2). At the level of intermediate **216** the S-goal strategy employing **219** (now available by enantioselective synthesis[71b]) became important. The full potential of this line of analysis remains untapped. For instance, the disconnection of bond *a* in **215** to give **218** is

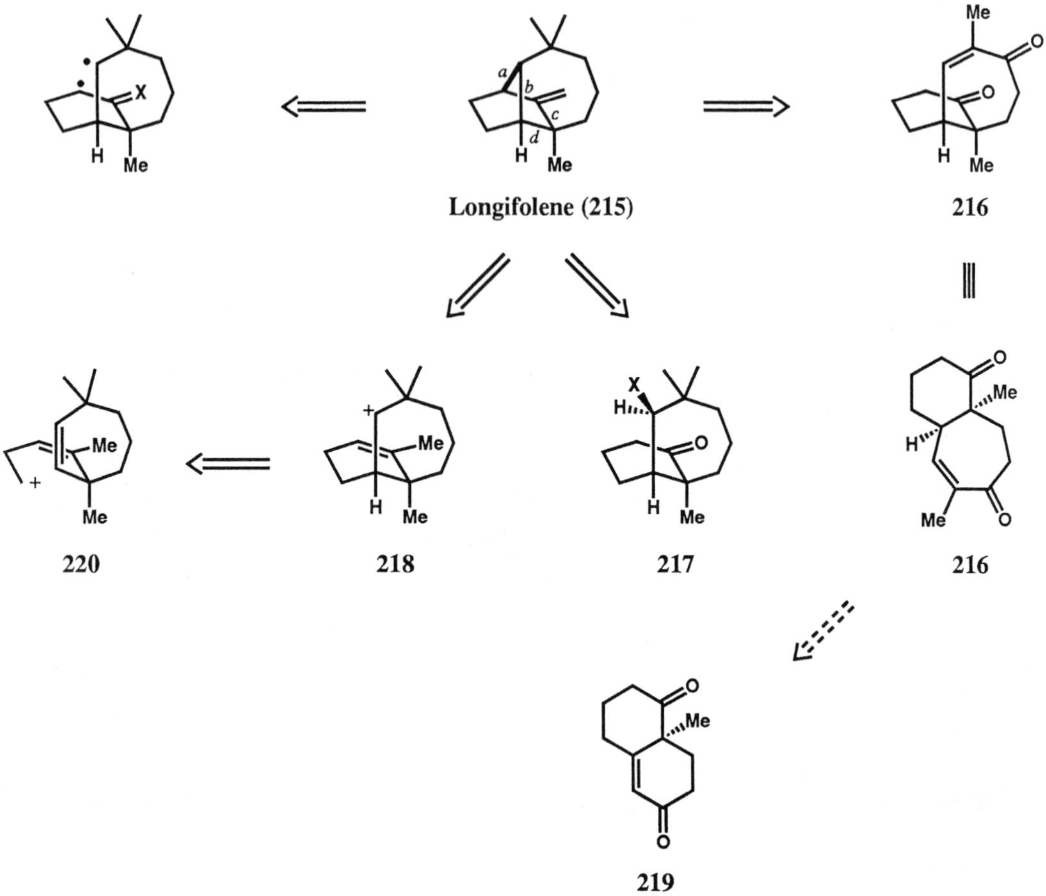
Longifolene (215)
216
216
220
218
217
219
X
H
Me
a
b
c
d
O
+
H''''

interesting because of the possibility of further simplification to monocyclic cation **220**. The strategic bond- pair (Diels-Alder) disconnection of longifolene to a precursor such as **221** is also appealing as a retrosynthetic step.

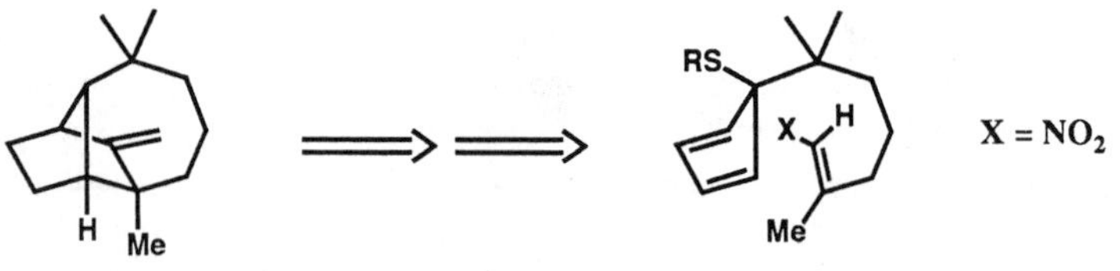
RS
X
H
Me
X = NO2

215 221

It was the problem of longifolene synthesis that suggested the idea of rigorously stepwise retrosynthetic planning and the use of independent topological strategies (in 1957). Widespread interest in the synthesis of target **215** as a purely academic problem (naturally produced longifolene is abundant and cheap) is evident from the subsequent development of several other syntheses.[72]

6.2 Multistrategic Retrosynthetic Analysis of Porantherine (222)

Porantherine (**222**) represents a typical problem in alkaloid synthesis and a good test of the effectiveness of multistrategic retrosynthetic analysis.[73] Bond *a* of this tetracyclic alkaloid is simultaneously strategic for disconnection, on a path between two functional groups, and attached to an unambiguously clearable stereocenter, and its disconnection is indicated for all three reasons. Further, it contains a partial retron (4-functionalized piperidine) for the doubly disconnective Mannich annulation transform, which is a strong candidate as T-goal. Functional group interchange of C=C in **222** for carbonyl as in **223** sets up the double Mannich retron which disconnects the strategic bond *a* in **223** and provides the reactive precursor **224** and its equivalent, the bicyclic intermediate **225**. Since **225** contains a full Mannich retron, it can be directly disconnected to monocyclic imine **226** which is equivalent to the acyclic diketo aldehyde **227**. Each of these retrosynthetic steps must be scrutinized in the synthetic direction to evaluate

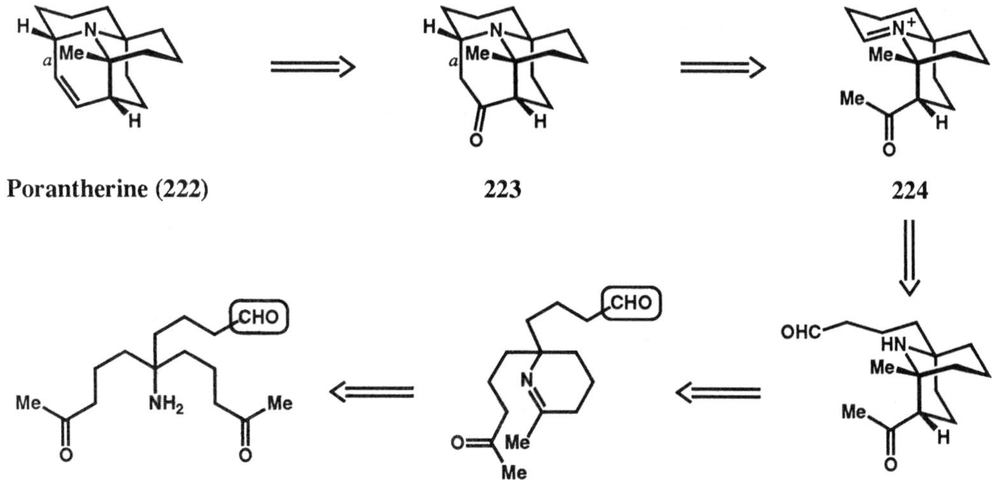
H
N
Me
a
H
O
N+
OHC
HN
CHO
NH2
Me

Porantherine (222) | **223** | **224** | **225** | **226** | **227**

the workability and merit of the transform in practice. From this evaluation it is clear that the aldehydic formyl group in **227** and **226** ought to be protected for the first two cyclization steps. During initial retrosynthetic analysis that protection can simply be indicated by boxing the functional group to be protected. The appropriate protecting group(s) and the optimum ordering of steps can be determined by subsequent analysis of the retrosynthetic sequence. This readily derived plan formed the basis of a simple synthesis of (±)-porantherine (Section 8.2).[73]

6.3 Multistrategic Retrosynthetic Analysis of Perhydrohistrionicotoxin (228)

Perhydrohistrionicotoxin (**228**) is a rare and highly bioactive alkaloid which is a useful tool in neuroscience for the study of ion transport.[74] Contributing to molecular complexity in this case are the four stereocenters (one being at a spiro origin), the amino and hydroxyl functions (*cis* about the cyclohexane ring), the two appendages, and the two spiro rings. A prominent strategic task is stereochemical, to render all stereocenters clearable, if possible in combination with functional group- and appendage-keyed transforms. The application of transforms which simultaneously remove stereocenters, appendages and functional groups would, if successful, probably lead to short and simple synthetic pathways. Stereocontrolled removal of the stereocenter common to the spiro rings is possible either by disconnection or by antithetic conversion of the piperidine ring to a symmetrical subunit, i.e. cyclopentane. The attractiveness of the latter possibility is signalled by the availability of an appropriate stereocontrolled T-goal, the Barton nitrite functionalization transform, in tactical combination with the Beckmann rearrangement transform. It also is suggested by the availability of the readily prepared spiroketone **235** as a possible S-goal. The combined T-goal and S-goal driven search then suggests the retrosynthetic sequence **228** ⇒ **235**. Disconnection of the amyl appendage in **228** with removal of a clearable stereocenter leads to **229**, which by functional group interchange (to **230**) and Beckmann transforms is converted to oxime **231**. Application of the Barton nitrite functionalization transform allows clearance of the spiro stereocenter to produce **233** via the nitrite ester **232**. Hydroboration and alkylmetal carbonyl addition transforms convert **233**

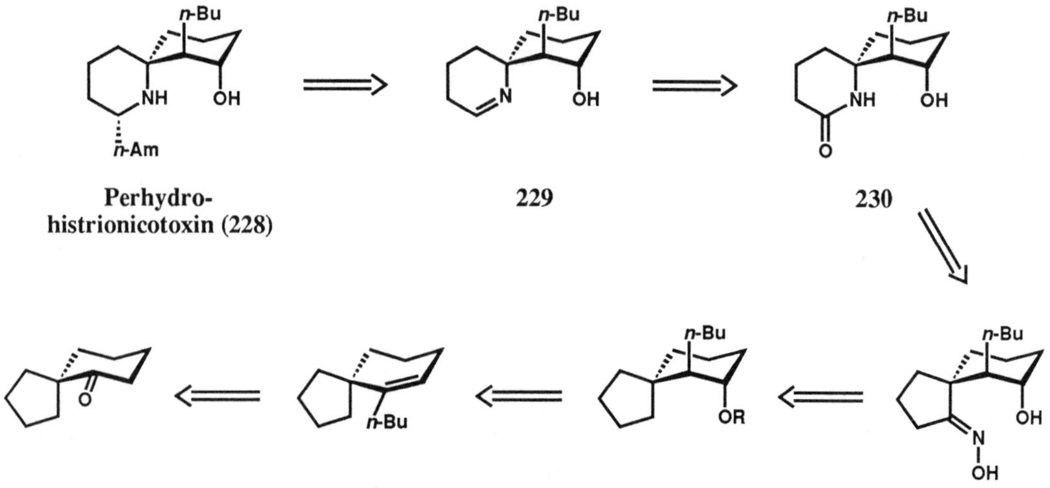
n-Bu
NH
OH
n-Am
n-Bu
N
OH
n-Bu
NH
OH
O
n-Bu
OH
N
OH
n-Bu
OR
n-Bu
O

Perhydro-histrionicotoxin (228) — **229** — **230** — **231** — **232** R = NO; **233** R = H — **234** — **235**

successively to **234** and S-goal **235** (made by pinacol coupling of cyclopentanone and acid-catalyzed rearrangement of the pinacol). A short synthesis of perhydrohistrionicotoxin by this retrosynthetic approach was demonstrated experimentally (Section 8.1).[75]

6.4 Multistrategic Retrosynthetic Analysis of Gibberellic Acid (236)

The plant bioregulator gibberellic acid (**236**) resisted synthesis for more than two decades because it abounds in all of the elements contributing to complexity, including reactive and dense functionality, in an usually forbidding arrangement. One contributing strategy in the retrosynthetic plan which led to the first synthesis of **236**[18] was the removal of those functional groups in ring A most responsible for chemical reactivity, with concomitant removal of two clearable stereocenters. These operations were also suggested by a transform-based strategy aimed at stereocontrolled disconnection of the A ring, which was identified as strategic for disassembly on the basis of combined topological and stereochemical considerations. The internal Diels-Alder transform was selected as T-goal because of the twin objectives of ring A disconnection and stereocenter clearance. With this combined guidance, halo- and hydroxylactonization transforms were applied to **236** to generate precursor **237** in five retrosynthetic steps. Further multistep search to apply the internal Diels-Alder transform produced **239** via a number of intermediate precursors including **238**. The inversion at C(6) of ring B was performed for purpose of stereocontrol and was justifiable because of the known 6α → 6β epimerization of the carboxyl function in the gibberellic acid series. A logical T-goal for the retrosynthetic simplification of **239** via the corresponding alcohol (*cis*-fused hydrindane) was the quinone-Diels-Alder transform. This transform could be applied successfully to **240** after 5 ⇒ 6 ring expansion, using the tactical combination of aldol and C=C oxidative cleavage transforms to generate **241**, and subsequent cleavage of the two-carbon bridge at the strategic bond to give **242**. The retron required for the quinone-Diels-Alder transform was established by the

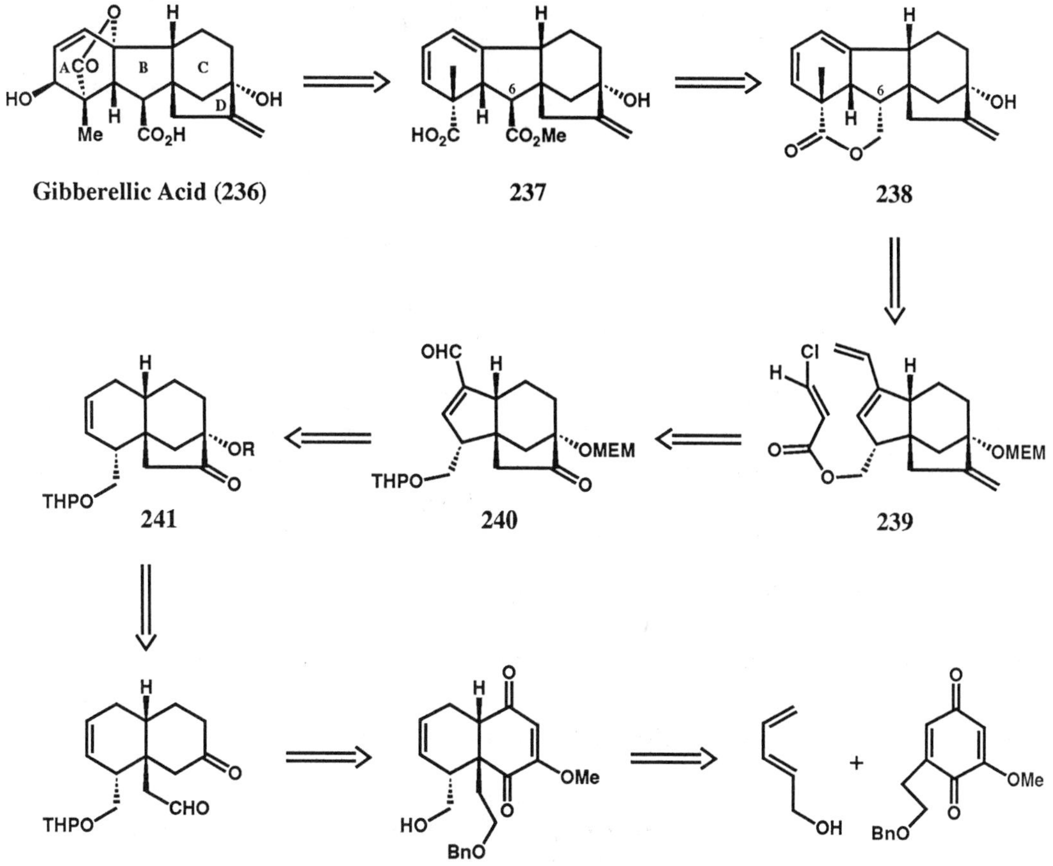

Gibberellic Acid (236)
237
238
239
240
241
242
243

application of FGA and FGI subgoals resulting in **243** which underwent disconnection to the diene and quinone starting materials. Detailed scrutiny of the individual steps and the overall retrosynthetic sequence provided data on possible protective groups and allowed a preliminary ordering of the 36 steps which were required for the total synthesis of **236** (see Section 10.8).[18] In subsequent work, a different and shorter route to the alcohol corresponding to **239** was developed from ketone **244**.[76,77] Strategic bond and functional group keyed disconnection of **244**

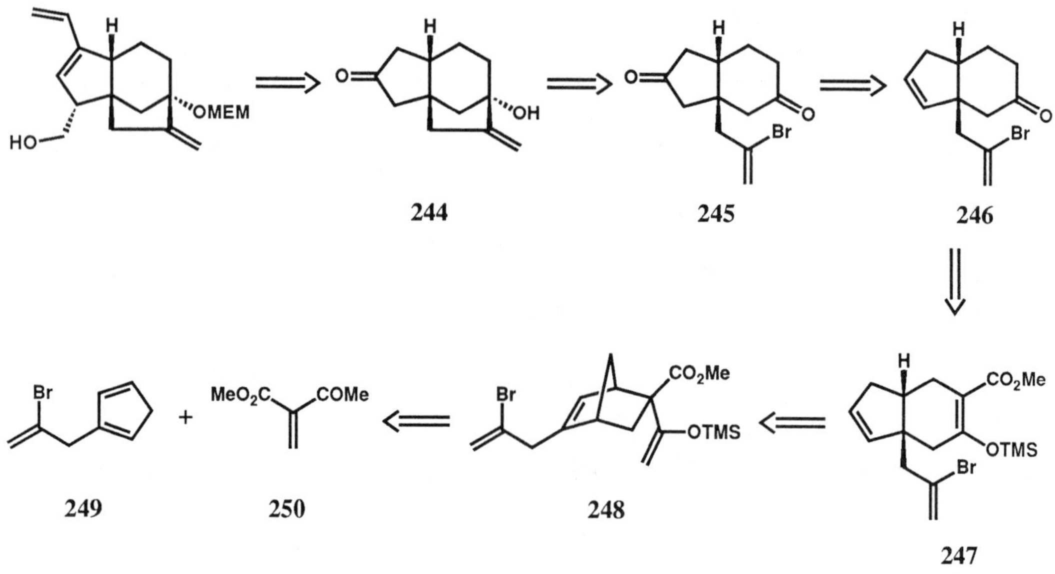
HO
OMEM
H
O
OH
244
Br
245
246
MeO2C
COMe
CO2Me
OTMS
249
250
248
247

led to the *cis*-hydrindanedione **245**. Functional group interchange of **245** to form **246** established the retron for the Cope rearrangement transform as T-goal. Application of this transform in tactical combination with the Diels-Alder transform produced the sequence **246** ⇒ **247** ⇒ **248** ⇒ **249** + **250** after control elements were added as a result of subsequent optimization.[76] In such a long and complex retrosynthetic analysis, it is invariably found that knowledge obtained from the study of the retrosynthetic steps at the lower levels of the EXTGT tree can provide information which influences the selection of steps and intermediates at the upper levels of the tree. This is usually the case either when an impasse is reached at the lower levels or when some particular pathway leads to an especially appealing lower-level sequence. Or, to state matters somewhat differently, there can be an important feedback of information from the lower levels to the upper levels of EXTGT tree analysis in the case of quite complex targets.

Still another synthetic route to gibberellic acid has been developed which differs from those above in the transforms used to disconnect the various rings, but which also depends on the initial disconnection of ring A (using aldol and carbonyl addition transforms).[78]

6.5 Multistrategic Retrosynthetic Analysis of Picrotoxinin (251)

The neurotoxin picrotoxinin (**251**) has been synthesized[32] by a plan which was derived retrosynthetically using the multistrategic approach. Antithetic removal of the two bridging lactone groups is a logical structural simplification for topological reasons, because it allows the generation of core functionality, and because it can be carried out with clearance of two stereocenters. The olefin bislactonization transform (or the halolactonization plus S_N2 displacement tactical combination) can be directly applied to generate precursor **252** from target **251**, except that the isopropenyl group would effectively interfere with the corresponding synthetic reaction. Internal protection of the spatially proximate hydroxyl and isopropenyl functions in the form of **253**, however, renders the bislactonization transform operable to produce

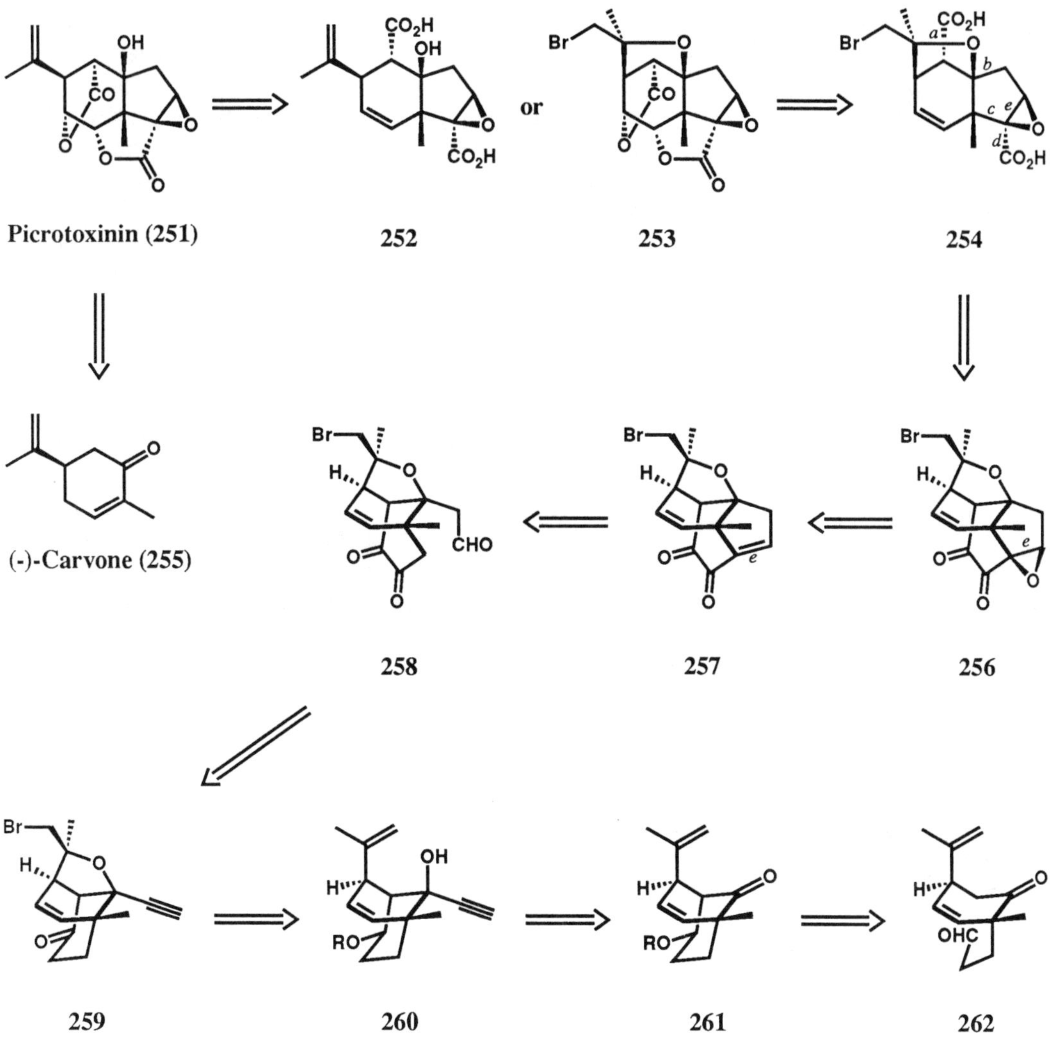

Picrotoxinin (251)
252
or
253
254
(-)-Carvone (255)
258
257
256
259
260
261
262

254. It can be seen that precursor **254** maps sufficiently well onto the structure of (-)-carvone (**255**) to suggest this as an eligible S-goal. That line of analysis dictates bonds *a, b, c,* and *d* of **254** to be strategic for disconnection whereas all of the bonds corresponding to the carvone framework are to be preserved. The proximity of the two carboxylic acid functions in **254** suggests their connection to **256** for the purpose of eventual stereocenter clearance and also to actuate bonds *a, d,* and *e* for retrosynthetic disconnection. After functional group interchange to **257**, bond *e* in this structure can be disconnected to afford **258**. From this point the remaining disconnections are keyed by the functionality of the intermediates which allow progressive antithetic simplification by the pathway shown to **262**, and eventually (-)-carvone (**255**). This line of analysis was supplemented by the addition of the required protective and control elements and optimization of the ordering of steps to provide a successful plan (see Section 9.17).[32] The synthesis of a related neurotoxin, picrotin, was accomplished in four steps from **251**.[79]

6.6 Multistep Retrosynthetic Analysis of Retigeranic Acid (263)

The novel pentacyclic sesterterpene retigeranic acid (**263**) possesses functionality in ring C, a ring which is also central and, hence, crucial for disassembly. In this regard, an attractive S-goal is exemplified by hydrindanone **264**, corresponding to rings A and B of **263**. Although ring C of **263** can be directly disconnected (T-goal strategy) at bond *a* by the vinylcyclopropane-cyclopentene rearrangement transform, the full retron being present, the synthetic stereoselectivity of this reaction (antithetic stereocenter clearability) was difficult to evaluate. Consequently another option for disconnection of ring C via **265** (generated from **263** by the tactical combination of aldol and C=C oxidative cleavage transform) was examined taking the Diels-Alder transform as T-goal. Direct disconnection of bonds *a* and *b* of **265** is not synthetically valid despite the presence of the Diels-Alder retron in that structure. Consequently, a more favorable precursor for Diels-Alder disconnection (i.e. suitably actuated), had to be generated by additional retrosynthetic steps. Specifically this approach called for the placement of electron withdrawing groups at the dienophile-related carbons of the Diels-Alder retron. In turn, this condition dictated the retrosynthetic cleavage of rings D and E, preferably in one step, by disconnection of the strategic bond pair designated *c* and *d* in **265**. Toward that end, **265** was converted retrosynthetically in several steps to **266** which could be antithetically cleaved to **267** by application of the ketene-olefin [2 + 2] addition transform since all stereocenters are definitely clearable. Application of C=C disconnective and FGI transforms converted **267** to a predecessor (**268**) which was suitably actuated for Diels-Alder transform application to provide precursors **269** and **270**. The first synthesis of (±)-retigeranic acid was accomplished stereoselectively by a route which corresponds to this retrosynthetic scheme (Section 10.10).[80]

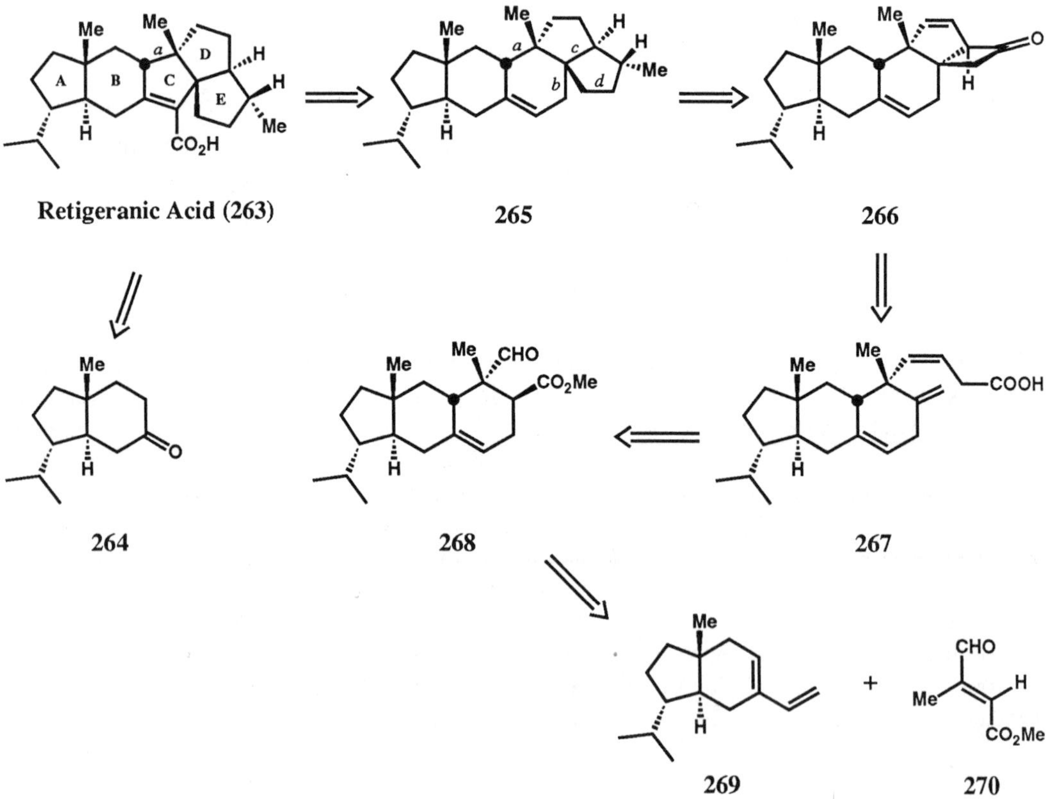
Me
Me
a
D
H
A
B
C
E
H
Me
H
CO_2H
Me
Me
a
c
H
H
Me
b
d
H
Me
Me
O
H
H
Me
O
H
Me
Me
CHO
CO_2Me
H
Me
Me
COOH
H
Me
H
CHO
H
Me
CO_2Me
+

Retigeranic Acid (263) 265 266 264 268 267 269 270

Subsequently, a study of the synthetic route corresponding to the direct disconnection of ring C, **263** ⇒ **271** was reported.[81] Although the critical synthetic conversion of **271** to **263** showed only modest stereoselectivity (*ca.* 3 : 1), the synthetic sequence is convergent.

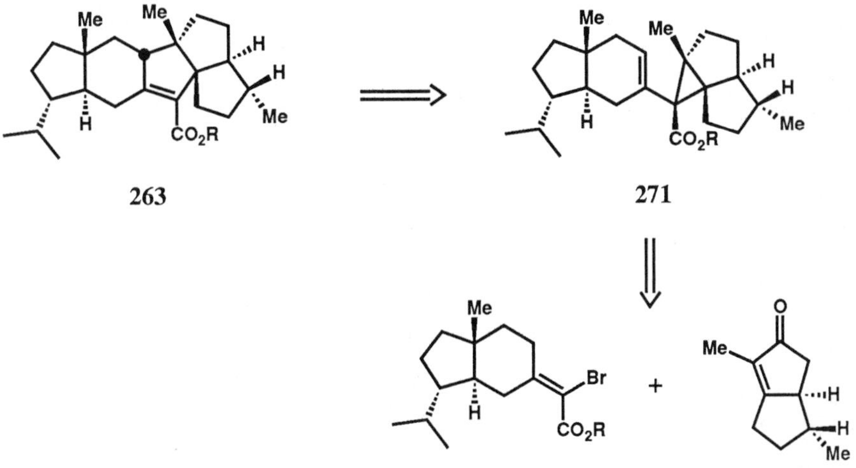
Me
Me
H
H
H
Me
CO2R
Me
Me
H
H
H
Me
CO2R
Me
Br
H
CO2R
+
O
Me
H
H
Me

263 271

6.7 Multistrategic Retrosynthetic Analysis of Ginkgolide B (272)

The ginkgo tree produces a family of polycyclic terpenoids of which ginkgolide B (**272**) is especially interesting as a therapeutic agent of potentially wide application. The problem of synthesis of **272** well illustrates not only the hierarchical structure of topological complexity but also the existence of different levels of functional-group and stereochemical-based complexity. Of the six rings which are present, the spiro carbocyclic rings, labeled A and B in **272**, are synthetically more critical than the four remaining oxacyclic units. The three oxacyclic rings (D, E and F) which are attached to the two spiro carbocyclic rings and to each other are next in order of topological import, followed by ring C which is essentially an appendage on ring A. The carbocyclic rings A and B are the most challenging for disassembly, clearly constituting a deeper topological problem than the remaining rings. Awareness in this case, as in others, of the *deeper problem* is frequently helpful in retrosynthetic analysis since it sharpens the focus of multistrategic attack. If rings A and B resist retrosynthetic disconnection, other rings become strategic units for disassembly as a means of simplifying the deeper problem. Retrosynthetic analysis of the ginkgolide B structure (**273**) was carried out by concurrent use of several different strategies which converged strongly on one particular synthetic approach. Although several variations are possible within this approach, one particular line of analysis was selected for experimental study. Its validity was demonstrated eventually by successful execution of the corresponding synthetic plan.[82-84] The individual strategies involved in the analysis are listed below.

1. Rings C and F are each peripheral to the cyclic network and also the sites of the most reactive functional groups. Reduction of reactivity and functionality at those locations is clearly strategic. Although atoms 10 and 11 of ring F are not directly disconnectible, the reactivity at these positions can be reduced, for example by functional group interchange to **273**. Ring C is potentially disconnectible by the tactical combination of hydroxylactonization and aldol transforms, since the required retrons are present.

Such a dissection, which would both remove functionality and lower reactivity, would also reduce the number of stereocenters by four. Simplification of this magnitude is

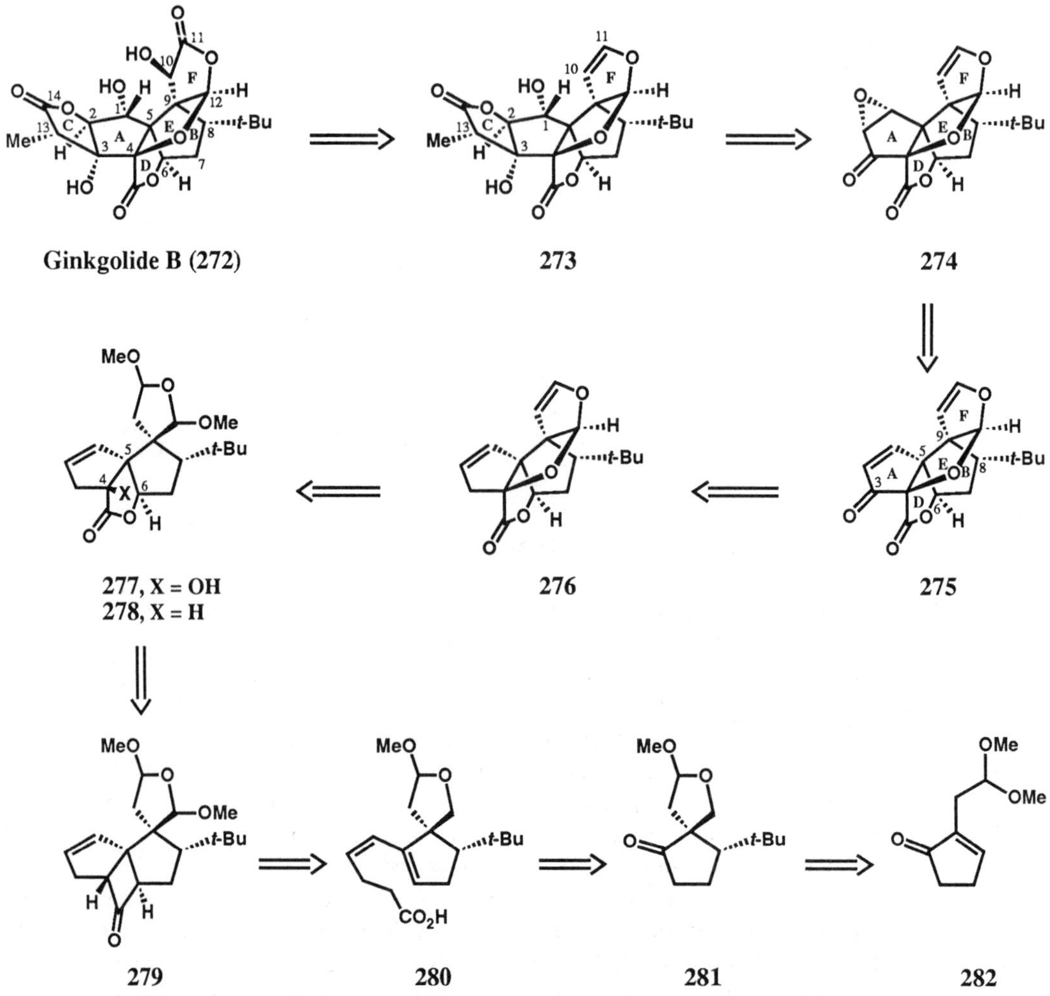
Ginkgolide B (272)
273
274
275
276
277, X = OH
278, X = H
279
280
281
282
t-Bu
MeO
OMe
CO2H

strategically indicated provided that the transforms are valid and the stereocenters are clearable. The transforms are not valid for **272** itself because of FG interference, but are acceptable for **273**, which is also so structured as to favor clearability of the stereocenters at C(1)-C(3). Clearability of C(13), which is presumably possible by use of a Z-propionate enolate as an aldol reagent, is favored by the trigonal geometry of carbons 10 and 11. Application of the hydroxy lactonization/aldol tactical combination of simplifying transforms to **273** provides successively **274** and **275**.

2. Of the two carbocyclic rings in **272** or in simpler precursors such as **275**, ring A, which has no independent stereocenters, is more strategic for disconnection. Indeed, ring B and the carbon stereocenter bearing the *t*-butyl (C-8) qualify for preservation as initial or *origin* structural units for the synthesis. The stereocenter at C(8) with its bulky

t-butyl substituent can potentially direct the stereoselective elaboration of the adjacent C(9), and subsequently the C(5) and C(6) stereocenters on the B ring.

3. The remaining oxacyclic rings of **274** or **275** (D, E and F) are topologically strategic for disconnection. Attractive procedures for such antithetic simplification are (a) coupled removal of rings E and F or (b) E and D. The tetrahydrofuran ring E is also crucial for disconnection because of its central relationship to the other rings (maximally fused ring). One approach to ring E disconnection is outlined by the sequence **275** ⇒ **276** ⇒ **277** ⇒ **278**.

4. Among the most logical disconnections of the carbocyclic rings of **272** (or the other precursors shown extending to **278**) is that involving the strategic C(4)-C(5), C(6)-O bond pair in **272** - **278**.

5. The highest priority ring disconnective T-goals for **272** are those which disconnect a cocyclic 5,5-fusion bond and offexendo bond pair. The internal ketene-olefin cycloaddition in tactical combination with the Baeyer-Villiger transform is well suited to the double disconnection of such a cyclopentane-γ-lactone ring pair.

These strategies guide the retrosynthetic conversion of **272** to **278** and the further conversion of **278** via **279** to **282**. The *t*-butyl substituent actuates the clearability of the stereocenters in **279**. Further retrosynthetic simplification as dictated by basic FG-, stereochemical and topological strategies then leads from **280** to **281** and to **282**, a previously described substance. The successful synthesis followed closely the above outlined retrosynthetic scheme. An enantioselective process was devised for the synthesis of **281** from **282** (see Section 10.12).[67, 83]

REFERENCES FOR PART ONE

1. E. J. Corey and W. T. Wipke, *Science,* **1969**, *166*, 178-192.

2. See, I. Fleming, *Selected Organic Syntheses* (J. Wiley, New York, 1972).

3. W. E. Bachmann, W. Cole, and A. L. Wilds, *J. Am. Chem. Soc.*, **1939**, *61*, 974-975.

4. (a) H. Fischer and A. Kirstahler, *Ann.,* **1928**, *466*, 178-188; (b) H. Fischer and K. Zeile, *Ann.,* **1929**, *468*, 98-116.

5. S. A. Harris and K. Folkers, *J. Am. Chem. Soc.*, **1939**, *61*, 1242, 1245, 3307.

6. R. B. Woodward and W. von E. Doering, *J. Am. Chem. Soc.,* **1945**, *67*, 860-874.

7. N. Anand, J. S. Bindra, and S. Ranganathan, *Art in Organic Synthesis* (Holden-Day, San Francisco, 1970).

8. E. J. Corey, A. K. Long, and S. D. Rubenstein, *Science,* **1985**, *228*, 408-418.

9. R. B. Woodward in *Perspectives in Organic Chemistry,* edited by A. R. Todd (Interscience, New York, 1956) pp. 155-184.

10. (a) E. J. Corey, *Quart. Rev. Chem. Soc.*, **1971**, *25*, 455-482; (b) E. J. Corey, *Pure and Applied Chem.*, **1971**, *2*, 45-68.

11. See also E. J. Corey, W. J. Howe, and D. A. Pensak, *J. Am. Chem. Soc.,* **1974**, *96*, 7724-7737.

12. See, E. J. Corey and B. B. Snider, *J. Am. Chem. Soc.*, **1972**, *94*, 2549-2550 for a synthesis of (±)-**37**.

13. D. Wagner, J. Verheyden, and J. G. Moffatt, *J. Org. Chem.*, **1974**, *39*, 24-30.

14. E. J. Corey and H. E. Ensley, *J. Am. Chem. Soc.*, **1975**, *97*, 6908-6909.

15. W. Oppolzer, *Angew. Chem. Int. Ed.*, **1984**, *23*, 876-889.

16. E. J. Corey and J. P. Dittami, *J. Am. Chem. Soc.*, **1985**, *107*, 256-257.

17. S. I. Salley, *J. Am. Chem. Soc.,* **1967**, *89*, 6762-6763.

18. (a) E. J. Corey, R. L. Danheiser, S. Chandrasekaran, P. Siret, G. E. Keck, and Jean-Louis Gras, *J. Am. Chem. Soc.*, **1978**, *100*, 8031-8034; (b) E. J. Corey, R. L. Danheiser, S. Chandra Sekaran, G. E. Keck, B. Gopalan, S. D. Larsen, P. Siret, and Jean-Louis Gras, *ibid.*, **1978**, *100*, 8034-8036.

19. T. Kametani and H. Nemoto, *Tetrahedron,* **1981**, *37*, 3-16.

20. E. J. Corey, A. P. Johnson, and A. K. Long, *J. Org. Chem.*, **1980**, *45*, 2051-2057.

21. (a) J. A. Marshall, G. L. Bundy, and W. I. Fanta, *J. Org. Chem.*, **1968**, *33*, 3913-3922; (b) E. Wenkert and D. A. Berges, *J. Am. Chem. Soc.*, **1967**, *89*, 2507-2509.

22. E. J. Corey and A. K. Long, *J. Org. Chem.*, **1978**, *43*, 2208-2216.

23. L. Werthemann and W. S. Johnson, *Proc. Nat. Acad. Sci. USA*, **1970**, *67*, 1465-1467.

24. See, J. D. Morrison, Editor, *Asymmetric Synthesis* (Academic Press, New York), Vols. 1-5.

25. See, B. E. Rossiter in *Asymmetric Synthesis*, edited by J. D. Morrison (Academic Press, New York) Vol. 5, 193-246.

26. D. A. Evans, S. L. Bender and J. Morris, *J. Am. Chem. Soc.*, **1988**, *110*, 2506-2526.

27. E. J. Corey, A. K. Long, J. Mulzer, H. W. Orf, A. P. Johnson, and A. P. Hewett, *J. Chem. Inf. Comp. Sci.*, **1980**, *20*, 221-230.

28. See, L. F. Fieser and M. Fieser, *Steroids*, (Reinhold Publishing, New York, 1959) pp. 645-659.

29. H. J. Dauben, Jr. and D. J. Bertelli, *J. Am. Chem. Soc.*, **1961**, *83*, 4657-4660.

30. G. Stork and F. H. Clarke, *J. Am. Chem. Soc.*, **1961**, *83*, 3114-3125.

31. E. J. Corey and S. Nozoe, *J. Am. Chem. Soc.*, **1965**, *87*, 5728-5733.

32. E. J. Corey and H. L. Pearce, *J. Am. Chem. Soc.*, **1979**, *101*, 5841-5843.

33. S. Hanessian, "*Total Synthesis of Natural Products:* The Chiron Approach" (Pergamon Press, Oxford, 1983).

34. (a) O. L. Chapman, M. R. Engel, J. P. Springer, and J. C. Clardy, *J. Am. Chem. Soc.*, **1971**, *93*, 6696-6697; (b) D. H. R. Barton, A. M. Deflorin, and O. E. Edwards, *J. Chem. Soc.*, **1956**, 530-534.

35. E. J. Corey, W. J. Howe, H. W. Orf, D. A. Pensak, and G. Petersson, *J. Am. Chem. Soc.*, **1975**, *97*, 6116-6124.

36. (a) E. J. Corey and T. Hase, *Tetrahedron Lett.*, **1979**, 335-338; (b) W. C. Still and J. C. Barrish, *J. Am. Chem. Soc.*, **1983**, *105*, 2487-2488; (c) S. L. Schreiber, T. S. Schreiber, and D. B. Smith, *J. Am. Chem. Soc.*, **1987**, *109*, 1525-1529; (d) S. L. Schreiber and M. T. Goulet, *J. Am. Chem. Soc.*, **1987**, *109*, 4718-4720.

37. See, S. Patai, *The Chemistry of Functional Groups* (J. Wiley-Interscience, New York 1964-1987, continuing series of over 50 volumes).

38. E. J. Corey, R. D. Cramer, III, and W. J. Howe, *J. Am. Chem. Soc.*, **1972**, *94*, 440-459.

39. E. J. Corey, H. W. Orf, and D. A. Pensak, *J. Am. Chem. Soc.*, **1976**, *98*, 210-221.

40. G. W. Gribble and C. F. Nutaitis, *Org. Prep. Proced. Int.*, **1985**, *17*, 317-384.

41. D. A. Evans and K. T. Chapman, *Tetrahedron Lett.*, **1986**, *27*, 5939-5942.

42. R. B. Woodward, *Pure and Applied Chem.*, **1968**, *17*, 519-547.

43. G. Giordano, G. Ribaldone, and G. Borsotti, *Synthesis,* **1971**, 92-95.

44. L. A. Paquette and K. S. Learn, *J. Am. Chem. Soc.,* **1986,** *108*, 7873-7875.

45. L. E. Overman and M. Kakimoto, *J. Am. Chem. Soc.*, **1979,** *101*, 1310-1312.

46. For a review of such synthetic equivalents see T. A. Hase, *Umpoled Synthons*, (J. Wiley, New York, 1987).

47. H. J. Reich and W. W. Willis, Jr., *J. Am. Chem. Soc.*, **1980,** *102*, 5967-5968.

48. E. J. Corey, M. Petrzilka, and Y. Ueda, *Tetrahedron Lett.,* **1975**, 4343-4346; *Helv. Chim. Acta,* **1977**, 60, 2294-2302.

49. A. Eschenmoser and C. E. Wintner, *Science*, **1977**, *196*, 1410-1420.

50. C. R. Hutchinson, K. C. Mattes, M. Nakone, J. J. Partridge, and M. R. Uskokovic, *Helv. Chim. Acta,* **1978,** *61*, 1221-1225.

51. K. J. Clark, G. I. Fray, R. H. Jaeger, and R. Robinson, *Tetrahedron*, **1959**, *6*, 217-224.

52. S. L. Schreiber, H. V. Meyers, and K. B. Wiberg, *J. Am. Chem. Soc.*, **1986**, *108,* 8274-8277.

53. S. E. Denmark and J. A. Sternberg, *J. Am. Chem. Soc.*, **1986**, *108*, 8277-8279.

54. S. L. Schreiber and H. V. Meyers, *J. Am. Chem. Soc.*, **1988**, *110*, 5198-5200.

55. R. D. Stipanovic and R. B. Turner, *J. Org. Chem., 1968, 33,* 3261-3263.

56. R. D. Clark and C. H. Heathcock, *Tetrahedron Lett.*, **1974**, 2027-2030.

57. G. H. Posner, *An Introduction to Synthesis Using Organocopper Reagents* (J. Wiley, New York, 1980).

58. E. E. van Tamelen, M. Shamma, A. W. Burgstahler, J. Wolinsky, R. Tamm, and P. E. Aldrich, *J. Am. Chem. Soc.*, **1958**, *80*, 5006-5007.

59. S. Masamune, S. K. Ang, C. Egli, N. Nakatsuka, S. K. Sakhar, and Y. Yasunari, *J. Am. Chem. Soc.*, **1967**, *89*, 2506-2507.

60. J. B. Hendrickson and R. K. Boeckman, Jr., *J. Org. Chem.*, **1971**, *36*, 2315-2319.

61. L. R. C. Barclay, C. E. Milligan, and D. N. Hall, *Can. J. Chem.*, **1962**, *40*, 1664-1671.

62. E. J. Corey and W. L. Jorgensen, *J. Am. Chem. Soc.*, **1976**, *98*, 189-203.

63. M. Suzuki, A. Yanagisawa, and R. Noyori, *J. Am. Chem. Soc.*, **1985**, *107*, 3348-3349.

64. (a) E. J. Corey, N. H. Andersen, R. M. Carlson, J. Paust, E. Vedejs, I. Vlattas, and R. E. K. Winter, *J. Am. Chem. Soc.,* **1968,** *90*, 3245-3247; (b) E. J. Corey, I. Vlattas, N. H. Andersen, and K. Harding, *J. Am. Chem. Soc.*, **1968**, *90*, 3247-3248; (c) E. J. Corey, I. Vlattas, and K. Harding, *J. Am. Chem. Soc.,* **1969**, *91*, 535-536; (d) E. J. Corey, *R. A. Welch Found. Res. Conf.,* **1968**, *XII*, 51-79; (e) E. J. Corey, *Ann. N. Y. Acad. Sci.*, **1971**, *180*, 24-37.

65. A. Bélanger, D. J. F. Berney, H.-J. Borschberg, R. Lapalme, D. M. Leturc, C.-C. Liao, F. N. MacLachlan, J.-P. Maffrand, F. Marazza, R. Martino, C. Moreau, L. Saint-Laurent, R. Saintonge, P. Soucy, L. Ruest, and P. Deslongchamps, *Can. J. Chem.*, **1979**, *57*, 3348-3354.

66. E. J. Corey, *Pure and Applied Chem.,* **1967**, *14*, 19-37.

67. E. J. Corey, *Chem. Soc. Rev.,* **1988**, *17*, 111-133.

68. E. J. Corey, A. K. Long, T. W. Greene, and J. W. Miller, *J. Org. Chem.,* **1985**, *50*, 1920-1927.

69. T. W. Greene, *Protective Groups in Organic Synthesis* (J. Wiley, New York, 1981).

70. E. J. Corey, M. Ohno, P. A. Vatakencherry, and R. B. Mitra, **1961**, *83*, 1251-1253; **1964**, *86*, 478-485.

71. (a) R. B. Woodward, F. I. Brutschy, and H. Baer, *J. Am. Chem. Soc.*, **1948**, *70*, 4216-4221; (b) P. Buchschacher and A. Fürst, *Organic Synth.*, **1985**, *63*, 37-43.

72. (a) J. E. McMurry and S. J. Isser, *J. Am. Chem. Soc.,* **1972**, *94*, 7132-7137; (b) R. A. Volkmann, G. C. Andrews, and W. S. Johnson, *J. Am Chem. Soc.,* **1975**, *97*, 4777-4778; (c) W. Oppolzer and T. Godel, *J. Am. Chem. Soc.,* **1978**, *100*, 2583-2584.

73. E. J. Corey and R. D. Balanson, *J. Am. Chem. Soc.*, **1974**, *96*, 6516-6517.

74. (a) B. Witkop, *Experientia*, **1971**, *27*, 1121-1138; (b) E. X. Albuquerque, E. A. Barnard, T. H. Chiu, A. J. Lapa, J. O. Dolly, S. Jansson, J. Daly, and B. Witkop, *Proc. Nat. Acad. Sci. USA*, **1973**, *70*, 949-953.

75. E. J. Corey, J. F. Arnett, and G. N. Widiger, *J. Am. Chem. Soc.*, **1975**, *97*, 430-431.

76. E. J. Corey and J. E. Munroe, *J. Am. Chem. Soc.,* **1982**, *104*, 6129-6130.

77. E. J. Corey and J. G. Smith, *J. Am. Chem. Soc.*, **1979**, *101*, 1038-1039.

78. L. Lombardo, L. N. Mander and J. V. Turner, *J. Am. Chem. Soc.*, **1980**, *102*, 6626-6628.

79. E. J. Corey and H. L. Pearce, *Tetrahedron Lett.,* **1980**, *21*, 1823-1824.

80. E. J. Corey, M. C. Desai, and T. A. Engler, *J. Am. Chem. Soc.*, **1985**, *107*, 4339-4341.

81. T. Hudlicky, L. Radesca-Kwart, L. Li, and T. Bryant, *Tetrahedron Lett.*, **1988**, *29*, 3283-3286.

82. E. J. Corey, M.-c. Kang, M. C. Desai, A. K. Ghosh, and I. N. Houpis, *J. Am. Chem. Soc.*, **1988**, *110*, 649-651.

83. E. J. Corey and A. V. Gavai, *Tetrahedron Lett.,* **1988**, *29*, 3201-3204.

84. E. J. Corey and A. K. Ghosh, *Tetrahedron Lett.*, **1988**, 29, 3205-3206.

GLOSSARY FOR PART ONE

Actuate. To make possible or enable transform function on a target in a manner analogous to that in which a structural subunit can *activate* a molecule for chemical reaction. The verb actuate is the retrosynthetic equivalent of activate in the synthetic direction.

Ancillary Keying Groups. Those structural subunits which provide keying of a particular transform over and above that associated with the retron itself.

Antithetic Analysis. (Synonymous with *Retrosynthetic Analysis)* A problem-solving technique for transforming the structure of a synthetic target molecule to a sequence of progressively simpler structures along a pathway which ultimately leads to simple or commercially available starting materials for a chemical synthesis.

APD. Appendage disconnection.

Appendage. A structural subunit consisting of one or more carbon atoms and their substituents which is bonded to a ring or functional group origin.

Carbogen. A member of the family of carbon-containing molecules.

Chiral Controller. (Synonymous with *Chiral Auxiliary)*. A chiral structural unit which when attached to a substrate enhances stereoselectivity in the formation of new stereocenter(s).

Clearable Stereocenter(s). Stereocenter(s) which can be eliminated retrosynthetically by application of a transform with stereocontrol (stereoselectivity).

Cocyclic Bonds. Endo bonds which are within the same ring.

Core Functional Group. A member of the most simple, fundamental, and synthetically versatile class of functional groups, such as carbonyl, hydroxyl, olefinic, acetylenic, and amino groups.

Endo Bond. A bond within a ring.

Equivalents of Reactive Intermediates. Reactive intermediates which are readily generated and which can be used in synthesis to produce indirectly structures that cannot be made directly.

Ex-Target Tree. (EXTGT Tree) A branching tree structure formed by retrosynthetic analysis of a target molecule (treetop). Such trees grow out from a target and consist of nodes which correspond to the structures of intermediates along a pathway of synthesis.

Exendo Bond. A bond which is directly attached to a ring and within another ring.

Exo Bond. A bond which is attached to a ring directly.

FGA. Functional group addition.

FGI. Functional group interchange.

FGR. Functional group removal.

Functional Group Interchange. (FGI) The replacement of one functional group by another.

Functional Group Origin. The atom to which a functional group is attached.

Functional group Equivalents. Functional groups from which a particular functional group can be produced chemically.

Functional Group-based Strategy. The use of functional groups to guide the retrosynthetic reduction of molecular complexity.

Mechanism-Control of Stereochemistry. Stereocontrol in a reaction or transform as a result of mechanistic factors rather than substrate structure alone.

Mechanistic Transform. A transform involving a sequence of reactive intermediates such as carbocations or carbon radicals which are generated in a stepwise mechanistic manner and which lead finally to stable predecessor structure(s).

Molecular Complexity. A measure of the combined effects of molecular size, element and functional-group content, internal connectedness, stereocenter content, reactivity, and instability that lead to difficulties of synthesis.

Multistrategic Analysis. Concurrent use of two or more strategies to guide retrosynthetic analysis.

Offexendo Bond. A bond connected to an exendo bond.

Partial Retron. An incomplete retron for a particular transform.

Perimeter Ring. (Synonymous with *Envelope Ring).* A ring which can be constructed by a logical OR operation on the atoms of two smaller rings which are fused or bridged to one another.

Positional Selectivity. Selective attachment of a group by a reaction at only one of several possible locations. Synonymous with "regioselectivity."

Preserve Bond. A bond which is not to be broken retrosynthetically.

Preserve Ring (or Other Substructure). A ring (or other substructure) which for strategic reasons is to be preserved during retrosynthetic simplification.

Primary Ring. A ring within a structure which cannot be expressed as a periphery (or perimeter) of two smaller rings which are fused or bridged together.

Retron. The minimal substructural element in a target structure which keys the direct application of a transform to generate a synthetic precursor.

Retrosynthetic Analysis. (Synonymous with *Antithetic Analysis)* A problem-solving technique for transforming the structure of a synthetic target molecule to a sequence of progressively simpler structures along a pathway which ultimately leads to simple or commercially available starting materials for a chemical synthesis.

Retrosynthetic. The direction of chemical change opposite to that of laboratory execution of a reaction (synthetic direction); the reverse-synthetic sense.

RGD. Ring disconnection.

S-Goal. A structure corresponding to a potential intermediate which can be used for retrosynthetic guidance.

SM-goal. A potential starting material for retrosynthetic guidance.

SS-Goal. A retrosynthetic intermediate generated from a target by the introduction of a substructure corresponding to the retron for a powerfully simplifying transform.

Stereoselectivity. Selectivity in the creation of one or more stereocenters during a reaction.

Structure-goal Strategy. The use of a particular structure corresponding to a potentially available starting material or synthetic intermediate as a guide for retrosynthetic search.

Supra Retron. A structural subunit in a target molecule which consists not only of a complete retron for a particular transform but also of additional structural elements which signal the applicability and effectiveness of that transform. A supra retron contains ancillary keying groups in addition to those of the basic or minimal retron.

Tactical Combination of Transforms. A standard combination of two or more transforms which can be used together to provide overall simplification of a target structure.

Target Molecule. (TGT) A molecule whose synthesis is under examination by retrosynthetic analysis.

Topological Strategy. The use of a particular bond, pair of bonds, set of bonds, or subunit as eligible for disconnection to guide retrosynthetic analysis; conversely the designation of bonds or cyclic subunits as ineligible for disconnection (i.e. to be preserved).

Transform-based Strategy. A strategic guide for retrosynthetic analysis in which the application of a particular powerfully simplifying transform becomes a goal.

Transform. The exact reverse of a synthetic reaction such that when applied to the structure of a reaction product it results in the generation of reactant structure(s).

PART TWO

Specific Pathways for the Synthesis of Complex Molecules

Introduction

This Part of the book could as well have been titled "Synthesis in Action" for it consists of specific multistep sequences of reactions which have been demonstrated by experiment to allow the synthesis of a variety of interesting target molecules. Graphical flowcharts for each synthesis define precisely the pathway of molecular construction in terms of individual reactions and reagents. Each synthetic sequence is accompanied by references to the original literature.

The syntheses documented in this Part were published over a span of more than three decades. Almost all were derived retrosynthetically using the basic logic outlined in Part One. Consequently, familiarity with the concepts and principles developed in Part One allows the identification of the strategies which were utilized in planning the synthetic schemes shown in Part Two. Together, Parts One and Two cover both the planning and execution phases of synthetic chemistry, from general logic to specific chemical methodology. Effective teaching (or learning) of synthesis requires emphasis on both sectors.

The organization of Part Two is according to structural type. The first section, Chapter Seven, is concerned with the synthesis of macrocyclic compounds. Syntheses of a number of heterocyclic target structures appear in Chapter Eight. Sesquiterpenoids and polycyclic higher isoprenoids are dealt with in Chapters Nine and Ten, respectively. The remainder of Part Two describes syntheses of prostanoids (Chapter Eleven) and biologically active acyclic polyenes including leukotrienes and other eicosanoids (Chapter Twelve).

There are a number of other books which provide concise graphical summaries of multistep synthetic sequences leading to complex naturally occurring molecules. This literature includes references 1-7 which follow.

1. N. Anand, J. S. Bindra, and S. Ranganathan, *Art in Organic Synthesis* (Holden-Day, Inc., San Francisco, first edition, 1970).
2. N. Anand, J. S. Bindra, and S. Ranganathan, *Art in Organic Synthesis* (J. Wiley, New York, second edition, 1987).
3. J. S. Bindra and R. Bindra, *Creativity in Organic Synthesis* (Academic Press, Inc., New York, 1975).
4. J. ApSimon, Ed., *The Total Synthesis of Natural Products*, Vols. 1-6 (J. Wiley, New York, 1973-1984).
5. I. Fleming, *Selected Organic Syntheses* (J. Wiley, New York, 1972).
6. S. Hanessian, *Total Synthesis of Natural Products: The 'Chiron' Approach* (Pergamon Press, London, 1983).
7. S. E. Danishefsky and S. Danishefsky, *Progress in Total Synthesis* (Appleton-Century-Crofts, New York, 1971).

The research leading to the syntheses which are outlined in Part Two was generously supported over the years by the National Institutes of Health, the National Science Foundation and Pfizer, Inc.

ABBREVIATIONS

Ac	Acetyl, acetate
acac	Acetylacetonate (as ligand)
AIBN	2,2'-Azobisisobutyronitrile
n-Am	*n*-Amyl
anh.	Anhydrous
aq.	Aqueous
atm	Atmosphere
9-BBN	9-Borabicyclo[3.3.1]nonane
Bn	Benzyl
BOP-Cl	*N, N*-Bis(2-oxo-3-oxazolidinyl)phosphinic chloride
bp	Boiling point
n-Bu	*n*-Butyl
s-Bu	*sec*-Butyl
t-Bu	*tert*-Butyl
Bz	Benzoyl
18-C-6	18-Crown-6
cat.	Catalytic amount
Cbz	Carbobenzyloxy (benzyloxycarbonyl)
Cp	Cyclopentadiene
CSA	Camphorsulfonic acid
Cy-Hex	Cyclohexyl
Dabco	1,4-Diazabicyclo[2.2.2]octane
DBN	1,5-Diazabicyclo[4.3.0]non-5-ene
DBU	1,8-Diazabicyclo[5.4.0]undec-7-ene
DCC	Dicyclohexylcarbodiimde
DDQ	2,3-Dichloro-5,6-dicyano-1,4-benzoquinone
DEAD	Diethyl azodicarboxylate
DET	Diethyl tartrate
DHP	Dihydropyran
Dibal-H	Diisobutylaluminum hydride
DIPT	Diisopropyl tartrate
DMAP	4-Dimethylaminopyridine
DME	1,2-Dimethoxyethane (glyme)
DMF	*N,N*-Dimethylformamide
DMSO	Dimethyl sulfoxide
DMT	Dimethyl tartrate
ee	Enantiomeric excess
en	Ethylenediamine (as ligand)
eq.	Equivalent
Et	Ethyl

GC	Gas chromatography
h	Hour
HETE	Hydroxy eicosatetraenoic acid
HMPA	Hexamethylphosphoric triamide
HMPT	Hexamethylphosphorous triamide
HPETE	Hydroperoxy eicosatetraenoic acid
HPLC	High performence liquid chromatography
HQ	Hydroquinone
hv	Light
imid	Imidazole
IPDMS	Isopropyldimethylsilyl
L-Selectride	Lithium tri-*sec*-butylborohydride
LAH	Lithium aluminum hydride
LDA	Lithium diisopropylamide
LICA	Lithium isopropylcyclohexylamide
LT	Leukotriene
MCPBA	*m*-Chloroperoxybenzoic acid
Me	Methyl
MEM	Methoxyethoxymethyl
Mes	Mesityl (2,4,6-trimethylbenzyl)
MICA	Magnesium isopropylcyclohexylamide
mol	Mole
mol. sieves	Molecular sieves, normally 4 Å
MOM	Methoxymethyl
mp	Melting point
Ms	Mesyl (methanesulfonyl)
MTM	Methylthiomethyl
MVK	Methyl vinyl ketone
NBS	*N*-Bromosuccinimide
NCS	*N*-Chlorosuccinimide
NMO	*N*-Methylmorpholine *N*-oxide
PBz	*p*-Phenylbenzoyl
PCC	Pyridinium chlorochromate
PDC	Pyridinium dichromate
PG	Prostaglandin
Ph	Phenyl
PMP	1,2,2,6,6-Pentamethylpiperidine
PPA	Polyphosphoric acid
PPTS	Pyridinium *p*-toluenesulfonate
n-Pr	*n*-Propyl
i-Pr	*iso*-Propyl
psi	Pounds per square inch
Py	Pyridine
PyCl	Pyridinium chloride
Ra Ni	Raney nickel, usually W-II type
Red-Al	Sodium dihydrobis(2-methoxyethoxy) aluminate

Ref.	Reference
SGC	Silica gel chromatography
Sia_2BH	Disiamylborane
TBDMS	*tert*-Butyldimethylsilyl
TBDPS	*tert*-Butyldiphenylsilyl
Tf	Trifluoromethanesulfonyl
TFA	Trifluoroacetic acid
TFAA	Trifluoroacetic anhydride
THF	Tetrahydrofuran
THP	Tetrahydropyranyl
Thx	Thexyl (Me_2CHMe_2C-)
TIPS	Triisopropylsilyl
TLC	Thin layer chromatography
TMEDA	*N, N, N', N'*-Tetramethylethylenediamine
TMS	Trimethylsilyl
Tol	*p*-Toluene or *p*-tolyl
Tr	Trityl (triphenylmethyl)
Ts	Tosyl (*p*-toluenesulfonyl)
TX	Thromboxane
Δ	Heat at reflux

CHAPTER SEVEN
Macrocyclic Structures

7.1

Erythronolide B

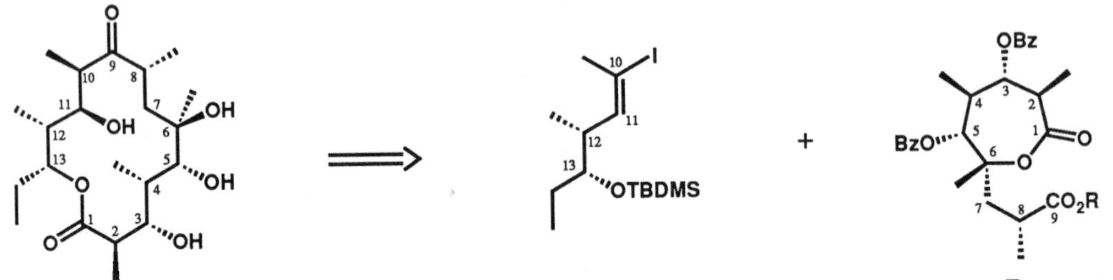

OH
OH
OH
OTBDMS
I
OBz
BzO
CO2R

A B

Erythronolide B, the biosynthetic progenitor of the erythromycin antibiotics, was synthesized for the first time, using as a key step a new method for macrolactone ring closure (double activation) which had been devised specifically for this problem.[1] Retrosynthetic simplification included the clearance of the stereocenters at carbons 10 and 11 and the disconnection of the 9,10-bond, leading to precursors **A** and **B**. Cyclic stereocontrol and especially the Baeyer-Villiger and halolactonization transforms played a major role in the retrosynthetic simplification of **B** which was synthesized starting from 2,4,6-trimethylphenol.

1. Synthesis of the fragment **A**:

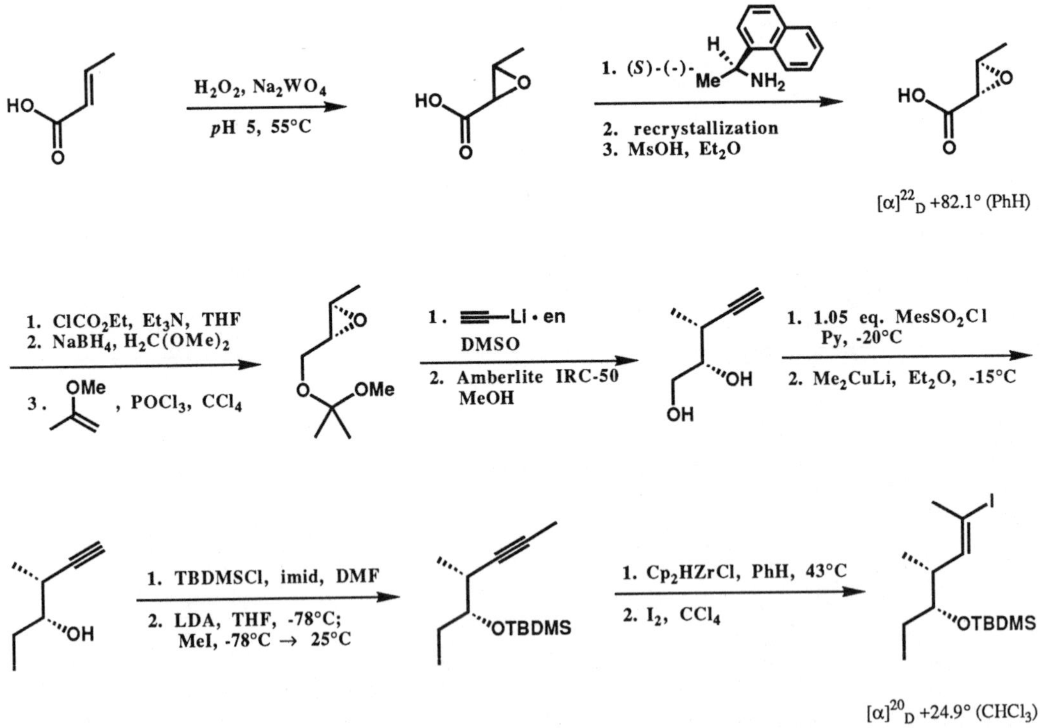

H2O2, Na2WO4
pH 5, 55°C
1. (S)-(-)-
Me NH2
2. recrystallization
3. MsOH, Et2O
[α]22D +82.1° (PhH)
1. ClCO2Et, Et3N, THF
2. NaBH4, H2C(OMe)2
3. OMe , POCl3, CCl4
1. Li • en
DMSO
2. Amberlite IRC-50
MeOH
1. 1.05 eq. MesSO2Cl
Py, -20°C
2. Me2CuLi, Et2O, -15°C
1. TBDMSCl, imid, DMF
2. LDA, THF, -78°C;
MeI, -78°C → 25°C
1. Cp2HZrCl, PhH, 43°C
2. I2, CCl4
OTBDMS
[α]20D +24.9° (CHCl3)

2. Synthesis of the fragment **B**:

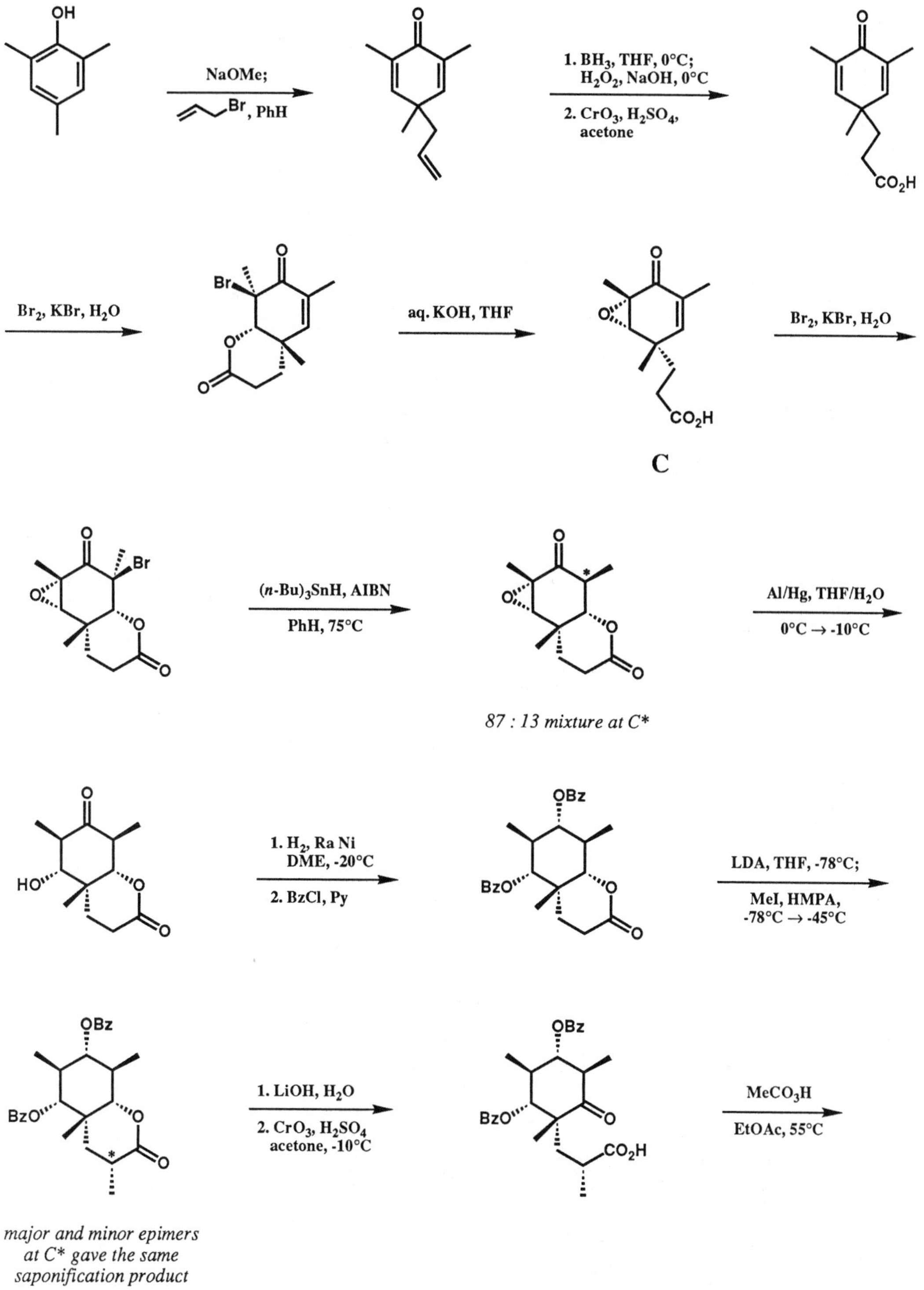

OH
NaOMe;
Br, PhH
O
1. BH3, THF, 0°C;
H2O2, NaOH, 0°C
2. CrO3, H2SO4,
acetone
CO2H
Br2, KBr, H2O
Br
aq. KOH, THF
C
(n-Bu)3SnH, AIBN
PhH, 75°C
87 : 13 mixture at C*
Al/Hg, THF/H2O
0°C → -10°C
HO
1. H2, Ra Ni
DME, -20°C
2. BzCl, Py
OBz
BzO
LDA, THF, -78°C;
MeI, HMPA,
-78°C → -45°C
major and minor epimers
at C* gave the same
saponification product
1. LiOH, H2O
2. CrO3, H2SO4
acetone, -10°C
MeCO3H
EtOAc, 55°C

7.1

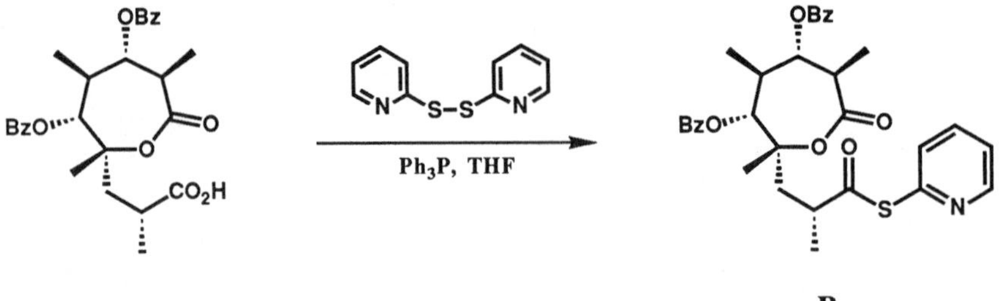
OBz
BzO
O
CO2H
N
S–S
Ph3P, THF

B

Fragment **B** was synthesized in optically active form with the required absolute configuration through resolution of the epoxy acid **C** as shown below.

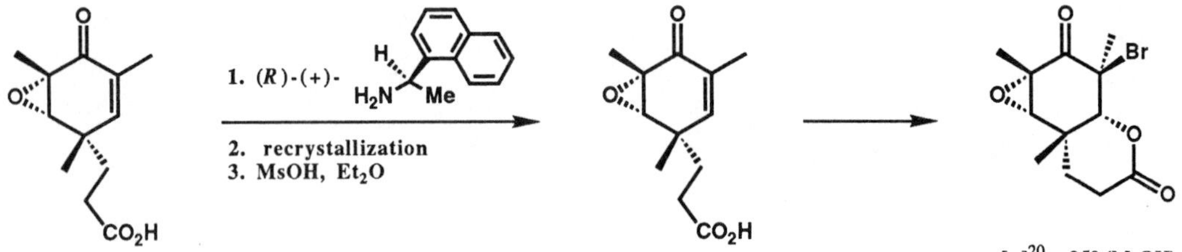
O
CO2H
1. (R)-(+)-
H
H2N
Me
2. recrystallization
3. MsOH, Et2O
Br

C

$[\alpha]^{28}_D$ -127° (MeOH)

$[\alpha]^{20}_D$ -25° (MeOH)

absolute configuration confirmed by X-ray crystallography .

3. Coupling of the fragments **A** and **B** and the completion of the synthesis:

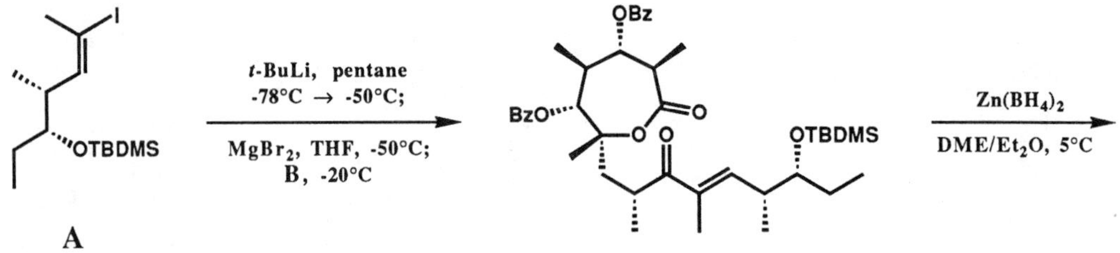
I
OTBDMS
t-BuLi, pentane
-78°C → -50°C;
MgBr2, THF, -50°C;
B, -20°C
OBz
BzO
O
OTBDMS
Zn(BH4)2
DME/Et2O, 5°C

A

$[\alpha]^{20}_D$ +24.9° ($CHCl_3$)

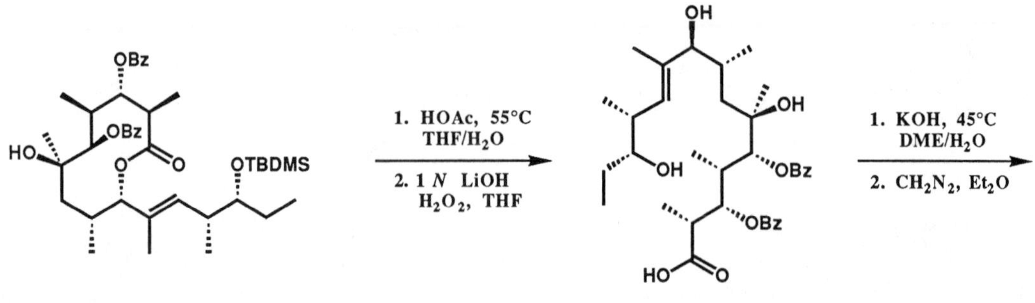
OBz
OBz
HO
O
OTBDMS
1. HOAc, 55°C
THF/H2O
2. 1 N LiOH
H2O2, THF
OH
OH
OH
OBz
OBz
HO
O
1. KOH, 45°C
DME/H2O
2. CH2N2, Et2O

7.1

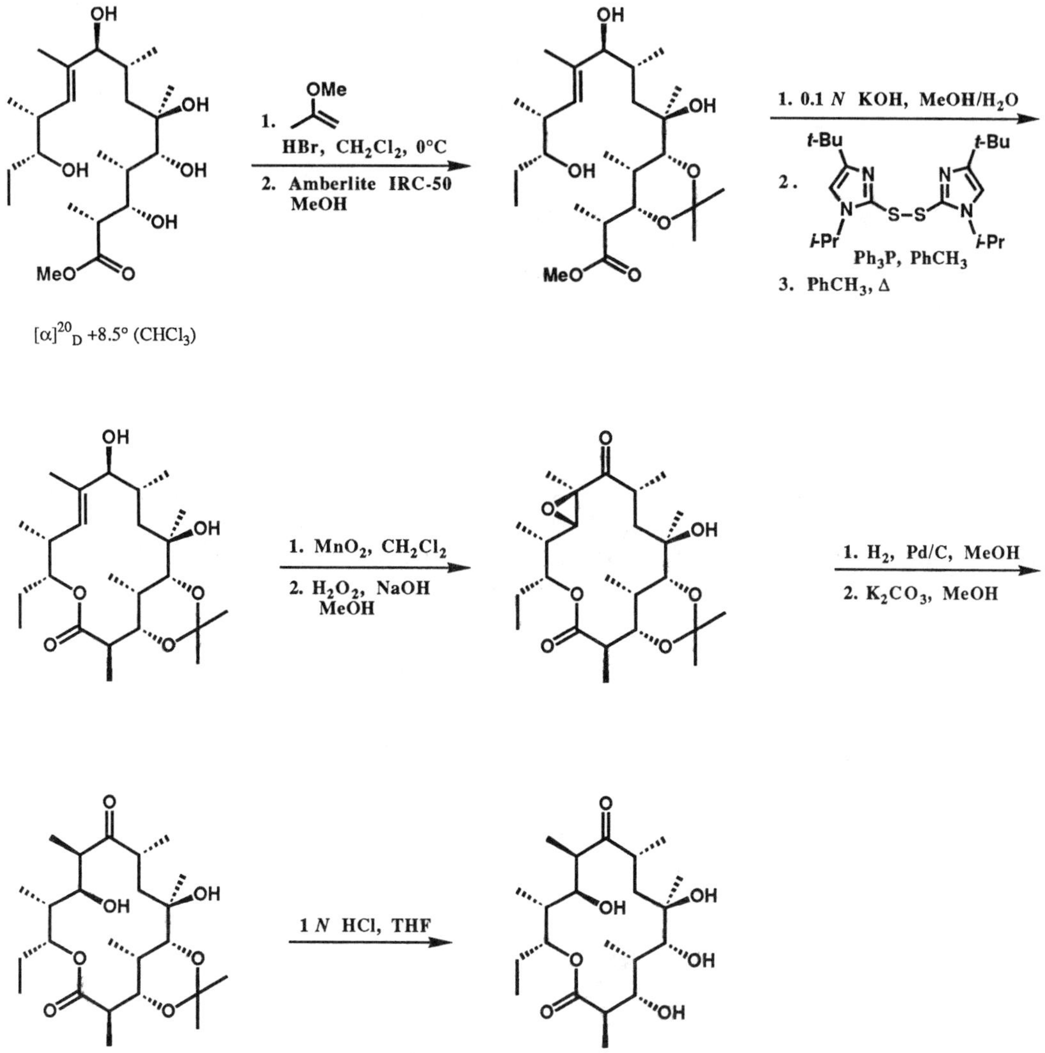
OH
OH
OH
OH
OH
MeO
O
[α]$^{20}_{D}$ +8.5° (CHCl$_3$)
1. OMe
HBr, CH$_2$Cl$_2$, 0°C
2. Amberlite IRC-50
MeOH
1. 0.1 *N* KOH, MeOH/H$_2$O
2. *t*-Bu N N S–S *t*-Bu *i*-Pr *i*-Pr
Ph$_3$P, PhCH$_3$
3. PhCH$_3$, Δ
1. MnO$_2$, CH$_2$Cl$_2$
2. H$_2$O$_2$, NaOH
MeOH
1. H$_2$, Pd/C, MeOH
2. K$_2$CO$_3$, MeOH
1 *N* HCl, THF

References:

1. E. J. Corey, K. C. Nicolaou, and L. S. Melvin, Jr., *J. Am. Chem. Soc.* **1975**, *97*, 654.

2. E. J. Corey, L. S. Melvin, Jr., and M. F. Haslanger, *Tetrahedron Lett.* **1975**, 3117.

3. E. J. Corey, E. J. Trybulski, L. S. Melvin, Jr., K. C. Nicolaou, J. A. Secrist, R. Lett, P. W. Sheldrake, J. R. Falck, D. J. Brunelle, M. F. Haslanger, S. Kim, and S.-e. Yoo, *J. Am. Chem. Soc.* **1978**, *100*, 4618.

4. E. J. Corey, S. Kim, S.-e. Yoo, K. C. Nicolaou, and L. S. Melvin, Jr., D. J. Brunelle, J. R. Falck, E. J. Trybulski, R. Lett, and P. W. Sheldrake, *J. Am. Chem. Soc.* **1978**, *100*, 4620.

Erythronolide A

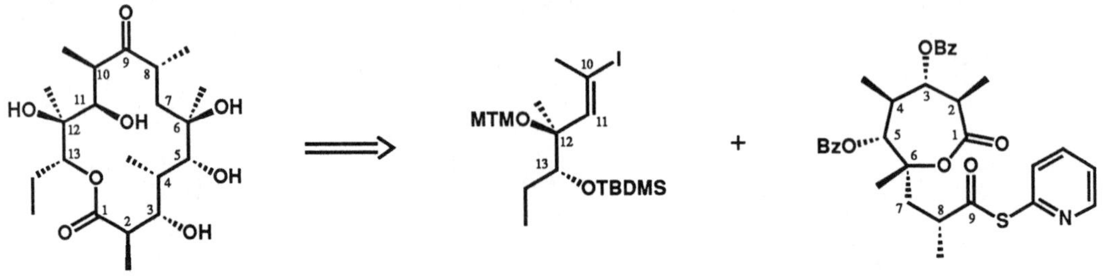

O
HO
OH
OH
OH
OH
O
MTMO
I
OTBDMS
+
OBz
BzO
O
O
O
S
N

A　　　　B

The first total synthesis of erythronolide A was accomplished from iodide **A** and lactone **B**, the same intermediate which had been used for the synthesis of erythronolide B. The pronounced acid sensitivity of erythronolide A necessitated a digression of the final steps of the synthesis from those used for the earlier synthesis of erythronolide B.

1. Synthesis of the fragment **A**:

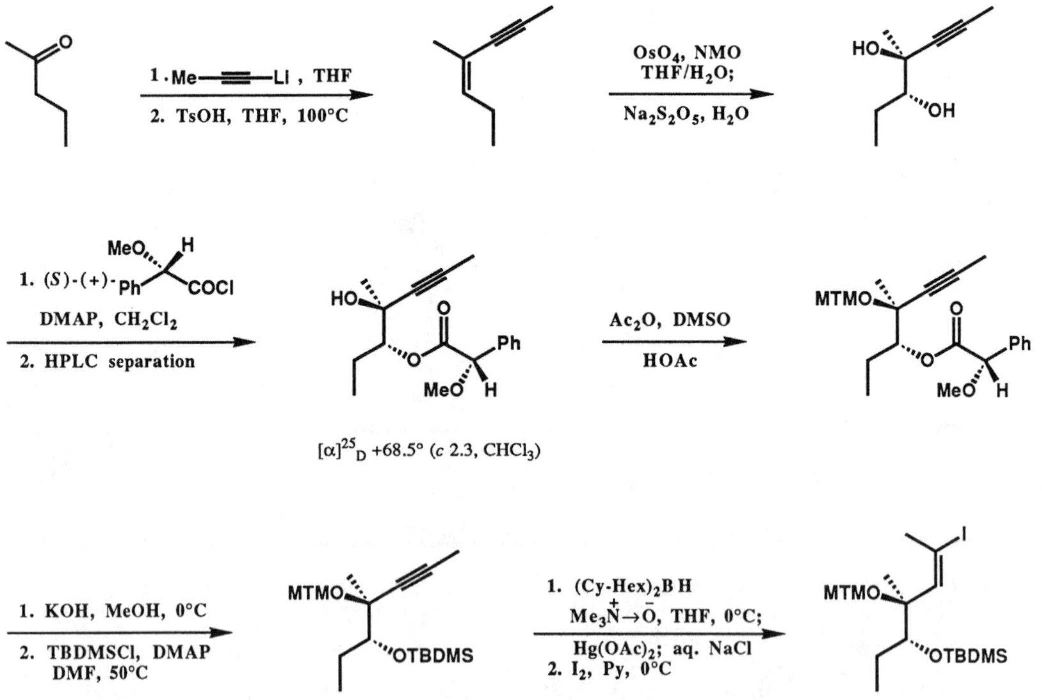

1. Me—≡—Li , THF
2. TsOH, THF, 100°C
OsO4, NMO
THF/H2O;
Na2S2O5, H2O
HO
OH
1. (S)-(+)-Ph(MeO)(H)COCl
DMAP, CH2Cl2
2. HPLC separation
[α]25D +68.5° (c 2.3, CHCl3)
Ac2O, DMSO
HOAc
MTMO
Ph
MeO H
1. KOH, MeOH, 0°C
2. TBDMSCl, DMAP
DMF, 50°C
OTBDMS
1. (Cy-Hex)2BH
Me3N→O, THF, 0°C;
Hg(OAc)2; aq. NaCl
2. I2, Py, 0°C
I

$[\alpha]^{25}_D$ +27.9° (*c* 1.04, EtOH)

A

2. Synthesis of the fragment **B**:

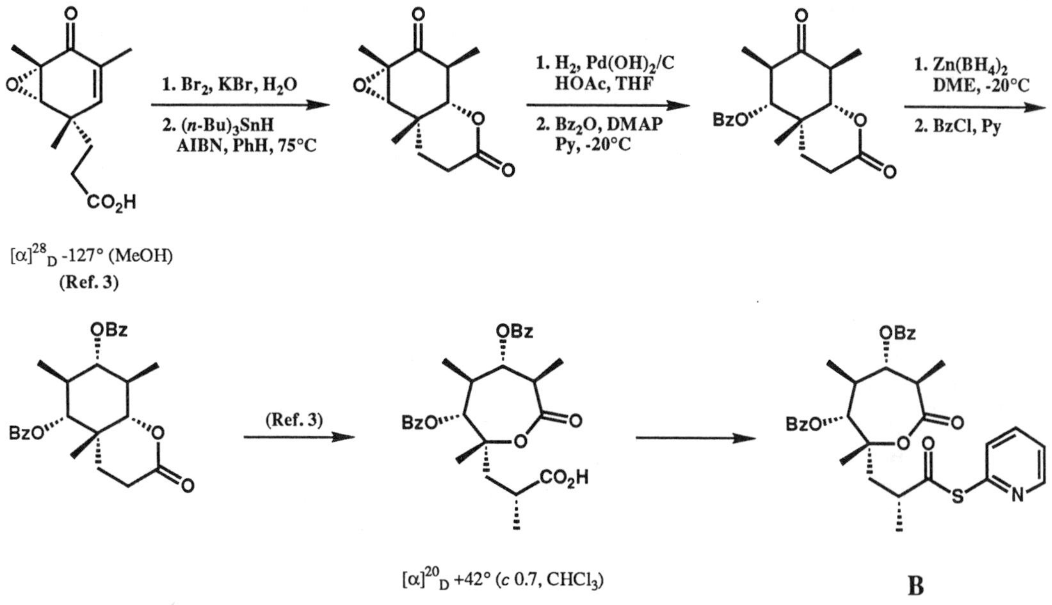
1. Br2, KBr, H2O
2. (n-Bu)3SnH
AIBN, PhH, 75°C
1. H2, Pd(OH)2/C
HOAc, THF
2. Bz2O, DMAP
Py, -20°C
1. Zn(BH4)2
DME, -20°C
2. BzCl, Py
CO2H
[α]28D -127° (MeOH)
(Ref. 3)
OBz
BzO
(Ref. 3)
[α]20D +42° (c 0.7, CHCl3)
B

3. Coupling of the fragments **A** and **B** and the completion of the synthesis:

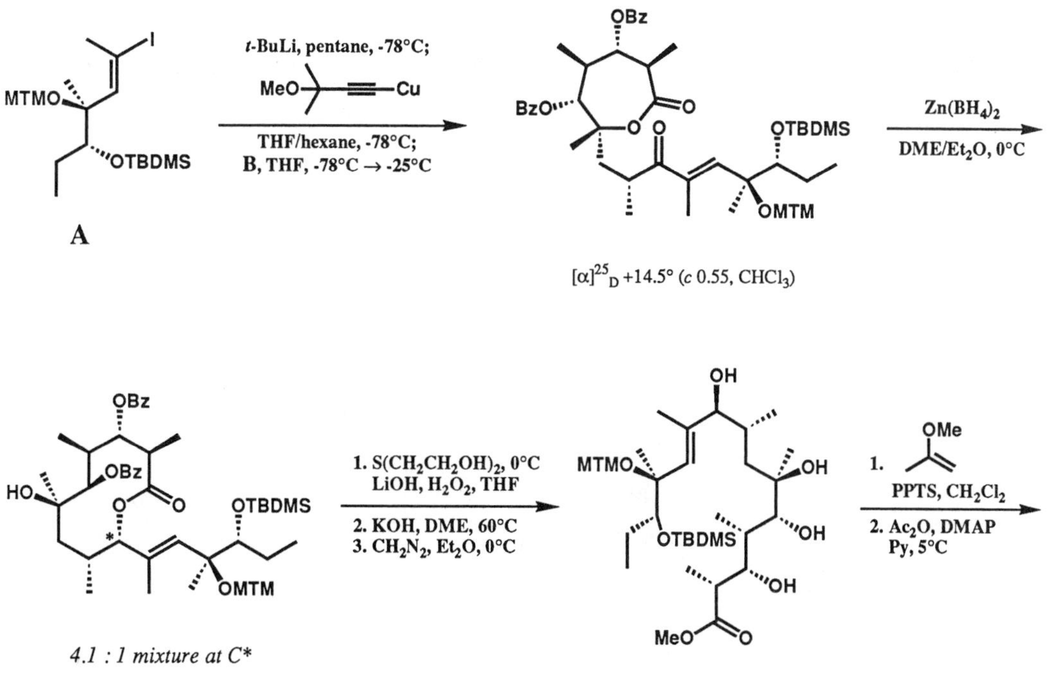
MTMO
OTBDMS
A
t-BuLi, pentane, -78°C;
MeO
Cu
THF/hexane, -78°C;
B, THF, -78°C → -25°C
OBz
BzO
OTBDMS
OMTM
Zn(BH4)2
DME/Et2O, 0°C
[α]25D +14.5° (c 0.55, CHCl3)
HO
4.1 : 1 mixture at C*
1. S(CH2CH2OH)2, 0°C
LiOH, H2O2, THF
2. KOH, DME, 60°C
3. CH2N2, Et2O, 0°C
OH
MeO
OMe
PPTS, CH2Cl2
2. Ac2O, DMAP
Py, 5°C

7.2

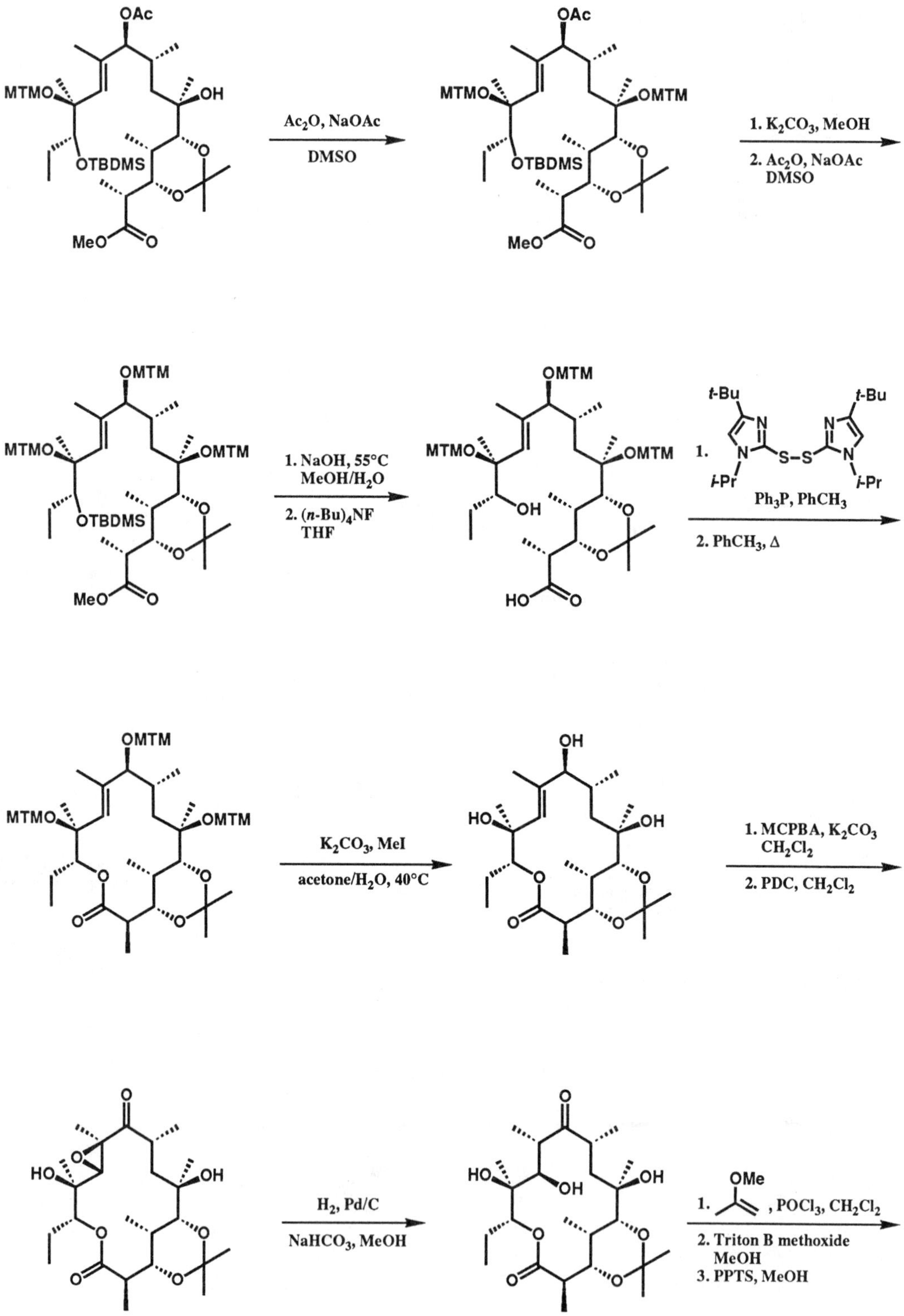
OAc
MTMO
OH
OTBDMS
MeO
O
Ac₂O, NaOAc
DMSO
OMTM
1. K₂CO₃, MeOH
2. Ac₂O, NaOAc
DMSO
1. NaOH, 55°C
MeOH/H₂O
2. (n-Bu)₄NF
THF
HO
t-Bu
i-Pr
S–S
Ph₃P, PhCH₃
2. PhCH₃, Δ
K₂CO₃, MeI
acetone/H₂O, 40°C
1. MCPBA, K₂CO₃
CH₂Cl₂
2. PDC, CH₂Cl₂
H₂, Pd/C
NaHCO₃, MeOH
OMe
1. , POCl₃, CH₂Cl₂
2. Triton B methoxide
MeOH
3. PPTS, MeOH

7.2

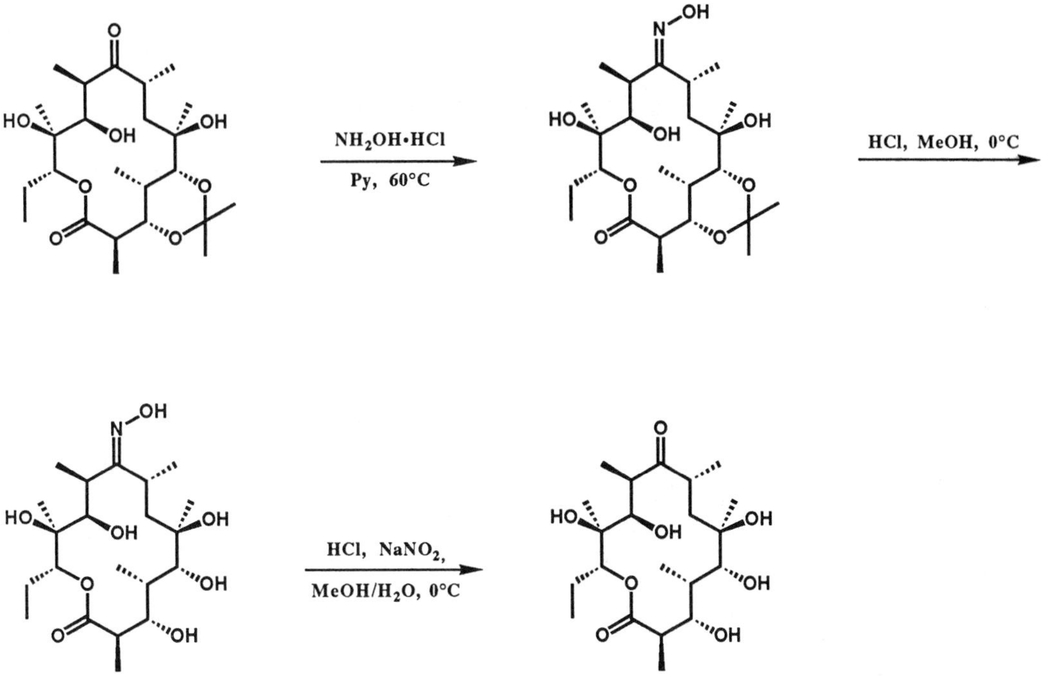

O
HO
OH
OH
O
O
O
O
NH2OH•HCl
Py, 60°C
N
OH
HCl, MeOH, 0°C
HCl, NaNO2,
MeOH/H2O, 0°C

$[\alpha]^{25}_D$ -37° (*c* 0.9, MeOH)

References:

1. E. J. Corey, P. B. Hopkins, S. Kim, S.-e. Yoo, K. P. Nambiar, and J. R. Falck, *J. Am. Chem. Soc.* **1979**, *101*, 7131.

2. D. Schomburg, P. B. Hopkins, W. N. Lipscomb, and E. J. Corey, *J. Org. Chem.* **1980**, *45*, 1544.

3. E. J. Corey, E. J. Trybulski, L. S. Melvin, Jr., K. C. Nicolaou, J. A. Secrist, R. Lett, P. W. Sheldrake, J. R. Falck, D. J. Brunelle, M. F. Haslanger, S. Kim, and S.-e. Yoo, *J. Am. Chem. Soc.* **1978**, *100*, 4618.

4. E. J. Corey, S. Kim, S.-e. Yoo, K. C. Nicolaou, and L. S. Melvin, Jr., D. J. Brunelle, J. R. Falck, E. J. Trybulski, R. Lett, and P. W. Sheldrake, *J. Am. Chem. Soc.* **1978**, *100*, 4620.

Recifeiolide, (±)-11-Hydroxy-*trans*-8-dodecenoic Acid Lactone a Natural Macrolide from *Cephalosporium Recifei*

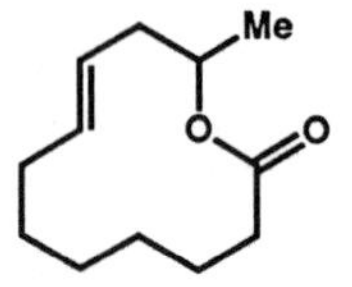
Me
O
O

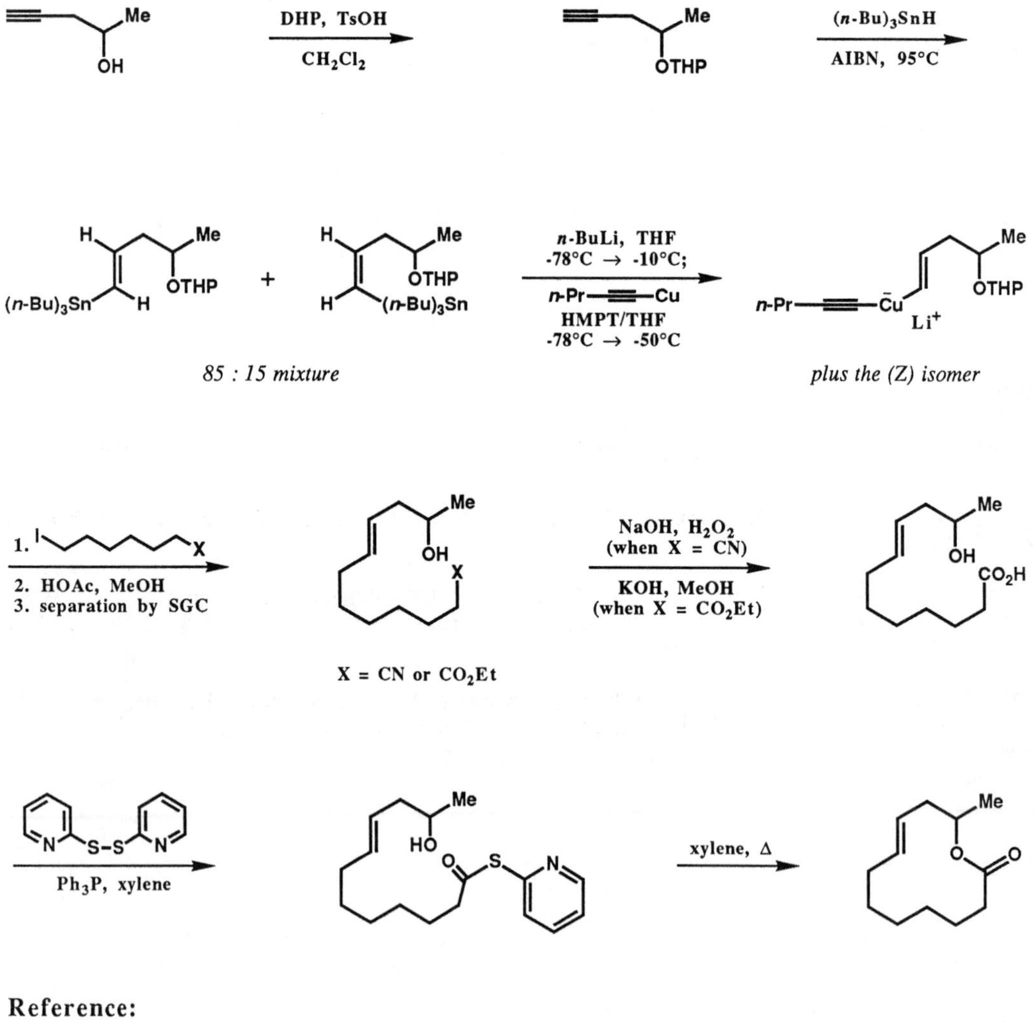
Me
OH
DHP, TsOH
CH2Cl2
Me
OTHP
(n-Bu)3SnH
AIBN, 95°C
H
Me
OTHP
(n-Bu)3Sn
H
+
H
Me
OTHP
H
(n-Bu)3Sn
85 : 15 mixture
n-BuLi, THF
-78°C → -10°C;
n-Pr—≡—Cu
HMPT/THF
-78°C → -50°C
Me
OTHP
n-Pr—≡—Cu
Li+
plus the (Z) isomer
1. I⁄⁄⁄X
2. HOAc, MeOH
3. separation by SGC
Me
OH
X
X = CN or CO2Et
NaOH, H2O2
(when X = CN)
KOH, MeOH
(when X = CO2Et)
Me
OH
CO2H
N S-S N
Ph3P, xylene
Me
HO
O
S
N
xylene, Δ
Me
O
O

Reference:

E. J. Corey, P. Ulrich, and J. M. Fitzpatrick, *J. Am. Chem. Soc.* **1976**, *98*, 222.

(±)-Vermiculine

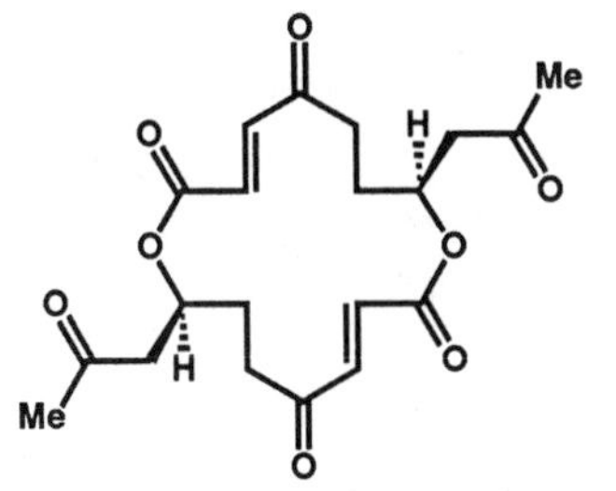
O
Me
H
O
O
O
O
O
H
O
Me
O

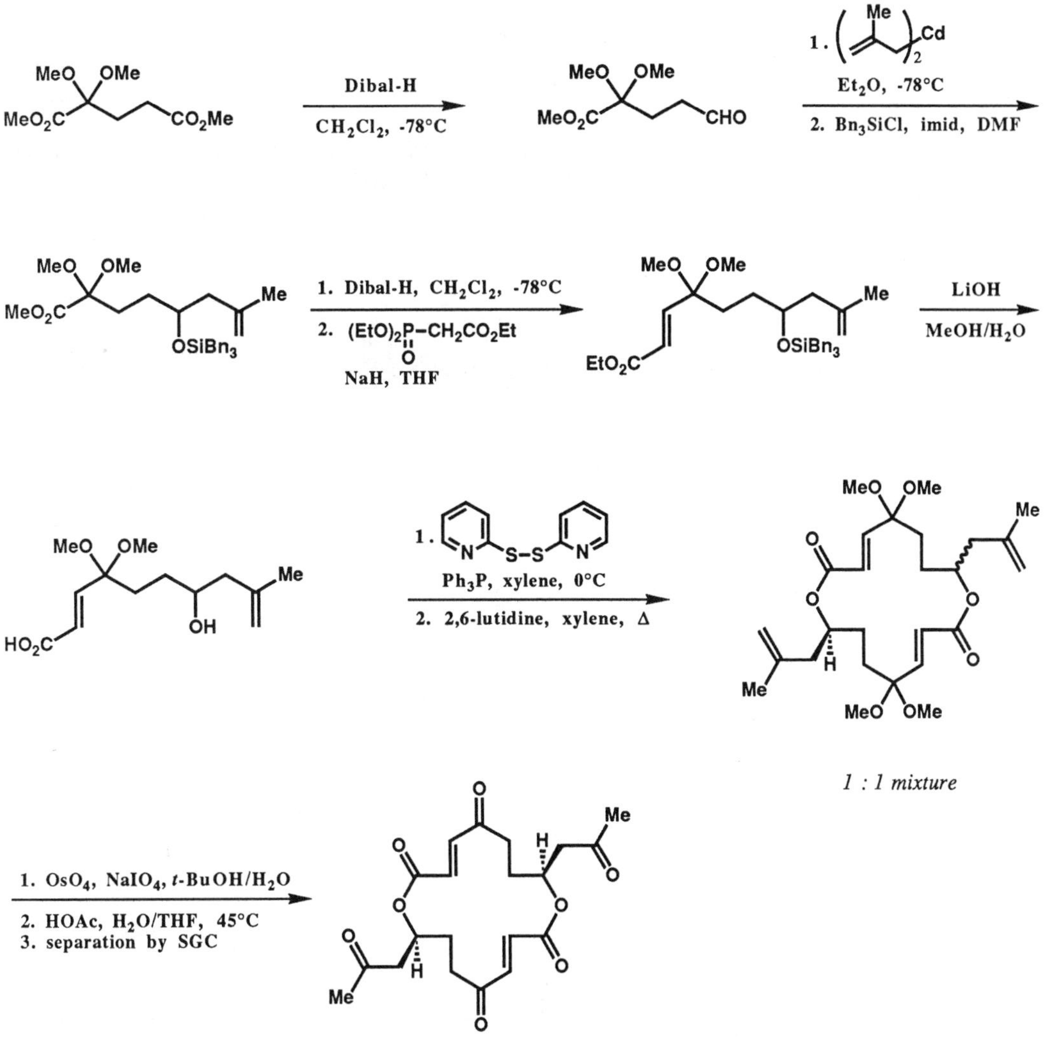
MeO OMe
MeO2C
CO2Me
Dibal-H
CH2Cl2, -78°C
MeO OMe
MeO2C
CHO
1. Me Cd 2
Et2O, -78°C
2. Bn3SiCl, imid, DMF
MeO OMe
MeO2C
Me
OSiBn3
1. Dibal-H, CH2Cl2, -78°C
2. (EtO)2P-CH2CO2Et
O
NaH, THF
MeO OMe
EtO2C
Me
OSiBn3
LiOH
MeOH/H2O
MeO OMe
HO2C
OH
Me
1. N S-S N
Ph3P, xylene, 0°C
2. 2,6-lutidine, xylene, Δ
MeO OMe
Me
O
O
O
O
H
Me
MeO OMe
1 : 1 mixture
1. OsO4, NaIO4, t-BuOH/H2O
2. HOAc, H2O/THF, 45°C
3. separation by SGC
O
Me
H
O
O
O
O
O
H
O
Me
O

Reference:

E. J. Corey, K. C. Nicolaou, and T. Toru, *J. Am. Chem. Soc.* **1975**, *97*, 2287.

7.5

Enterobactin

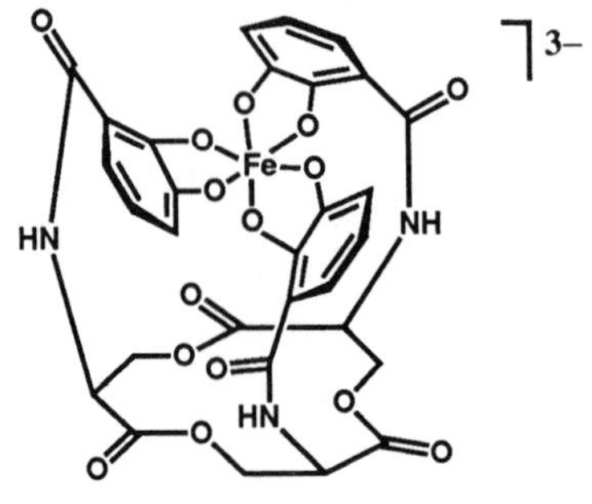

Fe
HN
NH
HN
3–

The first synthesis of enterobactin, a microbial chelator and transporter of environmental iron, was accomplished by the coupling of three protected *L*-serine units and macrocyclization by the double activation method.

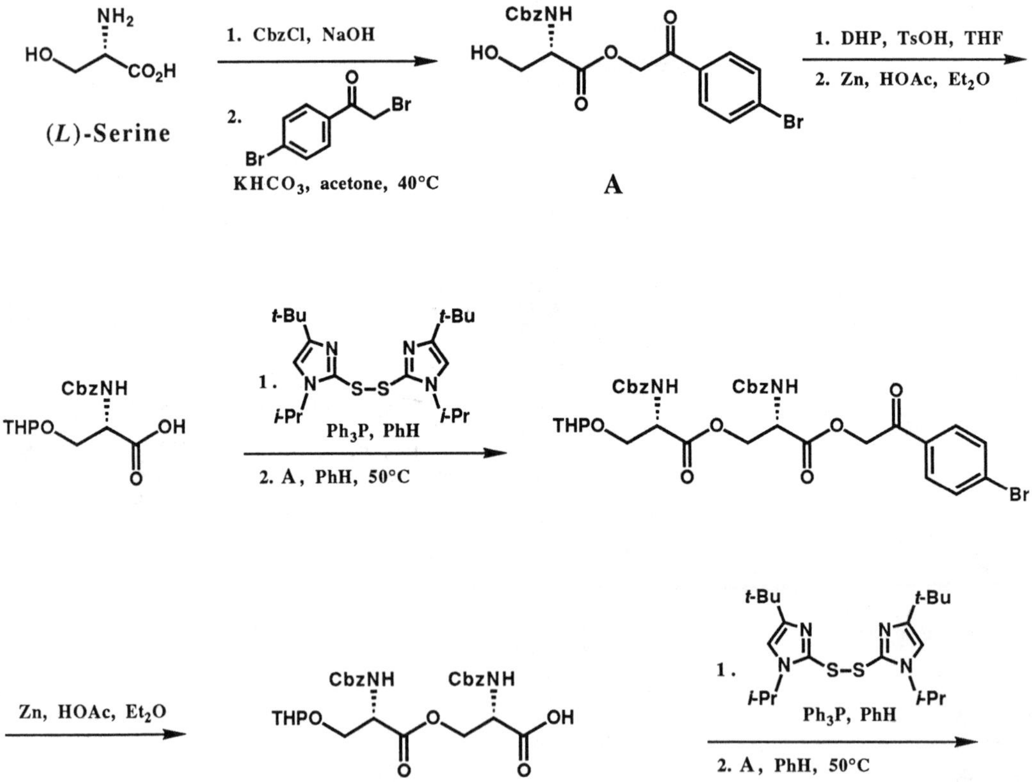

NH2
HO
CO2H
(L)-Serine
1. CbzCl, NaOH
2.
Br
Br
KHCO3, acetone, 40°C
CbzNH
HO
Br
A
1. DHP, TsOH, THF
2. Zn, HOAc, Et2O
CbzNH
THPO
OH
t-Bu
t-Bu
1.
S-S
i-Pr
i-Pr
Ph3P, PhH
2. A, PhH, 50°C
CbzNH
CbzNH
THPO
Br
Zn, HOAc, Et2O
CbzNH
CbzNH
THPO
OH
t-Bu
t-Bu
1.
S-S
i-Pr
i-Pr
Ph3P, PhH
2. A, PhH, 50°C

7.5

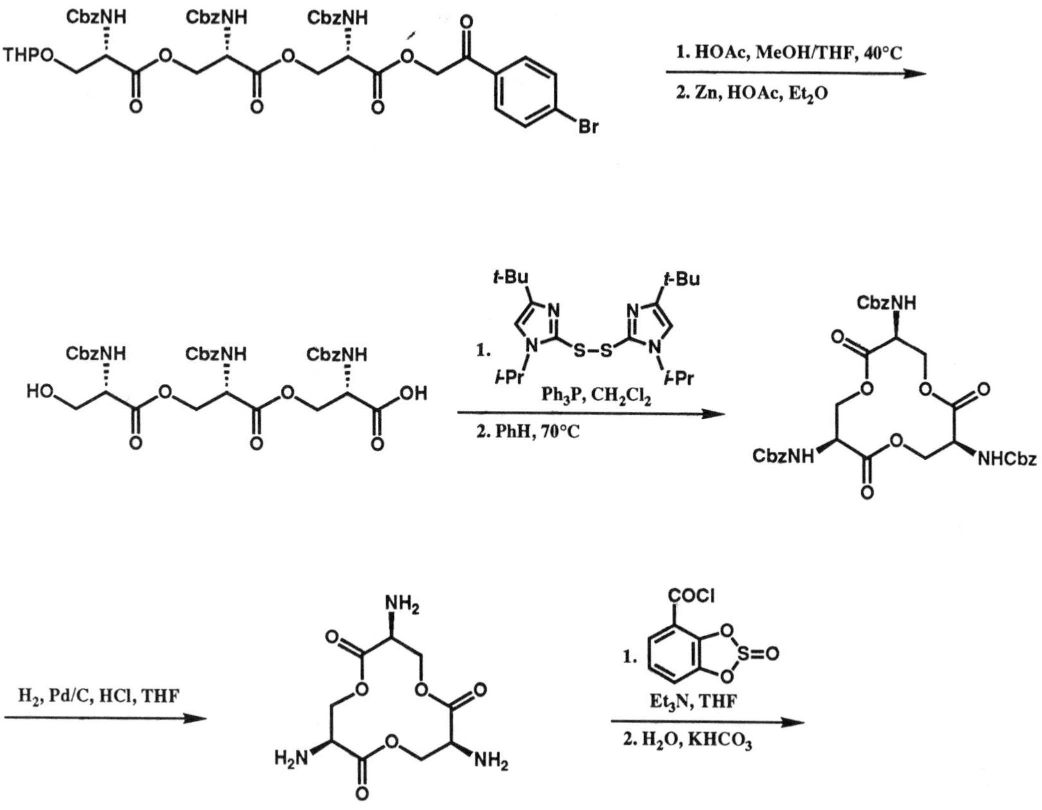

CbzNH
THPO
Br
1. HOAc, MeOH/THF, 40°C
2. Zn, HOAc, Et₂O
HO
OH
t-Bu
i-Pr
S–S
Ph₃P, CH₂Cl₂
2. PhH, 70°C
NHCbz
H₂, Pd/C, HCl, THF
NH₂
H₂N
COCl
S=O
Et₃N, THF
2. H₂O, KHCO₃

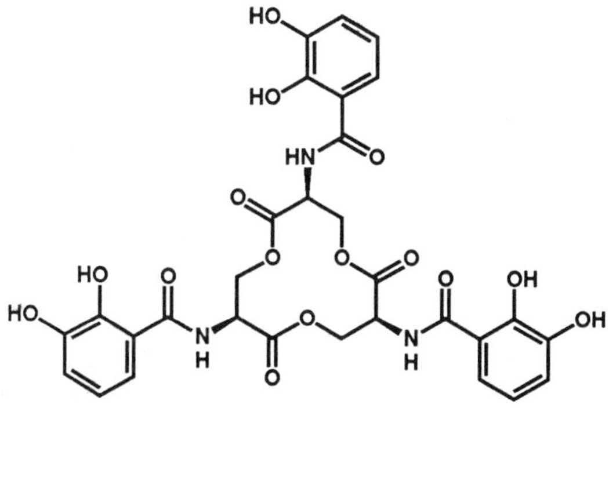

HO
HN
OH

synthetic sample: $[\alpha]^{25}_D$ +7.56° (*c* 1.0, EtOH)

natural ***Enterobactin***: $[\alpha]^{25}_D$ +7.40° (*c* 1.0, EtOH)

References:

1. E. J. Corey and S. Bhattacharyya, *Tetrahedron Lett.* **1977**, 3919.
2. E. J. Corey and S. D. Hurt, *Tetrahedron Lett.* **1977**, 3923.

(±)-*N*-Methylmaysenine

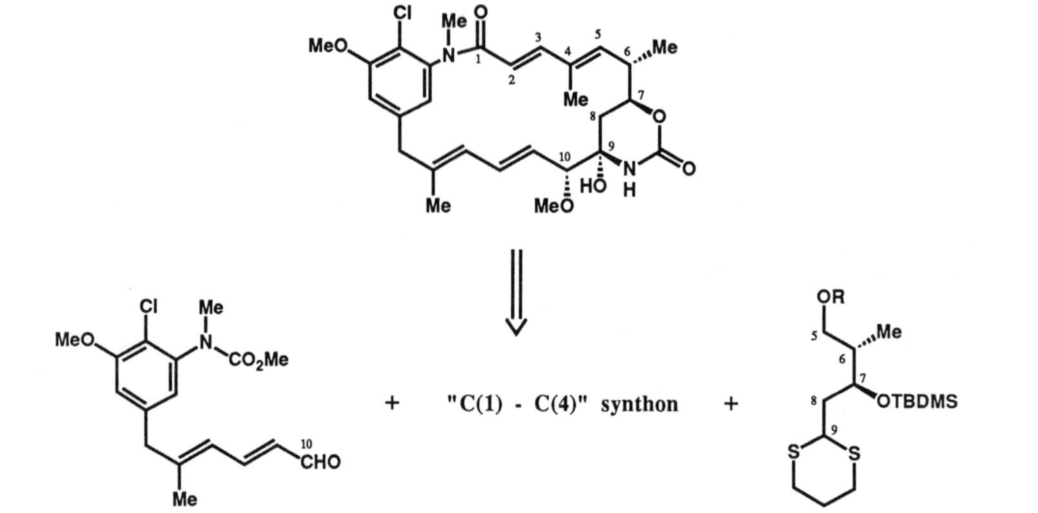

+ "C(1) - C(4)" synthon +

A B

The synthesis of (±)-*N*-methylmaysenine, a preliminary for the later synthesis of the antitumor agent maytansine, was accomplished by the joining of fragments **A** and **B**, chain extension and macrolactam closure using a mixed carboxylic-sulfonic acid anhydride.

1. Synthesis of the fragment **A** (Ref. 1, 2):

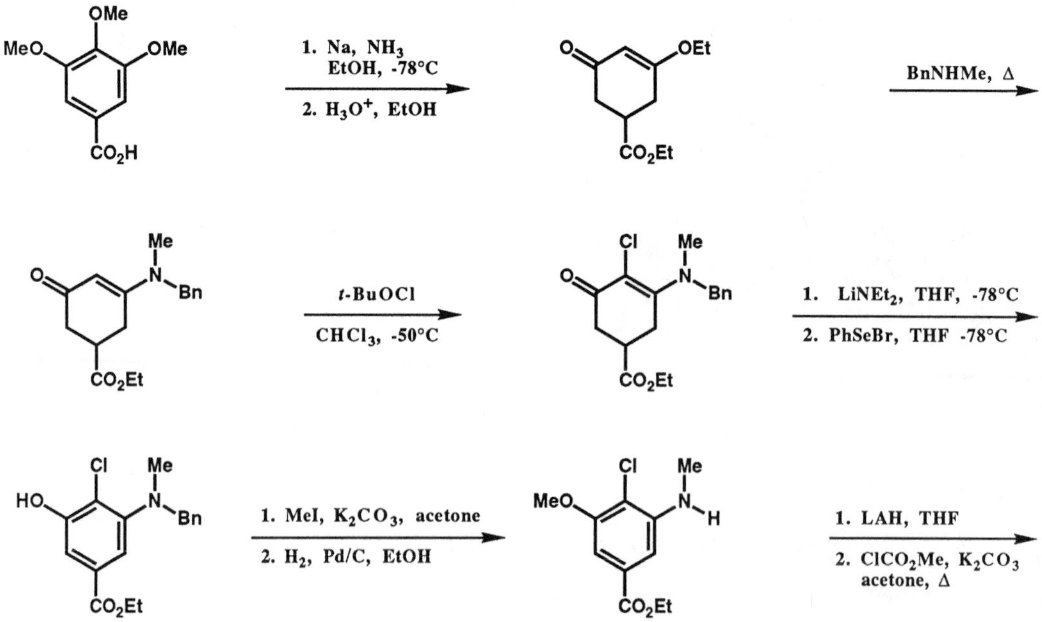

1. Na, NH3 EtOH, -78°C
2. H3O+, EtOH
BnNHMe, Δ
t-BuOCl
CHCl3, -50°C
1. LiNEt2, THF, -78°C
2. PhSeBr, THF -78°C
1. MeI, K2CO3, acetone
2. H2, Pd/C, EtOH
1. LAH, THF
2. ClCO2Me, K2CO3 acetone, Δ

7.6

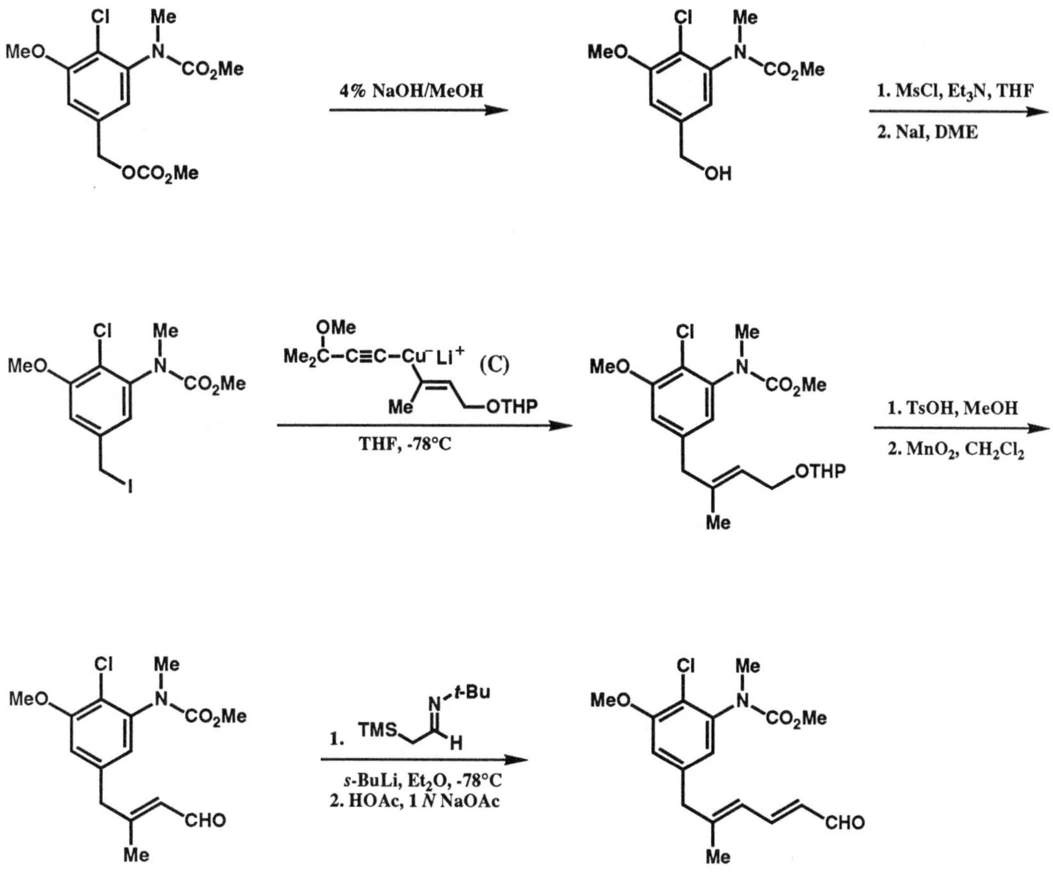
4% NaOH/MeOH
1. MsCl, Et3N, THF
2. NaI, DME
THF, -78°C
1. TsOH, MeOH
2. MnO2, CH2Cl2
s-BuLi, Et2O, -78°C
2. HOAc, 1 N NaOAc

A

Preparation of the reagent **C**:

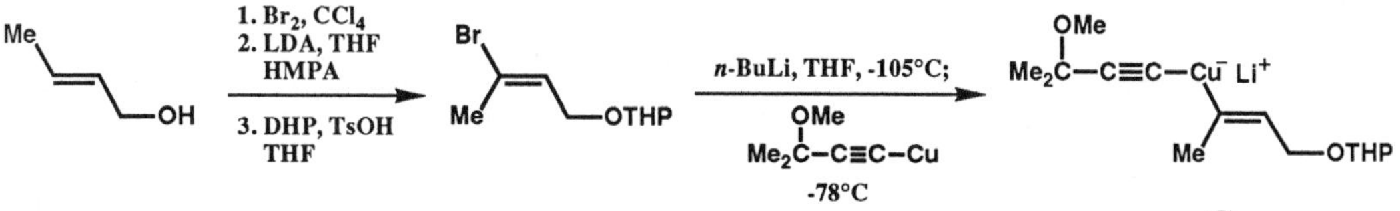
1. Br2, CCl4
2. LDA, THF
HMPA
3. DHP, TsOH
THF
n-BuLi, THF, -105°C;
-78°C

C

2. Synthesis of the fragment **B** (Ref. 3):

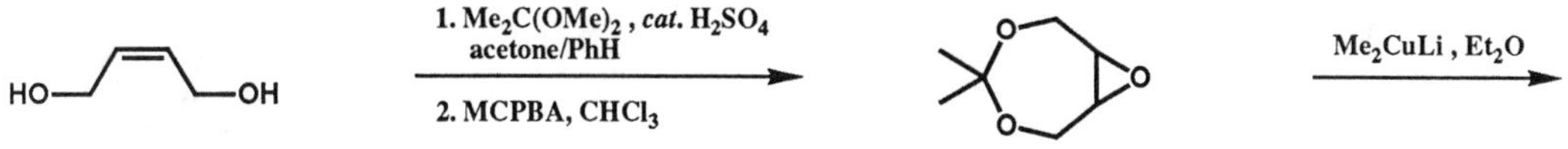
1. Me2C(OMe)2 , cat. H2SO4
acetone/PhH
2. MCPBA, CHCl3
Me2CuLi , Et2O

7.6

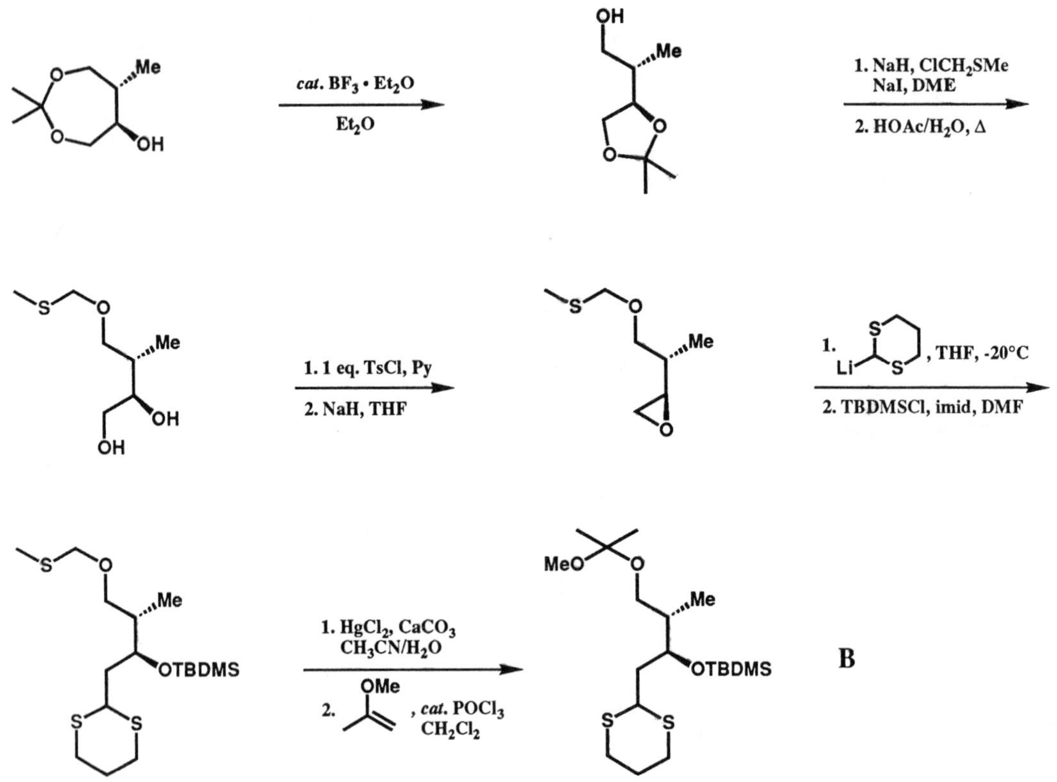
cat. BF3 • Et2O
Et2O
1. NaH, ClCH2SMe
NaI, DME
2. HOAc/H2O, Δ
1. 1 eq. TsCl, Py
2. NaH, THF
1. Li-dithiane, THF, -20°C
2. TBDMSCl, imid, DMF
1. HgCl2, CaCO3
CH3CN/H2O
2. cat. POCl3
CH2Cl2
B

3. Coupling of the fragments **A** and **B** and the completion of the synthesis (Ref. 4):

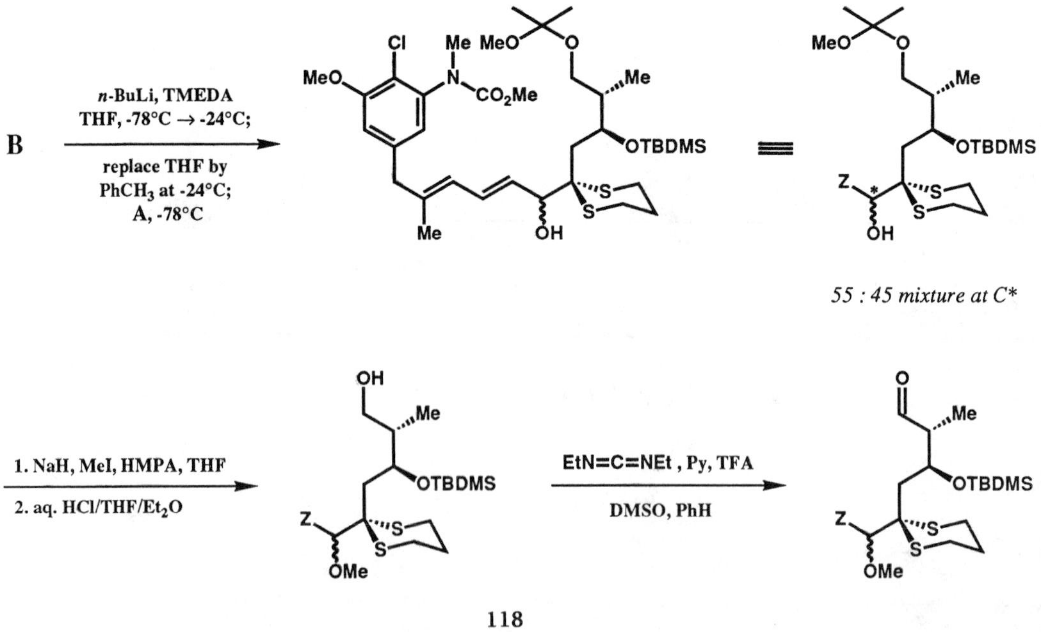
B
n-BuLi, TMEDA
THF, -78°C → -24°C;
replace THF by
PhCH3 at -24°C;
A, -78°C
55 : 45 mixture at C*
1. NaH, MeI, HMPA, THF
2. aq. HCl/THF/Et2O
EtN=C=NEt, Py, TFA
DMSO, PhH

7.6

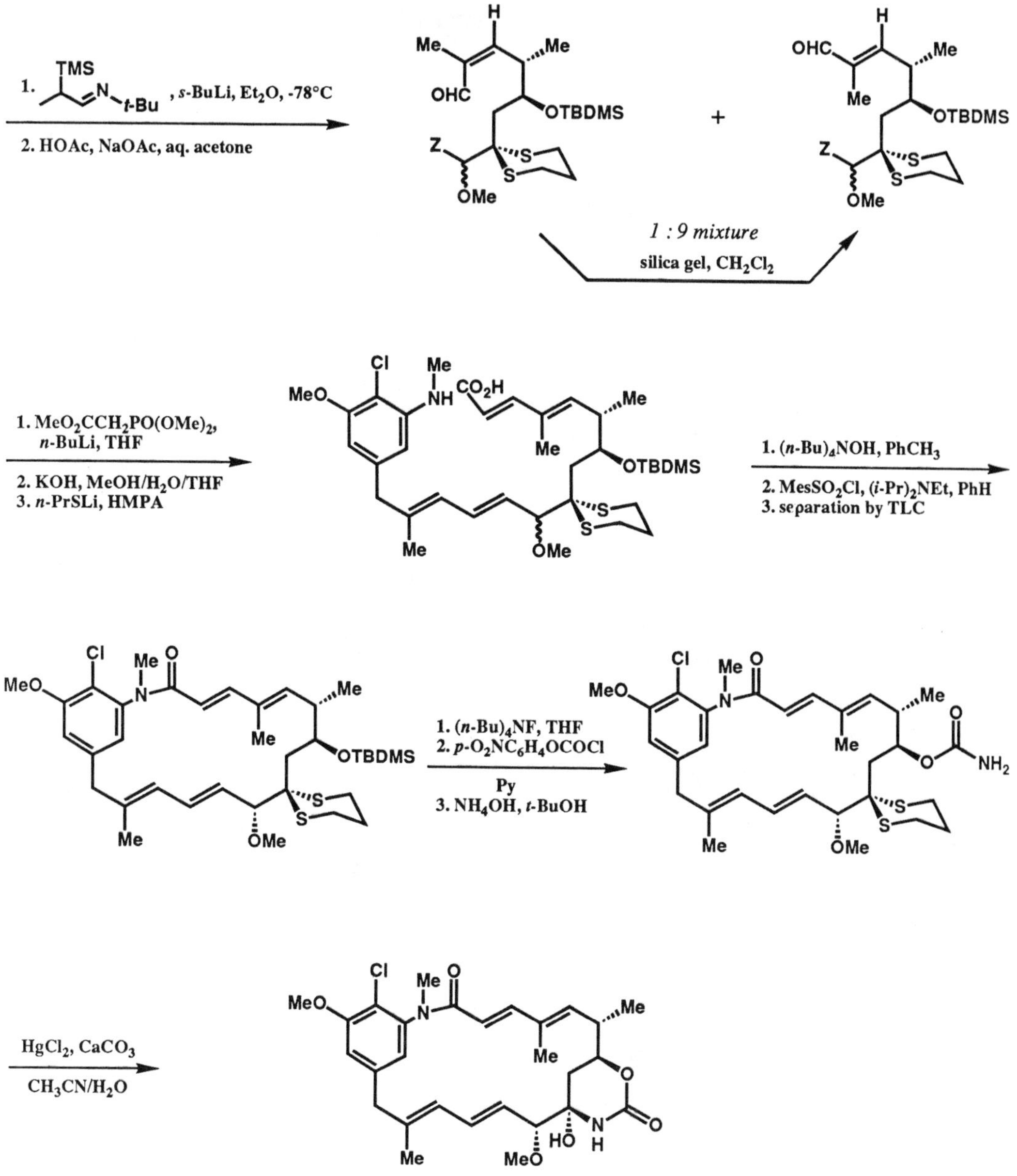
1. TMS N t-Bu , s-BuLi, Et2O, -78°C
2. HOAc, NaOAc, aq. acetone
1 : 9 mixture
silica gel, CH2Cl2
1. MeO2CCH2PO(OMe)2, n-BuLi, THF
2. KOH, MeOH/H2O/THF
3. n-PrSLi, HMPA
1. (n-Bu)4NOH, PhCH3
2. MesSO2Cl, (i-Pr)2NEt, PhH
3. separation by TLC
1. (n-Bu)4NF, THF
2. p-O2NC6H4OCOCl
Py
3. NH4OH, t-BuOH
HgCl2, CaCO3
CH3CN/H2O

References:

1. E. J. Corey, H. F. Wetter, A. P. Kozikowski, and A. V. R. Rao, *Tetrahedron Lett.* **1977**, 777.

2. E. J. Corey, M. G. Bock, A. P. Kozikowski, A. V. R. Rao, D. Floyd, and B. Lipshutz, *Tetrahedron Lett.* **1978**, 1051.

3. E. J. Corey and M. G. Bock, *Tetrahedron Lett.* **1975**, 2643.

4. E. J. Corey, L. O. Weigel, D. Floyd, and M. G. Bock, *J. Am. Chem. Soc.* **1978**, *100*, 2916.

(-)-*N*-Methylmaysenine

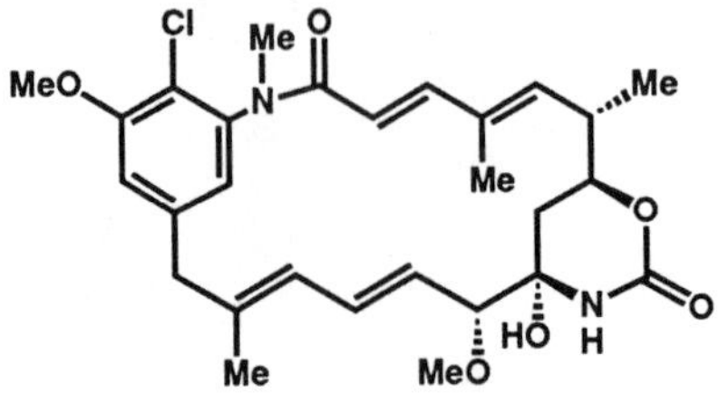

A simple modification of the foregoing route to (±)-*N*-methylmaysenine allowed the enantiospecific synthesis of (-)-*N*-methylmaysenine.

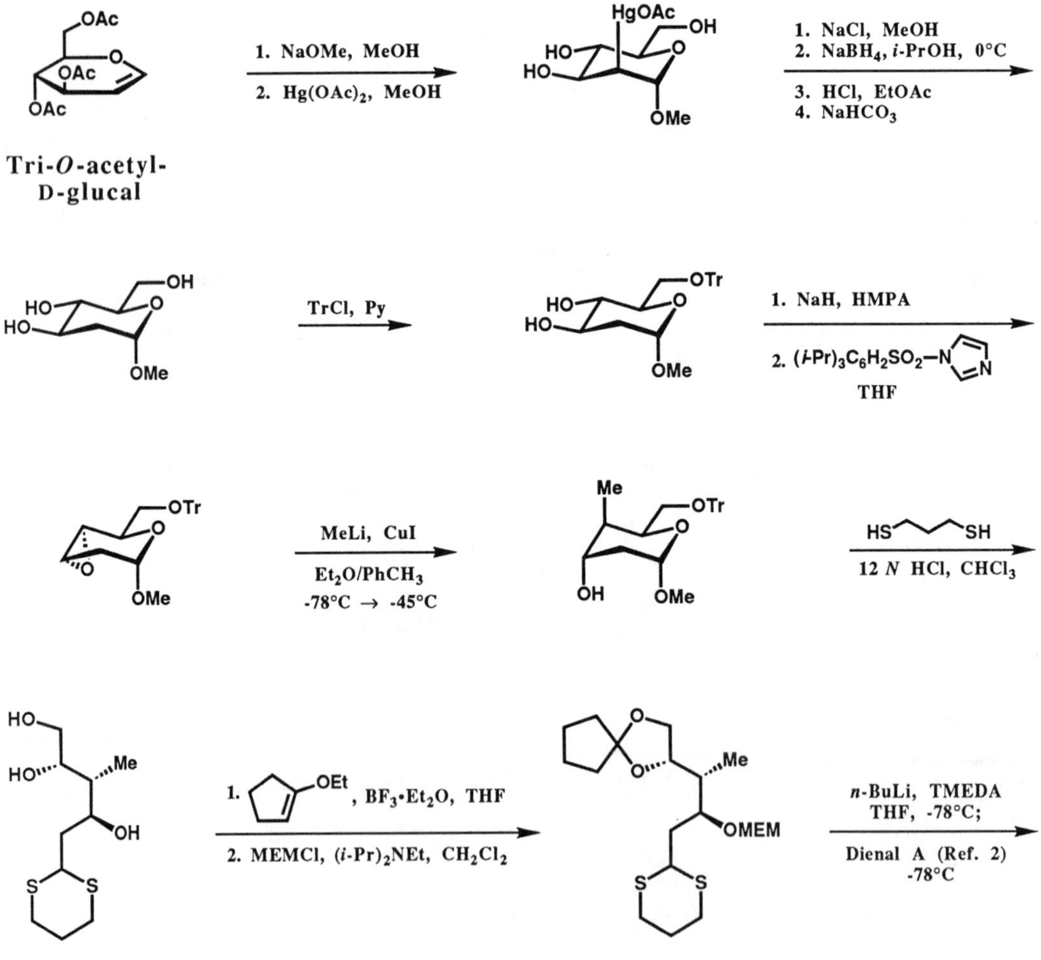

Tri-*O*-acetyl-D-glucal
1. NaOMe, MeOH
2. $Hg(OAc)_2$, MeOH
1. NaCl, MeOH
2. $NaBH_4$, *i*-PrOH, 0°C
3. HCl, EtOAc
4. $NaHCO_3$
TrCl, Py
1. NaH, HMPA
2. $(i\text{-Pr})_3C_6H_2SO_2$–N
THF
MeLi, CuI
$Et_2O/PhCH_3$
-78°C → -45°C
HS⁀SH
12 *N* HCl, $CHCl_3$
1. OEt, $BF_3 \cdot Et_2O$, THF
2. MEMCl, $(i\text{-Pr})_2NEt$, CH_2Cl_2
n-BuLi, TMEDA
THF, -78°C;
Dienal A (Ref. 2)
-78°C

$[\alpha]^{20}_D$ -16.8° (*c* 1.2, $CHCl_3$)

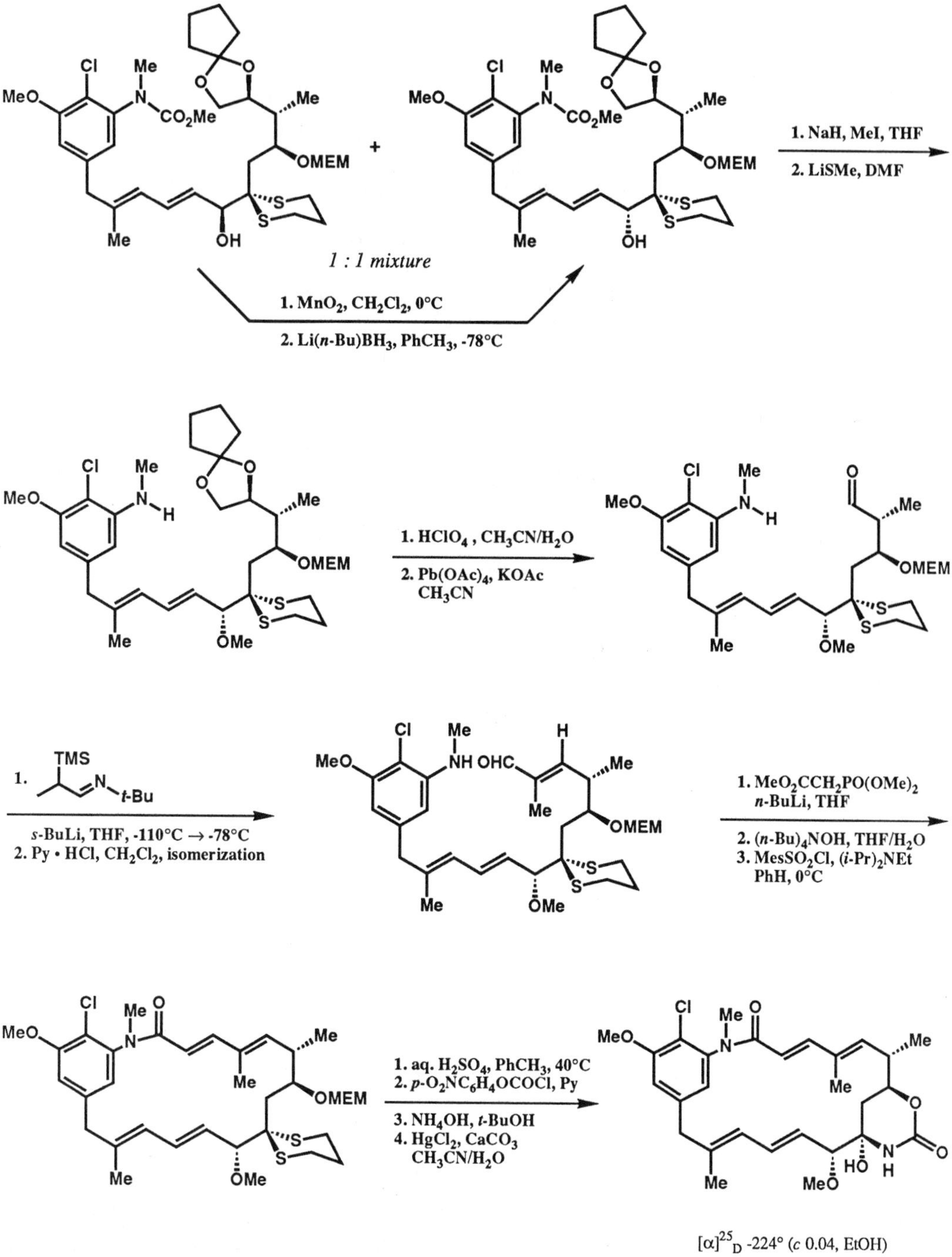

Cl
Me
MeO
N
CO2Me
O
O
Me
OMEM
S
S
Me
OH
+
1 : 1 mixture
1. MnO2, CH2Cl2, 0°C
2. Li(n-Bu)BH3, PhCH3, -78°C
1. NaH, MeI, THF
2. LiSMe, DMF
H
OMe
1. HClO4 , CH3CN/H2O
2. Pb(OAc)4, KOAc
CH3CN
TMS
N
t-Bu
s-BuLi, THF, -110°C → -78°C
2. Py • HCl, CH2Cl2, isomerization
NH OHC
1. MeO2CCH2PO(OMe)2
n-BuLi, THF
2. (n-Bu)4NOH, THF/H2O
3. MesSO2Cl, (i-Pr)2NEt
PhH, 0°C
1. aq. H2SO4, PhCH3, 40°C
2. p-O2NC6H4OCOCl, Py
3. NH4OH, t-BuOH
4. HgCl2, CaCO3
CH3CN/H2O
HO
[α]25D -224° (c 0.04, EtOH)

References:

1. E. J. Corey, L. O. Weigel, A. R. Chamberlin, and B. Lipshutz, *J. Am. Chem. Soc.* **1980**, *102*, 1439.
2. E. J. Corey, L. O. Weigel, D. Floyd, and M. G. Bock, *J. Am. Chem. Soc.* **1978**, *100*, 2916.

7.8

Maytansine

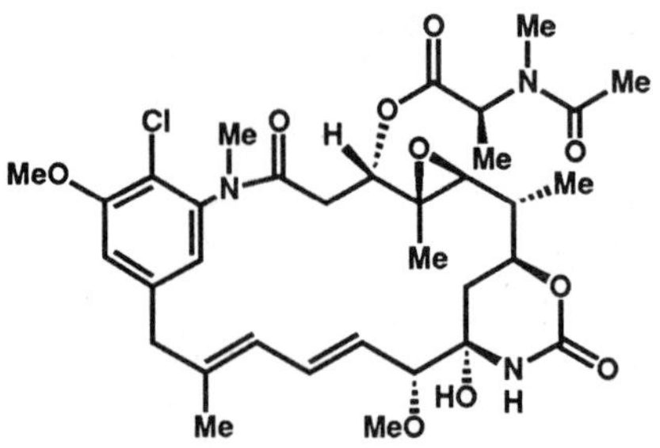

The first synthesis of the potent antitumor agent maytansine was carried out by the elaboration of aldehyde **D**, an intermediate in the enantioselective synthesis of (-)-*N*-methylmaysenine (Ref. 1,2), using enantioselective and diastereoselective steps.

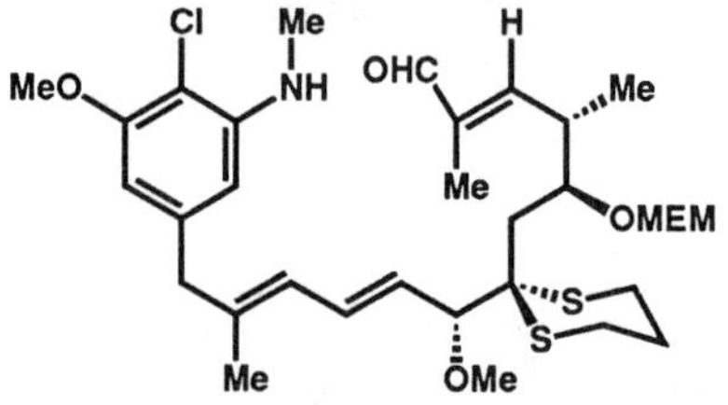

D

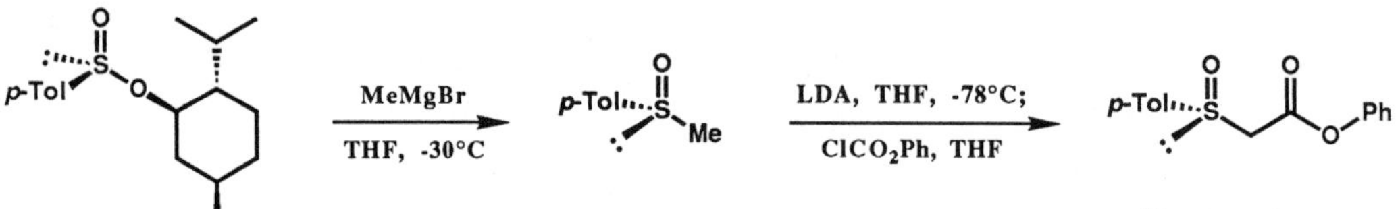
MeMgBr
THF, -30°C
LDA, THF, -78°C;
ClCO$_2$Ph, THF

$[\alpha]^{25}_D$ +87° (*c* 0.95, $CHCl_3$)

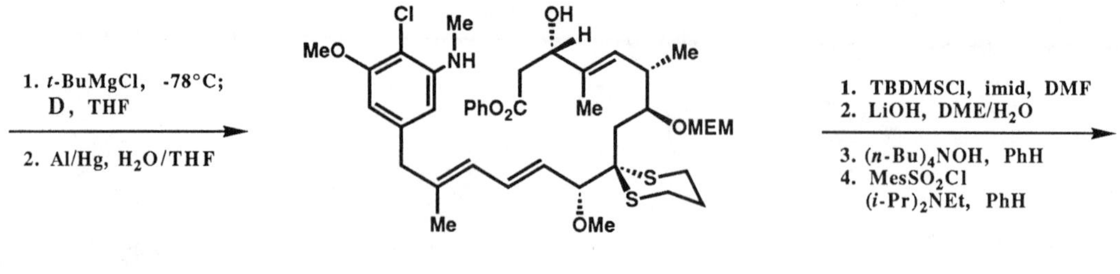
1. *t*-BuMgCl, -78°C; D, THF
2. Al/Hg, H$_2$O/THF
1. TBDMSCl, imid, DMF
2. LiOH, DME/H$_2$O
3. (*n*-Bu)$_4$NOH, PhH
4. MesSO$_2$Cl (*i*-Pr)$_2$NEt, PhH

diastereoselection 93 : 7

7.8

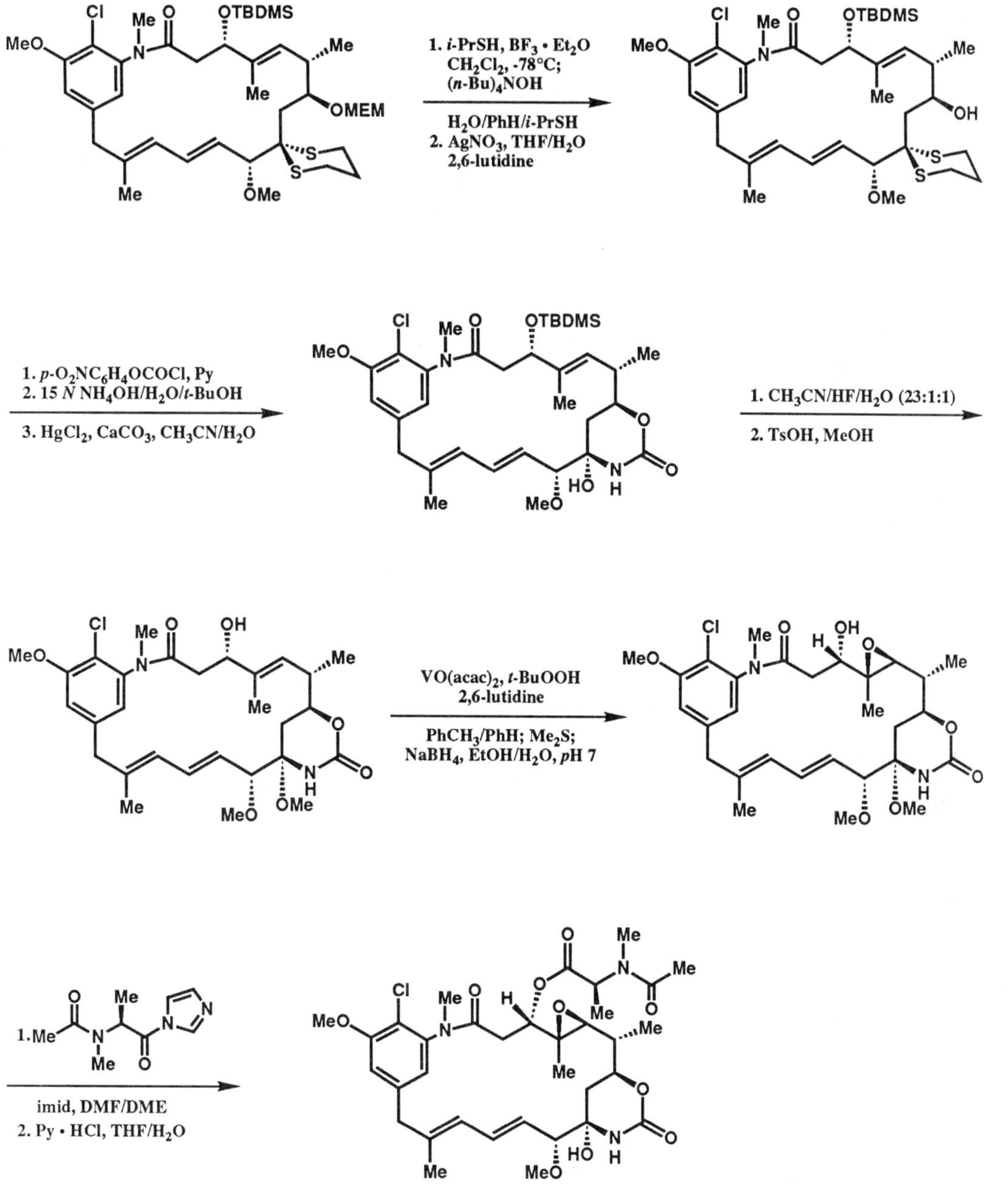
1. i-PrSH, BF3 • Et2O
CH2Cl2, -78°C;
(n-Bu)4NOH
H2O/PhH/i-PrSH
2. AgNO3, THF/H2O
2,6-lutidine
1. p-O2NC6H4OCOCl, Py
2. 15 N NH4OH/H2O/t-BuOH
3. HgCl2, CaCO3, CH3CN/H2O
1. CH3CN/HF/H2O (23:1:1)
2. TsOH, MeOH
VO(acac)2, t-BuOOH
2,6-lutidine
PhCH3/PhH; Me2S;
NaBH4, EtOH/H2O, pH 7
imid, DMF/DME
2. Py • HCl, THF/H2O

References:

1. E. J. Corey, L. O. Weigel, D. Floyd, and M. G. Bock, *J. Am. Chem. Soc.* **1978**, *100*, 2916.
2. E. J. Corey, L. O. Weigel, A. R. Chamberlin, and B. Lipshutz, *J. Am. Chem. Soc.* **1980**, *102*, 1439.
3. E. J. Corey, L. O. Weigel, A. R. Chamberlin, H. Cho, and D. H. Hua, *J. Am. Chem. Soc.* **1980**, *102*, 6613.

7.9

(±)-Brefeldin A

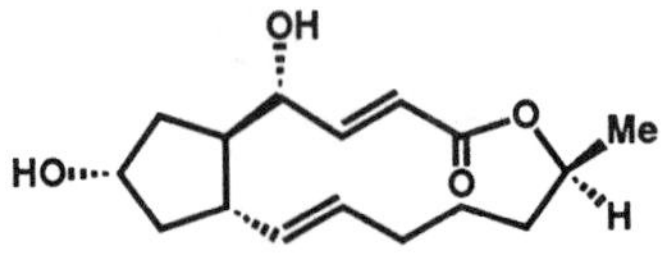
OH
HO
O
O
Me
H

Brefeldin A, an antiviral agent which impedes protein transport from the endoplasmic reticulum to the Golgi complex, was synthesized as the racemate using a number of interesting diastereoselective reactions.

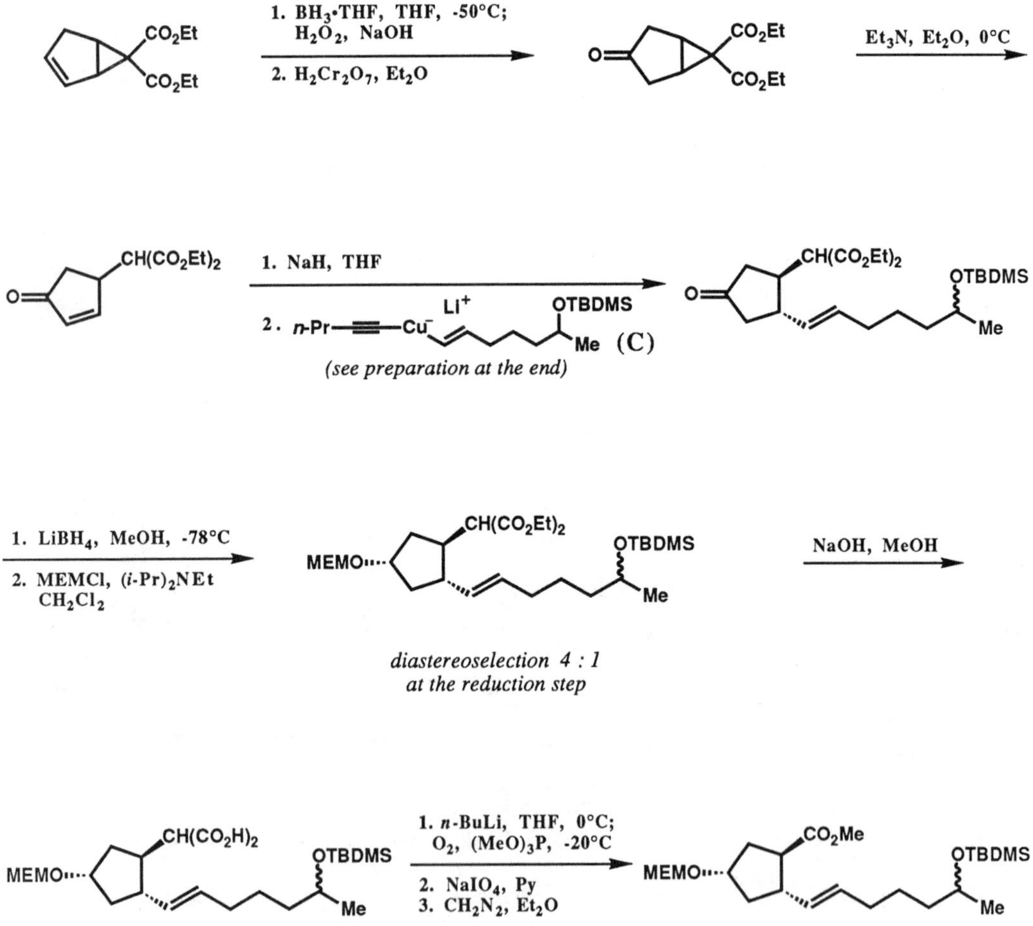
CO_2Et
CO_2Et
1. BH_3•THF, THF, -50°C; H_2O_2, NaOH
2. $H_2Cr_2O_7$, Et_2O
O
CO_2Et
CO_2Et
Et_3N, Et_2O, 0°C
O
$CH(CO_2Et)_2$
1. NaH, THF
2. n-Pr—≡—Cu⁻ Li⁺ OTBDMS Me (C)
(see preparation at the end)
O
$CH(CO_2Et)_2$
OTBDMS
Me
1. $LiBH_4$, MeOH, -78°C
2. MEMCl, $(i\text{-}Pr)_2NEt$ CH_2Cl_2
MEMO
$CH(CO_2Et)_2$
OTBDMS
Me
diastereoselection 4 : 1
at the reduction step
NaOH, MeOH
MEMO
$CH(CO_2H)_2$
OTBDMS
Me
1. n-BuLi, THF, 0°C; O_2, $(MeO)_3P$, -20°C
2. $NaIO_4$, Py
3. CH_2N_2, Et_2O
MEMO
CO_2Me
OTBDMS
Me

A

7.9

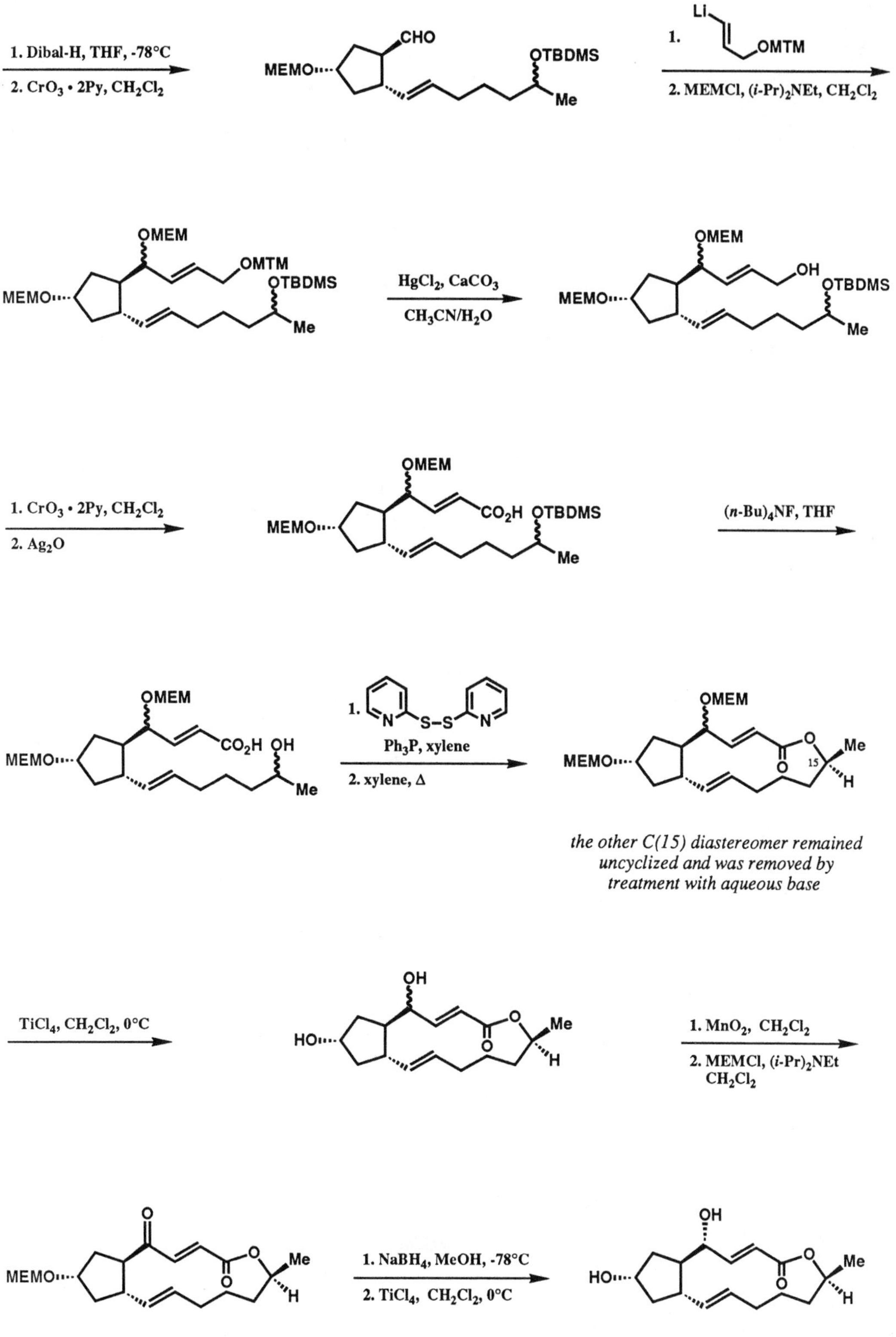
1. Dibal-H, THF, -78°C
2. CrO3 • 2Py, CH2Cl2
MEMO
CHO
OTBDMS
Me
1. Li OMTM
2. MEMCl, (i-Pr)2NEt, CH2Cl2
OMEM
OMTM
HgCl2, CaCO3
CH3CN/H2O
OH
1. CrO3 • 2Py, CH2Cl2
2. Ag2O
CO2H
(n-Bu)4NF, THF
1. Ph3P, xylene
2. xylene, Δ
15
H
the other C(15) diastereomer remained uncyclized and was removed by treatment with aqueous base
TiCl4, CH2Cl2, 0°C
HO
1. MnO2, CH2Cl2
2. MEMCl, (i-Pr)2NEt CH2Cl2
1. NaBH4, MeOH, -78°C
2. TiCl4, CH2Cl2, 0°C

7.9

The synthesis described above from intermediate **A** was later improved as following (Ref. 2):

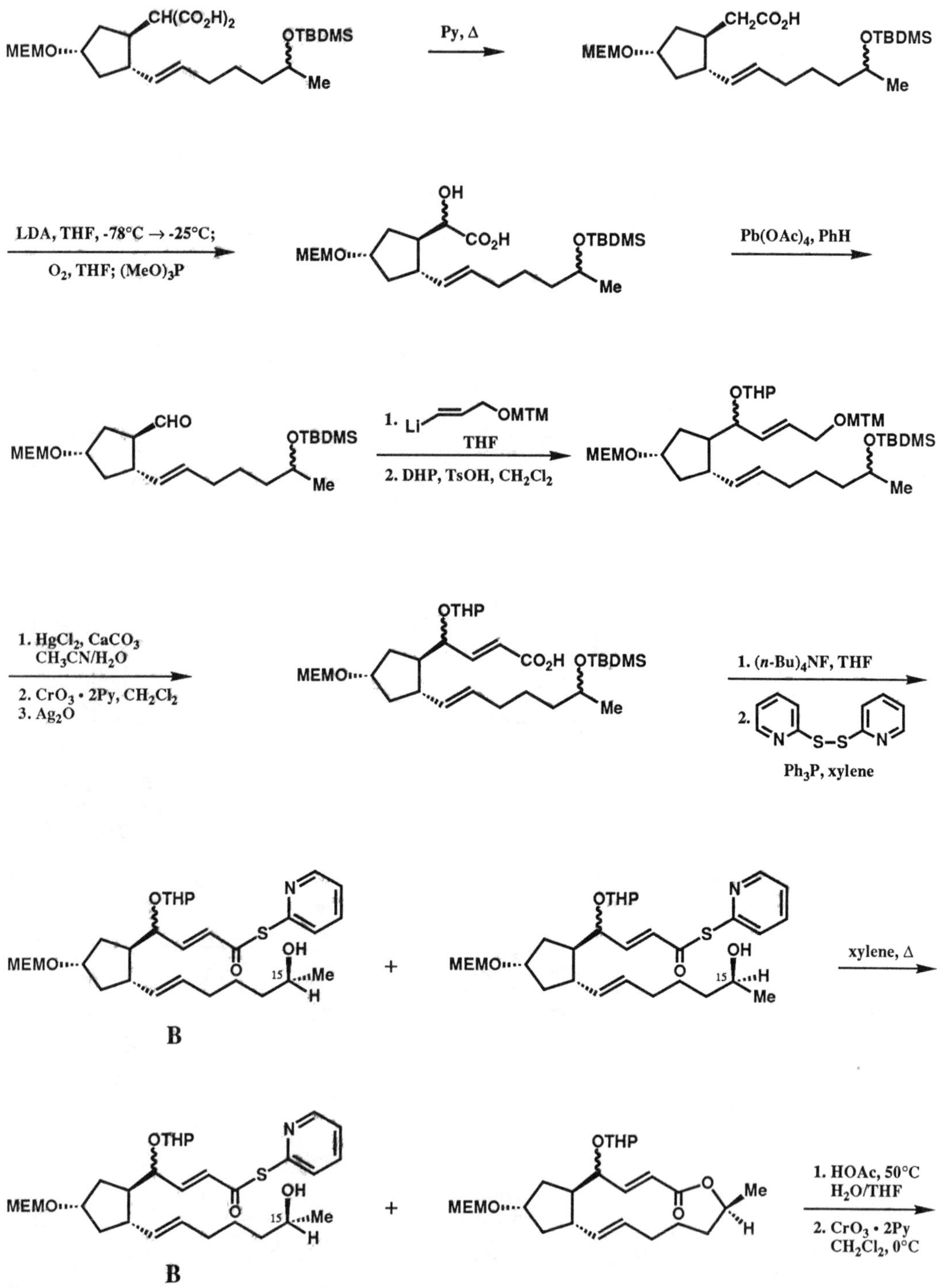

MEMO
CH(CO2H)2
OTBDMS
Me
Py, Δ
CH2CO2H
LDA, THF, -78°C → -25°C;
O2, THF; (MeO)3P
OH
CO2H
Pb(OAc)4, PhH
CHO
1. Li OMTM
THF
2. DHP, TsOH, CH2Cl2
OTHP
OMTM
1. HgCl2, CaCO3
CH3CN/H2O
2. CrO3 • 2Py, CH2Cl2
3. Ag2O
1. (n-Bu)4NF, THF
2. S–S N N
Ph3P, xylene
OH
15
H
N
S
O
B
+
xylene, Δ
1. HOAc, 50°C
H2O/THF
2. CrO3 • 2Py
CH2Cl2, 0°C

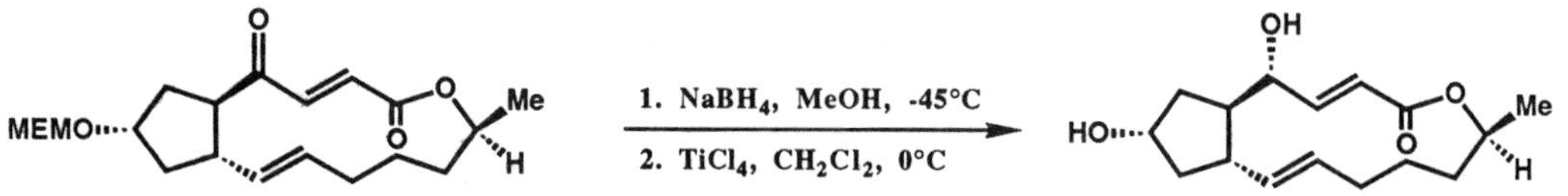
MEMO
Me
H
1. NaBH4, MeOH, -45°C
2. TiCl4, CH2Cl2, 0°C
OH
HO
Me
H

One of the C(15) epimeric thio esters (**B**) cyclizes more slowly than the other (by a factor of *ca.* 15) due to steric repulsions involving the methyl group at C(15). After lactonization, the uncyclized diastereomer was recovered and used for the synthesis as following.

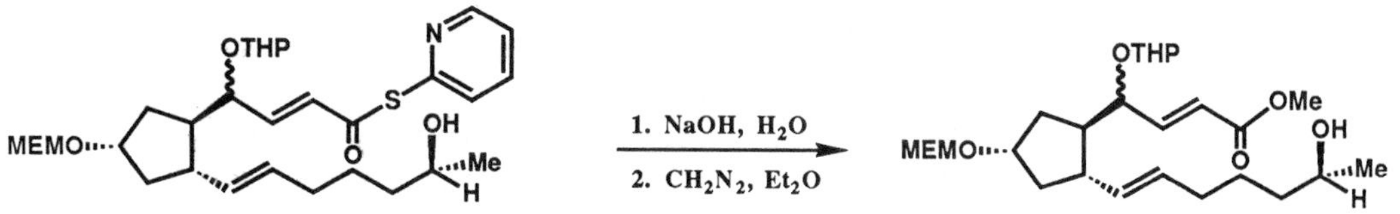
OTHP
N
S
OH
MEMO
Me
H
1. NaOH, H2O
2. CH2N2, Et2O
OTHP
OMe
OH
MEMO
Me
H

B

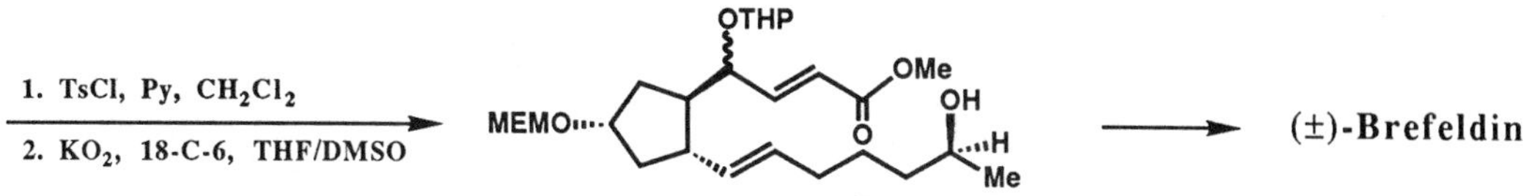
1. TsCl, Py, CH2Cl2
2. KO2, 18-C-6, THF/DMSO
OTHP
OMe
OH
MEMO
H
Me
(±)-Brefeldin

Preparation of the mixed cuprate (**C**):

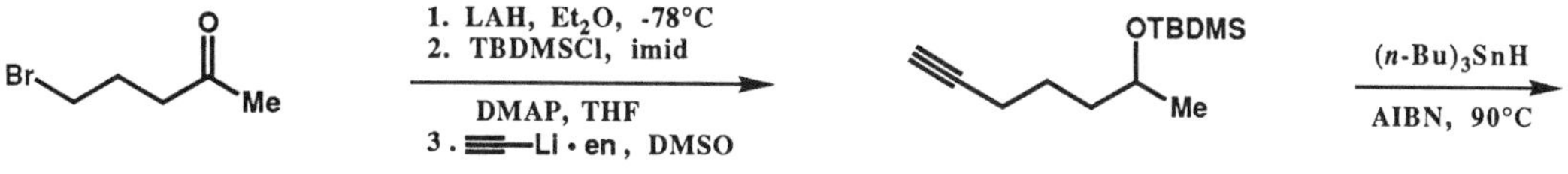
Br
Me
1. LAH, Et2O, -78°C
2. TBDMSCl, imid
DMAP, THF
3. Li • en, DMSO
OTBDMS
Me
(n-Bu)3SnH
AIBN, 90°C

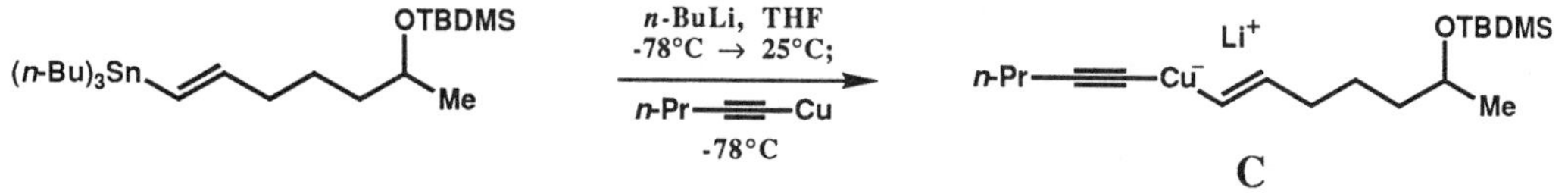
OTBDMS
(n-Bu)3Sn
Me
n-BuLi, THF
-78°C → 25°C;
n-Pr
Cu
-78°C
Li+
n-Pr
Cu
OTBDMS
Me

C

References:

1. E. J. Corey and R. H. Wollenberg, *Tetrahedron Lett.* **1976**, 4705.

2. E. J. Corey, R. H. Wollenberg, and D. R. Williams, *Tetrahedron Lett.* **1977**, 2243.

3. E. J. Corey and R. H. Wollenberg, *Tetrahedron Lett.* **1976**, 4701.

4. E. J. Corey, K. C. Nicolaou, and L. S. Melvin, Jr., *J. Am. Chem. Soc.* **1975**, *97*, 654.

7.10

Aplasmomycin

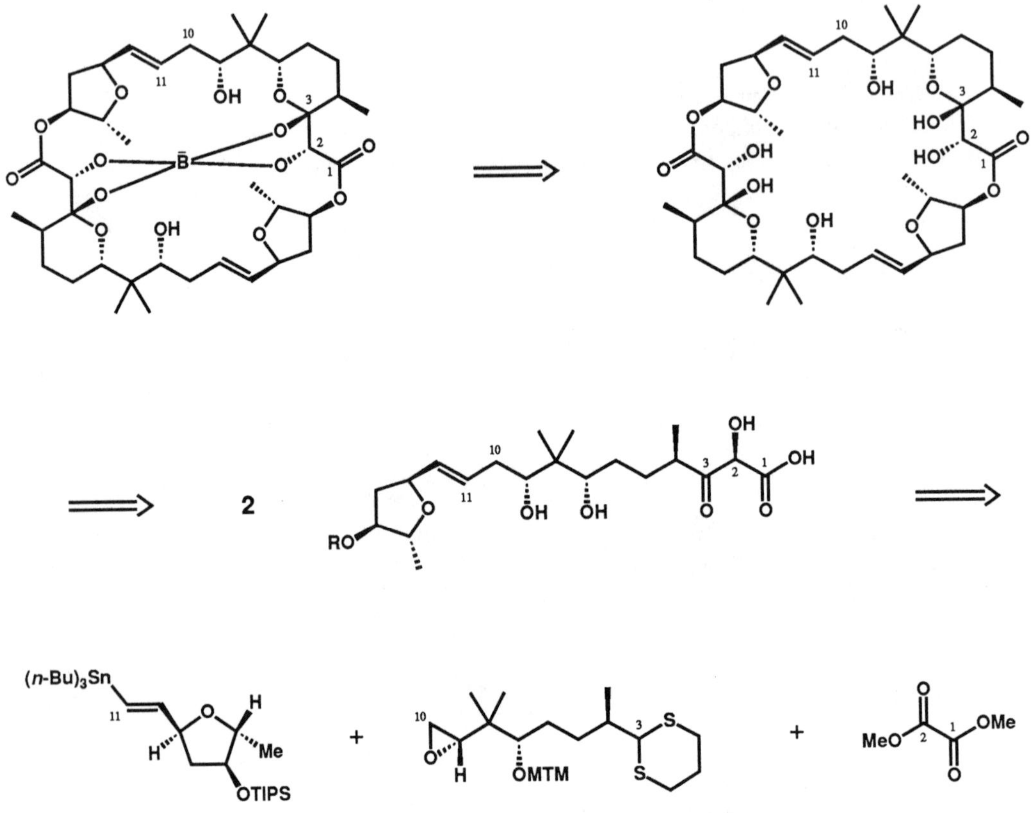

2
RO
(n-Bu)₃Sn
H
Me
OTIPS
+
OMTM
+
MeO
OMe

A

B

The antibiotic aplasmomycin serves as a receptor or transporter for borate. Retrosynthetic simplification of the boron-free macrocycle to identical hydroxy acid subunits is clearly appropriate. Further retrosynthetic dissection produced fragments **A**, **B** and oxalate and provided a workable synthetic plan.

1. Synthesis of the fragment **A**:

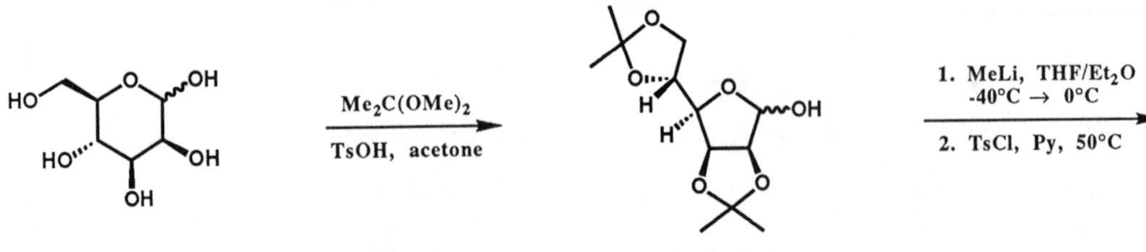

Me₂C(OMe)₂
TsOH, acetone
1. MeLi, THF/Et₂O
-40°C → 0°C
2. TsCl, Py, 50°C

(+)-D-Mannose

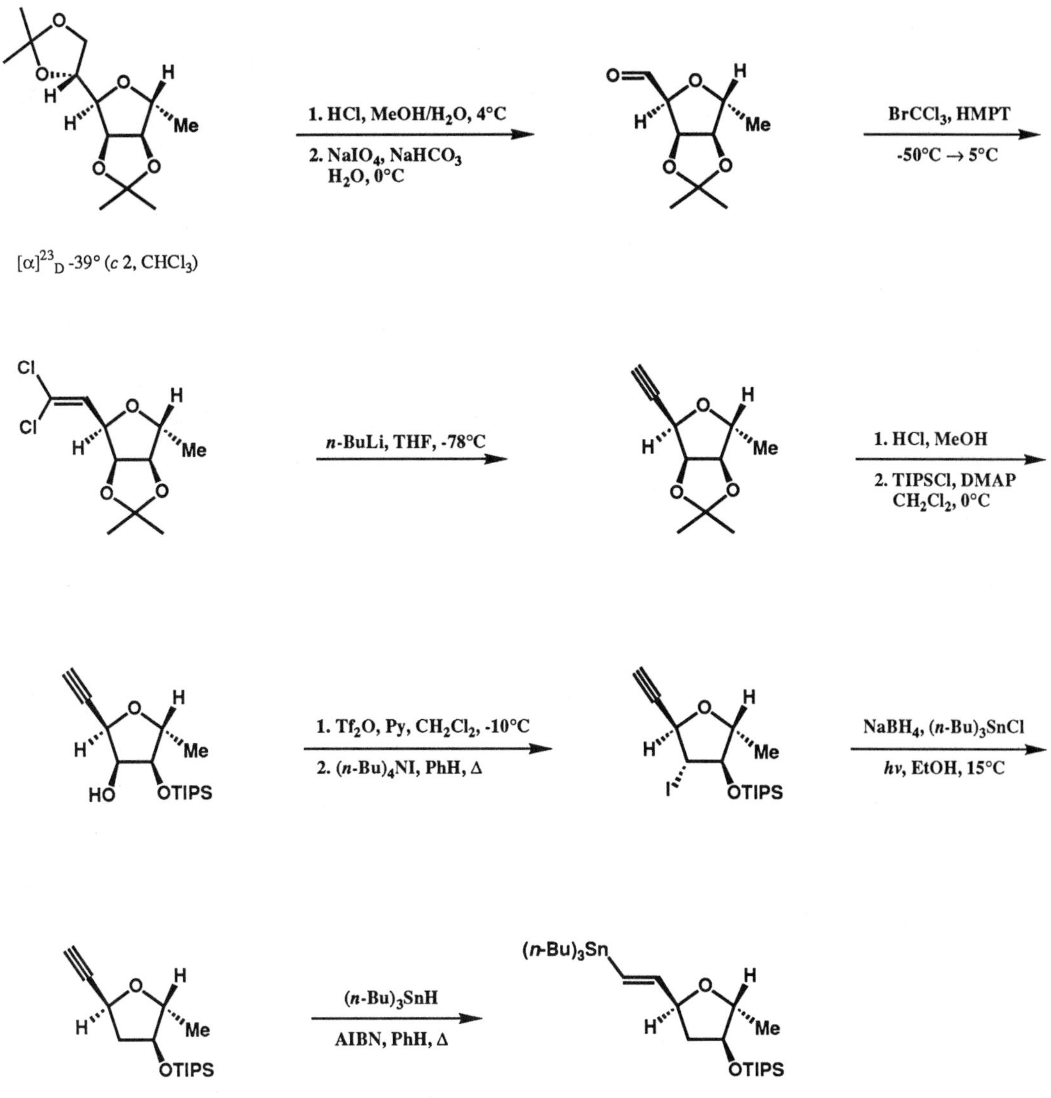

1. HCl, MeOH/H$_2$O, 4°C
2. NaIO$_4$, NaHCO$_3$ H$_2$O, 0°C
BrCCl$_3$, HMPT
-50°C → 5°C
[α]$^{23}_D$ -39° (c 2, CHCl$_3$)
n-BuLi, THF, -78°C
1. HCl, MeOH
2. TIPSCl, DMAP CH$_2$Cl$_2$, 0°C
1. Tf$_2$O, Py, CH$_2$Cl$_2$, -10°C
2. (n-Bu)$_4$NI, PhH, Δ
NaBH$_4$, (n-Bu)$_3$SnCl
hν, EtOH, 15°C
(n-Bu)$_3$SnH
AIBN, PhH, Δ

A

2. Synthesis of the fragment **B** :

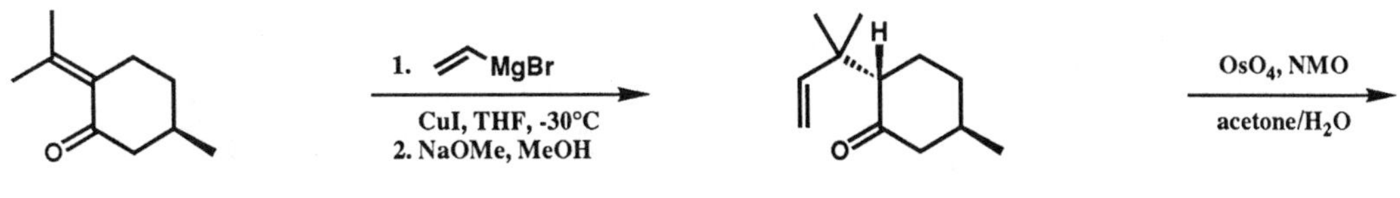

1. MgBr
CuI, THF, -30°C
2. NaOMe, MeOH
OsO$_4$, NMO
acetone/H$_2$O

(*R*)-(+)-Pulegone

85 : 15 mixture (equilibrated)

7.10

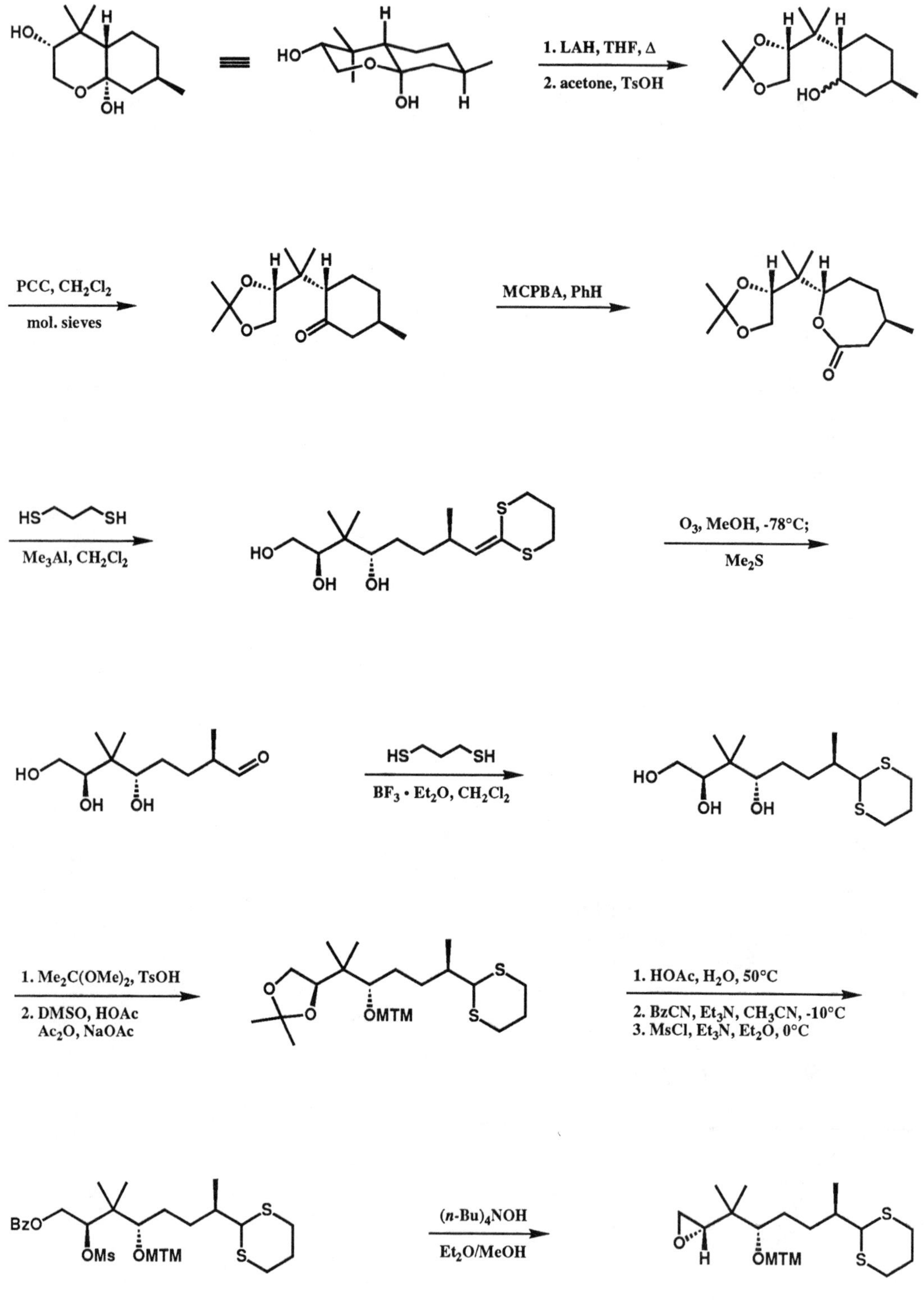
HO
H
OH
HO
O
OH
H
1. LAH, THF, Δ
2. acetone, TsOH
HO
PCC, CH_2Cl_2
mol. sieves
MCPBA, PhH
HS
SH
Me_3Al, CH_2Cl_2
O_3, MeOH, -78°C;
Me_2S
$BF_3 \cdot Et_2O$, CH_2Cl_2
1. $Me_2C(OMe)_2$, TsOH
2. DMSO, HOAc
Ac_2O, NaOAc
OMTM
1. HOAc, H_2O, 50°C
2. BzCN, Et_3N, CH_3CN, -10°C
3. MsCl, Et_3N, Et_2O, 0°C
BzO
OMs
(n-Bu)$_4$NOH
Et_2O/MeOH
S

B

3. Coupling of the fragments **A** and **B** and the completion of the synthesis:

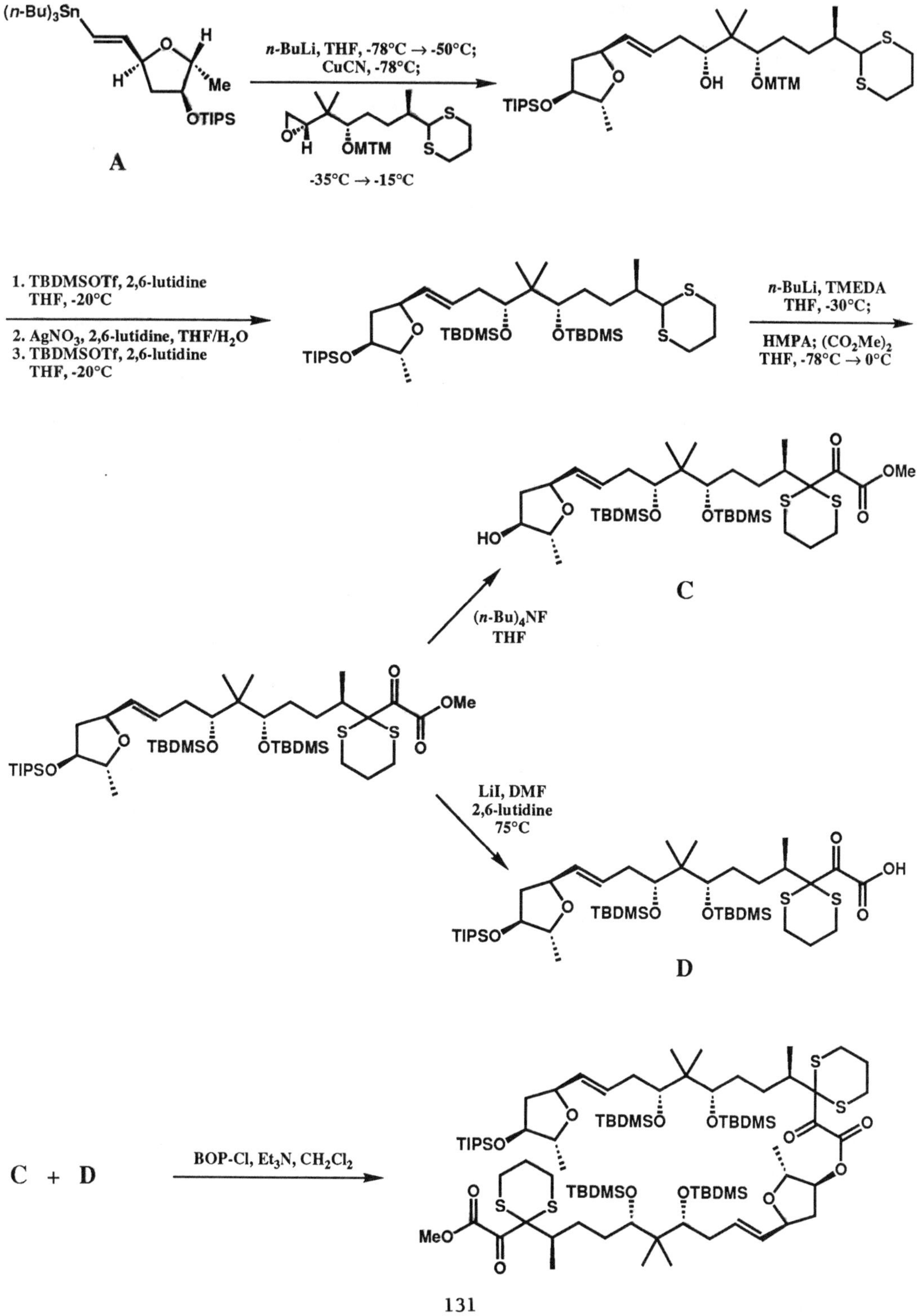

(n-Bu)3Sn
H
Me
OTIPS
A
n-BuLi, THF, -78°C → -50°C;
CuCN, -78°C;
H
OMTM
-35°C → -15°C
TIPSO
OH
OMTM
1. TBDMSOTf, 2,6-lutidine
THF, -20°C
2. AgNO3, 2,6-lutidine, THF/H2O
3. TBDMSOTf, 2,6-lutidine
THF, -20°C
TBDMSO
OTBDMS
n-BuLi, TMEDA
THF, -30°C;
HMPA; (CO2Me)2
THF, -78°C → 0°C
OMe
HO
C
(n-Bu)4NF
THF
LiI, DMF
2,6-lutidine
75°C
OH
D
C + D
BOP-Cl, Et3N, CH2Cl2
MeO

7.10

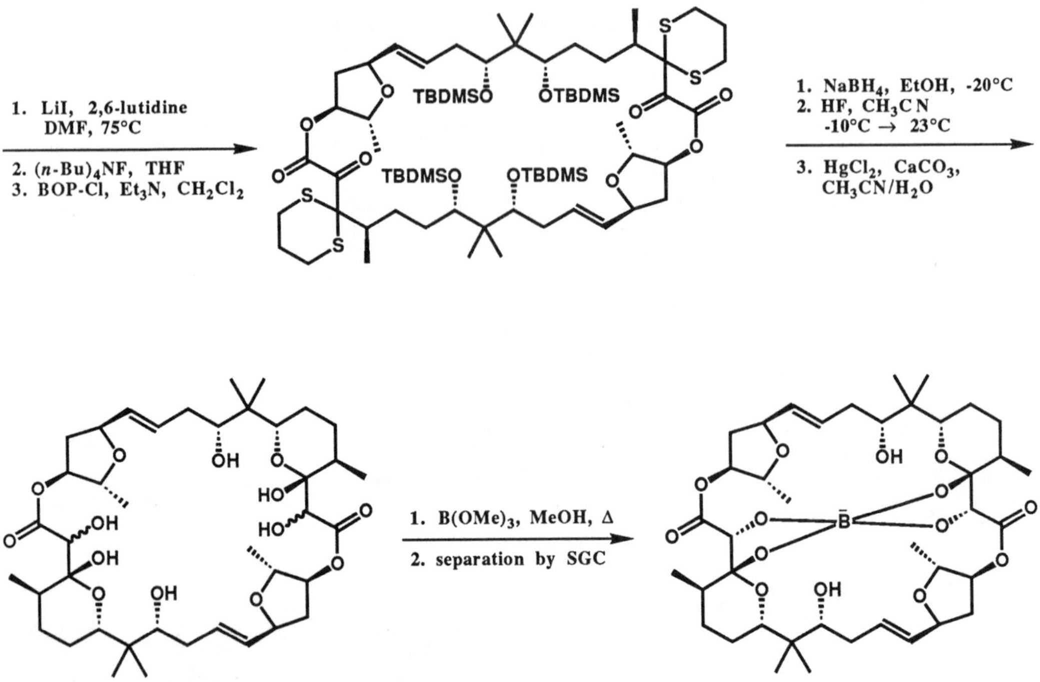
1. LiI, 2,6-lutidine
DMF, 75°C
2. (n-Bu)4NF, THF
3. BOP-Cl, Et3N, CH2Cl2
TBDMSO
OTBDMS
TBDMSO
OTBDMS
1. NaBH4, EtOH, -20°C
2. HF, CH3CN
-10°C → 23°C
3. HgCl2, CaCO3,
CH3CN/H2O
OH
HO
HO
OH
OH
OH
1. B(OMe)3, MeOH, Δ
2. separation by SGC
OH
OH

synthetic sample: $[\alpha]^{20}_{D}$ +202° (*c* 0.05, $CHCl_3$)
natural ***Aplasmomycin***: $[\alpha]^{23}_{D}$ +205° (*c* 0.05, $CHCl_3$)

The MOM group was also used for hydroxy protection to give fragment **E** as shown below:

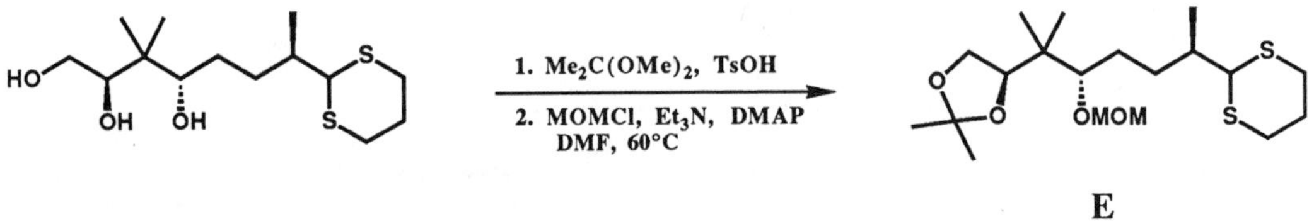
HO
OH
OH
1. Me2C(OMe)2, TsOH
2. MOMCl, Et3N, DMAP
DMF, 60°C
OMOM

E

Further elaboration of the MOM derivative **E** was carried out as described above for the MTM protected fragment **B** to afford fragment **F** which was converted to aplasmomycin by an 8-step sequence.

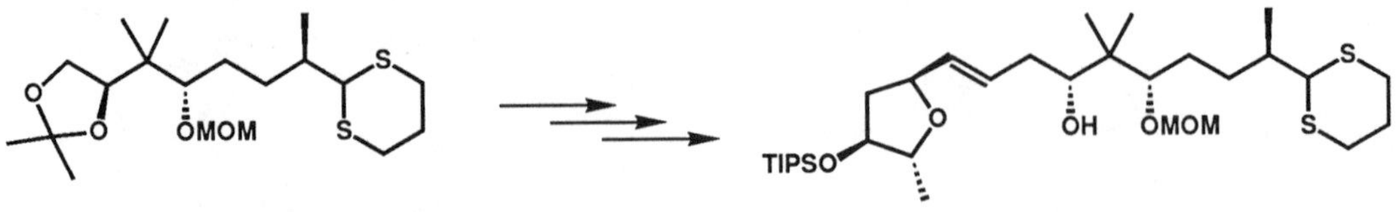
OMOM
TIPSO
OH
OMOM

E **F**

7.10

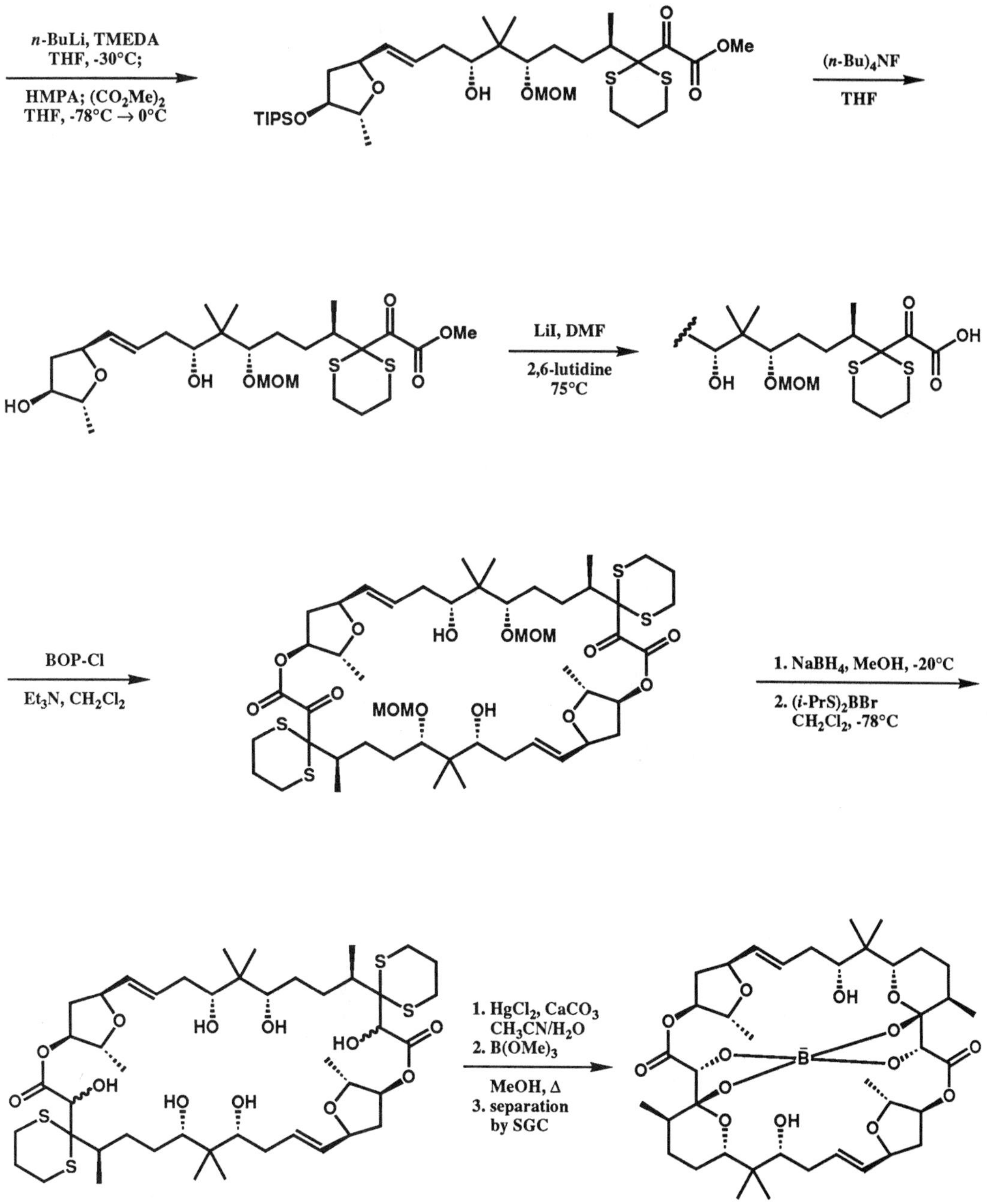
n-BuLi, TMEDA
THF, -30°C;
HMPA; (CO2Me)2
THF, -78°C → 0°C
TIPSO
OH
OMOM
OMe
(n-Bu)4NF
THF
HO
LiI, DMF
2,6-lutidine
75°C
OH
BOP-Cl
Et3N, CH2Cl2
MOMO
1. NaBH4, MeOH, -20°C
2. (i-PrS)2BBr
CH2Cl2, -78°C
1. HgCl2, CaCO3
CH3CN/H2O
2. B(OMe)3
MeOH, Δ
3. separation
by SGC

References:

1. E. J. Corey, B.-C. Pan, D. H. Hua, and D. R. Deardorff, *J. Am. Chem. Soc.* **1982**, *104*, 6816.
2. E. J. Corey, D. H. Hua, B.-C. Pan, and S. P. Seitz, *J. Am. Chem. Soc.* **1982**, *104*, 6818.

CHAPTER EIGHT
Heterocyclic Structures

8.1

(±)-Perhydrohistrionicotoxin

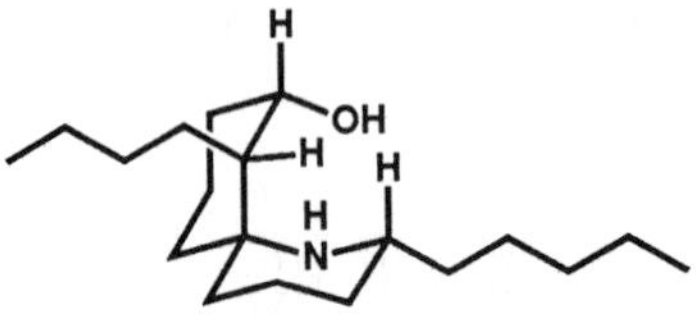
H
OH
H
H
H
N

The application of multistrategic retrosynthetic analysis to the target perhydro-histrionicotoxin is outlined in Section 6.3 of Part One.

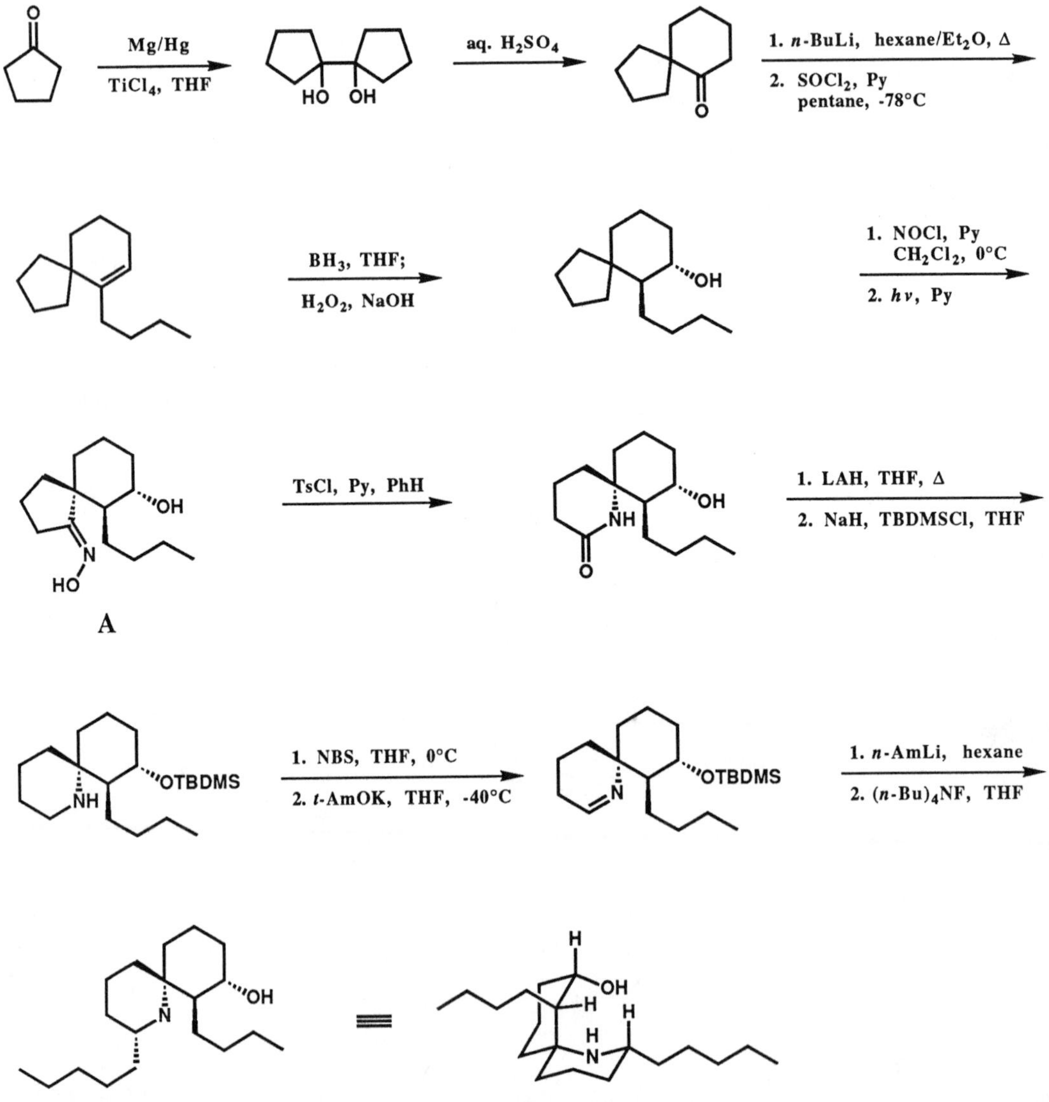
Mg/Hg
$TiCl_4$, THF
HO OH
aq. H_2SO_4
1. *n*-BuLi, hexane/Et_2O, Δ
2. $SOCl_2$, Py
pentane, -78°C
BH_3, THF;
H_2O_2, NaOH
OH
1. NOCl, Py
CH_2Cl_2, 0°C
2. *hν*, Py
OH
N
HO
A
TsCl, Py, PhH
OH
NH
1. LAH, THF, Δ
2. NaH, TBDMSCl, THF
OTBDMS
NH
1. NBS, THF, 0°C
2. *t*-AmOK, THF, -40°C
OTBDMS
N
1. *n*-AmLi, hexane
2. $(n\text{-Bu})_4NF$, THF
OH
N
H
OH
H
H
H
N

The intermediate **A** was also made by an alternative route using interesting new methodology for stereocontrol (Ref. 2).

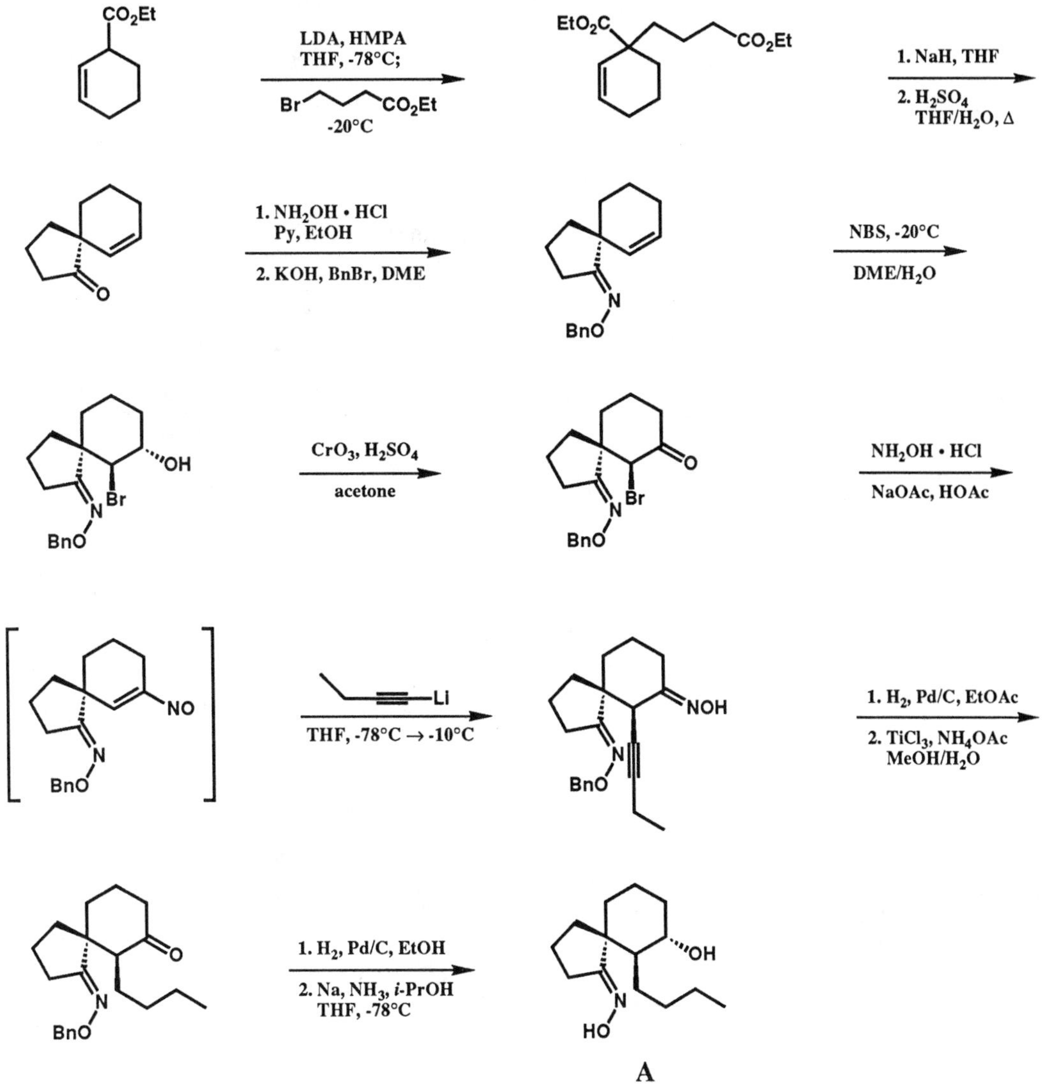
CO2Et
LDA, HMPA
THF, -78°C;
Br CO2Et
-20°C
EtO2C
CO2Et
1. NaH, THF
2. H2SO4
THF/H2O, Δ
O
1. NH2OH • HCl
Py, EtOH
2. KOH, BnBr, DME
N
BnO
NBS, -20°C
DME/H2O
OH
Br
CrO3, H2SO4
acetone
NH2OH • HCl
NaOAc, HOAc
NO
Li
THF, -78°C → -10°C
NOH
1. H2, Pd/C, EtOAc
2. TiCl3, NH4OAc
MeOH/H2O
1. H2, Pd/C, EtOH
2. Na, NH3, i-PrOH
THF, -78°C
HO
A

References:

1. E. J. Corey, J. F. Arnett, and G. N. Widiger, *J. Am. Chem. Soc.* **1975**, *97*, 430.
2. E. J. Corey, M. Petrzilka, and Y. Ueda, *Tetrahedron Lett.* **1975**, 4343.
3. E. J. Corey, Y. Ueda, and R. A. Ruden, *Tetrahedron Lett.* **1975**, 4347.
4. E. J. Corey and R. D. Balanson, *Heterocycles* **1976**, *5*, 445.
5. E. J. Corey, M. Petrzilka, and Y. Ueda, *Helv. Chim. Acta* **1977**, *60*, 2294.

8.2

(±)-Porantherine

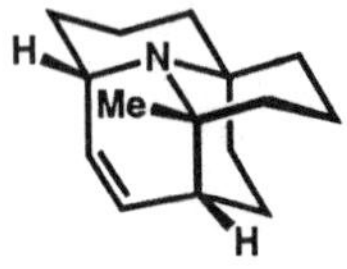
H
N
Me
H

A multistrategic retrosynthetic analysis of porantherine is outlined in Section 6.2 of Part One.

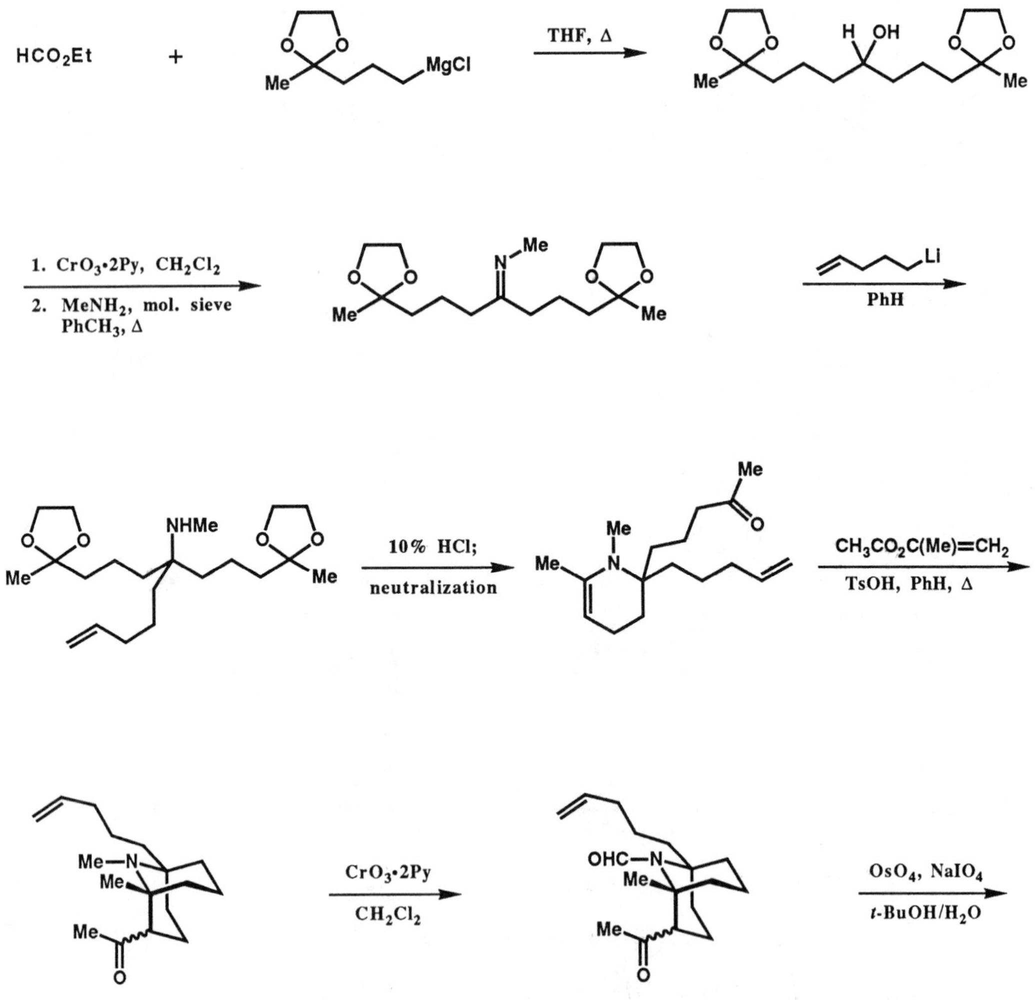
HCO2Et
+
MgCl
THF, Δ
H OH
1. CrO3•2Py, CH2Cl2
2. MeNH2, mol. sieve
PhCH3, Δ
Li
PhH
NHMe
10% HCl;
neutralization
CH3CO2C(Me)=CH2
TsOH, PhH, Δ
Me—N
CrO3•2Py
CH2Cl2
OHC—N
OsO4, NaIO4
t-BuOH/H2O

8.2

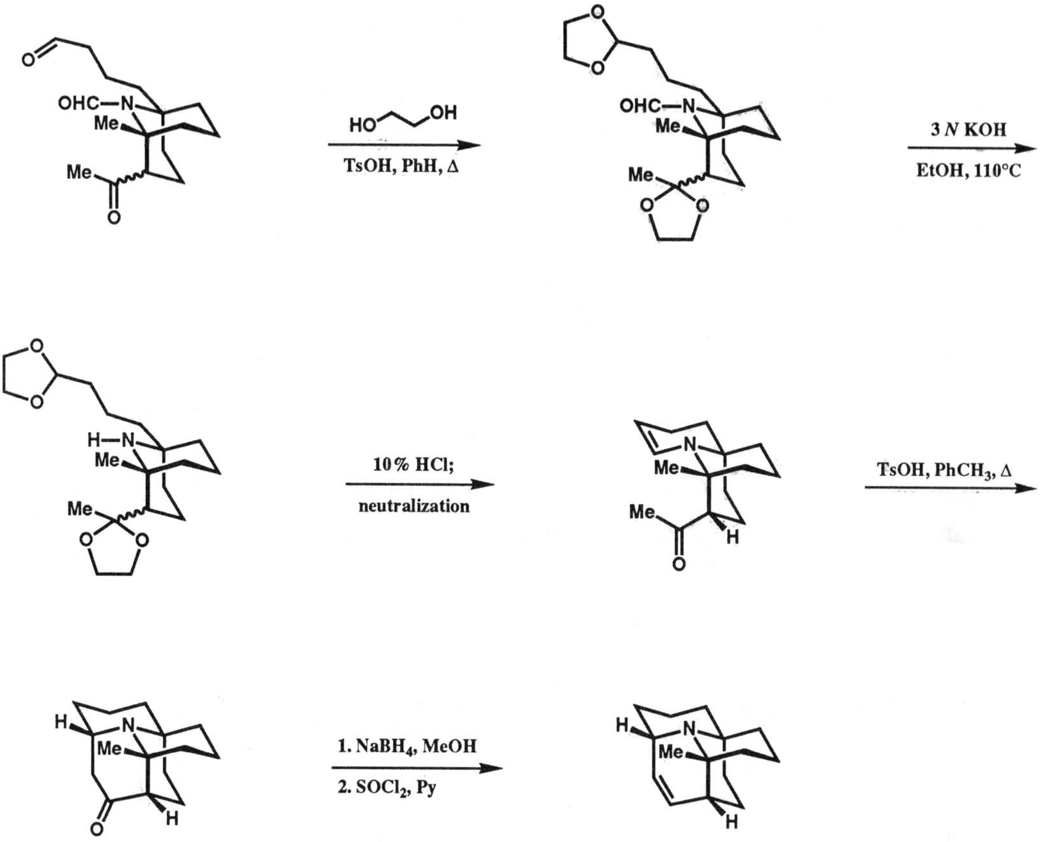
OHC—N
Me
Me
HO OH
TsOH, PhH, Δ
3 N KOH
EtOH, 110°C
H—N
10% HCl;
neutralization
TsOH, PhCH3, Δ
1. NaBH4, MeOH
2. SOCl2, Py
H

Reference:

E. J. Corey and R. D. Balanson, *J. Am. Chem. Soc.* **1974**, *96*, 6516.

8.3

(+)-Biotin

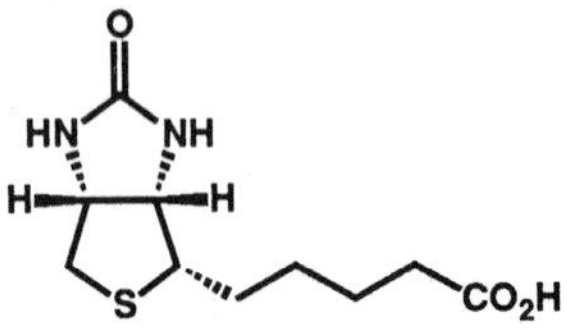

HN
NH
H
H
S
CO2H

The importance of biotin in nutrition and increasing commercial needs combine to suggest the need for short and economical synthesis. Retrosynthetic analysis using cysteine as SM goal suggested a number of synthetic pathways for study, one of which has been demonstrated as shown below.

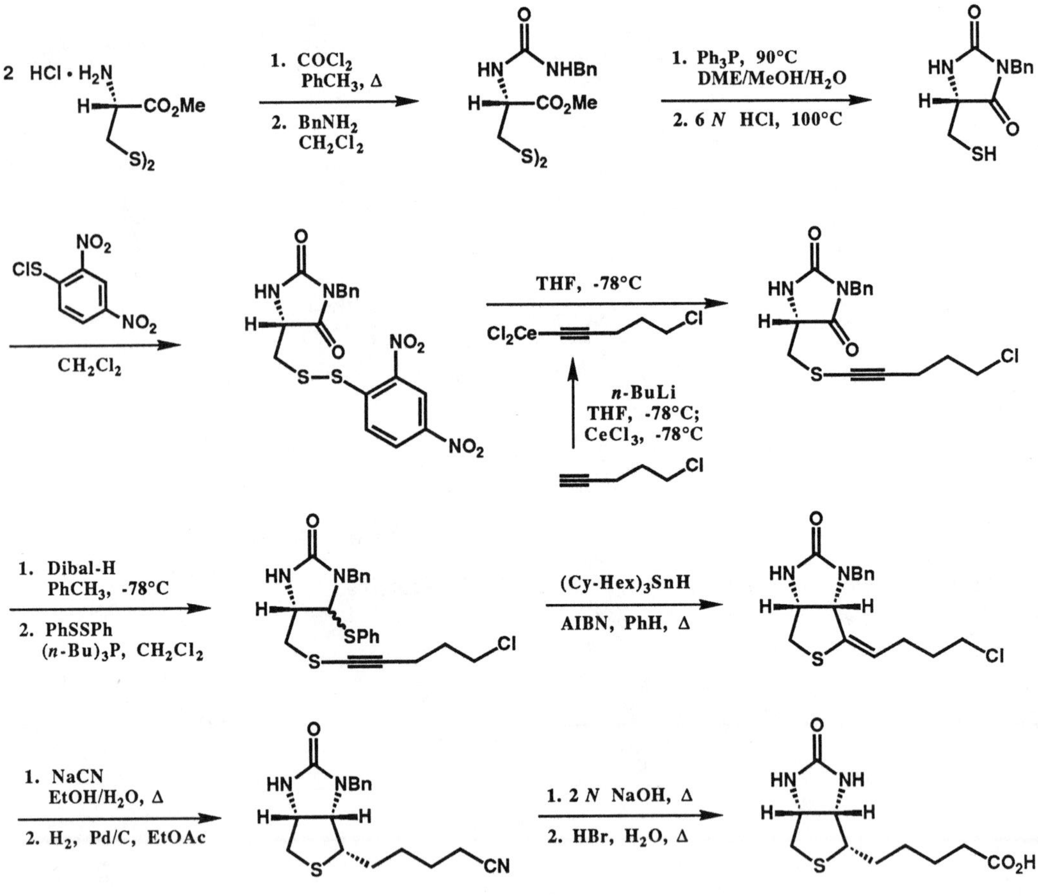

2 HCl • H2N
CO2Me
S)2
1. COCl2
PhCH3, Δ
2. BnNH2
CH2Cl2
HN
NHBn
1. Ph3P, 90°C
DME/MeOH/H2O
2. 6 N HCl, 100°C
NBn
SH
NO2
ClS
CH2Cl2
S–S
THF, -78°C
Cl2Ce
Cl
n-BuLi
THF, -78°C;
CeCl3, -78°C
1. Dibal-H
PhCH3, -78°C
2. PhSSPh
(n-Bu)3P, CH2Cl2
SPh
(Cy-Hex)3SnH
AIBN, PhH, Δ
1. NaCN
EtOH/H2O, Δ
2. H2, Pd/C, EtOAc
CN
1. 2 N NaOH, Δ
2. HBr, H2O, Δ
NH
CO2H

(+)-Biotin

synthetic methyl ester: $[\alpha]^{23}_{D}$ +81.0°
authentic methyl ester: $[\alpha]^{23}_{D}$ +80.7°

Reference:

E. J. Corey and M. M. Mehrotra, *Tetrahedron Lett.* **1988**, *29*, 57.

Methoxatin (Coenzyme PQQ)

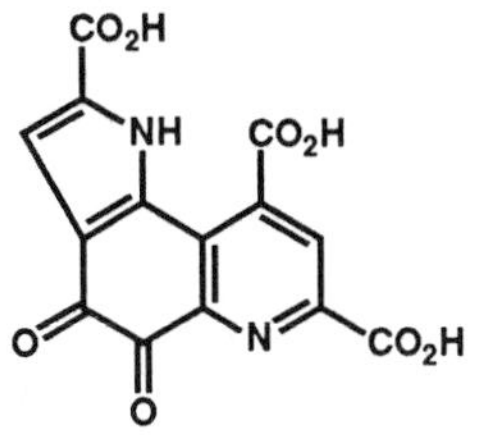
CO2H
NH
CO2H
N
CO2H
O
O

Methoxatin, now known as coenzyme PQQ, was originally obtained from methylotrophic bacteria but is now known to be a mammalian cofactor, for example, for lysyl oxidase and dopamine β-hydroxylase. The first synthesis of this rare compound was accomplished by the route outlined below. In the retrosynthetic analysis both of the heterocyclic rings were disconnected using directly keyed transforms.

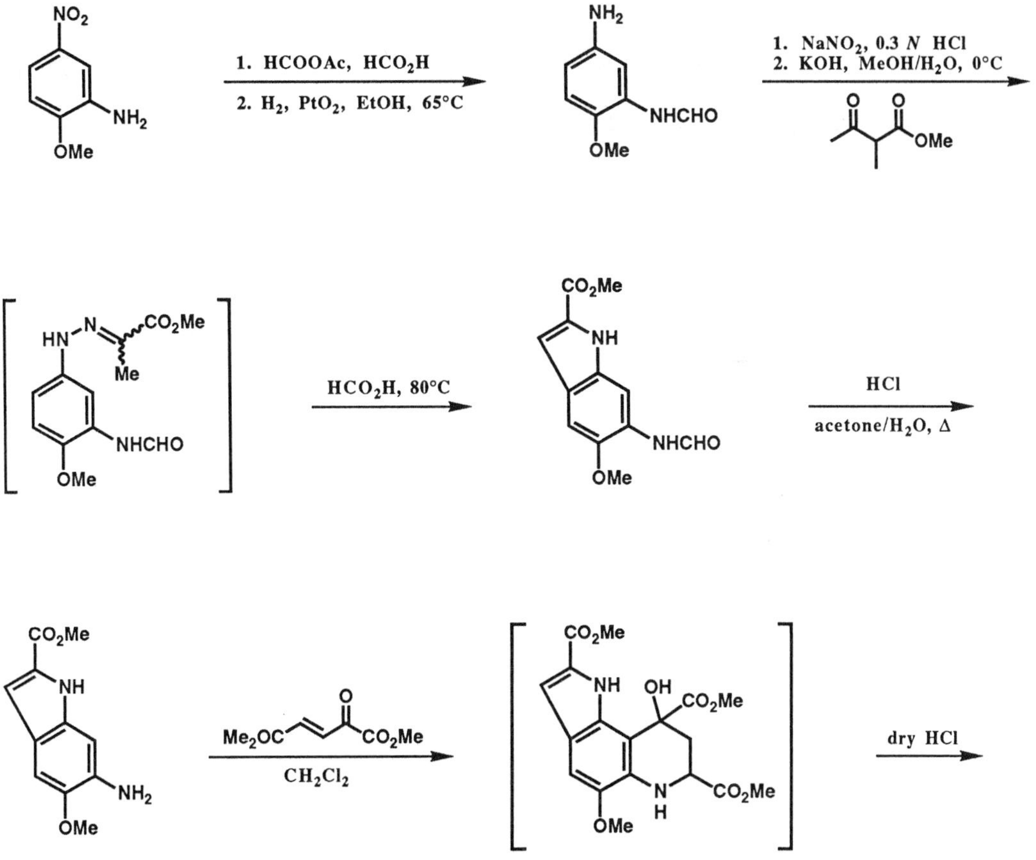
NO2
NH2
OMe
1. HCOOAc, HCO2H
2. H2, PtO2, EtOH, 65°C
NH2
NHCHO
OMe
1. NaNO2, 0.3 N HCl
2. KOH, MeOH/H2O, 0°C
O
O
OMe
HN
N
CO2Me
Me
NHCHO
OMe
HCO2H, 80°C
CO2Me
NH
NHCHO
OMe
HCl
acetone/H2O, Δ
CO2Me
NH
NH2
OMe
Me2OC
O
CO2Me
CH2Cl2
CO2Me
NH
OH
CO2Me
N
H
CO2Me
OMe
dry HCl

8.4

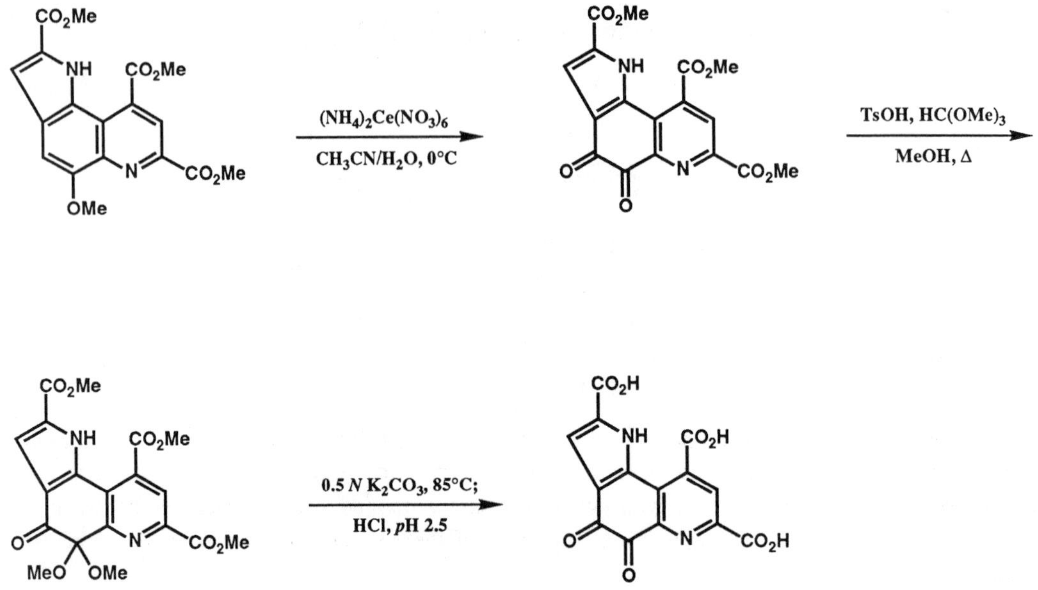

CO2Me
NH
CO2Me
N
CO2Me
OMe
(NH4)2Ce(NO3)6
CH3CN/H2O, 0°C
CO2Me
NH
CO2Me
O
O
N
CO2Me
TsOH, HC(OMe)3
MeOH, Δ
CO2Me
NH
CO2Me
O
N
CO2Me
MeO
OMe
0.5 N K2CO3, 85°C;
HCl, pH 2.5
CO2H
NH
CO2H
O
O
N
CO2H

Reference:

E. J. Corey and A. Tramontano, *J. Am. Chem. Soc.* **1981**, *103*, 5599.

20(*S*)-Camptothecin

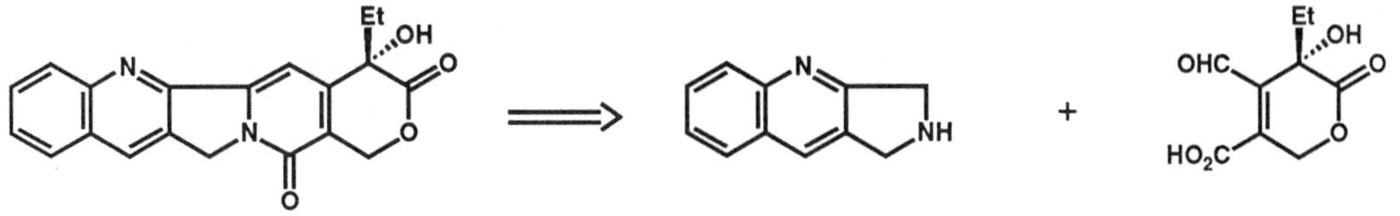
Et
OH
O
N
N
O
O
NH
+
OHC
HO2C

A B

The pentacyclic plant alkaloid camptothecin has been a popular synthetic target because of its antitumor activity. Retrosynthetic disconnection to tricyclic intermediate **A** and chiral lactone **B** followed from multistrategic planning.

1. Synthesis of the fragment **A**:

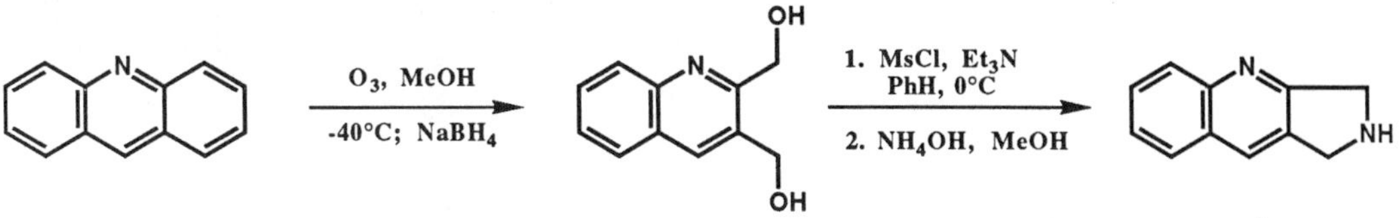
N
O3, MeOH
-40°C; NaBH4
OH
OH
1. MsCl, Et3N
PhH, 0°C
2. NH4OH, MeOH
NH

A

2. Synthesis of the structural equivalent of the fragment **B**:

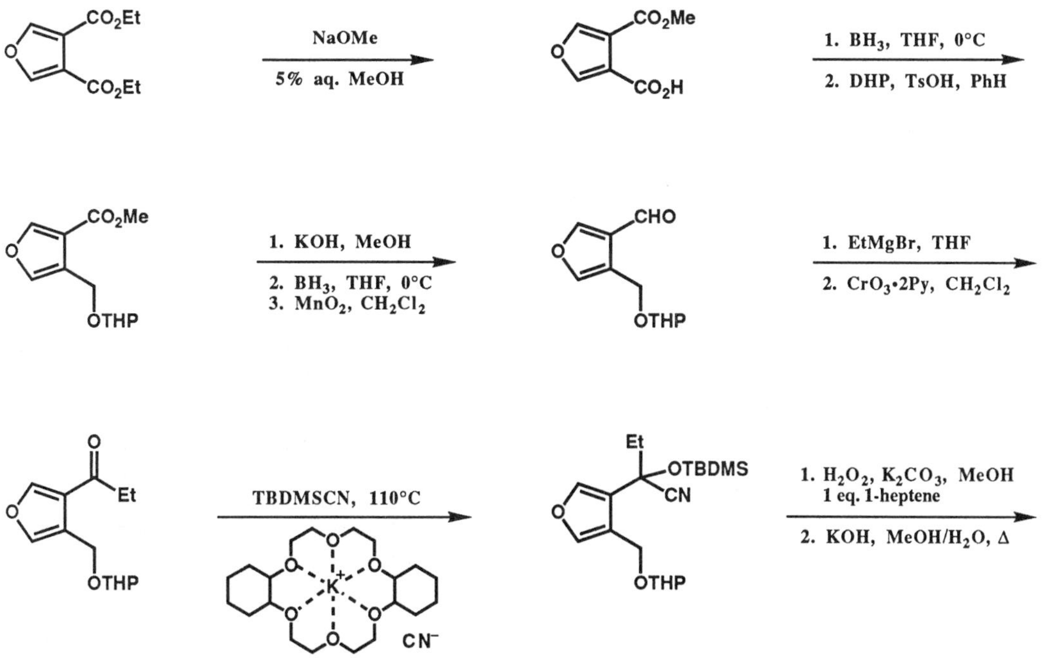
CO2Et
CO2Et
NaOMe
5% aq. MeOH
CO2Me
CO2H
1. BH3, THF, 0°C
2. DHP, TsOH, PhH
CO2Me
OTHP
1. KOH, MeOH
2. BH3, THF, 0°C
3. MnO2, CH2Cl2
CHO
OTHP
1. EtMgBr, THF
2. CrO3•2Py, CH2Cl2
Et
TBDMSCN, 110°C
CN−
OTBDMS
CN
1. H2O2, K2CO3, MeOH
1 eq. 1-heptene
2. KOH, MeOH/H2O, Δ

8.5

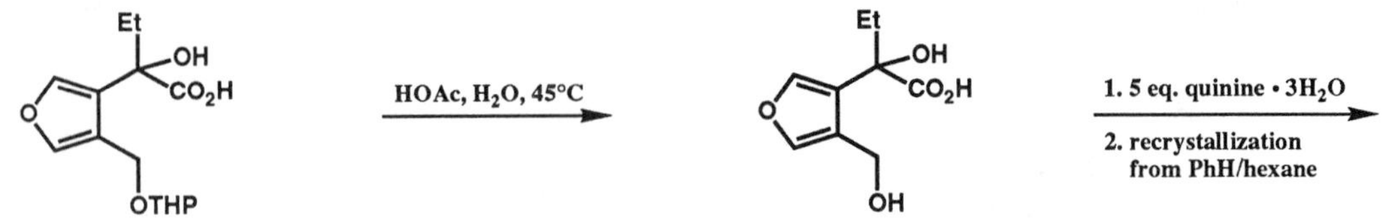
Et
OH
CO2H
OTHP
HOAc, H2O, 45°C
Et
OH
CO2H
OH
1. 5 eq. quinine • 3H2O
2. recrystallization from PhH/hexane

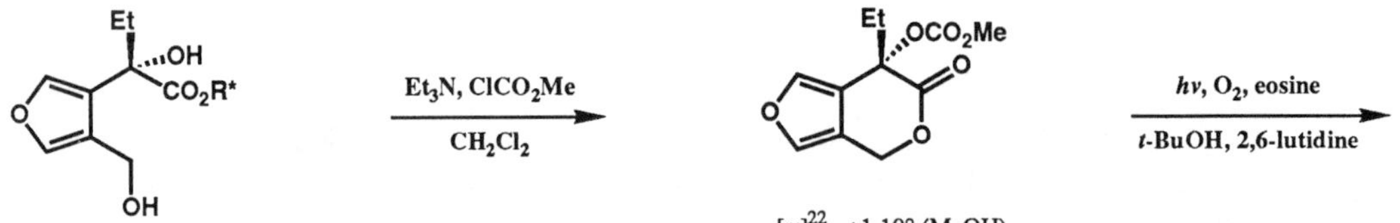
Et
OH
CO2R*
OH
Et3N, ClCO2Me
CH2Cl2
Et
OCO2Me
hv, O2, eosine
t-BuOH, 2,6-lutidine

R^* = quinine H^+

$[\alpha]^{22}_D$ +1.10° (MeOH)

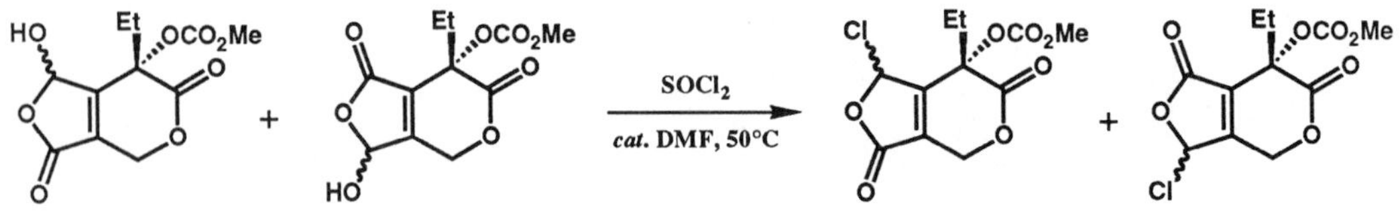
HO
Et
OCO2Me
Et
OCO2Me
HO
SOCl2
cat. DMF, 50°C
Cl
Et
OCO2Me
Et
OCO2Me
Cl

2.5 : 1 mixture

B

(equivalent form)

3. Coupling of the fragments **A** and **B** and the completion of the synthesis:

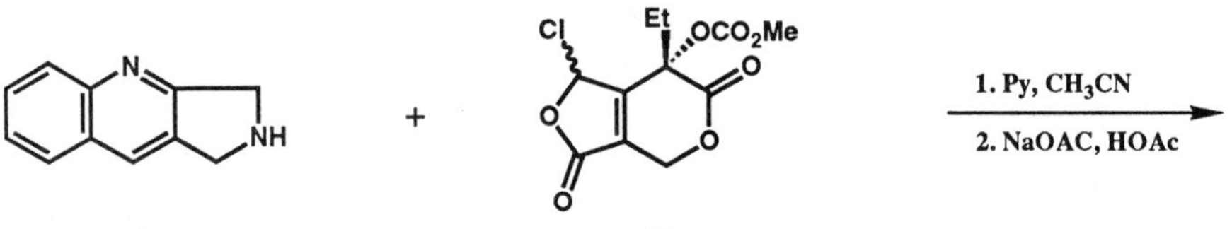
N
NH
Cl
Et
OCO2Me
1. Py, CH3CN
2. NaOAC, HOAc

A

B

(equivalent form)

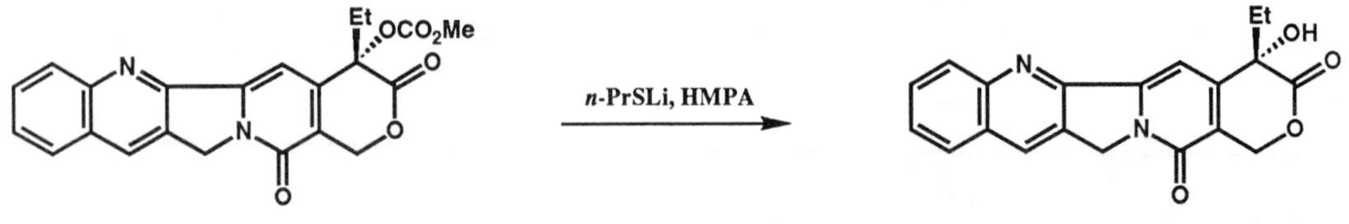
Et
OCO2Me
N
N
n-PrSLi, HMPA
Et
OH
N
N

synthetic sample: $[\alpha]^{22}_D$ +31.1° (20% MeOH in $CHCl_3$)

natural product: $[\alpha]^{22}_D$ +31.3° (20% MeOH in $CHCl_3$)

Reference:

E. J. Corey, D. N. Crouse, and J. E. Anderson, *J. Org. Chem.* **1975**, *40*, 2140.

CHAPTER NINE
Sesquiterpenoids

9.1

Stereospecific Synthesis of Trisubstituted Olefins from Acetylenes or Aldehydes Applications to the Total Synthesis of Cecropia Juvenile Hormones (JH) and Farnesol

The determination of the structure of the insect juvenile hormone of Cecropia (now called JH-I) (*Ang. Chem. Int.*, **1967**, *6*, 179) posed a challenge to synthesis because no methods then existed for the stereospecific synthesis of the type of trisubstituted olefinic linkages contained in JH-I. The first stereospecific synthesis, outlined below, depended on a number of new synthetic methods which now find general use. The synthesis of trisubstituted olefins from acetylenes by overall carbometallation can now be effected by a variety of reagents involving metals such as copper, titanium, zirconium, palladium, and nickel. Although JH-I is not a practical agent for insect control, the availability of synthetic material paved the way for the understanding of its action and the development of useful analogs.

1. Synthesis of trisubstituted olefins *via* LAH reduction of the propargylic alcohols and iodination as the key step (Ref. 1).

Propargylic alcohols are reduced by reaction with lithium aluminum hydride and subsequent hydrolysis to *(E)*-allylic alcohols *via* an organoaluminum intermediate (**A**) as shown below:

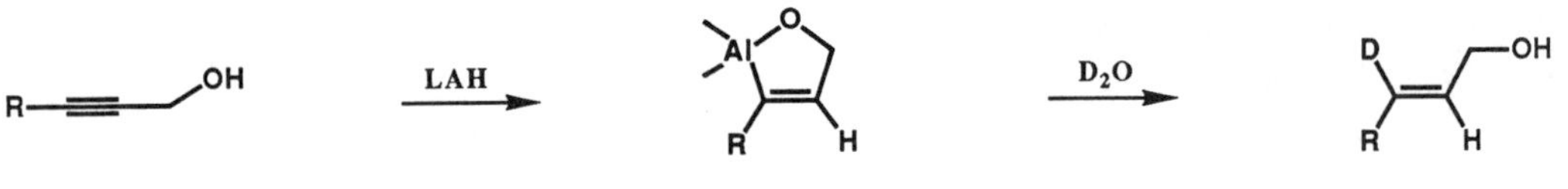

R
OH
LAH
Al
O
R
H
A
D_2O
D
OH
R
H

The reaction was adapted to the stereospecific synthesis of trisubstituted olefins in two ways as shown in the following scheme.

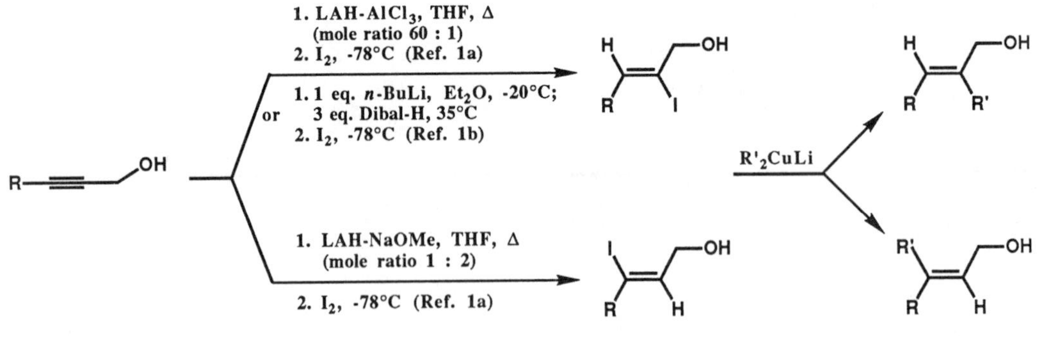

R
OH
1. LAH-AlCl$_3$, THF, Δ
(mole ratio 60 : 1)
2. I$_2$, -78°C (Ref. 1a)
or
1. 1 eq. *n*-BuLi, Et$_2$O, -20°C;
3 eq. Dibal-H, 35°C
2. I$_2$, -78°C (Ref. 1b)
1. LAH-NaOMe, THF, Δ
(mole ratio 1 : 2)
2. I$_2$, -78°C (Ref. 1a)
H
OH
R
I
I
OH
R
H
R'$_2$CuLi
H
OH
R
R'
R'
OH
R
H

The first synthesis of trisubstituted olefins from acetylenes was applied to the total synthesis of several natural products as outlined on the next page.

i. Total synthesis of *dl*-C_{18} Cecropia juvenile hormone (Ref. 2):

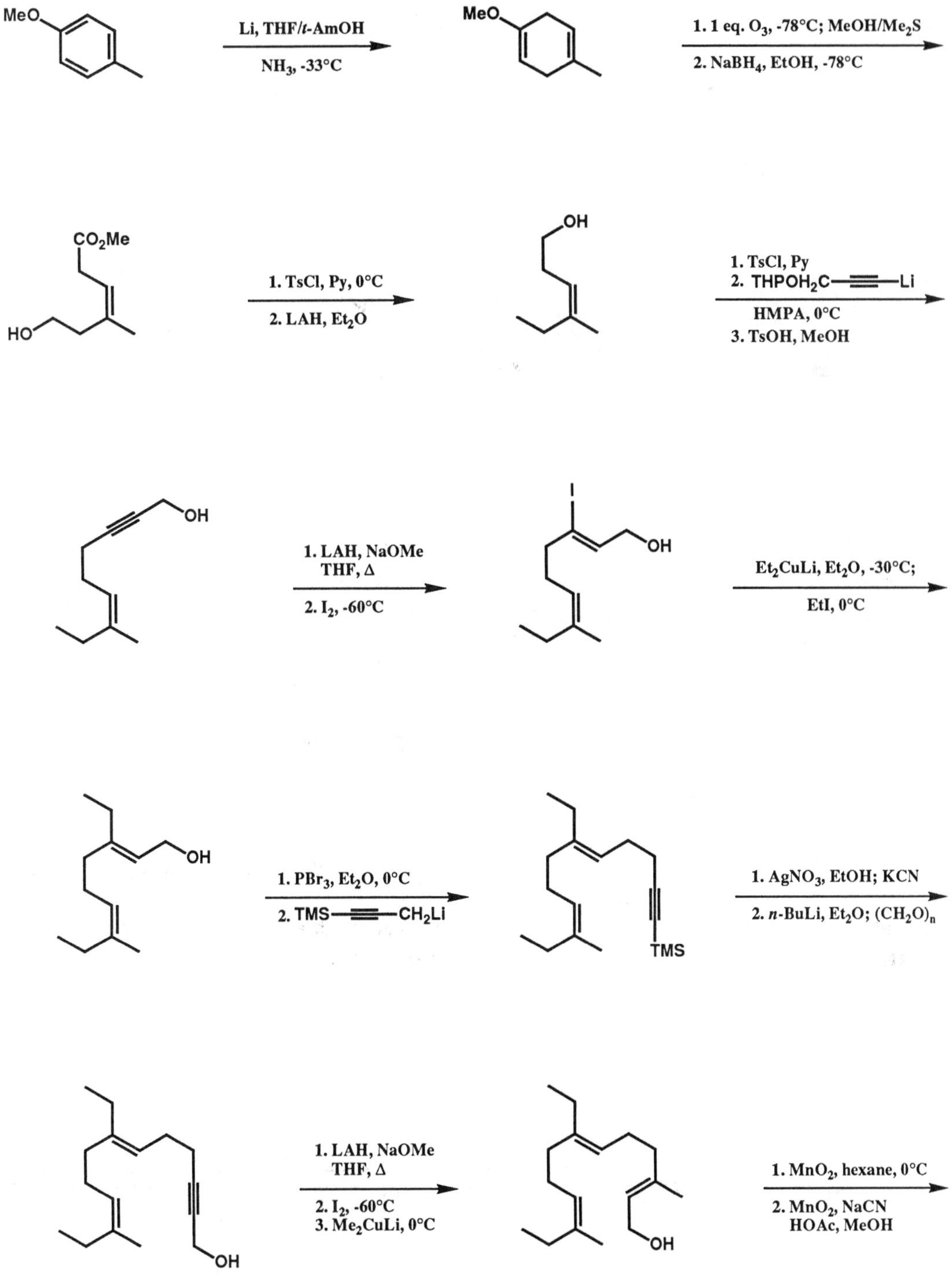
MeO
Li, THF/t-AmOH
NH3, -33°C
MeO
1. 1 eq. O3, -78°C; MeOH/Me2S
2. NaBH4, EtOH, -78°C
CO2Me
HO
1. TsCl, Py, 0°C
2. LAH, Et2O
OH
1. TsCl, Py
2. THPOH2C—≡—Li
HMPA, 0°C
3. TsOH, MeOH
OH
1. LAH, NaOMe
THF, Δ
2. I2, -60°C
I
OH
Et2CuLi, Et2O, -30°C;
EtI, 0°C
OH
1. PBr3, Et2O, 0°C
2. TMS—≡—CH2Li
TMS
1. AgNO3, EtOH; KCN
2. n-BuLi, Et2O; (CH2O)n
OH
1. LAH, NaOMe
THF, Δ
2. I2, -60°C
3. Me2CuLi, 0°C
OH
1. MnO2, hexane, 0°C
2. MnO2, NaCN
HOAc, MeOH

9.1

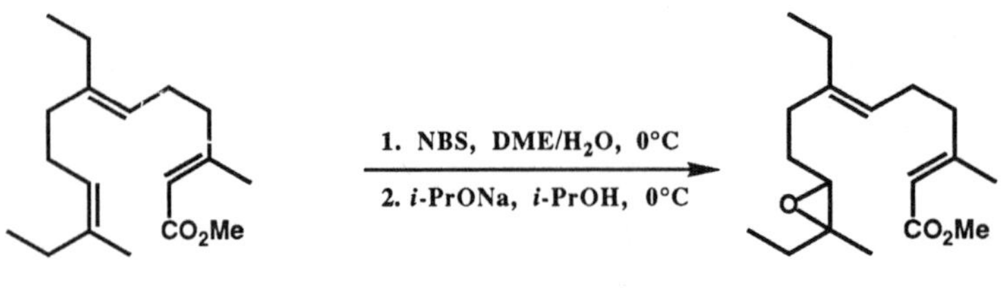

CO₂Me
1. NBS, DME/H₂O, 0°C
2. i-PrONa, i-PrOH, 0°C
O
CO₂Me

C_{18} **JH (JH-I)**

ii. Total synthesis of farnesol (Ref. 1a):

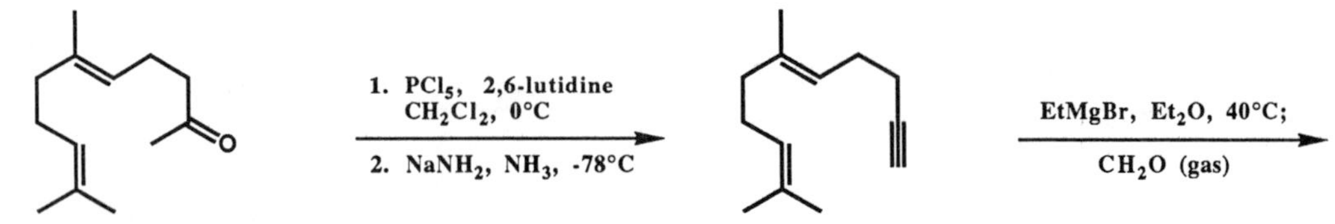

O
1. PCl₅, 2,6-lutidine
CH₂Cl₂, 0°C
2. NaNH₂, NH₃, -78°C
EtMgBr, Et₂O, 40°C;
CH₂O (gas)

Geranylacetone

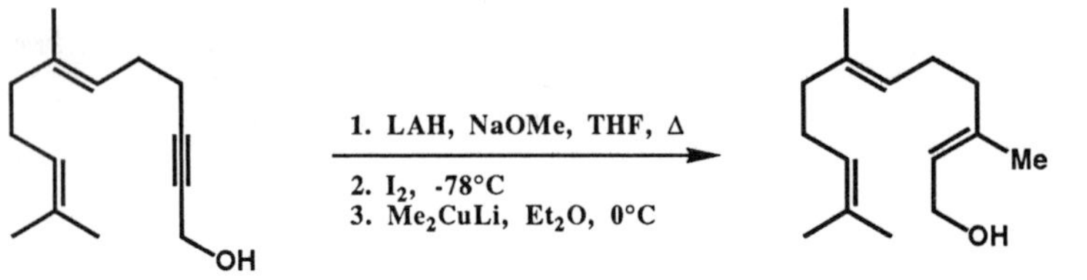

OH
1. LAH, NaOMe, THF, Δ
2. I₂, -78°C
3. Me₂CuLi, Et₂O, 0°C
Me
OH

(*E*, *E*)-Farnesol

Reaction of an alkyltriphenylphosphorane in tetrahydrofuran with an aldehyde produces the oxaphosphetane **B**, which can be further treated with 1 eq. of *s*-butyllithium to form the β-oxidophosphonium ylide **C**. This ylide can in turn react with another aldehyde, for instance, paraformaldehyde to give, after work-up, the trisubstituted olefin **D**.

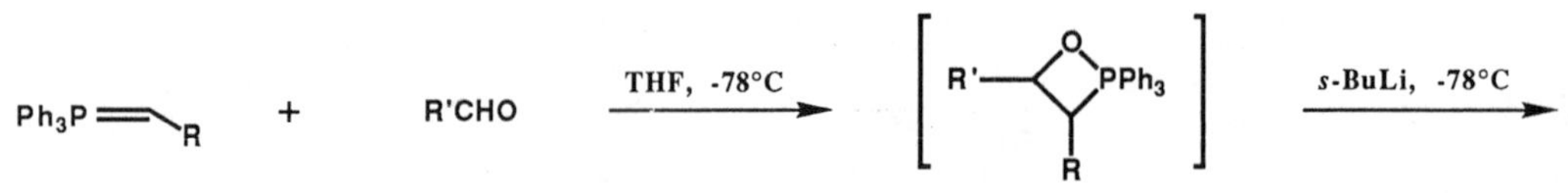

Ph₃P
R
+
R'CHO
THF, -78°C
R'
O
PPh₃
R
s-BuLi, -78°C

B

9.1

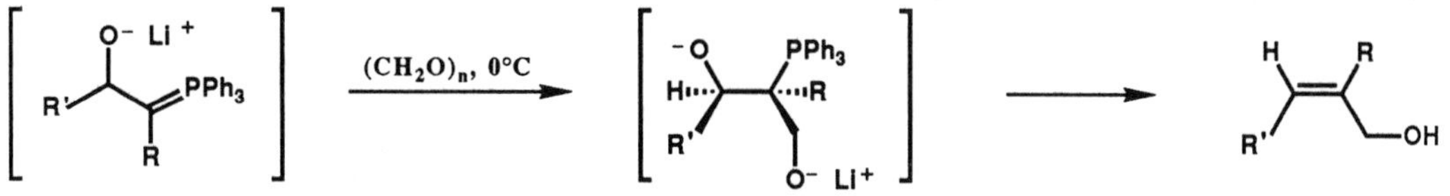

(CH₂O)n, 0°C

C D

This remarkable process allows the joining of three components in one synthetic step. This methodology was applied to the syntheses of several juvenile hormones and farnesol.

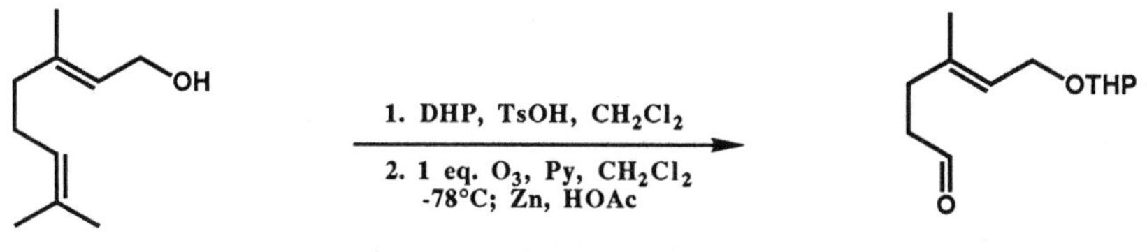

1. DHP, TsOH, CH₂Cl₂
2. 1 eq. O₃, Py, CH₂Cl₂ -78°C; Zn, HOAc

E

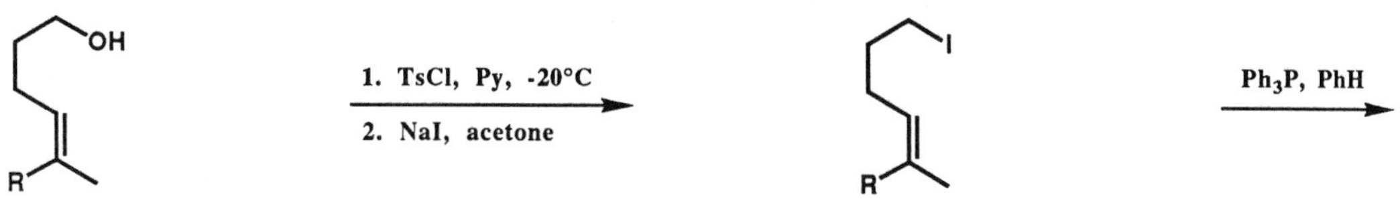

1. TsCl, Py, -20°C
2. NaI, acetone
Ph₃P, PhH

R = Me, Et.

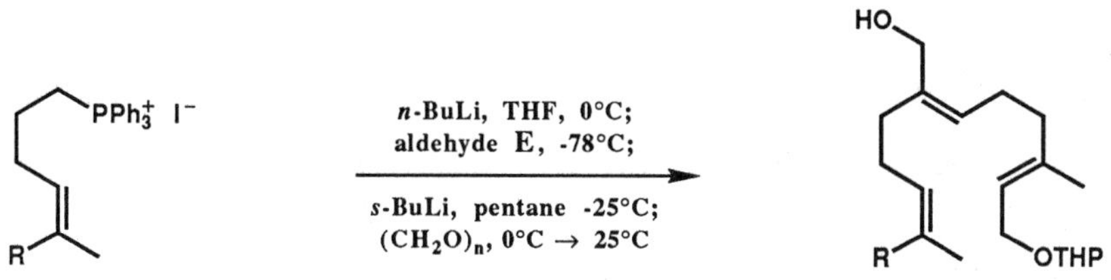

n-BuLi, THF, 0°C; aldehyde E, -78°C;
s-BuLi, pentane -25°C; (CH₂O)n, 0°C → 25°C

F

(Ref. 4)

9.1

Intermediate **F** was used in a series of syntheses as shown below:

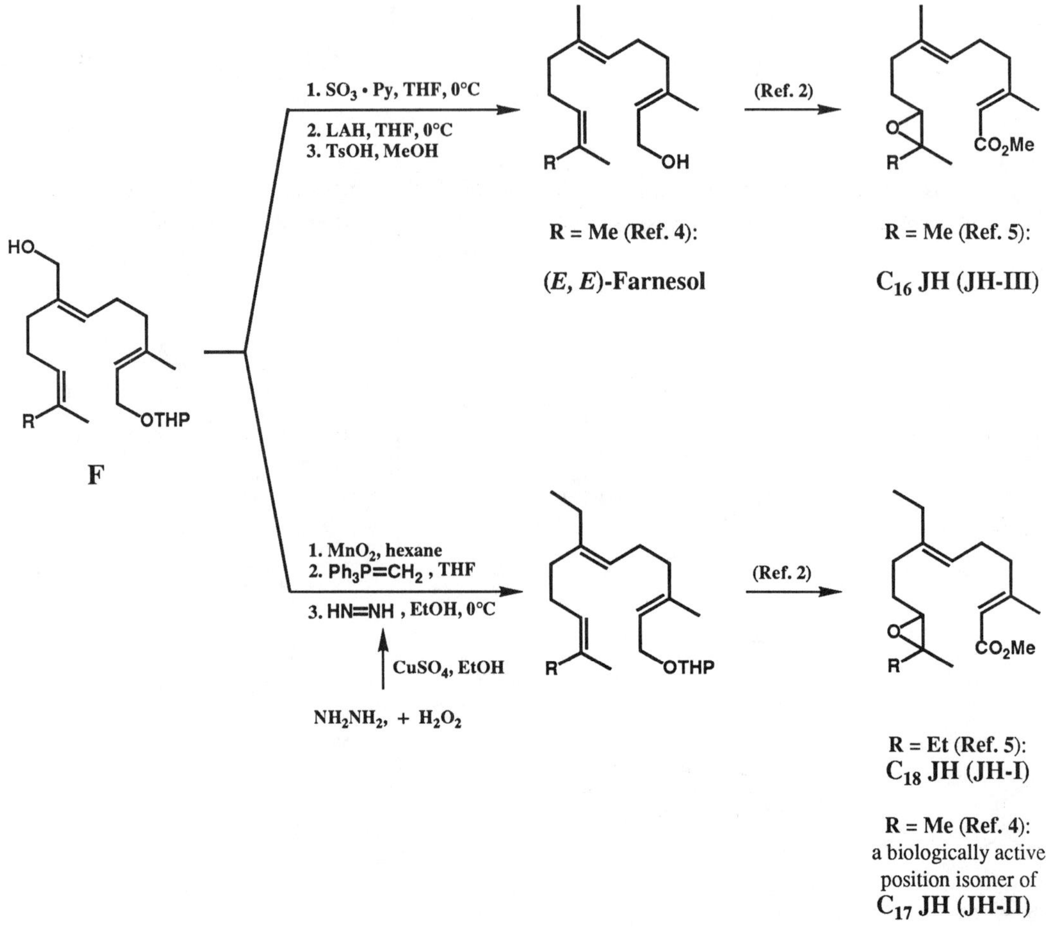

HO
R
OTHP
F
1. SO3 • Py, THF, 0°C
2. LAH, THF, 0°C
3. TsOH, MeOH
R
OH
(Ref. 2)
R
CO2Me
R = Me (Ref. 4):
(E, E)-Farnesol
R = Me (Ref. 5):
C16 JH (JH-III)
1. MnO2, hexane
2. Ph3P=CH2, THF
3. HN=NH, EtOH, 0°C
CuSO4, EtOH
NH2NH2, + H2O2
R
OTHP
(Ref. 2)
R
CO2Me
R = Et (Ref. 5):
C18 JH (JH-I)
R = Me (Ref. 4):
a biologically active
position isomer of
C17 JH (JH-II)

References:

1. (a). E. J. Corey, J. A. Katzenellenbogen, and G. H. Posner, *J. Am. Chem. Soc.* **1967**, *89*, 4245.
   (b). E. J. Corey, H. A. Kirst, and J. A. Katzenellenbogen, *J. Am. Chem. Soc.* **1970**, *92*, 6314.
2. E. J. Corey, J. A. Katzenellenbogen, N. W. Gilman, S. A. Roman, and B. W. Erickson, *J. Am. Chem. Soc.* **1968**, *90*, 5618.
3. E. J. Corey and H. Yamamoto, *J. Am. Chem. Soc.* **1970**, *92*, 226, 3523.
4. E. J. Corey and H. Yamamoto, *J. Am. Chem. Soc.* **1970**, *92*, 6637.
5. E. J. Corey and H. Yamamoto, *J. Am. Chem. Soc.* **1970**, *92*, 6636.
6. E. J. Corey, *Bull. Soc. Entomologique Suisse* **1971**, *44*, 87.

Longifolene

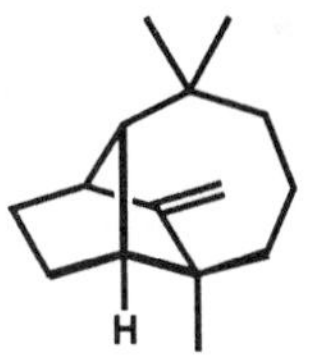
H

A discussion of the synthesis of longifolene appears in Section 6.1 of Part One.

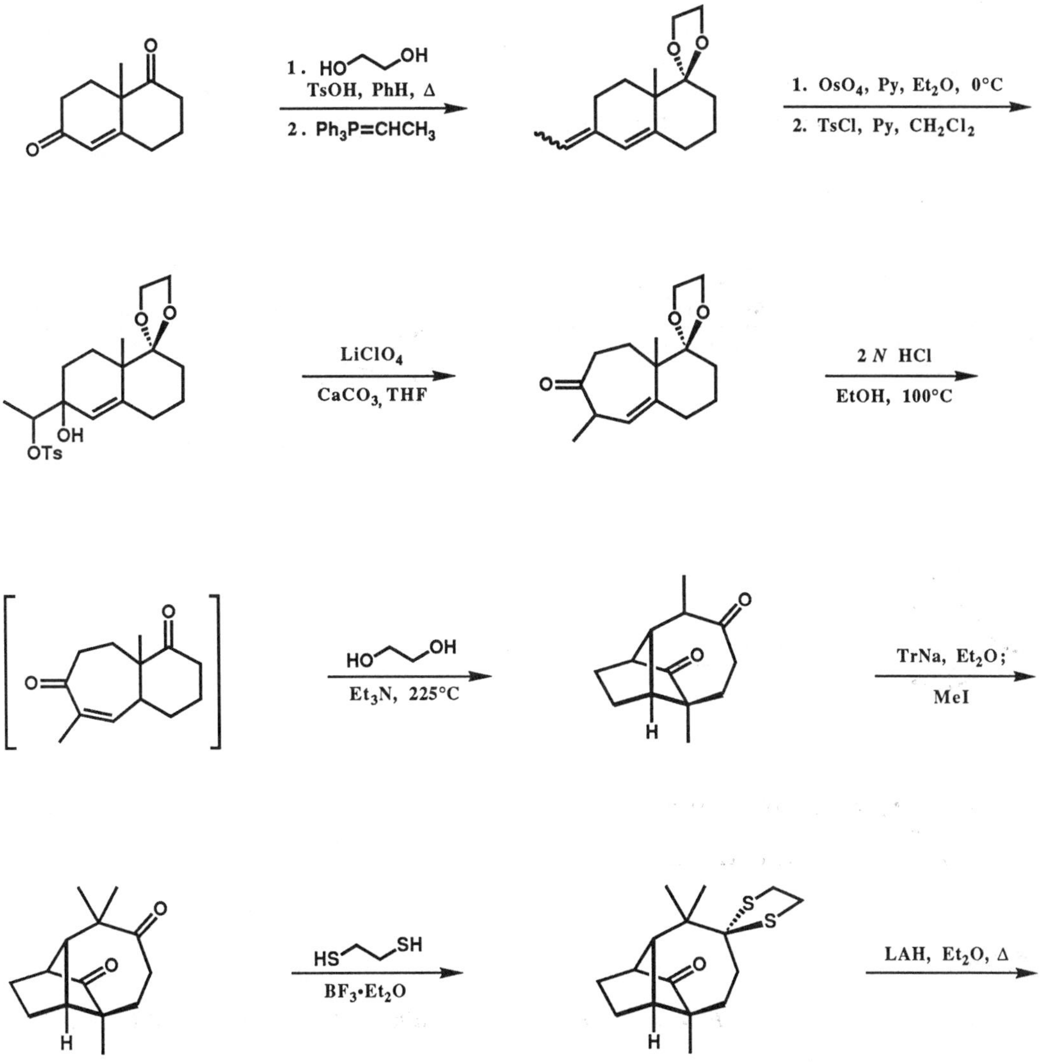
1. HO⌒⌒OH
TsOH, PhH, Δ
2. $Ph_3P{=}CHCH_3$
1. OsO_4, Py, Et_2O, 0°C
2. TsCl, Py, CH_2Cl_2
OH
OTs
$LiClO_4$
$CaCO_3$, THF
2 *N* HCl
EtOH, 100°C
HO⌒⌒OH
Et_3N, 225°C
H
TrNa, Et_2O;
MeI
H
HS⌒⌒SH
$BF_3{\cdot}Et_2O$
H
LAH, Et_2O, Δ

A

9.2

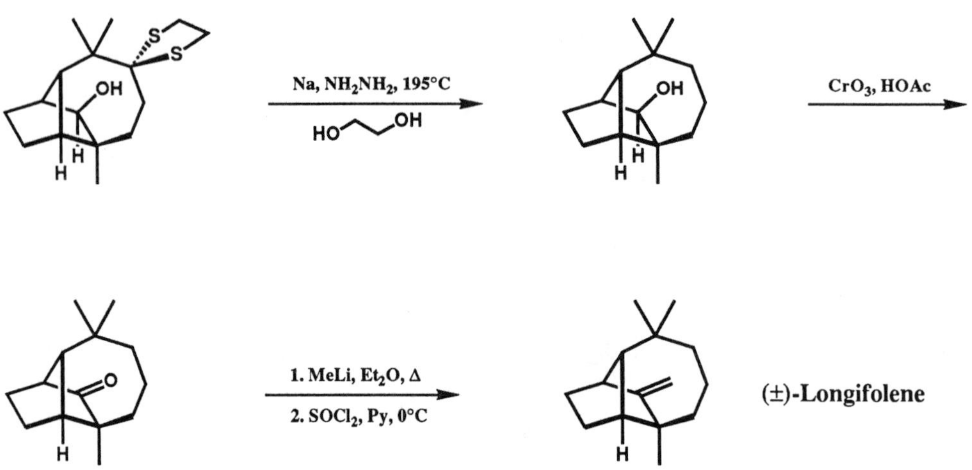
Na, NH$_2$NH$_2$, 195°C
HO⌒⌒OH
CrO$_3$, HOAc
1. MeLi, Et$_2$O, Δ
2. SOCl$_2$, Py, 0°C
(±)-Longifolene

The synthesis was modified to produce optically active longifolene:

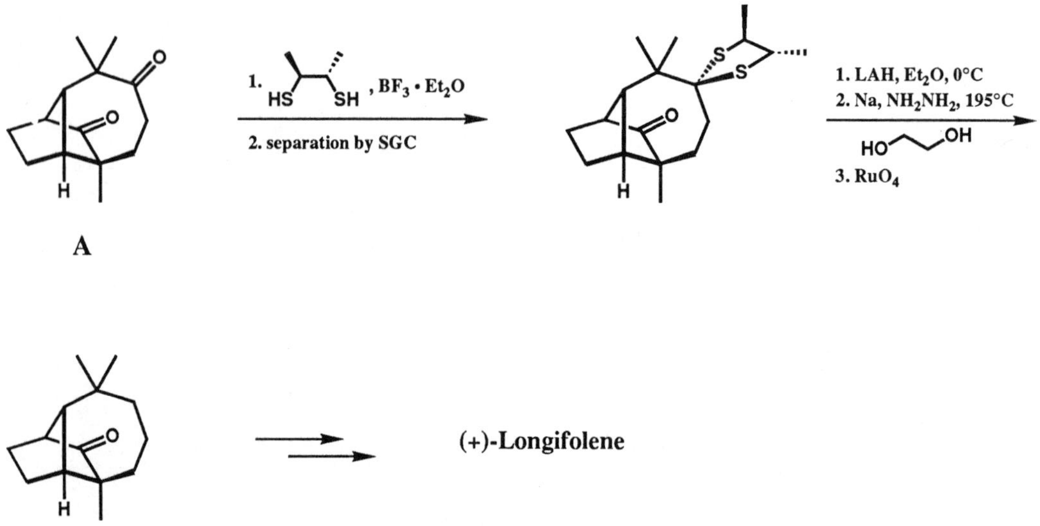
1. HS SH, BF$_3$ • Et$_2$O
2. separation by SGC
1. LAH, Et$_2$O, 0°C
2. Na, NH$_2$NH$_2$, 195°C
HO⌒⌒OH
3. RuO$_4$
A
(+)-Longifolene

synthetic sample: [α] = +587° at 296.6 nm
derivative from
natural ***Longifolene***: [α] = +610° at 296.6 nm

References:

1. E. J. Corey, M. Ohno, P. A. Vatakencherry, and R. B. Mitra, *J. Am. Chem. Soc.* **1961**, *83*, 1251.
2. E. J. Corey, M. Ohno, R. B. Mitra, and P. A. Vatakencherry, *J. Am. Chem. Soc.* **1964**, *86*, 478.

dl-Caryophyllene *dl*-Isocaryophyllene

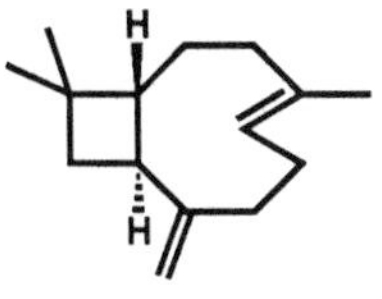

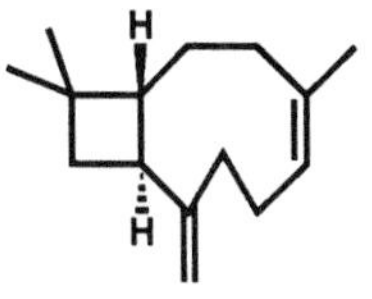

The total synthesis of caryophyllene and its *Z*-isomer involved a photochemical [2 + 2] cycloaddition reaction to generate the 4-membered ring and a fragmentation process (*Helv. Chim. Acta*, **1951**, *34*, 2338) to establish the 9-membered ring. Caryophyllene and various oxygenated derivatives protect plants against insects.

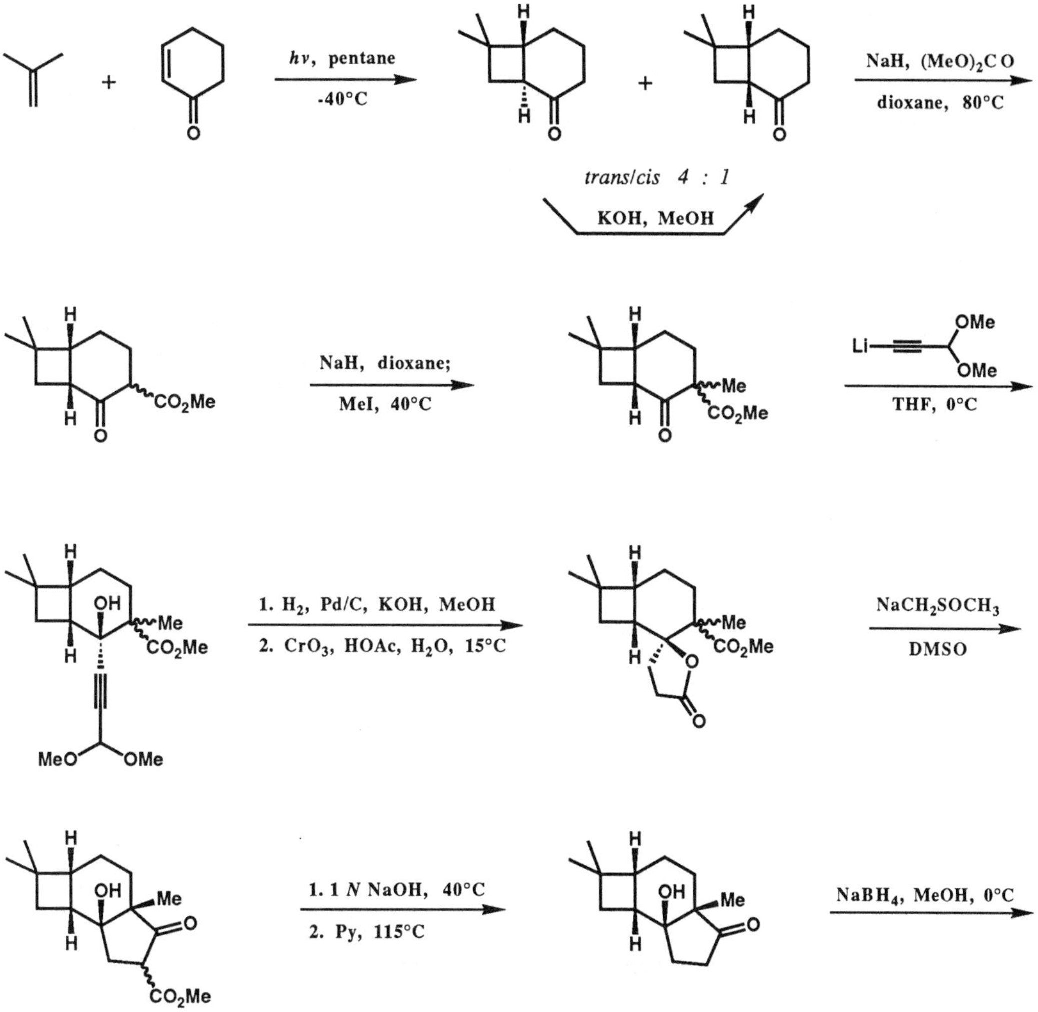
hν, pentane
-40°C
NaH, (MeO)2CO
dioxane, 80°C
trans/cis 4 : 1
KOH, MeOH
NaH, dioxane;
MeI, 40°C
THF, 0°C
1. H2, Pd/C, KOH, MeOH
2. CrO3, HOAc, H2O, 15°C
NaCH2SOCH3
DMSO
1. 1 N NaOH, 40°C
2. Py, 115°C
NaBH4, MeOH, 0°C

A

9.3

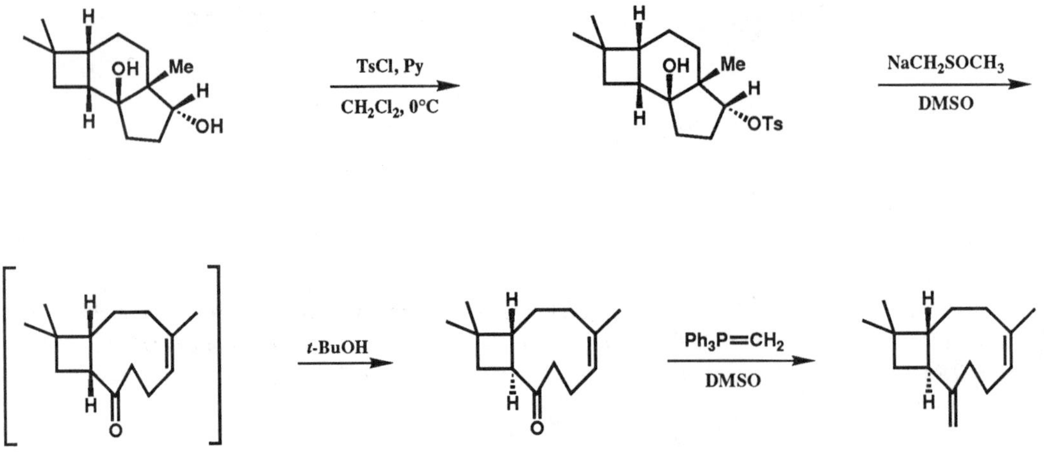
TsCl, Py
CH_2Cl_2, 0°C
$NaCH_2SOCH_3$
DMSO
t-BuOH
$Ph_3P{=}CH_2$
DMSO

Isocaryophyllene

The synthesis of *dl*-caryophyllene was also accomplished starting from the hydroxy ketone **A**, *via* a different reduction process as shown below:

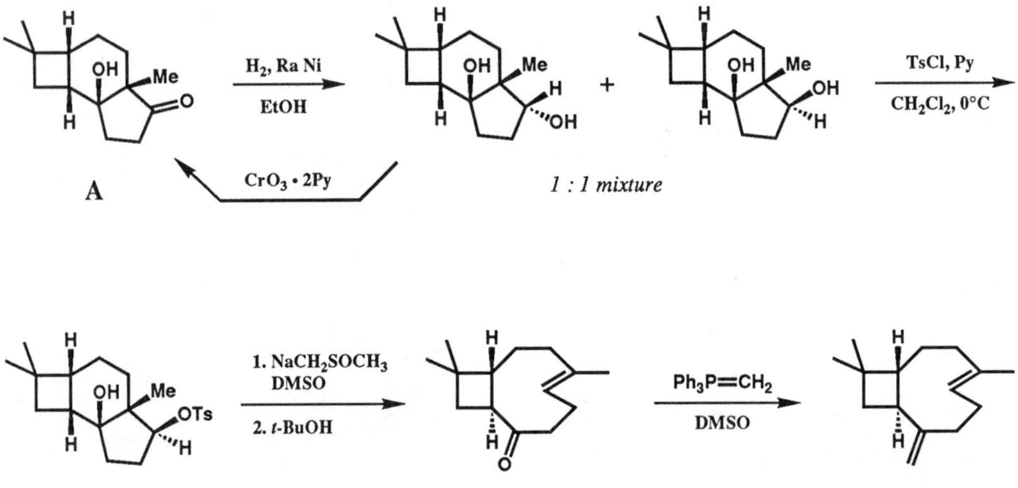
H_2, Ra Ni
EtOH
TsCl, Py
CH_2Cl_2, 0°C
A
$CrO_3 \cdot 2Py$
1 : 1 mixture
1. $NaCH_2SOCH_3$
DMSO
2. *t*-BuOH
$Ph_3P{=}CH_2$
DMSO

Caryophyllene

References:

1. E. J. Corey, R. B. Mitra, and H. Uda, *J. Am. Chem. Soc.* **1963**, *85*, 362.
2. E. J. Corey, R. B. Mitra, and H. Uda, *J. Am. Chem. Soc.* **1964**, *86*, 485.

9.4

α-Caryophyllene Alcohol

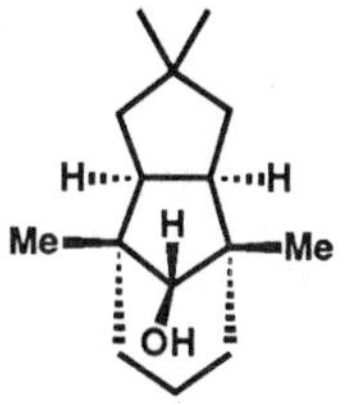
H
H
H
Me
Me
OH

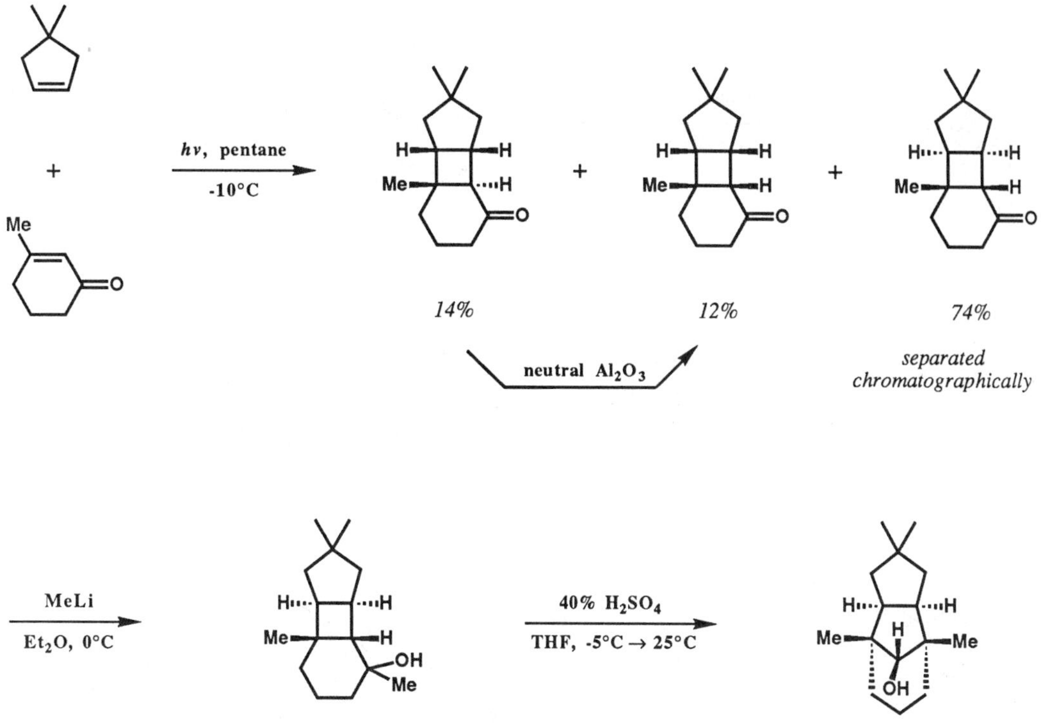
hv, pentane
-10°C
Me
O
14%
12%
74%
neutral Al2O3
separated chromatographically
MeLi
Et2O, 0°C
OH
Me
40% H2SO4
THF, -5°C → 25°C

References:

1. E. J. Corey and S. Nozoe, *J. Am. Chem. Soc.* **1964**, *86*, 1652.
2. E. J. Corey and S. Nozoe, *J. Am. Chem. Soc.* **1965**, *87*, 5733.

9.5

(±)-Cedrene **(±)-Cedrol**

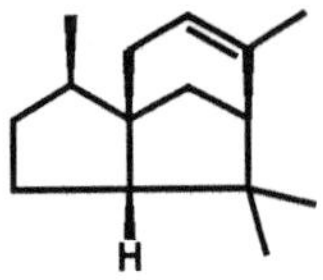

H

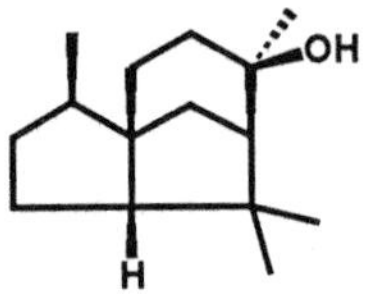

OH
H

The tricyclic fragrances of cedar wood, cedrene and cedrol, were synthesized by several routes involving cation-olefin cyclization.

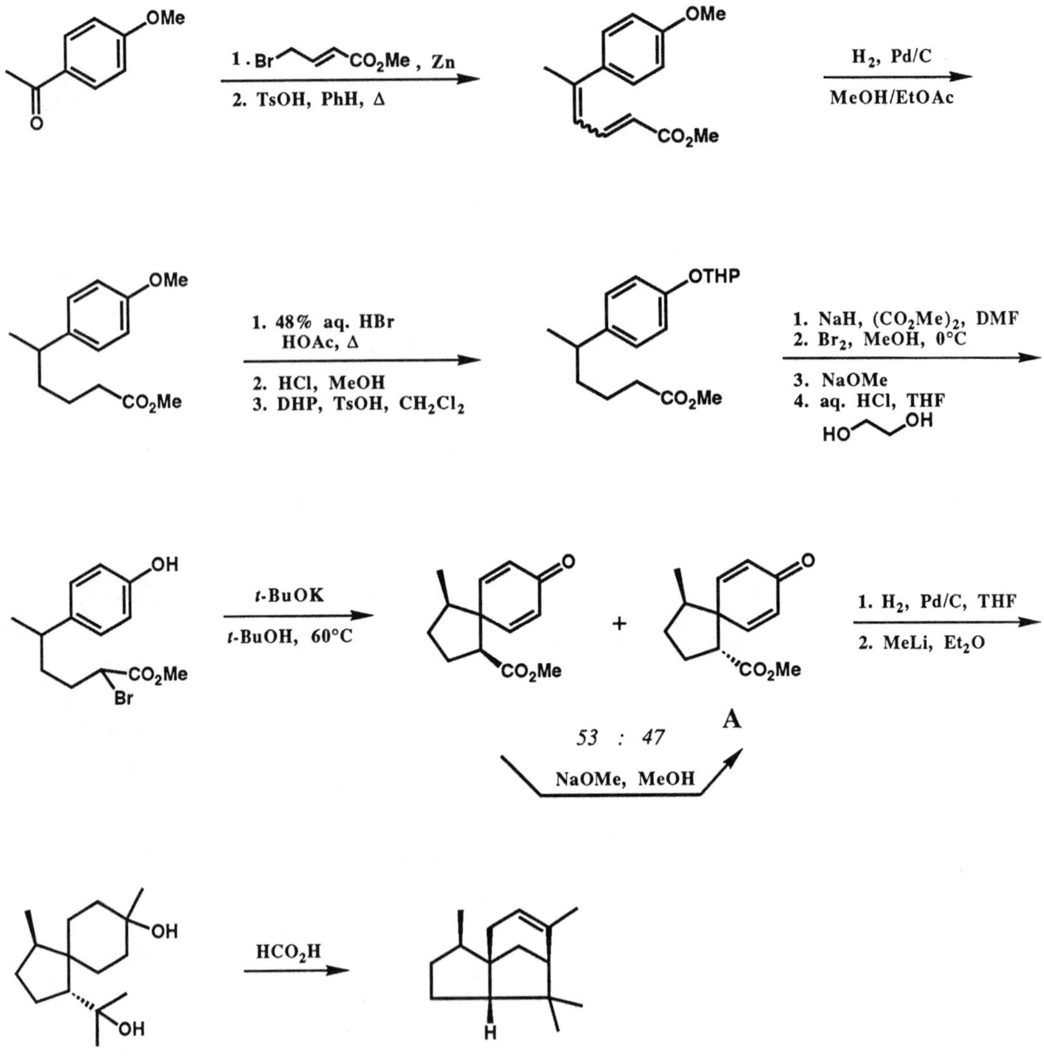

OMe
O
1. Br CO2Me , Zn
2. TsOH, PhH, Δ
OMe
CO2Me
H2, Pd/C
MeOH/EtOAc
OMe
CO2Me
1. 48% aq. HBr
HOAc, Δ
2. HCl, MeOH
3. DHP, TsOH, CH2Cl2
OTHP
CO2Me
1. NaH, (CO2Me)2, DMF
2. Br2, MeOH, 0°C
3. NaOMe
4. aq. HCl, THF
HO OH
OH
CO2Me
Br
t-BuOK
t-BuOH, 60°C
O
CO2Me
+
O
CO2Me
A
53 : 47
NaOMe, MeOH
1. H2, Pd/C, THF
2. MeLi, Et2O
OH
OH
HCO2H
H

(±)-Cedrene

9.5

From the intermediate **A**, cedrene was also made by an alternative route:

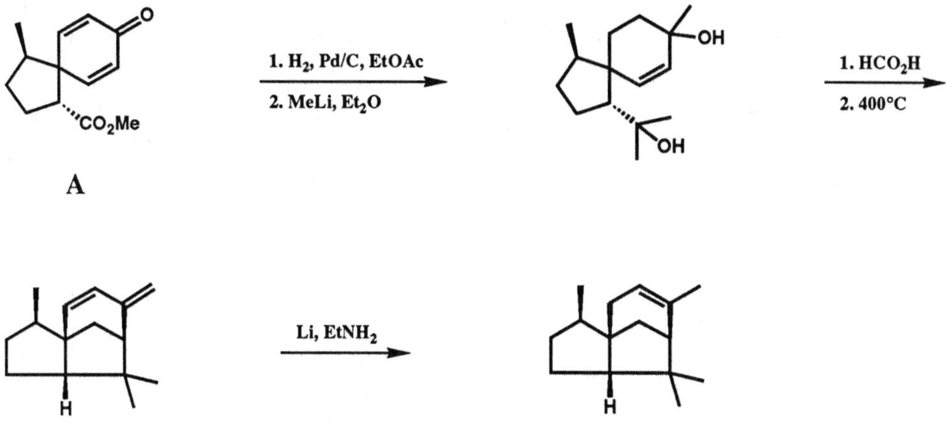

A
1. H_2, Pd/C, EtOAc
2. MeLi, Et_2O
1. HCO_2H
2. 400°C
Li, $EtNH_2$

(±)-Cedrene

Cedrol was also synthesized from intermediate **A**:

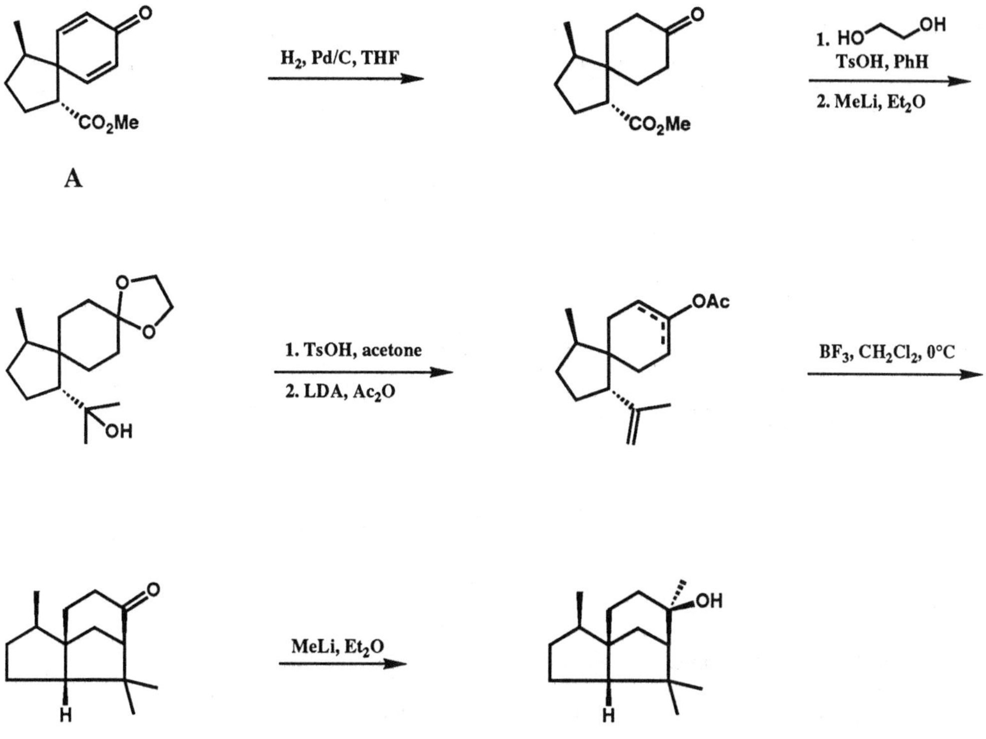

A
H_2, Pd/C, THF
1. HO⌒⌒OH
TsOH, PhH
2. MeLi, Et_2O
1. TsOH, acetone
2. LDA, Ac_2O
BF_3, CH_2Cl_2, 0°C
MeLi, Et_2O

(±)-Cedrol

9.5

Another synthesis of cedrene and of cedrol was achieved *via* the following sequence (Ref. 2):

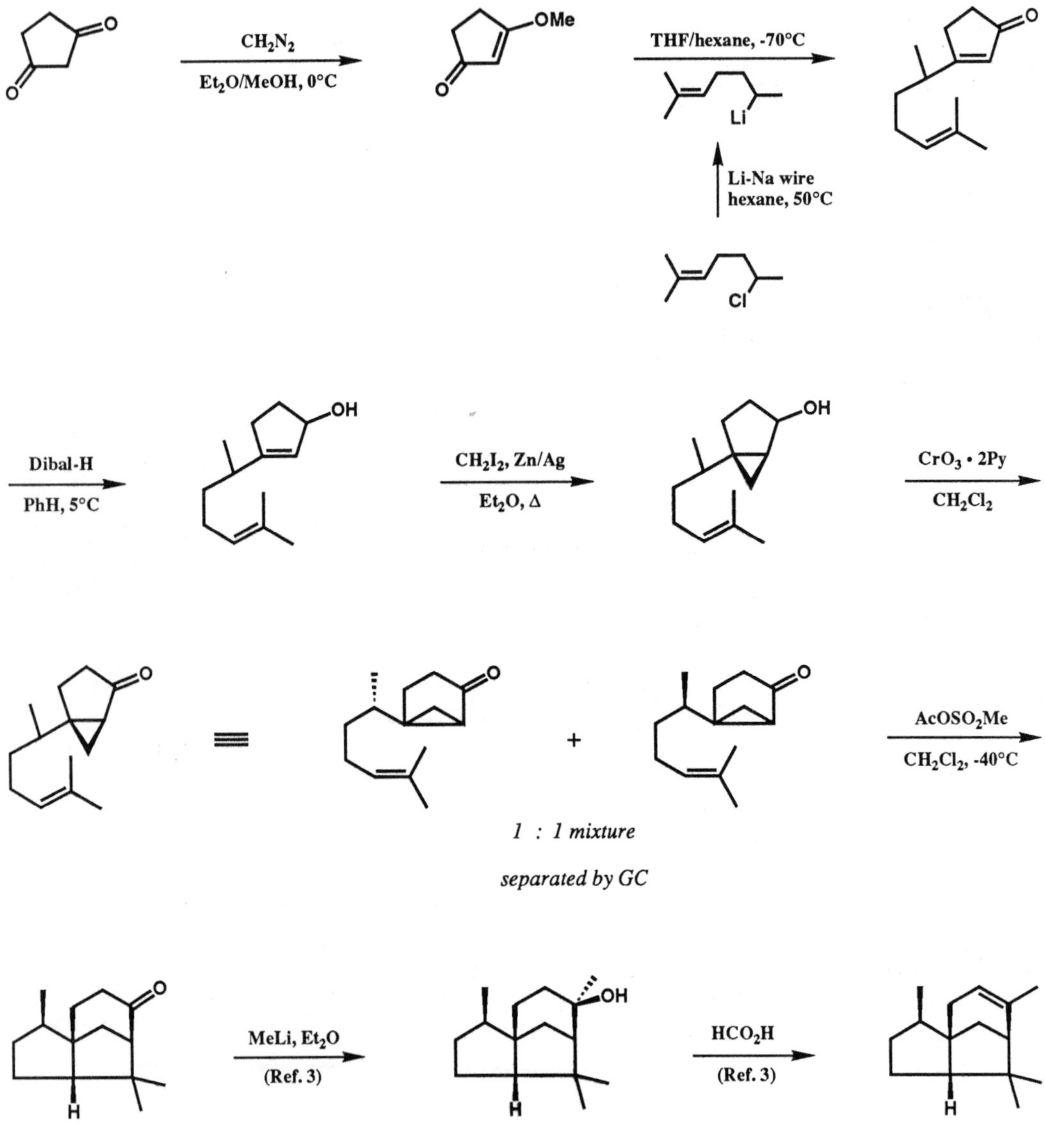
O
O
CH_2N_2
Et_2O/MeOH, 0°C
OMe
O
THF/hexane, -70°C
Li
Li-Na wire
hexane, 50°C
Cl
O
Dibal-H
PhH, 5°C
OH
CH_2I_2, Zn/Ag
Et_2O, Δ
OH
CrO_3 • 2Py
CH_2Cl_2
O
O
O
AcOSO$_2$Me
CH_2Cl_2, -40°C
1 : 1 mixture
separated by GC
O
H
MeLi, Et_2O
(Ref. 3)
OH
H
HCO_2H
(Ref. 3)
H

(±)-Cedrol

(±)-Cedrene

References:

1. E. J. Corey, N. N. Girotra, and C. T. Mathew, *J. Am. Chem. Soc.* **1969**, *91*, 1557.
2. E. J. Corey and R. D. Balanson, *Tetrahedron Lett.* **1973**, 3153.
3. G. Stork and F. H. Clarke, Jr., *J. Am. Chem. Soc.* **1955**, *77*, 1072; *ibid.* **1961**, *83*, 3114.

Humulene

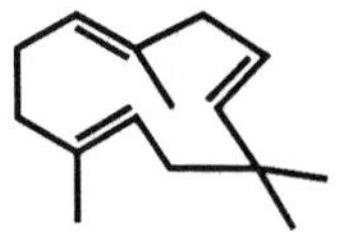

The 11-membered ring of humulene, a major impediment to synthesis, was constructed by a novel nickel-mediated cyclization.

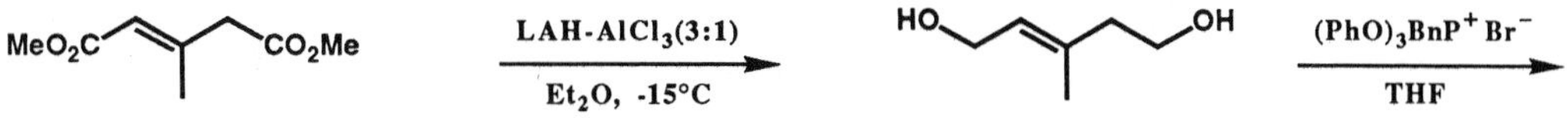
MeO2C
CO2Me
LAH-AlCl3(3:1)
Et2O, -15°C
HO
OH
(PhO)3BnP+ Br−
THF

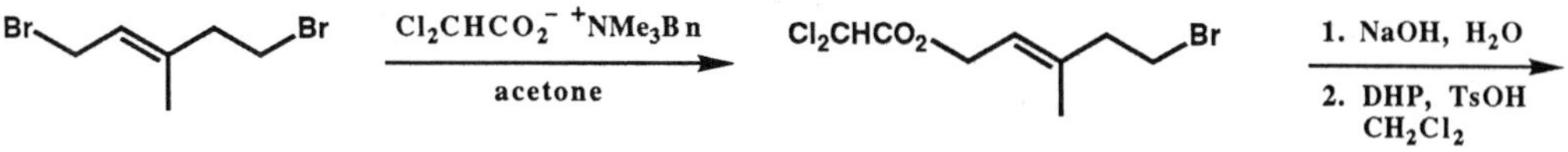
Br
Br
Cl2CHCO2− +NMe3Bn
acetone
Cl2CHCO2
Br
1. NaOH, H2O
2. DHP, TsOH
CH2Cl2

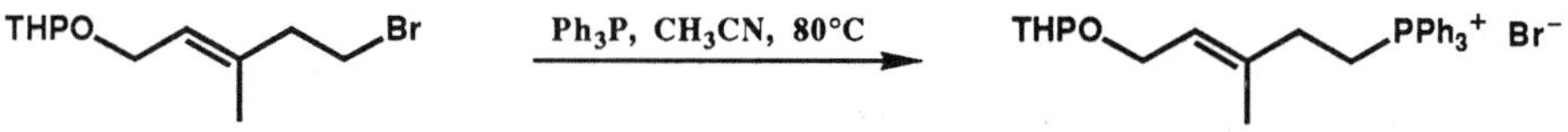
THPO
Br
Ph3P, CH3CN, 80°C
THPO
PPh3+ Br−

A

Cl
OAc
MesCO2− +NMe3Bn
EtOH, 80°C
MesCO2
OAc
1. NaOH, MeOH
2. PBr3, Et2O, 0°C

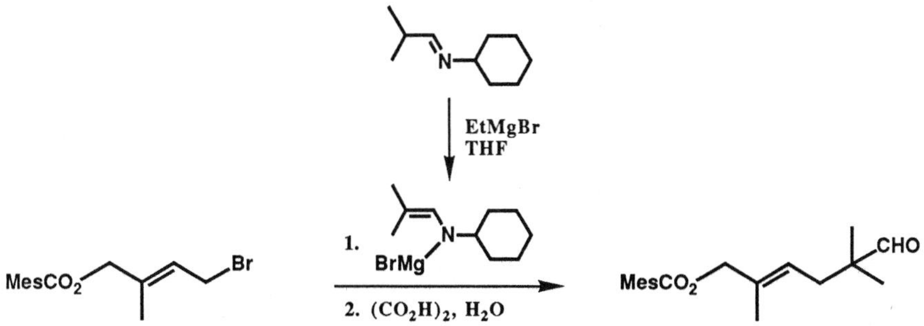
N
EtMgBr
THF
1.
N
BrMg
2. (CO2H)2, H2O
MesCO2
Br
MesCO2
CHO

B

9.6

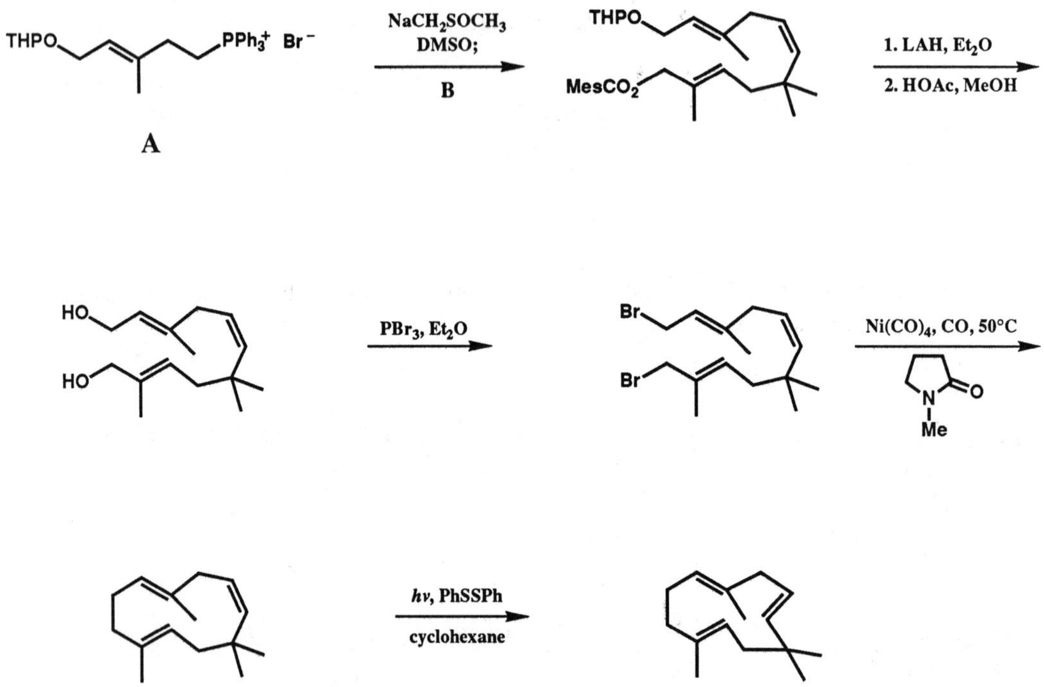
THPO
PPh3+ Br−
A
NaCH2SOCH3
DMSO;
B
THPO
MesCO2
1. LAH, Et2O
2. HOAc, MeOH
HO
HO
PBr3, Et2O
Br
Br
Ni(CO)4, CO, 50°C
N
O
Me
hν, PhSSPh
cyclohexane

Reference:

E. J. Corey and E. Hamanaka, *J. Am. Chem. Soc.* **1967**, *89*, 2758.

Dihydrocostunolide

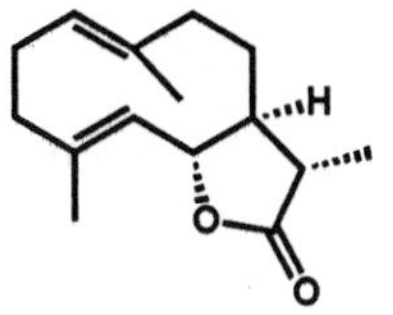

Dihydrocostunolide was synthesized from santonin as SM goal by a sequence in which the photochemical cleavage of a 6,6-fusion bond was a key step.

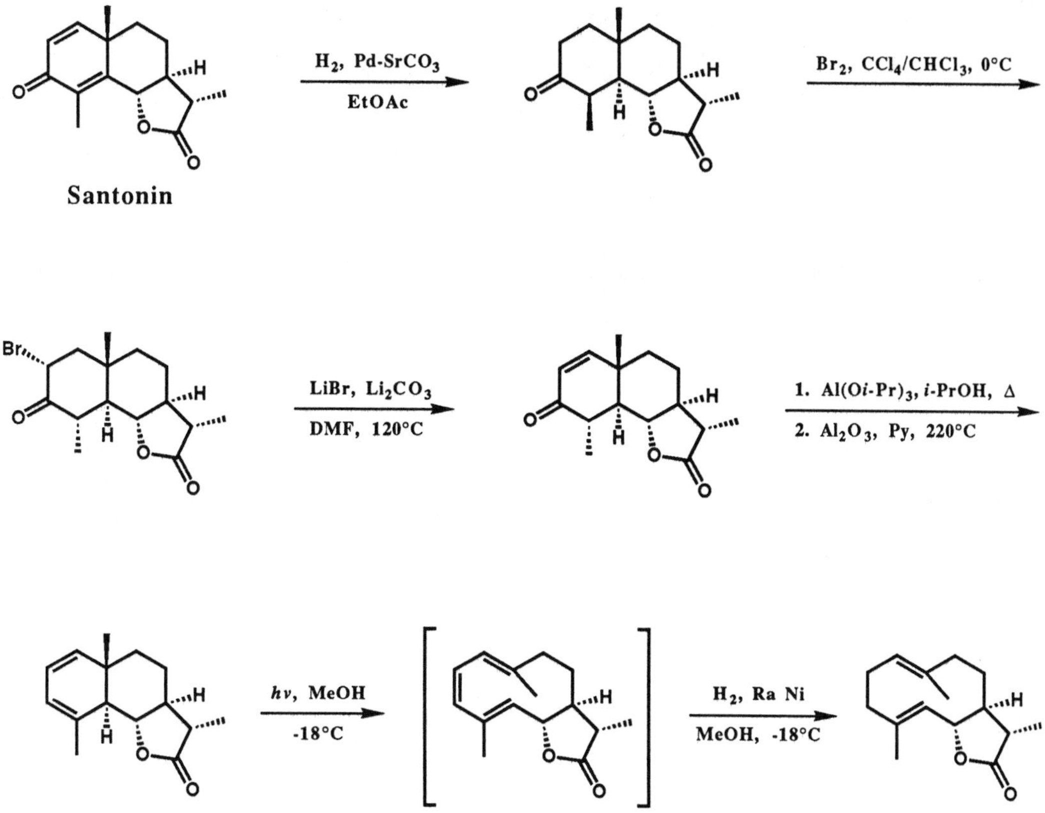
Santonin
H₂, Pd-SrCO₃
EtOAc
Br₂, CCl₄/CHCl₃, 0°C
LiBr, Li₂CO₃
DMF, 120°C
1. Al(O*i*-Pr)₃, *i*-PrOH, Δ
2. Al₂O₃, Py, 220°C
hv, MeOH
-18°C
H₂, Ra Ni
MeOH, -18°C

References:

1. E. J. Corey and A. G. Hortmann, *J. Am. Chem. Soc.* **1963**, *85*, 4033.
2. E. J. Corey and A. G. Hortmann, *J. Am. Chem. Soc.* **1965**, *87*, 5736.

9.8

dl-Elemol

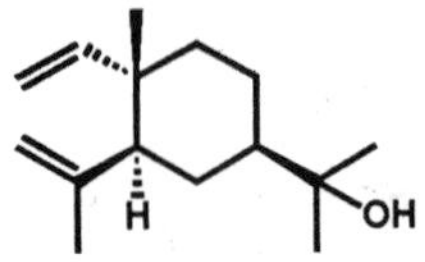
H
OH

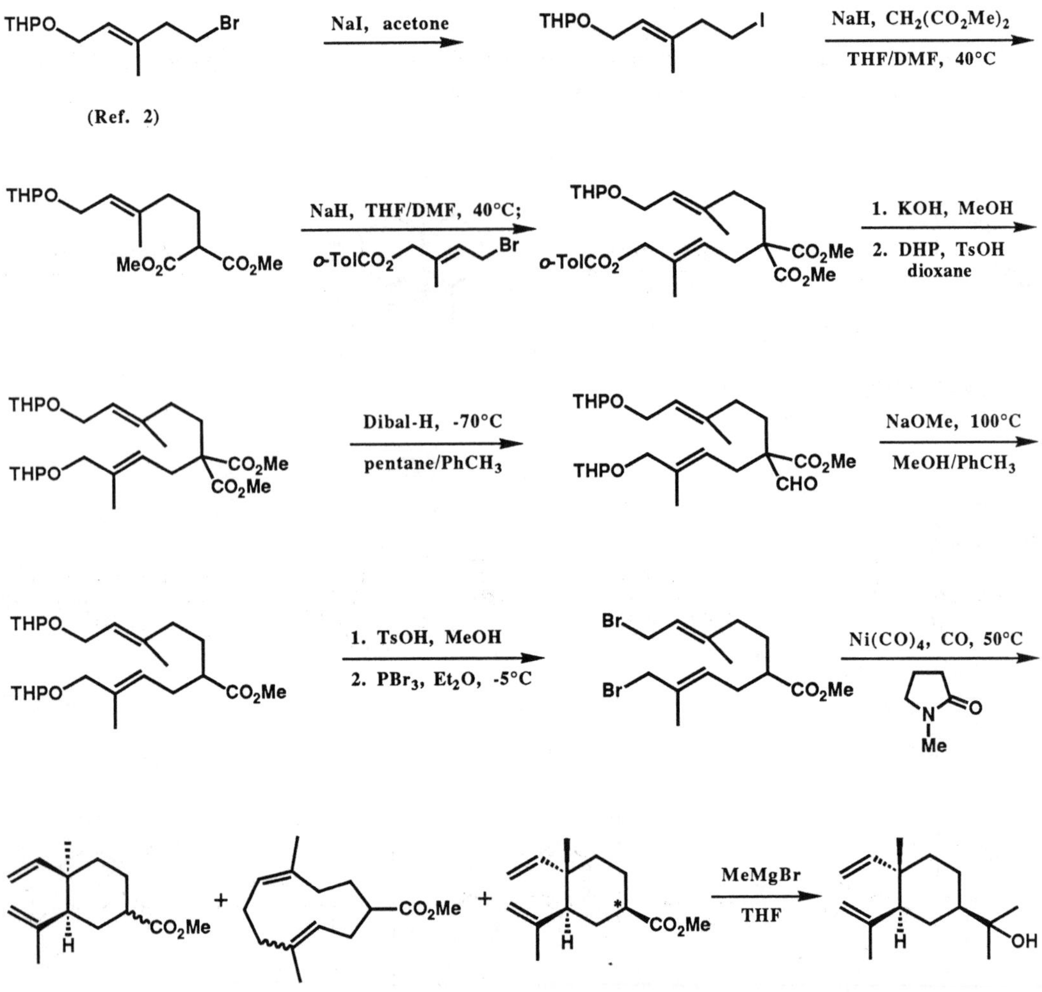
THPO
Br
(Ref. 2)
NaI, acetone
THPO
I
NaH, CH2(CO2Me)2
THF/DMF, 40°C
THPO
MeO2C
CO2Me
NaH, THF/DMF, 40°C;
o-TolCO2
Br
THPO
o-TolCO2
CO2Me
CO2Me
1. KOH, MeOH
2. DHP, TsOH
dioxane
THPO
THPO
CO2Me
CO2Me
Dibal-H, -70°C
pentane/PhCH3
THPO
THPO
CO2Me
CHO
NaOMe, 100°C
MeOH/PhCH3
THPO
THPO
CO2Me
1. TsOH, MeOH
2. PBr3, Et2O, -5°C
Br
Br
CO2Me
Ni(CO)4, CO, 50°C
N
O
Me
H
CO2Me
+
CO2Me
+
H
*
CO2Me
MeMgBr
THF
H
OH

27% | *11% E, Z mixture* | *32%, 9 : 1 mixture at C*, the minor diastereomer was removed by SGC*

References:

1. E. J. Corey and E. A. Broger, *Tetrahedron Lett.* **1969**, 1779.
2. E. J. Corey and E. Hamanaka, *J. Am. Chem. Soc.* **1967**, *89*, 2758.

Helminthosporal

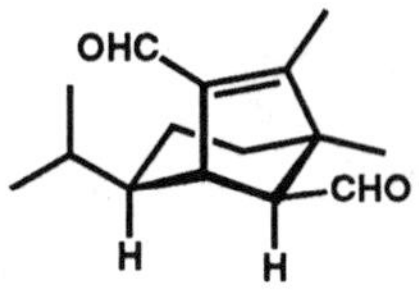
OHC
CHO
H
H

The synthesis of the fungal wheat plant toxin helminthosporal is referred to in Sections 3.1 and 5.2 of Part One.

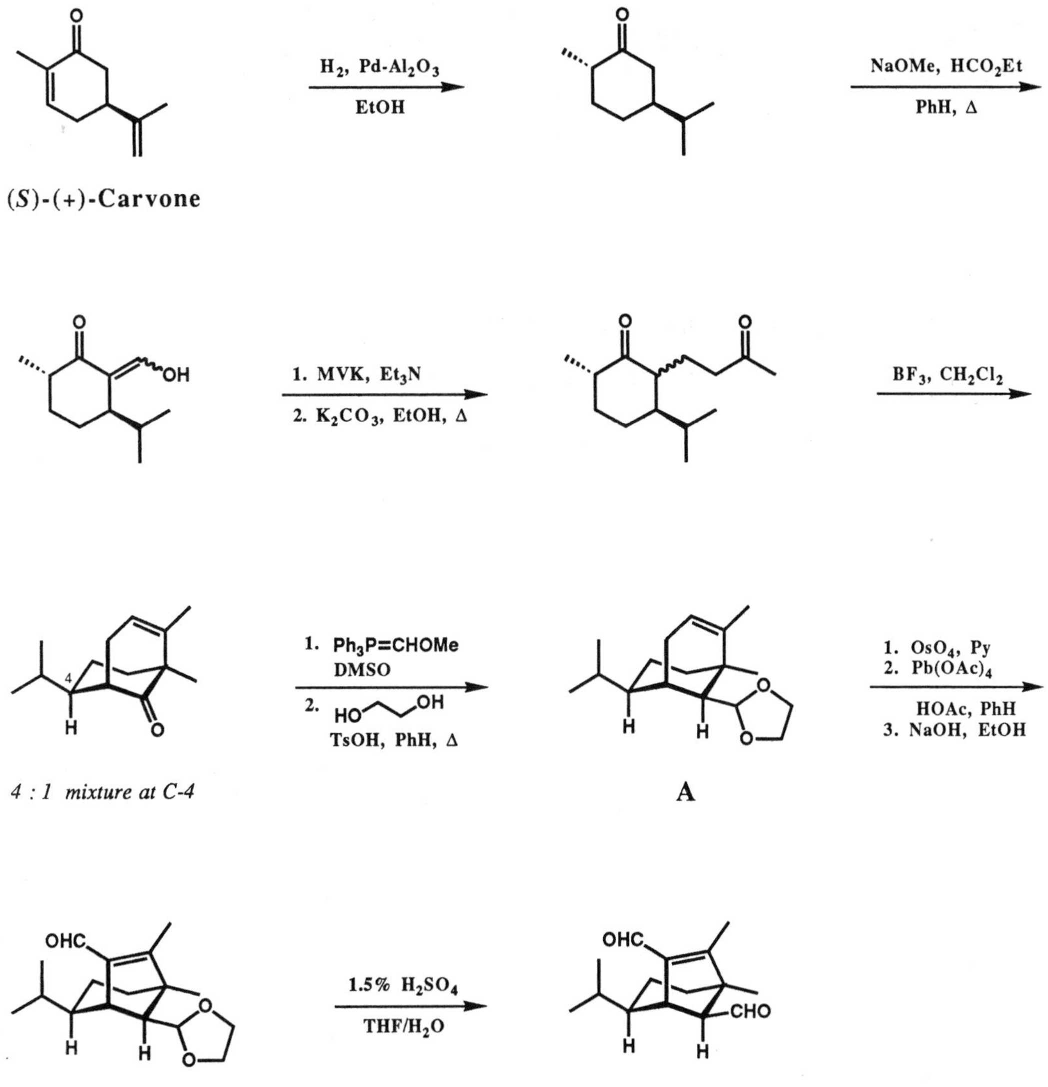
H₂, Pd-Al₂O₃
EtOH
NaOMe, HCO₂Et
PhH, Δ
(S)-(+)-Carvone
OH
1. MVK, Et₃N
2. K₂CO₃, EtOH, Δ
BF₃, CH₂Cl₂
1. Ph₃P=CHOMe
DMSO
2. HO OH
TsOH, PhH, Δ
1. OsO₄, Py
2. Pb(OAc)₄
HOAc, PhH
3. NaOH, EtOH
4 : 1 mixture at C-4
A
OHC
1.5% H₂SO₄
THF/H₂O
CHO

9.9

The ethylene acetal **A** was also prepared by an alternative approach.

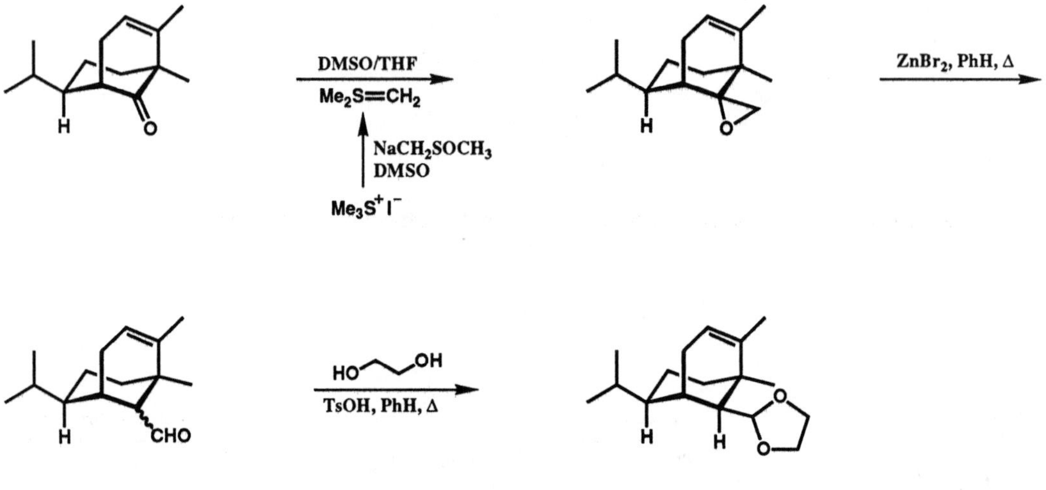
DMSO/THF
$Me_2S{=}CH_2$
$NaCH_2SOCH_3$
DMSO
$Me_3S^+ I^-$
H
O
$ZnBr_2$, PhH, Δ
CHO
HO
OH
TsOH, PhH, Δ

A

References:

1. E. J. Corey and S. Nozoe, *J. Am. Chem. Soc.* **1963**, *85*, 3527.

2. E. J. Corey and S. Nozoe, *J. Am. Chem. Soc.* **1965**, *87*, 5728.

dl-Sirenin

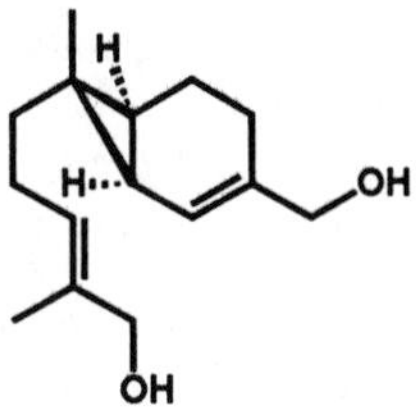
H
H
OH
OH

Sirenin is the sperm attractant produced by the female gametes of the water mold, *Allomyces*. Its synthesis has been accomplished by the use of an internal [2 + 1] cycloaddition step.

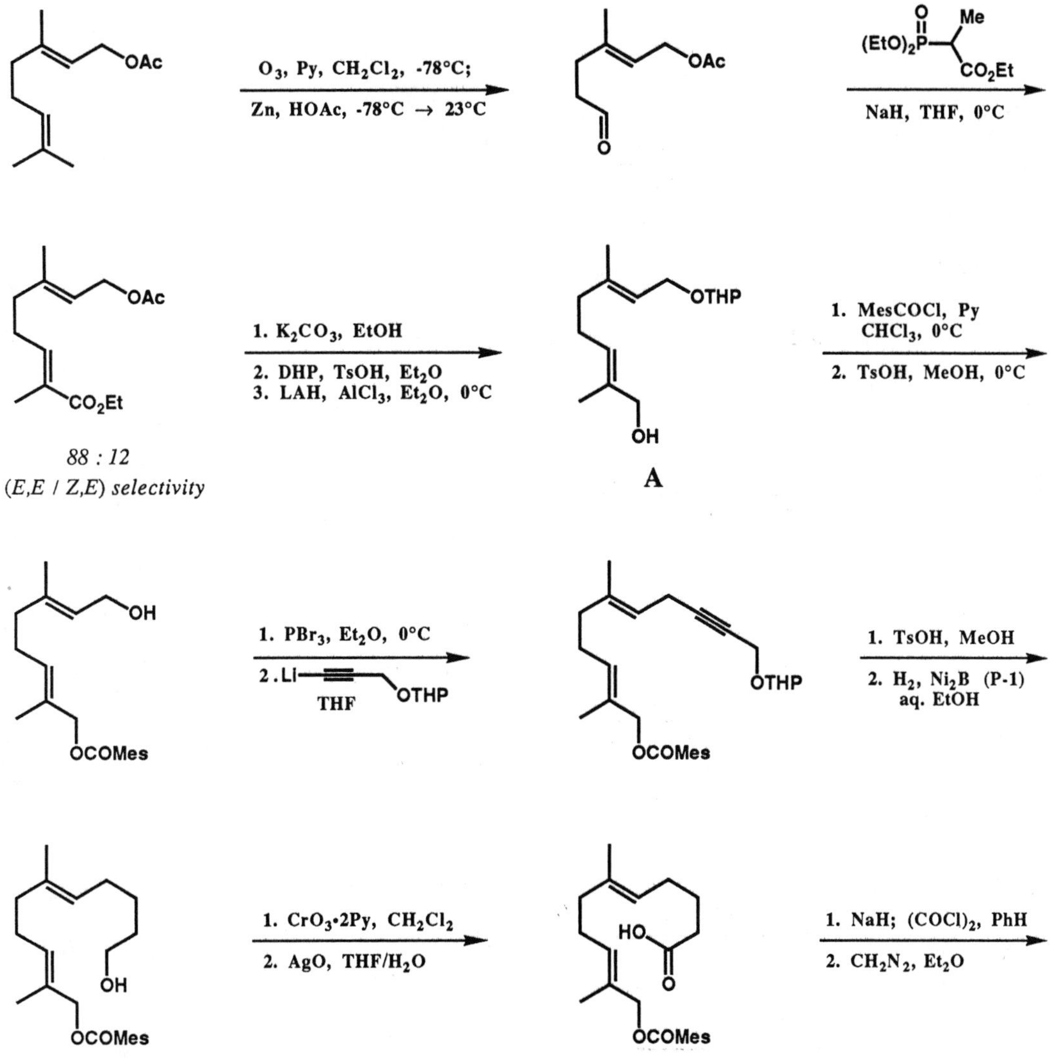
OAc
O3, Py, CH2Cl2, -78°C;
Zn, HOAc, -78°C → 23°C
OAc
O
O
(EtO)2P
Me
CO2Et
NaH, THF, 0°C
OAc
CO2Et
88 : 12
(E,E / Z,E) selectivity
1. K2CO3, EtOH
2. DHP, TsOH, Et2O
3. LAH, AlCl3, Et2O, 0°C
OTHP
OH
A
1. MesCOCl, Py
CHCl3, 0°C
2. TsOH, MeOH, 0°C
OH
OCOMes
1. PBr3, Et2O, 0°C
2. Li
OTHP
THF
OTHP
OCOMes
1. TsOH, MeOH
2. H2, Ni2B (P-1)
aq. EtOH
OH
OCOMes
1. CrO3•2Py, CH2Cl2
2. AgO, THF/H2O
HO
O
OCOMes
1. NaH; (COCl)2, PhH
2. CH2N2, Et2O

9.10

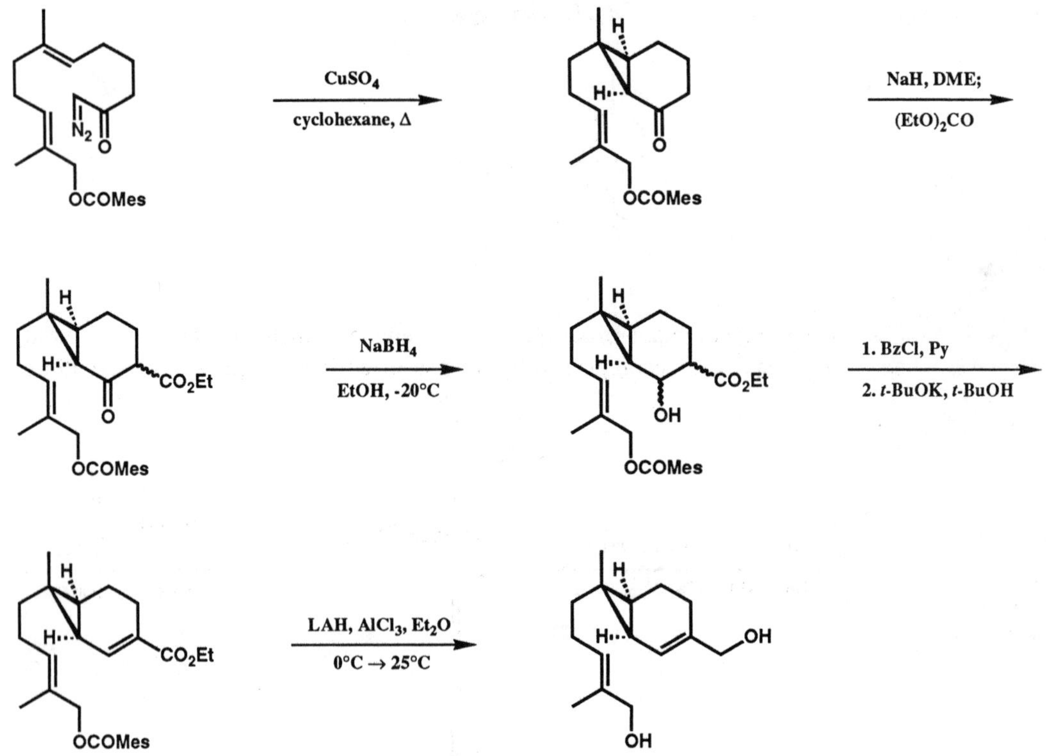
N2
OCOMes
CuSO4
cyclohexane, Δ
H
H
OCOMes
NaH, DME;
(EtO)2CO
H
H
CO2Et
OCOMes
NaBH4
EtOH, -20°C
H
H
CO2Et
OH
OCOMes
1. BzCl, Py
2. t-BuOK, t-BuOH
H
H
CO2Et
OCOMes
LAH, AlCl3, Et2O
0°C → 25°C
H
H
OH
OH

The intermediate **A** was also made by the following method:

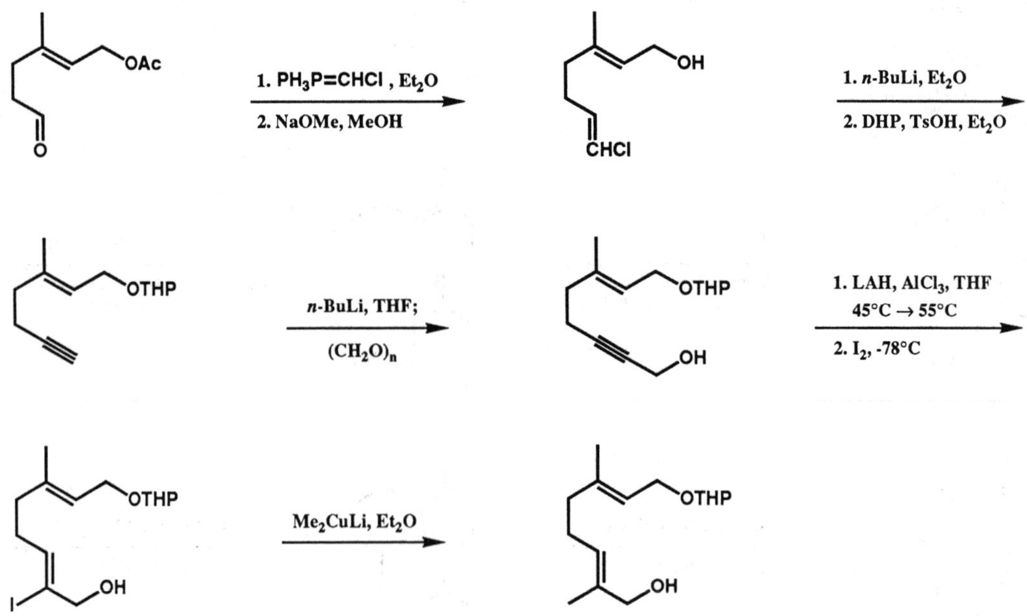
OAc
1. PH3P=CHCl , Et2O
2. NaOMe, MeOH
OH
CHCl
1. n-BuLi, Et2O
2. DHP, TsOH, Et2O
OTHP
n-BuLi, THF;
(CH2O)n
OTHP
OH
1. LAH, AlCl3, THF
45°C → 55°C
2. I2, -78°C
OTHP
I
OH
Me2CuLi, Et2O
OTHP
OH

A

A shorter synthesis of sirenin was achieved following the sequence shown below (Ref. 2):

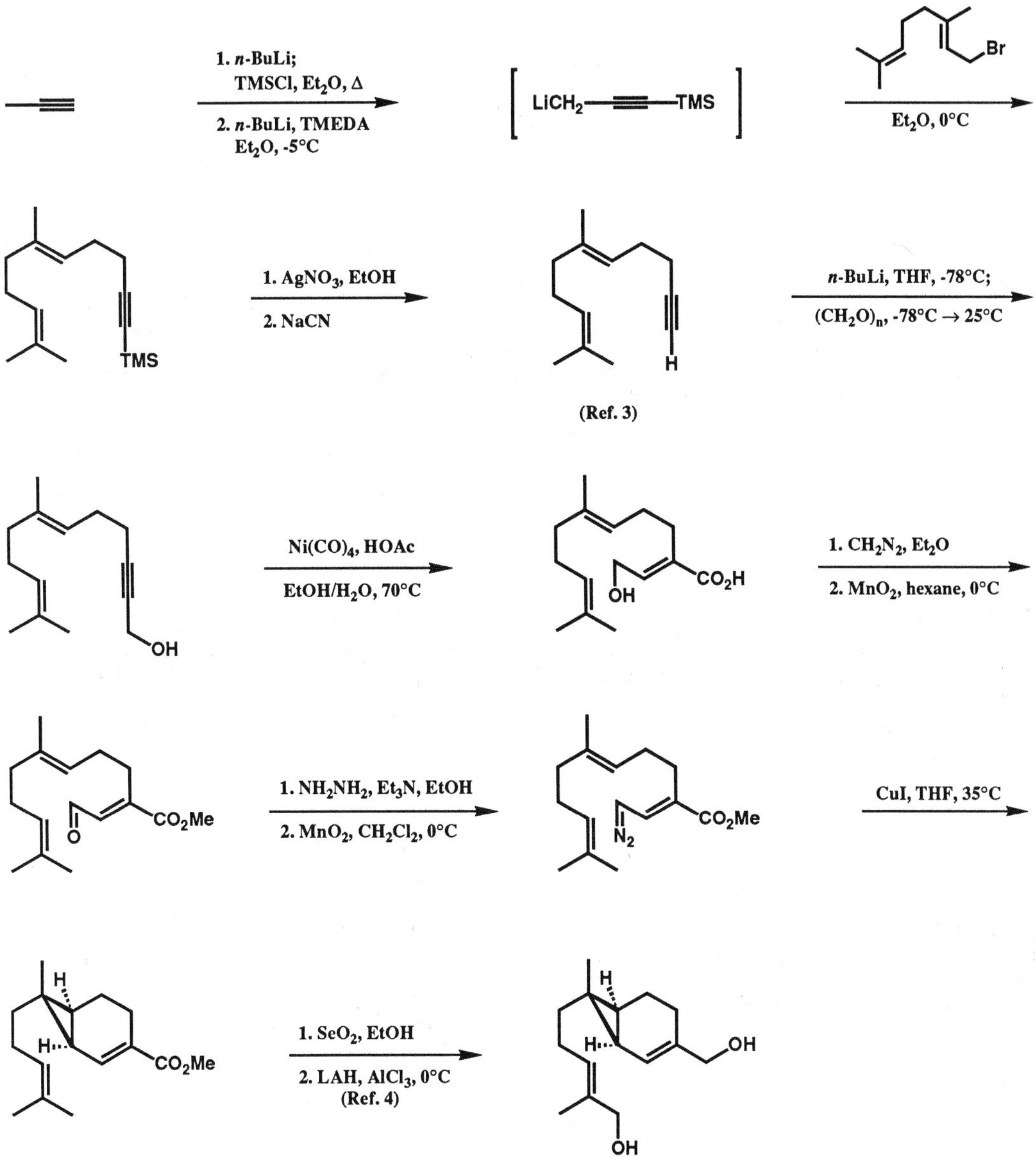
1. n-BuLi;
TMSCl, Et2O, Δ
2. n-BuLi, TMEDA
Et2O, -5°C
LiCH2—≡—TMS
Br
Et2O, 0°C
TMS
1. AgNO3, EtOH
2. NaCN
H
(Ref. 3)
n-BuLi, THF, -78°C;
(CH2O)n, -78°C → 25°C
OH
Ni(CO)4, HOAc
EtOH/H2O, 70°C
OH
CO2H
1. CH2N2, Et2O
2. MnO2, hexane, 0°C
O
CO2Me
1. NH2NH2, Et3N, EtOH
2. MnO2, CH2Cl2, 0°C
N2
CO2Me
CuI, THF, 35°C
H
H
CO2Me
1. SeO2, EtOH
2. LAH, AlCl3, 0°C
(Ref. 4)
H
H
OH
OH

References:

1. E. J. Corey, K. Achiwa, and J. A. Katzenellenbogen, *J. Am. Chem. Soc.* **1969**, *91*, 4318.
2. E. J. Corey and K. Achiwa, *Tetrahedron Lett.* **1970**, 2245.
3. E. J. Corey and H. A. Kirst, *Tetrahedron Lett.* **1968**, 5041.
4. J. J. Plattner, U. T. Bhalerao, and H. Rapoport, *J. Am. Chem. Soc.* **1969**, *91*, 4933.

9.11

dl-Sesquicarene

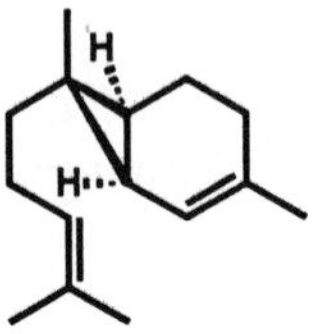
H
H

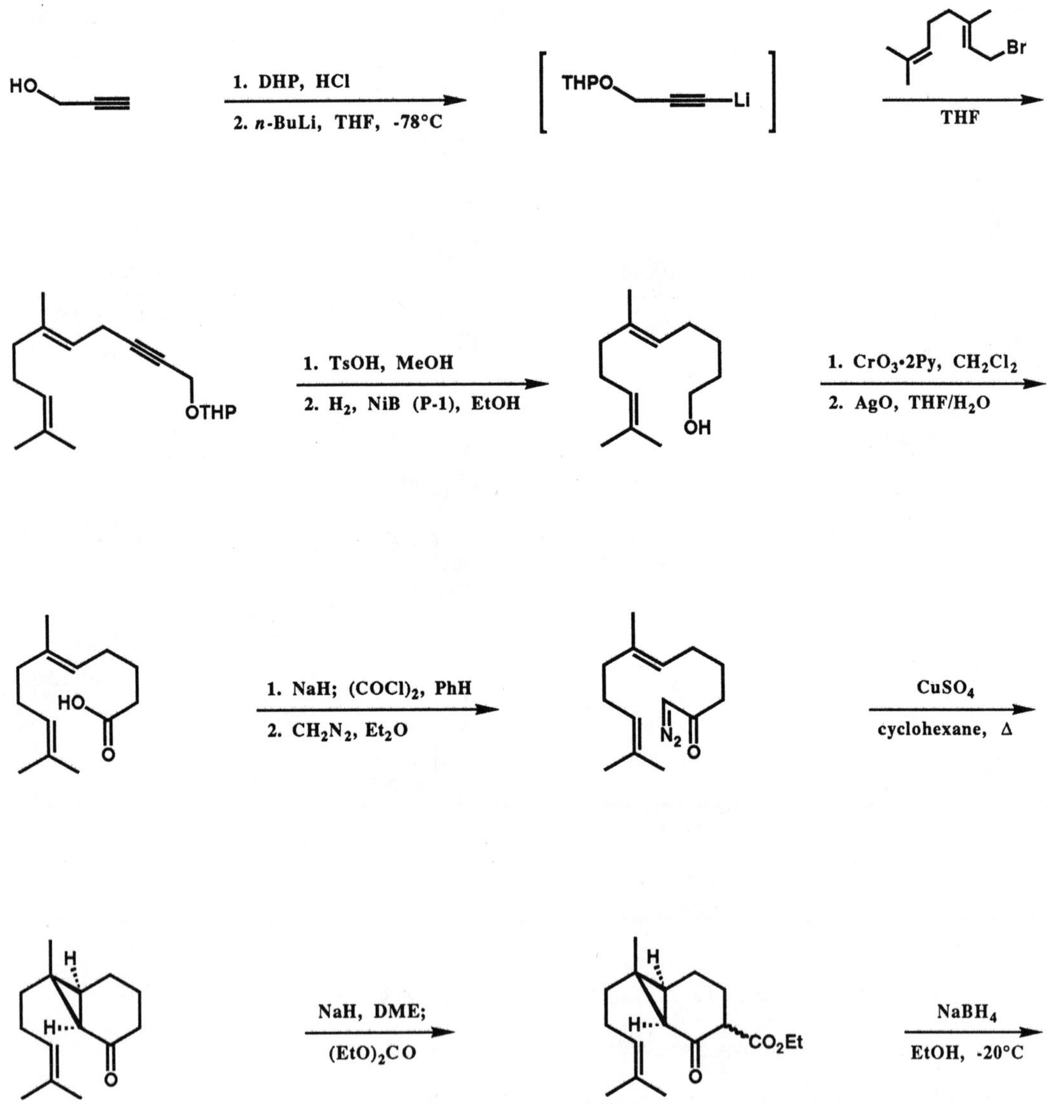
HO
1. DHP, HCl
2. n-BuLi, THF, -78°C
THPO
Li
Br
THF
OTHP
1. TsOH, MeOH
2. H2, NiB (P-1), EtOH
OH
1. CrO3•2Py, CH2Cl2
2. AgO, THF/H2O
HO
O
1. NaH; (COCl)2, PhH
2. CH2N2, Et2O
N2
O
CuSO4
cyclohexane, Δ
H
H
O
NaH, DME;
(EtO)2CO
CO2Et
NaBH4
EtOH, -20°C

9.11

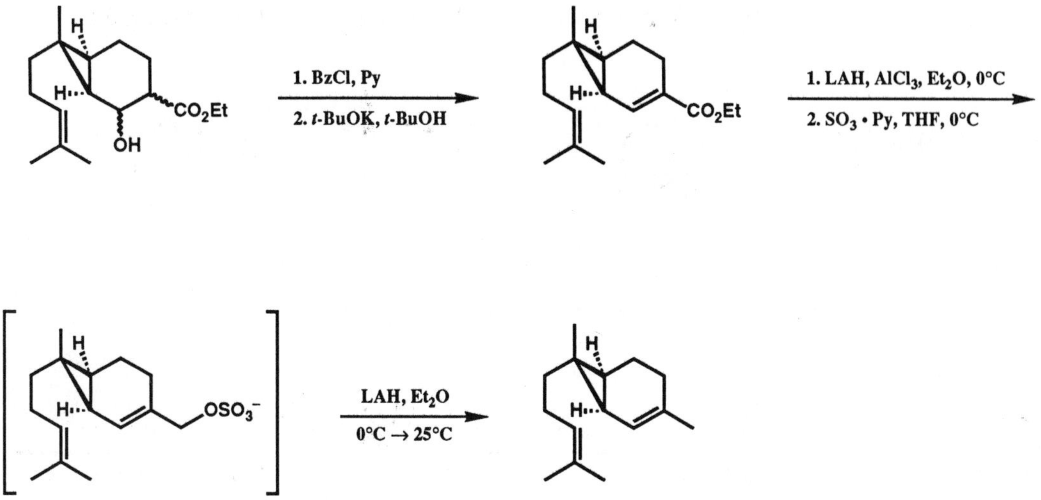
H
H
CO_2Et
OH
1. BzCl, Py
2. t-BuOK, t-BuOH
H
H
CO_2Et
1. LAH, $AlCl_3$, Et_2O, 0°C
2. SO_3 • Py, THF, 0°C
H
H
OSO_3^-
LAH, Et_2O
0°C → 25°C
H
H

Sesquicarene was also synthesized from (*Z, E*)-farnesol as shown below:

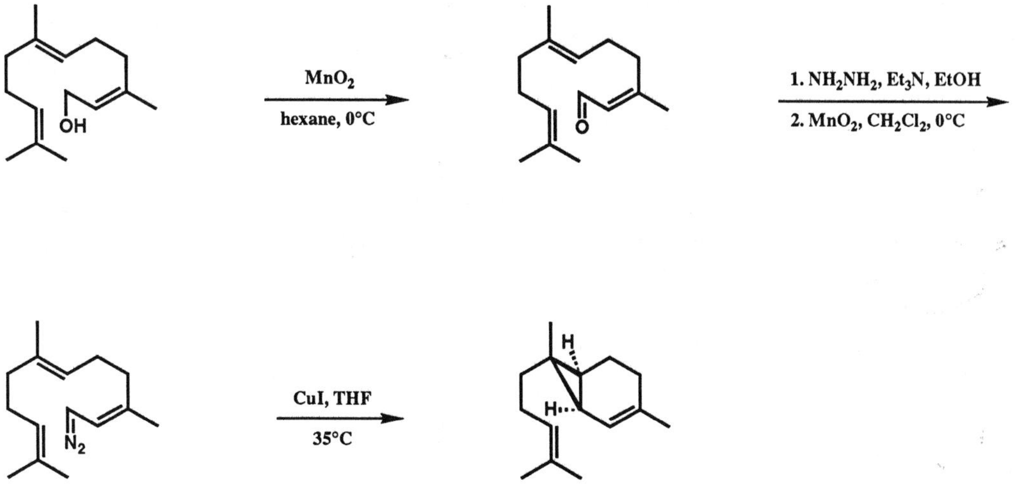
OH
MnO_2
hexane, 0°C
O
1. NH_2NH_2, Et_3N, EtOH
2. MnO_2, CH_2Cl_2, 0°C
N_2
CuI, THF
35°C
H
H

References:

1. E. J. Corey and K. Achiwa, *Tetrahedron Lett.* **1969**, 1837.

2. E. J. Corey and K. Achiwa, *Tetrahedron Lett.* **1969**, 3257.

9.12

(±)-α-Copaene (±)-α-Ylangene (±)-β-Copaene (±)-β-Ylangene

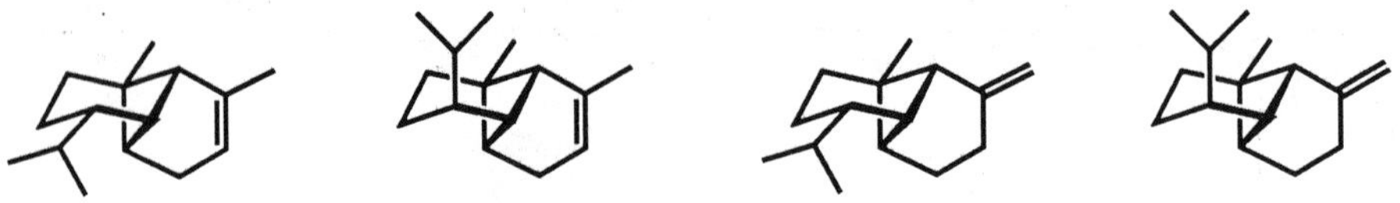

1. Total synthesis of (±)-α-copaene and (±)-α-ylangene:

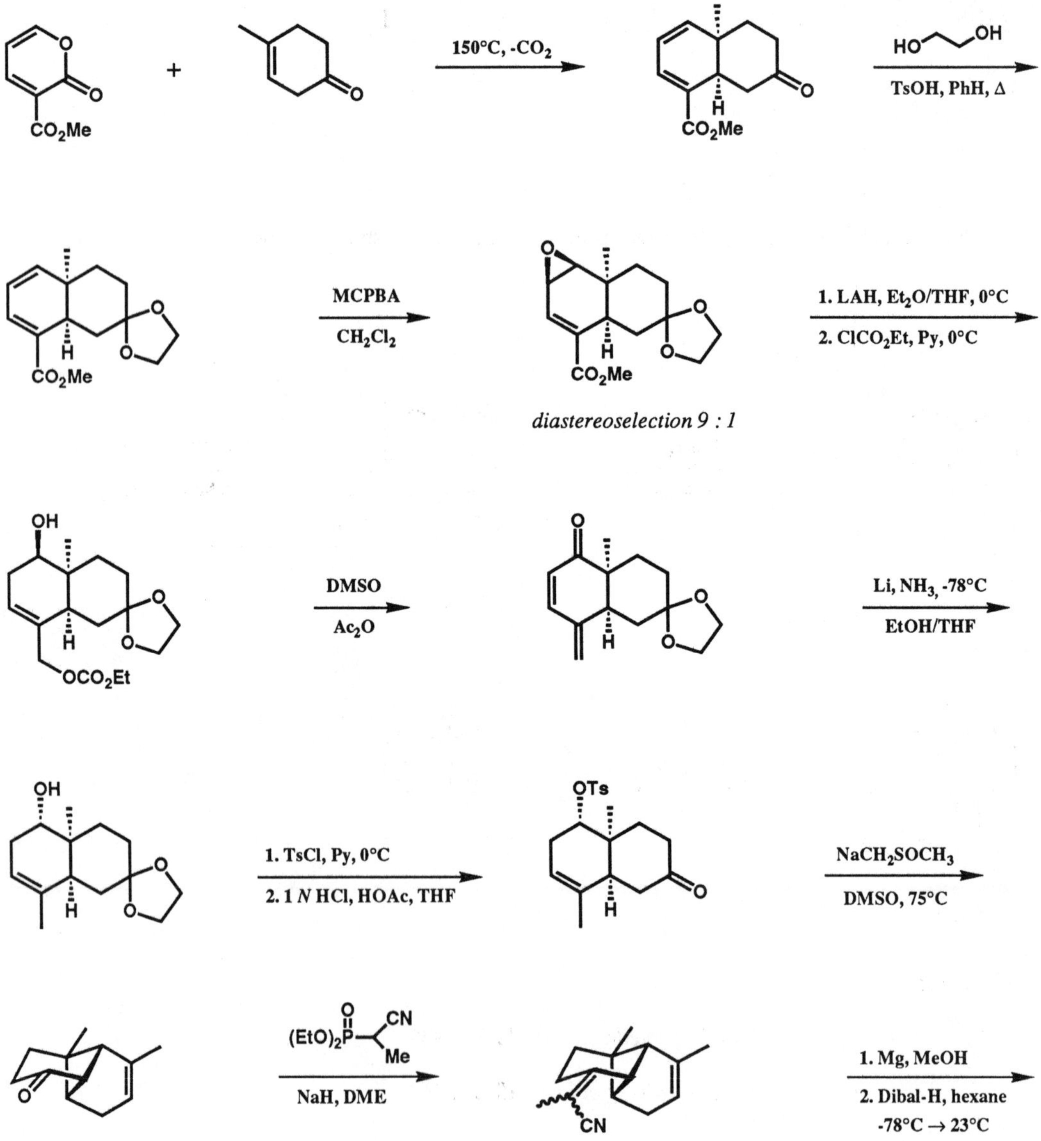

CO2Me
+
150°C, -CO2
H
CO2Me
HO OH
TsOH, PhH, Δ
MCPBA
CH2Cl2
diastereoselection 9 : 1
1. LAH, Et2O/THF, 0°C
2. ClCO2Et, Py, 0°C
OH
OCO2Et
DMSO
Ac2O
Li, NH3, -78°C
EtOH/THF
OTs
1. TsCl, Py, 0°C
2. 1 N HCl, HOAc, THF
NaCH2SOCH3
DMSO, 75°C
(EtO)2P
CN
Me
NaH, DME
1. Mg, MeOH
2. Dibal-H, hexane
-78°C → 23°C

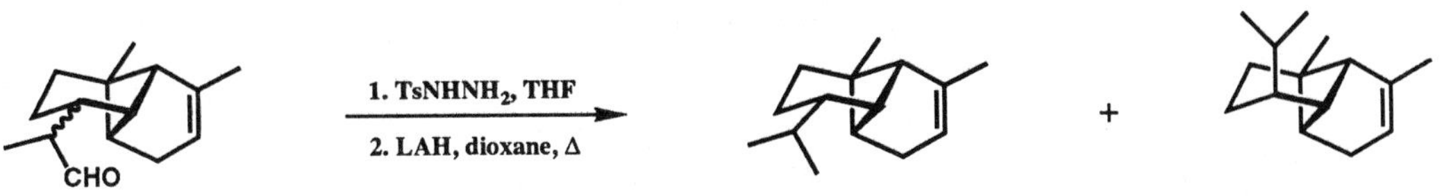
CHO
1. TsNHNH₂, THF
2. LAH, dioxane, Δ
+

(±)-α-Copaene (±)-α-Ylangene

2. Total synthesis of (±)-β-copaene and (±)-β-ylangene:

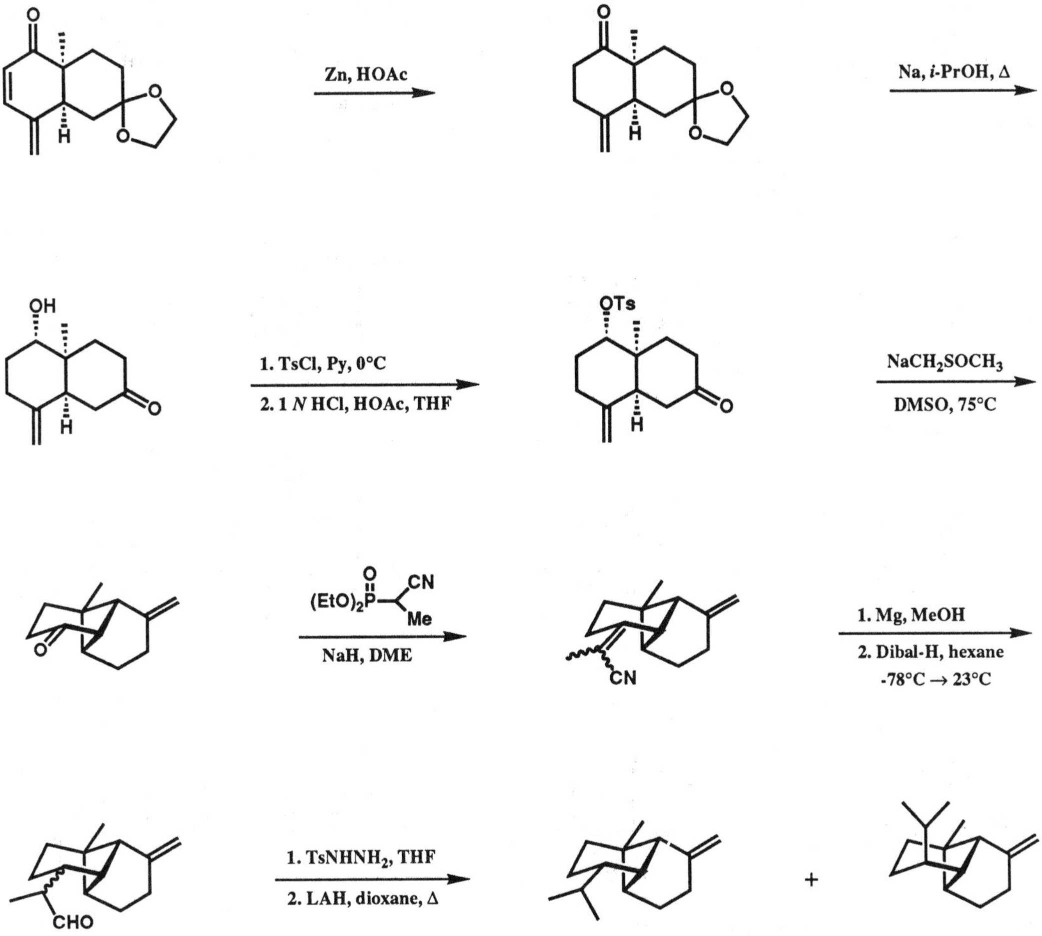
O
H
O
O
Zn, HOAc
Na, i-PrOH, Δ
OH
O
1. TsCl, Py, 0°C
2. 1 N HCl, HOAc, THF
OTs
NaCH₂SOCH₃
DMSO, 75°C
(EtO)₂P
CN
Me
NaH, DME
1. Mg, MeOH
2. Dibal-H, hexane
-78°C → 23°C
CHO
1. TsNHNH₂, THF
2. LAH, dioxane, Δ
+

(±)-β-Copaene (±)-β-Ylangene

Reference:

E. J. Corey and D. S. Watt, *J. Am. Chem. Soc.* **1973**, *95*, 2303.

9.13

(±)-Occidentalol

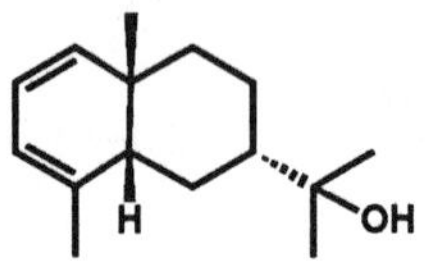
H
OH

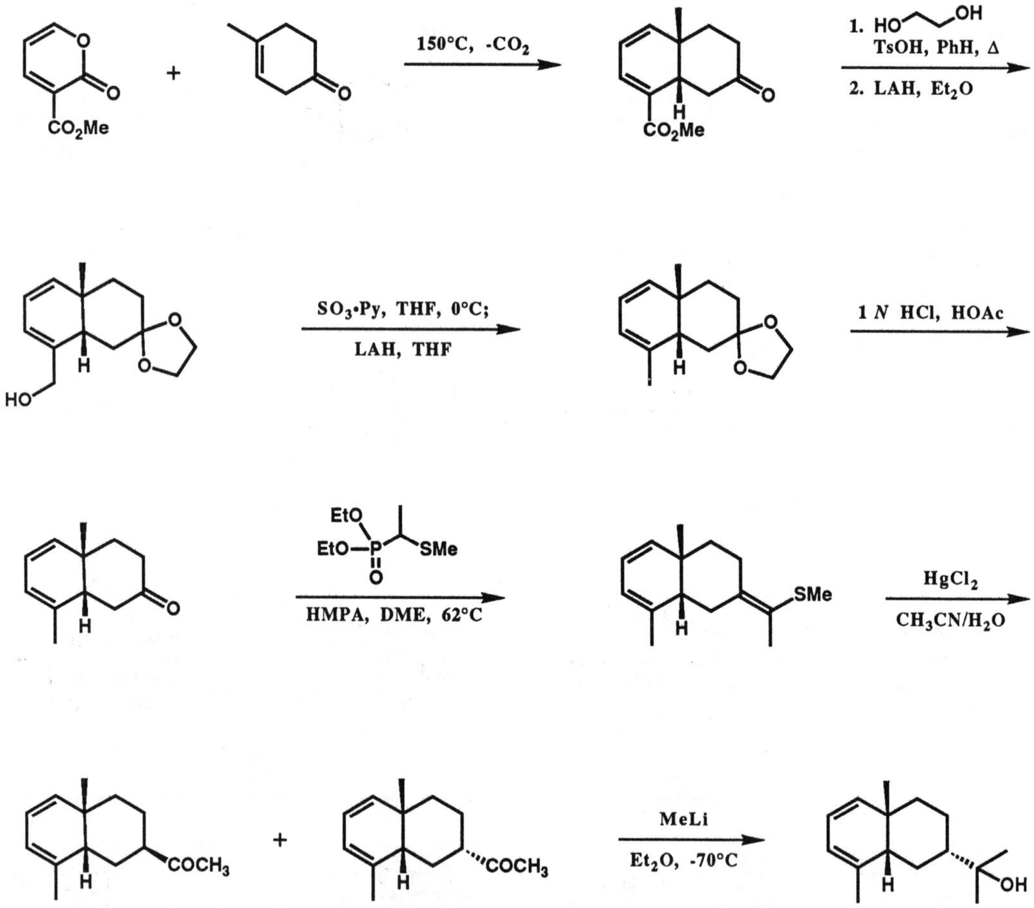
CO2Me
150°C, -CO2
1. HO OH
TsOH, PhH, Δ
2. LAH, Et2O
SO3•Py, THF, 0°C;
LAH, THF
1 N HCl, HOAc
EtO
EtO—P SMe
O
HMPA, DME, 62°C
SMe
HgCl2
CH3CN/H2O
COCH3
MeLi
Et2O, -70°C
OH

diastereoselection 1 : 2.9
the β-isomer was removed by SGC

Reference:

D. S. Watt and E. J. Corey, *Tetrahedron Lett.* **1972**, 4651.

(±)-*β*-*trans*-Bergamotene

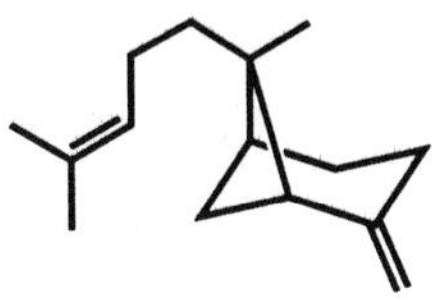

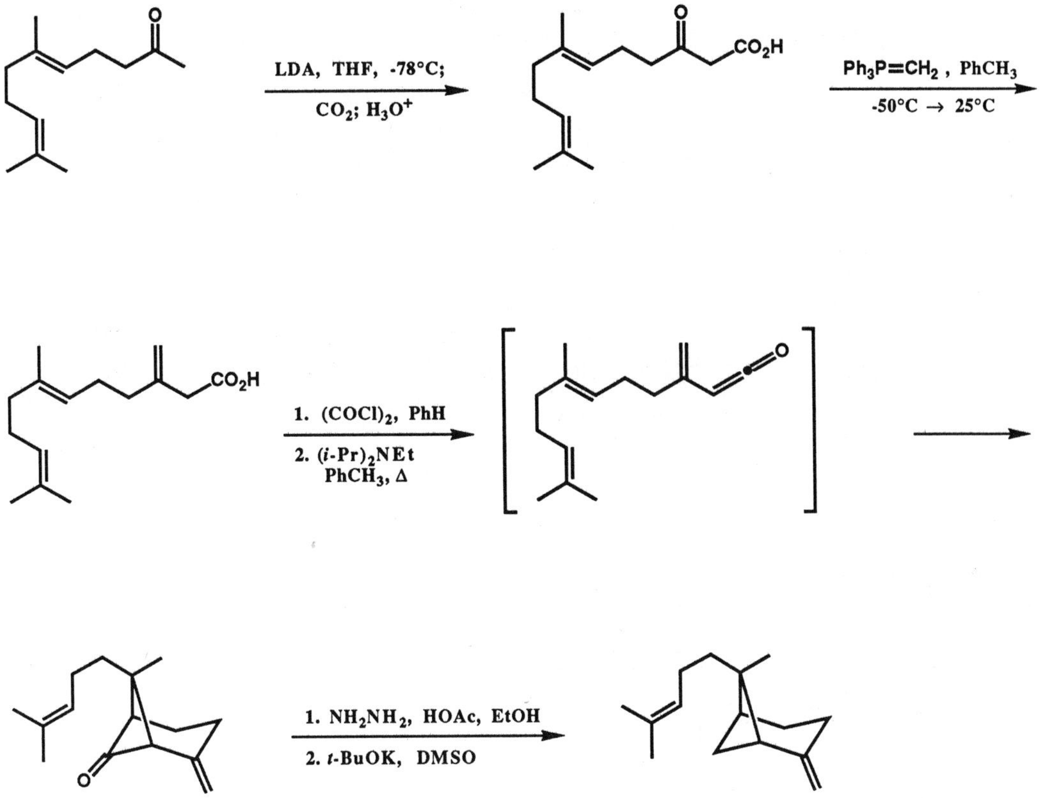
O
LDA, THF, -78°C;
CO2; H3O+
O
CO2H
Ph3P=CH2 , PhCH3
-50°C → 25°C
CO2H
1. (COCl)2, PhH
2. (i-Pr)2NEt
PhCH3, Δ
O
O
1. NH2NH2, HOAc, EtOH
2. t-BuOK, DMSO

Reference:

E. J. Corey and M. C. Desai, *Tetrahedron Lett.* **1985**, *26*, 3535.

9.15

(±)-Fumagillin

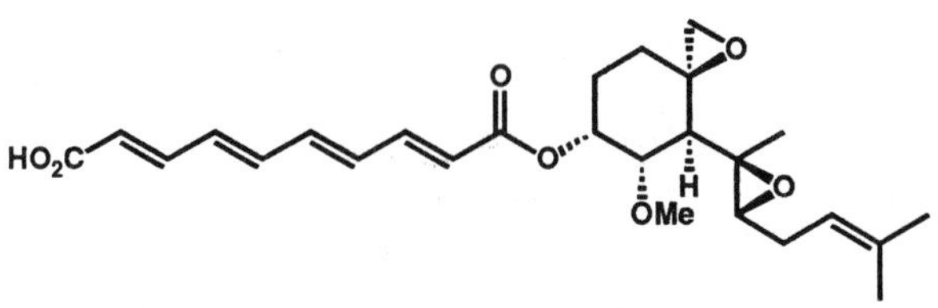

HO2C
O
O
H
OMe
O
O

The retrosynthetic analysis of fumagillol, the alcohol from which the antibiotic fumagillin is derived, has been outlined in Section 2.3. The experimentally demonstrated synthesis of fumagillol was derived by T-goal directed search to apply the Diels-Alder transform.

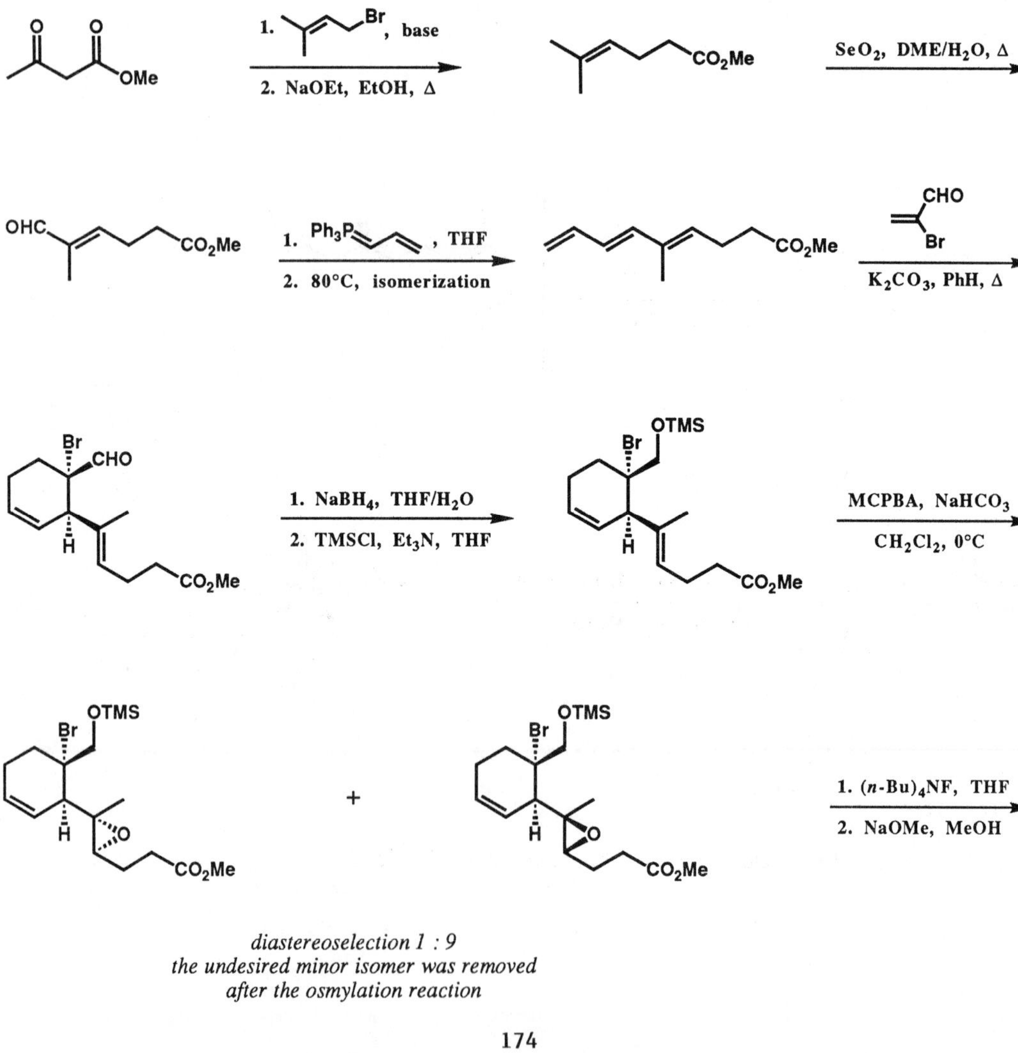

OMe
1. Br, base
2. NaOEt, EtOH, Δ
CO2Me
SeO2, DME/H2O, Δ
OHC
CO2Me
1. Ph3P, THF
2. 80°C, isomerization
CO2Me
CHO
Br
K2CO3, PhH, Δ
Br
CHO
H
CO2Me
1. NaBH4, THF/H2O
2. TMSCl, Et3N, THF
OTMS
Br
H
CO2Me
MCPBA, NaHCO3
CH2Cl2, 0°C
OTMS
Br
H
O
CO2Me
+
OTMS
Br
H
O
CO2Me
1. (n-Bu)4NF, THF
2. NaOMe, MeOH

diastereoselection 1 : 9
the undesired minor isomer was removed
after the osmylation reaction

9.15

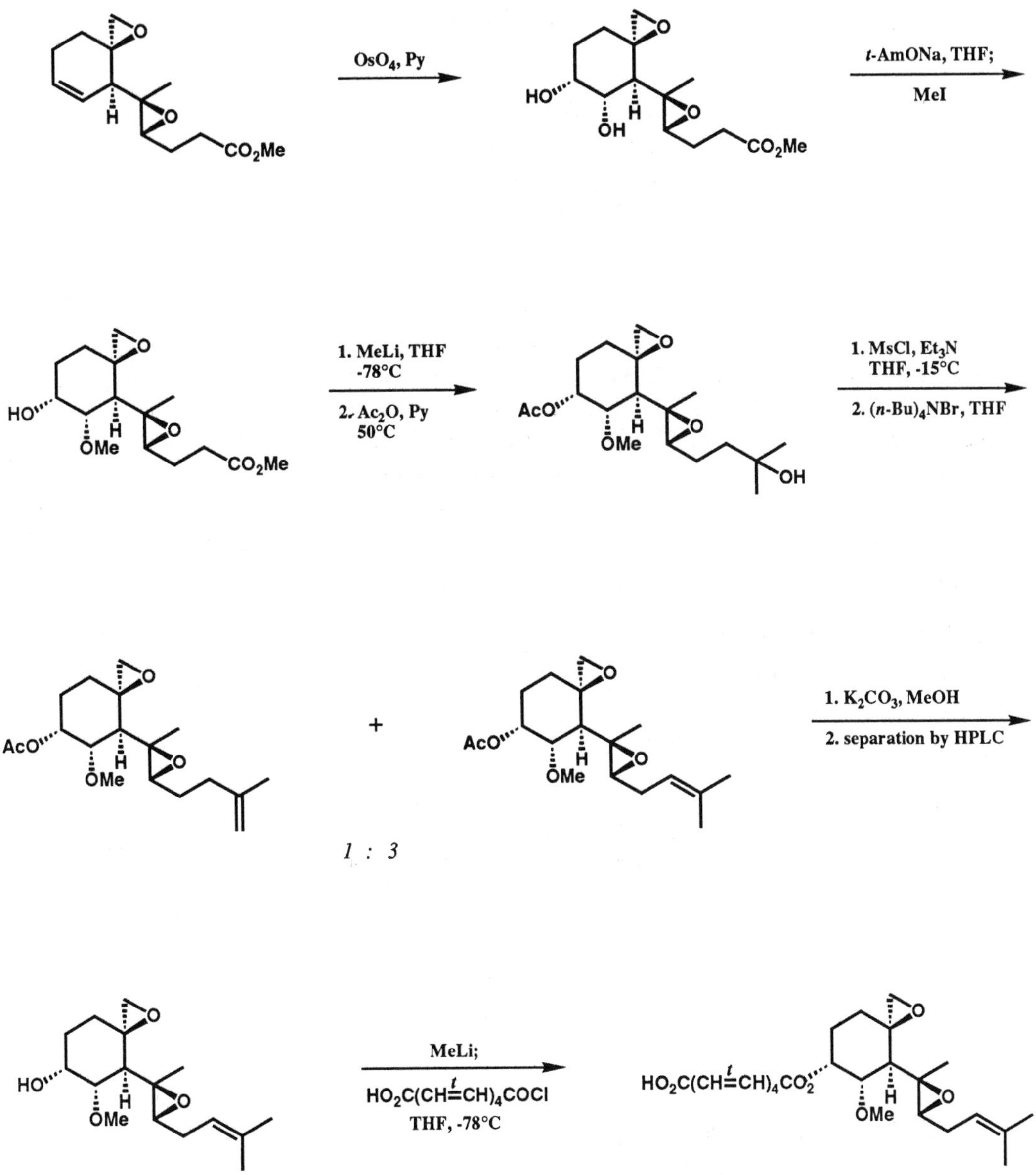
CO2Me
OsO4, Py
HO
OH
H
CO2Me
t-AmONa, THF;
MeI
HO
OMe
H
CO2Me
1. MeLi, THF
-78°C
2. Ac2O, Py
50°C
AcO
OMe
H
OH
1. MsCl, Et3N
THF, -15°C
2. (n-Bu)4NBr, THF
AcO
OMe
+
1 : 3
1. K2CO3, MeOH
2. separation by HPLC
HO
OMe
MeLi;
HO2C(CH=CH)4COCl
THF, -78°C
HO2C(CH=CH)4CO2
OMe

(±)-Fumagillol

(±)-Fumagillin

Reference:

E. J. Corey and B. B. Snider, *J. Am. Chem. Soc.* **1972**, *94*, 2549.

9.16

(±)-Ovalicin

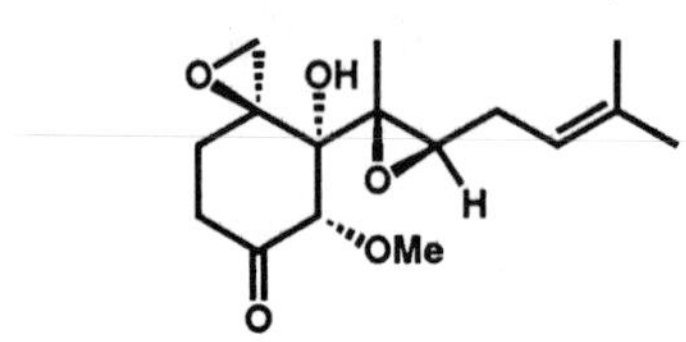

O
OH
O
H
OMe
O

The synthesis of ovalicin was accomplished following a line of analysis which was totally different from that employed for the synthesis of the structural relative fumagillol. The plan for ovalicin was based on S-goal, appendage, stereochemical and functional group derived strategies. A key requirement for the synthesis was the stereospecific construction of the *E*-1,4-pentadienyl subunit, which was achieved by a method of potentially wide utility.

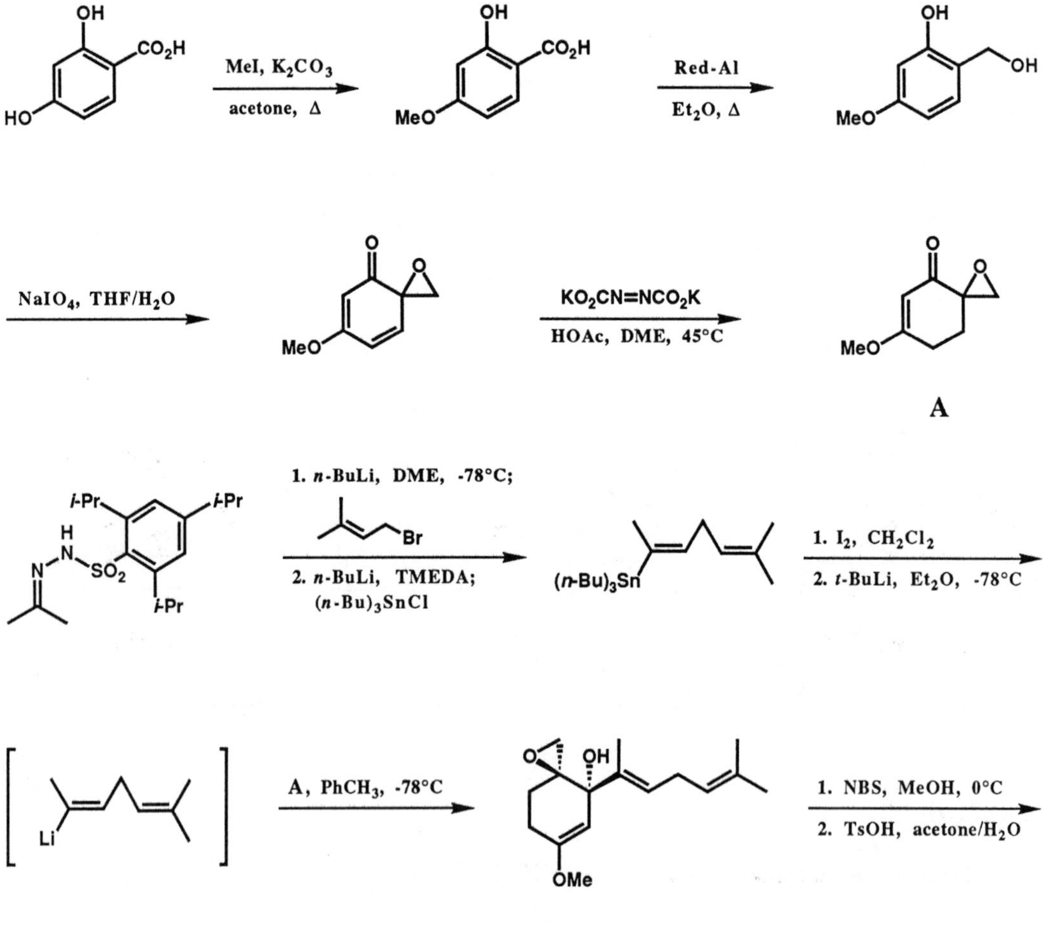

OH
CO2H
HO
MeI, K2CO3
acetone, Δ
MeO
Red-Al
Et2O, Δ
OH
NaIO4, THF/H2O
KO2CN=NCO2K
HOAc, DME, 45°C
A
i-Pr
H
N
SO2
1. n-BuLi, DME, -78°C;
Br
2. n-BuLi, TMEDA;
(n-Bu)3SnCl
(n-Bu)3Sn
1. I2, CH2Cl2
2. t-BuLi, Et2O, -78°C
Li
A, PhCH3, -78°C
1. NBS, MeOH, 0°C
2. TsOH, acetone/H2O

diastereoselection 17 : 1

9.16

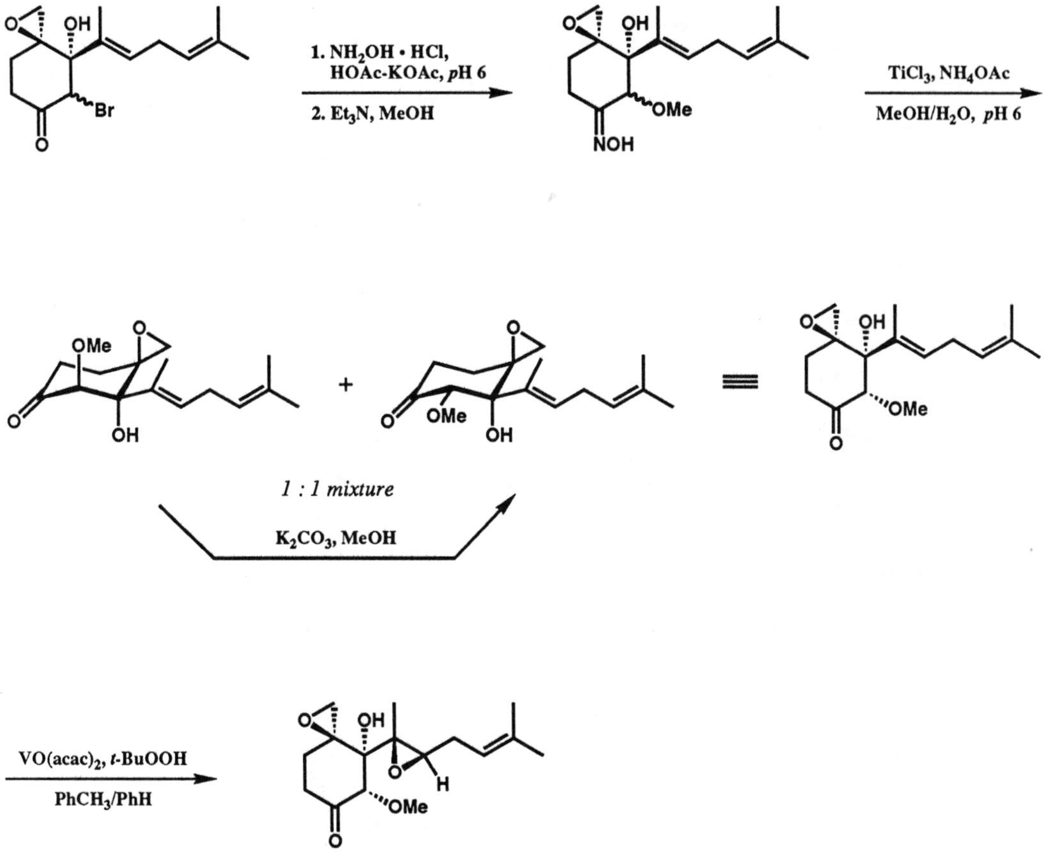
OH
Br
O
1. NH₂OH • HCl,
HOAc-KOAc, pH 6
2. Et₃N, MeOH
OH
OMe
NOH
TiCl₃, NH₄OAc
MeOH/H₂O, pH 6
OMe
OH
+
OMe
OH
≡
OH
OMe
1 : 1 mixture
K₂CO₃, MeOH
VO(acac)₂, t-BuOOH
PhCH₃/PhH
OH
H
OMe

Reference:

E. J. Corey and J. P. Dittami, *J. Am. Chem. Soc*. **1985**, *107*, 256.

9.17

Picrotoxinin **Picrotin**

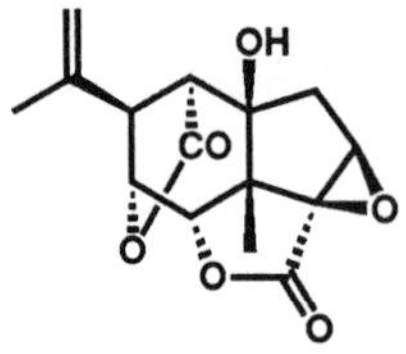

OH
CO
O
O
O
O

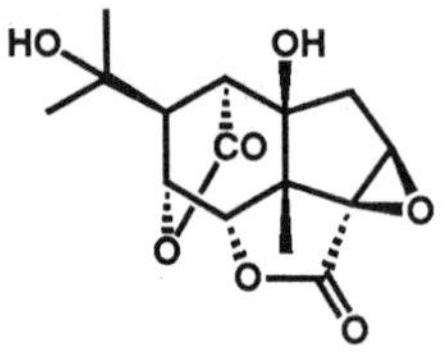

HO
OH
CO
O
O
O
O

Picrotoxin, a potent antagonist of γ–aminobutyric acid at neural synapses, has been synthesized from *(R)*-(-) carvone as SM-goal (Sections 3.1 and 6.5).

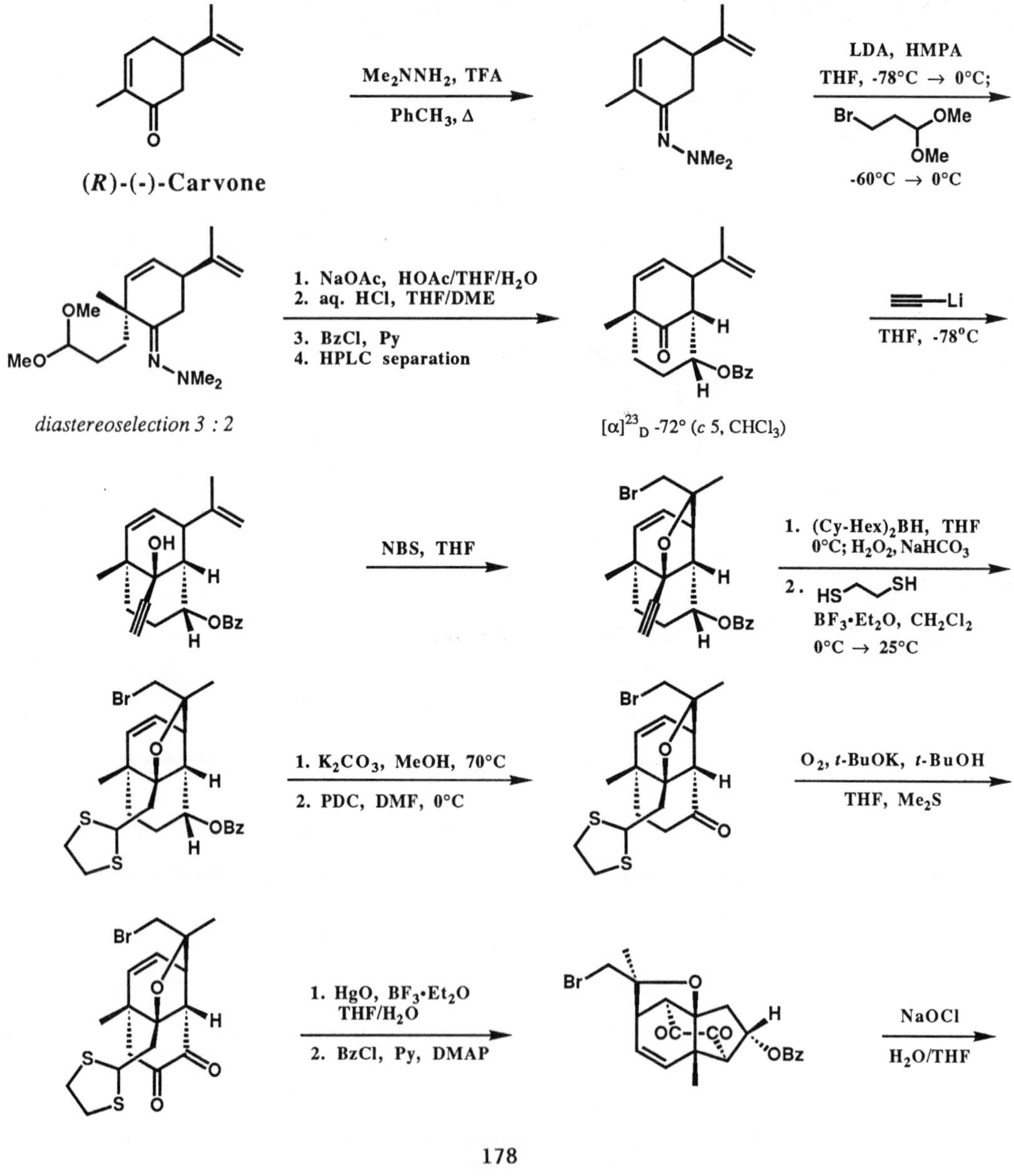

(R)-(-)-Carvone
Me2NNH2, TFA
PhCH3, Δ
N
NMe2
LDA, HMPA
THF, -78°C → 0°C;
Br OMe
OMe
-60°C → 0°C
OMe
MeO
diastereoselection 3 : 2
1. NaOAc, HOAc/THF/H2O
2. aq. HCl, THF/DME
3. BzCl, Py
4. HPLC separation
OBz
[α]23D -72° (c 5, CHCl3)
Li
THF, -78°C
OH
NBS, THF
Br
1. (Cy-Hex)2BH, THF
0°C; H2O2, NaHCO3
2. HS SH
BF3•Et2O, CH2Cl2
0°C → 25°C
1. K2CO3, MeOH, 70°C
2. PDC, DMF, 0°C
O2, t-BuOK, t-BuOH
THF, Me2S
1. HgO, BF3•Et2O
THF/H2O
2. BzCl, Py, DMAP
OC- -CO
NaOCl
H2O/THF

9.17

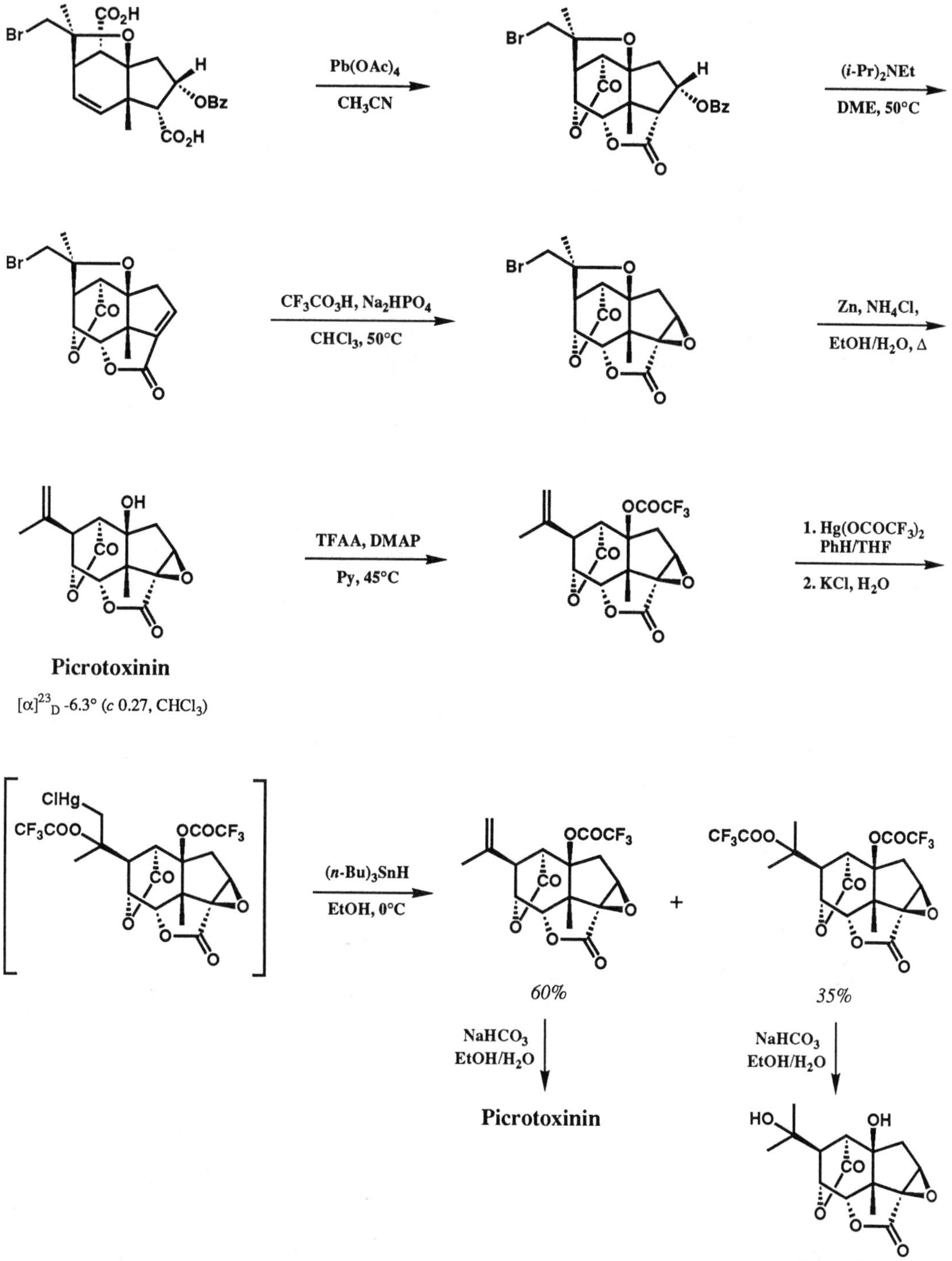
Pb(OAc)4
CH3CN
(i-Pr)2NEt
DME, 50°C
CF3CO3H, Na2HPO4
CHCl3, 50°C
Zn, NH4Cl,
EtOH/H2O, Δ
Picrotoxinin
[α]23D -6.3° (c 0.27, CHCl3)
TFAA, DMAP
Py, 45°C
1. Hg(OCOCF3)2
PhH/THF
2. KCl, H2O
(n-Bu)3SnH
EtOH, 0°C
60%
35%
NaHCO3
EtOH/H2O
NaHCO3
EtOH/H2O
Picrotoxinin

Picrotin

References:

1. E. J. Corey and H. L. Pearce, *J. Am, Chem. Soc.* **1979**, *101*, 5841.
2. E. J. Corey and H. L. Pearce, *Tetrahedron Lett.* **1980**, *21*, 1823.

9.18

9-Isocyanopupukeanane   2-Isocyanopupukeanane

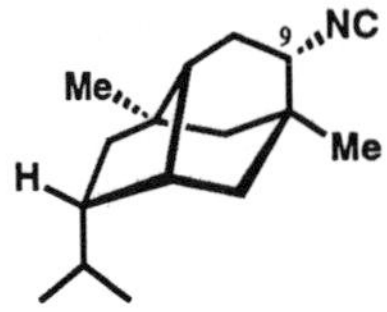

9
NC
Me
Me
H

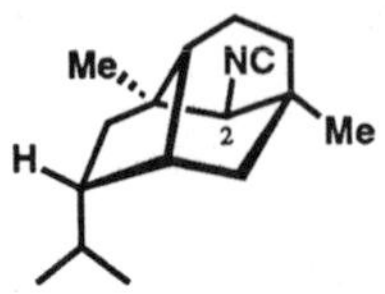

Me
NC
2
Me
H

The co-occurring marine allomones 2- and 9-isocyanopupukeanane have been synthesized from a common intermediate. This condition along with topologically based strategic disconnection had a major impact on the retrosynthetic analysis.

1. Total synthesis of 9-isocyanopupukeanane (Ref. 1):

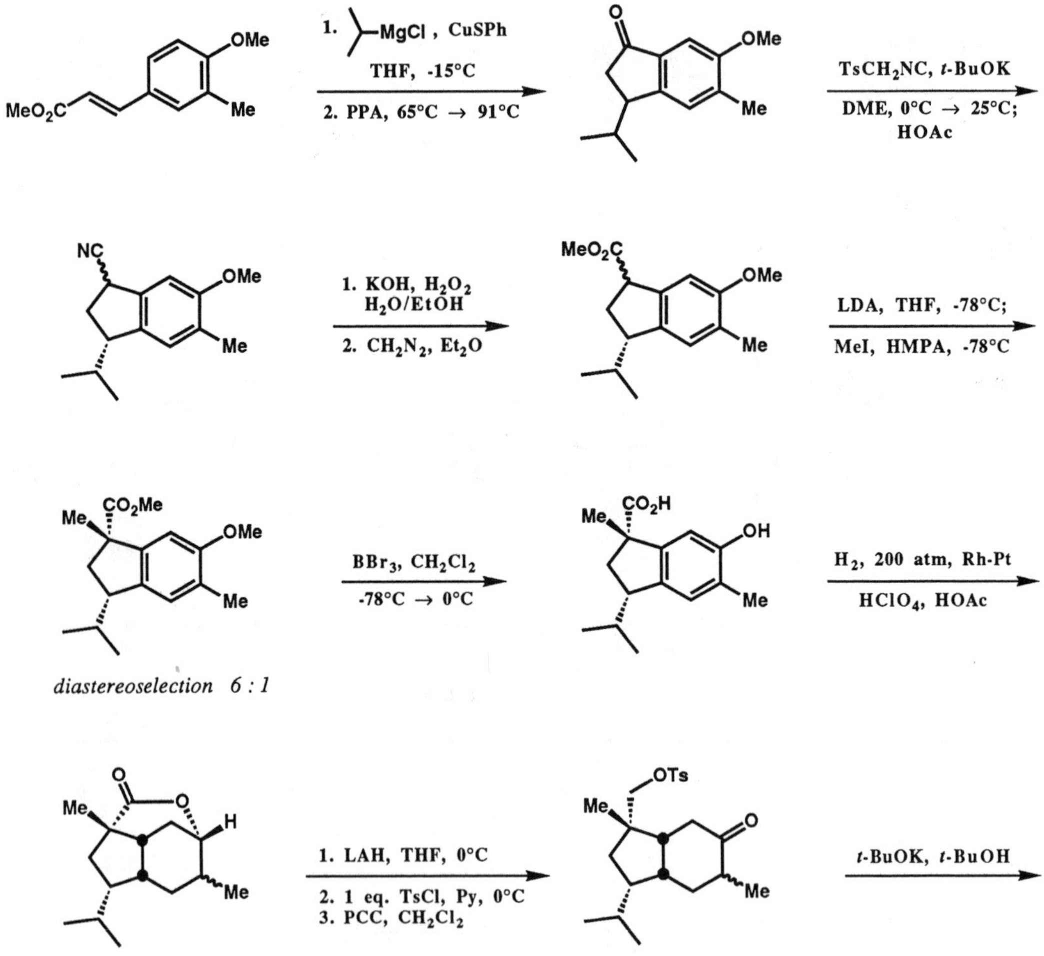

MeO2C
OMe
Me
1. MgCl, CuSPh
THF, -15°C
2. PPA, 65°C → 91°C
O
OMe
Me
TsCH2NC, t-BuOK
DME, 0°C → 25°C;
HOAc
NC
OMe
Me
1. KOH, H2O2
H2O/EtOH
2. CH2N2, Et2O
MeO2C
OMe
Me
LDA, THF, -78°C;
MeI, HMPA, -78°C
Me
CO2Me
OMe
Me
diastereoselection 6 : 1
BBr3, CH2Cl2
-78°C → 0°C
Me
CO2H
OH
Me
H2, 200 atm, Rh-Pt
HClO4, HOAc
O
Me
O
H
Me
1. LAH, THF, 0°C
2. 1 eq. TsCl, Py, 0°C
3. PCC, CH2Cl2
OTs
Me
O
Me
t-BuOK, t-BuOH

A

9.18

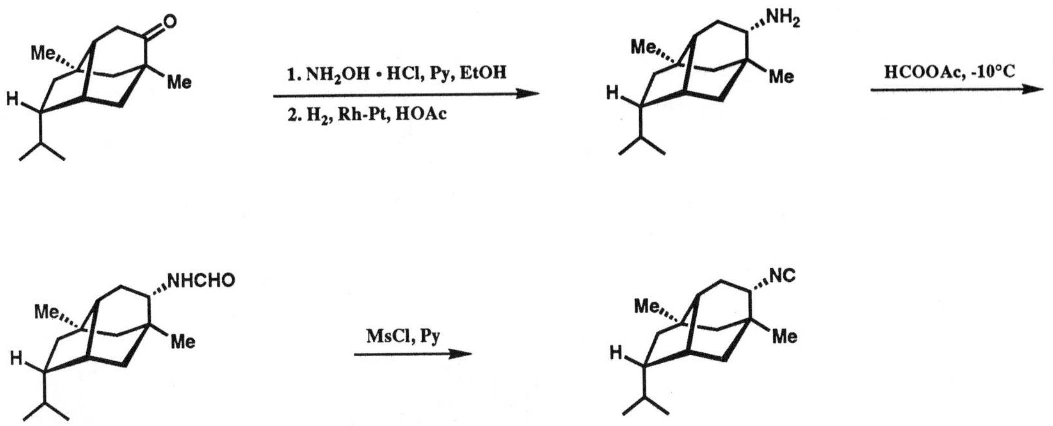
Me
H
O
Me
1. NH2OH • HCl, Py, EtOH
2. H2, Rh-Pt, HOAc
NH2
HCOOAc, -10°C
NHCHO
MsCl, Py
NC

2. Total synthesis of 2-isocyanopupukeanane (Ref. 2):

The total synthesis of 2-isocyanopupukeanane was accomplished starting from the same lactone **A** which was used for the synthesis of 9-isocyanopupukeanane.

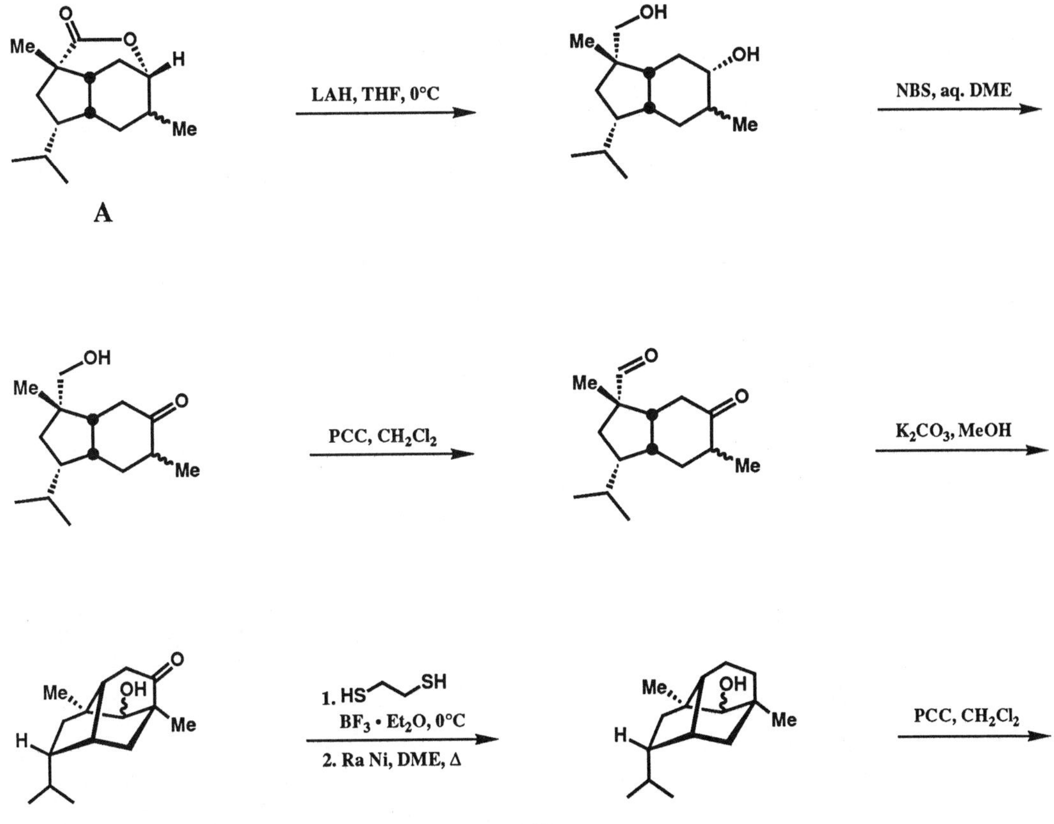
O
Me
H
A
LAH, THF, 0°C
OH
NBS, aq. DME
PCC, CH2Cl2
K2CO3, MeOH
1. HS SH
BF3 • Et2O, 0°C
2. Ra Ni, DME, Δ
PCC, CH2Cl2

9.18

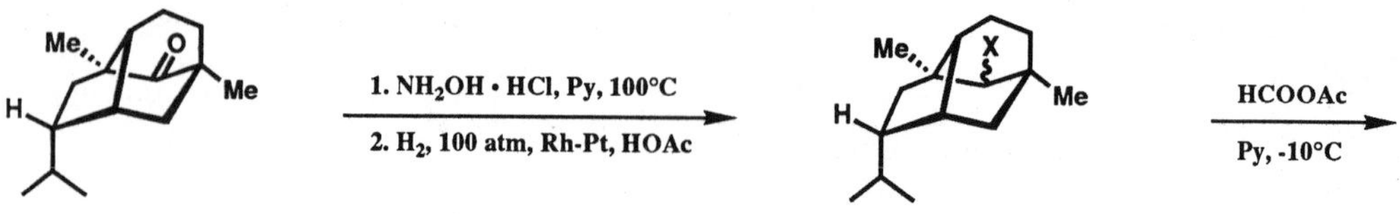
Me
O
Me
H
1. $NH_2OH \cdot HCl$, Py, 100°C
2. H_2, 100 atm, Rh-Pt, HOAc
X
HCOOAc
Py, -10°C

X = NH_2

1 : 1 mixture

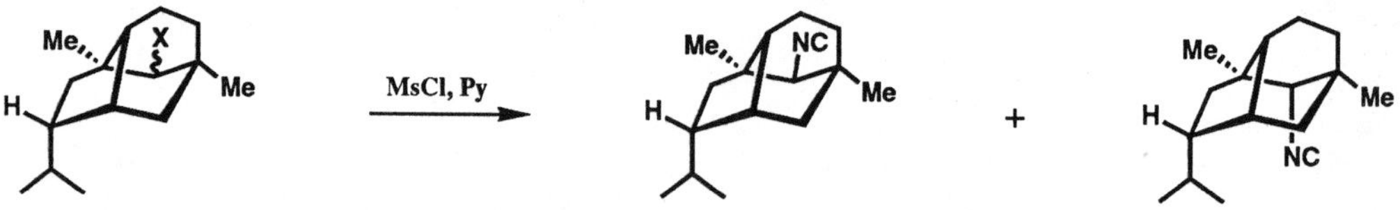
Me
X
Me
H
MsCl, Py
NC
+

X = NHCHO

2-Isocyano-pupukeanane

***epi*-2-Isocyano-pupukeanane**

References:

1. E. J. Corey, M. Behforouz, and M. Ishiguro, *J. Am. Chem. Soc.* **1979**, *101*, 1608.
2. E. J. Corey and M. Ishiguro, *Tetrahedron Lett.* **1979**, 2745.

(*E*)–γ-Bisabolene (*Z*)–γ-Bisabolene

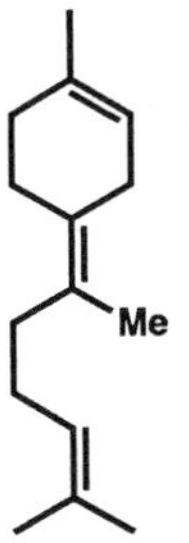
Me

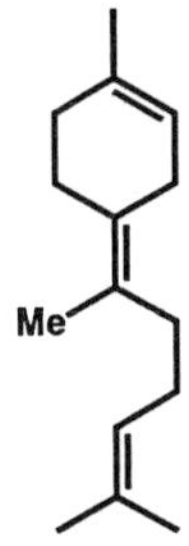
Me

The stereoselective synthesis of γ-bisabolenes was made possible by the development of a new method for the carbosilylation and double alkylation of an acetylenic function coupled with ring closure, overall addition of three carbon substituents to two acetylenic carbons.

1. Total synthesis of (*E*)-γ-bisabolene (Ref. 1):

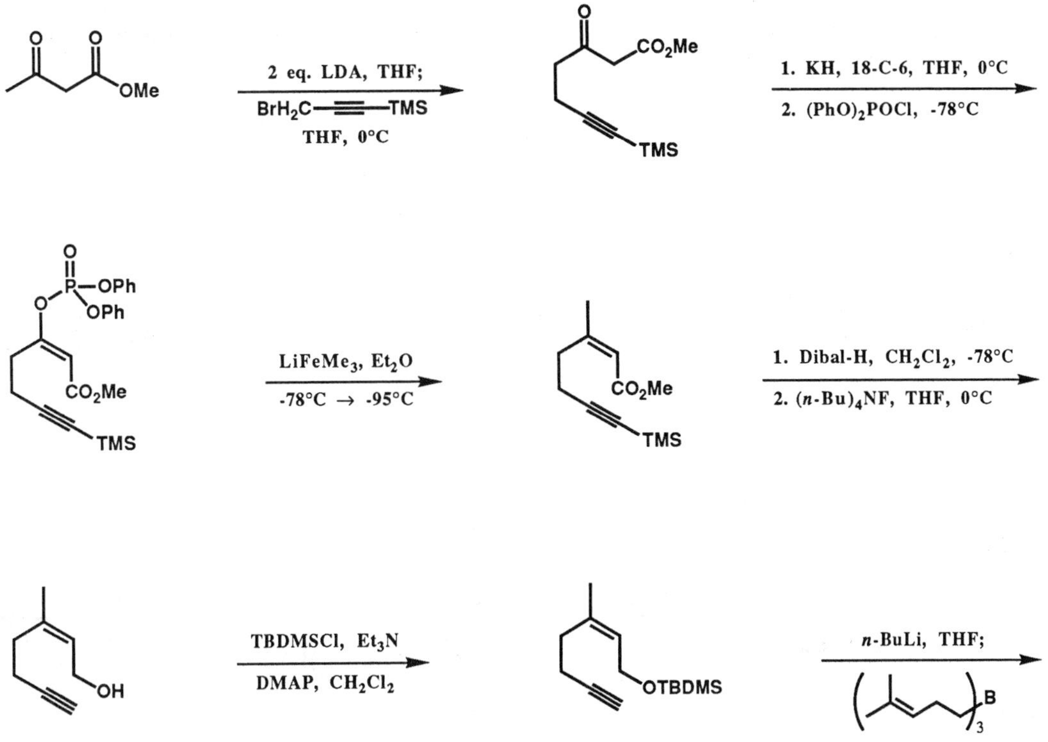
OMe
2 eq. LDA, THF;
BrH2C—≡—TMS
THF, 0°C
CO2Me
TMS
1. KH, 18-C-6, THF, 0°C
2. (PhO)2POCl, -78°C
OPh
OPh
CO2Me
TMS
LiFeMe3, Et2O
-78°C → -95°C
CO2Me
TMS
1. Dibal-H, CH2Cl2, -78°C
2. (n-Bu)4NF, THF, 0°C
OH
TBDMSCl, Et3N
DMAP, CH2Cl2
OTBDMS
n-BuLi, THF;
B
3

A

9.19

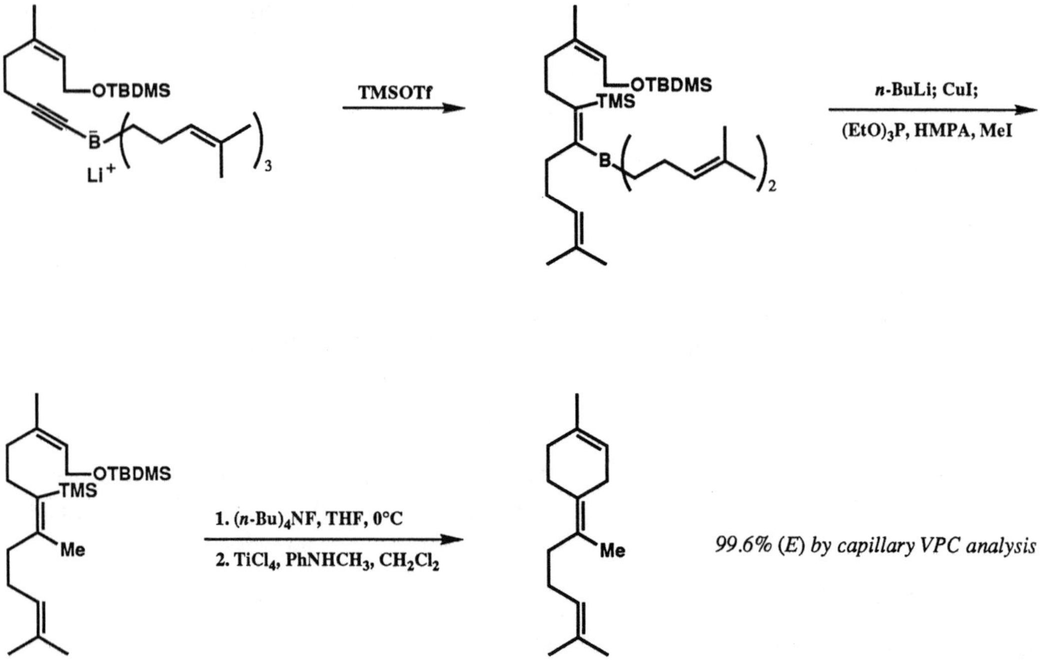
OTBDMS
B̄
Li+
3
TMSOTf
OTBDMS
TMS
B
2
n-BuLi; CuI;
(EtO)3P, HMPA, MeI
OTBDMS
TMS
Me
1. (n-Bu)4NF, THF, 0°C
2. TiCl4, PhNHCH3, CH2Cl2
Me
99.6% (E) by capillary VPC analysis

An alternative route to (*E*)-γ-bisabolene was demonstrated starting from the acetylene **B**, which was made from 5-hexyn-2-one as shown:

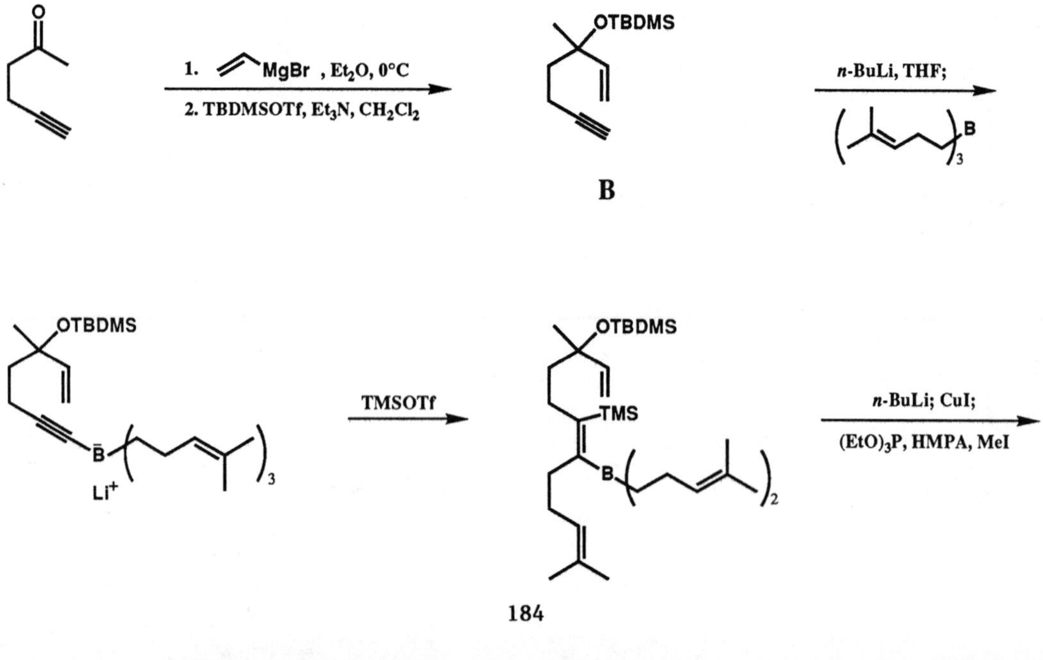
O
1. MgBr , Et2O, 0°C
2. TBDMSOTf, Et3N, CH2Cl2
OTBDMS
B
n-BuLi, THF;
B
3
OTBDMS
B̄
Li+
3
TMSOTf
OTBDMS
TMS
B
2
n-BuLi; CuI;
(EtO)3P, HMPA, MeI

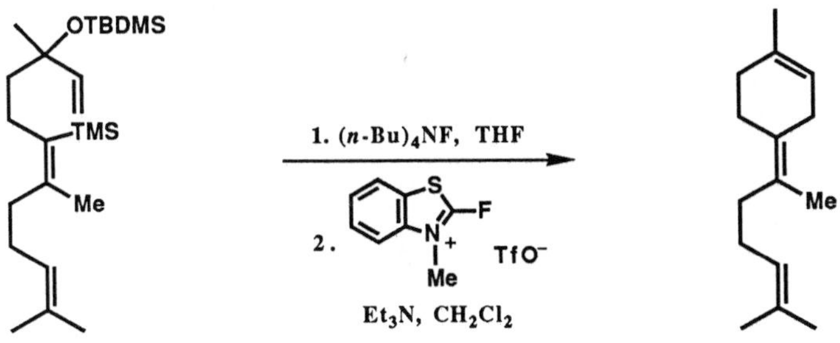
OTBDMS
TMS
Me
1. (n-Bu)$_4$NF, THF
2.
S
F
N+
Me
TfO$^-$
Et$_3$N, CH$_2$Cl$_2$
Me

97.5% (E) by capillary VPC analysis

2. Total synthesis of (Z)-γ-bisabolene (Ref. 2):

The acetylenic alcohol **A**, which was used as a key intermediate in the synthesis of (*E*)-γ-bisabolene, served as the starting point of this synthesis.

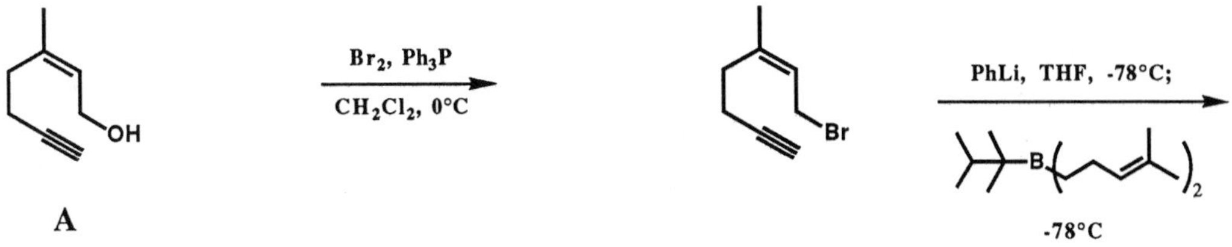
OH
Br$_2$, Ph$_3$P
CH$_2$Cl$_2$, 0°C
Br
PhLi, THF, -78°C;
B
2
-78°C

A

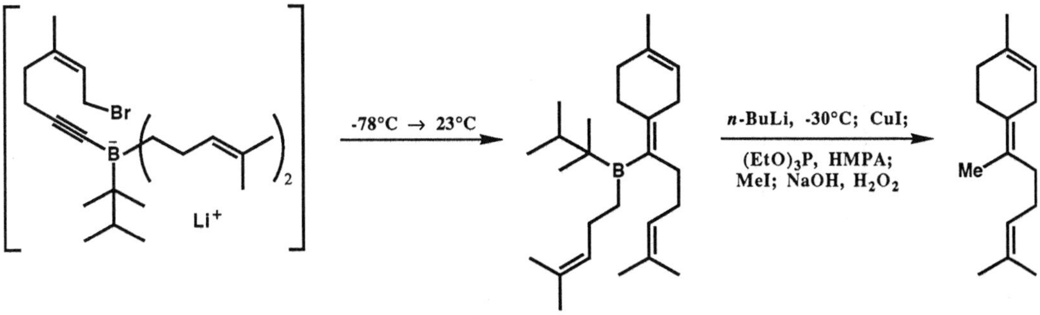
Br
B
2
Li$^+$
-78°C → 23°C
B
n-BuLi, -30°C; CuI;
(EtO)$_3$P, HMPA;
MeI; NaOH, H$_2$O$_2$
Me

79 : 21 mixture of
(Z)- and (E)-
Bisabolenes

References:

1. E. J. Corey and W. L. Seibel, *Tetrahedron Lett.* **1986**, *27*, 905.
2. E. J. Corey and W. L. Seibel, *Tetrahedron Lett.* **1986**, *27*, 909.

CHAPTER TEN

Polycyclic Isoprenoids

10.1

(±)-Aphidicolin

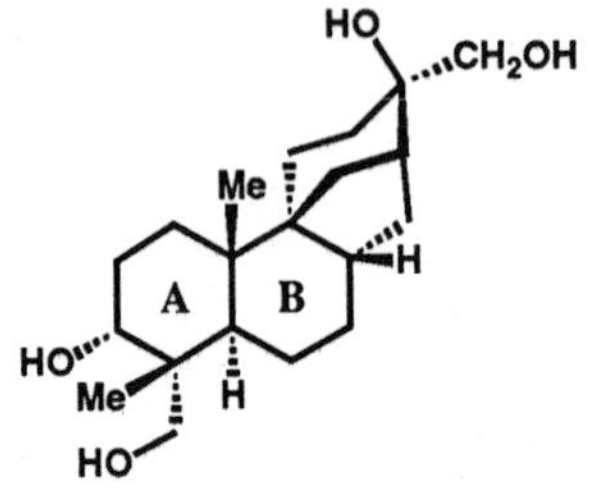

HO
CH2OH
Me
H
A
B
HO
Me
H
HO

Aphidicolin, a potent antiviral and antimitotic agent, possesses an interesting arrangement of fused, spiro and bridged rings. The synthesis shown below followed from the concurrent application of topological, transform-based, stereochemical and FG-based strategies. Noteworthy steps include stereospecific double annulation to form the A/B ring pair, introduction of the CH_2OH group at a highly hindered carbonyl carbon, and position-selective enolate generation.

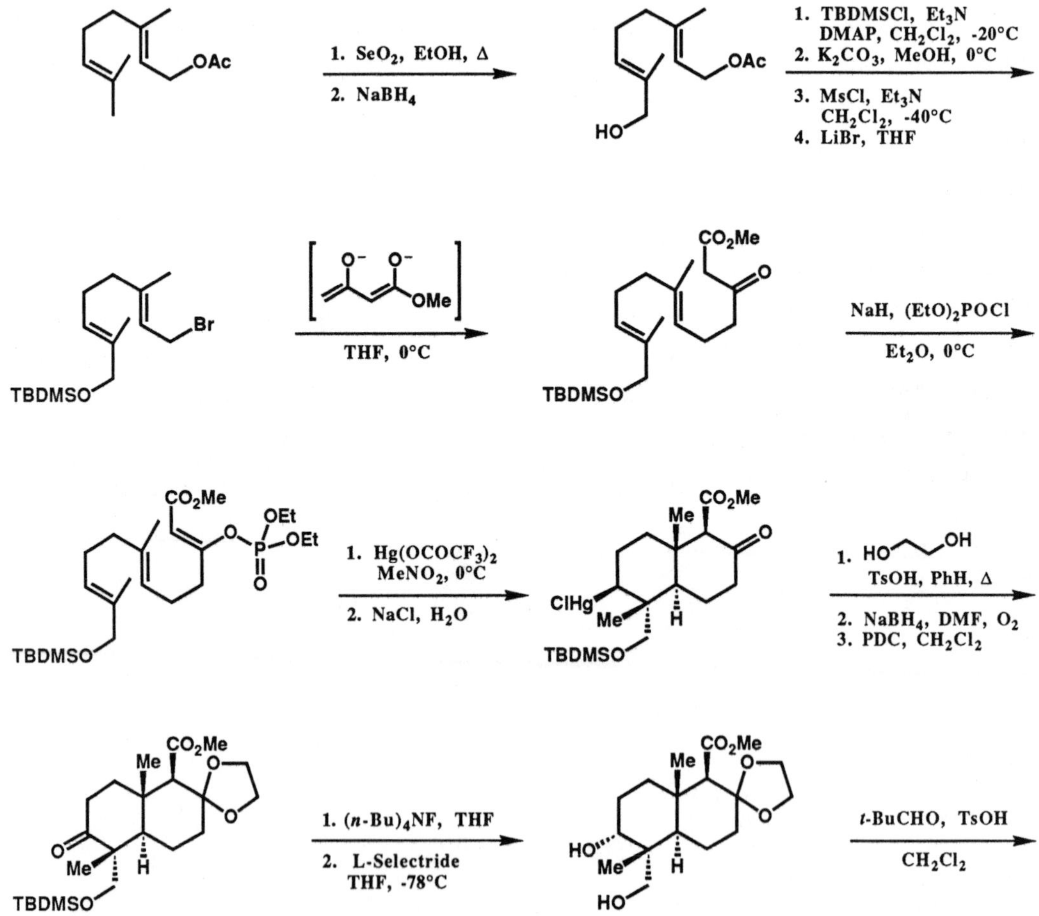

OAc
1. SeO2, EtOH, Δ
2. NaBH4
HO
OAc
1. TBDMSCl, Et3N DMAP, CH2Cl2, -20°C
2. K2CO3, MeOH, 0°C
3. MsCl, Et3N CH2Cl2, -40°C
4. LiBr, THF
TBDMSO
Br
O- O- OMe
THF, 0°C
CO2Me
TBDMSO
NaH, (EtO)2POCl
Et2O, 0°C
CO2Me
OEt
OEt
TBDMSO
1. Hg(OCOCF3)2 MeNO2, 0°C
2. NaCl, H2O
CO2Me
Me
ClHg
Me
H
TBDMSO
1. HO OH TsOH, PhH, Δ
2. NaBH4, DMF, O2
3. PDC, CH2Cl2
CO2Me
Me
Me
H
TBDMSO
1. (n-Bu)4NF, THF
2. L-Selectride THF, -78°C
CO2Me
Me
HO
Me
H
HO
t-BuCHO, TsOH
CH2Cl2

10.1

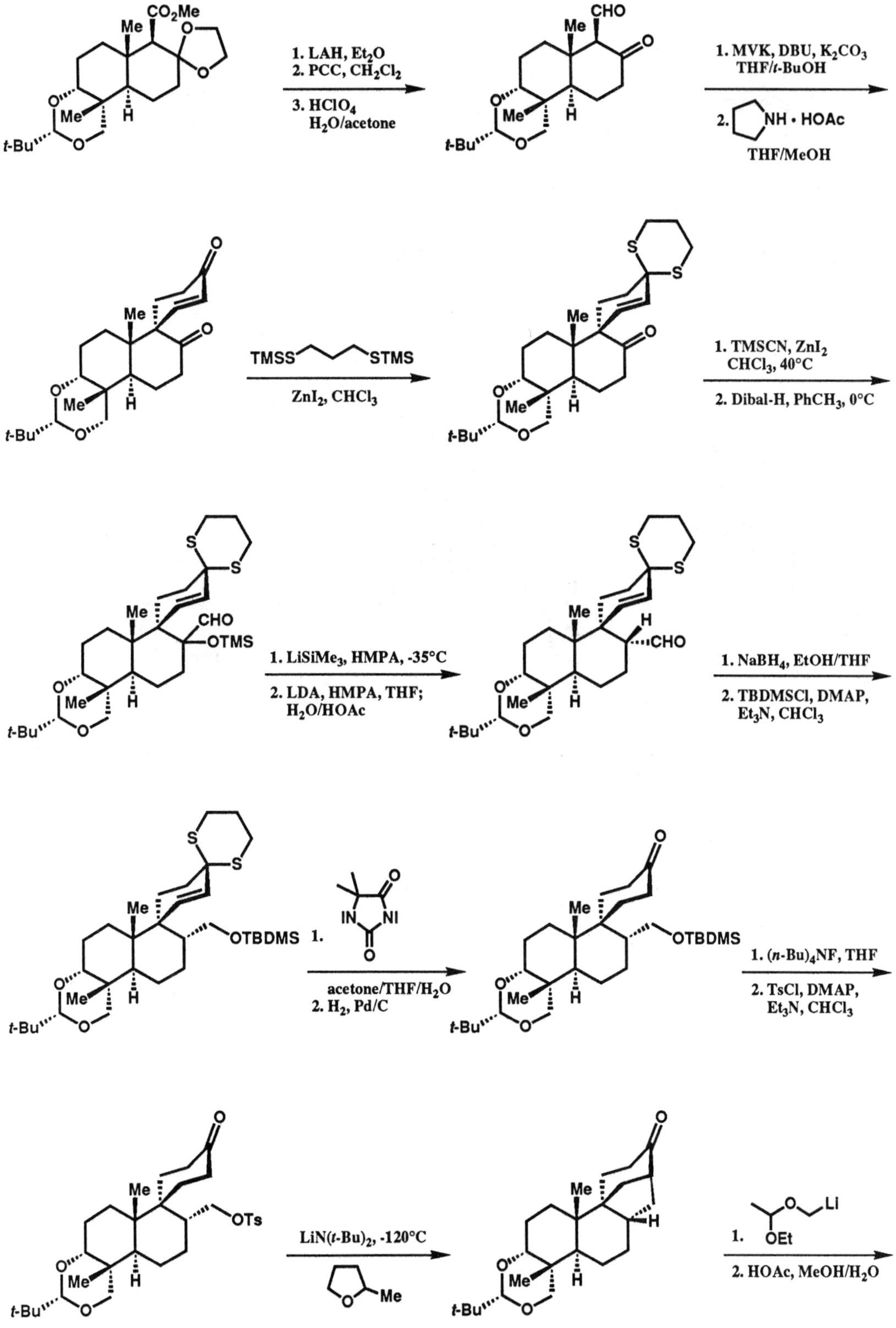

1. LAH, Et2O
2. PCC, CH2Cl2
3. HClO4
H2O/acetone
1. MVK, DBU, K2CO3
THF/t-BuOH
2. NH • HOAc
THF/MeOH
TMSS STMS
ZnI2, CHCl3
1. TMSCN, ZnI2
CHCl3, 40°C
2. Dibal-H, PhCH3, 0°C
1. LiSiMe3, HMPA, -35°C
2. LDA, HMPA, THF;
H2O/HOAc
1. NaBH4, EtOH/THF
2. TBDMSCl, DMAP,
Et3N, CHCl3
acetone/THF/H2O
2. H2, Pd/C
1. (n-Bu)4NF, THF
2. TsCl, DMAP,
Et3N, CHCl3
LiN(t-Bu)2, -120°C
1. OEt
2. HOAc, MeOH/H2O

10.1

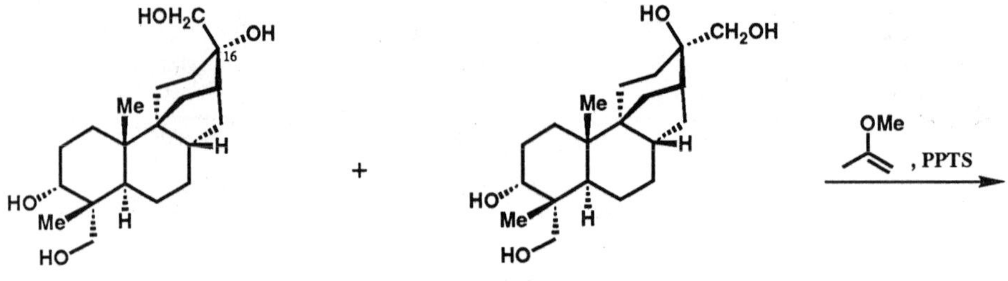
HOH$_2$C
OH
16
Me
H
HO
Me
H
HO
+
HO
CH$_2$OH
Me
H
HO
Me
H
HO
OMe
, PPTS

C$_{16}$-*epi*-(±)-**Aphidicolin** **(±)-Aphidicolin**

1 : 1 mixture, inseparable by chromatography.

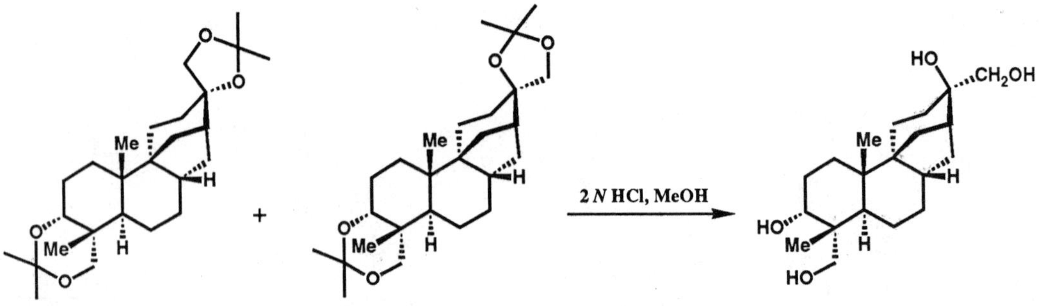
O
O
Me
H
O
Me
H
O
+
O
O
Me
H
O
Me
H
O
2 *N* HCl, MeOH
HO
CH$_2$OH
Me
H
HO
Me
H
HO

removed by chromatography **(±)-Aphidicolin**

Reference:

E. J. Corey, M. A. Tius, and J. Das, *J. Am. Chem. Soc.* **1980,** *102*, 1742.

(±)-Stemodinone (±)-Stemodin

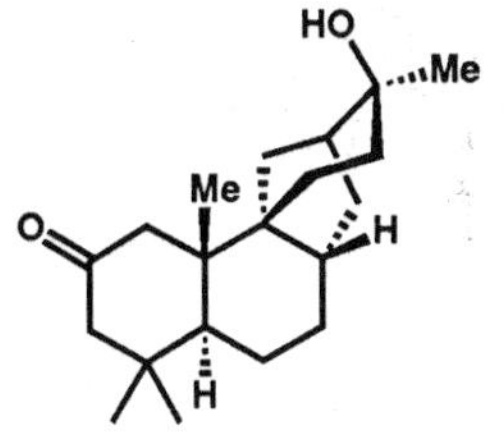
HO
Me
Me
O
H
H

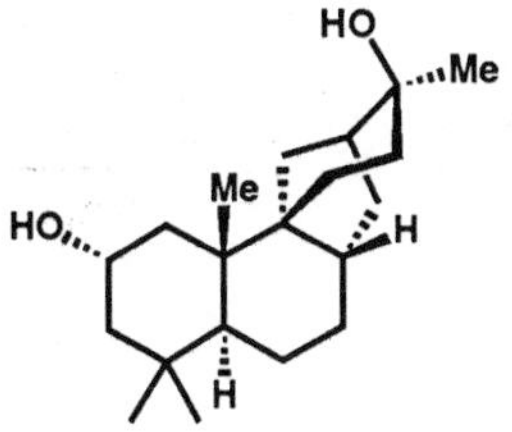
HO
Me
Me
HO
H
H

The syntheses of stemodin and stemodinone, structural relatives of aphidicolin, were accomplished using the A/B double annulation and B/C spiro annulation processes developed for the assembly of aphidicolin.

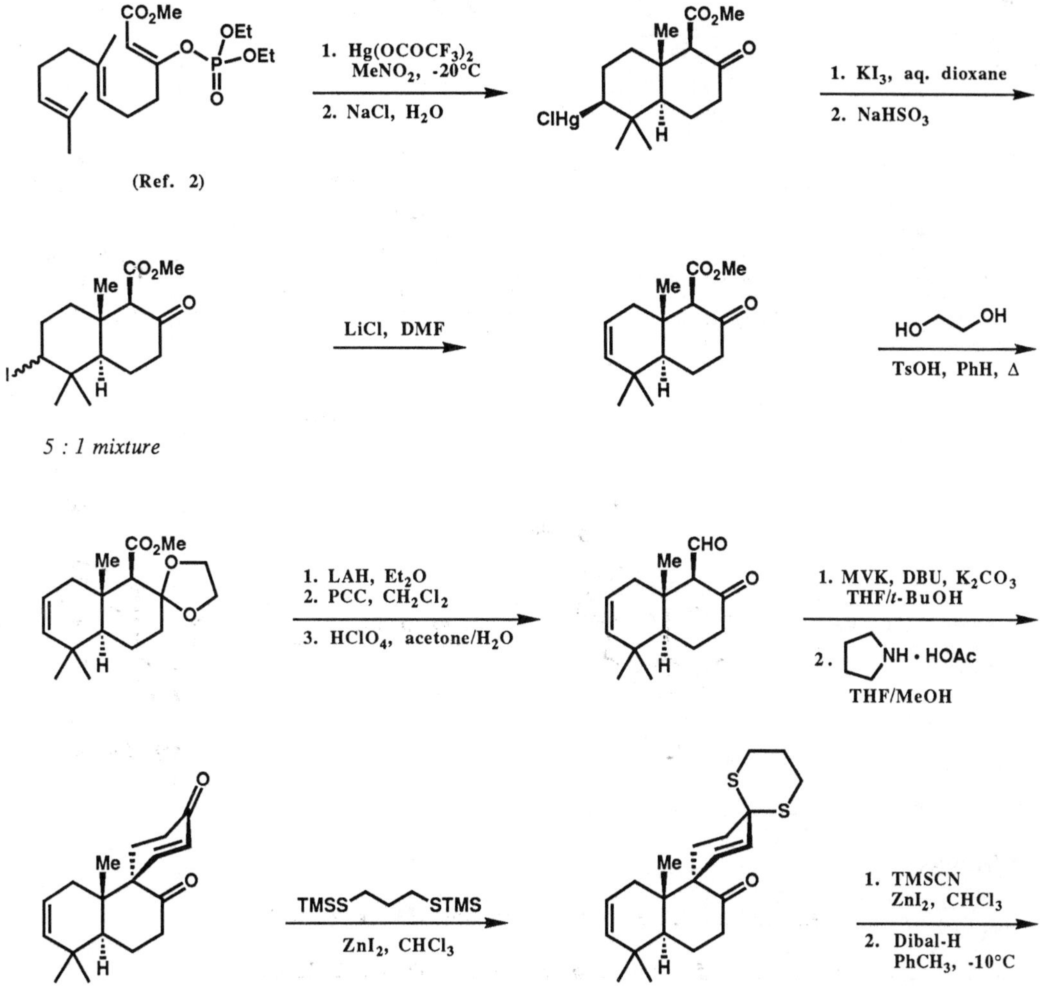
(Ref. 2)
1. Hg(OCOCF3)2 MeNO2, -20°C
2. NaCl, H2O
1. KI3, aq. dioxane
2. NaHSO3
5 : 1 mixture
LiCl, DMF
TsOH, PhH, Δ
1. LAH, Et2O
2. PCC, CH2Cl2
3. HClO4, acetone/H2O
1. MVK, DBU, K2CO3 THF/t-BuOH
NH • HOAc
THF/MeOH
TMSS
STMS
ZnI2, CHCl3
1. TMSCN ZnI2, CHCl3
2. Dibal-H PhCH3, -10°C

10.2

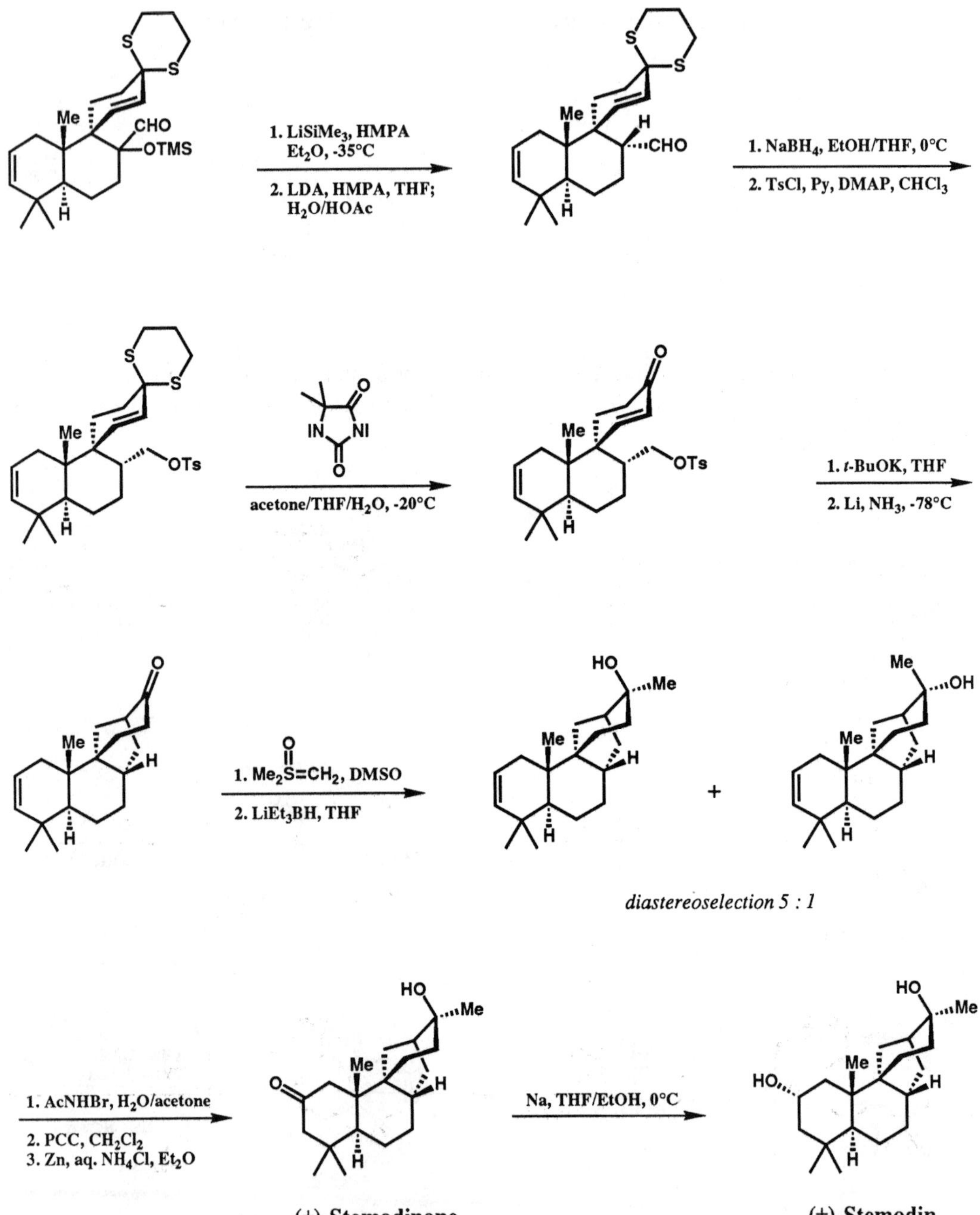
1. LiSiMe3, HMPA Et2O, -35°C
2. LDA, HMPA, THF; H2O/HOAc
1. NaBH4, EtOH/THF, 0°C
2. TsCl, Py, DMAP, CHCl3
acetone/THF/H2O, -20°C
1. t-BuOK, THF
2. Li, NH3, -78°C
1. Me2S=CH2, DMSO
2. LiEt3BH, THF
diastereoselection 5 : 1
1. AcNHBr, H2O/acetone
2. PCC, CH2Cl2
3. Zn, aq. NH4Cl, Et2O
Na, THF/EtOH, 0°C

(±)-Stemodinone **(±)-Stemodin**

References:

1. E. J. Corey, M. A. Tius, and J. Das, *J. Am. Chem. Soc.* **1980**, *102*, 7612.
2. E. J. Corey, M. A. Tius, and J. Das, *J. Am. Chem. Soc.* **1980**, *102*, 1742.

The Complement Inhibitor *K*-76

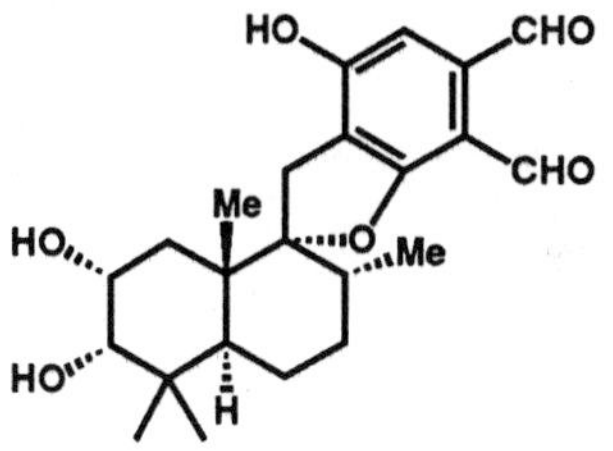

Otsuka K-76, a fungal product with strong anticomplement activity, was synthesized from the A/B bicyclic precursor of stemodin. The aromatic subunit was retrosynthetically disconnected to a symmetrical precursor. A surprising non-selectivity of olefinic hydroxylation by osmium tetroxide was noted.

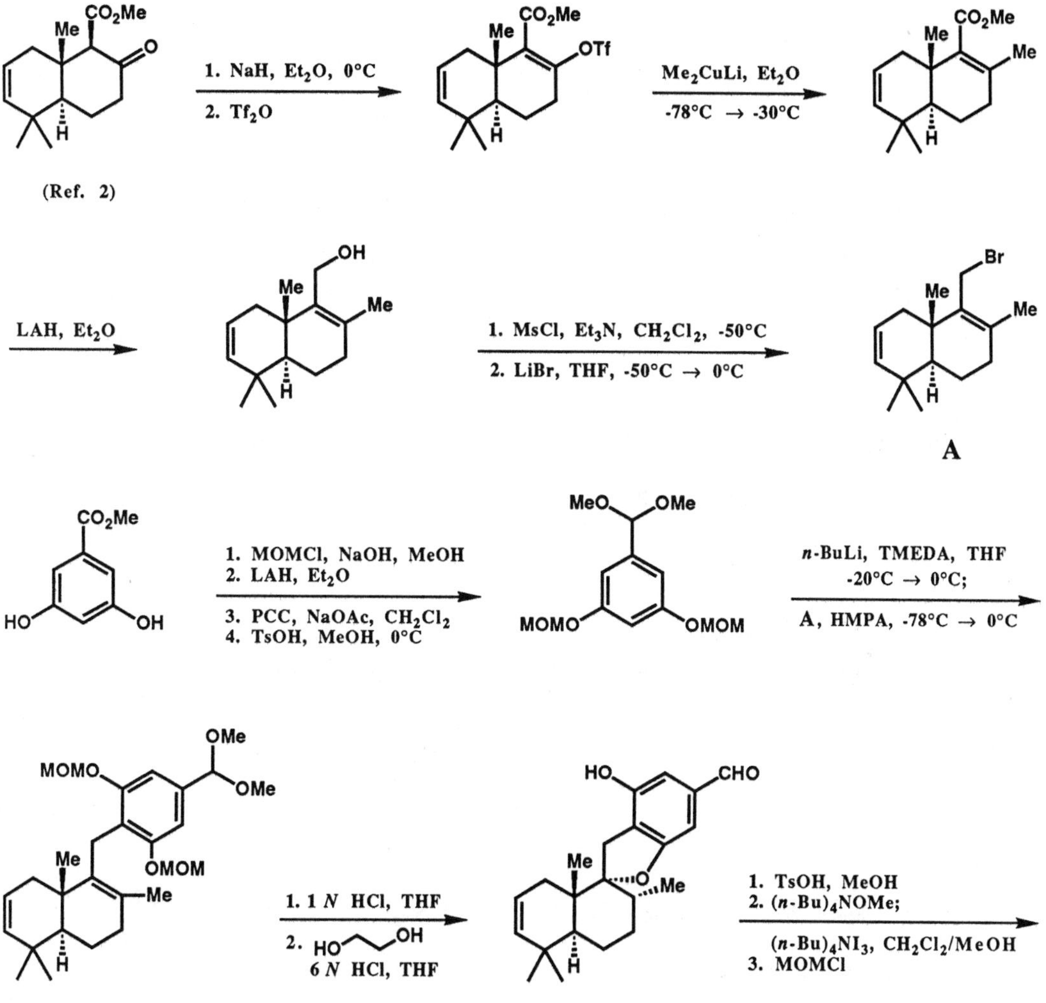

CO2Me
Me
O
1. NaH, Et2O, 0°C
2. Tf2O
OTf
Me2CuLi, Et2O
-78°C → -30°C
H
(Ref. 2)
LAH, Et2O
OH
1. MsCl, Et3N, CH2Cl2, -50°C
2. LiBr, THF, -50°C → 0°C
Br
A
HO
OH
1. MOMCl, NaOH, MeOH
2. LAH, Et2O
3. PCC, NaOAc, CH2Cl2
4. TsOH, MeOH, 0°C
MeO
OMe
MOMO
OMOM
n-BuLi, TMEDA, THF
-20°C → 0°C;
A, HMPA, -78°C → 0°C
1. 1 N HCl, THF
2. HO OH
6 N HCl, THF
CHO
1. TsOH, MeOH
2. (n-Bu)4NOMe;
(n-Bu)4NI3, CH2Cl2/MeOH
3. MOMCl

10.3

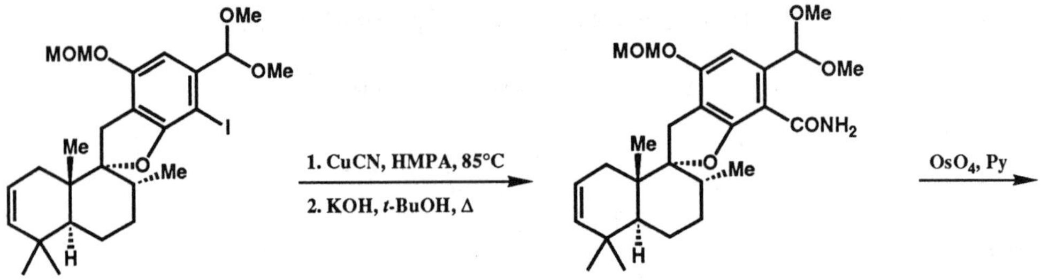
1. CuCN, HMPA, 85°C
2. KOH, t-BuOH, Δ
OsO4, Py

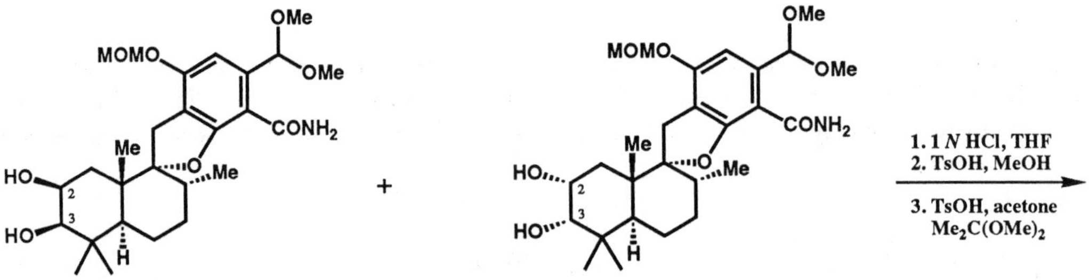
1. 1 N HCl, THF
2. TsOH, MeOH
3. TsOH, acetone
Me2C(OMe)2

1 : 1 mixture of 2β, 3β and 2α, 3α diols which were used to carry out the following reactions.

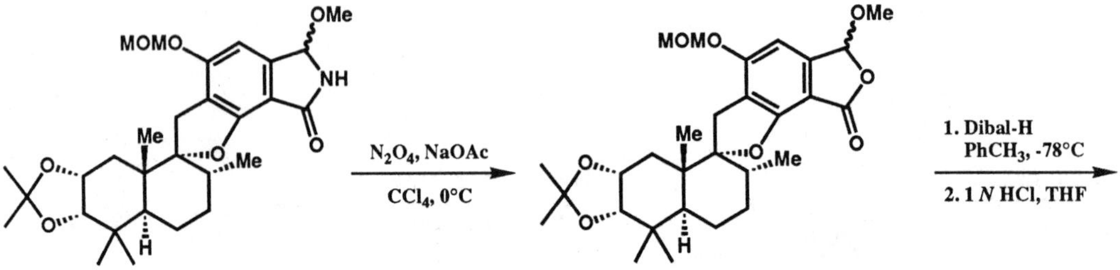
N2O4, NaOAc
CCl4, 0°C
1. Dibal-H
PhCH3, -78°C
2. 1 N HCl, THF

plus the 2β, 3β isomer

The 2β, 3β isomer was separated easily at this stage

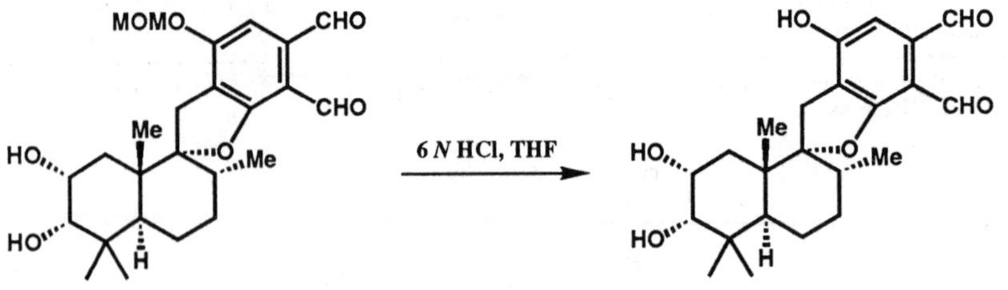
6 N HCl, THF

References:

1. E. J. Corey and J. Das, *J. Am. Chem. Soc.* **1982**, *104*, 5551.
2. E. J. Corey, M. A. Tius, and J. Das, *J. Am. Chem. Soc.* **1980**, *102*, 7612.

(±)-Tricyclohexaprenol

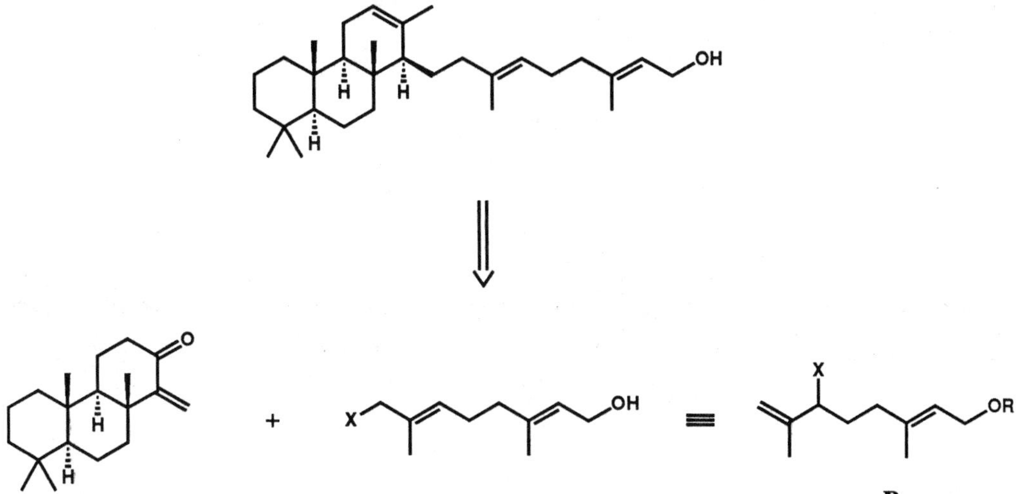
OH
H
H
H
O
H
H
X
OH
X
OR

A

B

Tricyclohexaprenol, a possible forerunner of sterols in the evolution of biomembranes, was synthesized by construction of the cyclic network in one step using cation-olefin tricyclization and subsequent stereocontrolled attachment of the C_{10} appendage to ring C.

1. Synthesis of the fragment **A**:

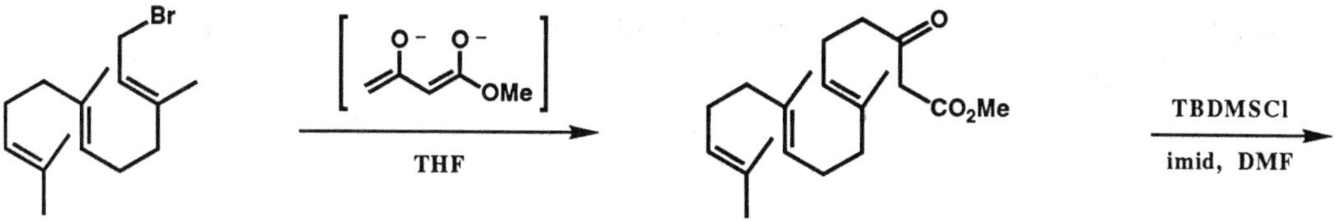
Br
O⁻ O⁻
OMe
THF
O
CO₂Me
TBDMSCl
imid, DMF

(Ref. 2)

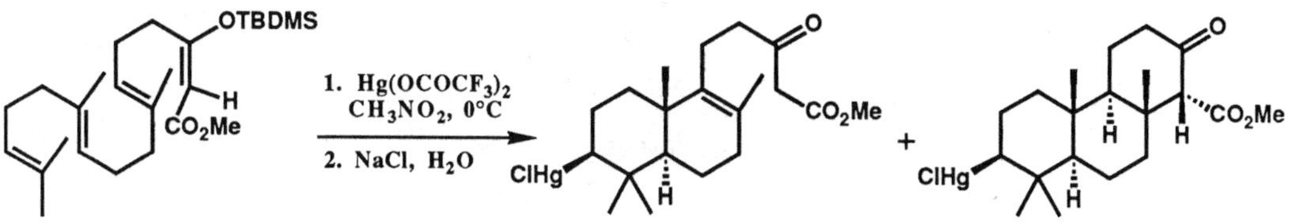
OTBDMS
H
CO₂Me
1. Hg(OCOCF₃)₂
CH₃NO₂, 0°C
2. NaCl, H₂O
O
CO₂Me
ClHg
H
+
O
CO₂Me
H
H
ClHg
H

10.4

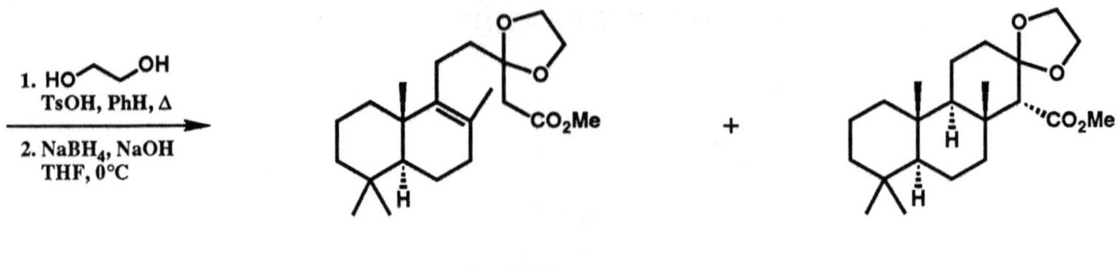
1. HO OH
TsOH, PhH, Δ
2. NaBH4, NaOH
THF, 0°C
CO2Me
+
H
CO2Me
H
H

removed by SGC

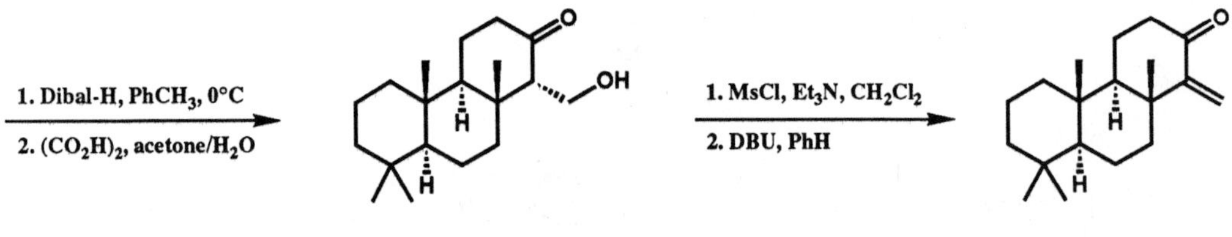
1. Dibal-H, PhCH3, 0°C
2. (CO2H)2, acetone/H2O
OH
H
H
1. MsCl, Et3N, CH2Cl2
2. DBU, PhH
O
H
H

A

2. Synthesis of the fragment **B**:

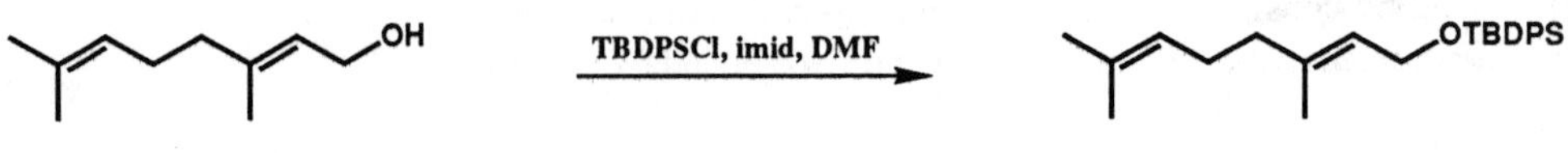
OH
TBDPSCl, imid, DMF
OTBDPS

Geraniol

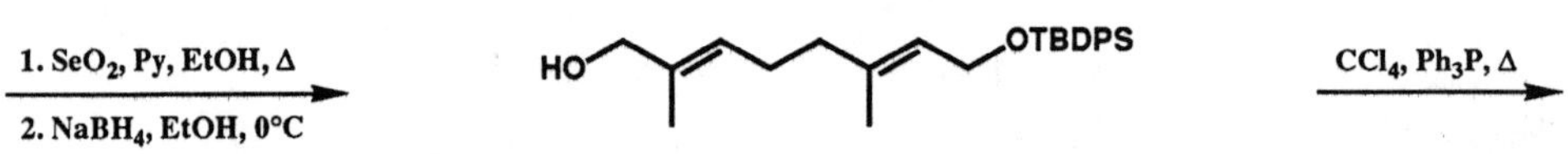
1. SeO2, Py, EtOH, Δ
2. NaBH4, EtOH, 0°C
HO
OTBDPS
CCl4, Ph3P, Δ

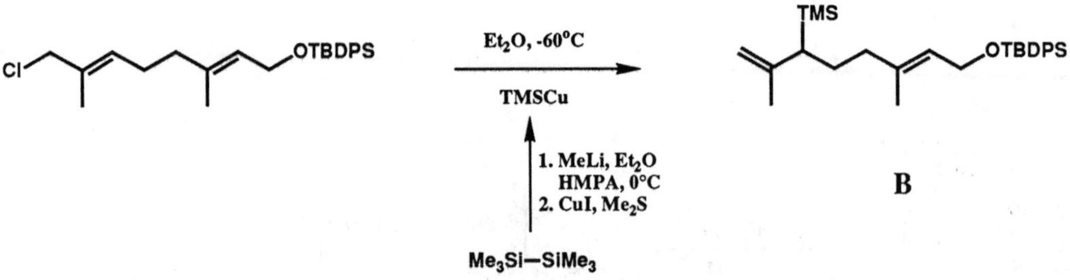
Cl
OTBDPS
Et2O, -60°C
TMSCu
1. MeLi, Et2O
HMPA, 0°C
2. CuI, Me2S
Me3Si—SiMe3
TMS
OTBDPS

B

10.4

3. Coupling of the fragments **A** and **B** and completion of the synthesis:

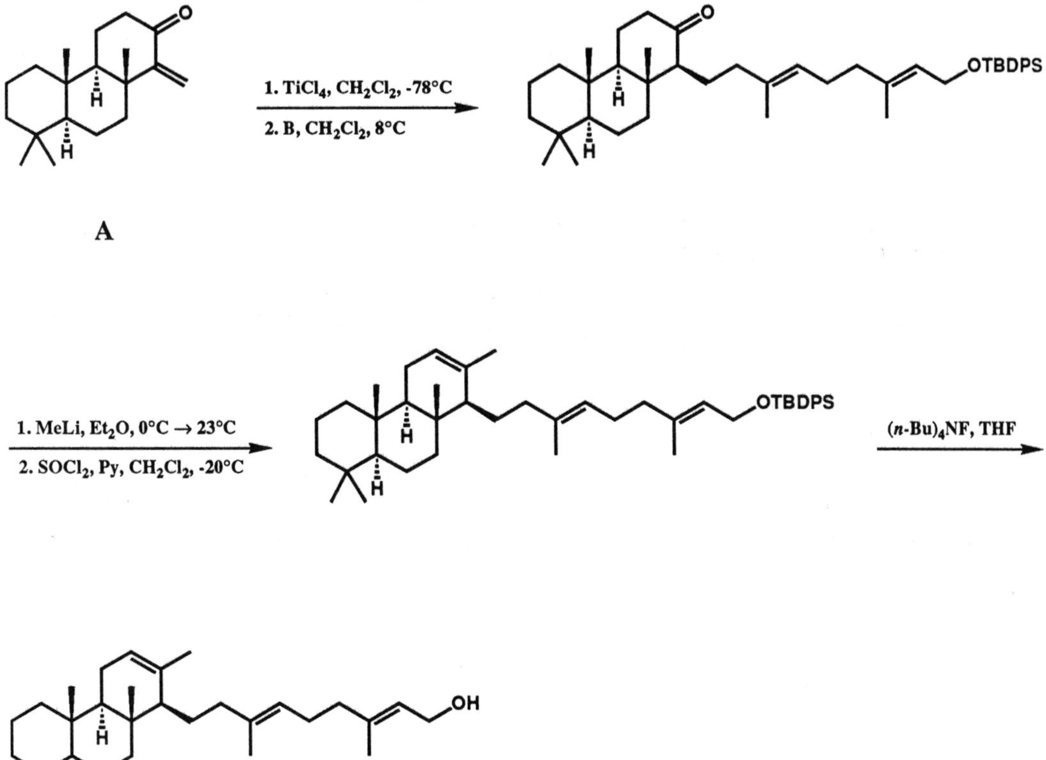
O
H
H
A
1. TiCl4, CH2Cl2, -78°C
2. B, CH2Cl2, 8°C
OTBDPS
1. MeLi, Et2O, 0°C → 23°C
2. SOCl2, Py, CH2Cl2, -20°C
(n-Bu)4NF, THF
OH

References:

1. E. J. Corey and R. M. Burk, *Tetrahedron Lett.* **1987**, *28*, 6413.
2. E. J. Corey, J. G. Reid, A. G. Myers, and R. W. Hahl, *J. Am. Chem. Soc.* **1987**, *109*, 918.

10.5

(±)-Atractyligenin

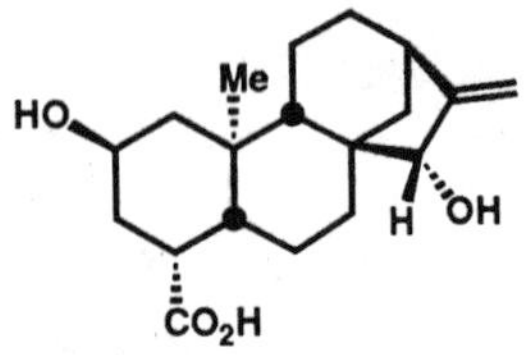
Me
HO
H
OH
CO2H

Atractyligenin and its sulfated glucoside (actractyloside) are toxins which block the transport of ADP into mitochondria and which occur in the coffee bean. Atractyligenin was synthesized following a multistrategic retrosynthetic plan in which the disconnection of ring B was a major objective. Novel stereocontrolled processes were employed for the critical cyclization to form the tetracarbocyclic network and for introduction of the carboxylic substituent.

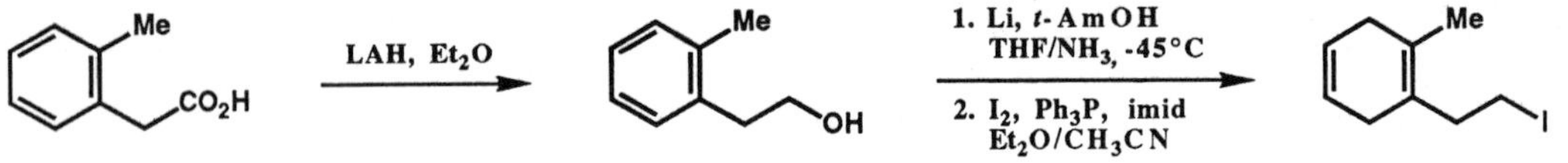
Me
CO2H
LAH, Et2O
Me
OH
1. Li, t-AmOH
THF/NH3, -45°C
2. I2, Ph3P, imid
Et2O/CH3CN
Me
I

A

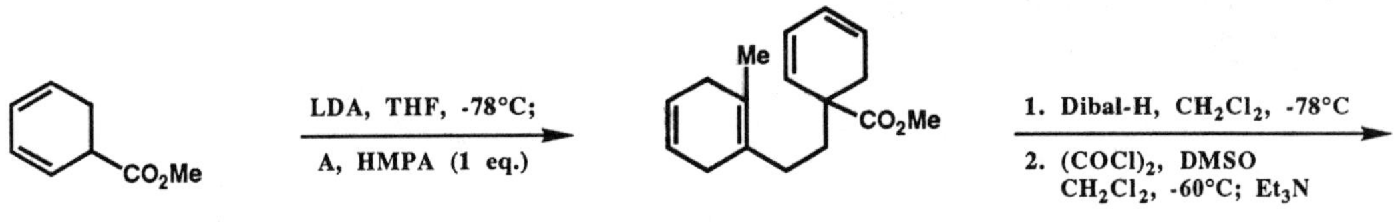
CO2Me
LDA, THF, -78°C;
A, HMPA (1 eq.)
Me
CO2Me
1. Dibal-H, CH2Cl2, -78°C
2. (COCl)2, DMSO
CH2Cl2, -60°C; Et3N

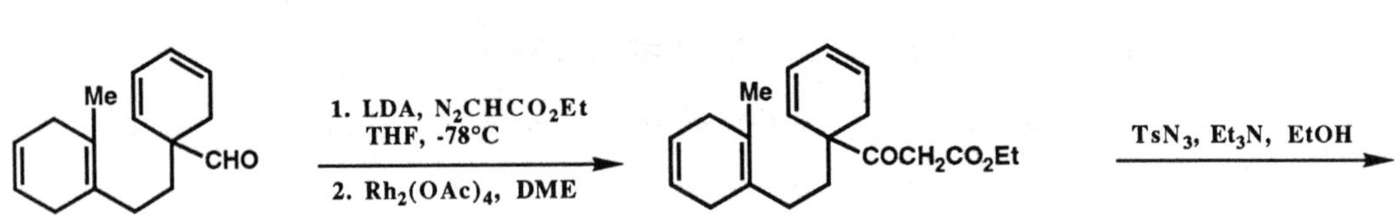
Me
CHO
1. LDA, N2CHCO2Et
THF, -78°C
2. Rh2(OAc)4, DME
Me
COCH2CO2Et
TsN3, Et3N, EtOH

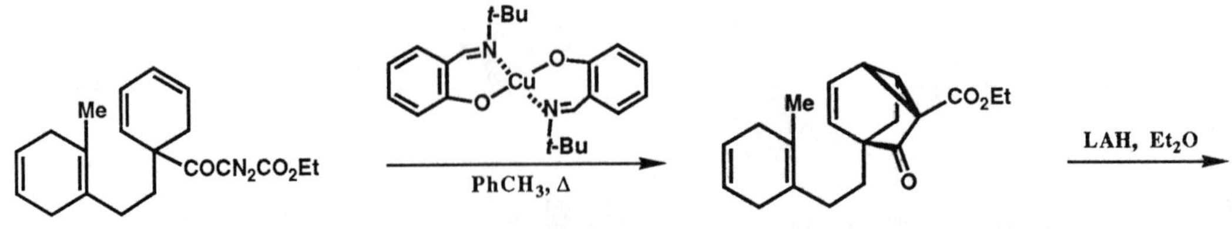
Me
COCN2CO2Et
t-Bu
N
Cu
O
O
N
t-Bu
PhCH3, Δ
Me
CO2Et
O
LAH, Et2O

10.5

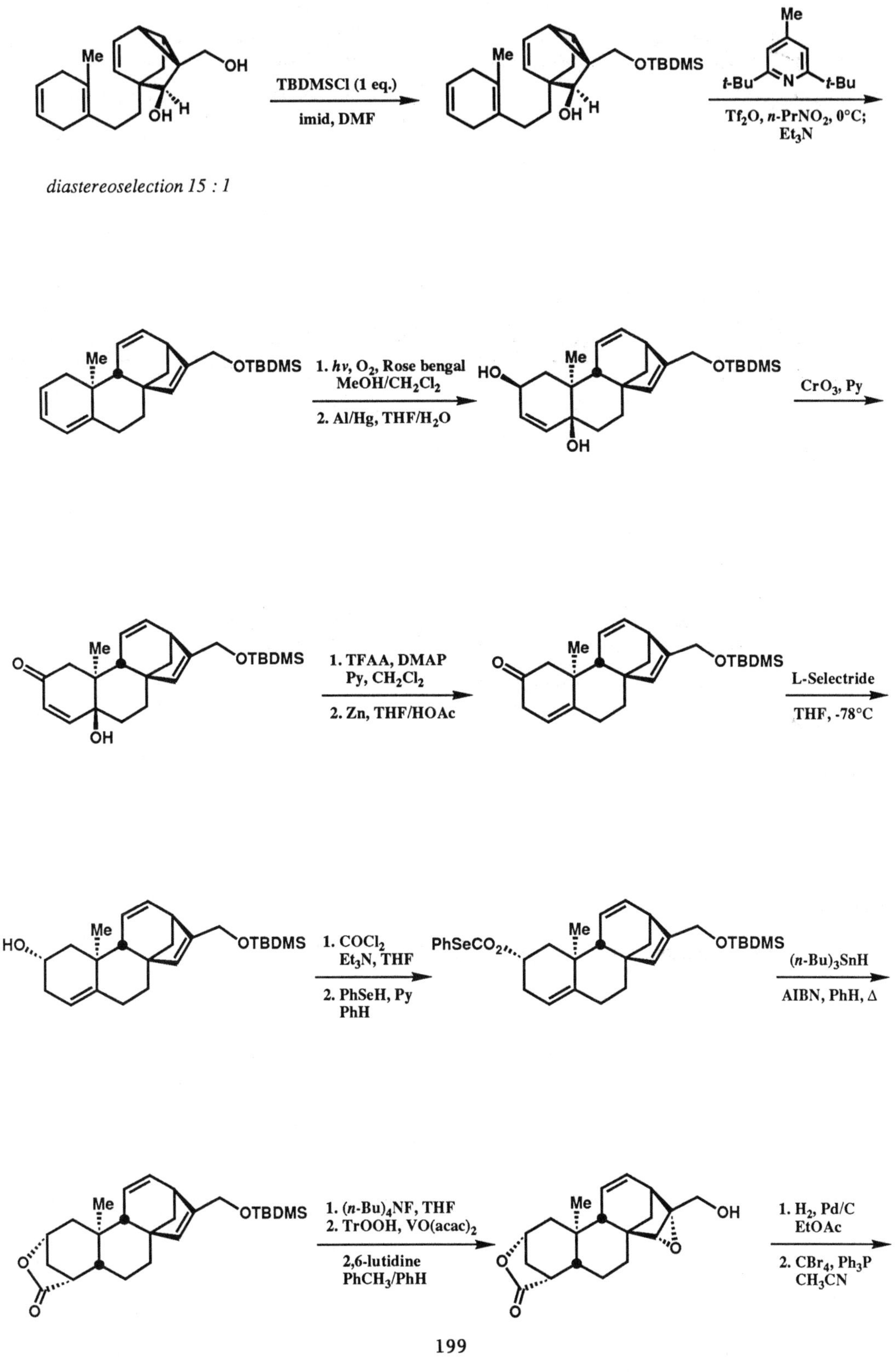
TBDMSCl (1 eq.)
imid, DMF
t-Bu
t-Bu
Tf2O, n-PrNO2, 0°C; Et3N
diastereoselection 15 : 1
1. hν, O2, Rose bengal MeOH/CH2Cl2
2. Al/Hg, THF/H2O
CrO3, Py
1. TFAA, DMAP Py, CH2Cl2
2. Zn, THF/HOAc
L-Selectride
THF, -78°C
1. COCl2 Et3N, THF
2. PhSeH, Py PhH
(n-Bu)3SnH
AIBN, PhH, Δ
1. (n-Bu)4NF, THF
2. TrOOH, VO(acac)2
2,6-lutidine PhCH3/PhH
1. H2, Pd/C EtOAc
2. CBr4, Ph3P CH3CN

10.5

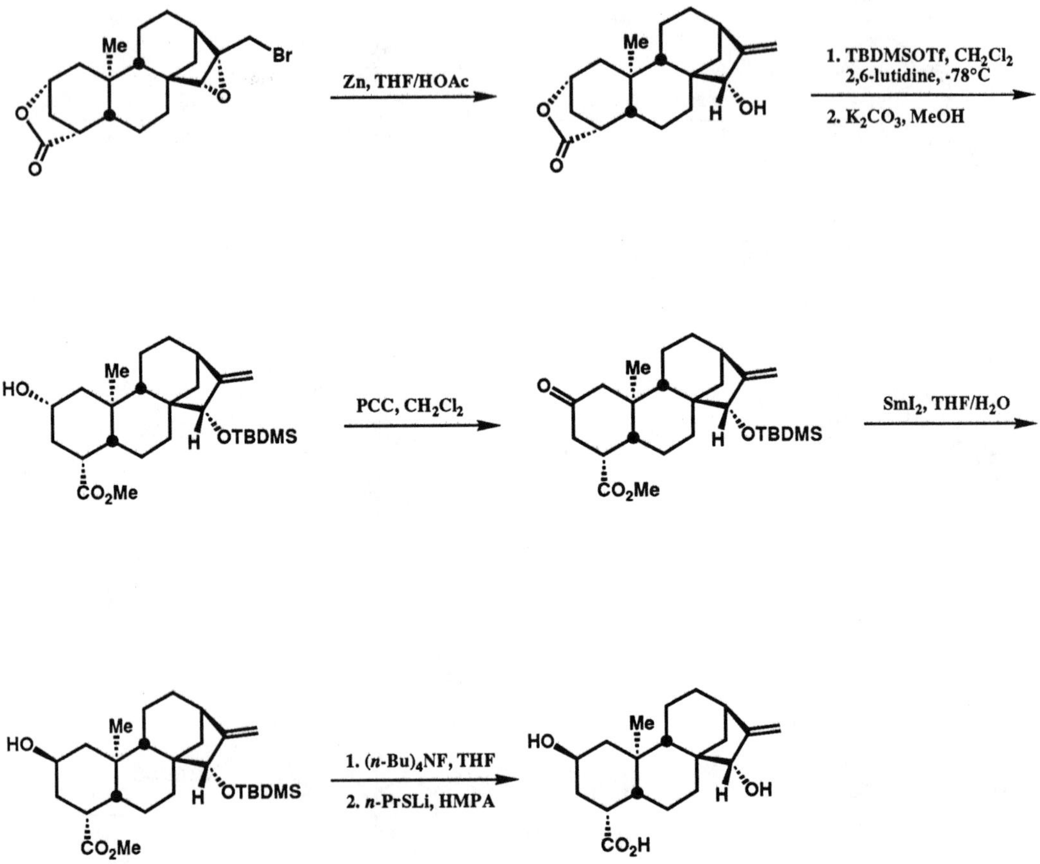
Me
Br
O
O
O
Zn, THF/HOAc
H
OH
1. TBDMSOTf, CH_2Cl_2
2,6-lutidine, -78°C
2. K_2CO_3, MeOH
HO
OTBDMS
CO_2Me
PCC, CH_2Cl_2
SmI_2, THF/H_2O
1. $(n\text{-Bu})_4NF$, THF
2. n-PrSLi, HMPA
CO_2H

diastereoselection 13 : 1

References:

A. K. Singh, R. K. Bakshi, and E. J. Corey, *J. Am. Chem. Soc.* **1987**, *109*, 6187.

(±)-Cafestol

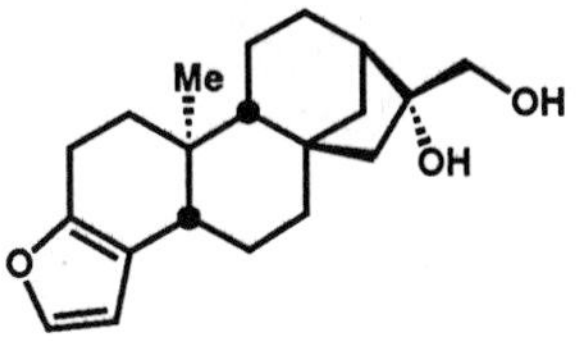

The synthesis of cafestol, an antiinflammatory agent which occurs in coffee beans along with related diterpenoids such as actractyloside and kahweol, was accomplished by the same strategic approach which was applied to its companion atractyligenin.

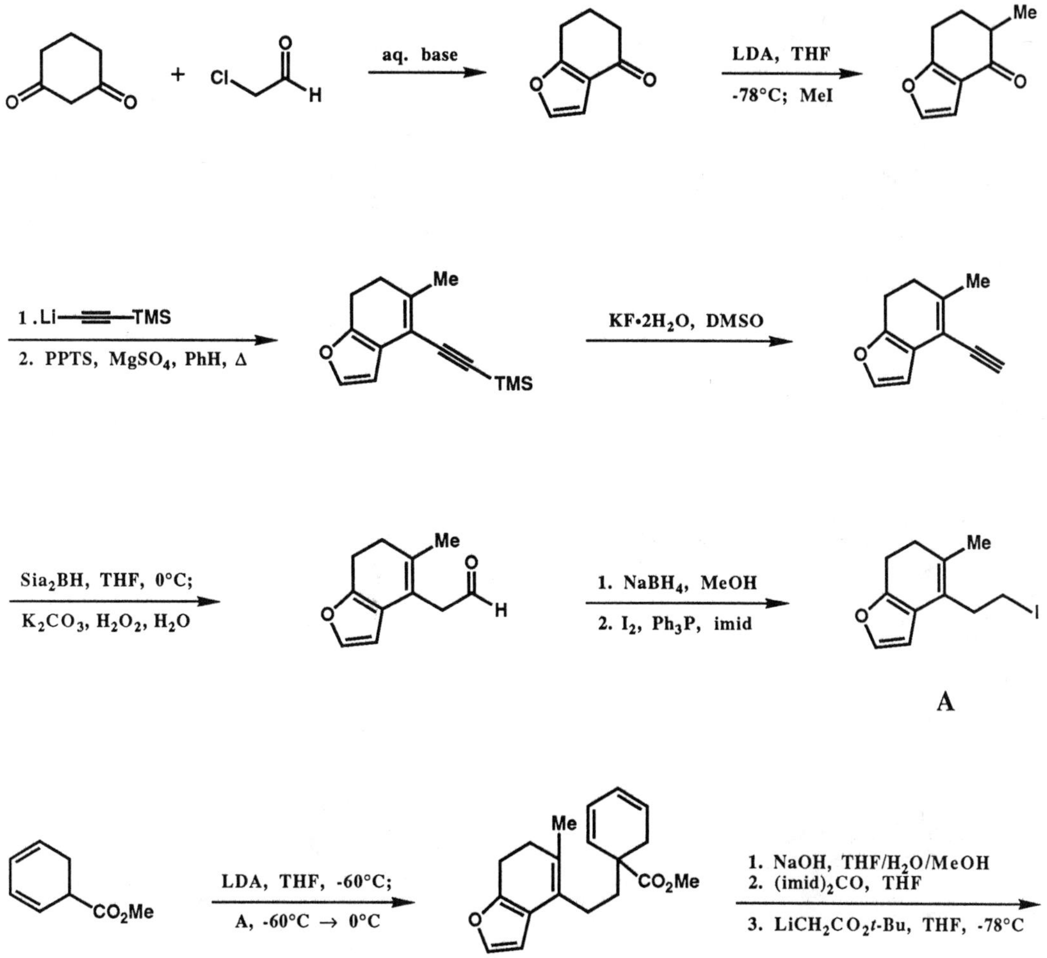
aq. base
LDA, THF
-78°C; MeI
Me
1. Li—≡—TMS
2. PPTS, MgSO4, PhH, Δ
TMS
KF•2H2O, DMSO
Sia2BH, THF, 0°C;
K2CO3, H2O2, H2O
1. NaBH4, MeOH
2. I2, Ph3P, imid
A
CO2Me
LDA, THF, -60°C;
A, -60°C → 0°C
1. NaOH, THF/H2O/MeOH
2. (imid)2CO, THF
3. LiCH2CO2t-Bu, THF, -78°C

10.6

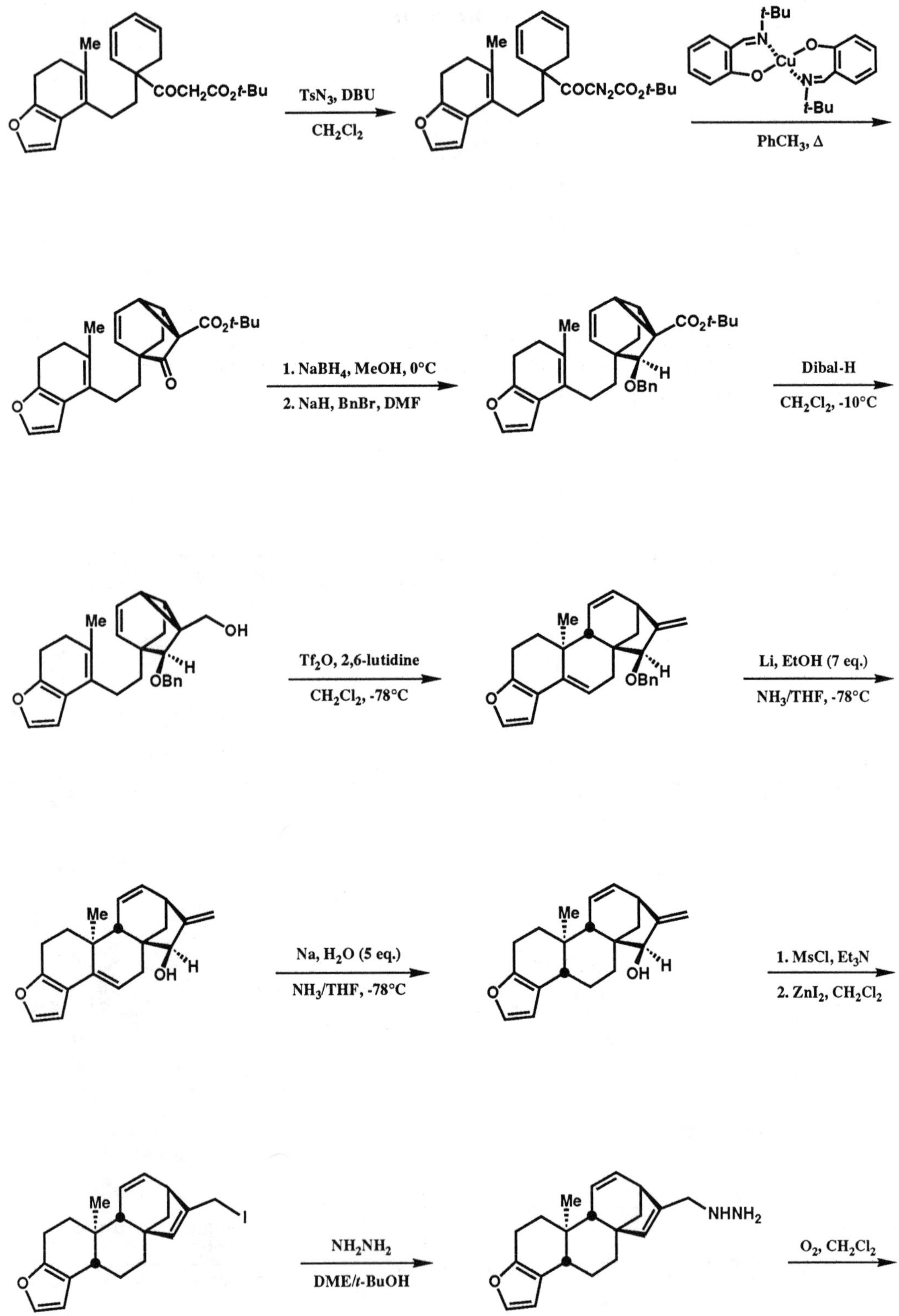
Me
COCH2CO2t-Bu
TsN3, DBU
CH2Cl2
COCN2CO2t-Bu
t-Bu
N
Cu
O
t-Bu
PhCH3, Δ
CO2t-Bu
1. NaBH4, MeOH, 0°C
2. NaH, BnBr, DMF
H
OBn
Dibal-H
CH2Cl2, -10°C
OH
Tf2O, 2,6-lutidine
CH2Cl2, -78°C
Li, EtOH (7 eq.)
NH3/THF, -78°C
Na, H2O (5 eq.)
NH3/THF, -78°C
1. MsCl, Et3N
2. ZnI2, CH2Cl2
I
NH2NH2
DME/t-BuOH
NHNH2
O2, CH2Cl2

10.6

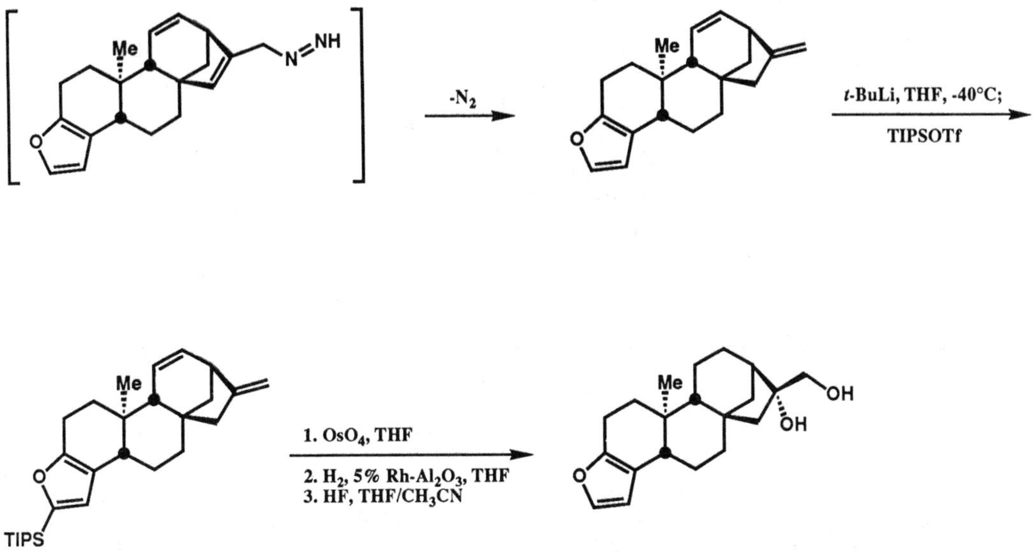

Me
N=NH
-N₂
Me
t-BuLi, THF, -40°C;
TIPSOTf
Me
TIPS
1. OsO₄, THF
2. H₂, 5% Rh-Al₂O₃, THF
3. HF, THF/CH₃CN
Me
OH
OH

Reference:

E. J. Corey, G. Wess, Y. B. Xiang, and A. K. Singh, *J. Am. Chem. Soc.* **1987**, *109*, 4717.

10.7

Kahweol

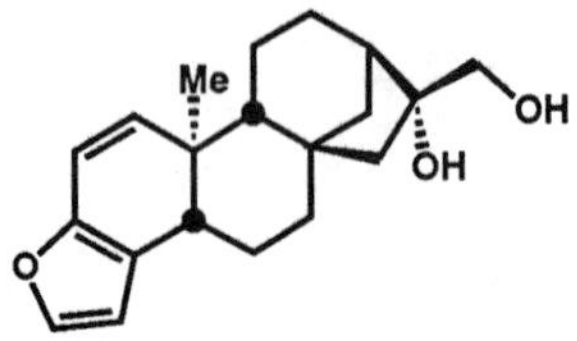
Me
OH
OH
O

Kahweol, a "coffee" diterpenoid, was synthesized from the co-occurring natural product cafestol.

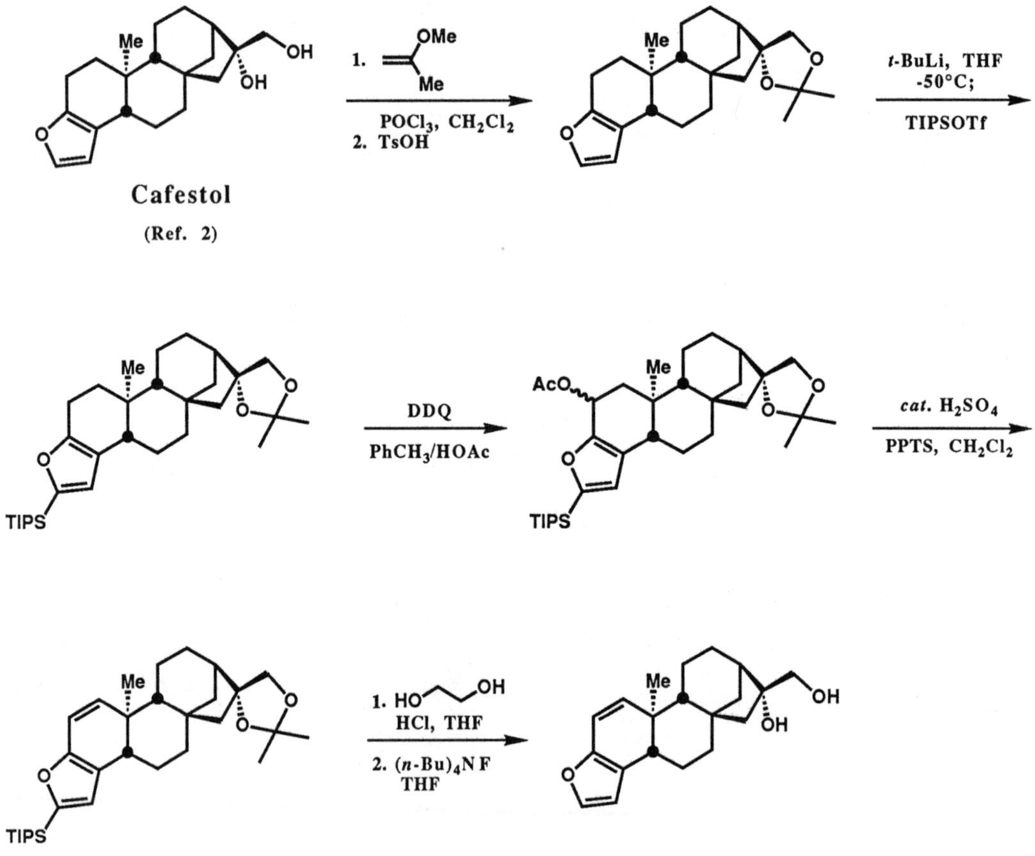
Cafestol
(Ref. 2)
1. OMe Me
POCl3, CH2Cl2
2. TsOH
t-BuLi, THF
-50°C;
TIPSOTf
DDQ
PhCH3/HOAc
AcO
TIPS
cat. H2SO4
PPTS, CH2Cl2
1. HO OH
HCl, THF
2. (n-Bu)4NF
THF
Me
OH
O

References:

1. E. J. Corey and Y. B. Xiang, *Tetrahedron Lett.* **1987**, *28*, 5403.

2. E. J. Corey, G. Wess, Y. B. Xiang, and A. K. Singh, *J. Am. Chem. Soc.* **1987**, *109*, 4717.

Gibberellic Acid

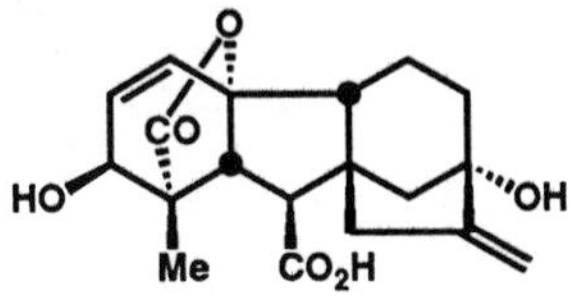

O
CO
HO
OH
Me
CO_2H

The plant growth regulator gibberellic acid was synthesized along the lines of the plan discussed in Section 6.4 of Part One.

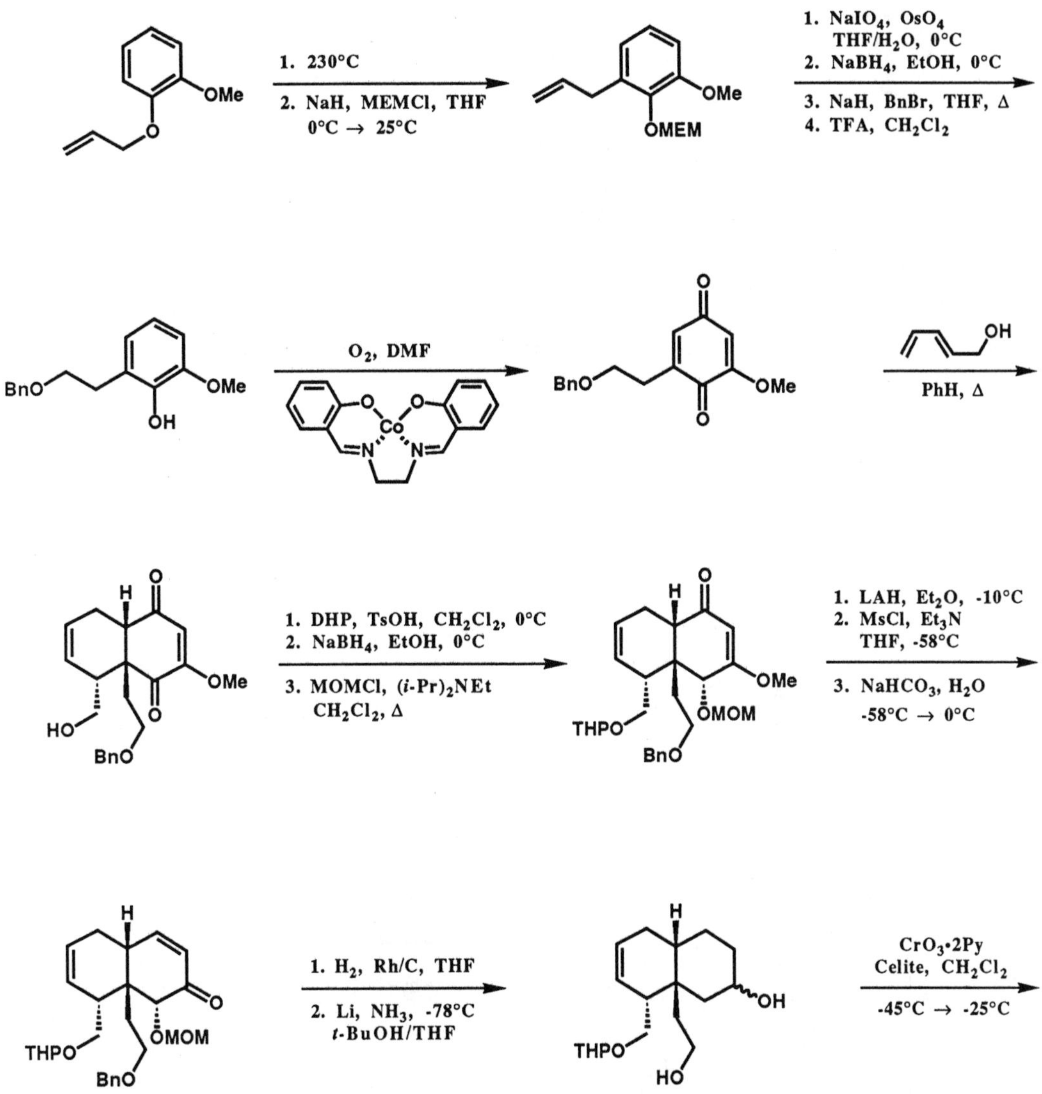

OMe
O
1. 230°C
2. NaH, MEMCl, THF
0°C → 25°C
OMe
OMEM
1. $NaIO_4$, OsO_4
THF/H_2O, 0°C
2. $NaBH_4$, EtOH, 0°C
3. NaH, BnBr, THF, Δ
4. TFA, CH_2Cl_2
BnO
OMe
OH
O_2, DMF
Co
BnO
OMe
OH
PhH, Δ
H
HO
BnO
OMe
1. DHP, TsOH, CH_2Cl_2, 0°C
2. $NaBH_4$, EtOH, 0°C
3. MOMCl, $(i\text{-Pr})_2NEt$
CH_2Cl_2, Δ
THPO
BnO
OMOM
OMe
1. LAH, Et_2O, -10°C
2. MsCl, Et_3N
THF, -58°C
3. $NaHCO_3$, H_2O
-58°C → 0°C
THPO
BnO
OMOM
1. H_2, Rh/C, THF
2. Li, NH_3, -78°C
t-BuOH/THF
THPO
HO
OH
CrO_3•2Py
Celite, CH_2Cl_2
-45°C → -25°C

10.8

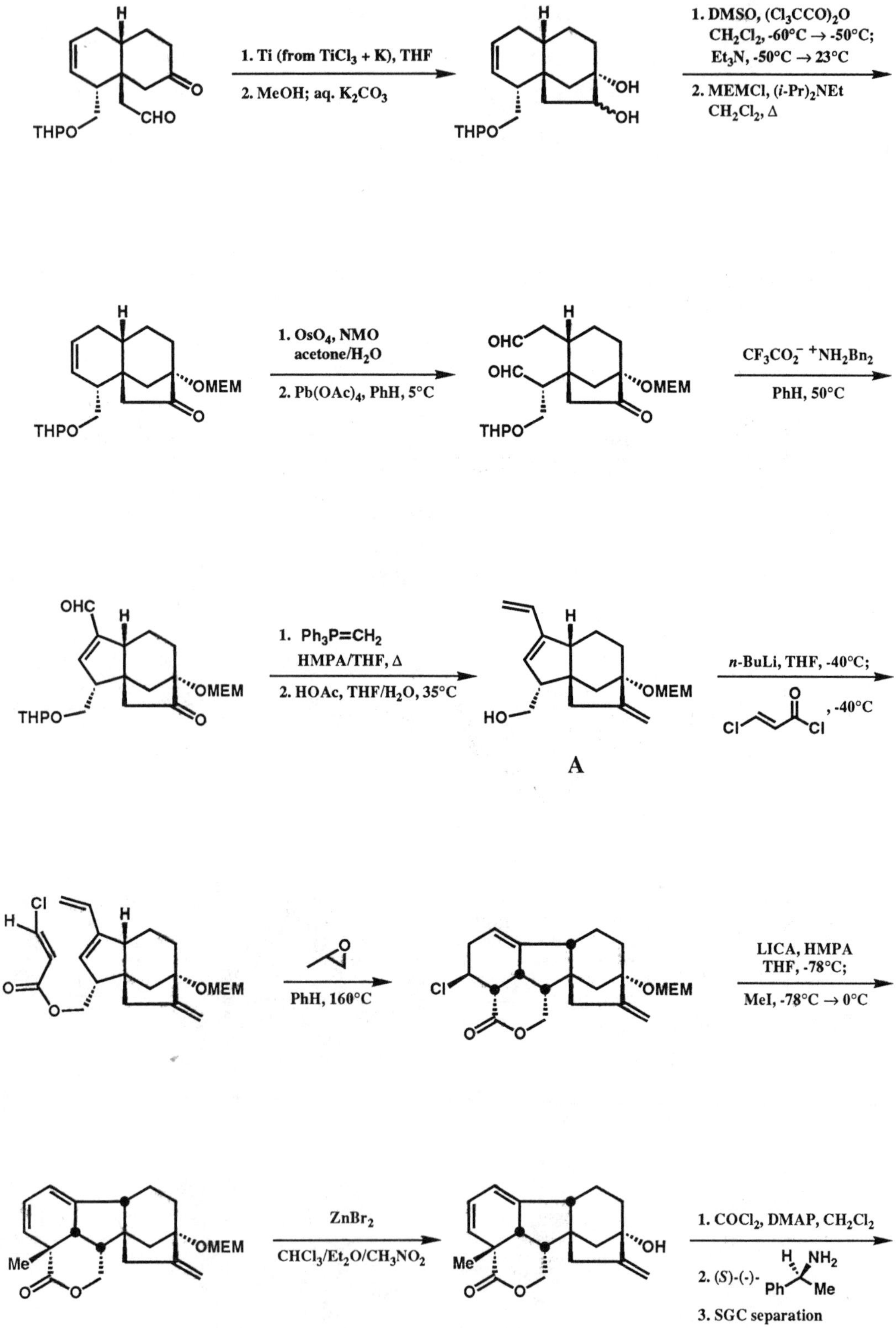
H
THPO
CHO
O
1. Ti (from TiCl3 + K), THF
2. MeOH; aq. K2CO3
OH
OH
1. DMSO, (Cl3CCO)2O
CH2Cl2, -60°C → -50°C;
Et3N, -50°C → 23°C
2. MEMCl, (i-Pr)2NEt
CH2Cl2, Δ
OMEM
1. OsO4, NMO
acetone/H2O
2. Pb(OAc)4, PhH, 5°C
OHC
OHC
CF3CO2− +NH2Bn2
PhH, 50°C
1. Ph3P=CH2
HMPA/THF, Δ
2. HOAc, THF/H2O, 35°C
HO
A
n-BuLi, THF, -40°C;
Cl
, -40°C
PhH, 160°C
LICA, HMPA
THF, -78°C;
MeI, -78°C → 0°C
Me
ZnBr2
CHCl3/Et2O/CH3NO2
1. COCl2, DMAP, CH2Cl2
2. (S)-(-)-
NH2
Ph
3. SGC separation

10.8

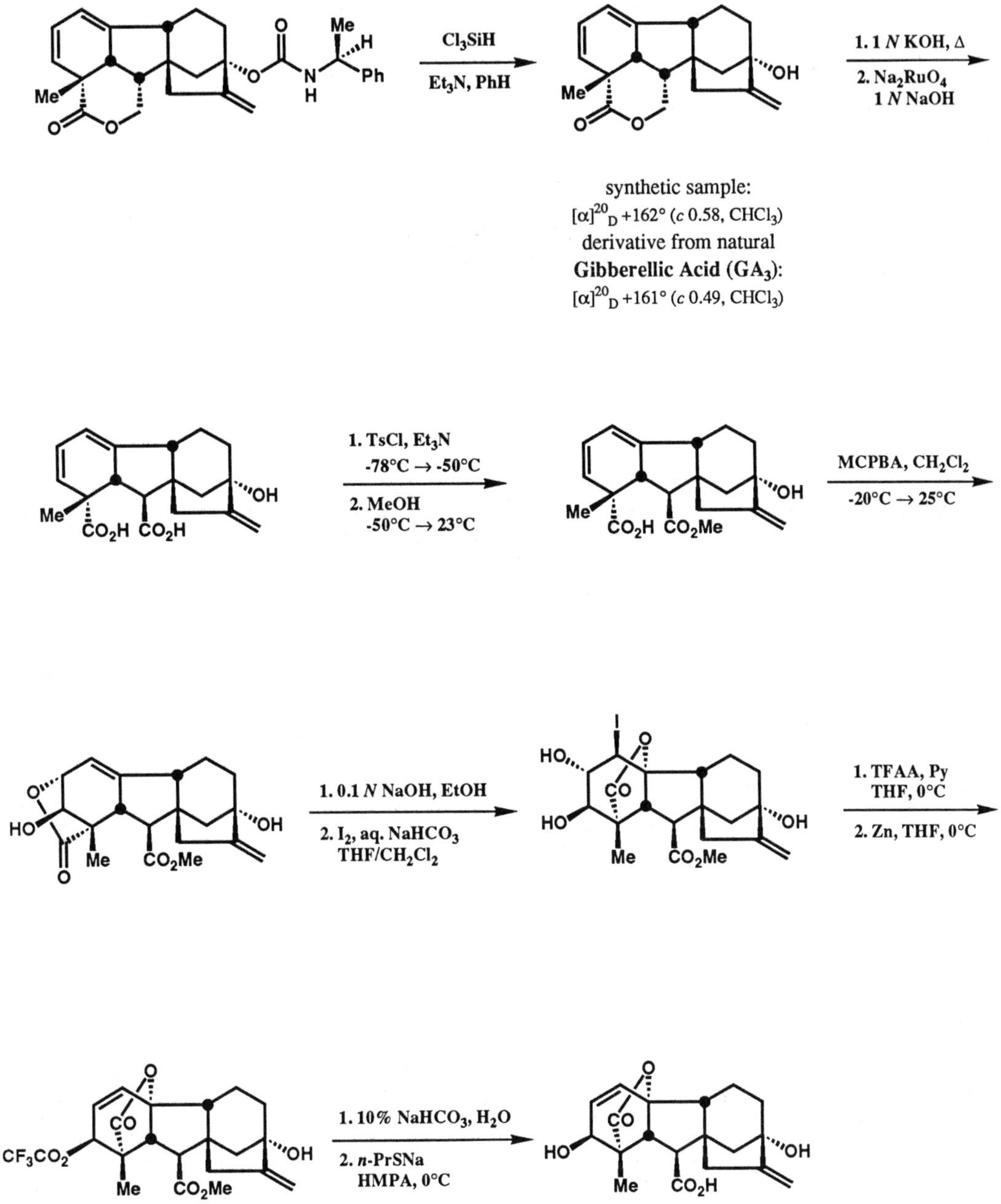
Cl3SiH
Et3N, PhH
1. 1 N KOH, Δ
2. Na2RuO4
1 N NaOH
synthetic sample:
[α]20D +162° (c 0.58, CHCl3)
derivative from natural
Gibberellic Acid (GA3):
[α]20D +161° (c 0.49, CHCl3)
1. TsCl, Et3N
-78°C → -50°C
2. MeOH
-50°C → 23°C
MCPBA, CH2Cl2
-20°C → 25°C
1. 0.1 N NaOH, EtOH
2. I2, aq. NaHCO3
THF/CH2Cl2
1. TFAA, Py
THF, 0°C
2. Zn, THF, 0°C
1. 10% NaHCO3, H2O
2. n-PrSNa
HMPA, 0°C

Gibberellic Acid (GA_3)

10.8

A second route to the key tricyclic intermediate **A** for the synthesis of gibberellic acid was also developed (Ref. 8):

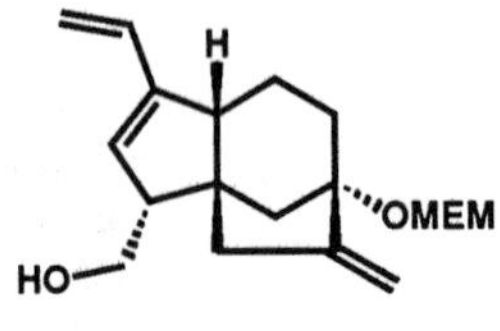
H
OMEM
HO

A

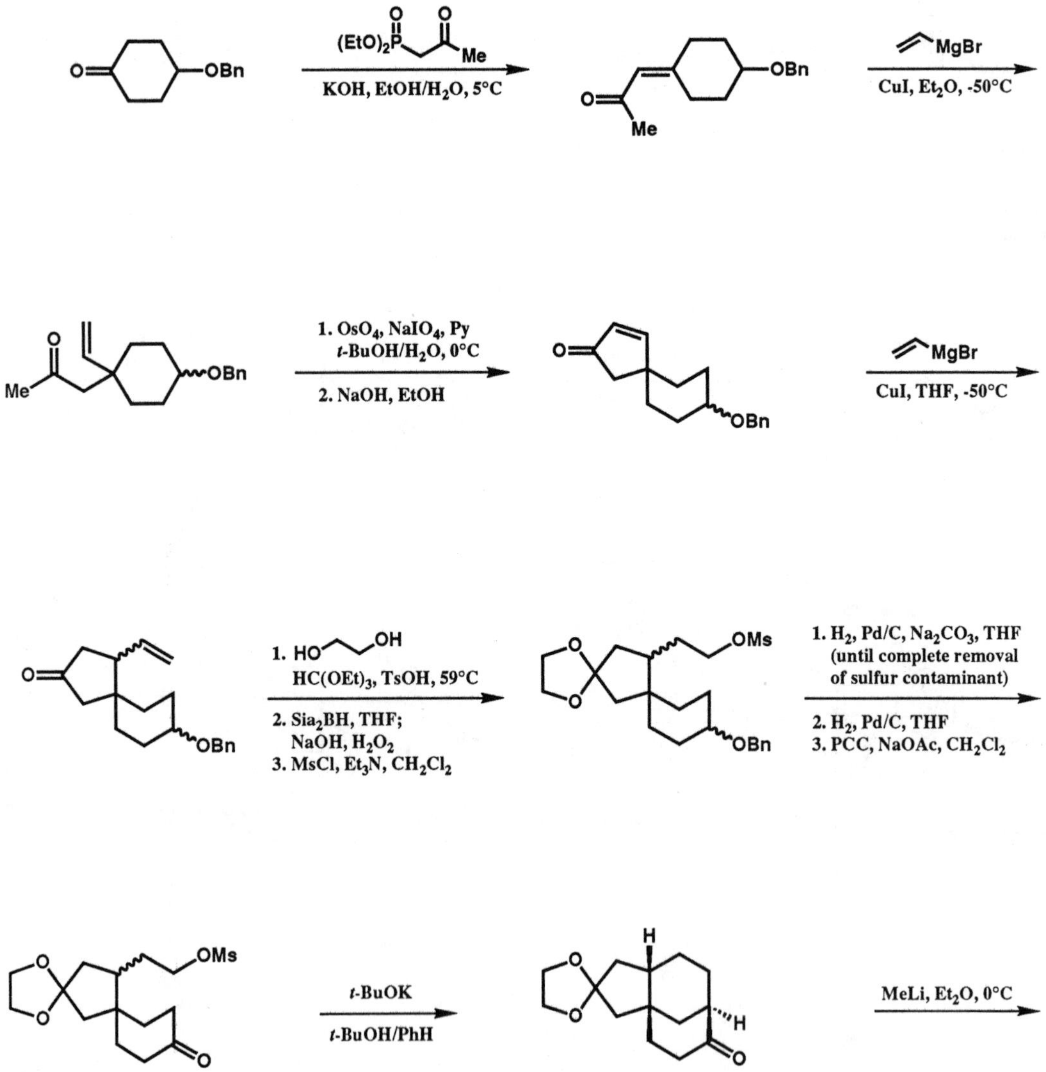
OBn
(EtO)2P
Me
KOH, EtOH/H2O, 5°C
O
Me
MgBr
CuI, Et2O, -50°C
1. OsO4, NaIO4, Py
t-BuOH/H2O, 0°C
2. NaOH, EtOH
CuI, THF, -50°C
1. HO OH
HC(OEt)3, TsOH, 59°C
2. Sia2BH, THF;
NaOH, H2O2
3. MsCl, Et3N, CH2Cl2
OMs
1. H2, Pd/C, Na2CO3, THF
(until complete removal
of sulfur contaminant)
2. H2, Pd/C, THF
3. PCC, NaOAc, CH2Cl2
t-BuOK
t-BuOH/PhH
MeLi, Et2O, 0°C

10.8

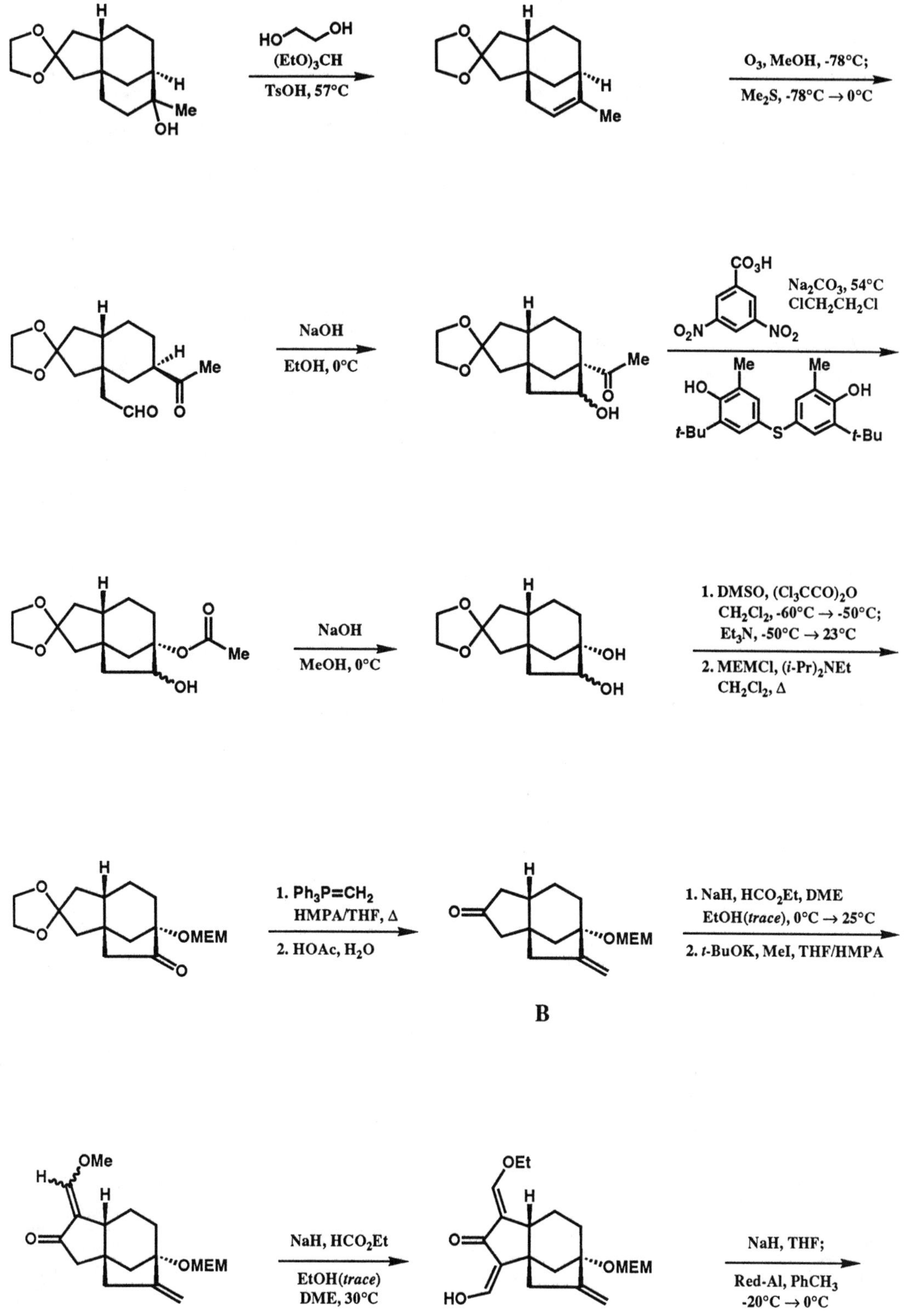

HO OH
(EtO)3CH
TsOH, 57°C
O3, MeOH, -78°C;
Me2S, -78°C → 0°C
NaOH
EtOH, 0°C
CO3H
O2N NO2
Na2CO3, 54°C
ClCH2CH2Cl
HO OH
t-Bu S t-Bu
NaOH
MeOH, 0°C
1. DMSO, (Cl3CCO)2O
CH2Cl2, -60°C → -50°C;
Et3N, -50°C → 23°C
2. MEMCl, (i-Pr)2NEt
CH2Cl2, Δ
1. Ph3P=CH2
HMPA/THF, Δ
2. HOAc, H2O
B
1. NaH, HCO2Et, DME
EtOH(trace), 0°C → 25°C
2. t-BuOK, MeI, THF/HMPA
NaH, HCO2Et
EtOH(trace)
DME, 30°C
NaH, THF;
Red-Al, PhCH3
-20°C → 0°C

10.8

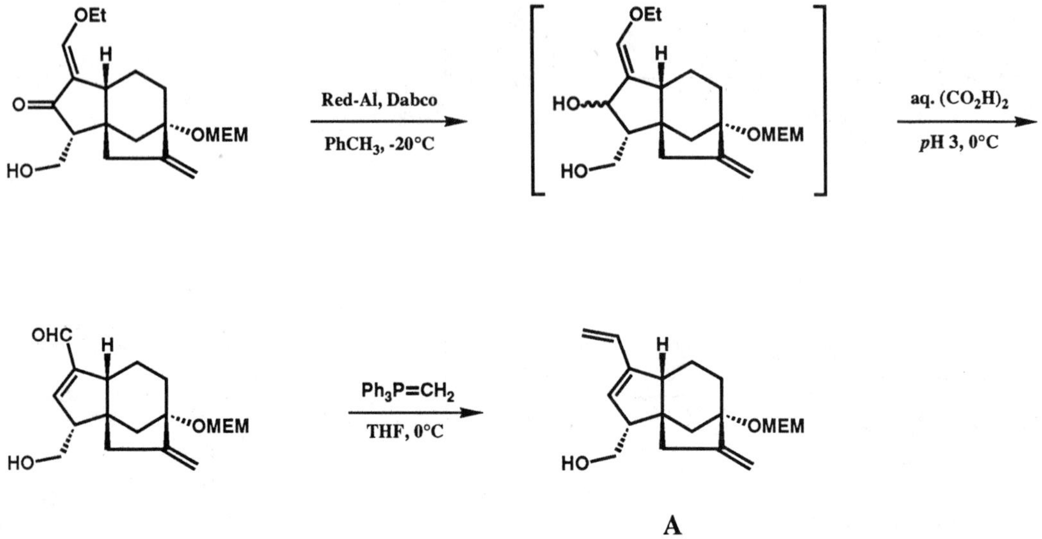

OEt
H
O
OMEM
HO
Red-Al, Dabco
PhCH3, -20°C
HO
aq. (CO2H)2
pH 3, 0°C
OHC
Ph3P=CH2
THF, 0°C

A

The synthesis of the tricyclic intermediate **A** was further improved by the development of a short and stereocontrolled synthesis of compound **B** (Ref. 9):

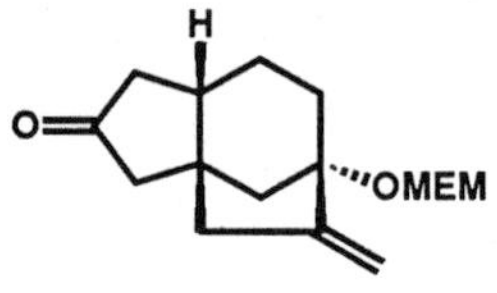

H
O
OMEM

B

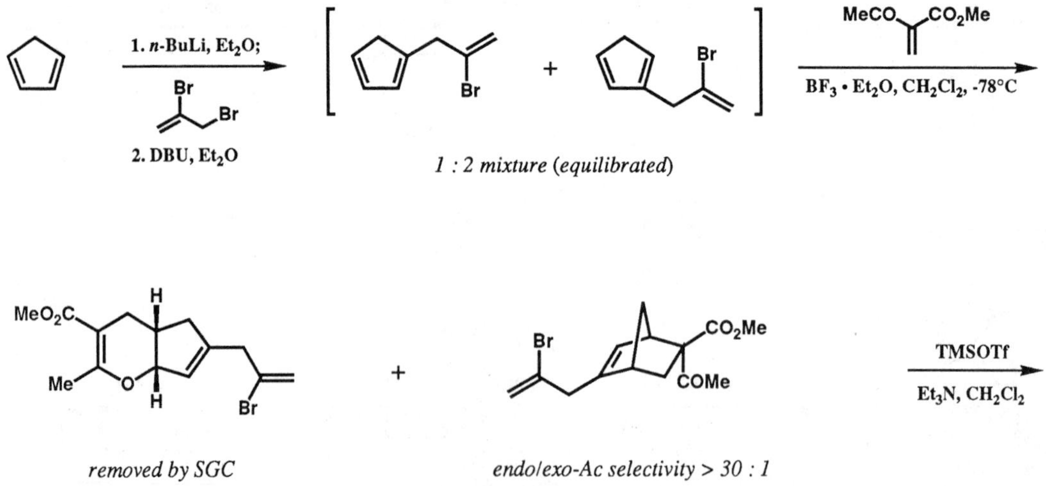

1. n-BuLi, Et2O;
Br
Br
2. DBU, Et2O
Br
+
Br
1 : 2 mixture (equilibrated)
MeCO
CO2Me
BF3 • Et2O, CH2Cl2, -78°C
MeO2C
H
Me
O
H
Br
+
Br
CO2Me
COMe
TMSOTf
Et3N, CH2Cl2
removed by SGC
endo/exo-Ac selectivity > 30 : 1

10.8

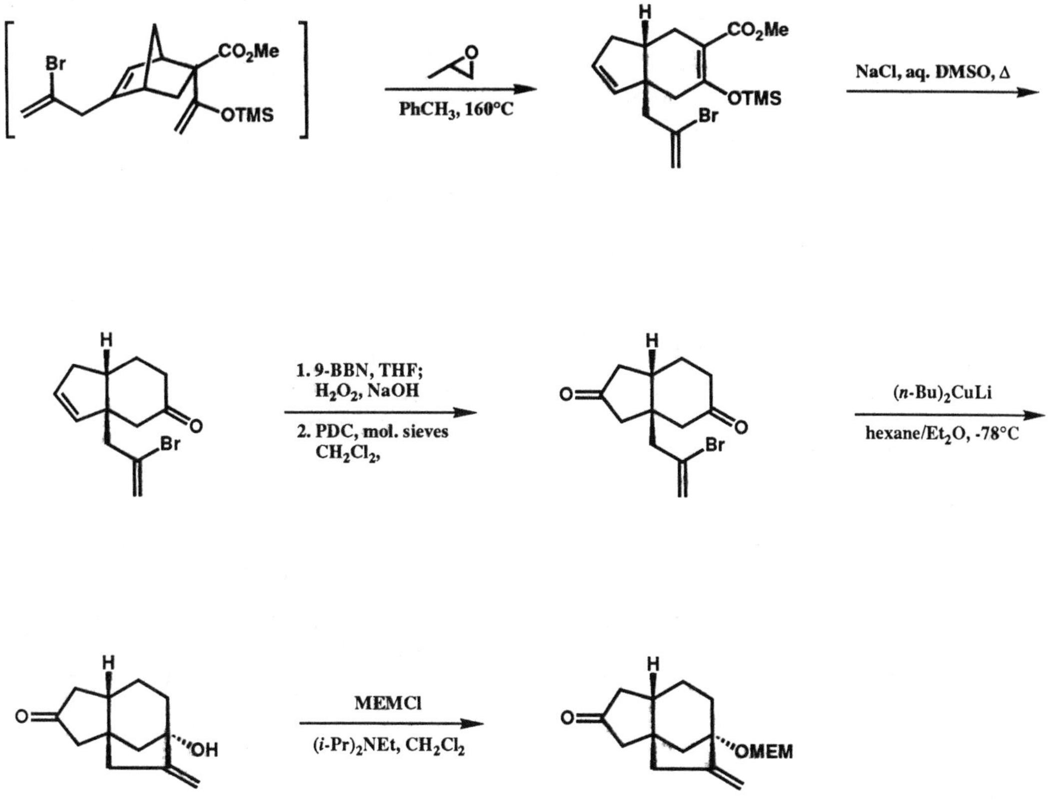
Br
CO2Me
OTMS
PhCH3, 160°C
H
CO2Me
OTMS
Br
NaCl, aq. DMSO, Δ
H
O
Br
1. 9-BBN, THF; H2O2, NaOH
2. PDC, mol. sieves CH2Cl2,
H
O
O
Br
(n-Bu)2CuLi
hexane/Et2O, -78°C
H
O
OH
MEMCl
(i-Pr)2NEt, CH2Cl2
H
O
OMEM

B

References:

1. E. J. Corey, M. Narisada, T. Hiraoka, and R. A. Ellison, *J. Am. Chem. Soc.* **1970**, *92*, 396.
2. E. J. Corey, T. M. Brennan, and R. L. Carney, *J. Am. Chem. Soc.* **1971**, *93*, 7316.
3. E. J. Corey and Carney, R. L. *J. Am. Chem. Soc.* **1971**, *93*, 7318.
4. E. J. Corey and Danheiser, R. L. *Tetrahedron Lett.* **1973**, 4477.
5. E. J. Corey, R. L. Danheiser, and S. Chandrasekaran, *J. Org. Chem.* **1976**, *41*, 260.
6. E. J. Corey, R. L. Danheiser, S. Chandrasekaran, P. Siret, G. E. Keck, and J.-L. Gras, *J. Am. Chem. Soc.* **1978**, *100*, 8031.
7. E. J. Corey, R. L. Danheiser, S. Chandrasekaran, G. E. Keck, B. Gopalan, S. D. Larsen, P. Siret, and J.-L. Gras, *J. Am. Chem. Soc.* **1978**, *100*, 8034.
8. E. J. Corey and J. G. Smith, *J. Am. Chem. Soc.* **1979**, *101*, 1038.
9. E. J. Corey and J. E. Munroe, *J. Am. Chem. Soc.* **1982**, *104*, 6129.

10.9

(±)-Antheridium-Inducing Factor (A_{An}) — Antheridic Acid

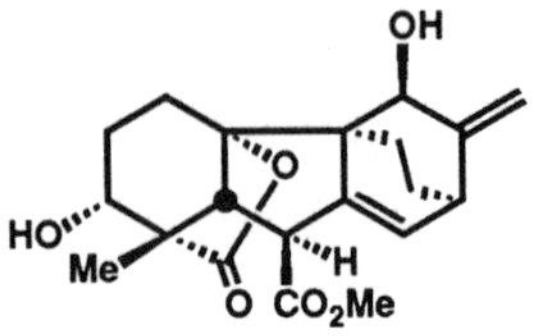

Retrosynthetic analysis of antheridic acid produced a totally different plan of synthesis from that which had been employed for the structurally related target gibberellic acid. The synthesis of antheridic acid, which included a number of novel steps, allowed definitive assignment of structure and revised stereochemistry at C(3).

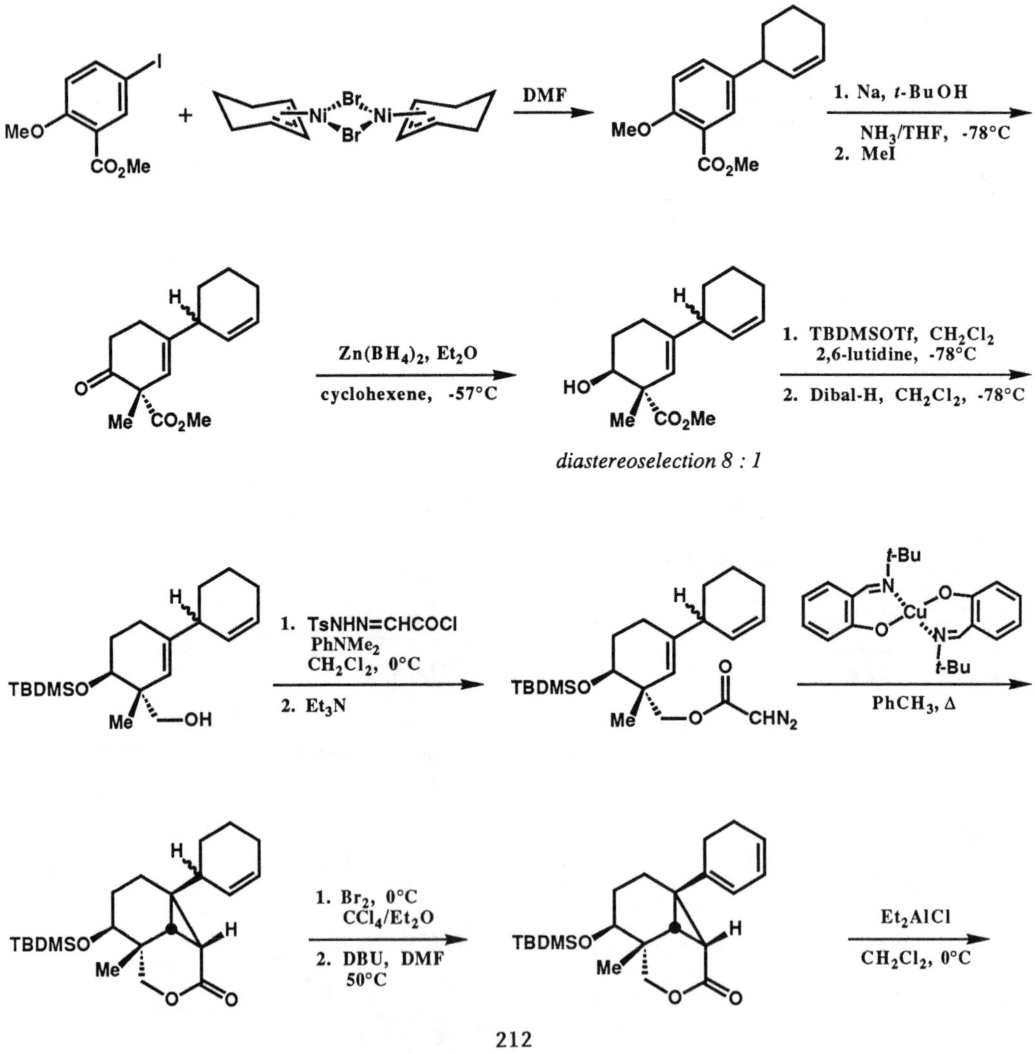

MeO
CO_2Me
DMF
1. Na, t-BuOH
NH_3/THF, -78°C
2. MeI
Me
$Zn(BH_4)_2$, Et_2O
cyclohexene, -57°C
HO
diastereoselection 8 : 1
1. TBDMSOTf, CH_2Cl_2
2,6-lutidine, -78°C
2. Dibal-H, CH_2Cl_2, -78°C
TBDMSO
OH
1. TsNHN=CHCOCl
$PhNMe_2$
CH_2Cl_2, 0°C
2. Et_3N
CHN_2
t-Bu
Cu
$PhCH_3$, Δ
1. Br_2, 0°C
CCl_4/Et_2O
2. DBU, DMF
50°C
Et_2AlCl
CH_2Cl_2, 0°C

10.9

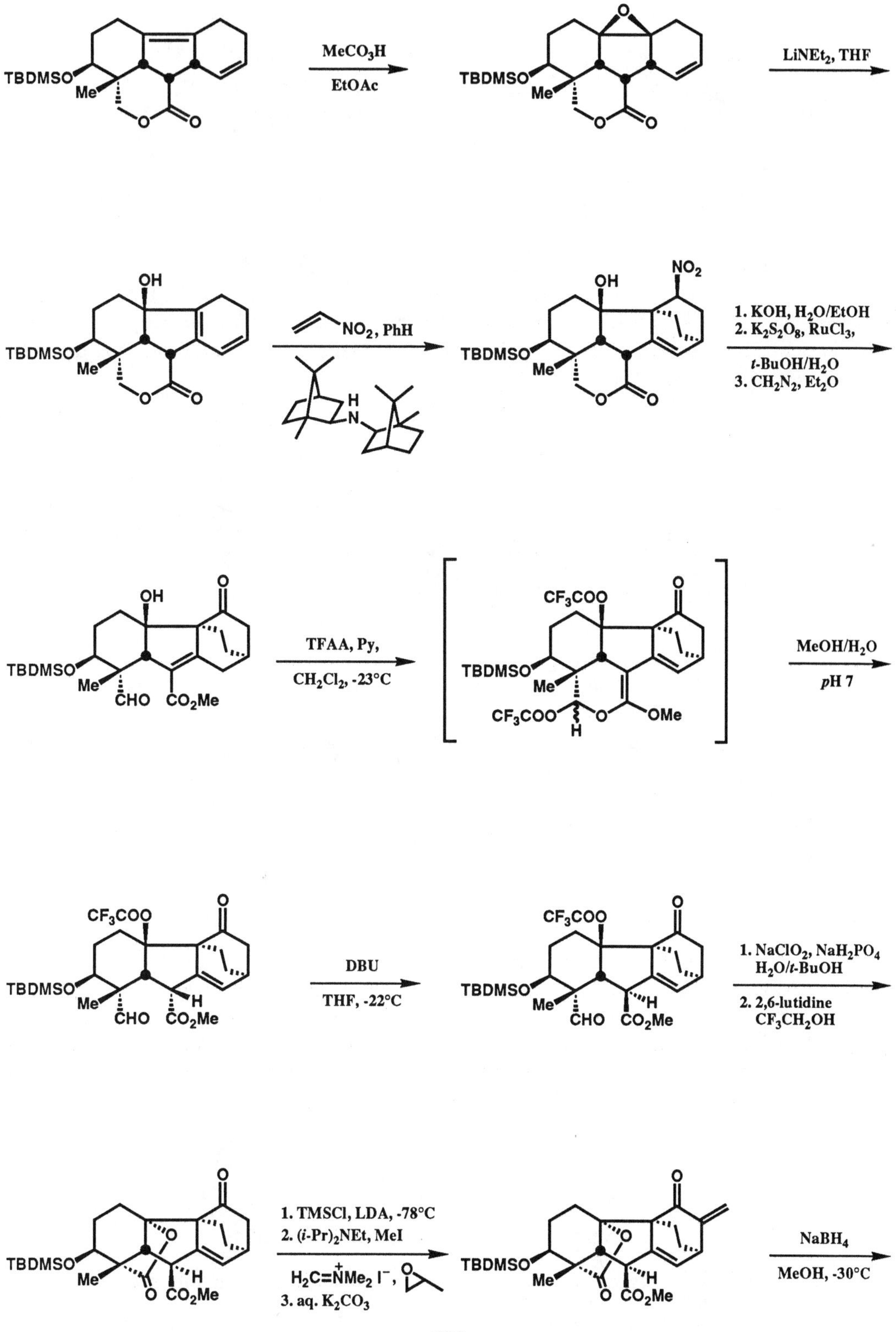
TBDMSO
Me
O
MeCO3H
EtOAc
LiNEt2, THF
OH
NO2, PhH
H
N
NO2
1. KOH, H2O/EtOH
2. K2S2O8, RuCl3,
t-BuOH/H2O
3. CH2N2, Et2O
CHO
CO2Me
TFAA, Py,
CH2Cl2, -23°C
CF3COO
OMe
H
MeOH/H2O
pH 7
DBU
THF, -22°C
1. NaClO2, NaH2PO4
H2O/t-BuOH
2. 2,6-lutidine
CF3CH2OH
1. TMSCl, LDA, -78°C
2. (i-Pr)2NEt, MeI
H2C=N+Me2 I−,
3. aq. K2CO3
NaBH4
MeOH, -30°C

10.9

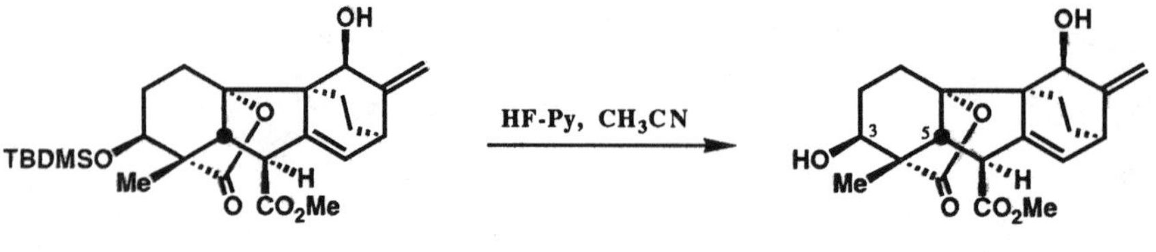

OH
TBDMSO
Me
O
H
O CO2Me
HF-Py, CH3CN
OH
HO
3
5
Me
O
H
O CO2Me

A

Compound **A** possesses the structure originally proposed for the methyl ester of A_{An} (see K. Nakanishi, M. Endo, U. Näf, and L. F. Johnson, *J. Am. Chem. Soc,* **1971**, *93*, 5579). However, the 270-MHz ^{1}H NMR spectrum of synthetic **A** was different from that reported for A_{An} methyl ester with respect to the protons at C(3) and C(5). Therefore, the 3α-alcohol, compound **B**, was synthesized as shown below.

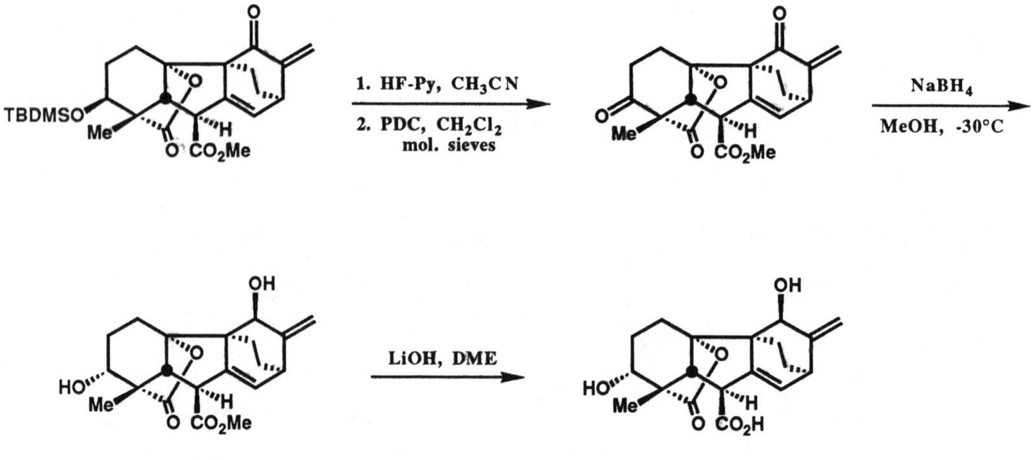

O
TBDMSO
Me
O
H
O CO2Me
1. HF-Py, CH3CN
2. PDC, CH2Cl2
mol. sieves
O
O
Me
O
H
O CO2Me
NaBH4
MeOH, -30°C
OH
HO
Me
O
H
O CO2Me
LiOH, DME
OH
HO
Me
O
H
O CO2H

B

^{1}H NMR and infrared spectral data of (±)-**B**, (±)-**B** methyl ester, and (±)-**B** methyl ester 3-benzoate were identical with those of A_{An} and the corresponding derivatives. Mass spectra of the methyl esters of (±)-**B** and A_{An} were identical. Chromatographic mobility of (±)-**B** relative to gibberellic acid (GA_3) (as standard) was identical with that reported for A_{An}. Therefore, this synthesis also proved that antheridium inducing factor, A_{An}, must be regarded as possessing stereostructure **B** rather than **A** as originally supposed.

References:

1. E. J. Corey and A. G. Myers, *J. Am. Chem. Soc.* **1985**, *107*, 5574.

2. E. J. Corey, A. G. Myers, N. Takahashi, H. Yamane, and H. Schraudolf, *Tetrahedron Lett.* **1986**, *27*, 5083.

10.10

(±)-Retigeranic Acid

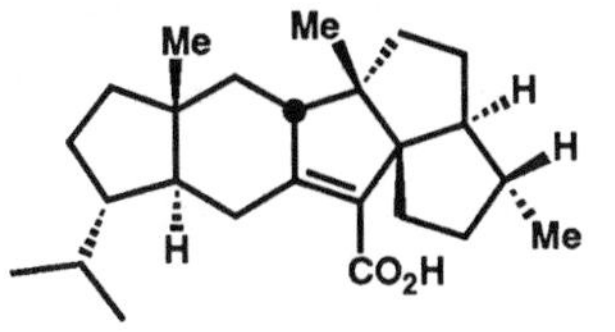

Me
Me
H
H
Me
H
CO_2H

The multistrategic retrosynthetic analysis of retigeranic acid, which led to the synthesis outlined below, has been described in Section 6.6 of Part One.

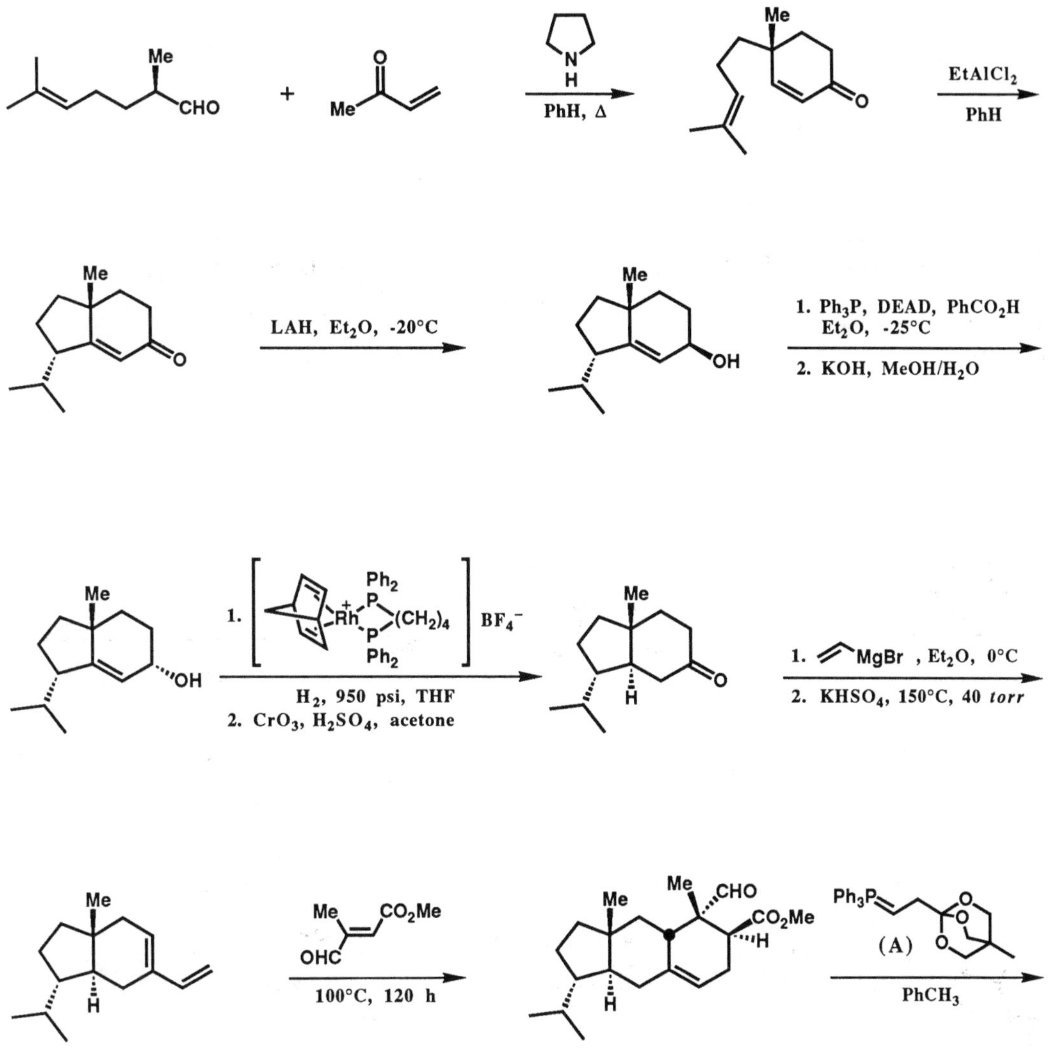

Me
CHO
+
Me
O
N
H
PhH, Δ
Me
O
$EtAlCl_2$
PhH
Me
O
LAH, Et_2O, -20°C
Me
OH
1. Ph_3P, DEAD, $PhCO_2H$
Et_2O, -25°C
2. KOH, $MeOH/H_2O$
Me
OH
1.
Ph_2
P
Rh
P
$(CH_2)_4$
Ph_2
BF_4^-
H_2, 950 psi, THF
2. CrO_3, H_2SO_4, acetone
Me
H
O
1. MgBr, Et_2O, 0°C
2. $KHSO_4$, 150°C, 40 torr
Me
H
Me
CO_2Me
OHC
100°C, 120 h
Me
Me
CHO
CO_2Me
H
H
Ph_3P
O
O
O
(A)
$PhCH_3$

10.10

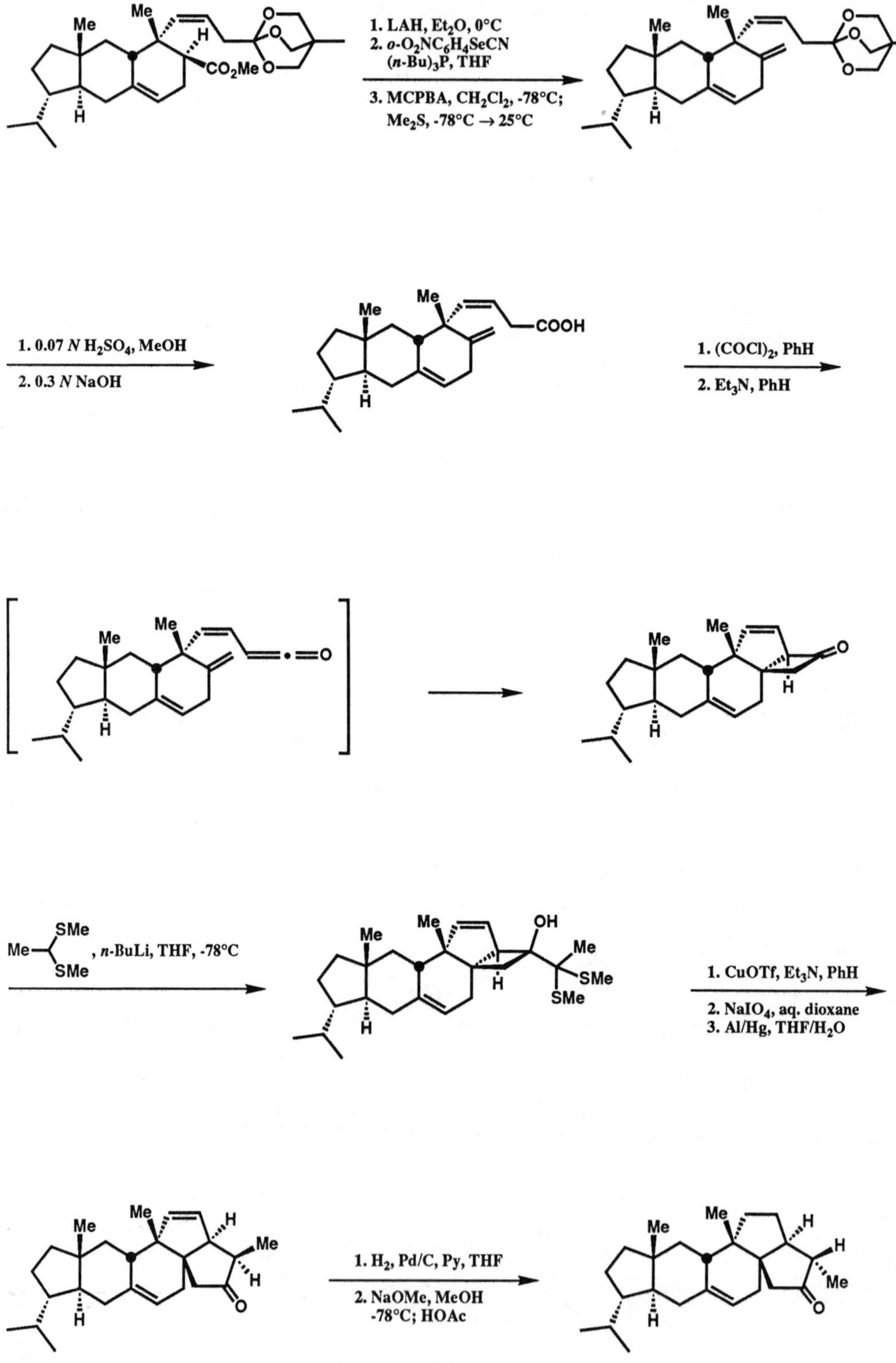
Me
Me
H
CO2Me
1. LAH, Et2O, 0°C
2. o-O2NC6H4SeCN
(n-Bu)3P, THF
3. MCPBA, CH2Cl2, -78°C;
Me2S, -78°C → 25°C
1. 0.07 N H2SO4, MeOH
2. 0.3 N NaOH
COOH
1. (COCl)2, PhH
2. Et3N, PhH
SMe
SMe
, n-BuLi, THF, -78°C
OH
SMe
SMe
1. CuOTf, Et3N, PhH
2. NaIO4, aq. dioxane
3. Al/Hg, THF/H2O
1. H2, Pd/C, Py, THF
2. NaOMe, MeOH
-78°C; HOAc

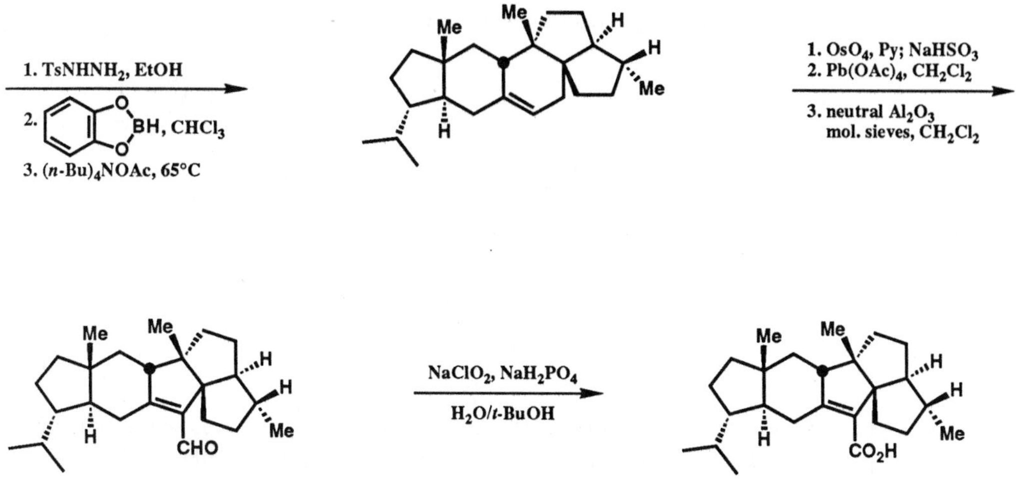
1. TsNHNH$_2$, EtOH
2. BH, CHCl$_3$
3. (n-Bu)$_4$NOAc, 65°C
Me
Me
H
H
Me
H
1. OsO$_4$, Py; NaHSO$_3$
2. Pb(OAc)$_4$, CH$_2$Cl$_2$
3. neutral Al$_2$O$_3$
mol. sieves, CH$_2$Cl$_2$
CHO
NaClO$_2$, NaH$_2$PO$_4$
H$_2$O/t-BuOH
CO$_2$H

Preparation of the ylide **A** (Ref. 3):

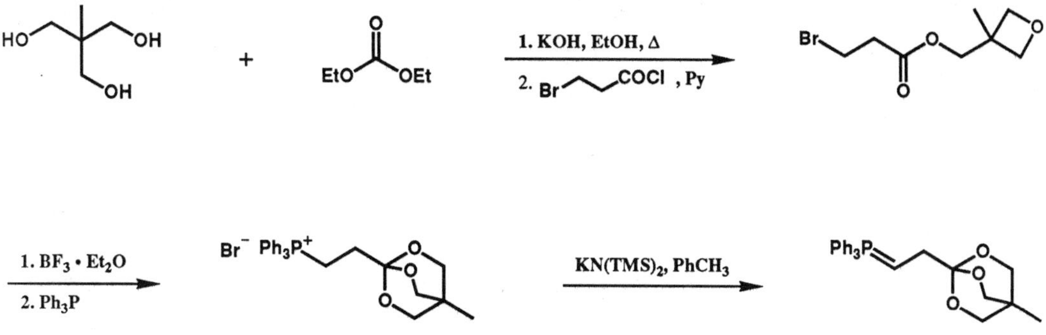
HO
OH
OH
EtO
OEt
1. KOH, EtOH, Δ
2. Br COCl , Py
Br
1. BF$_3$ • Et$_2$O
2. Ph$_3$P
Br$^-$ Ph$_3$P$^+$
KN(TMS)$_2$, PhCH$_3$
Ph$_3$P

A

References:

1. E. J. Corey, M. C. Desai, and T. A. Engler, *J. Am. Chem. Soc.* **1985**, *107*, 4339.
2. B. B. Snider, D. J. Rodini, and J. van Straten, *J. Am. Chem. Soc.* **1980**, *102*, 5872.
3. E. J. Corey and N. Raju, *Tetrahedron Lett.* **1983**, *24*, 5571.

10.11

7,20-Diisocyanoadociane

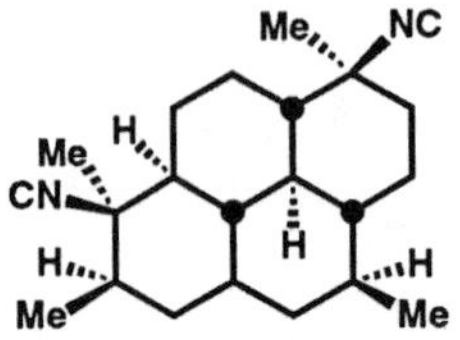

Me
NC
H
Me
CN
H
H
H
Me
Me

7,20-Diisocyanoadociane, a novel marine-derived diterpenoid, was analyzed retrosynthetically using the intramolecular Diels-Alder transform as T-goal concurrently with topological and stereochemical guidance. The enantioselective synthesis outlined below allowed assignment of absolute configuration.

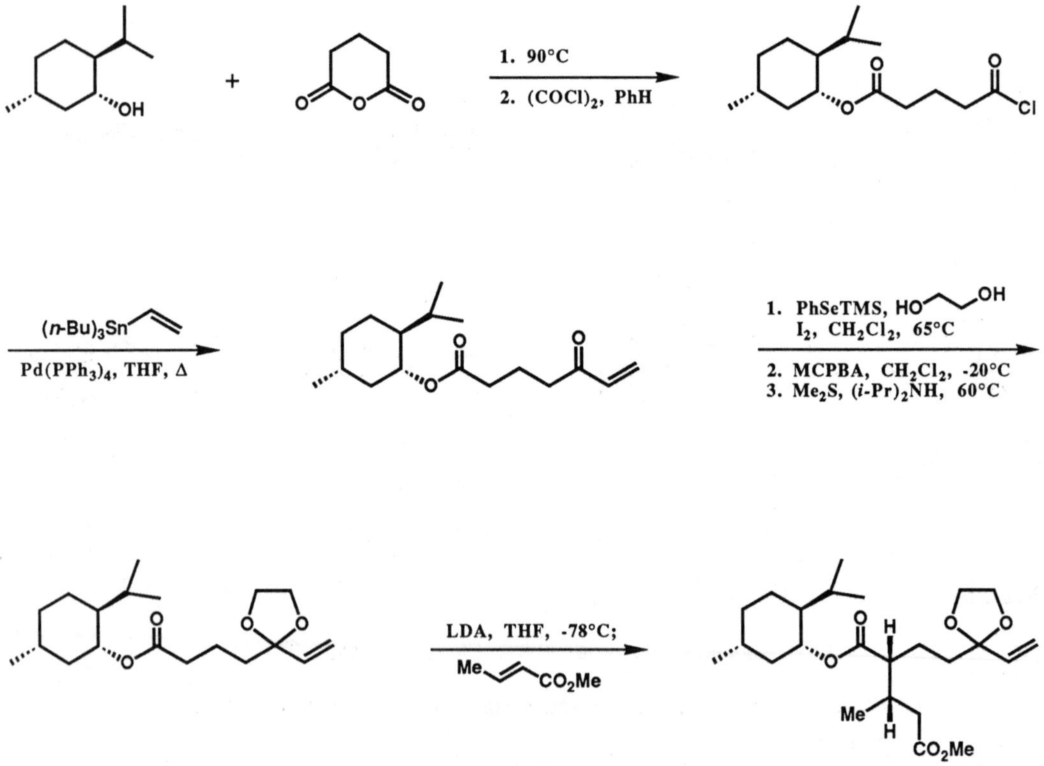

OH
+
1. 90°C
2. (COCl)$_2$, PhH
Cl
(n-Bu)$_3$Sn
Pd(PPh$_3$)$_4$, THF, Δ
1. PhSeTMS, HO OH
I$_2$, CH$_2$Cl$_2$, 65°C
2. MCPBA, CH$_2$Cl$_2$, -20°C
3. Me$_2$S, (i-Pr)$_2$NH, 60°C
LDA, THF, -78°C;
Me CO$_2$Me
H
Me
H
CO$_2$Me

threo/erythro-diastereoselection 8 : 1

ca. 60% ee

10.11

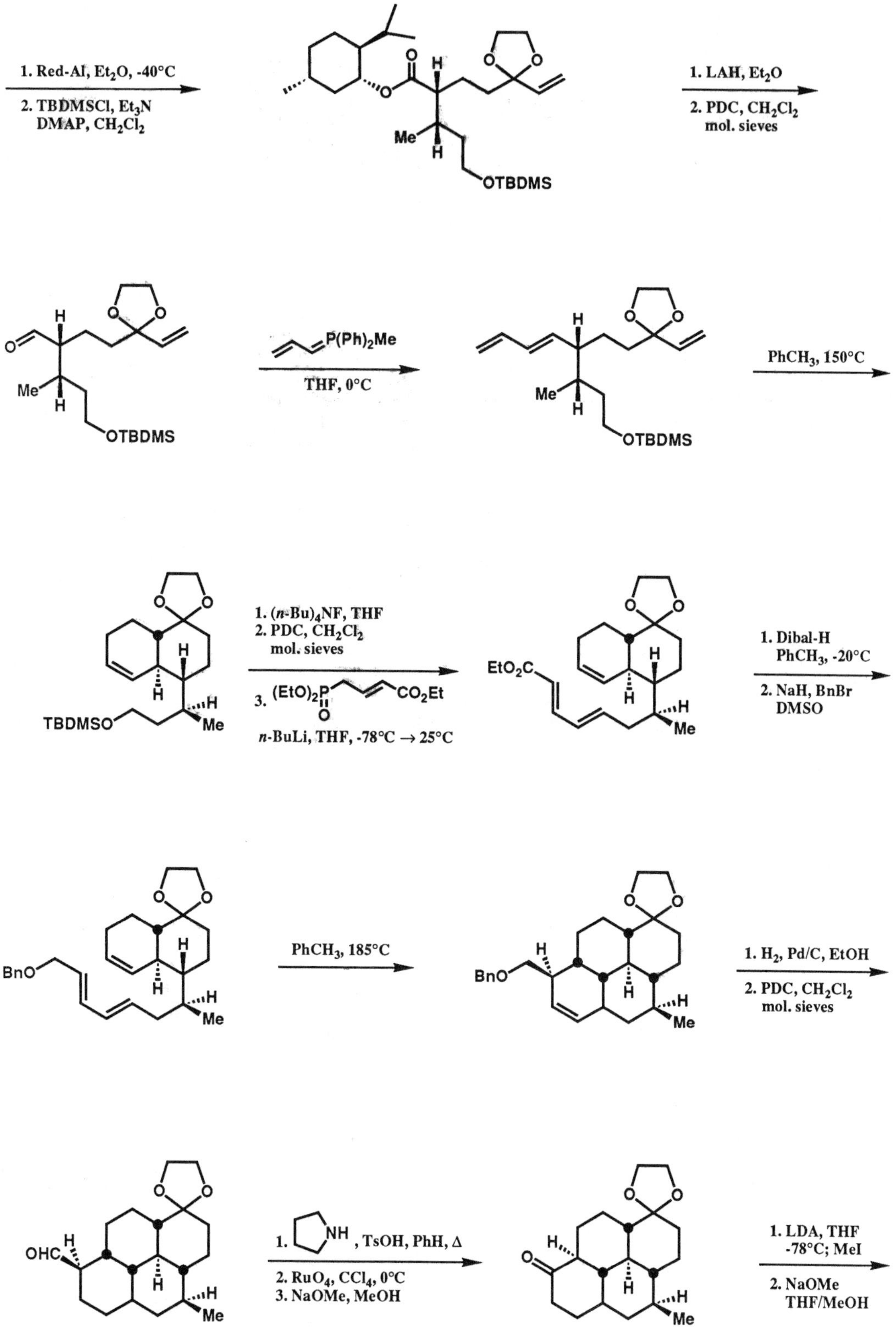
1. Red-Al, Et2O, -40°C
2. TBDMSCl, Et3N DMAP, CH2Cl2
OTBDMS
Me
1. LAH, Et2O
2. PDC, CH2Cl2 mol. sieves
OTBDMS
P(Ph)2Me
THF, 0°C
OTBDMS
PhCH3, 150°C
TBDMSO
1. (n-Bu)4NF, THF
2. PDC, CH2Cl2 mol. sieves
3. (EtO)2P CO2Et
n-BuLi, THF, -78°C → 25°C
EtO2C
1. Dibal-H PhCH3, -20°C
2. NaH, BnBr DMSO
BnO
PhCH3, 185°C
BnO
1. H2, Pd/C, EtOH
2. PDC, CH2Cl2 mol. sieves
OHC
1. NH, TsOH, PhH, Δ
2. RuO4, CCl4, 0°C
3. NaOMe, MeOH
1. LDA, THF -78°C; MeI
2. NaOMe THF/MeOH

10.11

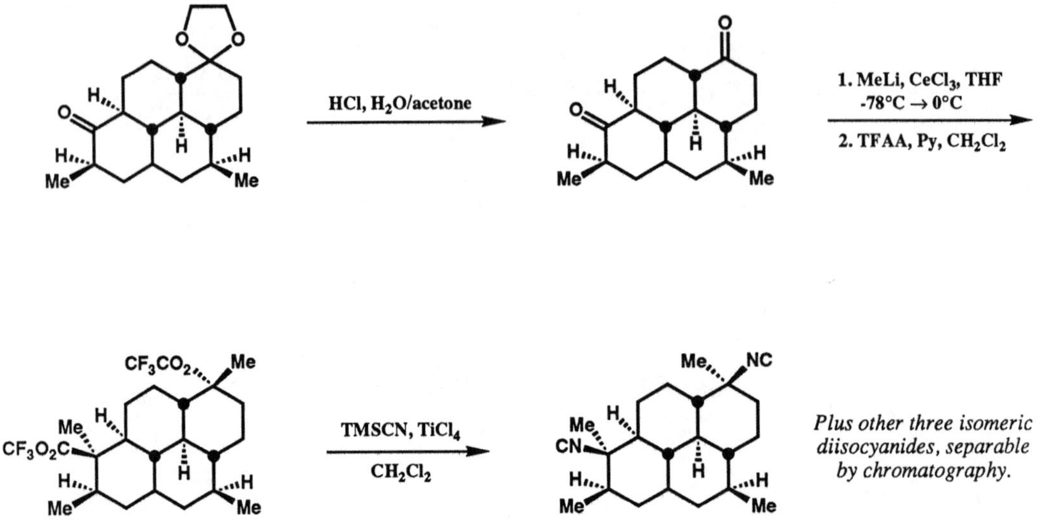
HCl, H_2O/acetone
1. MeLi, $CeCl_3$, THF
-78°C → 0°C
2. TFAA, Py, CH_2Cl_2
TMSCN, $TiCl_4$
CH_2Cl_2
CF_3CO_2
CF_3O_2C
Me
H
O
CN
NC
Plus other three isomeric diisocyanides, separable by chromatography.

synthetic sample: $[\alpha]^{23}_D$ +23.0° (*c* 0.27, $CHCl_3$), *ca.* *60% ee*

natural product: $[\alpha]^{23}_D$ +47.8° (*c* 0.23, $CHCl_3$)

Reference:

E. J. Corey and P. A. Magriotis, *J. Am. Chem. Soc.* **1987**, *109*, 287.

Ginkgolide B

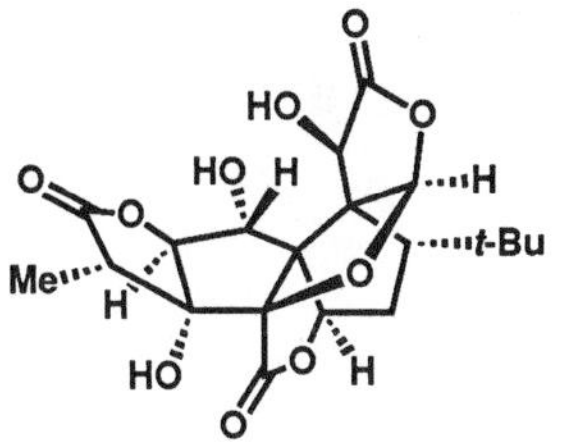
HO
HO H
O
Me
H
HO
O
O
H
H
t-Bu

Ginkgolide A

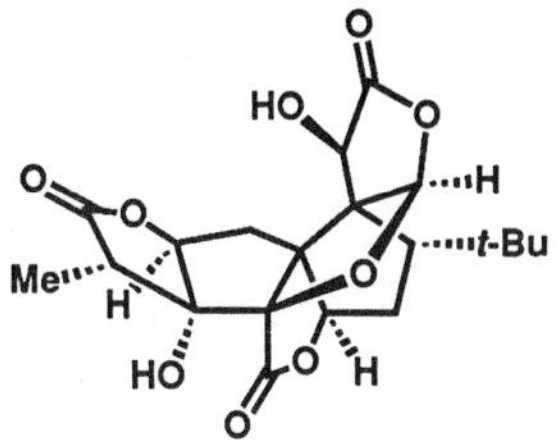
HO
O
Me
H
HO
O
O
H
H
t-Bu

The multistrategic retrosynthetic analysis of ginkgolide B, which led to the synthesis outlined below, is presented in Section 6.7 of Part One. The synthetic route, an enantioselective version of which was demonstrated, also led to ginkgolide A.

1. Total synthesis of (±)-ginkgolide B (Ref. 1):

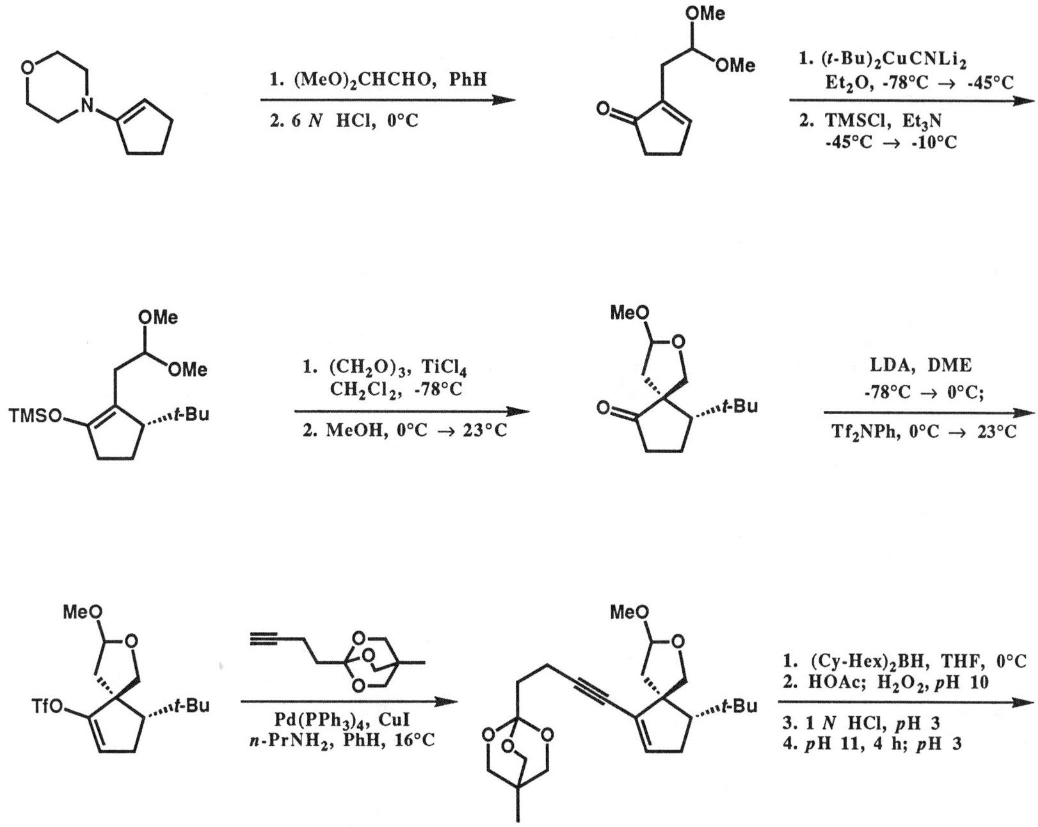
1. (MeO)2CHCHO, PhH
2. 6 N HCl, 0°C
OMe
OMe
1. (t-Bu)2CuCNLi2
Et2O, -78°C → -45°C
2. TMSCl, Et3N
-45°C → -10°C
TMSO
t-Bu
1. (CH2O)3, TiCl4
CH2Cl2, -78°C
2. MeOH, 0°C → 23°C
MeO
LDA, DME
-78°C → 0°C;
Tf2NPh, 0°C → 23°C
TfO
Pd(PPh3)4, CuI
n-PrNH2, PhH, 16°C
1. (Cy-Hex)2BH, THF, 0°C
2. HOAc; H2O2, pH 10
3. 1 N HCl, pH 3
4. pH 11, 4 h; pH 3

10.12

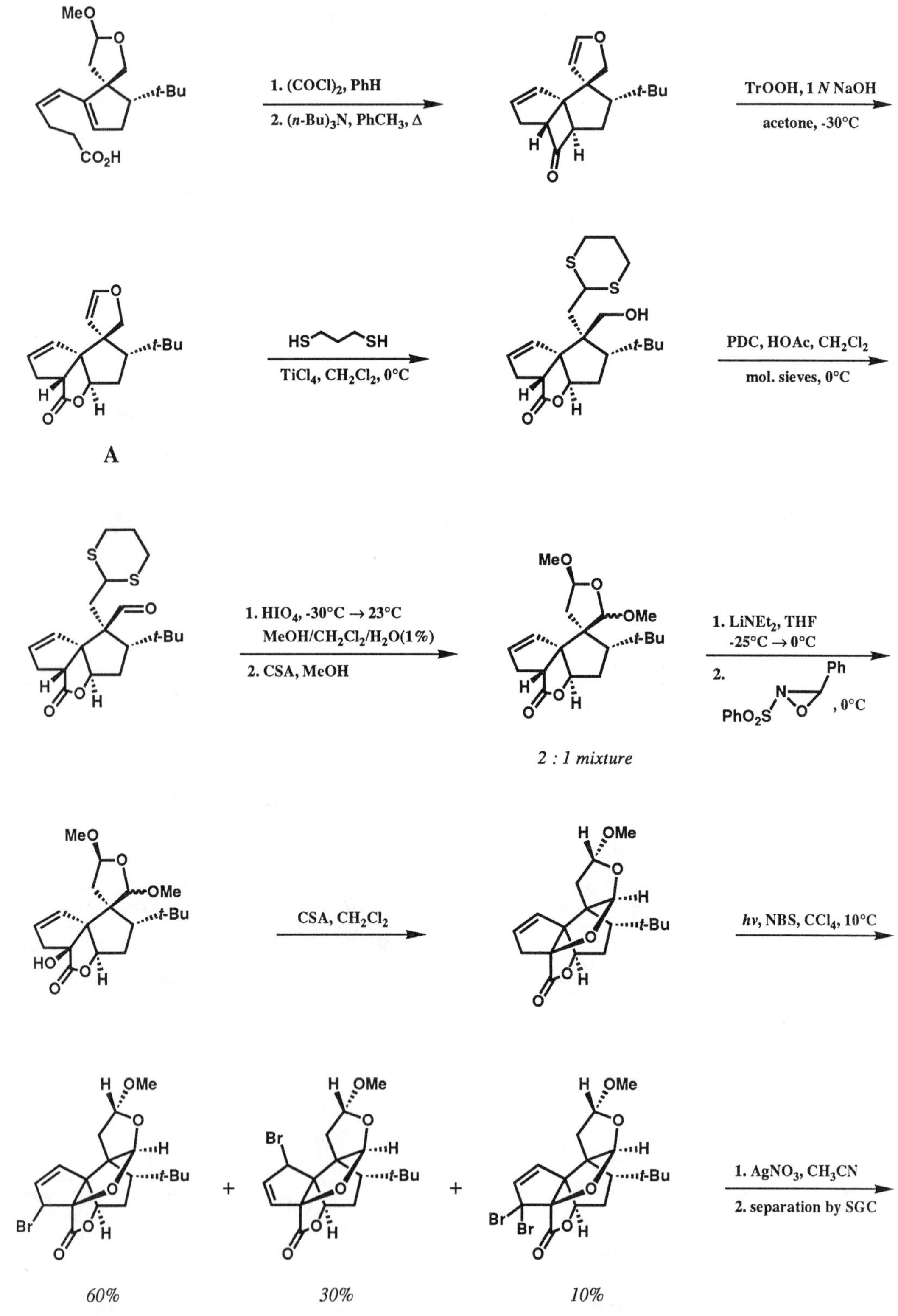
1. (COCl)2, PhH
2. (n-Bu)3N, PhCH3, Δ
TrOOH, 1 N NaOH
acetone, -30°C
A
HS SH
TiCl4, CH2Cl2, 0°C
PDC, HOAc, CH2Cl2
mol. sieves, 0°C
1. HIO4, -30°C → 23°C
MeOH/CH2Cl2/H2O(1%)
2. CSA, MeOH
2 : 1 mixture
1. LiNEt2, THF
-25°C → 0°C
2.
, 0°C
CSA, CH2Cl2
hv, NBS, CCl4, 10°C
1. AgNO3, CH3CN
2. separation by SGC
60%
30%
10%

10.12

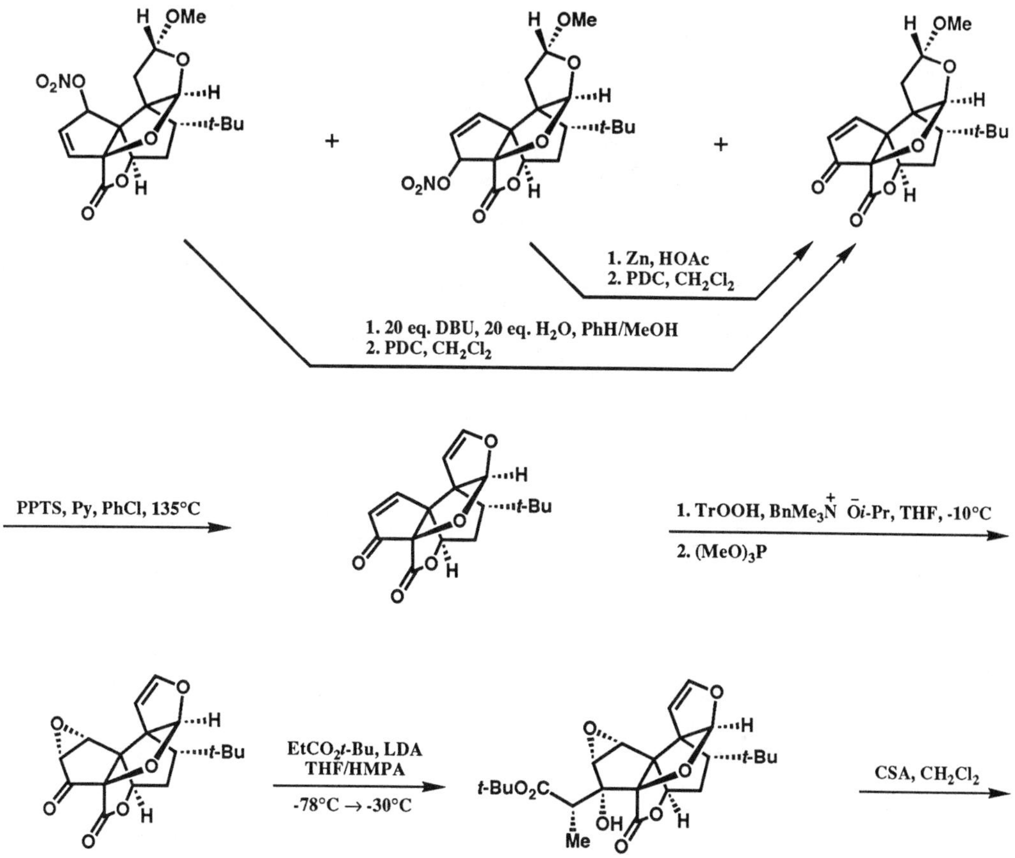

H OMe
O_2NO
H
t-Bu
O
H
+
1. Zn, HOAc
2. PDC, CH_2Cl_2
1. 20 eq. DBU, 20 eq. H_2O, PhH/MeOH
2. PDC, CH_2Cl_2
PPTS, Py, PhCl, 135°C
1. TrOOH, $BnMe_3\overset{+}{N}\ \overset{-}{O}i$-Pr, THF, -10°C
2. $(MeO)_3P$
$EtCO_2t$-Bu, LDA
THF/HMPA
-78°C → -30°C
t-BuO_2C
OH
Me
CSA, CH_2Cl_2

10.12

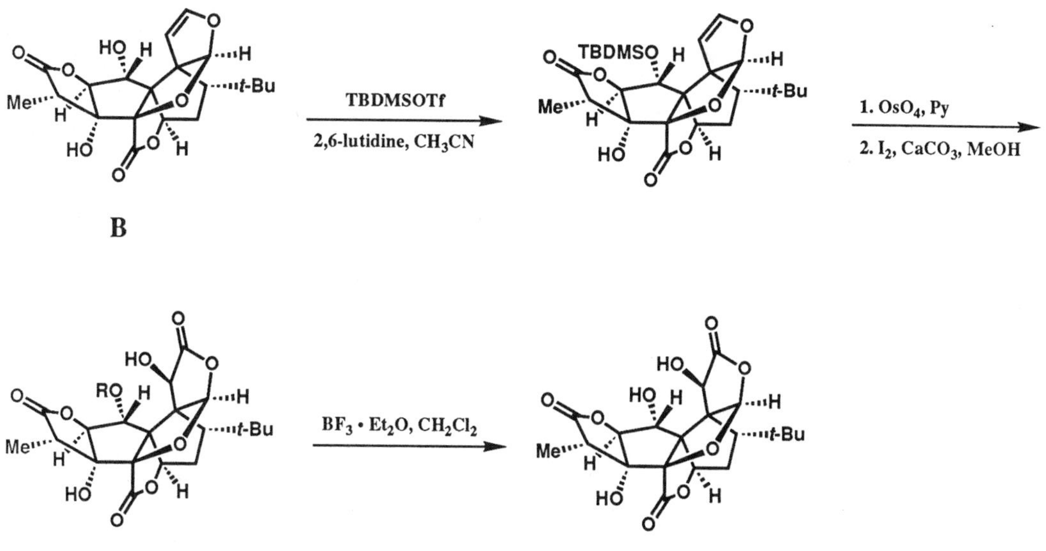

HO H
Me
HO
TBDMSOTf
2,6-lutidine, CH_3CN
TBDMSO H
1. OsO_4, Py
2. I_2, $CaCO_3$, MeOH
RO H
$BF_3 \cdot Et_2O$, CH_2Cl_2

B

R = TBDMS

10.12

Preparation of the acetylenic OBO ester: 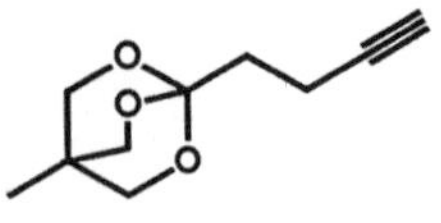

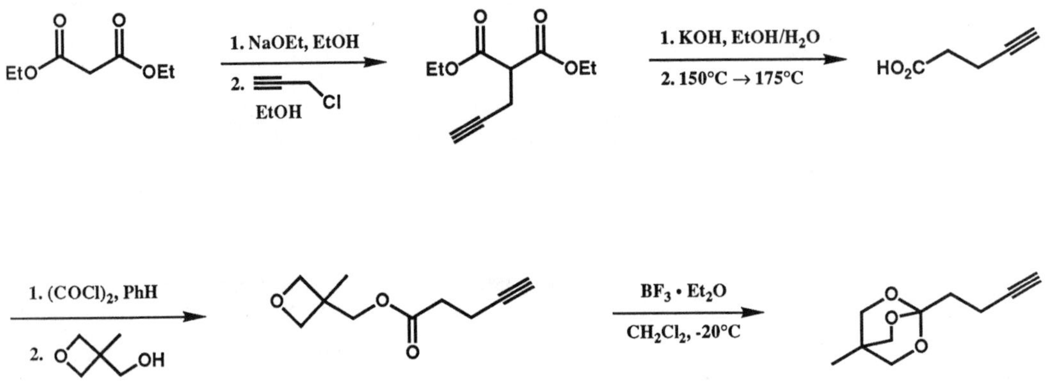

An efficient enantioselective route for the total synthesis of ginkgolide B has been established by synthesizing the key intermediate **A** in an enantiomerically pure form (Ref. 2).

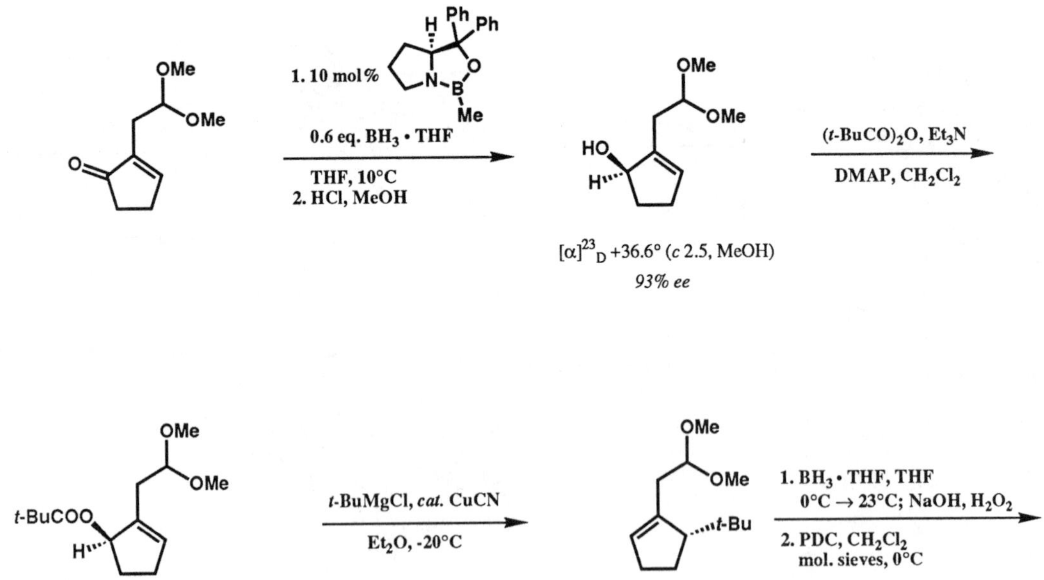

OMe
OMe
t-Bu
KH, THF;
TBDMSCl, 0°C
TBDMSO
1. (CH2O)3, TiCl4
CH2Cl2, -78°C
2. MeOH, 0°C → 23°C

2 : 1 mixture

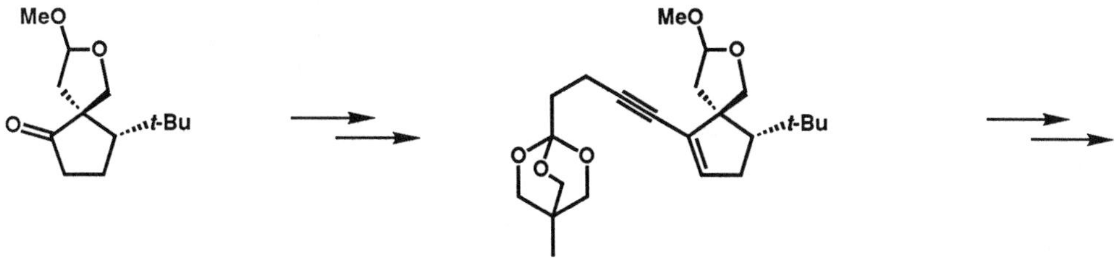
MeO
t-Bu

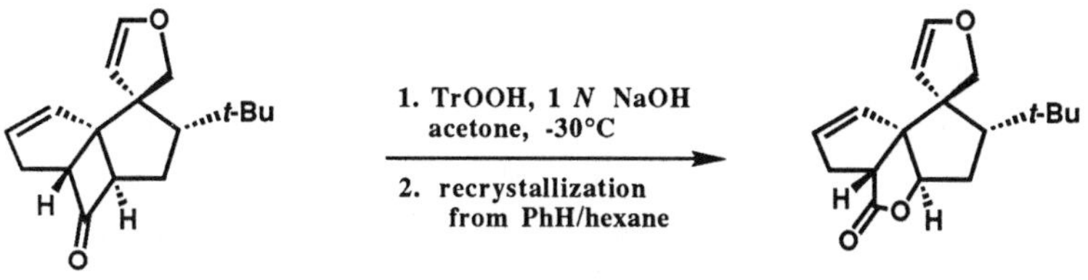
t-Bu
H
1. TrOOH, 1 N NaOH
acetone, -30°C
2. recrystallization
from PhH/hexane

A

$[\alpha]^{23}_{D}$ -151.9° (*c* 0.9, MeOH)
mp 111°C, 100% *ee*

2. Total synthesis of (±)-ginkgolide A (Ref. 3):

Based on the successful total synthesis of ginkgolide B, ginkgolide A was made by two different routes.

i. From the bislactone **B**:

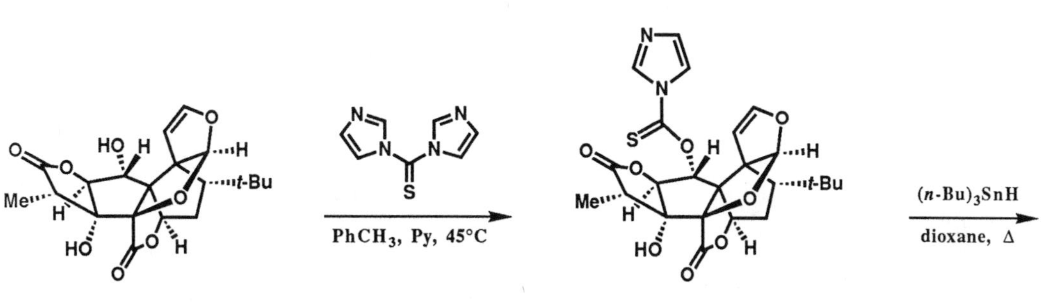
HO
H
Me
t-Bu
PhCH3, Py, 45°C
(n-Bu)3SnH
dioxane, Δ

B

10.12

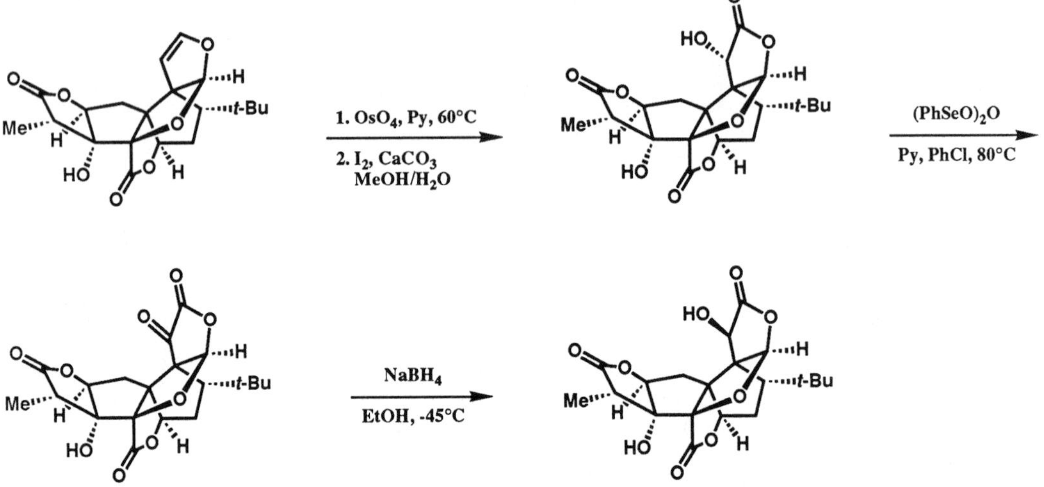

1. OsO_4, Py, 60°C
2. I_2, $CaCO_3$ $MeOH/H_2O$
$(PhSeO)_2O$
Py, PhCl, 80°C
$NaBH_4$
EtOH, -45°C

Ginkgolide A

ii. From ginkgolide B:

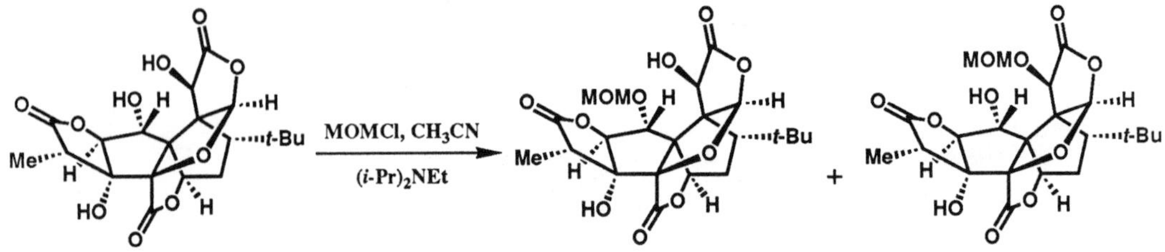

MOMCl, CH_3CN
$(i\text{-}Pr)_2NEt$

Ginkgolide B

1 : 3 mixture, separated by SGC

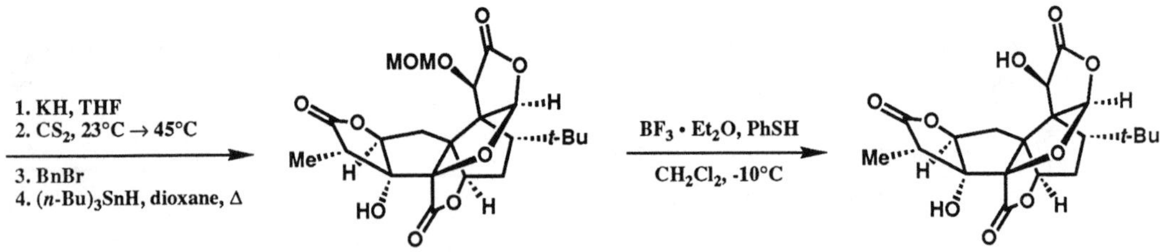

1. KH, THF
2. CS_2, 23°C → 45°C
3. BnBr
4. $(n\text{-}Bu)_3SnH$, dioxane, Δ
$BF_3 \cdot Et_2O$, PhSH
CH_2Cl_2, -10°C

Ginkgolide A

References:

1. E. J. Corey, M.-c. Kang, M. C. Desai, A. K. Ghosh, and I. N. Houpis, *J. Am. Chem. Soc.* **1988**, *110*, 649.

2. E. J. Corey and A. V. Gavai, *Tetrahedron Lett.* **1988**, *29*, 3201.

3. E. J. Corey and A. K. Ghosh, *Tetrahedron Lett.* **1988**, *29*, 3205.

Bilobalide

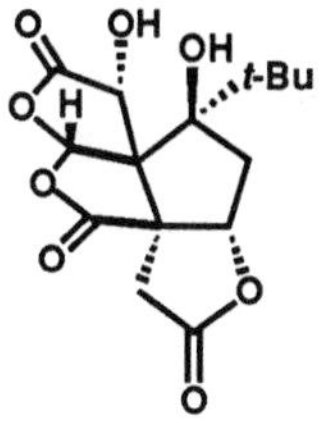

OH
OH
t-Bu
H
O

Despite the structural relationship between ginkgolide B and bilobalide, retrosynthetic analysis of the latter produced a totally different collection of sequences. A successful synthesis of bilobalide was implemented using a plan which depended on stereochemical and FG-based strategies. A process for enantioselective synthesis was based on an initial enantioselective Diels-Alder step in combination with a novel annulation method.

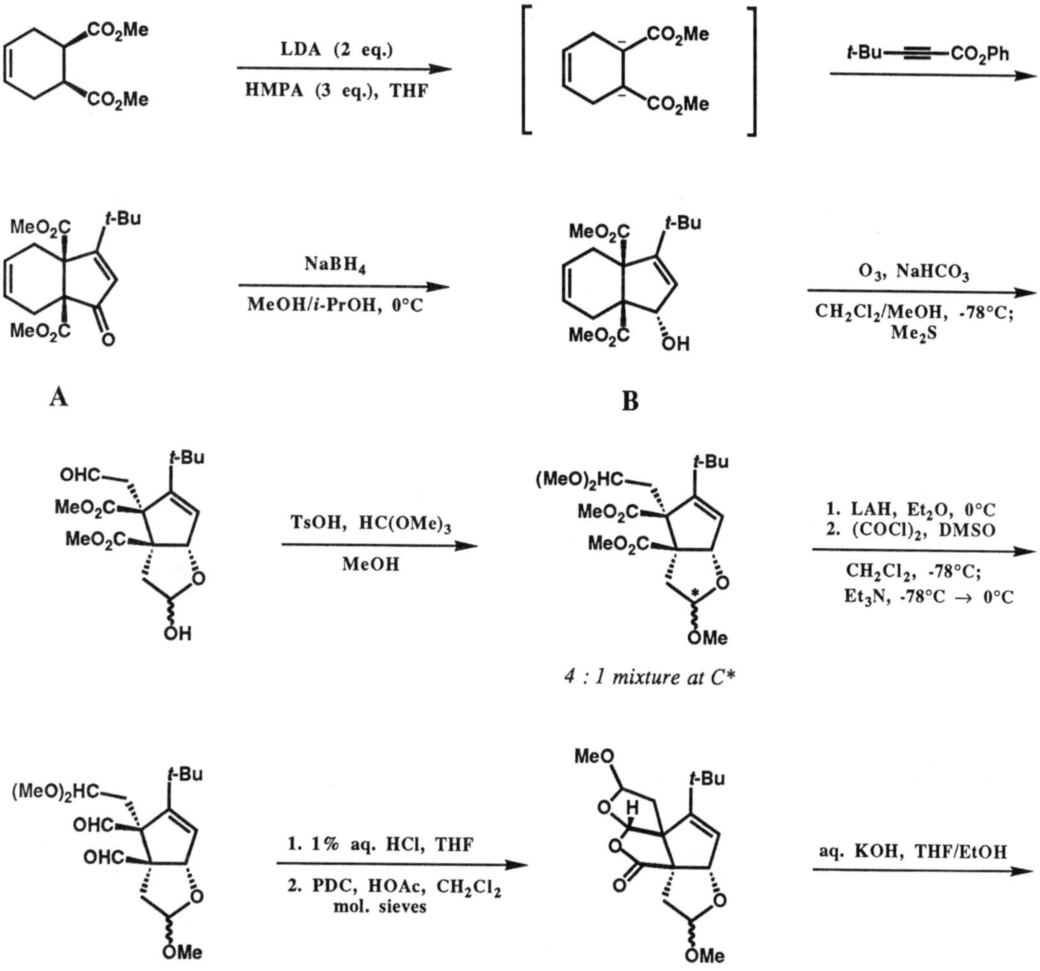

CO2Me
LDA (2 eq.)
HMPA (3 eq.), THF
t-Bu—≡—CO2Ph
MeO2C
t-Bu
NaBH4
MeOH/i-PrOH, 0°C
OH
O3, NaHCO3
CH2Cl2/MeOH, -78°C; Me2S
A
B
OHC
TsOH, HC(OMe)3
MeOH
(MeO)2HC
OMe
4 : 1 mixture at C*
1. LAH, Et2O, 0°C
2. (COCl)2, DMSO
CH2Cl2, -78°C; Et3N, -78°C → 0°C
1. 1% aq. HCl, THF
2. PDC, HOAc, CH2Cl2 mol. sieves
MeO
H
aq. KOH, THF/EtOH

10.13

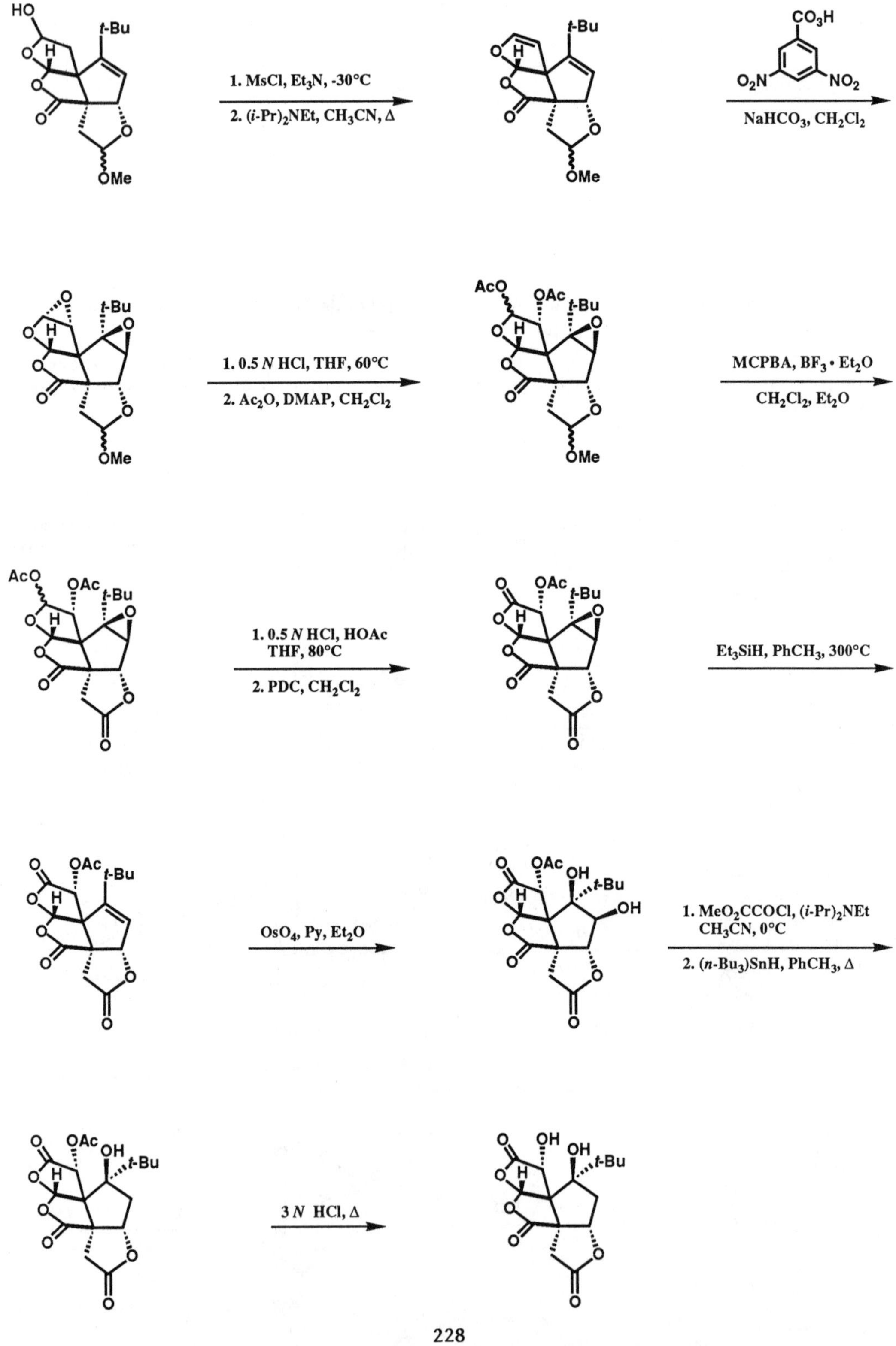
HO
t-Bu
H
OMe
1. MsCl, Et3N, -30°C
2. (i-Pr)2NEt, CH3CN, Δ
CO3H
O2N
NO2
NaHCO3, CH2Cl2
1. 0.5 N HCl, THF, 60°C
2. Ac2O, DMAP, CH2Cl2
AcO
OAc
MCPBA, BF3 • Et2O
CH2Cl2, Et2O
1. 0.5 N HCl, HOAc
THF, 80°C
2. PDC, CH2Cl2
Et3SiH, PhCH3, 300°C
OsO4, Py, Et2O
OH
1. MeO2CCOCl, (i-Pr)2NEt
CH3CN, 0°C
2. (n-Bu3)SnH, PhCH3, Δ
3 N HCl, Δ

The enantioselective synthesis of (-)-bilobalide was achieved based on successful synthesis of the chiral enone **A** and the highly stereoselective reduction of enone **A** to the desired α-alcohol **B**. Further transformation to (-)-bilobalide was accomplished following the route used for racemic bilobalide (Ref. 2).

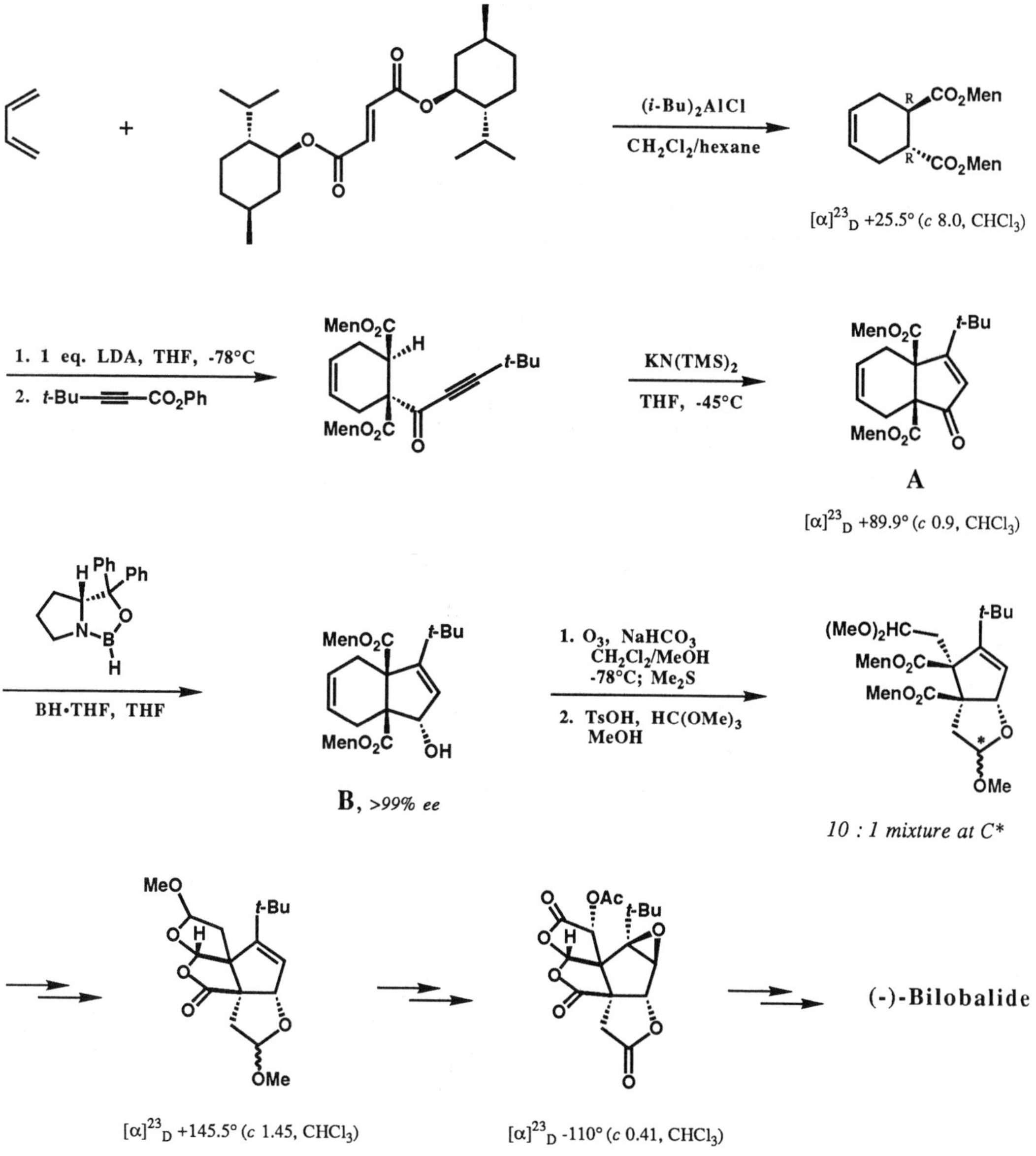

(i-Bu)2AlCl
CH2Cl2/hexane
CO2Men
CO2Men
[α]23D +25.5° (c 8.0, CHCl3)
1. 1 eq. LDA, THF, -78°C
2. t-Bu—≡—CO2Ph
MenO2C
MenO2C
t-Bu
KN(TMS)2
THF, -45°C
A
[α]23D +89.9° (c 0.9, CHCl3)
BH•THF, THF
B, >99% ee
1. O3, NaHCO3 CH2Cl2/MeOH -78°C; Me2S
2. TsOH, HC(OMe)3 MeOH
(MeO)2HC
OMe
10 : 1 mixture at C*
MeO
OAc
[α]23D +145.5° (c 1.45, CHCl3)
[α]23D -110° (c 0.41, CHCl3)
(-)-Bilobalide

References:

1. E. J. Corey and W.-g. Su, *J. Am. Chem. Soc.* **1987**, *109*, 7534.
2. E. J. Corey and W.-g. Su, *Tetrahedron Lett.* **1988**, *29*, 3423.

10.14

Forskolin

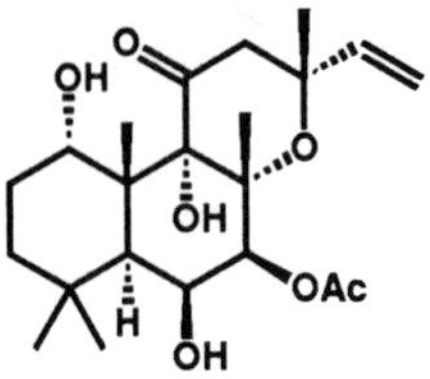

OH
O
O
OH
OAc
H
OH

Forskolin is an activator of the enzyme adenylate cyclase which has therapeutic utility. Outlined below are stereocontrolled routes to racemic and natural chiral forms of forskolin derived by multistrategic retrosynthetic analysis.

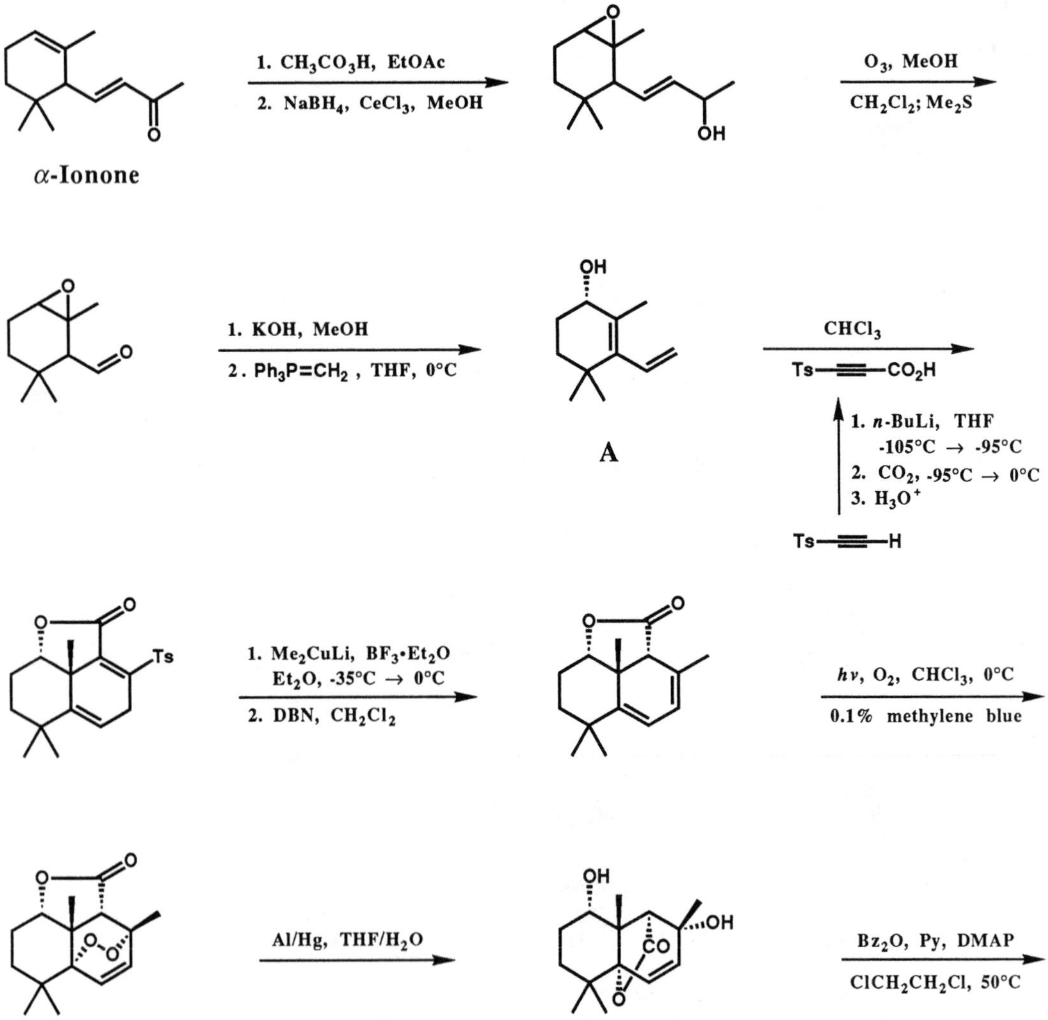

α-Ionone
1. CH_3CO_3H, EtOAc
2. $NaBH_4$, $CeCl_3$, MeOH
O_3, MeOH
CH_2Cl_2; Me_2S
1. KOH, MeOH
2. $Ph_3P{=}CH_2$, THF, 0°C
A
$CHCl_3$
Ts—≡—CO_2H
1. *n*-BuLi, THF -105°C → -95°C
2. CO_2, -95°C → 0°C
3. H_3O^+
Ts—≡—H
1. Me_2CuLi, $BF_3{\cdot}Et_2O$ Et_2O, -35°C → 0°C
2. DBN, CH_2Cl_2
hν, O_2, $CHCl_3$, 0°C
0.1% methylene blue
Al/Hg, THF/H_2O
Bz_2O, Py, DMAP
$ClCH_2CH_2Cl$, 50°C

10.14

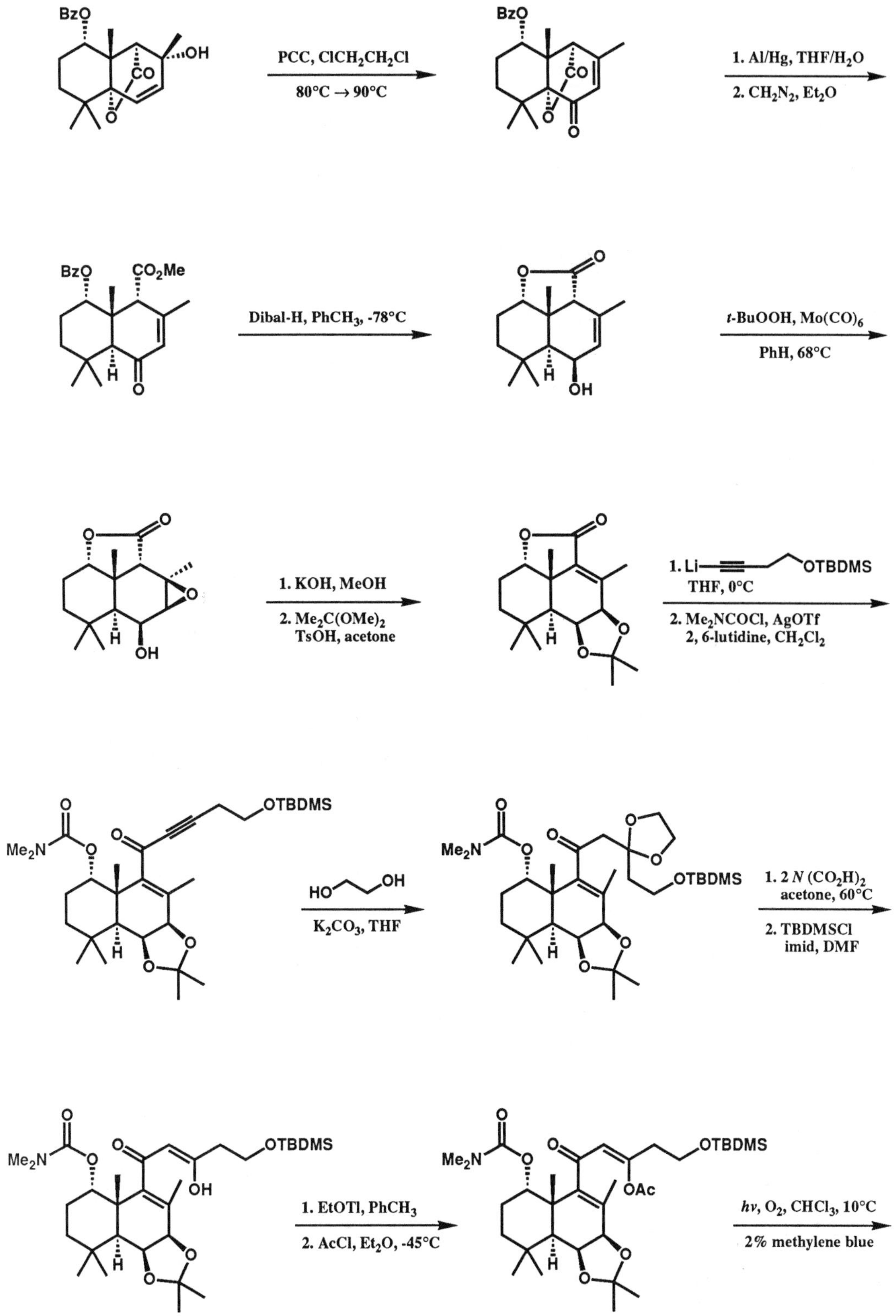
BzO
OH
CO
O
PCC, ClCH2CH2Cl
80°C → 90°C
1. Al/Hg, THF/H2O
2. CH2N2, Et2O
CO2Me
H
Dibal-H, PhCH3, -78°C
t-BuOOH, Mo(CO)6
PhH, 68°C
1. KOH, MeOH
2. Me2C(OMe)2
TsOH, acetone
1. Li—≡—OTBDMS
THF, 0°C
2. Me2NCOCl, AgOTf
2, 6-lutidine, CH2Cl2
Me2N
OTBDMS
HO—OH
K2CO3, THF
1. 2 N (CO2H)2
acetone, 60°C
2. TBDMSCl
imid, DMF
1. EtOTl, PhCH3
2. AcCl, Et2O, -45°C
OAc
hv, O2, CHCl3, 10°C
2% methylene blue

10.14

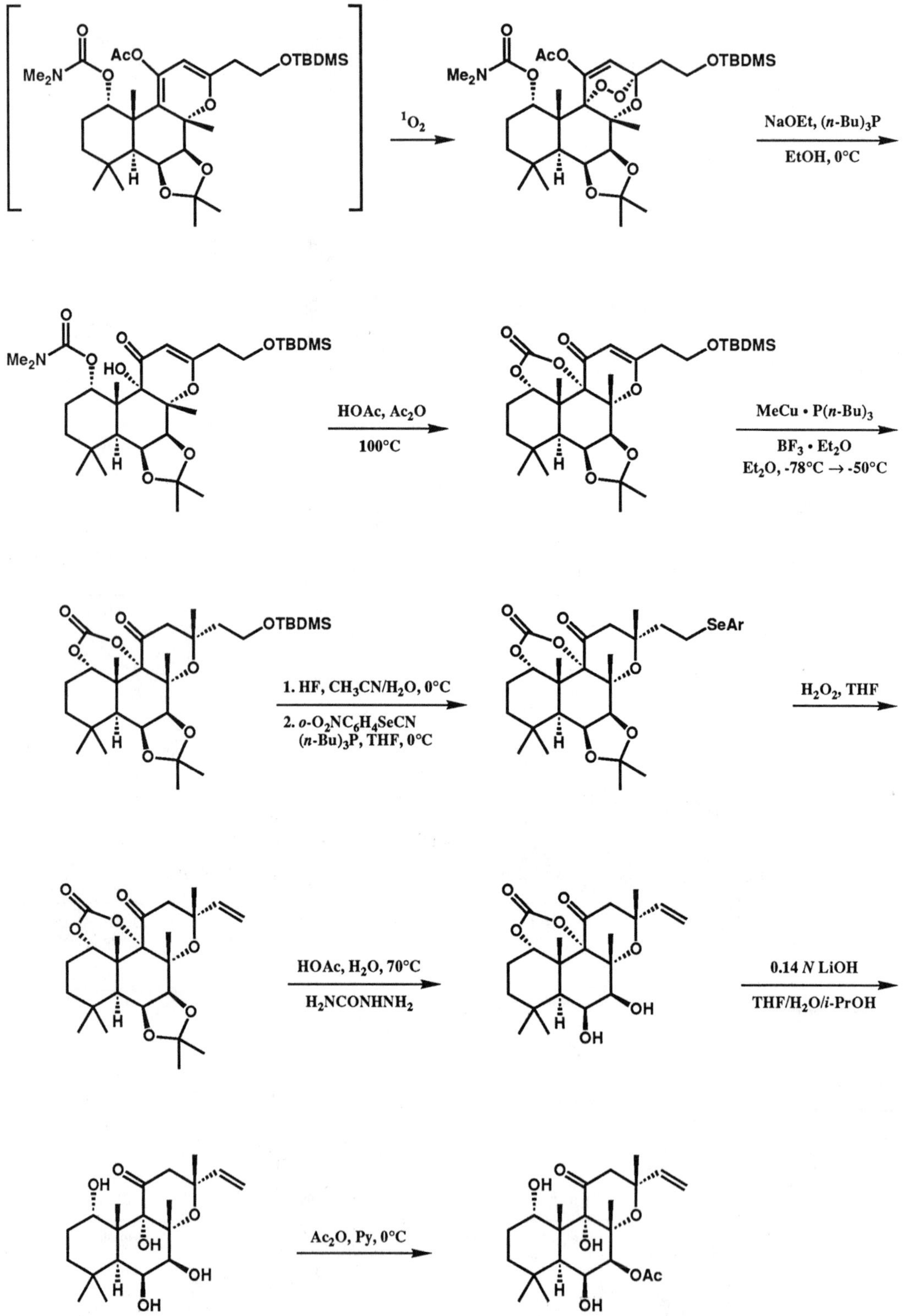

Me2N
O
AcO
OTBDMS
H
1O2
NaOEt, (n-Bu)3P
EtOH, 0°C
HO
HOAc, Ac2O
100°C
MeCu • P(n-Bu)3
BF3 • Et2O
Et2O, -78°C → -50°C
1. HF, CH3CN/H2O, 0°C
2. o-O2NC6H4SeCN
(n-Bu)3P, THF, 0°C
SeAr
H2O2, THF
HOAc, H2O, 70°C
H2NCONHNH2
OH
0.14 N LiOH
THF/H2O/i-PrOH
Ac2O, Py, 0°C
OAc

The hydroxydiene **A** could be obtained by enantioselective CBS reduction of dienone **B** in 90% *ee*, which led to an enantioselective synthesis of the natural occurring form of forskolin.

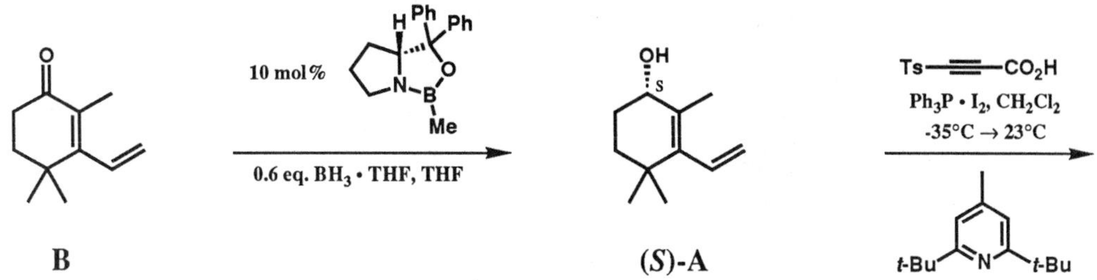
10 mol%
0.6 eq. $BH_3 \cdot THF$, THF
Ts—≡—CO_2H
$Ph_3P \cdot I_2$, CH_2Cl_2
-35°C → 23°C
t-Bu
t-Bu

B

(*S*)-A

$[\alpha]^{23}_D$ -53° (*c* 1.0, $CHCl_3$)
enantioselectivity 95 : 5

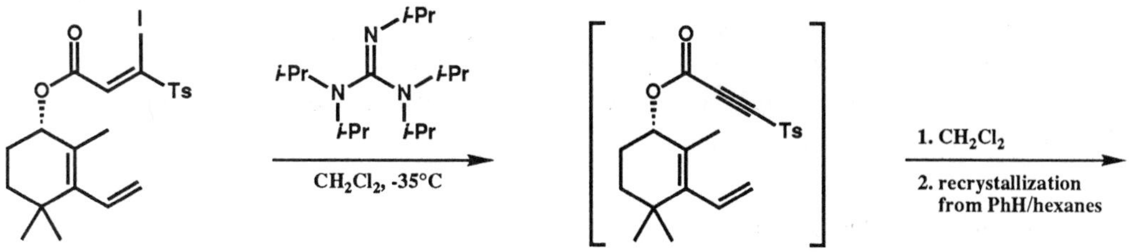
i-Pr
i-Pr
i-Pr
i-Pr
i-Pr
Ts
CH_2Cl_2, -35°C
Ts
1. CH_2Cl_2
2. recrystallization from PhH/hexanes

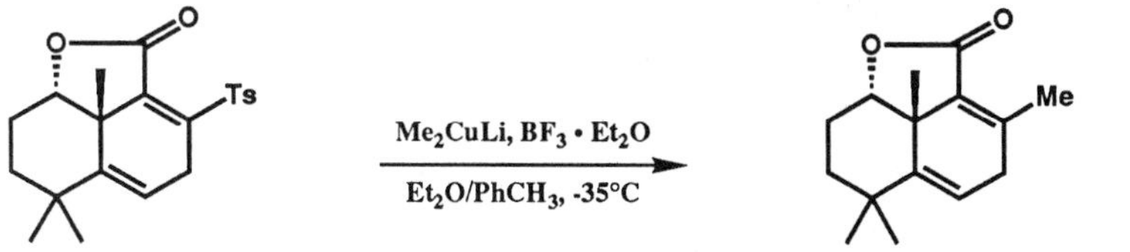
Ts
Me_2CuLi, $BF_3 \cdot Et_2O$
$Et_2O/PhCH_3$, -35°C
Me

$[\alpha]^{23}_D$ +8.7° (*c* 1.0, $CHCl_3$)
mp 153°C - 154°C, 100% *ee*

$[\alpha]^{23}_D$ -70.7° (*c* 1.0, $CHCl_3$)

Reference:

1. E. J. Corey, P. Da Silva Jardine, and J. C. Rohloff, *J. Am. Chem. Soc.* **1988**, *110*, 3672.

2. E. J. Corey, P. Da Silva Jardine, and T. Mohri, *Tetrahedron Lett.* **1988**, *29*, in press.

10.15

Venustatriol

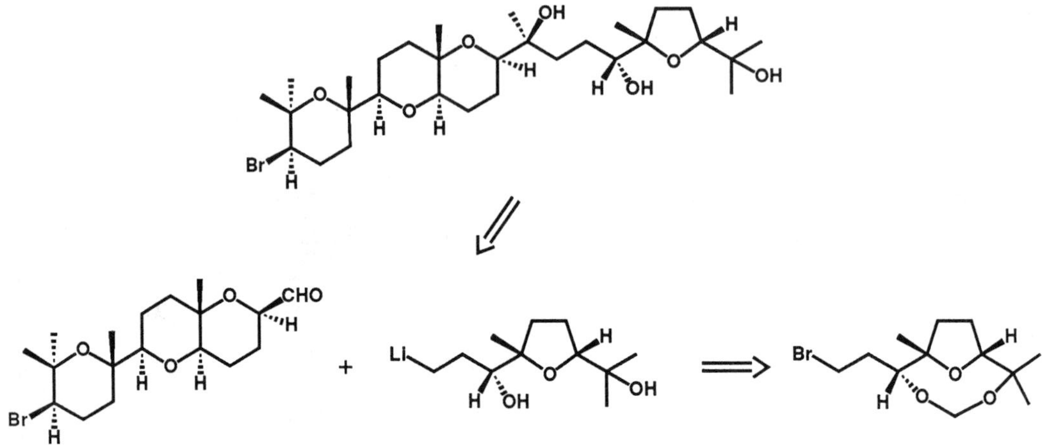

OH
H
H
OH
OH
O
O
O
O
H
H
Br
H
CHO
H
Li
H
OH
OH
+
Br
H
H
O
O

A B

Venustatriol, a marine-derived antiviral agent, as with many polyether structures, is a straightforward problem for retrosynthetic analysis. The major issues, clearance of stereocenters and topologically strategic disconnection, were readily resolved to generate the pathway of synthesis described below.

1. Synthesis of the fragment **A**:

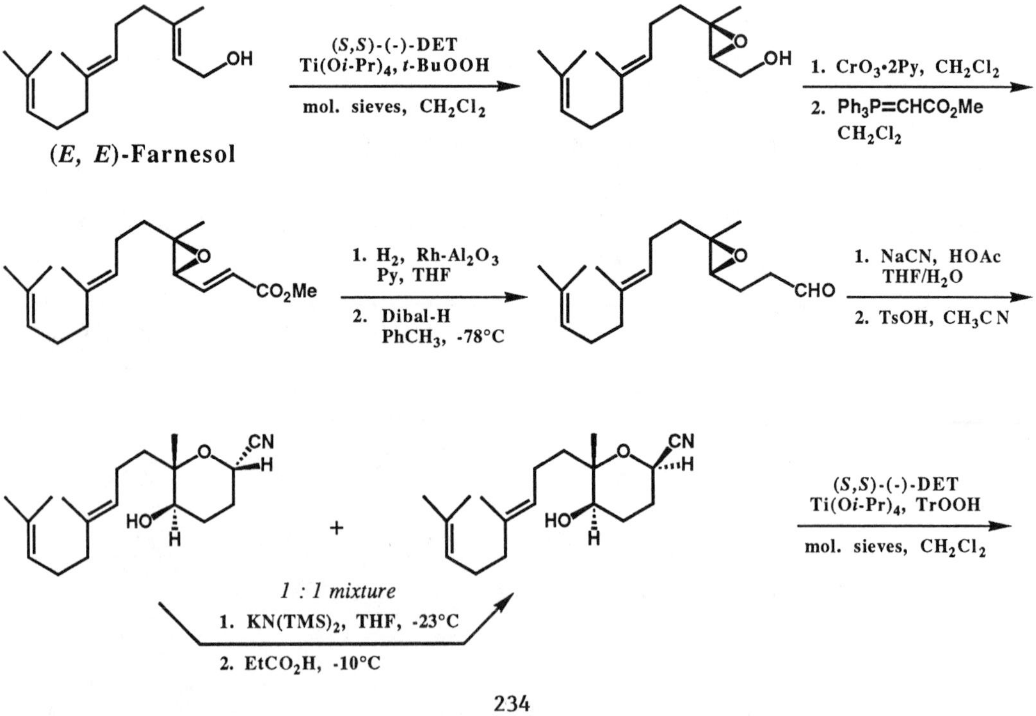

OH
(S,S)-(-)-DET
Ti(Oi-Pr)4, t-BuOOH
mol. sieves, CH2Cl2
O
OH
1. CrO3•2Py, CH2Cl2
2. Ph3P=CHCO2Me
CH2Cl2
(E, E)-Farnesol
O
CO2Me
1. H2, Rh-Al2O3
Py, THF
2. Dibal-H
PhCH3, -78°C
O
CHO
1. NaCN, HOAc
THF/H2O
2. TsOH, CH3CN
O
CN
H
HO
H
+
O
CN
H
HO
H
(S,S)-(-)-DET
Ti(Oi-Pr)4, TrOOH
mol. sieves, CH2Cl2
1 : 1 mixture
1. KN(TMS)2, THF, -23°C
2. EtCO2H, -10°C

10.15

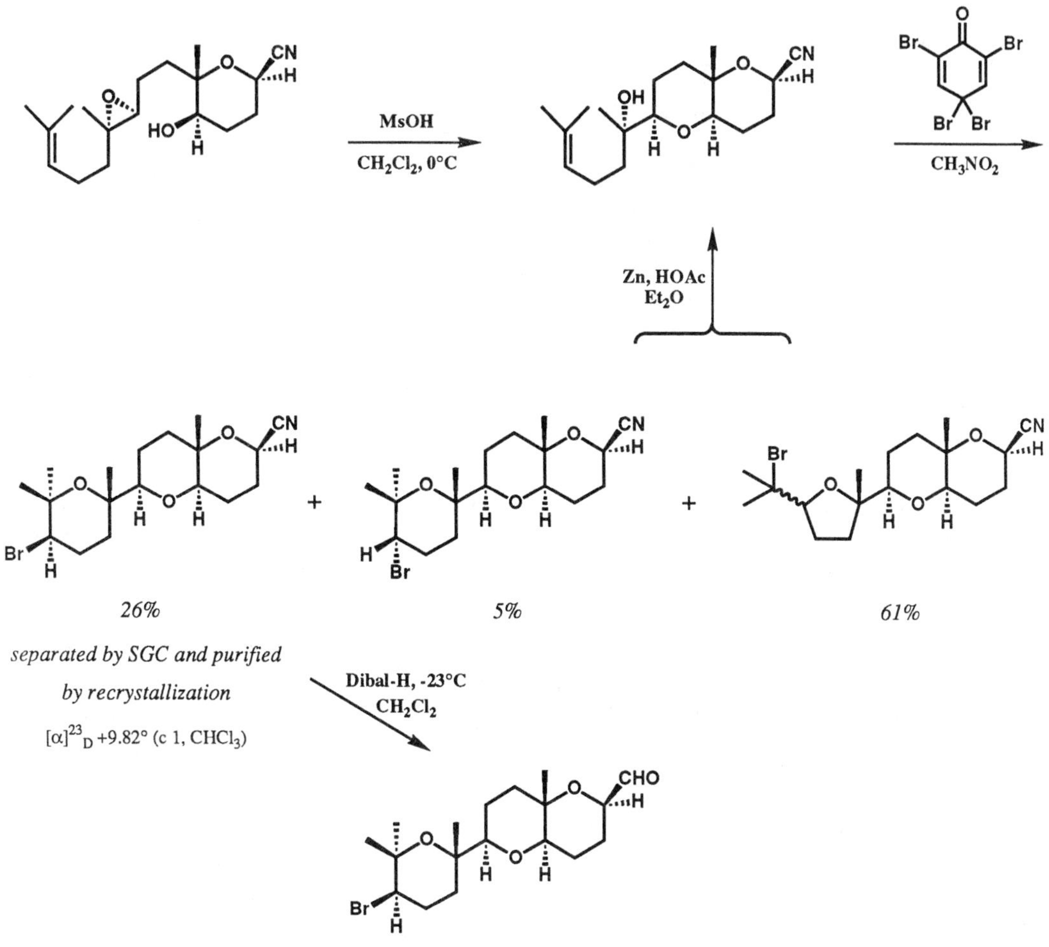
MsOH
CH2Cl2, 0°C
CH3NO2
Zn, HOAc
Et2O
26%
5%
61%
separated by SGC and purified by recrystallization
[α]23D +9.82° (c 1, CHCl3)
Dibal-H, -23°C
CH2Cl2

A

2. Synthesis of the fragment **B**:

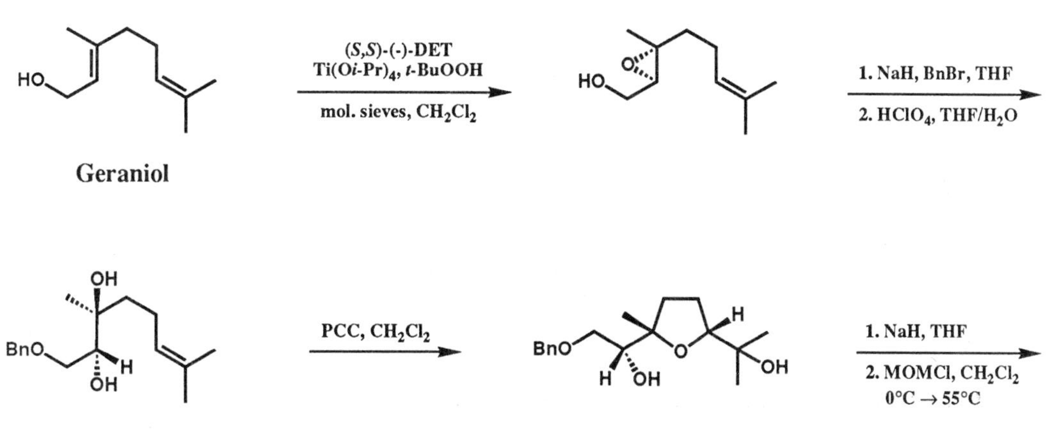
(S,S)-(-)-DET
Ti(Oi-Pr)4, t-BuOOH
mol. sieves, CH2Cl2
1. NaH, BnBr, THF
2. HClO4, THF/H2O
Geraniol
PCC, CH2Cl2
1. NaH, THF
2. MOMCl, CH2Cl2
0°C → 55°C

10.15

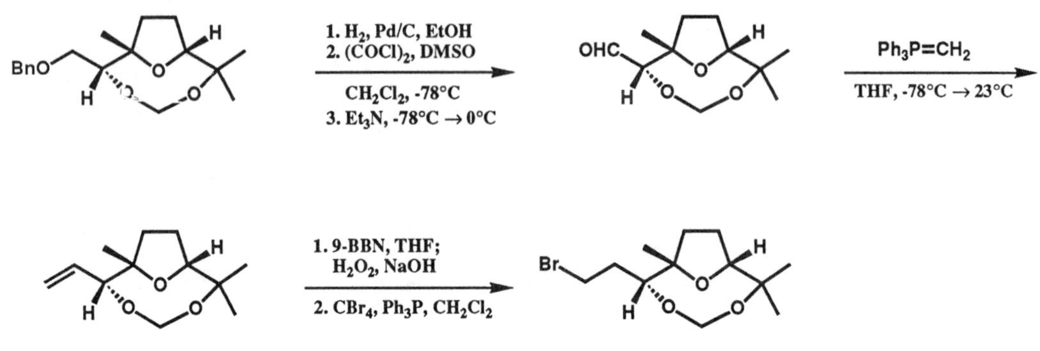

B

$[\alpha]^{23}_{D}$ +79.8° (*c* 2.5, $CHCl_3$)

3. Coupling of the fragments **A** and **B** and the completion of the synthesis:

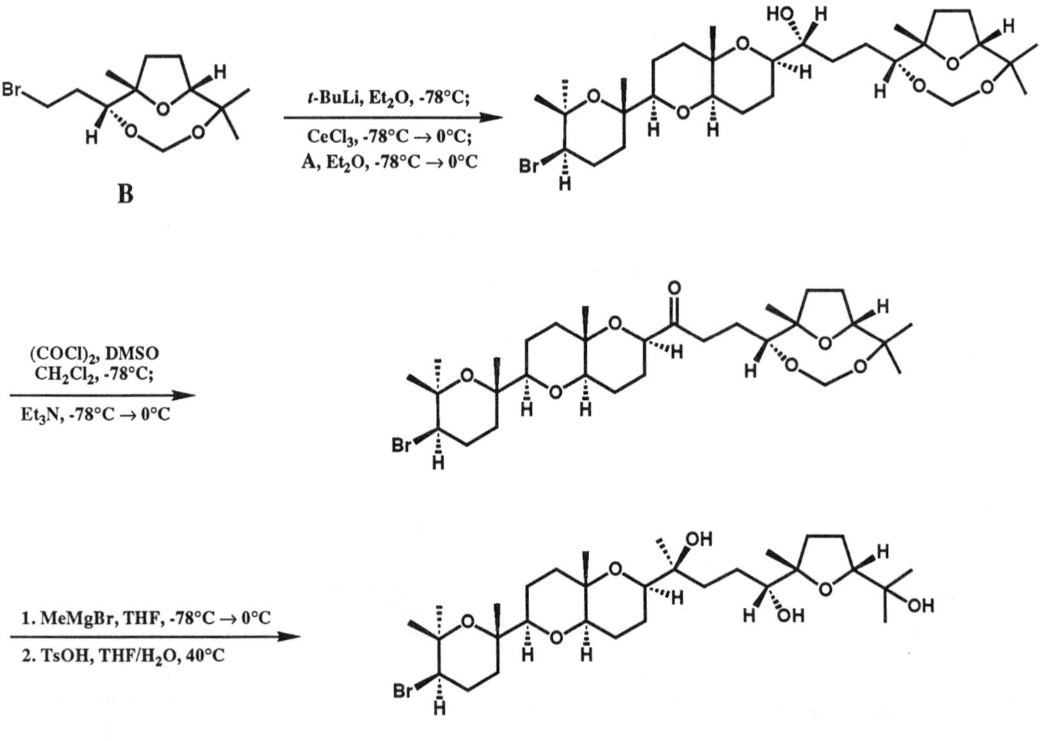

synthetic sample: $[\alpha]^{23}_{D}$ +9.3° (*c* 0.6, $CHCl_3$)

natural ***Venustatriol***: $[\alpha]^{23}_{D}$ +9.4° (*c* 3.2, $CHCl_3$)

Reference:

E. J. Corey and D.-C. Ha, *Tetrahedron Lett.* **1988**, *29*, 3171.

Pseudopterosin A

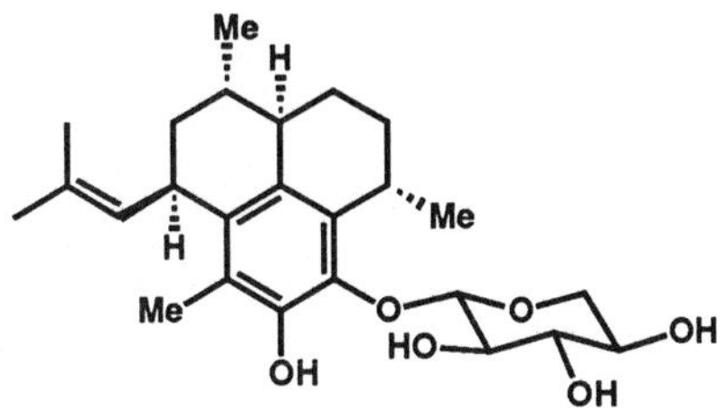
Me
H
Me
H
Me
O
O
OH
HO
OH
OH

Pseudopterosin A is a member of a group of marine natural products which show potent antiinflammatory properties, but which are not prostaglandin biosynthesis inhibitors. Structurally similar to phosphatidyl inositol, they may function as phospholipase inhibitors, and, as such, may be the forerunners of a new class of therapeutic agents.

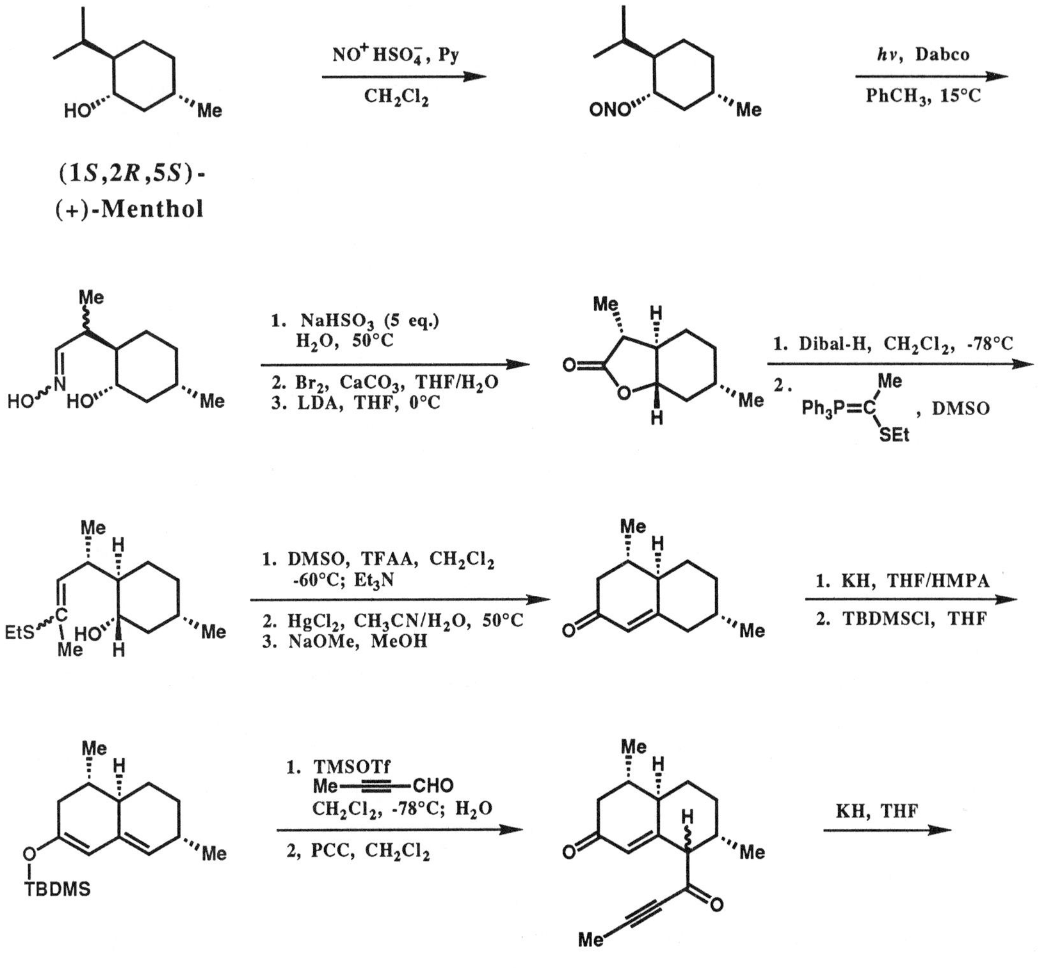
(1S,2R,5S)-
(+)-Menthol
NO$^+$ HSO$_4^-$, Py
CH$_2$Cl$_2$
ONO
hν, Dabco
PhCH$_3$, 15°C
1. NaHSO$_3$ (5 eq.) H$_2$O, 50°C
2. Br$_2$, CaCO$_3$, THF/H$_2$O
3. LDA, THF, 0°C
1. Dibal-H, CH$_2$Cl$_2$, -78°C
2. Ph$_3$P=C(Me)SEt, DMSO
1. DMSO, TFAA, CH$_2$Cl$_2$ -60°C; Et$_3$N
2. HgCl$_2$, CH$_3$CN/H$_2$O, 50°C
3. NaOMe, MeOH
1. KH, THF/HMPA
2. TBDMSCl, THF
TBDMS
1. TMSOTf Me—≡—CHO CH$_2$Cl$_2$, -78°C; H$_2$O
2, PCC, CH$_2$Cl$_2$
KH, THF

10.16

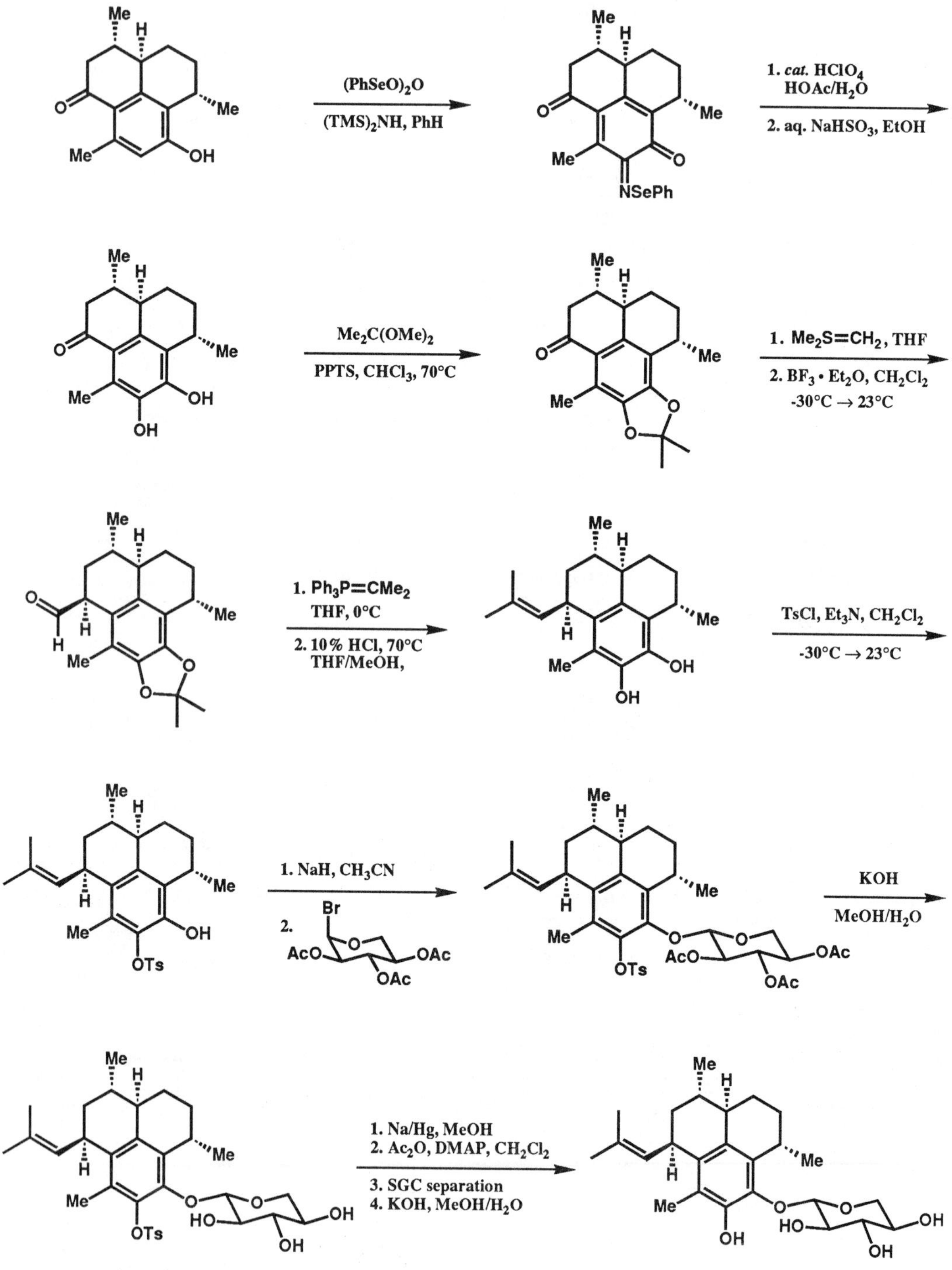

(PhSeO)2O
(TMS)2NH, PhH
1. cat. HClO4 HOAc/H2O
2. aq. NaHSO3, EtOH
Me2C(OMe)2
PPTS, CHCl3, 70°C
1. Me2S=CH2, THF
2. BF3 • Et2O, CH2Cl2 -30°C → 23°C
1. Ph3P=CMe2 THF, 0°C
2. 10% HCl, 70°C THF/MeOH,
TsCl, Et3N, CH2Cl2
-30°C → 23°C
1. NaH, CH3CN
2.
KOH
MeOH/H2O
1. Na/Hg, MeOH
2. Ac2O, DMAP, CH2Cl2
3. SGC separation
4. KOH, MeOH/H2O

Reference:

E. J. Corey and P. Carpino, Submitted for publication.

Partial Synthesis of α-Amyrin Acetate
Proof of the Structure and Stereochemistry

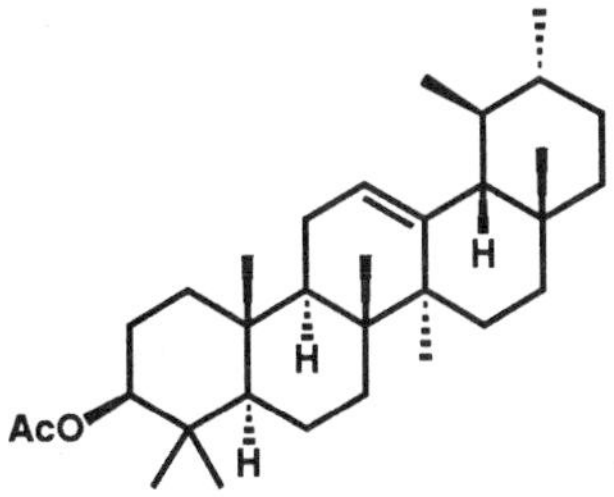
AcO
H
H
H

The comparison of degradation products from α- and β-amyrin led to the conclusion that these large classes of triterpenoids differed in absolute configuration. That such was not the case was established by synthesis of α-amyrin from a β-amyrin derivative, glycyrrhetic acid, an obvious SM goal. The synthesis of α-amyrin also demonstrated the relative stereochemistry beyond doubt. This *cis* D/E ring of α-amyrin is thermodynamically more stable than the *trans* arrangement mainly because of anchoring of ring E by the two equatorial non-angular methyl substituents on that ring.

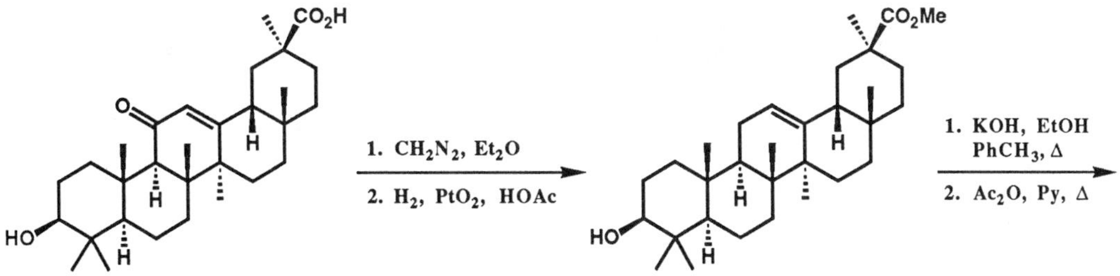
CO2H
O
HO
H
H
H
1. CH2N2, Et2O
2. H2, PtO2, HOAc
CO2Me
HO
H
H
H
1. KOH, EtOH
PhCH3, Δ
2. Ac2O, Py, Δ

Glycyrrhetic Acid
(a known derivative
of **β-Amyrin**)

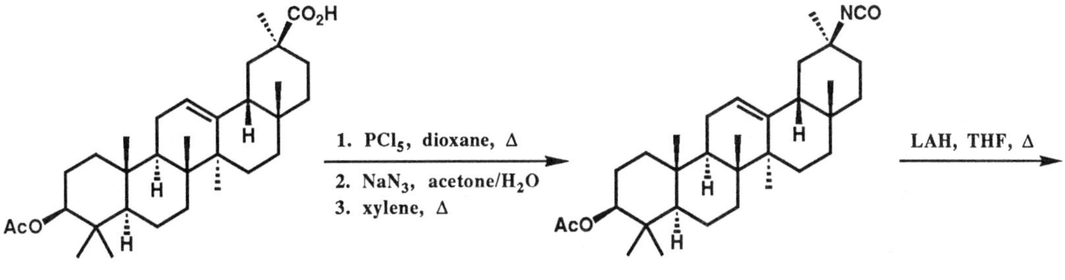
CO2H
AcO
H
H
H
1. PCl5, dioxane, Δ
2. NaN3, acetone/H2O
3. xylene, Δ
NCO
AcO
H
H
H
LAH, THF, Δ

10.17

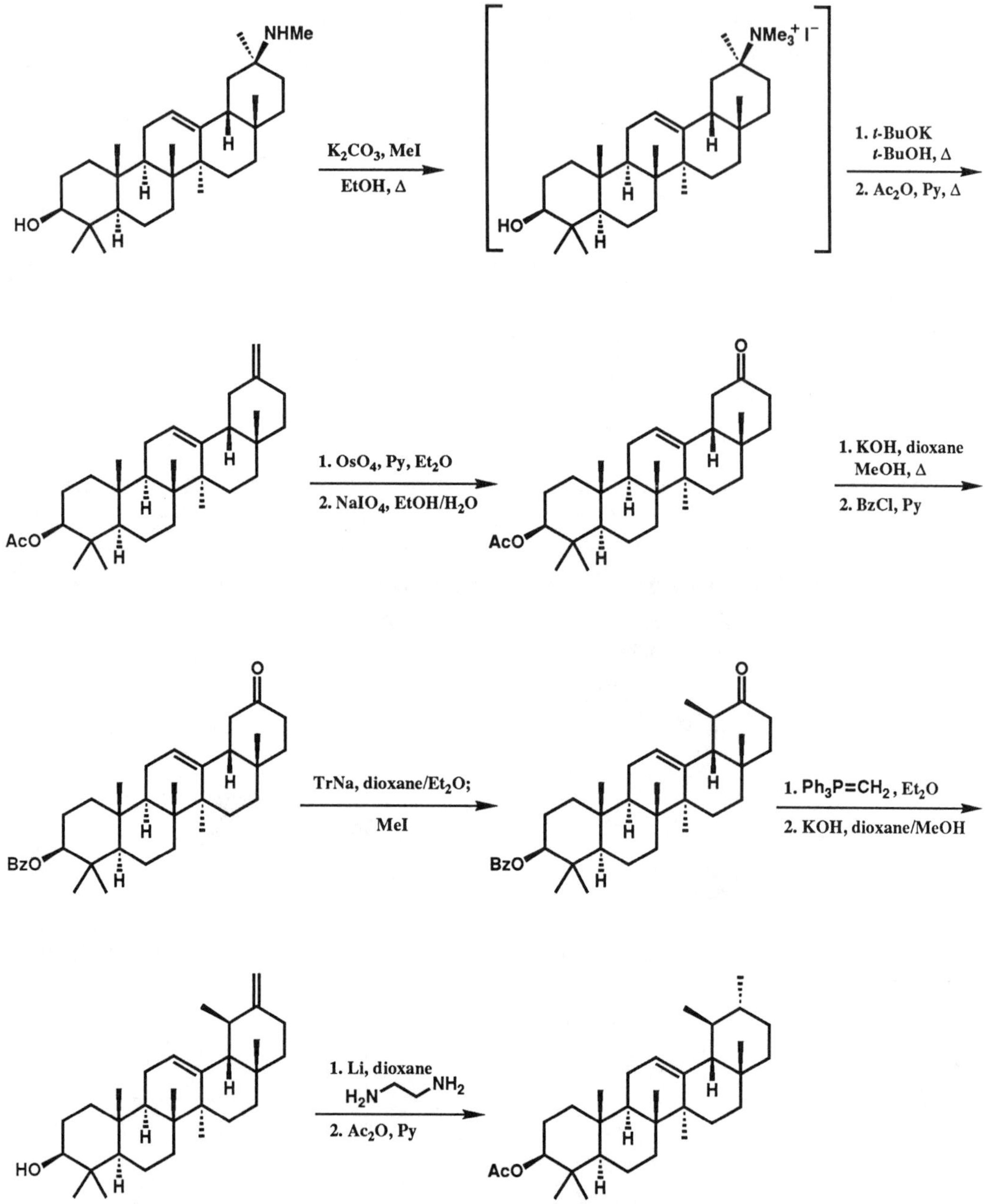

NHMe
K₂CO₃, MeI
EtOH, Δ
NMe₃⁺ I⁻
1. t-BuOK
t-BuOH, Δ
2. Ac₂O, Py, Δ
1. OsO₄, Py, Et₂O
2. NaIO₄, EtOH/H₂O
1. KOH, dioxane
MeOH, Δ
2. BzCl, Py
TrNa, dioxane/Et₂O;
MeI
1. Ph₃P=CH₂, Et₂O
2. KOH, dioxane/MeOH
1. Li, dioxane
H₂N NH₂
2. Ac₂O, Py
HO
AcO
BzO
H
O

α-Amyrin acetate

References:

1. E. J. Corey and E. W. Cantrall, *J. Am. Chem. Soc.* **1958**, *80*, 499.
2. E. J. Corey and E. W. Cantrall, *J. Am. Chem. Soc.* **1959**, *81*, 1745.

Olean-11,12;13,18-Diene, a β-Amyrin Derivative

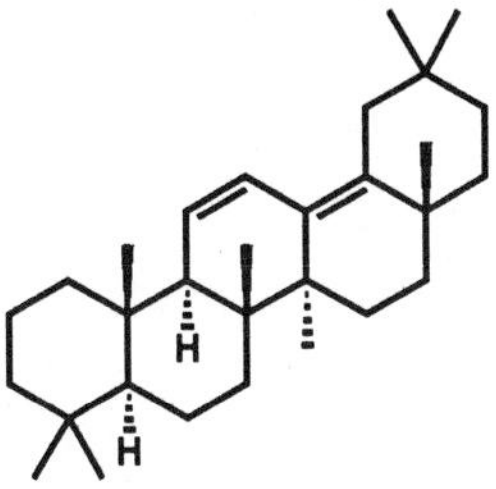

The first synthesis of a β-amyrin derivative was accomplished by a convergent route which depended on cation-olefin cyclization to form the critical central ring. The plan of synthesis was largely guided by the selection of SM goals for the A/B and E ring portions of the target.

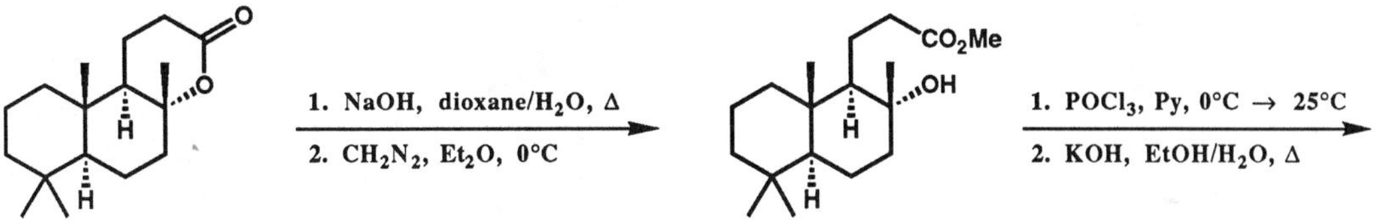
O
O
1. NaOH, dioxane/H_2O, Δ
2. CH_2N_2, Et_2O, 0°C
CO_2Me
OH
1. $POCl_3$, Py, 0°C → 25°C
2. KOH, EtOH/H_2O, Δ
H
H

(+)-Ambreinolide

(Ref. 3)

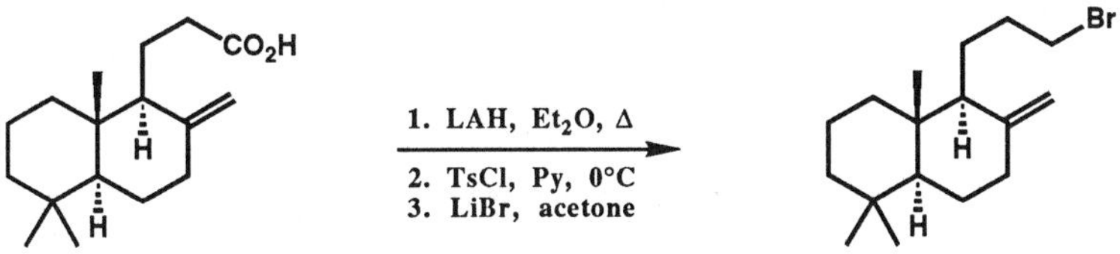
CO_2H
1. LAH, Et_2O, Δ
2. TsCl, Py, 0°C
3. LiBr, acetone
Br
H
H

A

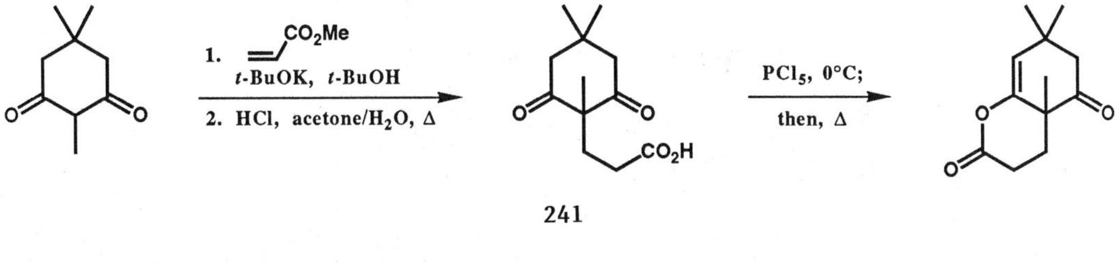
1. CO_2Me
t-BuOK, *t*-BuOH
2. HCl, acetone/H_2O, Δ
CO_2H
PCl_5, 0°C;
then, Δ
O
O

10.18

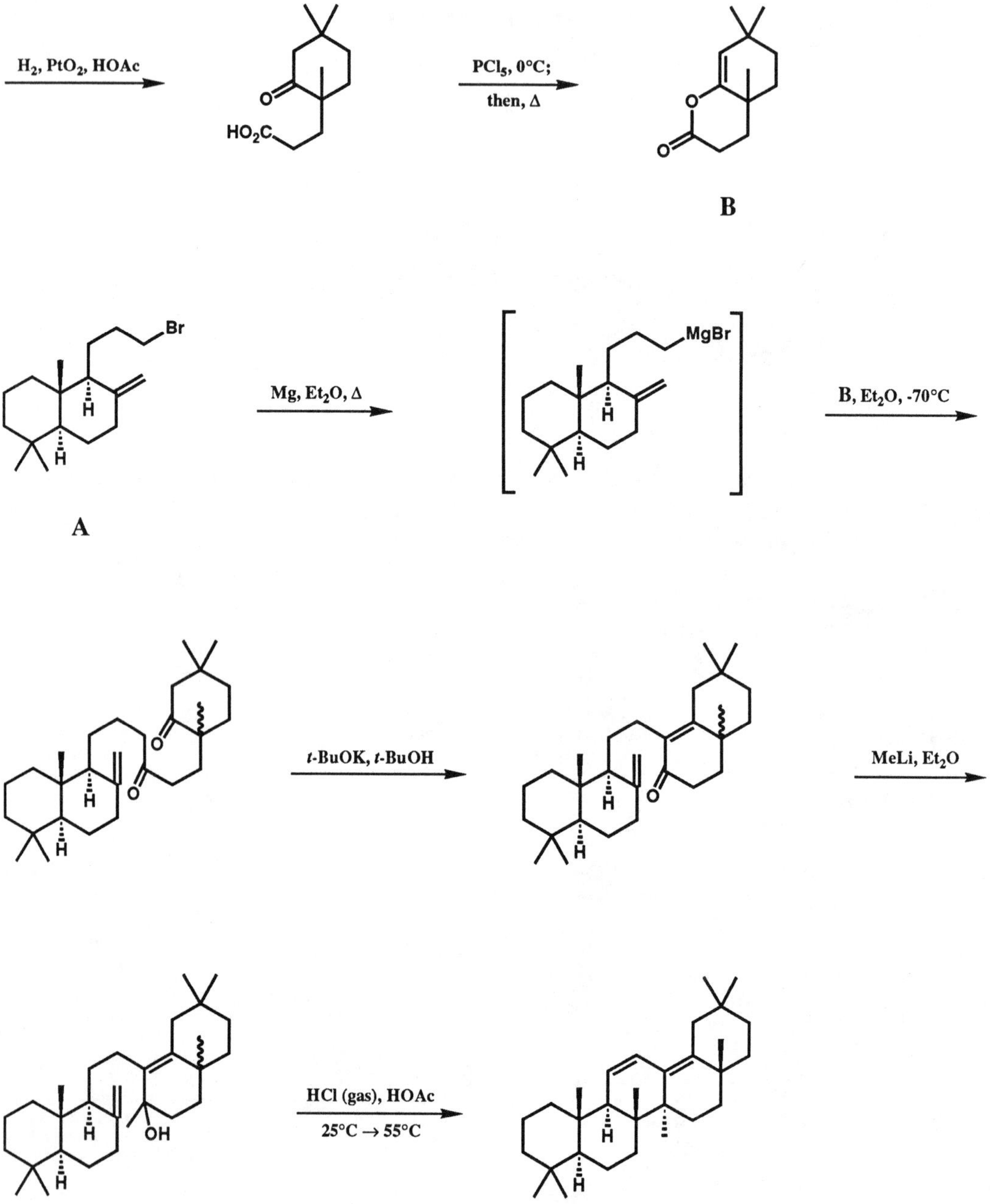
H2, PtO2, HOAc
HO2C
O
PCl5, 0°C;
then, Δ
O
O
B
Br
H
H
A
Mg, Et2O, Δ
MgBr
H
H
B, Et2O, -70°C
O
O
H
H
t-BuOK, t-BuOH
O
H
H
MeLi, Et2O
OH
H
H
HCl (gas), HOAc
25°C → 55°C
H
H

References:

1. E. J. Corey, H.-J. Hess, and S. Proskow, *J. Am. Chem. Soc.* **1959**, *81*, 5258.
2. E. J. Corey, H.-J. Hess, and S. Proskow, *J. Am. Chem. Soc.* **1963**, *85*, 3979.
3. P. Dietrich and E. Lederer, *Helv. Chim. Acta* **1952**, *35*, 1148.

10.19

Pentacyclosqualene

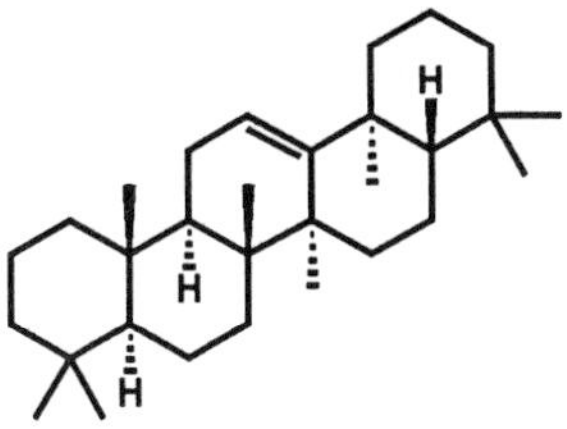
H
H
H

Pentacyclosqualene, the symmetrical hydropicene corresponding to squalene, has not been made by acid-induced cation-olefin cyclization of squalene, despite considerable experimental study. A simple, convergent synthesis of pentacyclosqualene using cation-olefin cyclization to generate ring C was developed. The C_{30}-framework was constructed by radical coupling to a tetracyclic intermediate that was also used for the synthesis of onoceradiene.

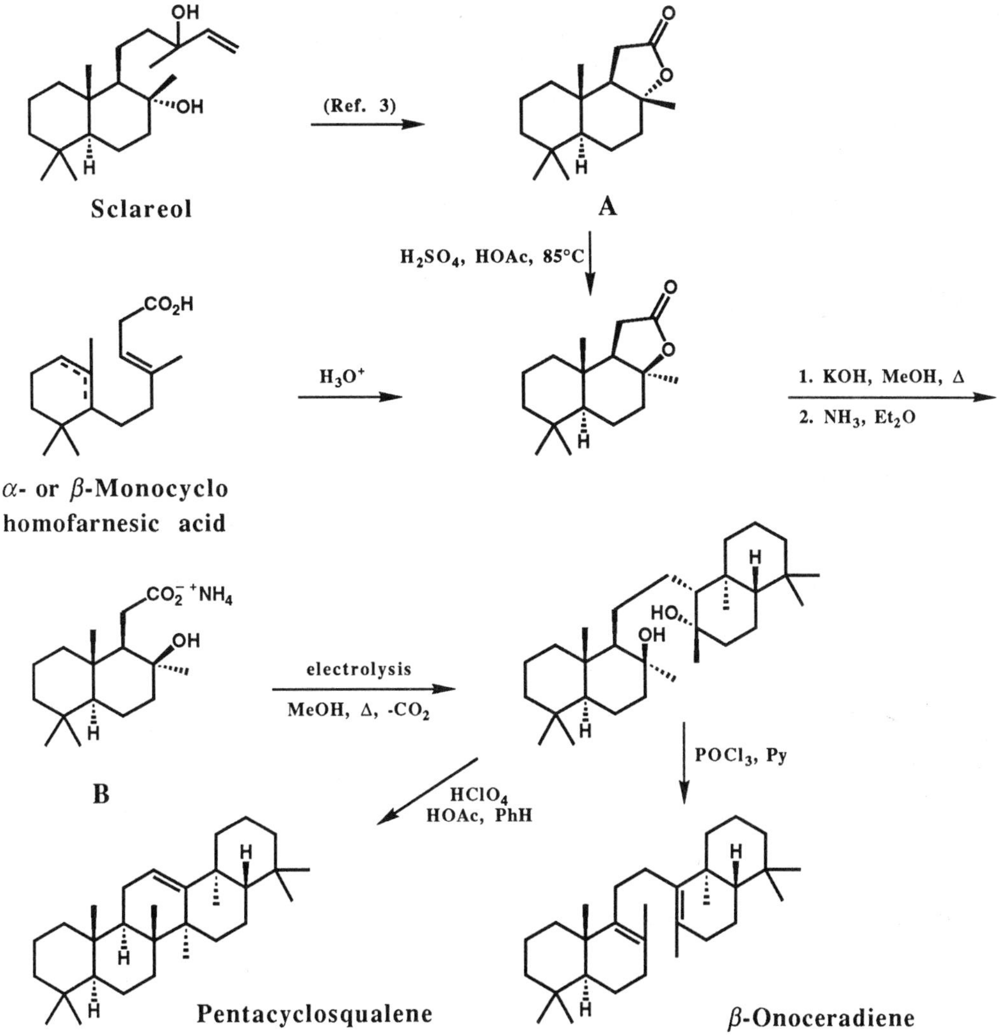
Sclareol
(Ref. 3)
A
H_2SO_4, HOAc, 85°C
α- or β-Monocyclo homofarnesic acid
H_3O^+
1. KOH, MeOH, Δ
2. NH_3, Et_2O
B
electrolysis
MeOH, Δ, $-CO_2$
POCl$_3$, Py
$HClO_4$
HOAc, PhH
Pentacyclosqualene
β-Onoceradiene

10.19

Pentacyclosqualene was also obtained from the isomeric lactone **A** in a similar fashion as shown below.

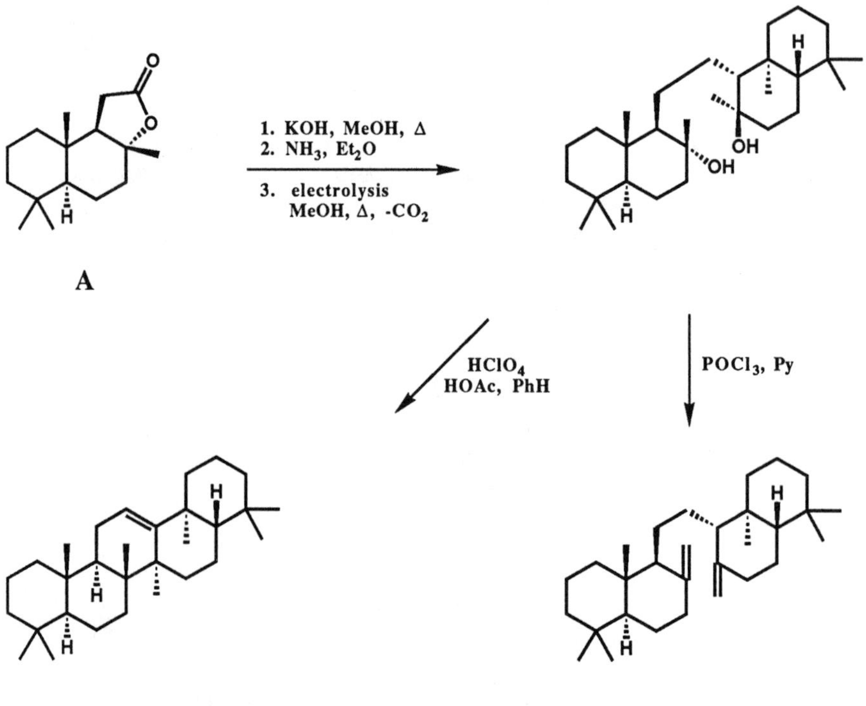

1. KOH, MeOH, Δ
2. NH_3, Et_2O
3. electrolysis
MeOH, Δ, $-CO_2$
OH
OH
H
$HClO_4$
HOAc, PhH
$POCl_3$, Py

A

Pentacyclosqualene **α-Onoceradiene**

It was found later that the electrolytic coupling reaction gave better yield with the acetate corresponding to **B**, since fragmentation was a major side reaction of the γ-hydroxy acid **B** (Ref. 2).

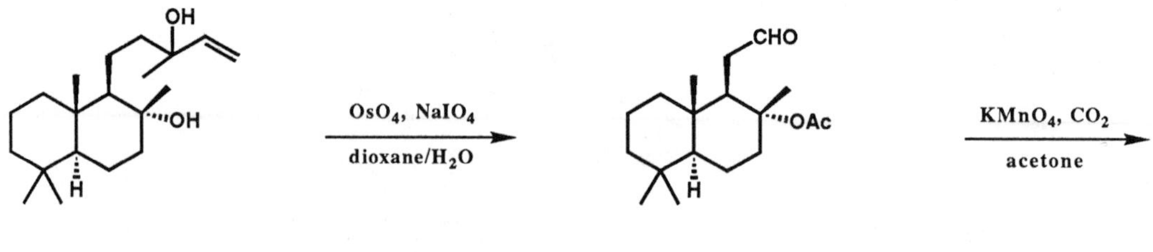

OH
OH
OsO_4, $NaIO_4$
dioxane/H_2O
CHO
OAc
$KMnO_4$, CO_2
acetone

Sclareol

10.19

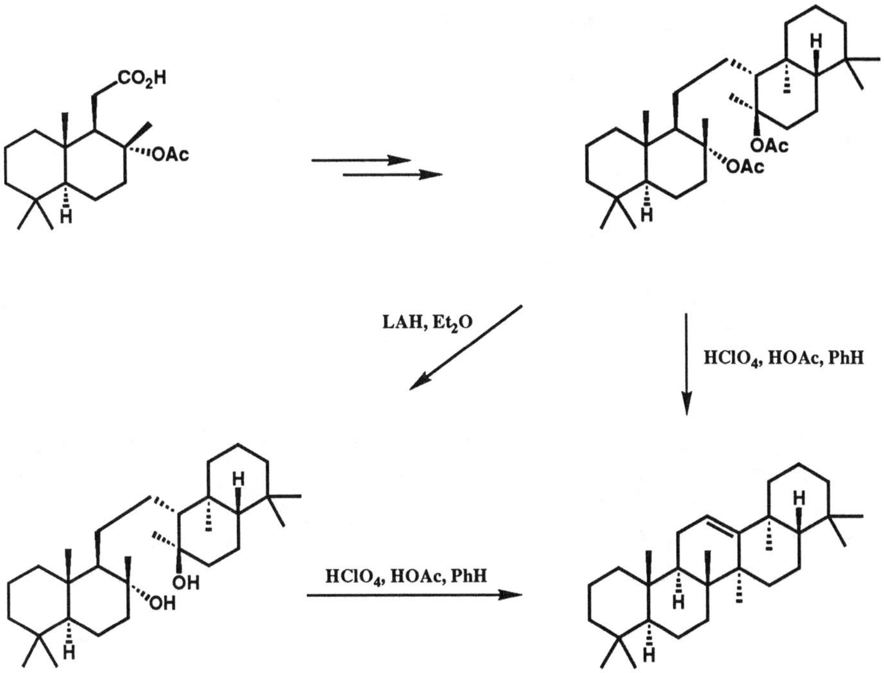
CO2H
OAc
H
OAc
OAc
H
H
LAH, Et2O
HClO4, HOAc, PhH
OH
OH
H
H
HClO4, HOAc, PhH
H
H
H

References:

1. E. J. Corey and R. R. Sauers, *J. Am. Chem. Soc.* **1957**, *79*, 3925.
2. E. J. Corey and R. R. Sauers, *J. Am. Chem. Soc.* **1959**, *81*, 1739.
3. L. Ruzicka and M. M. Janot, *Helv. Chim. Acta* **1931**, *14*, 645.

10.20

Dihydroconessine

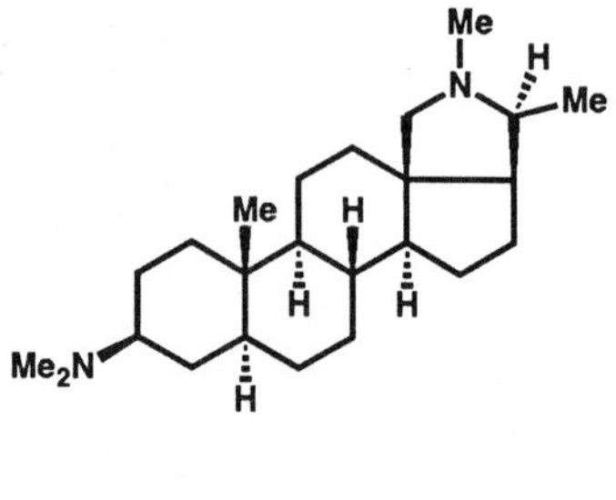

Me
N
H
Me
Me
H
H
H
Me2N
H

A simple synthesis of dihydroconessine from the SM goal 3β-acetoxybisnorcholenic acid was devised by applying concurrently the Hoffmann-Loeffler-Freitag transform to effect retrosynthetic removal of the C(18) functional group of the target structure. In the course of demonstrating the synthesis, a mild new method for the generation of isocyanides was discovered and the first functionalization of a steroidal angular methyl group was achieved.

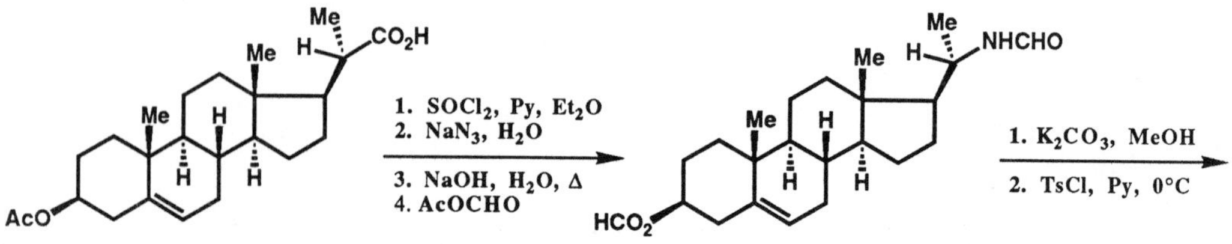

1. SOCl2, Py, Et2O
2. NaN3, H2O
3. NaOH, H2O, Δ
4. AcOCHO
1. K2CO3, MeOH
2. TsCl, Py, 0°C
CO2H
NHCHO
AcO
HCO2

3β-Acetoxybisnorcholenic Acid

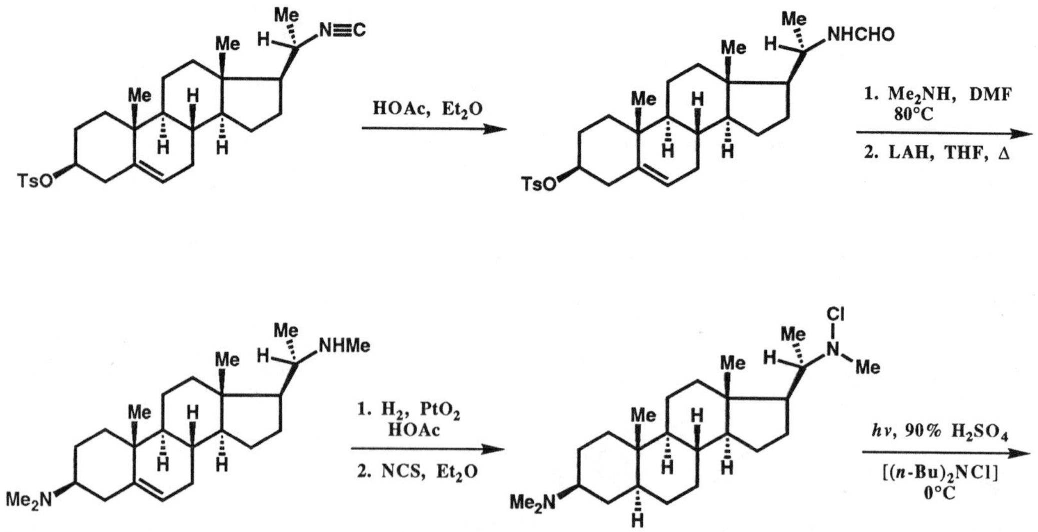

N≡C
HOAc, Et2O
NHCHO
1. Me2NH, DMF
80°C
2. LAH, THF, Δ
TsO
TsO
NHMe
1. H2, PtO2
HOAc
2. NCS, Et2O
Me2N
Me2N
Cl
N
hν, 90% H2SO4
[(n-Bu)2NCl]
0°C

10.20

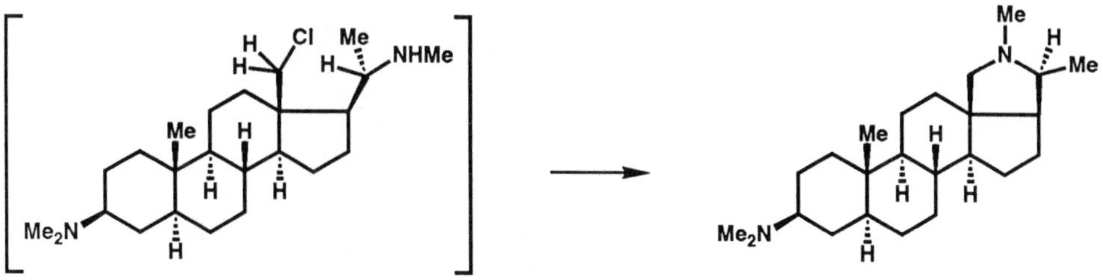
H
Cl
Me
H
H
NHMe
Me
H
H
H
H
Me2N
H
Me
N
H
Me
Me
H
H
H
Me2N
H

References:

1. E. J. Corey and W. R. Hertler, *J. Am. Chem. Soc.* **1958**, *80*, 2903.
2. E. J. Corey and W. R. Hertler, *J. Am. Chem. Soc.* **1959**, *81*, 5209.

CHAPTER ELEVEN

Prostanoids

11.1

Structures of Prostaglandins (PG's)

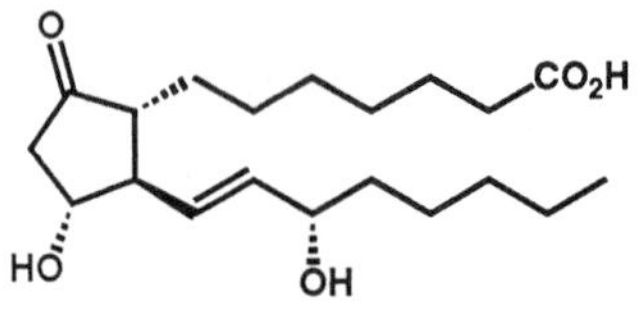
O
CO2H
HO
OH

PGE$_1$

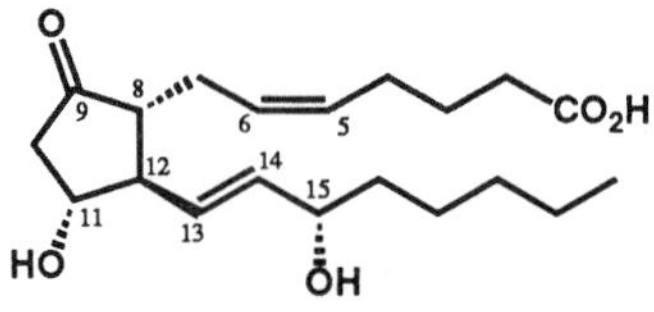
O
8
9
6
5
CO2H
12
14
15
11
13
HO
OH

PGE$_2$

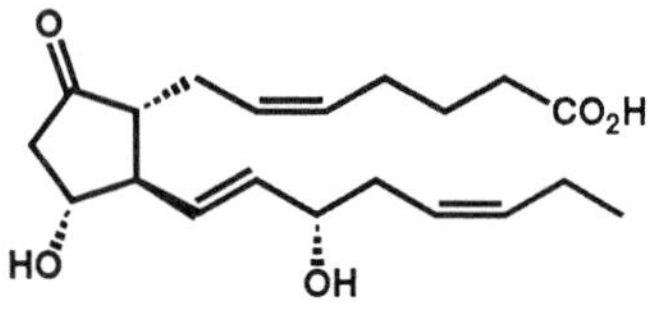
O
CO2H
HO
OH

PGE$_3$

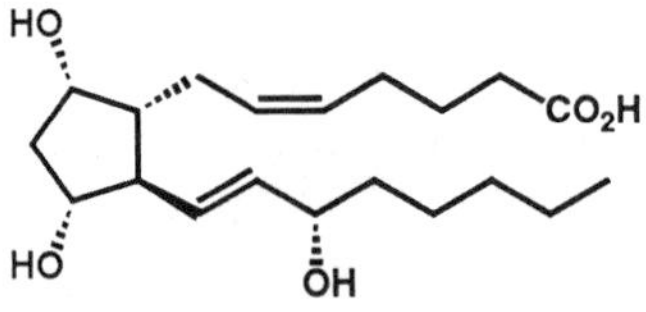
HO
CO2H
HO
OH

PGF$_{2\alpha}$

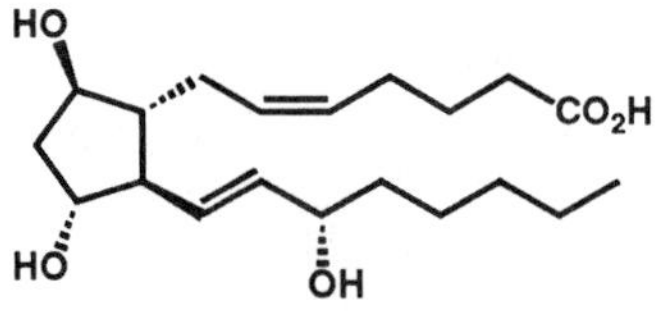
HO
CO2H
HO
OH

PGF$_{2\beta}$

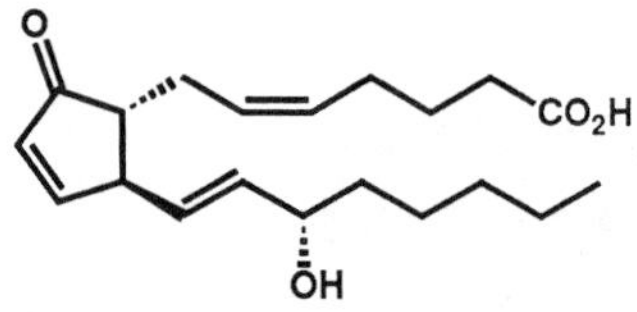
O
CO2H
OH

PGA$_2$

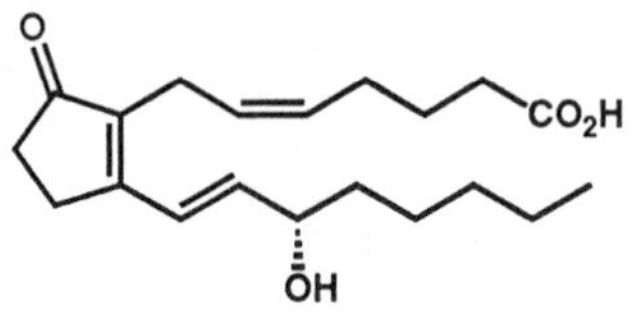
O
CO2H
OH

PGB$_2$

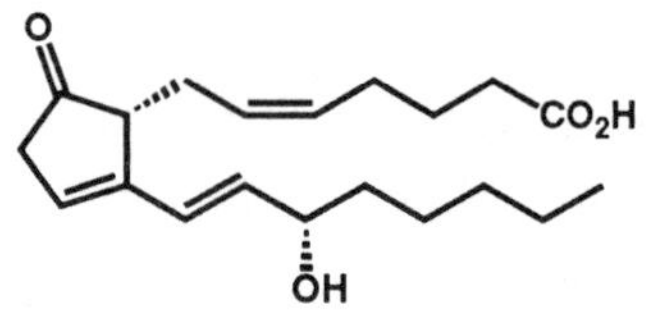
O
CO2H
OH

PGC$_2$

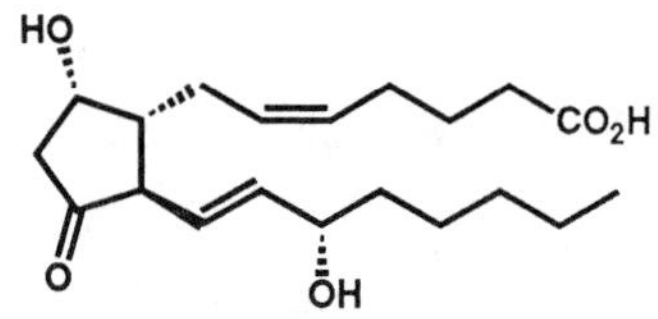
HO
CO2H
O
OH

PGD$_2$

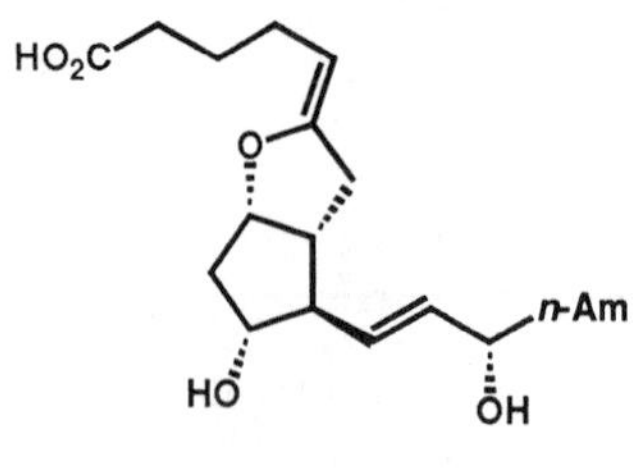
HO2C
O
n-Am
HO
OH

PGI$_2$

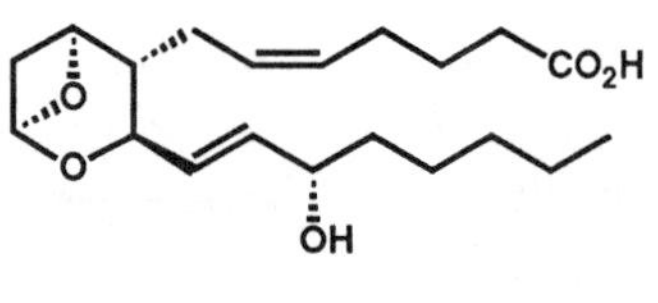
O
CO2H
O
OH

TXA$_2$

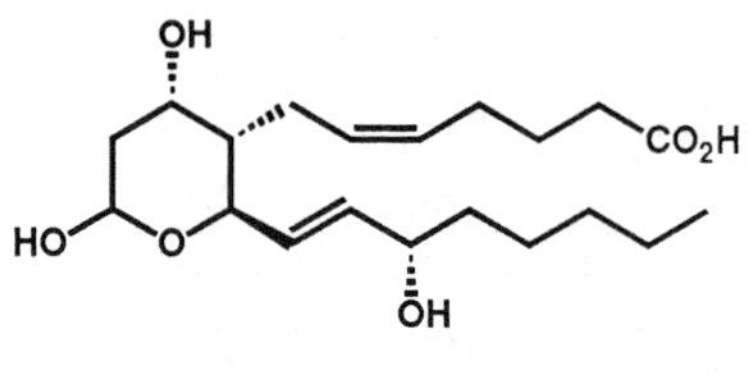
OH
CO2H
HO
O
OH

TXB$_2$

References:

1. A. Mitra, *The Synthesis of Prostaglandins*; J. Wiley-Interscience: New York, 1977.

2. J. S. Bindra and R. Bindra, *Prostaglandin Synthesis*; Academic Press: London, 1977.

3. *New Synthetic Routes to Prostaglandins and Thromboxanes*; S. M. Roberts and F. Scheinmann Eds.; Academic Press: London, 1982.

Total Synthesis of (±)-Prostaglandins E_1, $F_{1\alpha}$, $F_{1\beta}$, A_1, and B_1

The first synthesis of pure primary prostaglandins was designed retrosynthetically to allow the direct synthesis of E prostaglandins (PG's) under the mildest possible conditions to ensure survival of the sensitive β-ketol subunit (see Section 5.9 of Part One). The synthesis involved a number of novel processes including mild methods for transketalization, aldol cyclization, dehydration of β-ketols, amine oxidation, and cleavage of tetrahydropyranyl ethers. An important result leading to second generation syntheses was the finding that PGE's survive tetrahydropyranyl ether cleavage.

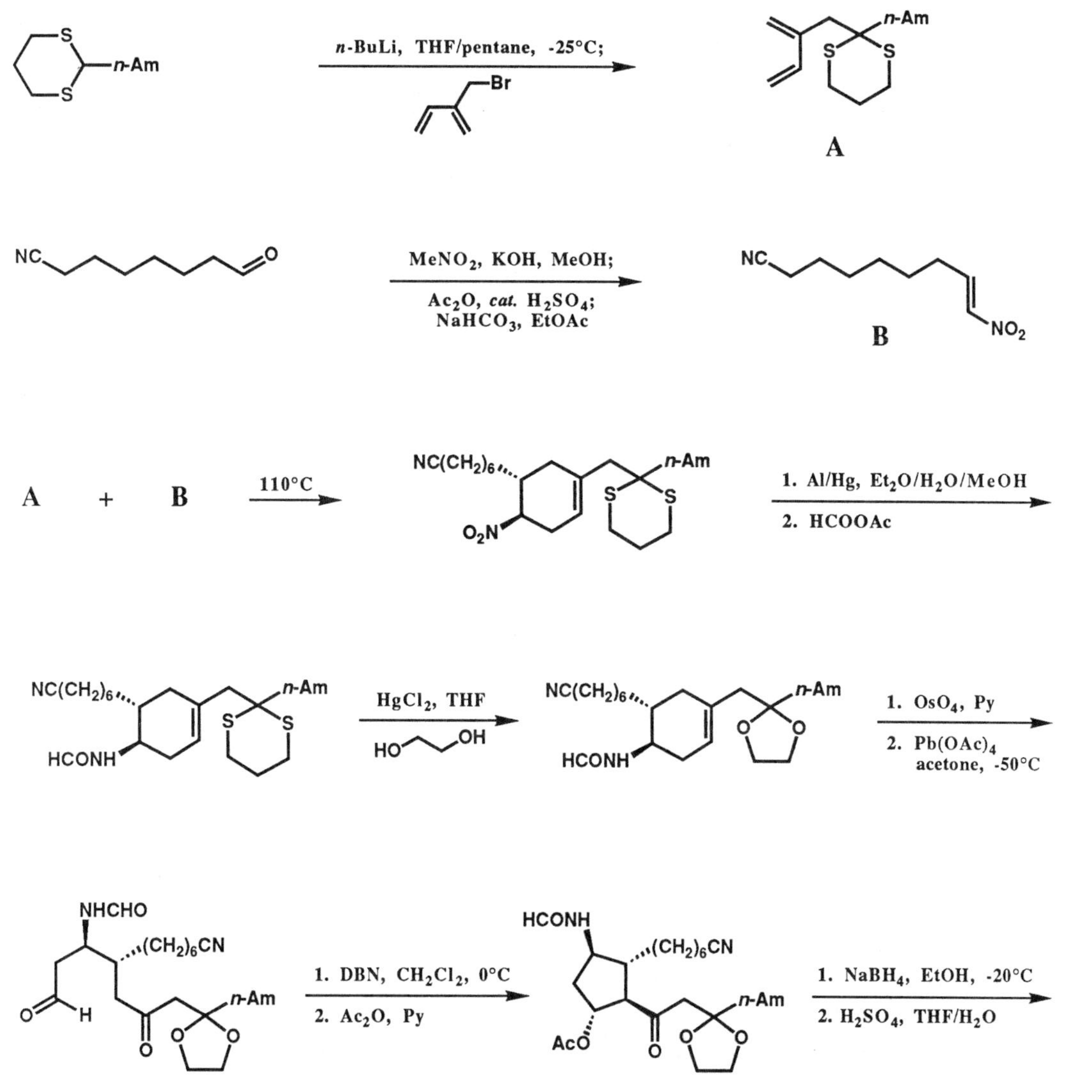
n-Am
n-BuLi, THF/pentane, -25°C;
Br
A
NC
O
MeNO2, KOH, MeOH;
Ac2O, cat. H2SO4;
NaHCO3, EtOAc
B
NO2
A + B
110°C
NC(CH2)6
O2N
1. Al/Hg, Et2O/H2O/MeOH
2. HCOOAc
HCONH
HgCl2, THF
HO OH
1. OsO4, Py
2. Pb(OAc)4
acetone, -50°C
NHCHO
(CH2)6CN
H
1. DBN, CH2Cl2, 0°C
2. Ac2O, Py
AcO
1. NaBH4, EtOH, -20°C
2. H2SO4, THF/H2O

11.2

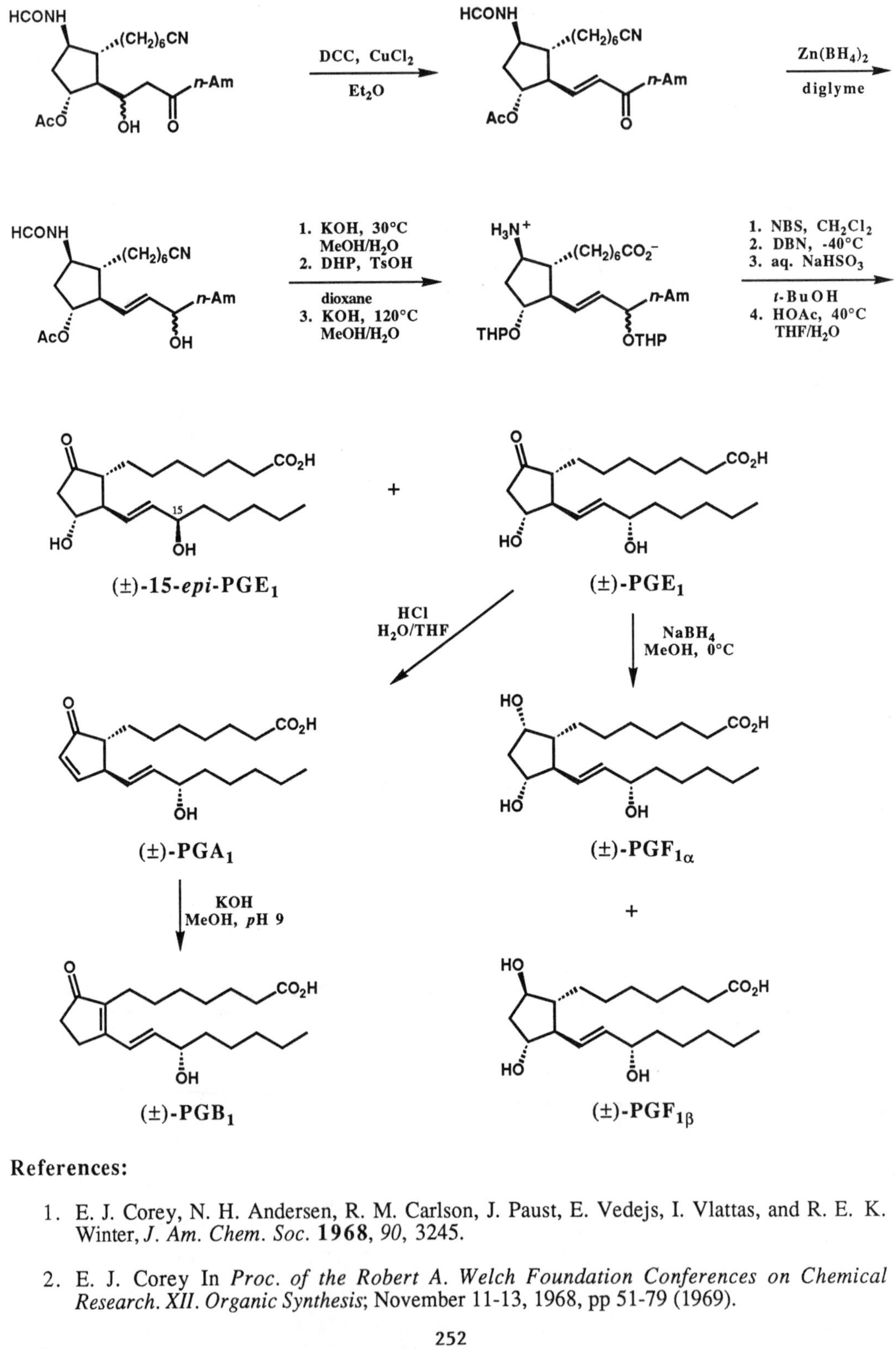
HCONH
(CH$_2$)$_6$CN
n-Am
AcO
OH
O
DCC, CuCl$_2$
Et$_2$O
Zn(BH$_4$)$_2$
diglyme
1. KOH, 30°C
MeOH/H$_2$O
2. DHP, TsOH
dioxane
3. KOH, 120°C
MeOH/H$_2$O
H$_3$N$^+$
(CH$_2$)$_6$CO$_2^-$
THPO
OTHP
1. NBS, CH$_2$Cl$_2$
2. DBN, -40°C
3. aq. NaHSO$_3$
t-BuOH
4. HOAc, 40°C
THF/H$_2$O
CO$_2$H
15
HO
(±)-15-*epi*-PGE$_1$
+
(±)-PGE$_1$
HCl
H$_2$O/THF
NaBH$_4$
MeOH, 0°C
(±)-PGA$_1$
(±)-PGF$_{1\alpha}$
KOH
MeOH, *p*H 9
(±)-PGB$_1$
(±)-PGF$_{1\beta}$

References:

1. E. J. Corey, N. H. Andersen, R. M. Carlson, J. Paust, E. Vedejs, I. Vlattas, and R. E. K. Winter, *J. Am. Chem. Soc.* **1968**, *90*, 3245.

2. E. J. Corey In *Proc. of the Robert A. Welch Foundation Conferences on Chemical Research. XII. Organic Synthesis*; November 11-13, 1968, pp 51-79 (1969).

11.3

Total Synthesis of Prostaglandins E_1, $F_{1\alpha}$ and Their 11-Epimers

The approach used for the initial synthesis of PG's *via* cyclopentylamine intermediates was shortened and broadened to include the synthesis of 11-*epi*-PG's and also chiral PG's by resolution.

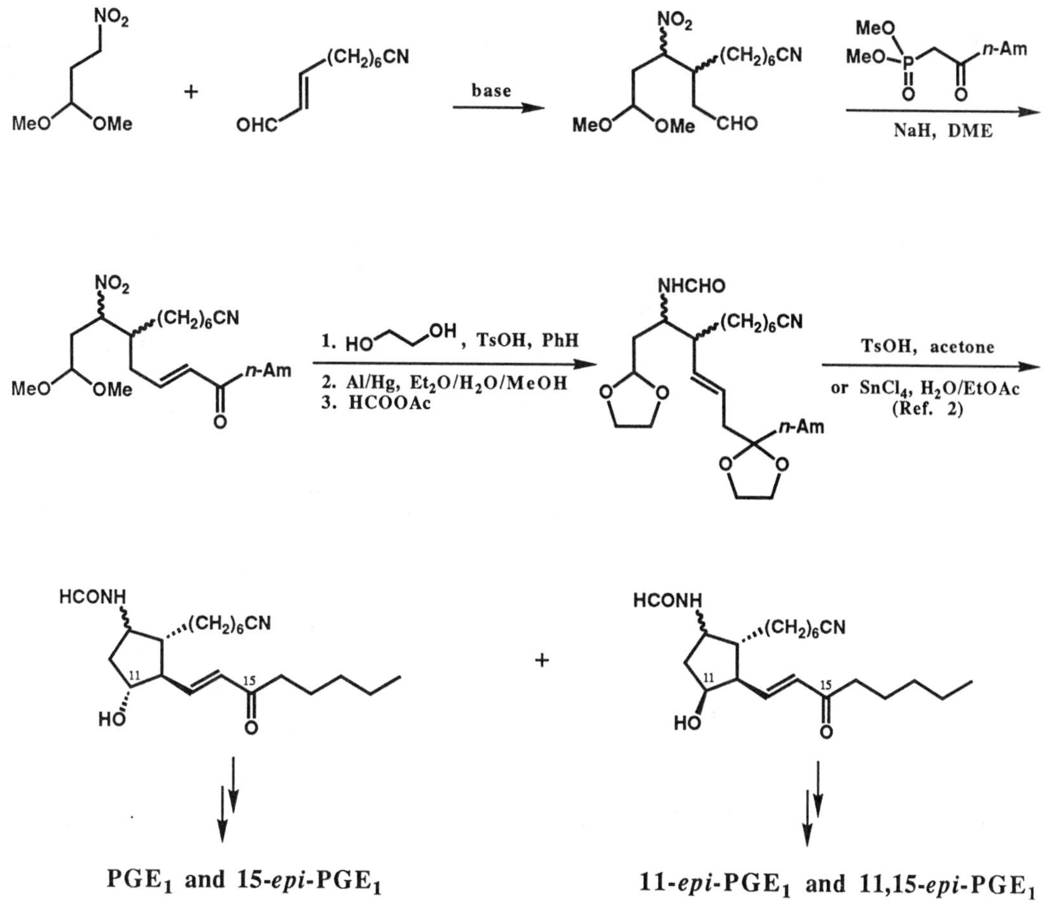

NO2
MeO
OMe
(CH2)6CN
OHC
base
CHO
MeO
MeO–P
n-Am
NaH, DME
1. HO⌒OH, TsOH, PhH
2. Al/Hg, Et2O/H2O/MeOH
3. HCOOAc
NHCHO
TsOH, acetone
or SnCl4, H2O/EtOAc
(Ref. 2)
HCONH
11
15
HO
PGE1 and 15-epi-PGE1
11-epi-PGE1 and 11,15-epi-PGE1

The cyclization reaction could also be carried out before the reduction of the nitro group. The resulting nitro alcohols were then converted to PGE_1 and 15-*epi*- and 11,15-*epi*-PGE_1's.

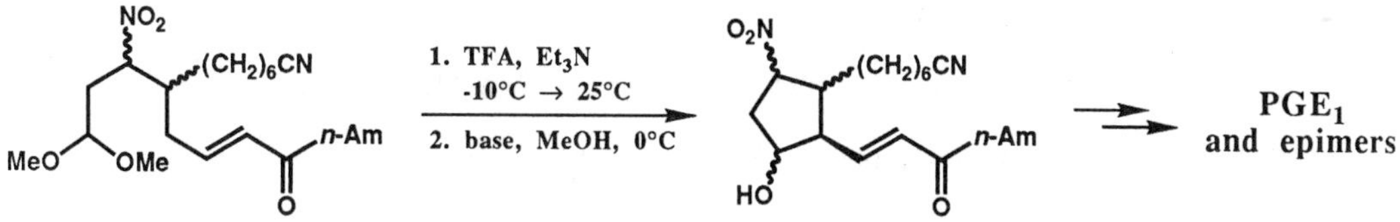

NO2
(CH2)6CN
n-Am
MeO
OMe
1. TFA, Et3N
-10°C → 25°C
2. base, MeOH, 0°C
O2N
HO
PGE1
and epimers

11.3

The synthesis of prostaglandin E_1 in the naturally occurring *levo* form was achieved by a modification of the sequence described above in which the racemic amine **A** was resolved with (-)-α-bromocamphor-π-sulfonic acid. The enantiomer of the natural PGE_1 ($[\alpha]_{578}$ +57° (*c* 0.5, THF)) was also synthesized (Ref. 3).

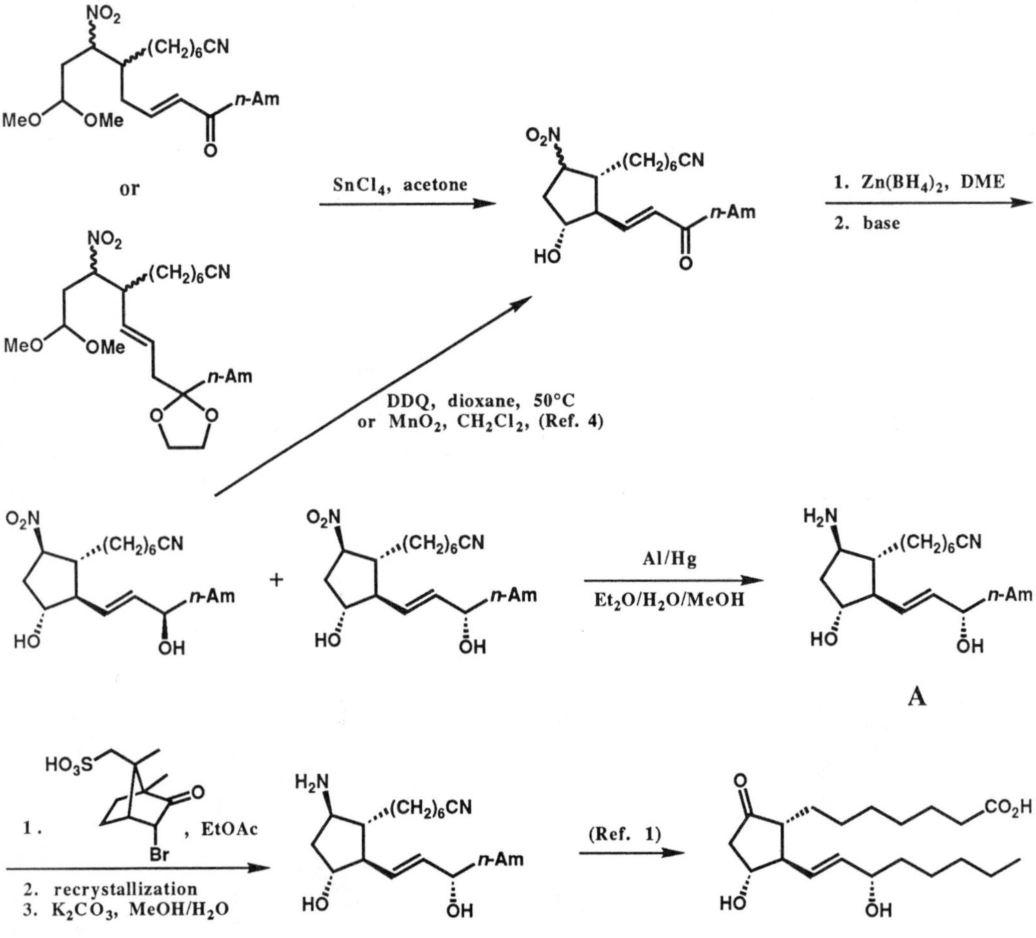

NO2
(CH2)6CN
n-Am
MeO
OMe
O
or
NO2
(CH2)6CN
MeO
OMe
n-Am
O
O
SnCl4, acetone
O2N
(CH2)6CN
n-Am
HO
O
1. Zn(BH4)2, DME
2. base
DDQ, dioxane, 50°C
or MnO2, CH2Cl2, (Ref. 4)
O2N
(CH2)6CN
n-Am
HO
OH
+
Al/Hg
Et2O/H2O/MeOH
H2N
(CH2)6CN
n-Am
HO
OH
A
HO3S
O
1.
, EtOAc
Br
2. recrystallization
3. K2CO3, MeOH/H2O
H2N
(CH2)6CN
n-Am
HO
OH
(Ref. 1)
O
CO2H
HO
OH

$[\alpha]_{578}$ -21° (*c* 1.7, MeOH)

$[\alpha]_{578}$ -61.6° (*c* 0.56, THF)

PGE_1

References:

1. E. J. Corey, I. Vlattas, N. H. Andersen, and K. Harding, *J. Am. Chem. Soc.* **1968**, *90*, 3247.

2. E. J. Corey, "Prostaglandins", In *Annals of the New York Academy of Sciences*; P. Ramwell and J. E. Shaw, Eds. **1971**, *180*, 24.

3. E. J. Corey, I. Vlattas, and K. Harding, *J. Am. Chem. Soc.* **1969**, *91*, 535.

4. E. J. Corey, N. M. Weinshenker, T. K. Schaaf, and W. Huber, *J. Am. Chem. Soc.* **1969**, *91*, 5675.

11.4

General Synthesis of Prostaglandins

The first general synthetic route to all the known prostaglandins was developed by way of bicyclo[2.2.1]heptene intermediates. The design was guided by the requirements that the route be versatile enough to allow the synthesis of many analogs and also allow early resolution. This synthesis has been used on a large scale and in laboratories throughout the world; it has been applied to the production of countless PG analogs.

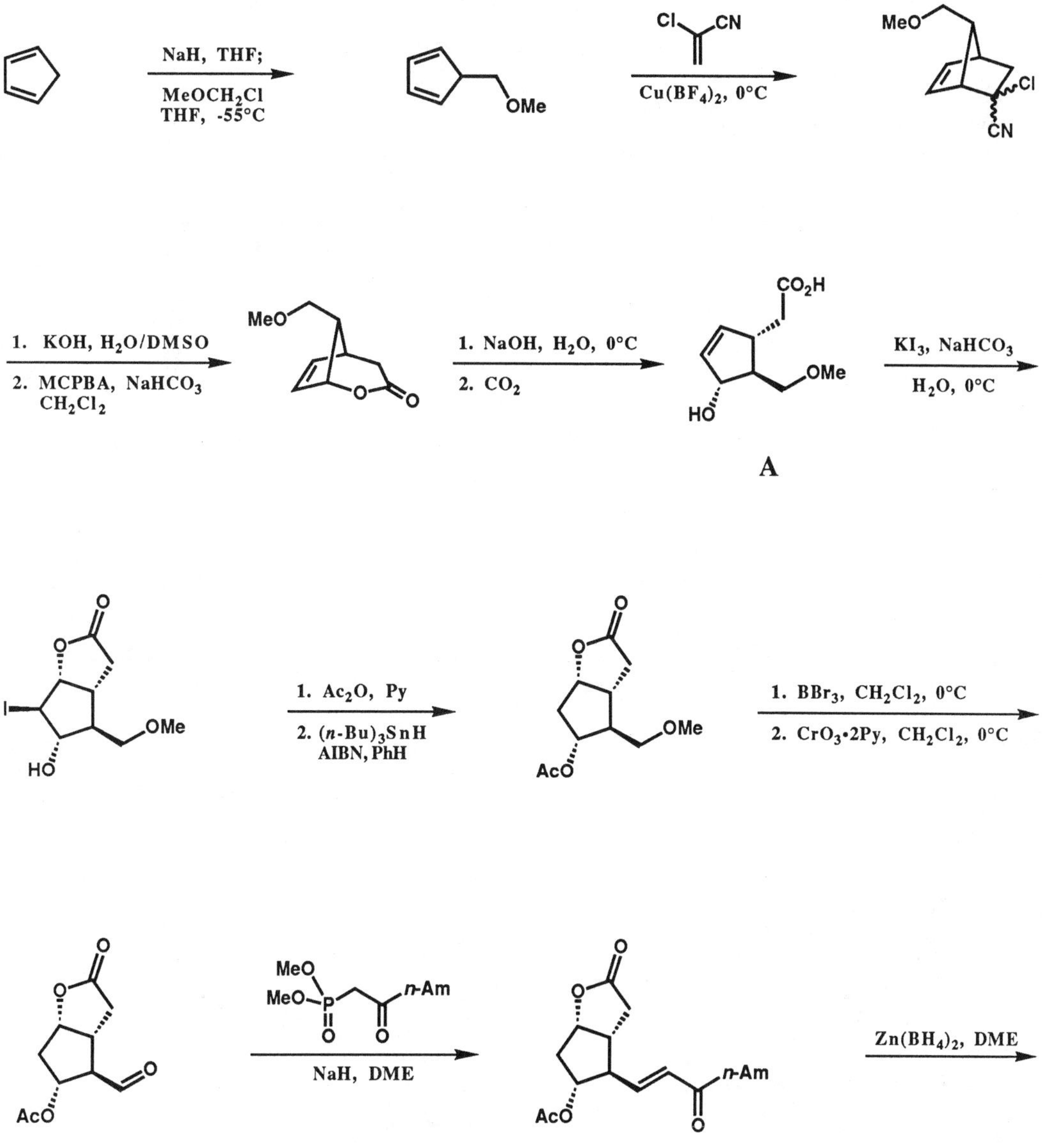

NaH, THF;
MeOCH2Cl
THF, -55°C
OMe
Cl
CN
Cu(BF4)2, 0°C
MeO
Cl
CN
1. KOH, H2O/DMSO
2. MCPBA, NaHCO3
CH2Cl2
MeO
O
O
1. NaOH, H2O, 0°C
2. CO2
CO2H
OMe
HO
A
KI3, NaHCO3
H2O, 0°C
O
O
I
OMe
HO
1. Ac2O, Py
2. (n-Bu)3SnH
AIBN, PhH
O
O
OMe
AcO
1. BBr3, CH2Cl2, 0°C
2. CrO3•2Py, CH2Cl2, 0°C
O
O
O
AcO
MeO
MeO
P
O
O
n-Am
NaH, DME
O
O
n-Am
O
AcO
Zn(BH4)2, DME

11.4

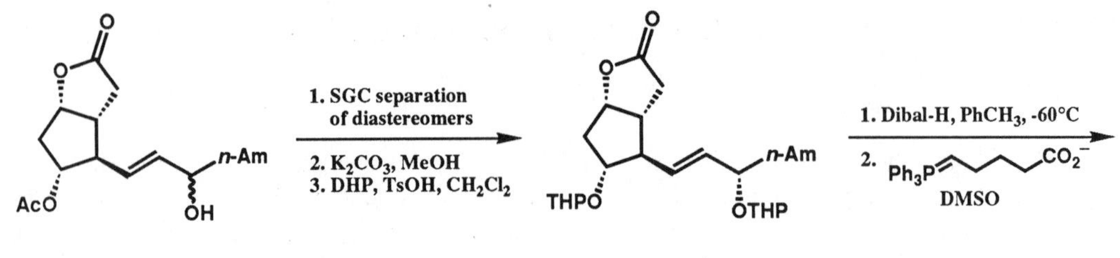

1 : 1 mixture

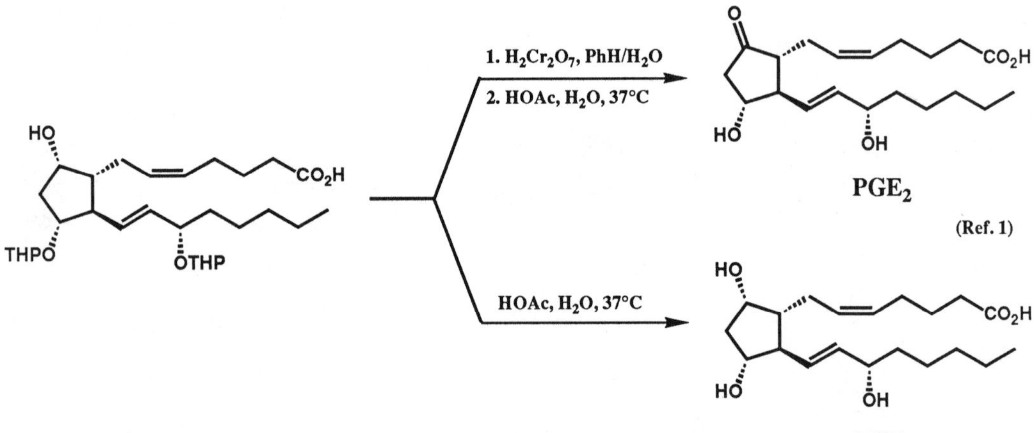

PGE_2

(Ref. 1)

$PGF_{2\alpha}$

The hydroxy acid **A** was resolved with (+)-ephedrine and converted to optically active $PGF_{2\alpha}$ and PGE_2 (Ref. 2).

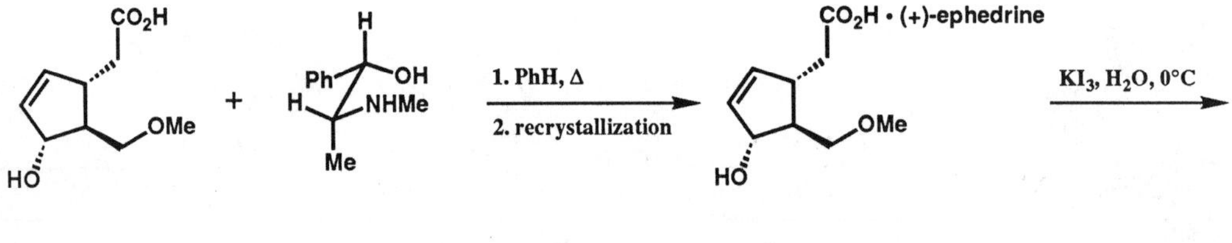

$[\alpha]^{23}_D$ +37.2° (*c* 1.0, MeOH)

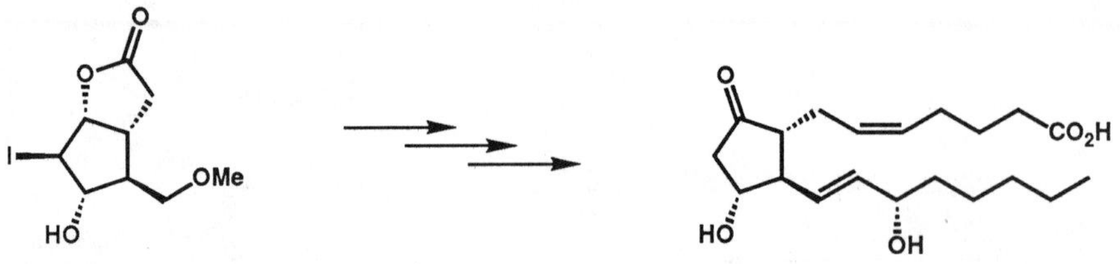

PGE_2

$[\alpha]^{26}_D$ -45.8° (*c* 1.0, MeOH)

synthetic sample: $[\alpha]^{25}_D$ -61° (*c* 1.0, THF)

natural product: $[\alpha]^{25}_D$ -61° (*c* 1.0, THF)

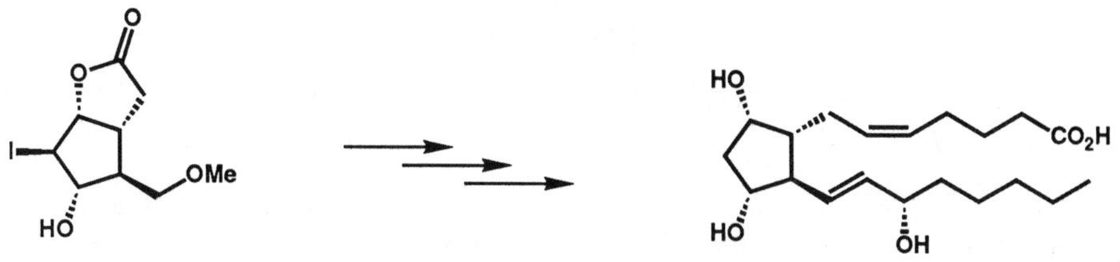
O
O
I
OMe
HO
HO
CO₂H
HO
OH

PGF$_{2\alpha}$

$[\alpha]^{26}_{D}$ -45.8° (*c* 1.0, MeOH)

synthetic sample: $[\alpha]^{25}_{D}$ +23.8° (*c* 1.0, THF)
natural product: $[\alpha]^{25}_{D}$ +23.5° (*c* 1.0, THF)

The (dextrorotatory) 11,15-bis-THP ether of PGF$_{2\alpha}$ was also transformed into prostaglandins of the first series by selective hydrogenation of the Z-Δ^5 bond (Ref. 3).

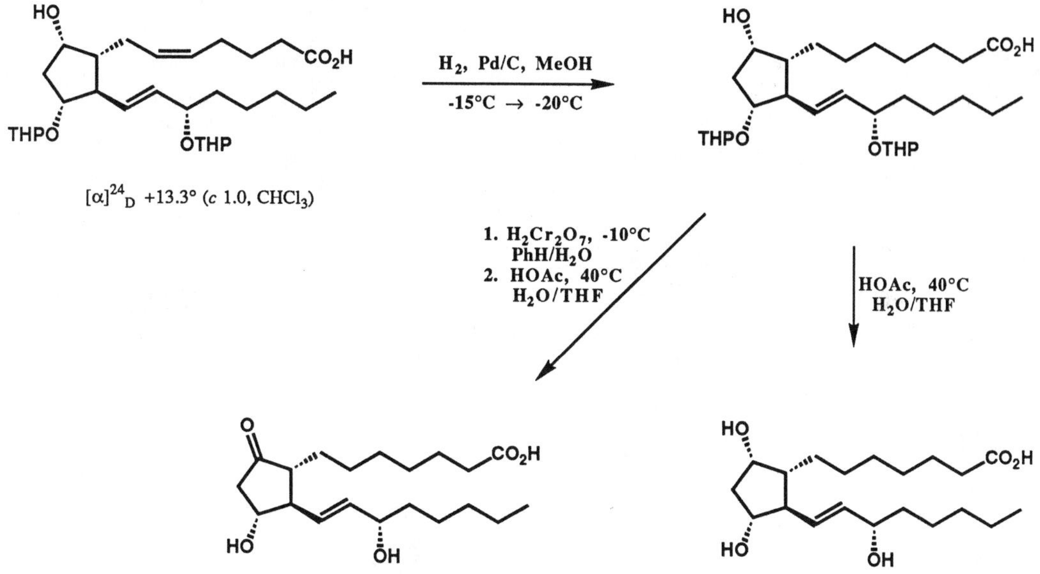
HO
CO₂H
THPO
OTHP
H₂, Pd/C, MeOH
-15°C → -20°C
HO
CO₂H
THPO
OTHP
1. H₂Cr₂O₇, -10°C
PhH/H₂O
2. HOAc, 40°C
H₂O/THF
HOAc, 40°C
H₂O/THF
O
CO₂H
HO
OH
HO
CO₂H
HO
OH

$[\alpha]^{24}_{D}$ +13.3° (*c* 1.0, $CHCl_3$)

PGE$_1$

$[\alpha]^{26}_{D}$ -61.1° (*c* 0.256, THF)

PGF$_{1\alpha}$

$[\alpha]^{24}_{D}$ +24° (*c* 0.87, THF)

References:

1. E. J. Corey, N. M. Weinshenker, T. K. Schaaf, and W. Huber, *J. Am. Chem. Soc.* **1969**, *91*, 5675.

2. E. J. Corey, T. K. Schaaf, W. Huber, U. Koelliker, and N. M. Weinshenker, *J. Am. Chem. Soc.* **1970**, *92*, 397.

3. E. J. Corey, R. Noyori, and T. K. Schaaf, *J. Am. Chem. Soc.* **1970**, *92*, 2586.

11.5

Refinements of the General Synthesis of Prostaglandins

The necessity for producing large amounts of synthetic prostaglandins and analogs provided the impetus for a number of improvements in the bicyclo[2.2.1]heptene approach. Especially important was the development of an enantioselective modification for the synthesis of chiral prostanoids without resolution (1975) and the invention of a chiral catalyst for the stereocontrolled conversion of 15-keto prostanoids to either 15(*S*)- or 15(*R*)- alcohols.

1. Improved route to:

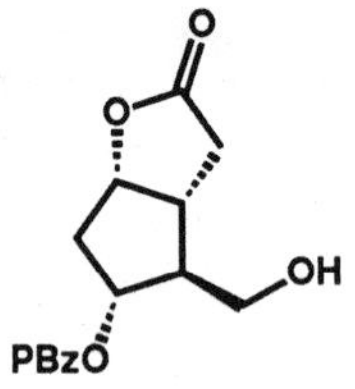

O
O
OH
PBzO

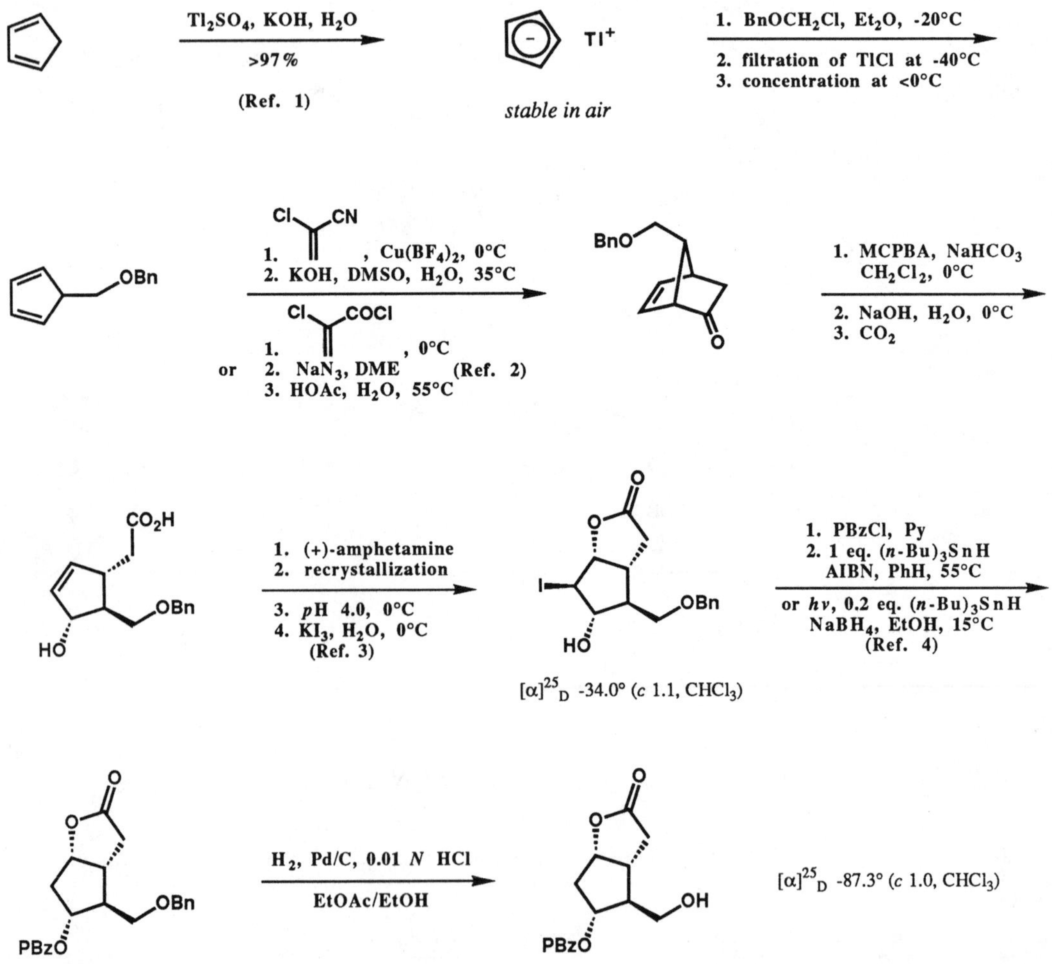

Tl2SO4, KOH, H2O
>97%
(Ref. 1)
Tl+
stable in air
1. BnOCH2Cl, Et2O, -20°C
2. filtration of TlCl at -40°C
3. concentration at <0°C
OBn
Cl CN
1. , Cu(BF4)2, 0°C
2. KOH, DMSO, H2O, 35°C
Cl COCl
1. , 0°C
or 2. NaN3, DME (Ref. 2)
3. HOAc, H2O, 55°C
BnO
O
1. MCPBA, NaHCO3
CH2Cl2, 0°C
2. NaOH, H2O, 0°C
3. CO2
CO2H
OBn
HO
1. (+)-amphetamine
2. recrystallization
3. pH 4.0, 0°C
4. KI3, H2O, 0°C
(Ref. 3)
I
OBn
HO
[α]25D -34.0° (c 1.1, CHCl3)
1. PBzCl, Py
2. 1 eq. (n-Bu)3SnH
AIBN, PhH, 55°C
or hv, 0.2 eq. (n-Bu)3SnH
NaBH4, EtOH, 15°C
(Ref. 4)
OBn
PBzO
H2, Pd/C, 0.01 N HCl
EtOAc/EtOH
OH
PBzO
[α]25D -87.3° (c 1.0, CHCl3)

2. Enantioselective route to: 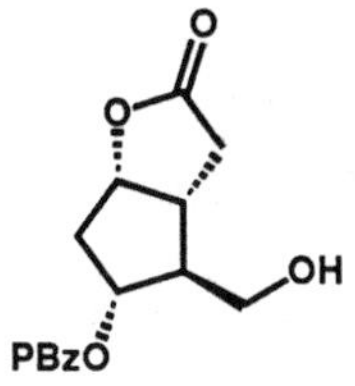 (Ref. 5).

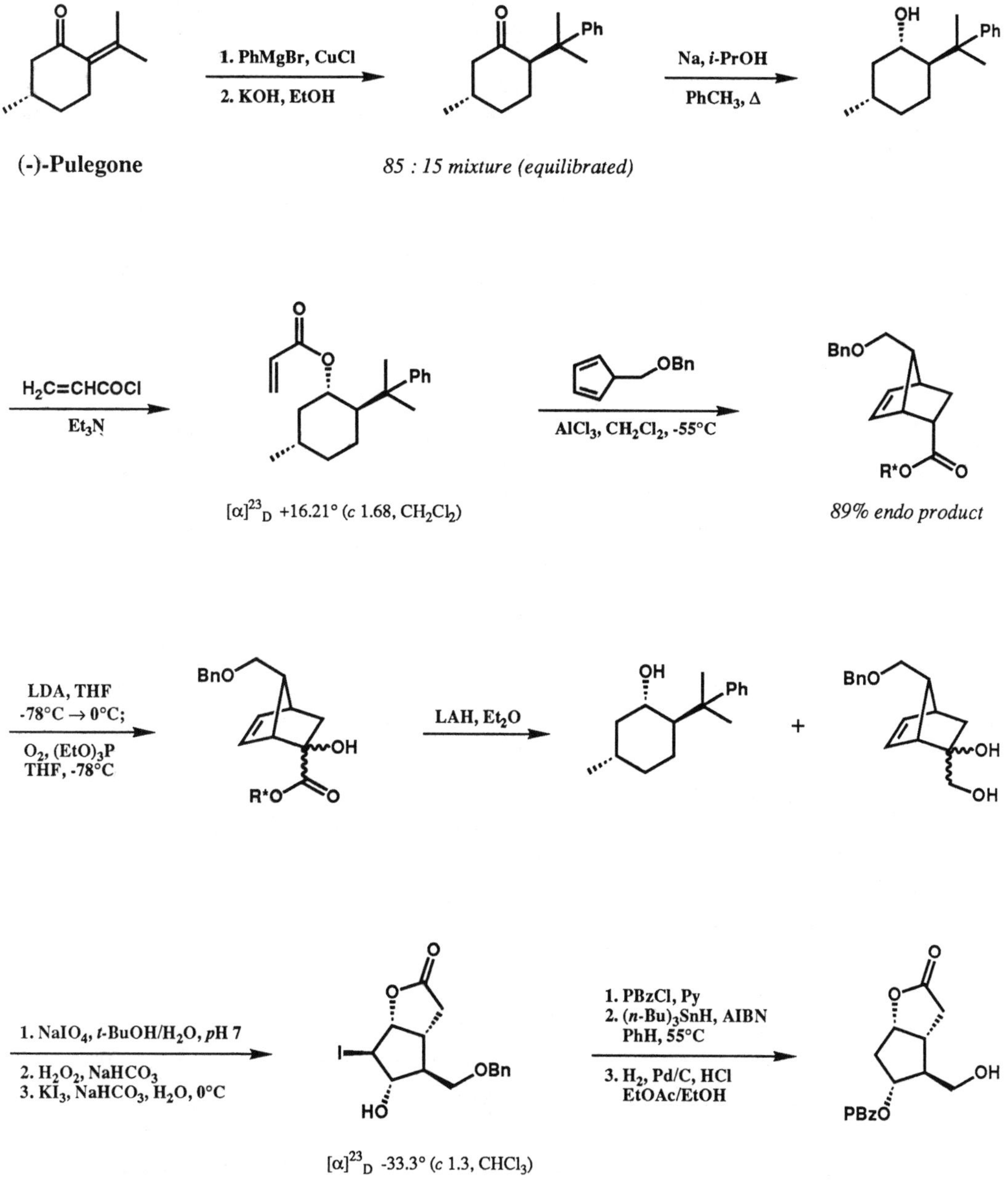
(-)-Pulegone
1. PhMgBr, CuCl
2. KOH, EtOH
85 : 15 mixture (equilibrated)
Na, i-PrOH
PhCH3, Δ
H2C=CHCOCl
Et3N
[α]23D +16.21° (c 1.68, CH2Cl2)
OBn
AlCl3, CH2Cl2, -55°C
89% endo product
LDA, THF
-78°C → 0°C;
O2, (EtO)3P
THF, -78°C
LAH, Et2O
1. NaIO4, t-BuOH/H2O, pH 7
2. H2O2, NaHCO3
3. KI3, NaHCO3, H2O, 0°C
[α]23D -33.3° (c 1.3, CHCl3)
1. PBzCl, Py
2. (n-Bu)3SnH, AIBN
PhH, 55°C
3. H2, Pd/C, HCl
EtOAc/EtOH

3. Stereoselective methods to the 15(*S*)-alcohol (**IV**):

In the early work on the synthesis of prostaglandins, zinc borohydride was used for the reduction of the 15-ketone function and a 1 : 1 mixture of epimeric 15(*S*)- and 15(*R*)-alcohols was generally obtained. Subsequent studies led to reaction conditions for highly selective reduction to the desired 15(*S*)-alcohol. Some of the results are summarized in the following table. The most practical method is **E** which utilizes borane as the stoichiometric reductant and a chiral, enzyme-like catalyst which is shown.

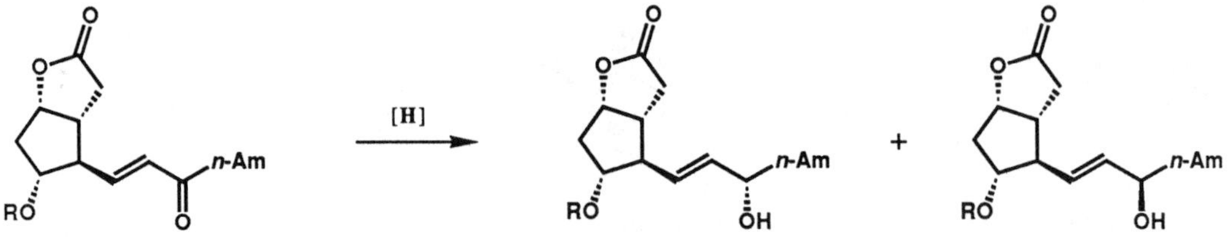
[H]
n-Am
RO
O
OH
+

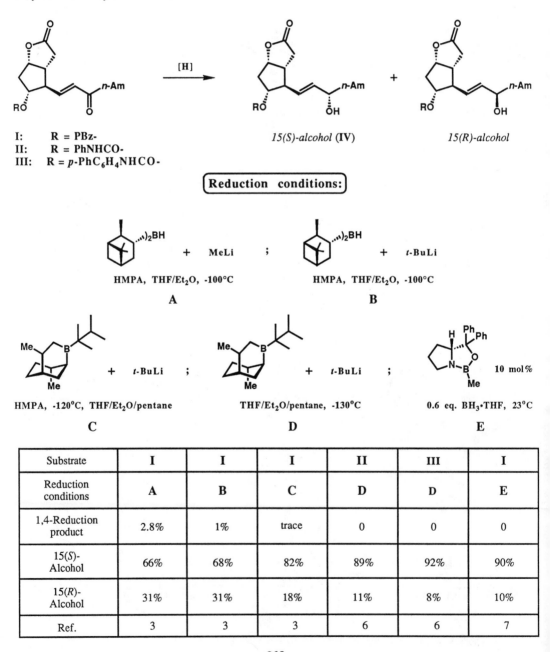

I: R = PBz-
II: R = PhNHCO-
III: R = *p*-PhC$_6$H$_4$NHCO-

15(S)-alcohol (**IV**)

15(R)-alcohol

Reduction conditions:

)2BH
+ MeLi
;
+ t-BuLi

HMPA, THF/Et$_2$O, -100°C — **A**

HMPA, THF/Et$_2$O, -100°C — **B**

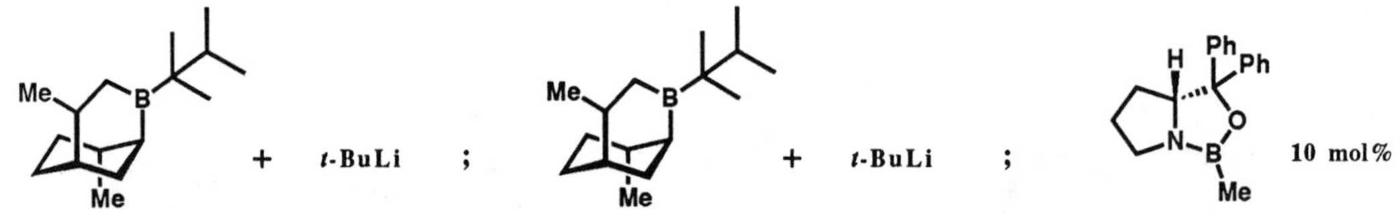
Me
B
+ t-BuLi
;
H
Ph
Ph
O
N
B
Me
10 mol%

HMPA, -120°C, THF/Et$_2$O/pentane — **C**

THF/Et$_2$O/pentane, -130°C — **D**

0.6 eq. BH$_3$•THF, 23°C — **E**

Substrate	I	I	I	II	III	I
Reduction conditions	A	B	C	D	D	E
1,4-Reduction product	2.8%	1%	trace	0	0	0
15(*S*)-Alcohol	66%	68%	82%	89%	92%	90%
15(*R*)-Alcohol	31%	31%	18%	11%	8%	10%
Ref.	3	3	3	6	6	7

4. Conversion of 15(*R*)- to 15(*S*)-alcohols (Ref. 8):

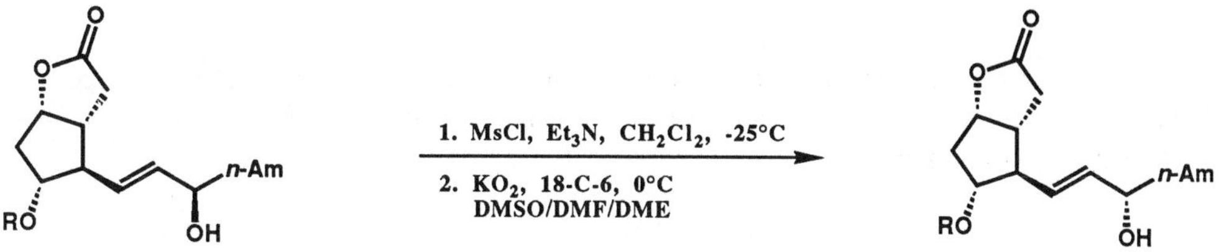

n-Am
RO
OH
1. MsCl, Et3N, CH2Cl2, -25°C
2. KO2, 18-C-6, 0°C
DMSO/DMF/DME
n-Am
RO
OH

15(R)-alcohol *15(S)-alcohol*

References:

1. E. J. Corey, U. Koelliker, and J. Neuffer, *J. Am. Chem. Soc.* **1971**, *93*, 1489.
2. E. J. Corey, T. Ravindranathan, and S. Terashima, *J. Am. Chem. Soc.* **1971**, *93*, 4326.
3. E. J. Corey, S. M. Albonico, U. Koelliker, T. K. Schaaf, and R. K. Varma, *J. Am. Chem. Soc.* **1971**, *93*, 1491.
4. E. J. Corey and J. W. Suggs, *J. Org. Chem.* **1975**, *40*, 2554.
5. E. J. Corey and H. E. Ensley, *J. Am. Chem. Soc.* **1975**, *97*, 6908.
6. E. J. Corey, K. B. Becker, and R. K. Varma, *J. Am. Chem. Soc.* **1972**, *94*, 8616.
7. E. J. Corey, R. K. Bakshi, S. Shibata, C.-P. Chen, and V. K. Singh, *J. Am. Chem. Soc.* **1987**, *109*, 7925.
8. (a) E. J. Corey, K. C. Nicolaou, and M. Shibasaki, *J. Chem. Soc., Chem. Commun.* **1975**, 658; (b) E. J. Corey, K. C. Nicolaou, M. Shibasaki, Y. Machida, and C. S. Shiner, *Tetrahedron Lett.* **1975**, 3183.

11.6

Total Synthesis of Prostaglandins E_3 and $F_{3\alpha}$

The third family of prostaglandins, based on the marine fatty acid precursor eicosapentaenoic acid, was synthesized with the help of a number of new methods designed for the introduction of the doubly unsaturated omega chain.

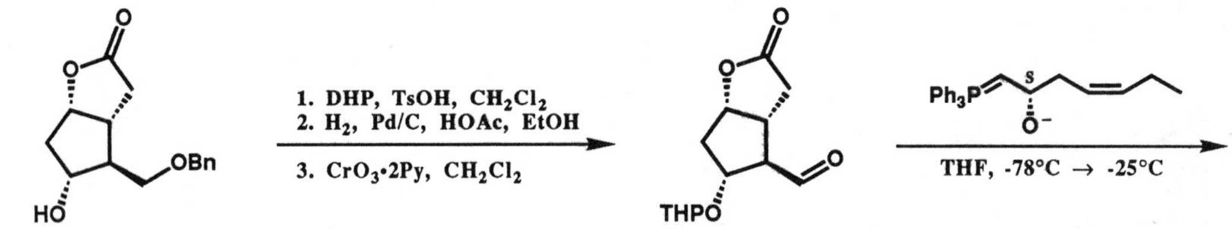
OBn
HO
1. DHP, TsOH, CH2Cl2
2. H2, Pd/C, HOAc, EtOH
3. CrO3•2Py, CH2Cl2
THPO
Ph3P
S
O⁻
THF, -78°C → -25°C

$[\alpha]^{25}_D$ -10° ($CHCl_3$)

(Ref. 2)

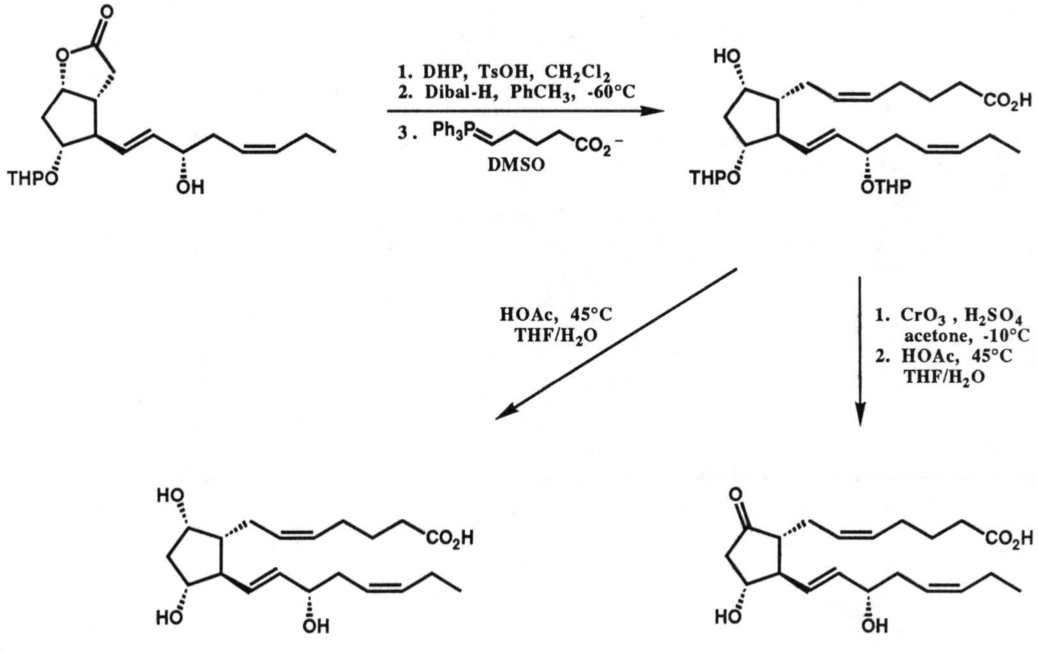
THPO
OH
1. DHP, TsOH, CH2Cl2
2. Dibal-H, PhCH3, -60°C
3. Ph3P CO2⁻
DMSO
HO
CO2H
THPO
OTHP
HOAc, 45°C
THF/H2O
1. CrO3 , H2SO4
acetone, -10°C
2. HOAc, 45°C
THF/H2O
HO
CO2H
HO
OH
O
CO2H
HO
OH

$[\alpha]^{26}_D$ +29.6° (*c* 0.54, THF)

$PGF_{3\alpha}$

$[\alpha]^{24}_D$ -48.9° (*c* 1.2, THF)

PGE_3

Preparation of the Wittig salt:

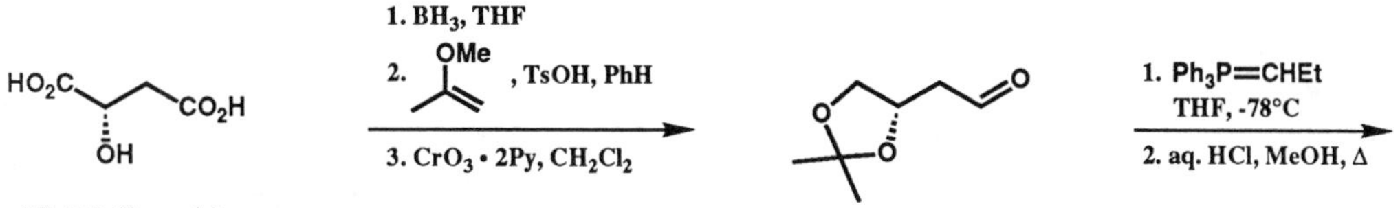
HO2C
CO2H
OH
1. BH3, THF
2. OMe , TsOH, PhH
3. CrO3 • 2Py, CH2Cl2
O
O
O
1. Ph3P=CHEt
THF, -78°C
2. aq. HCl, MeOH, Δ

(*S*)-Malic acid

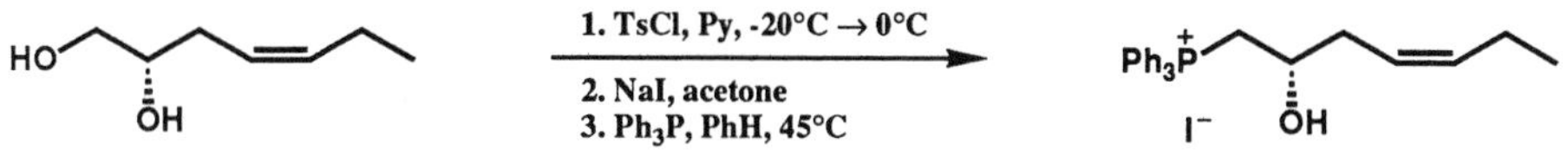
HO
OH
1. TsCl, Py, -20°C → 0°C
2. NaI, acetone
3. Ph3P, PhH, 45°C
Ph3P+
I−
OH

$[\alpha]^{25}_{D}$ +45.8° (*c* 1.0, $CHCl_3$)

The *ω*-appendage of the PG_3's can also be generated by the following methods (Ref. 3).

Method 1:

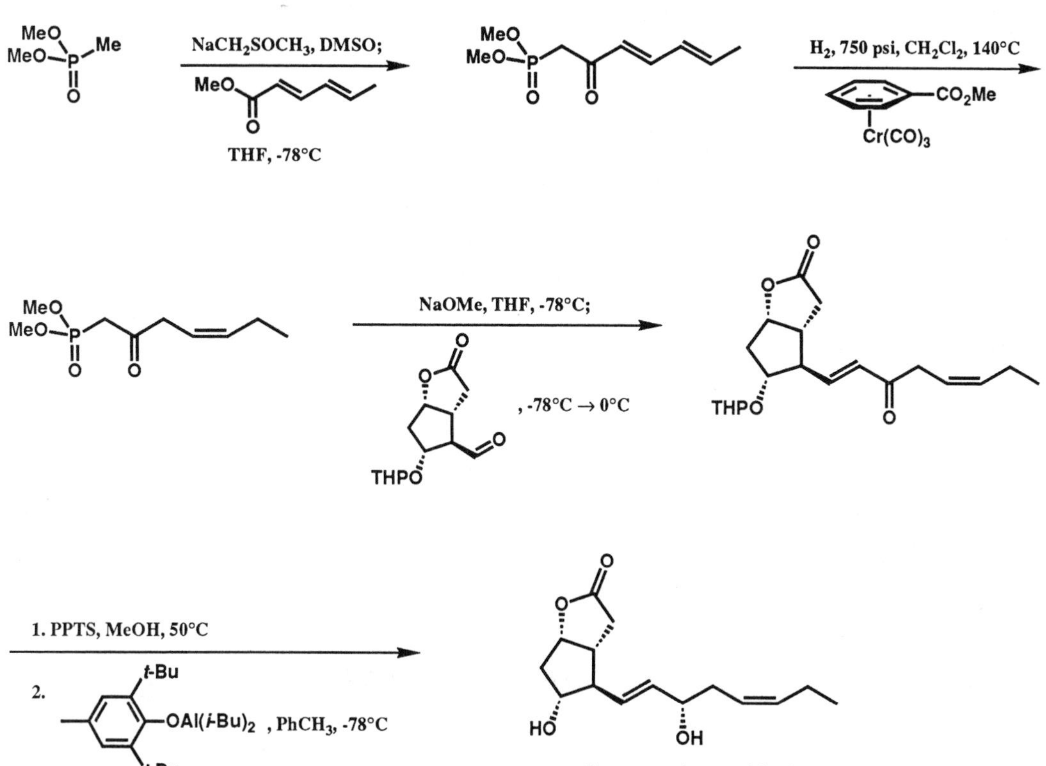
MeO
MeO
P
Me
O
NaCH2SOCH3, DMSO;
MeO
O
THF, -78°C
MeO
MeO
P
O
O
H2, 750 psi, CH2Cl2, 140°C
CO2Me
Cr(CO)3
MeO
MeO
P
O
O
NaOMe, THF, -78°C;
O
O
THPO
O
, -78°C → 0°C
O
O
THPO
O
1. PPTS, MeOH, 50°C
2.
t-Bu
OAl(i-Bu)2 , PhCH3, -78°C
t-Bu
O
O
HO
OH

diastereoselection 91 : 9

11.6

Method 2:

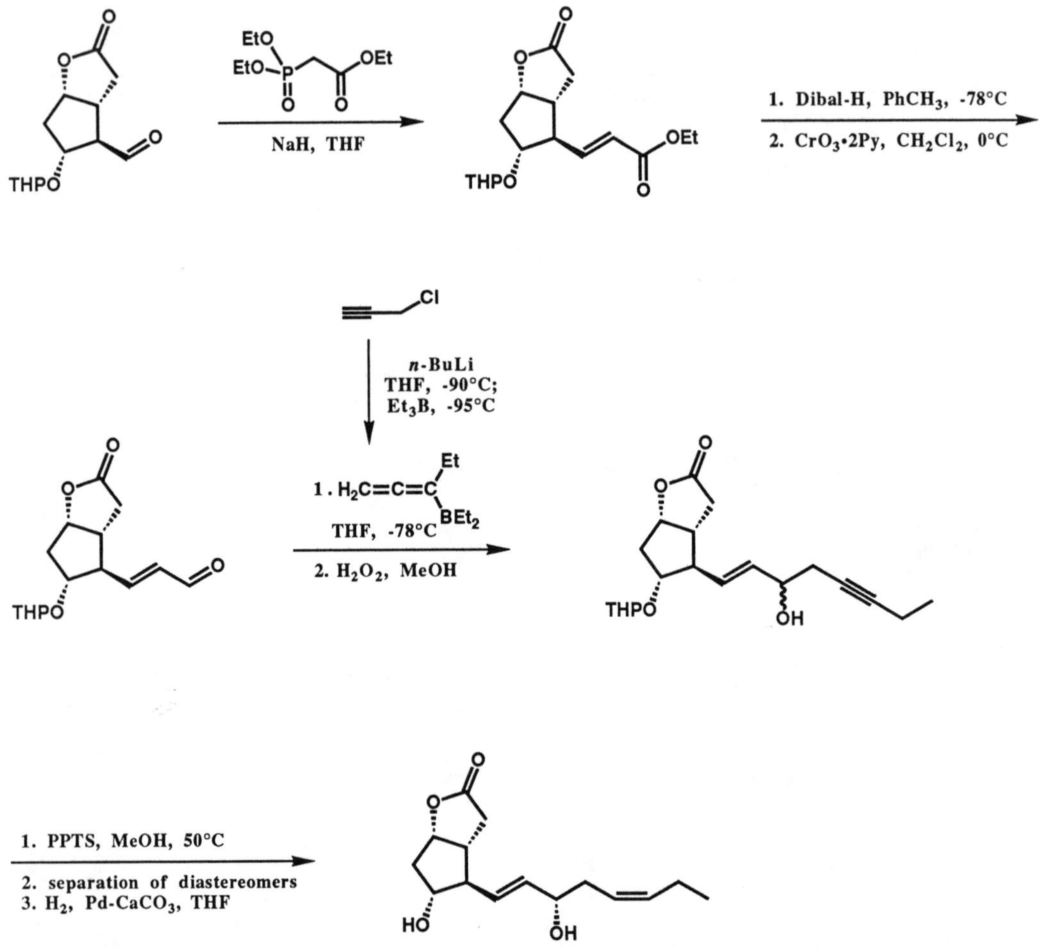
THPO
EtO
EtO–P
OEt
NaH, THF
OEt
THPO
1. Dibal-H, PhCH3, -78°C
2. CrO3•2Py, CH2Cl2, 0°C
Cl
n-BuLi
THF, -90°C;
Et3B, -95°C
Et
1. H2C=C=C
BEt2
THF, -78°C
2. H2O2, MeOH
THPO
THPO
OH
1. PPTS, MeOH, 50°C
2. separation of diastereomers
3. H2, Pd-CaCO3, THF
HO
OH

References:

1. E. J. Corey, H. Shirahama, H. Yamamoto, S. Terashima, A. Venkateswarlu, and T. K. Schaaf, *J. Am. Chem. Soc.* **1971**, *93*, 1490.

2. E. J. Corey, S. M. Albonico, U. Koelliker, T. K. Schaaf, and R. K. Varma, *J. Am. Chem. Soc.* **1971**, *93*, 1491.

3. E. J. Corey, S. Ohuchida, and R. Hahl, *J. Am. Chem. Soc.* **1984**, *106*, 3875.

Modified Bicyclo[2.2.1]heptane Routes to Prostaglandins

Route 1:

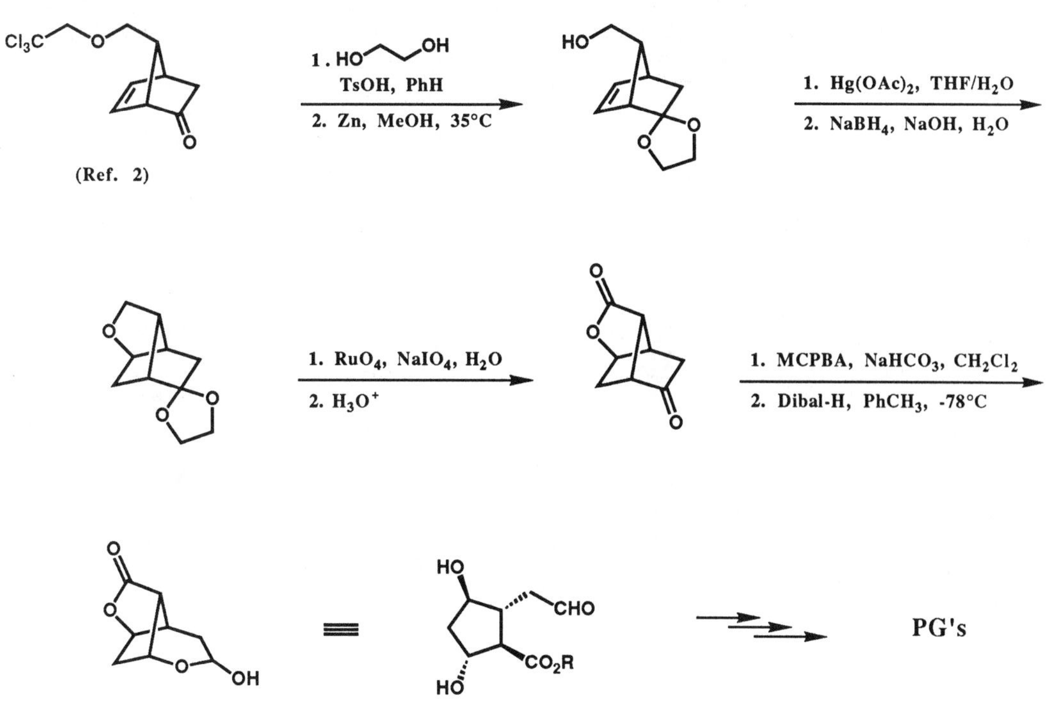
Cl3C
O
O
(Ref. 2)
1. HO OH
TsOH, PhH
2. Zn, MeOH, 35°C
HO
O
O
1. Hg(OAc)2, THF/H2O
2. NaBH4, NaOH, H2O
1. RuO4, NaIO4, H2O
2. H3O+
1. MCPBA, NaHCO3, CH2Cl2
2. Dibal-H, PhCH3, -78°C
OH
HO
CHO
CO2R
HO
PG's

Route 2:

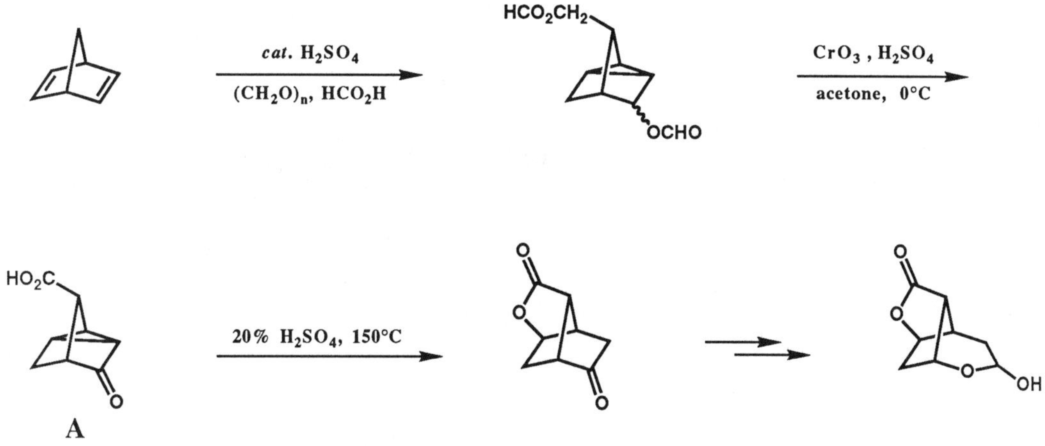
cat. H2SO4
(CH2O)n, HCO2H
HCO2CH2
OCHO
CrO3 , H2SO4
acetone, 0°C
HO2C
A
20% H2SO4, 150°C
OH

11.7

The chiral keto acid **A** could be obtained by resolution with (S)-(-)-α-methylbenzylamine. Another route from **A** to PG's is shown below:

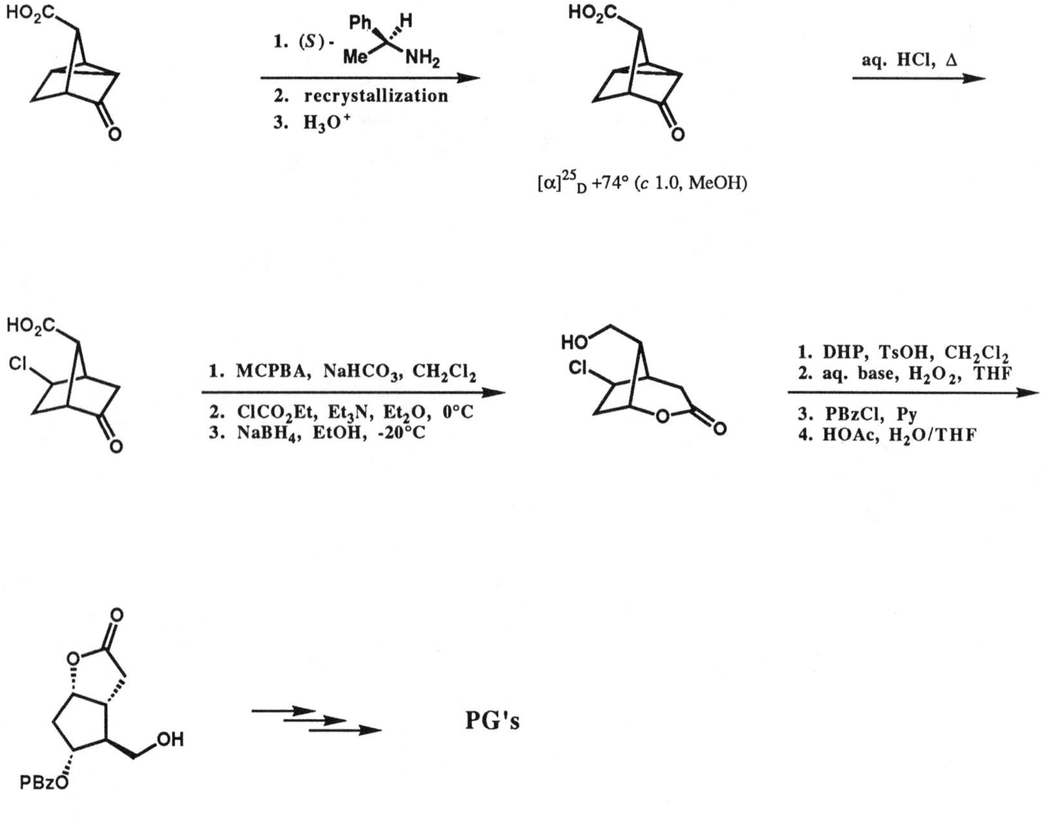

HO2C
1. (S)-
Ph
H
Me
NH2
2. recrystallization
3. H3O+
HO2C
[α]25D +74° (c 1.0, MeOH)
aq. HCl, Δ
HO2C
Cl
1. MCPBA, NaHCO3, CH2Cl2
2. ClCO2Et, Et3N, Et2O, 0°C
3. NaBH4, EtOH, -20°C
HO
Cl
1. DHP, TsOH, CH2Cl2
2. aq. base, H2O2, THF
3. PBzCl, Py
4. HOAc, H2O/THF
OH
PBzO
PG's

References:

1. J. S. Bindra, A. Grodski, T. K. Schaaf, and E. J. Corey, *J. Am. Chem. Soc.* **1973**, *95*, 7522.

2. (a) E. J. Corey, S. M. Albonico, U. Koelliker, T. K. Schaaf, and R. K. Varma, *J. Am. Chem. Soc.* **1971**, *93*, 1491; (b) E. J. Corey, T. Ravindranathan, and S. Terashima, *ibid*, **1971**, *93*, 4326.

Total Synthesis of Prostaglandin A_2 and Conversion to Other Prostaglandins

A number of direct routes to PGA_2, which were shorter than those *via* PGE_2, were developed. In addition a highly diastereoselective conversion of PGA_2 to PGE_2 employing a controller group at C(15) was achieved in an early example of unconventional acyclic diastereoselection. Finally two different syntheses of the highly sensitive PGC_2 were devised.

1. Total synthesis of PGA_2:

Scheme 1 (Ref. 1):

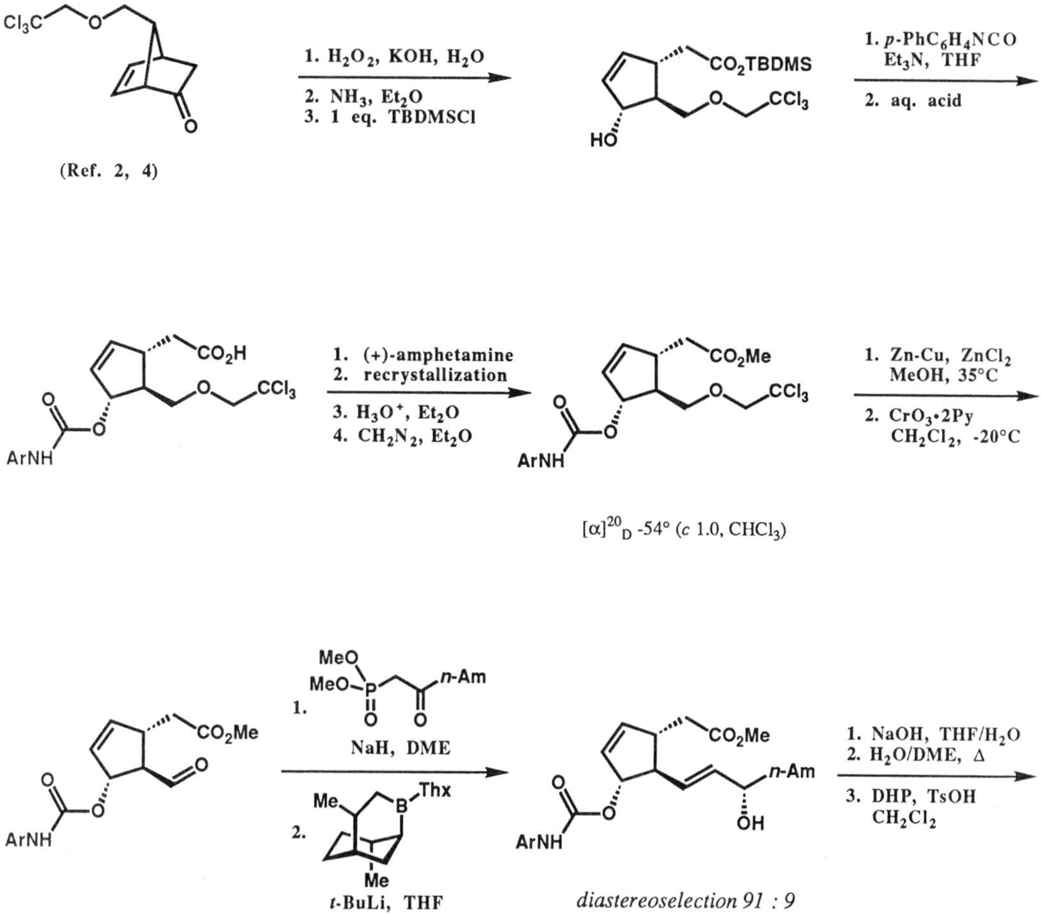

Cl3C
O
O
(Ref. 2, 4)
1. H2O2, KOH, H2O
2. NH3, Et2O
3. 1 eq. TBDMSCl
CO2TBDMS
O
CCl3
HO
1. p-PhC6H4NCO
Et3N, THF
2. aq. acid
CO2H
O
CCl3
O
O
ArNH
1. (+)-amphetamine
2. recrystallization
3. H3O+, Et2O
4. CH2N2, Et2O
CO2Me
O
CCl3
O
O
ArNH
[α]20D -54° (c 1.0, CHCl3)
1. Zn-Cu, ZnCl2
MeOH, 35°C
2. CrO3•2Py
CH2Cl2, -20°C
CO2Me
O
O
O
ArNH
MeO
MeO
P
n-Am
O
O
1.
NaH, DME
Me
B
Thx
Me
2.
t-BuLi, THF
CO2Me
n-Am
OH
O
O
ArNH
diastereoselection 91 : 9
1. NaOH, THF/H2O
2. H2O/DME, Δ
3. DHP, TsOH
CH2Cl2

11.8

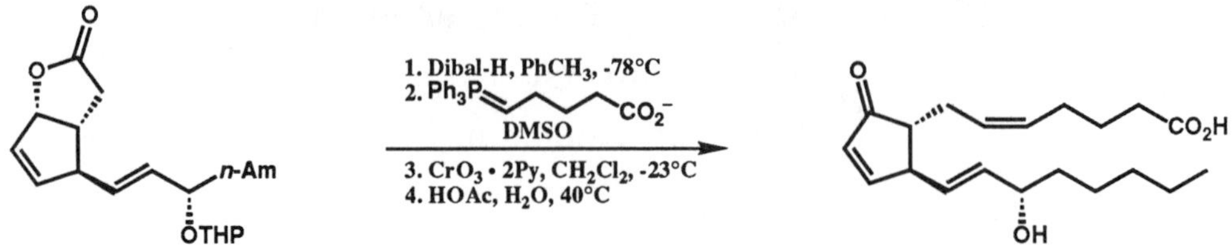

O
O
n-Am
OTHP
1. Dibal-H, PhCH3, -78°C
2. Ph3P=
CO2⁻
DMSO
3. CrO3 • 2Py, CH2Cl2, -23°C
4. HOAc, H2O, 40°C
O
CO2H
OH

PGA$_2$

synthetic sample: $[\alpha]^{20}_{D}$ +140° (*c* 1.15, CHCl$_3$)

natural product: $[\alpha]^{20}_{D}$ +131° (*c* 1.26, CHCl$_3$)

Scheme 2 (Ref. 3):

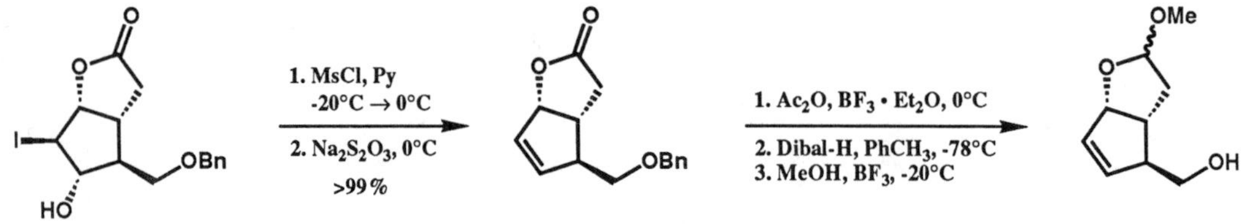

O
O
I
OBn
HO
1. MsCl, Py
-20°C → 0°C
2. Na2S2O3, 0°C
>99%
OBn
1. Ac2O, BF3 • Et2O, 0°C
2. Dibal-H, PhCH3, -78°C
3. MeOH, BF3, -20°C
OMe
OH

$[\alpha]^{28}_{D}$ -39.3° (*c* 0.975, MeOH)

(Ref. 4)

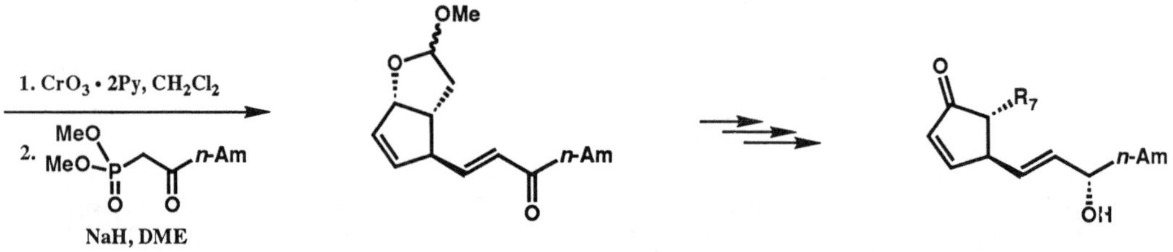

1. CrO3 • 2Py, CH2Cl2
2. MeO
MeO-P
O
O
n-Am
NaH, DME
OMe
O
n-Am
O
O
R7
n-Am
OH

PGA

Scheme 3 (Ref. 5):

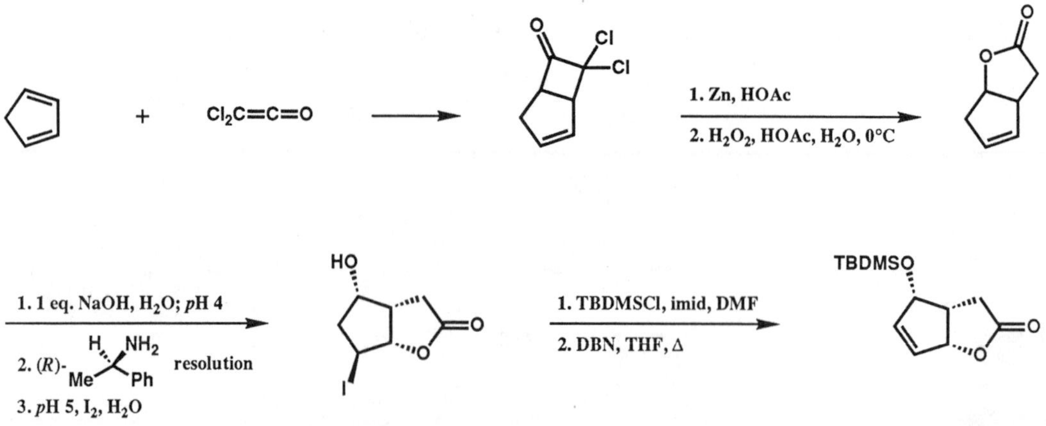

+
Cl2C=C=O
O
Cl
Cl
1. Zn, HOAc
2. H2O2, HOAc, H2O, 0°C
O
O
1. 1 eq. NaOH, H2O; pH 4
2. (R)-
H
NH2
Me
Ph
resolution
3. pH 5, I2, H2O
HO
O
O
I
1. TBDMSCl, imid, DMF
2. DBN, THF, Δ
TBDMSO
O
O

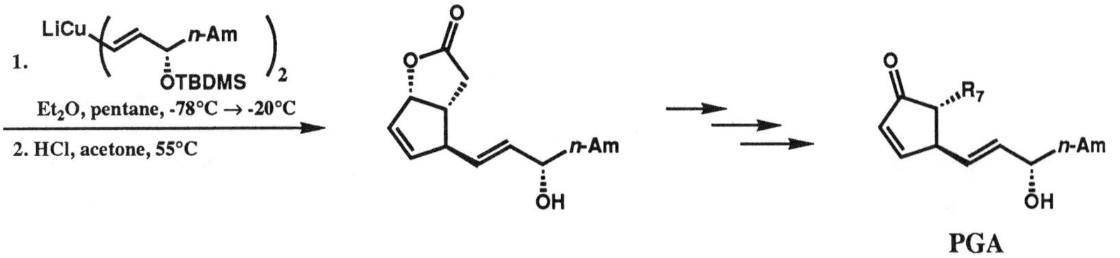
1. LiCu(n-Am, OTBDMS)2
Et2O, pentane, -78°C → -20°C
2. HCl, acetone, 55°C
n-Am
OH
R7
n-Am
OH
PGA

Preparation of the Gilman reagent (Ref. 6):

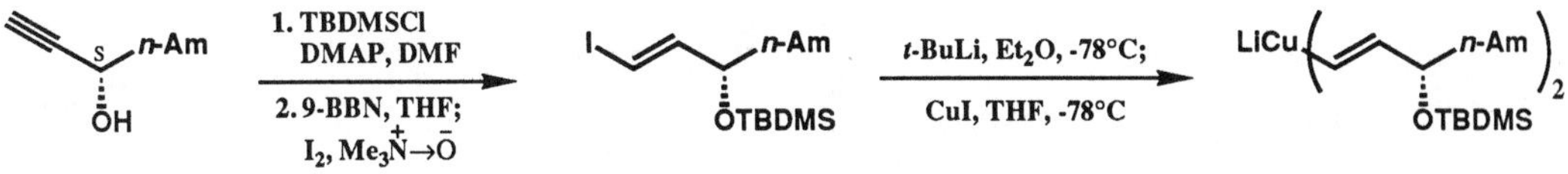
S
n-Am
OH
1. TBDMSCl
DMAP, DMF
2. 9-BBN, THF;
I2, Me3N→O
I
n-Am
OTBDMS
t-BuLi, Et2O, -78°C;
CuI, THF, -78°C
LiCu
n-Am
OTBDMS
2

2. Conversion of PGA_2 to PGE_2, $PGF_{2\alpha}$ and PGE_1, $PGF_{1\alpha}$ (Ref. 7):

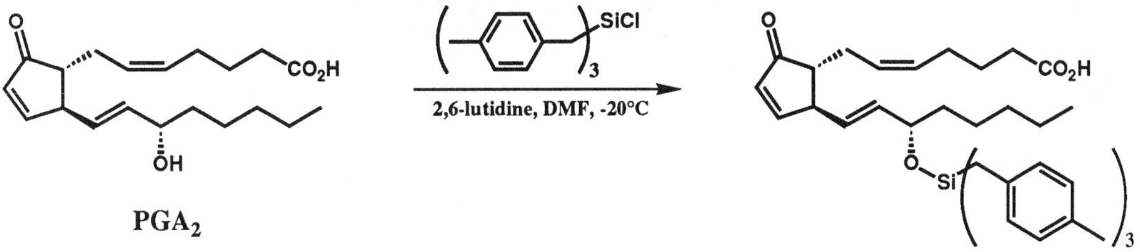
CO2H
OH
PGA2
SiCl
3
2,6-lutidine, DMF, -20°C
CO2H
Si
3

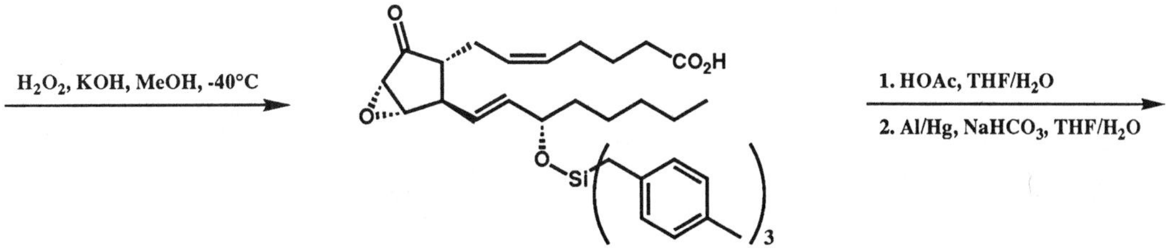
H2O2, KOH, MeOH, -40°C
CO2H
Si
3
1. HOAc, THF/H2O
2. Al/Hg, NaHCO3, THF/H2O

diastereoselection 94 : 6

11.8

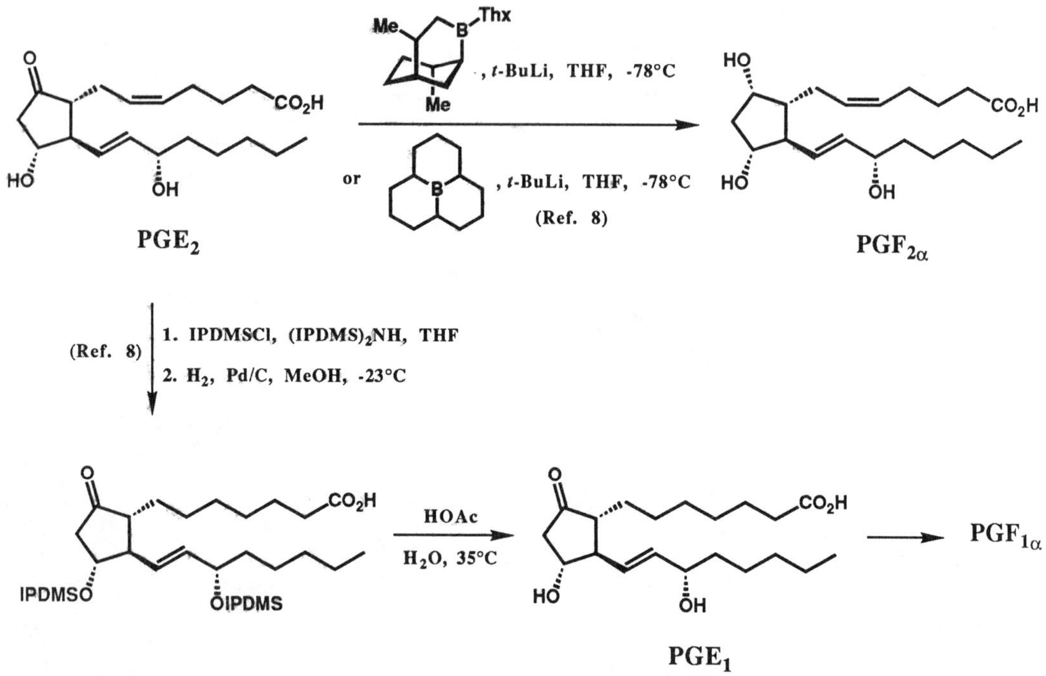

Thx
Me
B
Me
, t-BuLi, THF, -78°C
or
B
, t-BuLi, THF, -78°C
(Ref. 8)
O
CO2H
HO
OH
PGE2
HO
CO2H
HO
OH
PGF2α
(Ref. 8)
1. IPDMSCl, (IPDMS)2NH, THF
2. H2, Pd/C, MeOH, -23°C
O
CO2H
IPDMSO
OIPDMS
HOAc
H2O, 35°C
O
CO2H
HO
OH
PGE1
PGF1α

3. Conversion of PGA_2 to PGC_2:

Conversion of PGA_2 to the highly sensitive PGC_2 was accomplished by deconjugation of the enone system by formation of the γ-extended enolate using *tert*-alkoxide as base and α-protonation by *p*H 4 buffer.

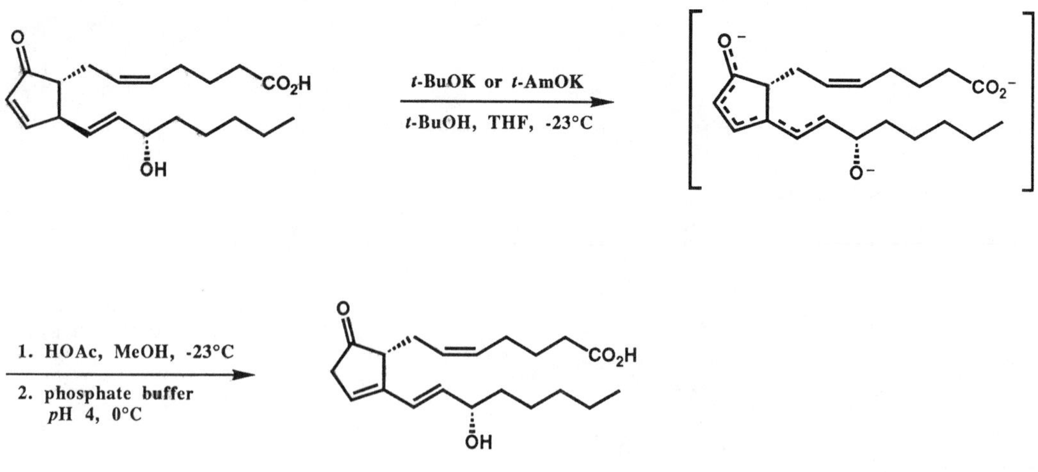

O
CO2H
OH
t-BuOK or t-AmOK
t-BuOH, THF, -23°C
O−
CO2−
O−
1. HOAc, MeOH, -23°C
2. phosphate buffer
pH 4, 0°C
O
CO2H
OH

PGC_2

The double bond transposition could also be achieved by the conversion of an intermediate for PGA_2 synthesis into a 1,3-diene iron tricarbonyl complex from which PGC_2 was synthesized in four steps. The $Fe(CO)_3$•diene complex which survived the Wittig reaction was cleanly removed by Collins reagent in the subsequent step (Ref. 10).

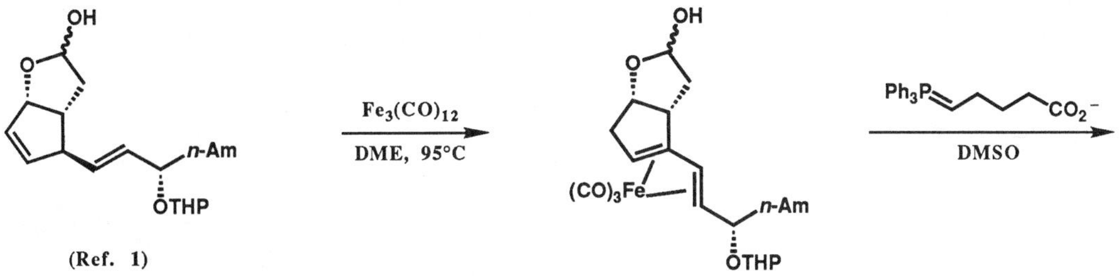
OH
O
n-Am
OTHP
Fe3(CO)12
DME, 95°C
OH
O
(CO)3Fe
n-Am
OTHP
Ph3P
CO2⁻
DMSO

(Ref. 1)

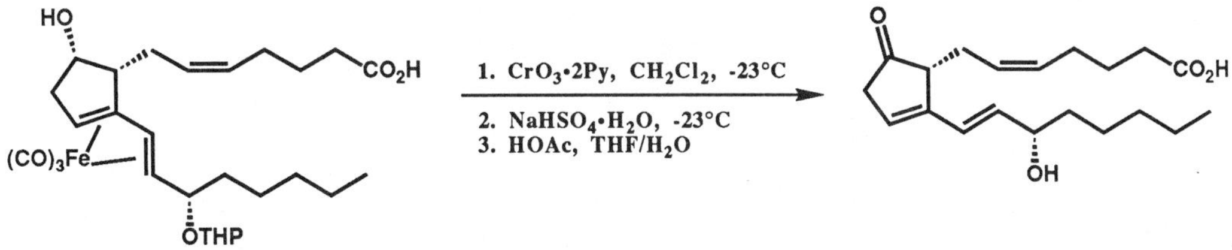
HO
CO2H
(CO)3Fe
OTHP
1. CrO3•2Py, CH2Cl2, -23°C
2. NaHSO4•H2O, -23°C
3. HOAc, THF/H2O
O
CO2H
OH

References:

1. E. J. Corey and G. Moinet, *J. Am. Chem. Soc.* **1973**, *95*, 6831.
2. E. J. Corey, S. M. Albonico, U. Koelliker, T. K. Schaaf, and R. K. Varma, *J. Am. Chem. Soc.* **1971**, *93*, 1491.
3. E. J. Corey and P. A. Grieco, *Tetrahedron Lett.* **1972**, 107.
4. E. J. Corey, T. Ravindranathan, and S. Terashima, *J. Am. Chem. Soc*, **1971**, *93*, 4326 and references cited therein.
5. E. J. Corey and J. Mann, *J. Am. Chem. Soc.* **1973**, *95*, 6832 and references cited therein.
6. E. J. Corey and D. J. Beames, *J. Am. Chem. Soc.* **1972**, *94*, 7210.
7. E. J. Corey and H. E. Ensley, *J. Org. Chem.* **1973**, *38*, 3187.
8. E. J. Corey and R. K. Varma, *J. Am. Chem. Soc.* **1971**, *93*, 7319.
9. E. J. Corey and C. R. Cyr, *Tetrahedron Lett.* **1974**, 1761.
10. E. J. Corey and G. Moinet, *J. Am. Chem. Soc.* **1973**, *95*, 7185.

11.9

Alternative Synthesis of Prostaglandins $F_{1\alpha}$ and E_1

A modification of the bicyclo[2.2.1]heptene route to PG's was developed in which the omega chain was introduced at the end of the synthesis, for the purpose of facilitating the preparation of large numbers of omega-chain-differentiated PG analogs of either the first or second PG family.

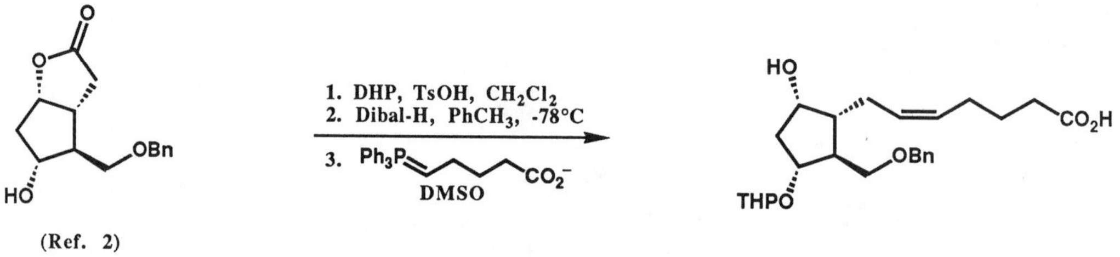

O
O
OBn
HO
(Ref. 2)
1. DHP, TsOH, CH_2Cl_2
2. Dibal-H, $PhCH_3$, -78°C
3. Ph_3P
CO_2^-
DMSO
HO
CO_2H
OBn
THPO

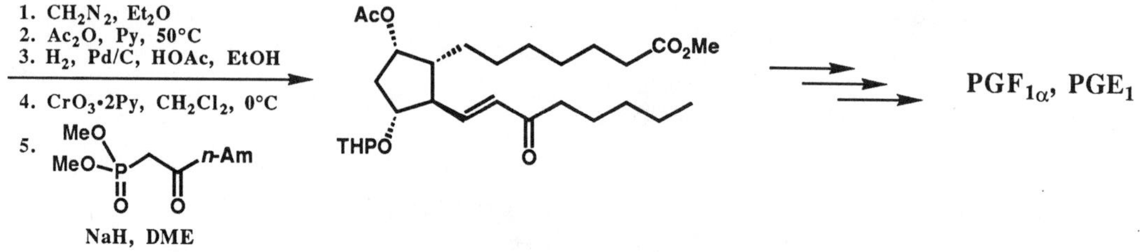

1. CH_2N_2, Et_2O
2. Ac_2O, Py, 50°C
3. H_2, Pd/C, HOAc, EtOH
4. CrO_3•2Py, CH_2Cl_2, 0°C
5.
MeO
MeO
P
n-Am
O
O
NaH, DME
AcO
CO_2Me
THPO
O
$PGF_{1\alpha}$, PGE_1

References:

1. T. K. Schaaf and E. J. Corey, *J. Org. Chem.* **1972**, *37*, 2921.

2. (a) E. J. Corey, H. Shirahama, H. Yamamoto, S. Terashima, A. Venkateswarlu, and T. K. Schaaf, *J. Am. Chem. Soc.* **1971**, *93*, 1490; (b) E. J. Corey, T. Ravindranathan, and S. Terashima, *ibid*, **1971**, *93*, 4326.

Conjugated Addition-Alkylation Route to Prostaglandins

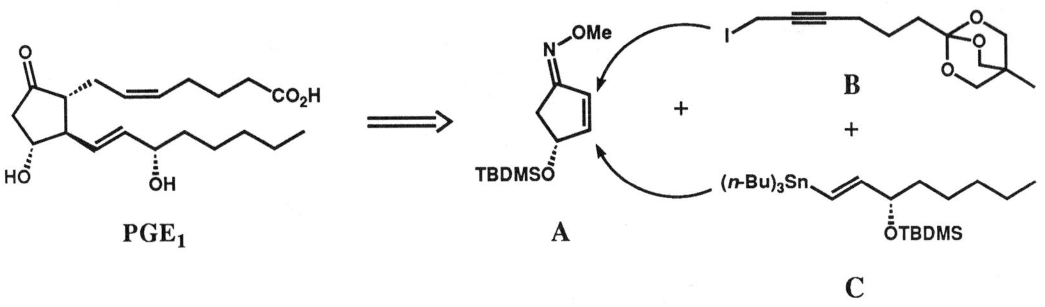

O
CO2H
HO
OH
PGE1
N
OMe
TBDMSO
A
I
O
O
O
B
+
+
(n-Bu)3Sn
OTBDMS
C

A short three-component route to prostaglandins was developed involving a number of novel steps, several of which were based on special properties of the *O*-methyloxime function.

1. Synthesis of the fragment **A**:

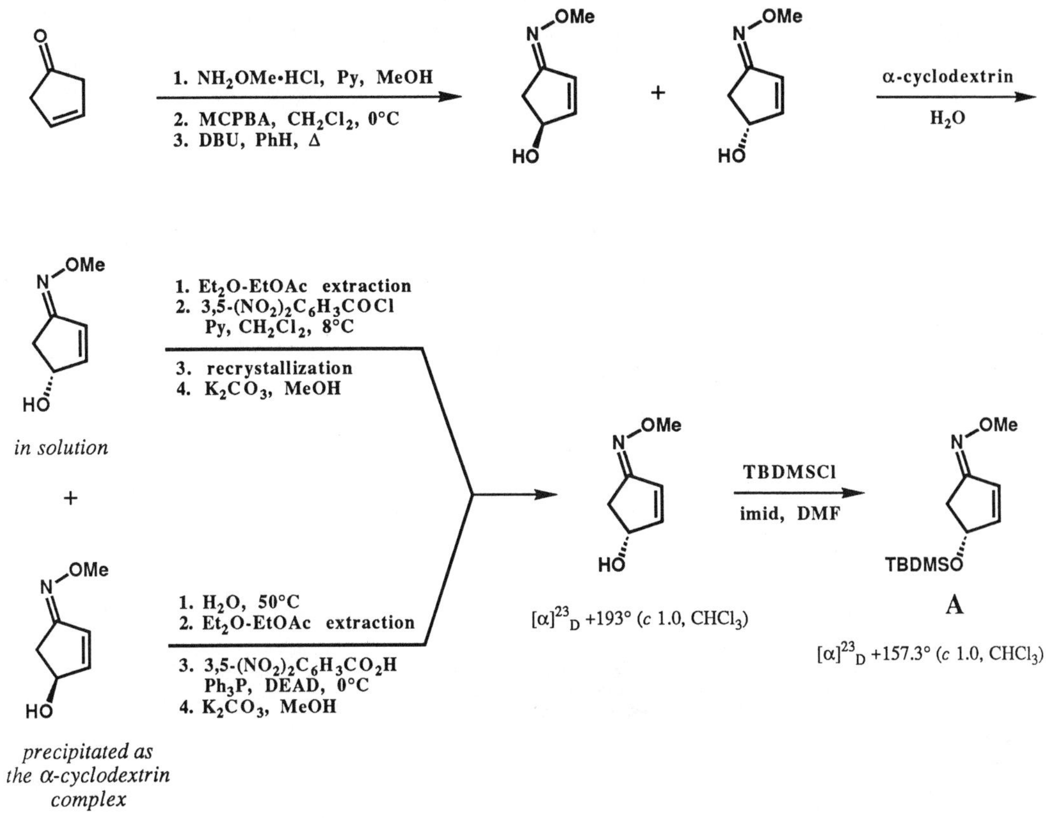

O
1. NH2OMe•HCl, Py, MeOH
2. MCPBA, CH2Cl2, 0°C
3. DBU, PhH, Δ
N
OMe
HO
+
α-cyclodextrin
H2O
in solution
precipitated as the α-cyclodextrin complex
1. Et2O-EtOAc extraction
2. 3,5-(NO2)2C6H3COCl
Py, CH2Cl2, 8°C
3. recrystallization
4. K2CO3, MeOH
1. H2O, 50°C
2. Et2O-EtOAc extraction
3. 3,5-(NO2)2C6H3CO2H
Ph3P, DEAD, 0°C
4. K2CO3, MeOH
[α]23D +193° (c 1.0, CHCl3)
TBDMSCl
imid, DMF
TBDMSO
A
[α]23D +157.3° (c 1.0, CHCl3)

11.10

A direct synthesis of the fragment **A** was also achieved starting from diethyl (*S*,*S*)-tartrate.

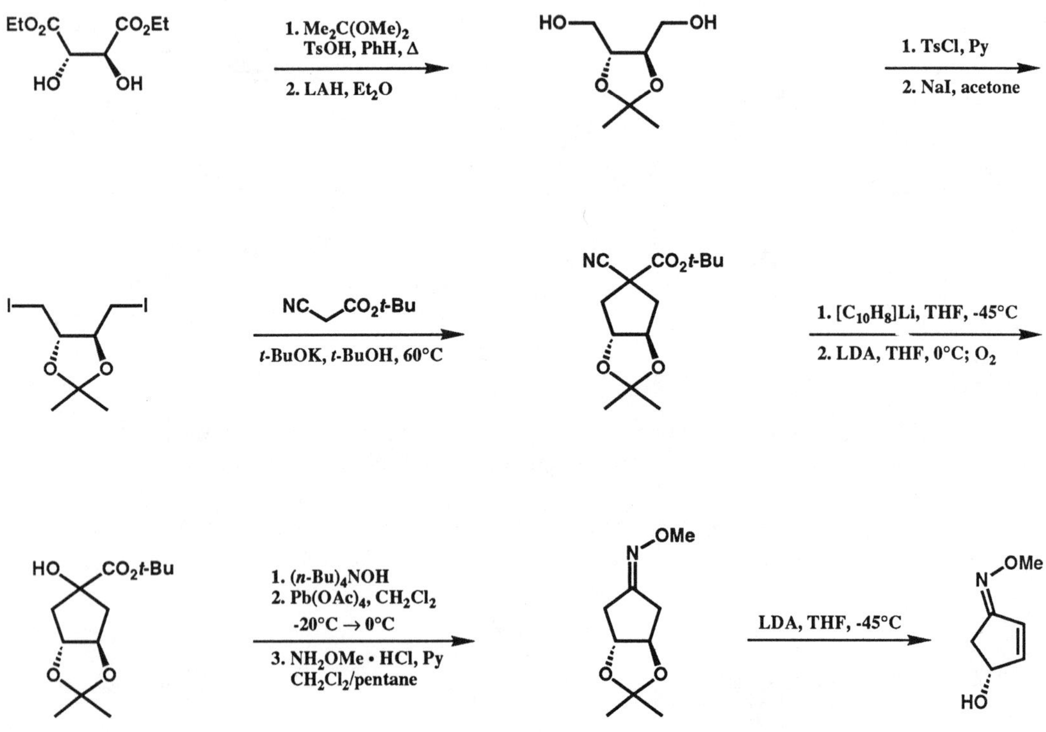
EtO2C
CO2Et
HO
OH
1. Me2C(OMe)2
TsOH, PhH, Δ
2. LAH, Et2O
HO
OH
1. TsCl, Py
2. NaI, acetone
I
I
NC CO2t-Bu
t-BuOK, t-BuOH, 60°C
NC
CO2t-Bu
1. [C10H8]Li, THF, -45°C
2. LDA, THF, 0°C; O2
HO
CO2t-Bu
1. (n-Bu)4NOH
2. Pb(OAc)4, CH2Cl2
-20°C → 0°C
3. NH2OMe • HCl, Py
CH2Cl2/pentane
N
OMe
LDA, THF, -45°C
N
OMe
HO

2. Synthesis of the fragment **B**:

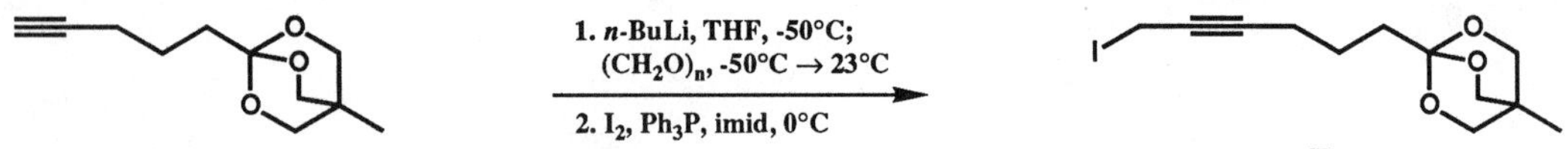
1. n-BuLi, THF, -50°C;
(CH2O)n, -50°C → 23°C
2. I2, Ph3P, imid, 0°C
I

(Ref. 2)

B

3. Synthesis of the fragment **C**:

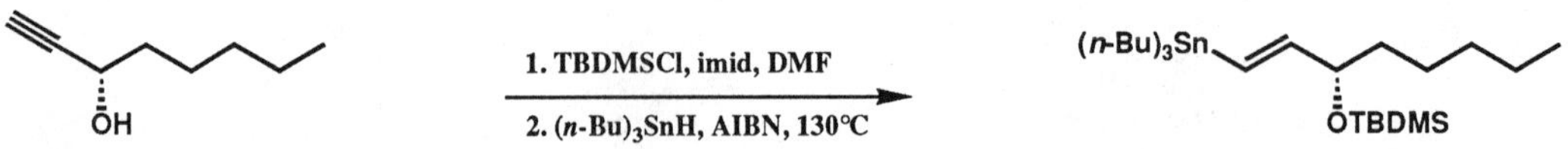
OH
1. TBDMSCl, imid, DMF
2. (n-Bu)3SnH, AIBN, 130°C
(n-Bu)3Sn
OTBDMS

C

$[\alpha]^{23}_{D}$ -12.0°(*c* 1.0, $CHCl_3$)

4. Three-component synthesis of PGE_2:

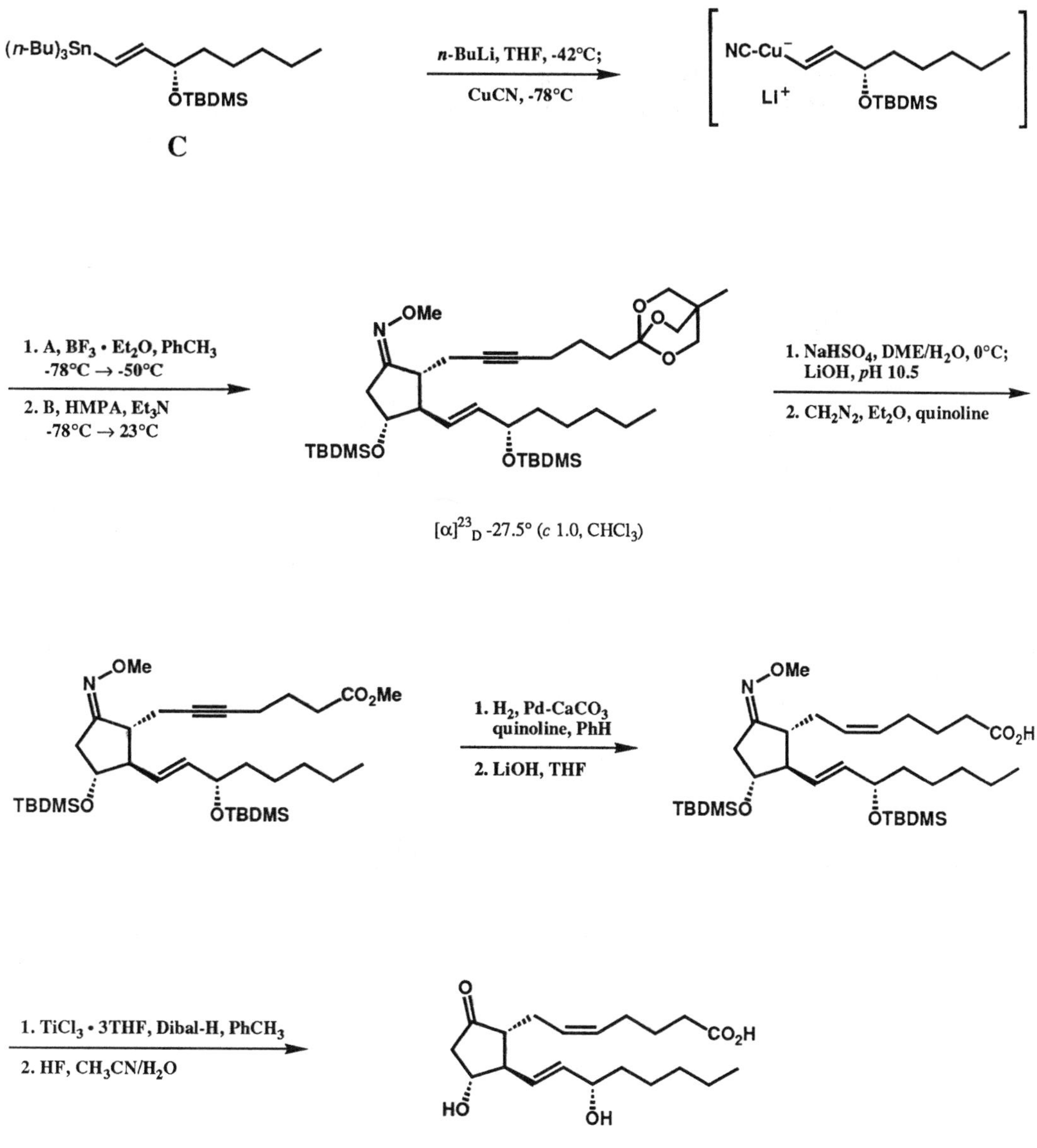
(n-Bu)3Sn
OTBDMS
C
n-BuLi, THF, -42°C;
CuCN, -78°C
NC-Cu
Li+
OTBDMS
1. A, BF3 • Et2O, PhCH3
-78°C → -50°C
2. B, HMPA, Et3N
-78°C → 23°C
OMe
N
TBDMSO
OTBDMS
[α]23D -27.5° (c 1.0, CHCl3)
1. NaHSO4, DME/H2O, 0°C;
LiOH, pH 10.5
2. CH2N2, Et2O, quinoline
CO2Me
1. H2, Pd-CaCO3
quinoline, PhH
2. LiOH, THF
CO2H
1. TiCl3 • 3THF, Dibal-H, PhCH3
2. HF, CH3CN/H2O
HO
OH

PGE_2

References:

1. E. J. Corey, K. Niimura, Y. Konishi, S. Hashimoto, and Y. Hamada, *Tetrahedron Lett.* **1986**, *27*, 2199, 3556.

2. E. J. Corey and N. Raju, *Tetrahedron Lett.* **1983**, *24*, 5571.

11.11

Bicyclo[3.1.0]hexane Routes to Prostaglandins

Two different routes to PG's *via* bicyclo[3.1.0]hexane intermediates are shown. In route 1 stereo- and position-specific addition of dichloroketene to a bicyclo[3.1.0]hexene provided the framework for elaboration to prostanoids. Route 2 featured stereospecific internal cyclopropanation and stereospecific S_N2 displacement of carbon to establish the prostanoid nucleus.

Route 1 (Ref. 1):

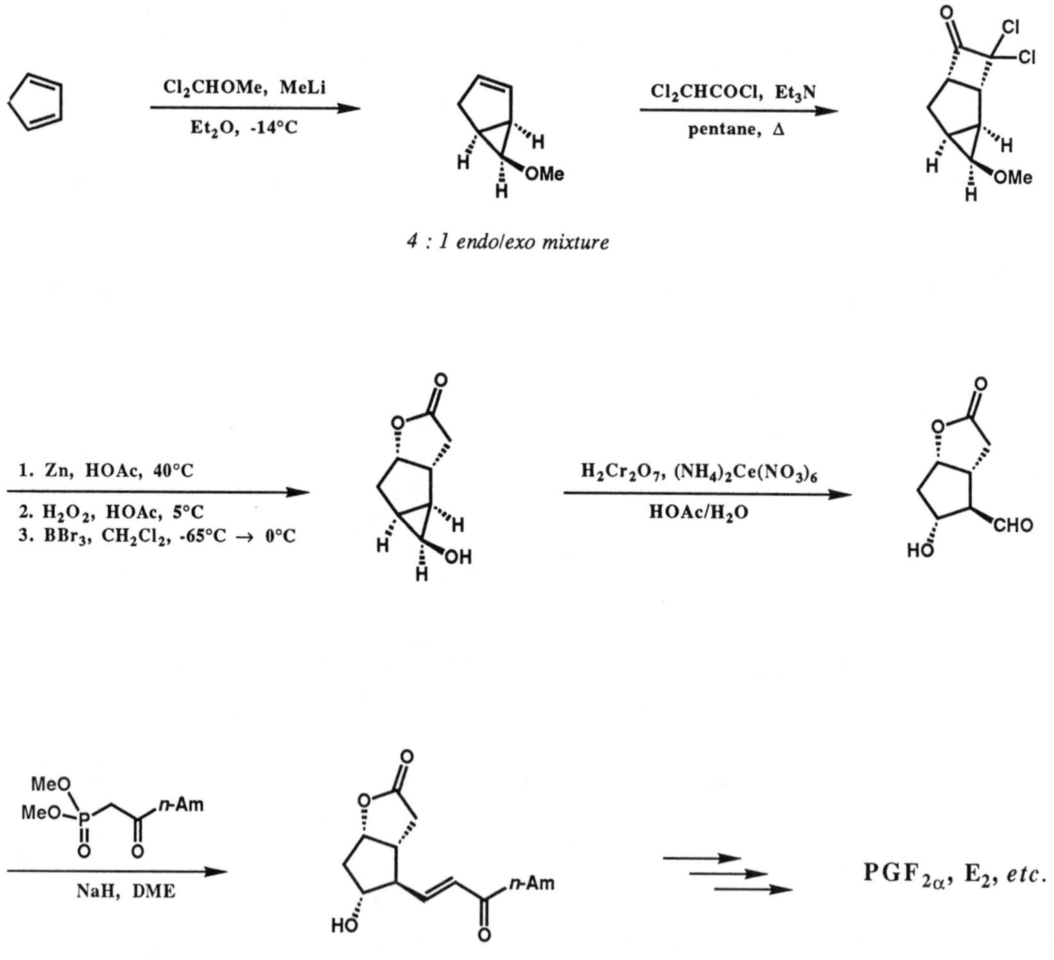

Cl2CHOMe, MeLi
Et2O, -14°C
4 : 1 endo/exo mixture
Cl2CHCOCl, Et3N
pentane, Δ
1. Zn, HOAc, 40°C
2. H2O2, HOAc, 5°C
3. BBr3, CH2Cl2, -65°C → 0°C
H2Cr2O7, (NH4)2Ce(NO3)6
HOAc/H2O
CHO
HO
MeO
MeO
n-Am
NaH, DME
PGF2α, E2, etc.

Route 2 (Ref. 2):

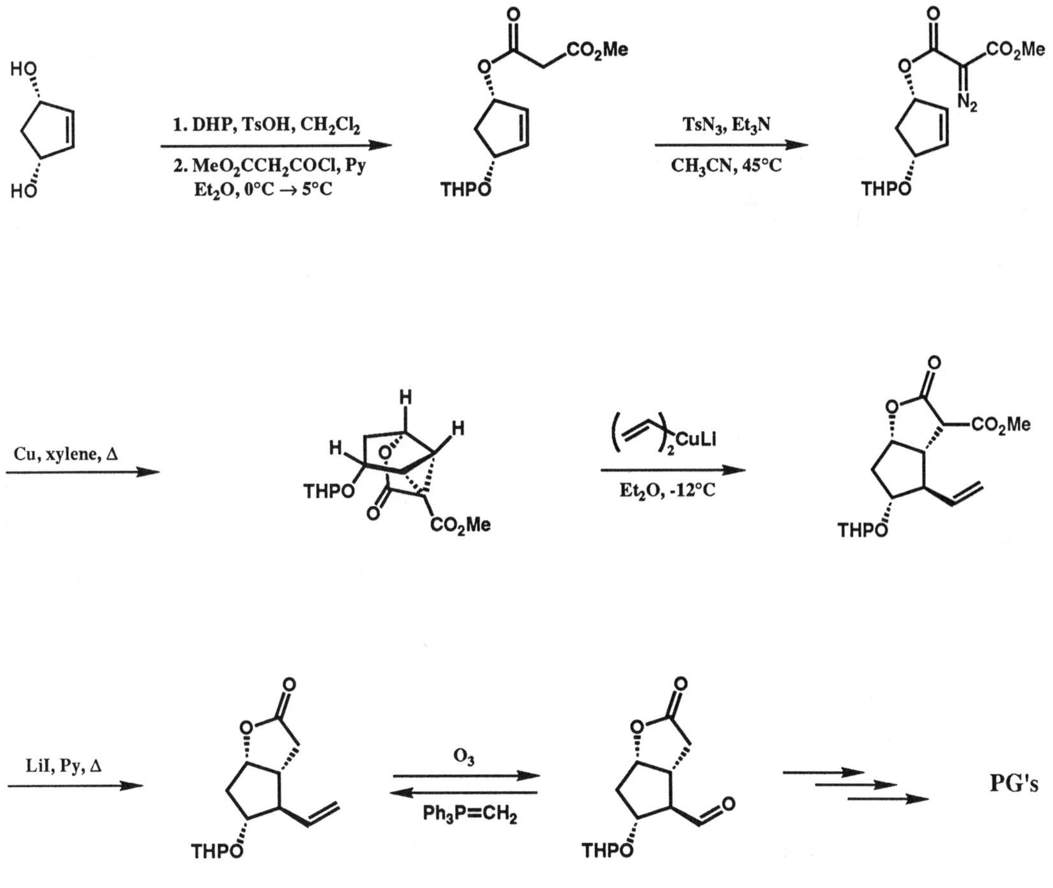
HO
HO
1. DHP, TsOH, CH2Cl2
2. MeO2CCH2COCl, Py
Et2O, 0°C → 5°C
O
CO2Me
THPO
TsN3, Et3N
CH3CN, 45°C
N2
Cu, xylene, Δ
H
CuLi
Et2O, -12°C
LiI, Py, Δ
O3
Ph3P=CH2
PG's

(Ref. 3)

References:

1. E. J. Corey, Z. Arnold, and J. Hutton, *Tetrahedron Lett.* **1970**, 307.

2. E. J. Corey and P. L. Fuchs, *J. Am. Chem. Soc.* **1972**, *94*, 4014.

3. E. J. Corey, H. Shirahama, H. Yamamoto, S. Terashima, A. Venkateswarlu, and T. K. Schaaf, *J. Am. Chem. Soc.* **1971**, *93*, 1490.

11.12

Synthesis of Prostaglandin $F_{2\alpha}$ from a 2-Oxabicyclo[3.3.0]octenone

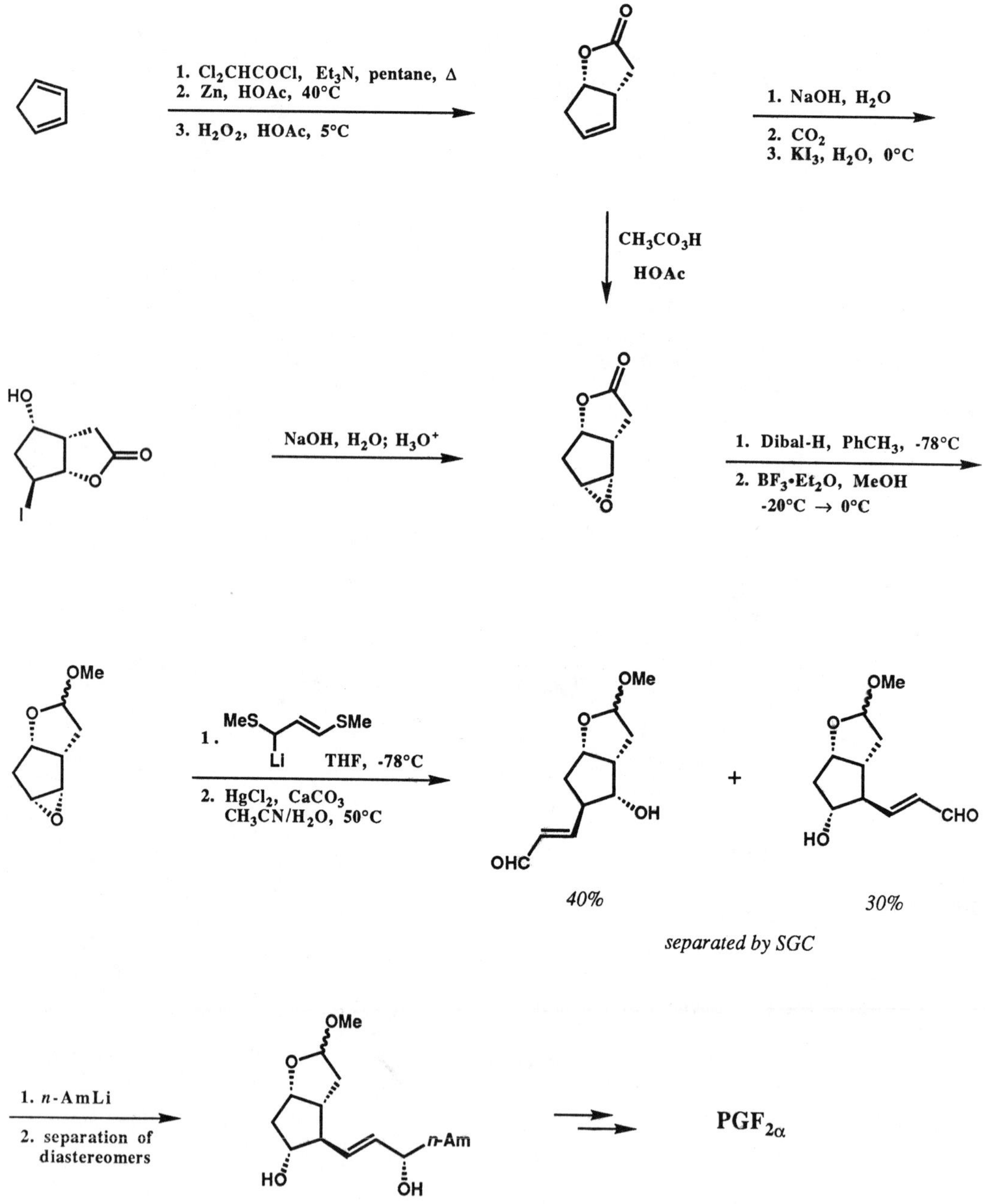

1. $Cl_2CHCOCl$, Et_3N, pentane, Δ
2. Zn, HOAc, 40°C
3. H_2O_2, HOAc, 5°C
1. NaOH, H_2O
2. CO_2
3. KI_3, H_2O, 0°C
CH_3CO_3H
HOAc
HO
I
NaOH, H_2O; H_3O^+
1. Dibal-H, $PhCH_3$, -78°C
2. BF_3•Et_2O, MeOH
-20°C → 0°C
OMe
MeS
SMe
Li
THF, -78°C
2. $HgCl_2$, $CaCO_3$
CH_3CN/H_2O, 50°C
OH
OHC
CHO
40%
30%
separated by SGC
1. n-AmLi
2. separation of diastereomers
n-Am
PGF$_{2\alpha}$

11.12

The low position selectivity in the epoxide opening step of this early synthesis was improved in a later study (Ref. 2).

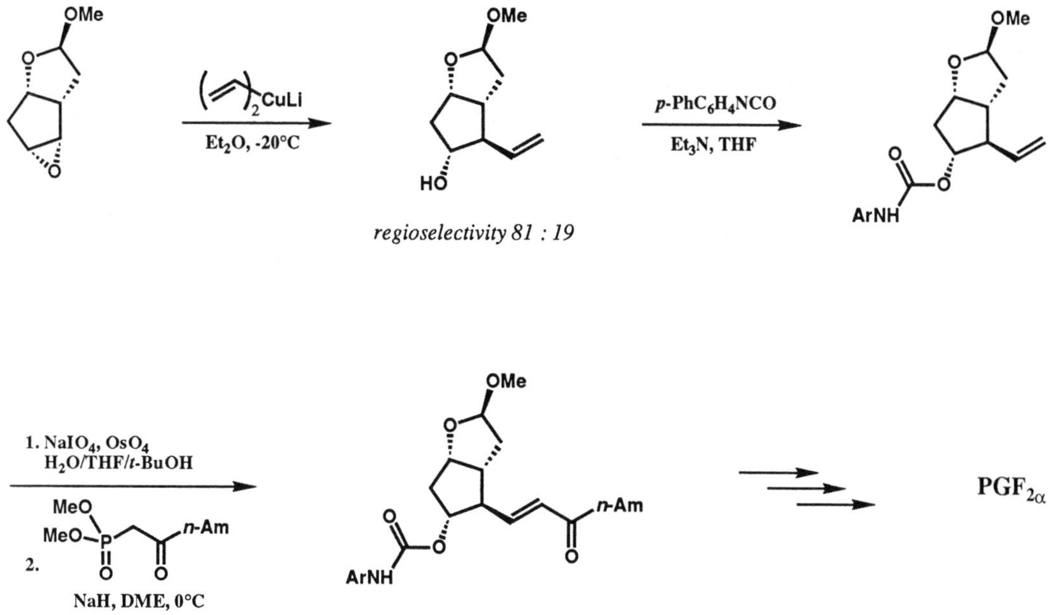

OMe
O
O
CuLi
2
Et2O, -20°C
HO
p-PhC6H4NCO
Et3N, THF
ArNH
regioselectivity 81 : 19
1. NaIO4, OsO4
H2O/THF/t-BuOH
MeO
MeO
P
n-Am
2.
NaH, DME, 0°C
PGF2α

References:

1. E. J. Corey and R. Noyori, *Tetrahedron Lett.* **1970**, 311.
2. E. J. Corey, K. C. Nicolaou, and D. J. Beames, *Tetrahedron Lett.* **1974**, 2439.

11.13

Synthesis of 11-Desoxyprostaglandins

The possibility that 11-desoxyprostaglandins might exhibit tissue selectivity in therapeutic applications provided an incentive to develop a short synthetic route to this series (Ref. 1).

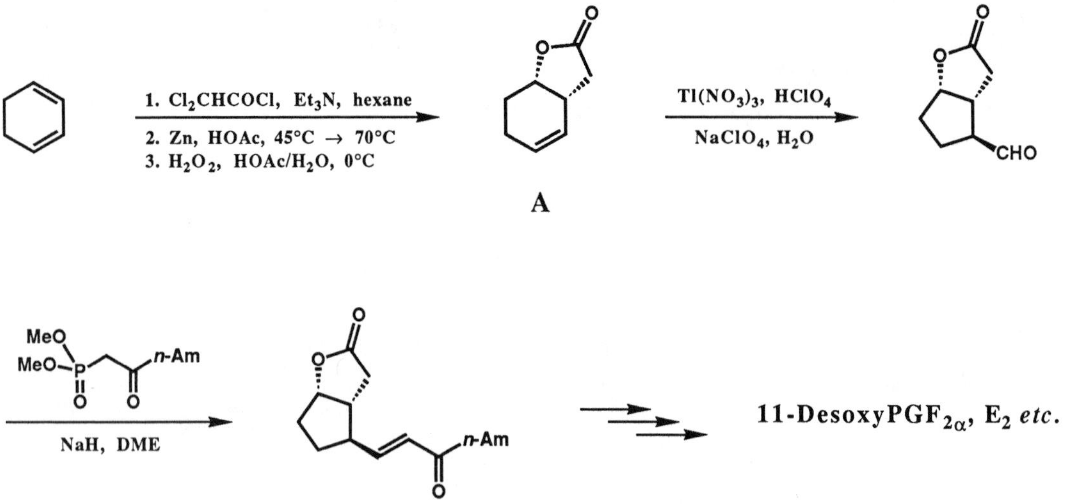

1. Cl2CHCOCl, Et3N, hexane
2. Zn, HOAc, 45°C → 70°C
3. H2O2, HOAc/H2O, 0°C
A
Tl(NO3)3, HClO4
NaClO4, H2O
CHO
MeO
MeO
P
n-Am
NaH, DME
n-Am
11-DesoxyPGF2α, E2 etc.

The lactone **A** was also used as starting material in the synthesis of the primary prostaglandins *via* an allylic substitution-semi-pinacolic rearrangement sequence (Ref. 2).

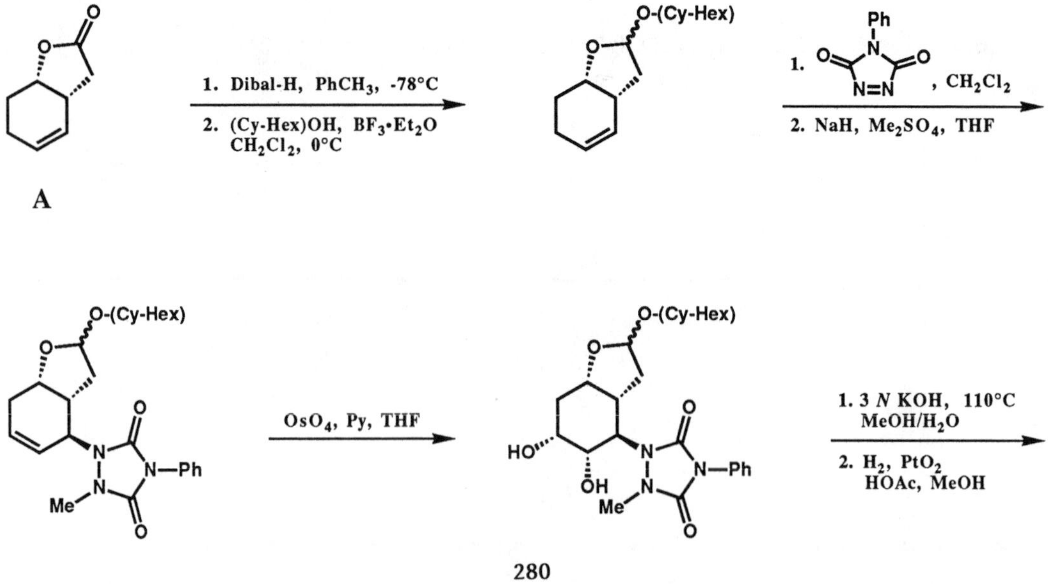

A
1. Dibal-H, PhCH3, -78°C
2. (Cy-Hex)OH, BF3•Et2O
CH2Cl2, 0°C
O-(Cy-Hex)
Ph
, CH2Cl2
2. NaH, Me2SO4, THF
N—Ph
Me
OsO4, Py, THF
HO
OH
1. 3 N KOH, 110°C
MeOH/H2O
2. H2, PtO2
HOAc, MeOH

11.13

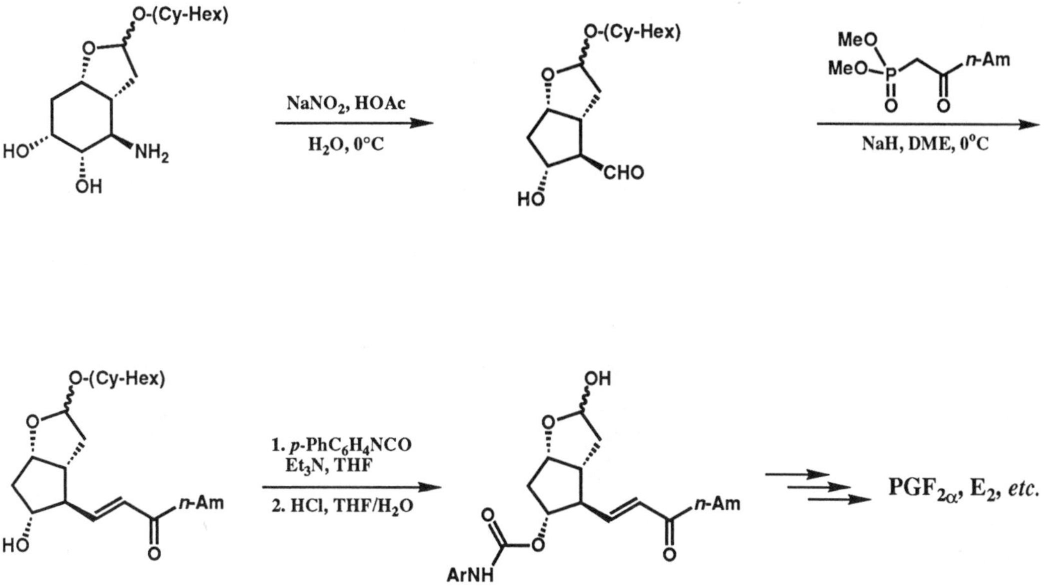

O-(Cy-Hex)
HO
NH2
OH
NaNO2, HOAc
H2O, 0°C
O-(Cy-Hex)
CHO
HO
MeO
MeO
P
O
O
n-Am
NaH, DME, 0°C
O-(Cy-Hex)
n-Am
HO
1. p-PhC6H4NCO
Et3N, THF
2. HCl, THF/H2O
OH
ArNH
n-Am
PGF2α, E2, etc.

The *cis*-fused lactone **A** was resolved using (+)-1-(1-naphthyl)ethylamine to give the enantiomer required for the synthesis of prostanoids (Ref. 3).

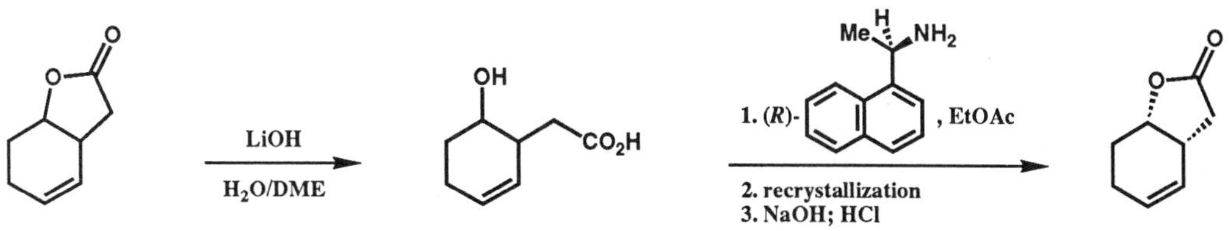

LiOH
H2O/DME
OH
CO2H
Me
H
NH2
1. (R)-
, EtOAc
2. recrystallization
3. NaOH; HCl

A

$[\alpha]^{27}_{D}$ -28° (*c* 0.83, MeOH)

References:

1. E. J. Corey and T. Ravindranathan, *Tetrahedron Lett.* **1971**, 4753.
2. E. J. Corey and B. B. Snider, *Tetrahedron Lett.* **1973**, 3091.
3. E. J. Corey and B. B. Snider, *J. Org. Chem.* **1974**, *39*, 256.

11.14

Synthesis of Vane's Prostaglandin X (Prostacycline, PGI_2)

Vane and co-workers isolated a new prostaglandin (initially called PGX) from microsomal fraction of stomach. The structure was established by chemical synthesis from $PGF_{2\alpha}$ (Ref. 1).

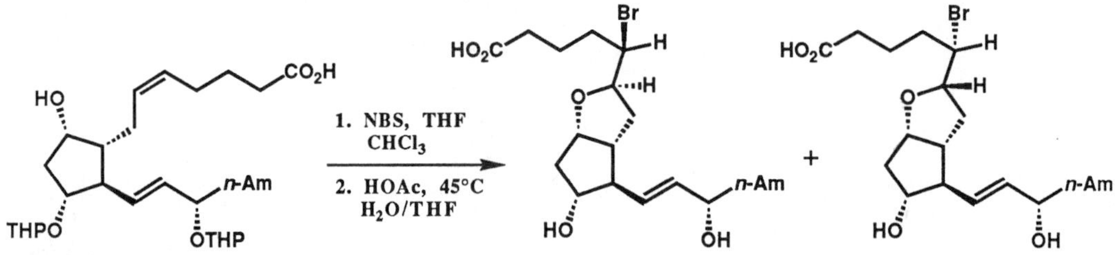

1. NBS, THF
CHCl3
2. HOAc, 45°C
H2O/THF

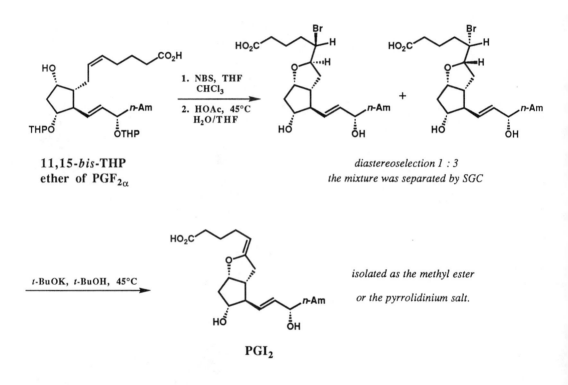

11,15-*bis*-THP ether of $PGF_{2\alpha}$

diastereoselection 1 : 3
the mixture was separated by SGC

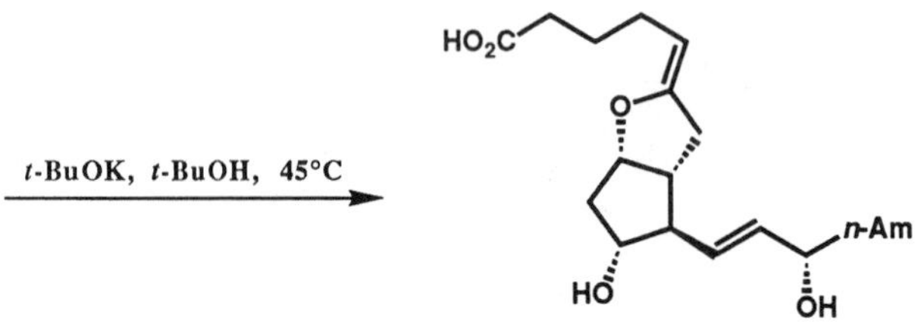

t-BuOK, t-BuOH, 45°C

isolated as the methyl ester

or the pyrrolidinium salt.

PGI_2

The 5-(*E*)-isomer of PGI_2 was also unambiguously synthesized in two different approaches (Ref. 2, 3).

Route 1:

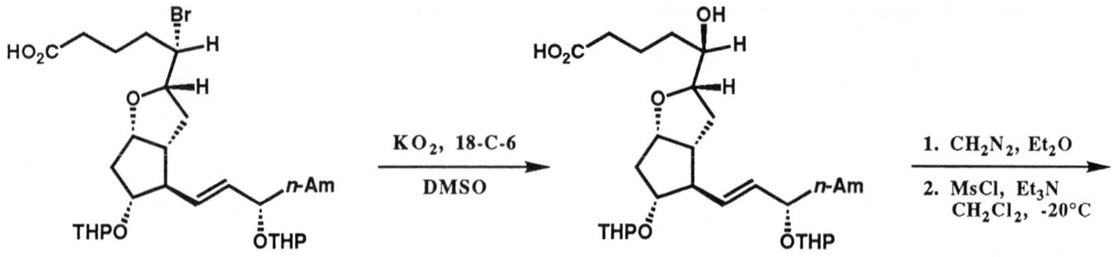

KO2, 18-C-6
DMSO
1. CH2N2, Et2O
2. MsCl, Et3N
CH2Cl2, -20°C

(Ref. 1)

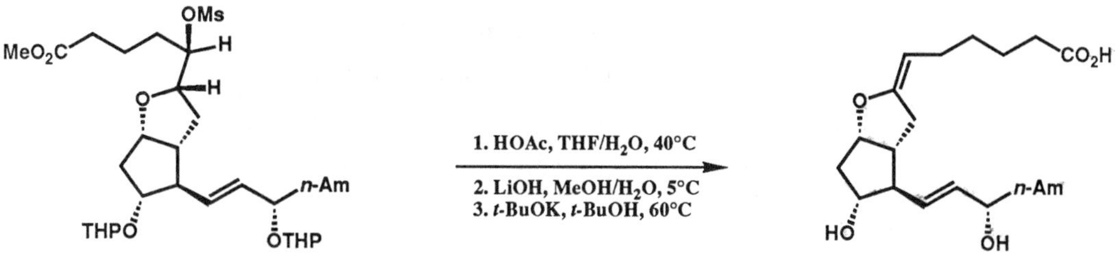
OMs
H
MeO2C
H
THPO
OTHP
n-Am
1. HOAc, THF/H2O, 40°C
2. LiOH, MeOH/H2O, 5°C
3. t-BuOK, t-BuOH, 60°C
CO2H
HO
OH
n-Am

isolated as the methyl ester
or the pyrrolidinium salt

Route 2:

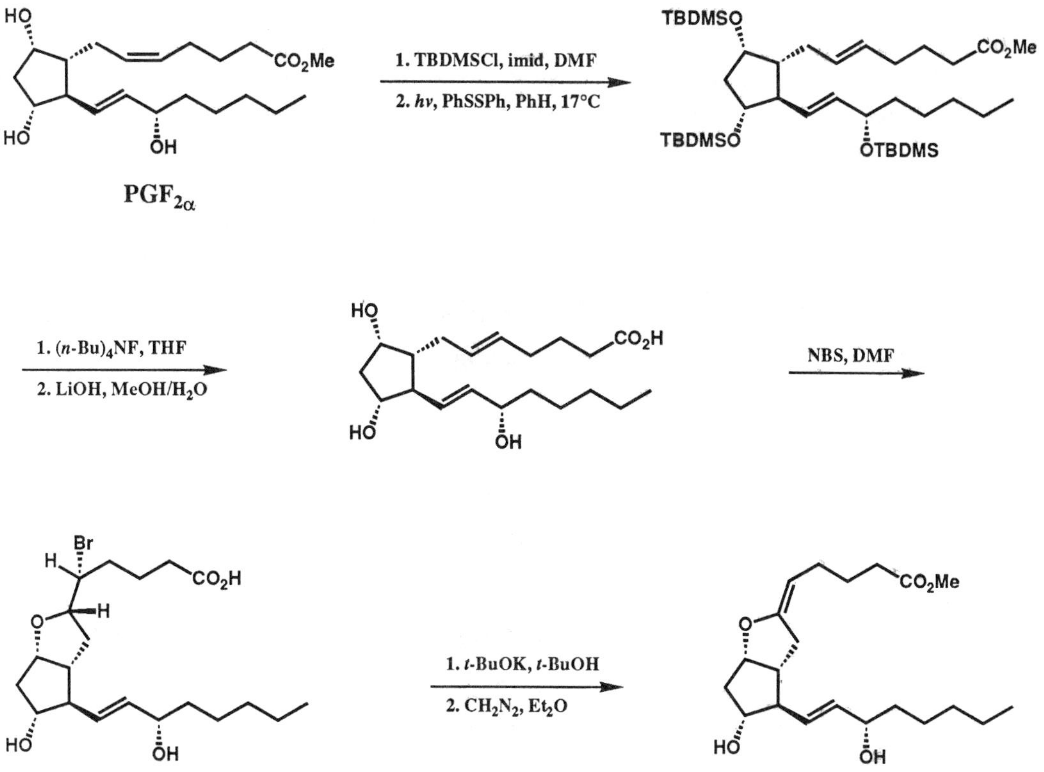
HO
CO2Me
HO
OH
PGF2α
1. TBDMSCl, imid, DMF
2. hν, PhSSPh, PhH, 17°C
TBDMSO
CO2Me
TBDMSO
OTBDMS
1. (n-Bu)4NF, THF
2. LiOH, MeOH/H2O
HO
CO2H
HO
OH
NBS, DMF
Br
H
CO2H
H
HO
OH
1. t-BuOK, t-BuOH
2. CH2N2, Et2O
CO2Me
HO
OH

major isomer

References:

1. E. J. Corey, G. E. Keck, and I. Székely, *J. Am. Chem. Soc.* **1977**, *99*, 2006.
2. E. J. Corey, I. Székely, and C. S. Shiner, *Tetrahedron Lett.* **1977**, 3529.
3. E. J. Corey, H. L. Pearce, I. Székely, and M. Ishiguro, *Tetrahedron Lett.* **1978**, 1023.

11.15

Synthesis of the Major Human Urinary Metabolite of Prostaglandin D_2

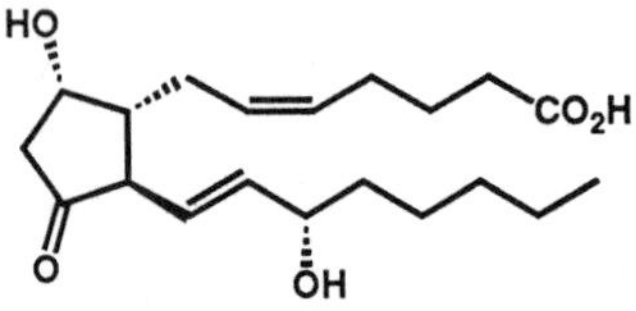
HO
CO_2H
O
OH

PGD_2

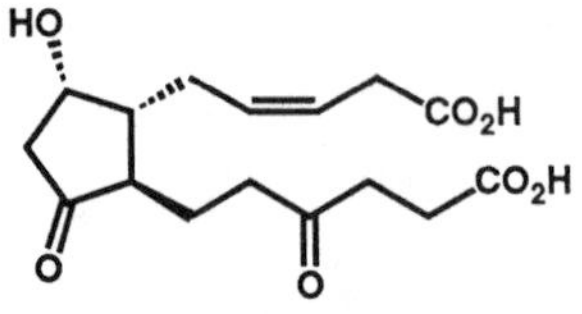
HO
CO_2H
CO_2H
O
O

major metabolite of PGD_2

The clinical measurement of prostaglandin D_2, produced in the body by activation of immune-active cells, required the synthesis of the major urinary metabolite. A synthesis was developed starting with the standard intermediate for PG synthesis.

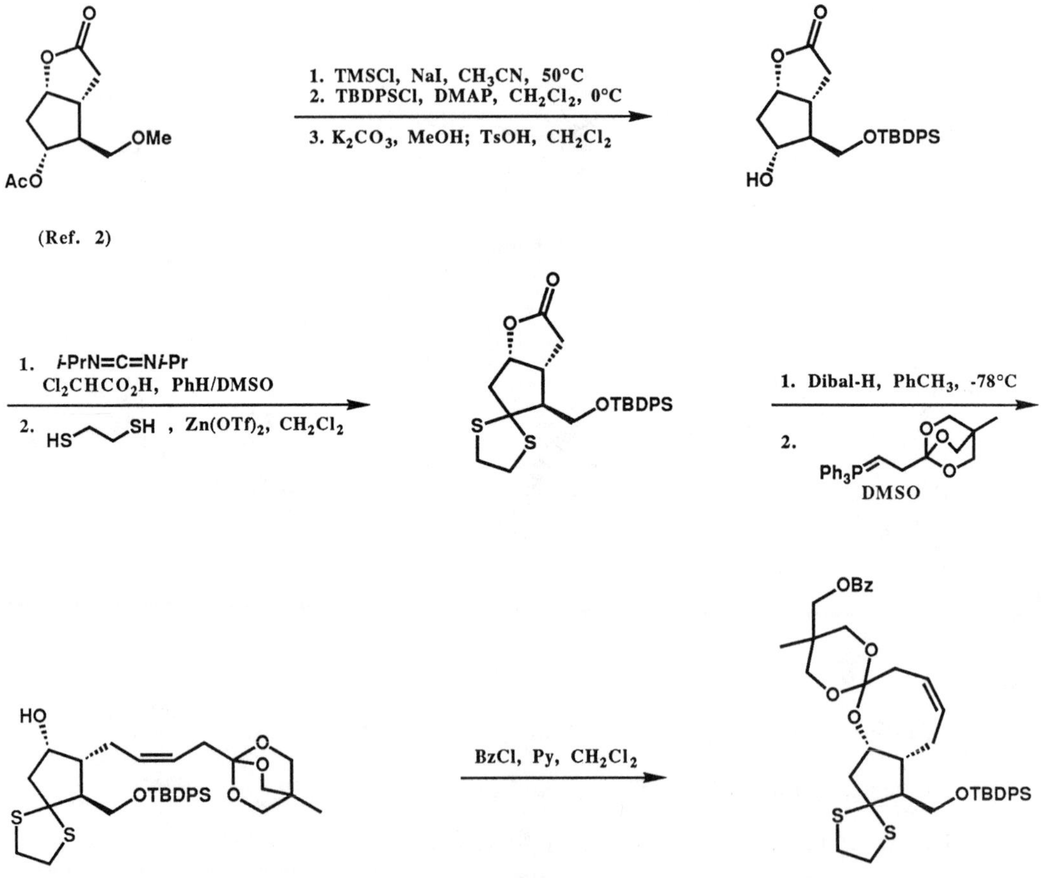
AcO
OMe
(Ref. 2)
1. TMSCl, NaI, CH_3CN, 50°C
2. TBDPSCl, DMAP, CH_2Cl_2, 0°C
3. K_2CO_3, MeOH; TsOH, CH_2Cl_2
HO
OTBDPS
1. i-PrN=C=Ni-Pr
Cl_2CHCO_2H, PhH/DMSO
2. HS SH, $Zn(OTf)_2$, CH_2Cl_2
OTBDPS
1. Dibal-H, $PhCH_3$, -78°C
2. Ph_3P
DMSO
HO
OTBDPS
BzCl, Py, CH_2Cl_2
OBz
OTBDPS

11.15

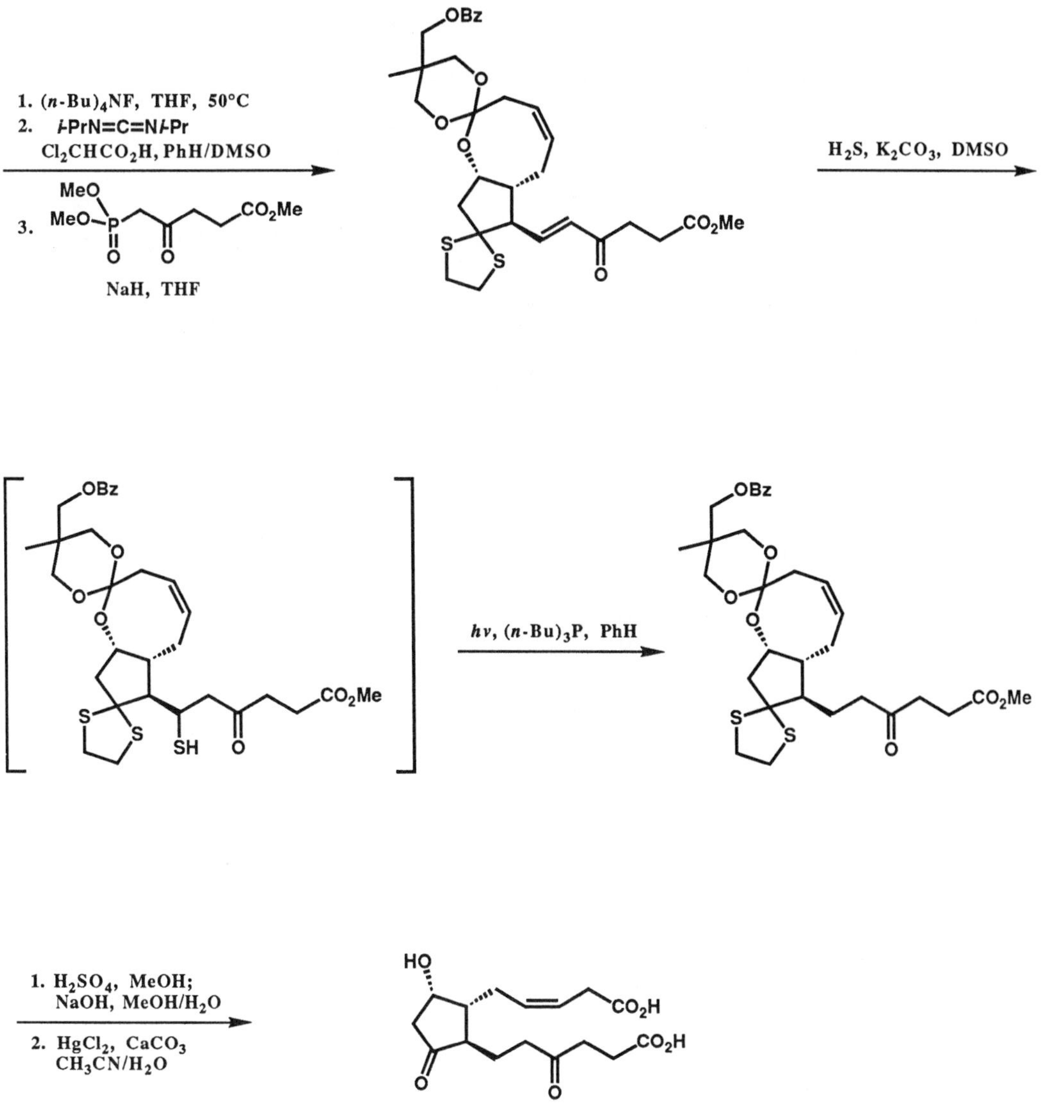

OBz
O
O
O
S
S
CO2Me
O
1. (n-Bu)4NF, THF, 50°C
2. i-PrN=C=Ni-Pr
Cl2CHCO2H, PhH/DMSO
3. MeO MeO P O O CO2Me
NaH, THF
H2S, K2CO3, DMSO
SH
hν, (n-Bu)3P, PhH
1. H2SO4, MeOH;
NaOH, MeOH/H2O
2. HgCl2, CaCO3
CH3CN/H2O
HO
CO2H
CO2H

References:

1. E. J. Corey and K. Shimoji, *J. Am. Chem. Soc.* **1983**, *105*, 1662.

2. (a) E. J. Corey, N. M. Weinshenker, T. K. Schaaf, and W. Huber, *J. Am. Chem. Soc.* **1969**, *91*, 5675; (b) E. J. Corey, T. K. Schaaf, W. Huber, U. Koelliker, and N. M. Weinshenker, *ibid,* **1970**, *92*, 397.

11.16

Synthesis of Several Analogs of the Prostaglandin Endoperoxide PGH_2

1. Synthesis of the 9,11-azo analog **A** (Ref. 1):

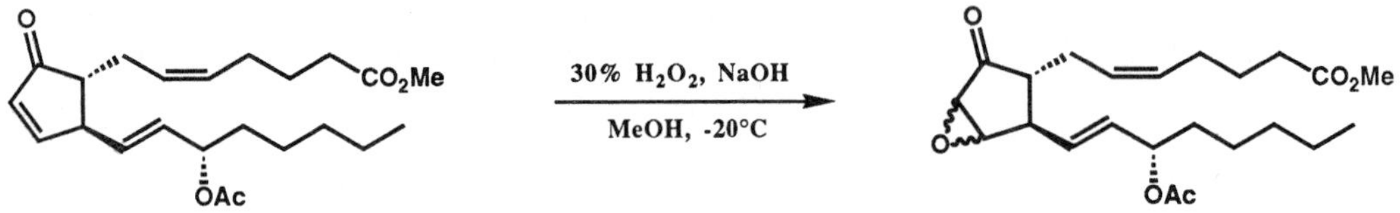

O
CO2Me
OAc
30% H2O2, NaOH
MeOH, -20°C
O
CO2Me
O
OAc

(Ref. 2, 3)

1.25 : 1 mixture of diastereomeric epoxides

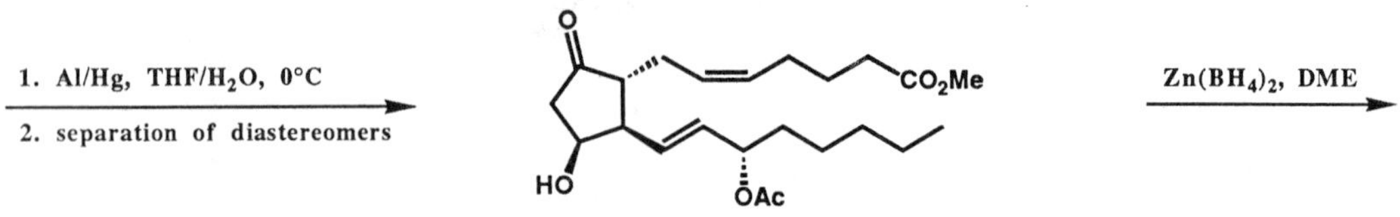

1. Al/Hg, THF/H2O, 0°C
2. separation of diastereomers
O
CO2Me
HO
OAc
Zn(BH4)2, DME

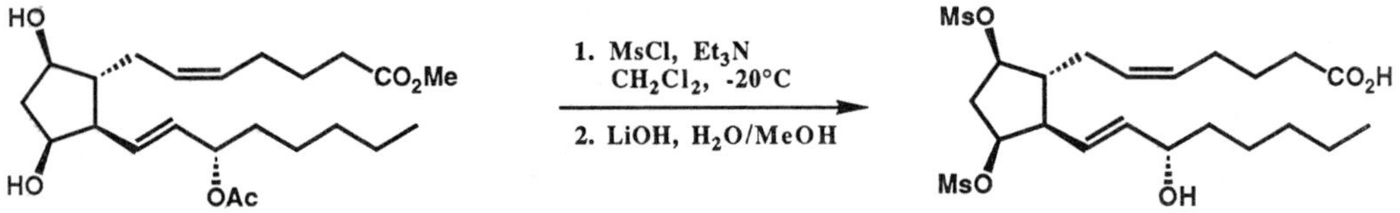

HO
CO2Me
HO
OAc
1. MsCl, Et3N
CH2Cl2, -20°C
2. LiOH, H2O/MeOH
MsO
CO2H
MsO
OH

diastereoselection 9 : 1

1. NH2NH2, t-BuOH/EtOH, Δ
2. air, Cu(OAc)2, MeOH/Et2O
N
N
CO2H
OH

$[\alpha]^{22}_D$ -17.7° ($CHCl_3$)

A

The 9,11-azo analog of PGH_2, compound **A**, is stable indefinitely and is substantially more active than PGH_2 in stimulating muscle contraction in the isolated rabbit aorta strip.

11.16

Compound **A** was also made by a direct, stereocontrolled total synthesis (Ref. 4).

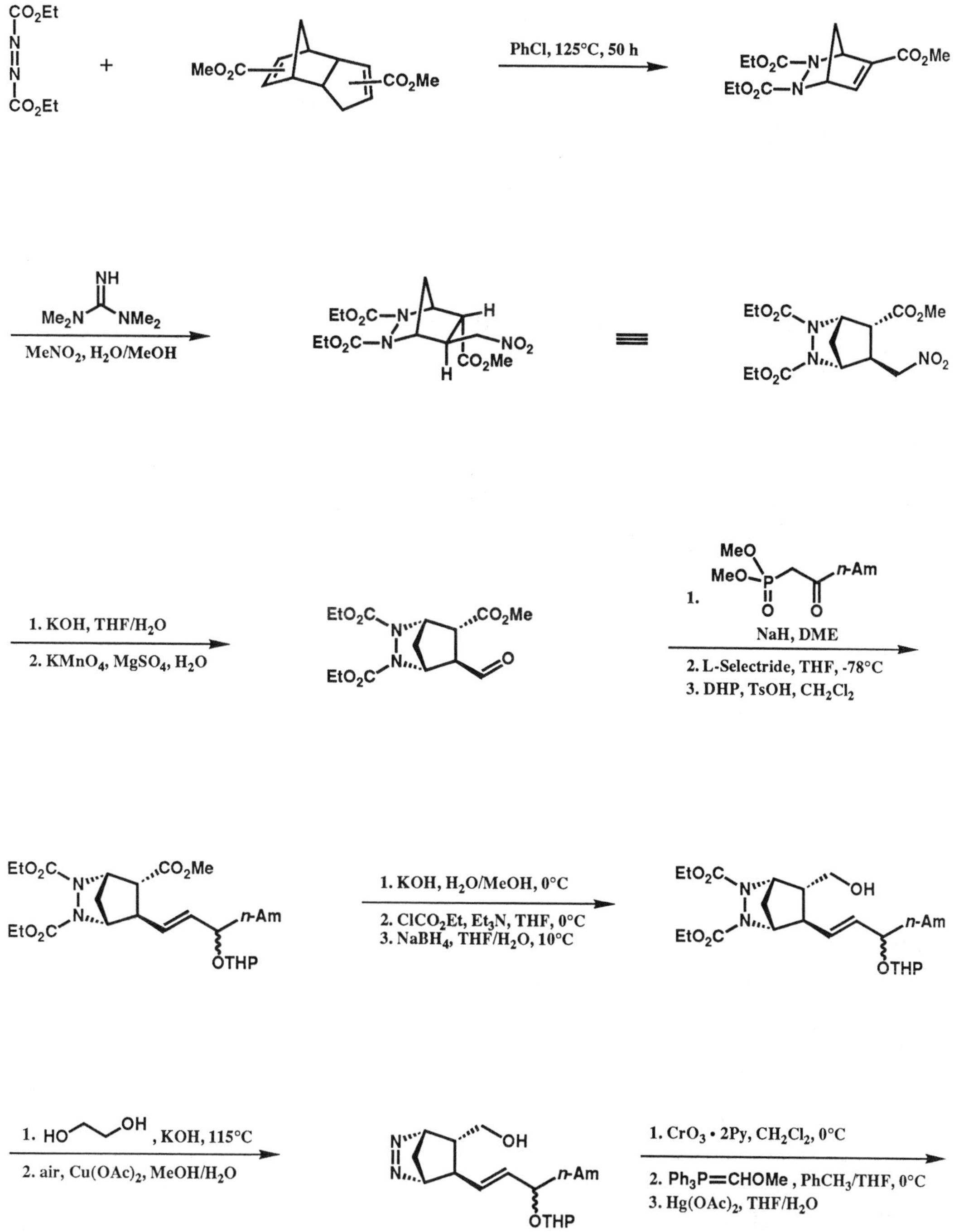
CO2Et
N
N
CO2Et
+
MeO2C
CO2Me
PhCl, 125°C, 50 h
EtO2C—N
EtO2C—N
CO2Me
NH
Me2N
NMe2
MeNO2, H2O/MeOH
EtO2C—N
EtO2C—N
H
NO2
H
CO2Me
EtO2C
N
N
EtO2C
CO2Me
NO2
1. KOH, THF/H2O
2. KMnO4, MgSO4, H2O
EtO2C
N
N
EtO2C
CO2Me
O
1.
MeO
MeO
P
O
O
n-Am
NaH, DME
2. L-Selectride, THF, -78°C
3. DHP, TsOH, CH2Cl2
EtO2C
N
N
EtO2C
CO2Me
n-Am
OTHP
1. KOH, H2O/MeOH, 0°C
2. ClCO2Et, Et3N, THF, 0°C
3. NaBH4, THF/H2O, 10°C
EtO2C
N
N
EtO2C
OH
n-Am
OTHP
1. HO
OH
, KOH, 115°C
2. air, Cu(OAc)2, MeOH/H2O
N
N
OH
n-Am
OTHP
1. CrO3 • 2Py, CH2Cl2, 0°C
2. Ph3P=CHOMe, PhCH3/THF, 0°C
3. Hg(OAc)2, THF/H2O

11.16

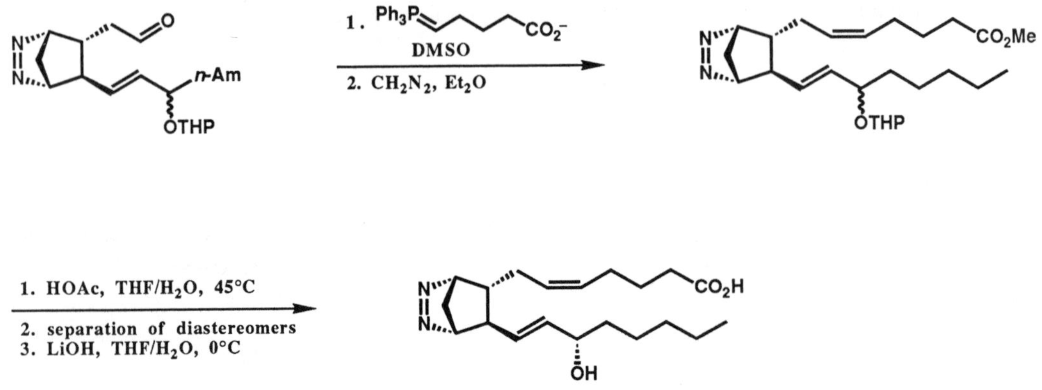

n-Am
OTHP
1. Ph3P=CH(CH2)3CO2⁻
DMSO
2. CH2N2, Et2O
CO2Me
OTHP
1. HOAc, THF/H2O, 45°C
2. separation of diastereomers
3. LiOH, THF/H2O, 0°C
CO2H
OH

A

2. Synthesis of the 9,11-etheno analog **B** (Ref. 5):

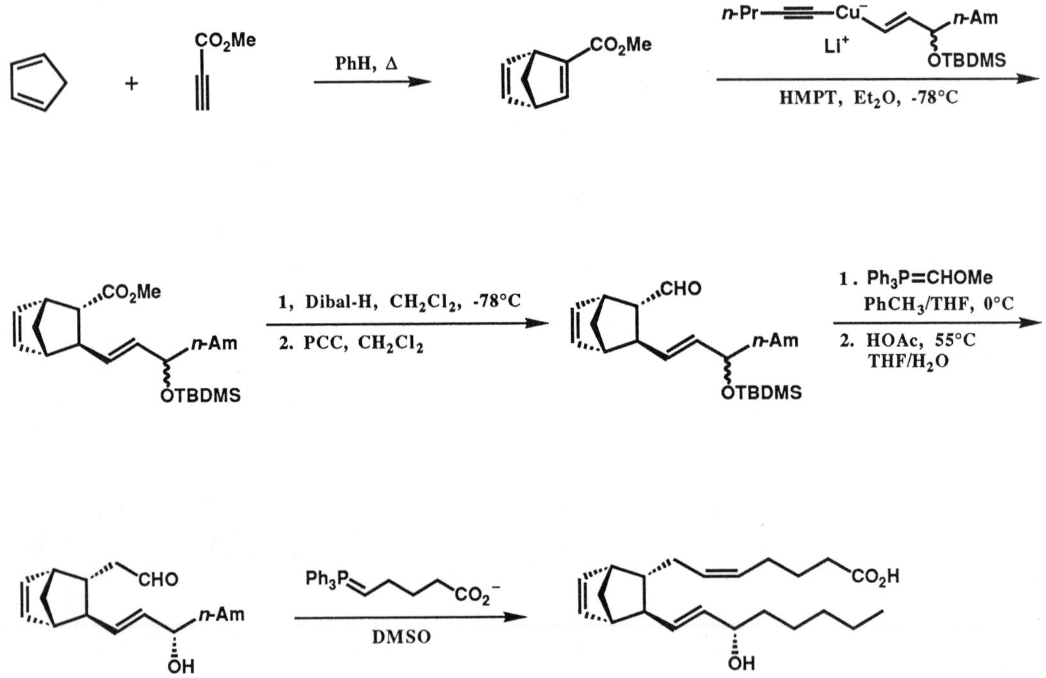

CO2Me
PhH, Δ
CO2Me
n-Pr—≡—Cu⁻
Li⁺
n-Am
OTBDMS
HMPT, Et2O, -78°C
CO2Me
n-Am
OTBDMS
1, Dibal-H, CH2Cl2, -78°C
2. PCC, CH2Cl2
CHO
n-Am
OTBDMS
1. Ph3P=CHOMe
PhCH3/THF, 0°C
2. HOAc, 55°C
THF/H2O
CHO
n-Am
OH
Ph3P=CH(CH2)3CO2⁻
DMSO
CO2H
OH

separated from the 15-β-isomer

B

The 9,11-etheno analog of PGH_2, compound **B**, induces aggregation in platelet-rich plasma, although somewhat less effectively than PGH_2.

Earlier, the 9,11-etheno analog of PGH_1, compound **C**, had been synthesized by the route shown below. It was found that this compound inhibits the PGH isomerase which forms PGE_1 but not the PGH reductase which catalyzes the formation of $PGF_{1\alpha}$ (Ref. 6).

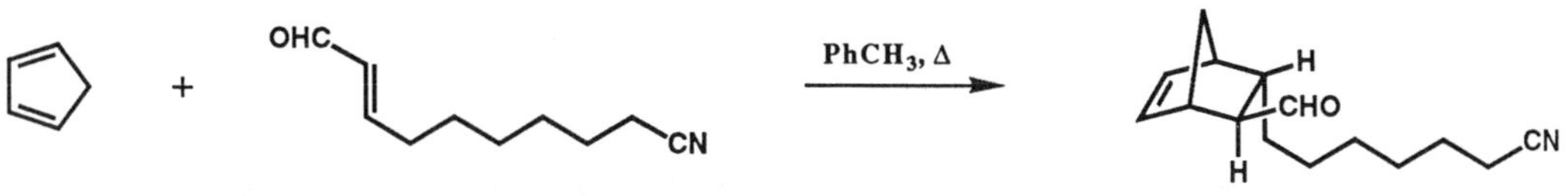
OHC
CN
PhCH3, Δ
H
CHO
CN
H

1.5 / 1 exo/endo-CHO mixture but exo- specific in the presence of $BF_3 \cdot Et_2O$

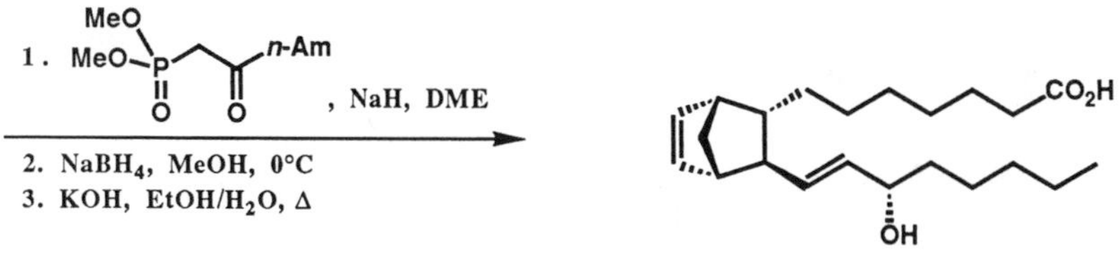
MeO
1. MeO–P
n-Am
, NaH, DME
2. NaBH4, MeOH, 0°C
3. KOH, EtOH/H2O, Δ
CO2H
OH

C

3. Synthesis of the 9-azo-11-methylene analog **D** (Ref. 7).

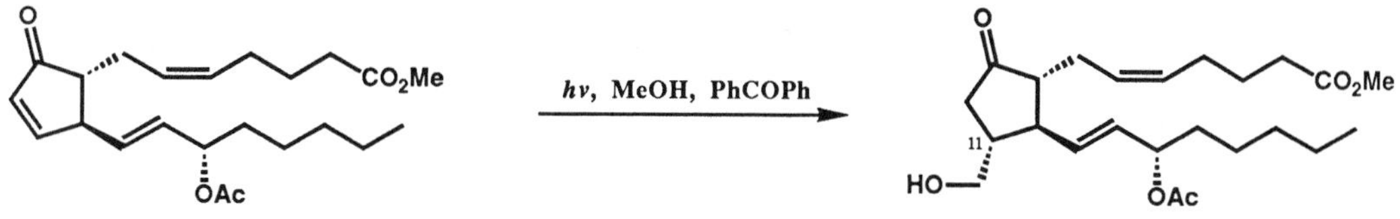
O
CO2Me
OAc
hν, MeOH, PhCOPh
O
CO2Me
11
HO
OAc

(Ref. 2, 3)

plus the 11-epimer

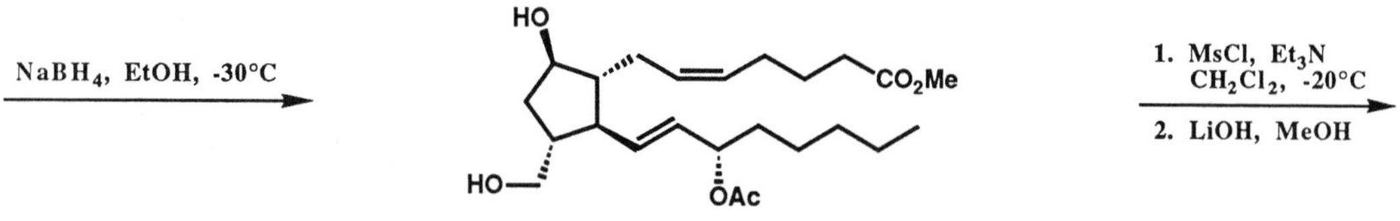
NaBH4, EtOH, -30°C
HO
CO2Me
HO
OAc
1. MsCl, Et3N
CH2Cl2, -20°C
2. LiOH, MeOH

11.16

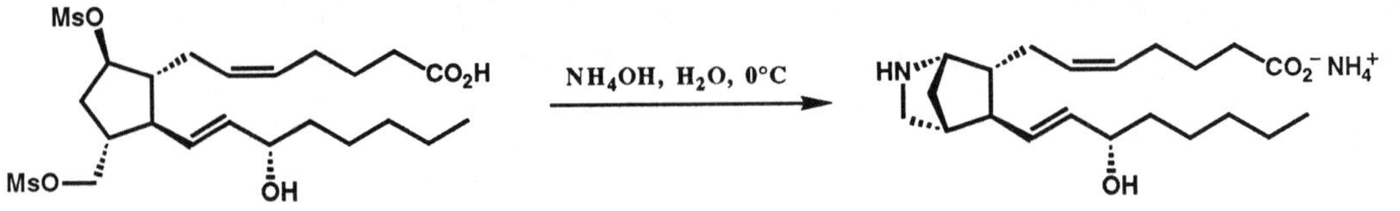
MsO
CO2H
MsO
OH
NH4OH, H2O, 0°C
HN
CO2⁻ NH4⁺
OH

$[\alpha]^{25}_{D}$ -12.5° (*c* 1.7, MeOH)

D

Compound **D** is an effective inhibitor of TXA_2 biosynthesis in human blood platelets at the micromolar level.

References:

1. E. J. Corey , K. C. Nicolaou, Y. Machida, C. L. Malmsten, and B. Samuelsson, *Proc. Nat. Acad. Sci. USA* **1975**, *72*, 3355.
2. (a) E. J. Corey, T. K. Schaaf, W. Huber, U. Koelliker, and N. M. Weinshenker, *J. Am. Chem. Soc.* **1970**, *92*, 397; (b) E. J. Corey and G. Moinet, *ibid.* **1973**, *95*, 6831; (c) E. J. Corey and J. Mann, *ibid.* **1973**, *95*, 6832.
3. A. J. Weinheimer and R. L. Spraggins, *Tetrahedron Lett.* **1969**, 5185.
4. E. J. Corey, K. Narasaka, and M. Shibasaki, *J. Am. Chem. Soc.* **1976**, *98*, 6417.
5. E. J. Corey, M. Shibasaki, K. C. Nicolaou, C. L. Malmsten, and B. Samuelsson, *Tetrahedron Lett.* **1976**, 737.
6. P. Wlodawer, B. Samuelsson, S. M. Albonico, and E. J. Corey, *J. Am. Chem. Soc.* **1971**, *93*, 2815.
7. E. J. Corey, H. Niwa, M. Bloom and P. W. Ramwell, *Tetrahedron Lett.* **1979**, 671.

Total Synthesis of 12-Methylprostaglandin A_2 and 8-Methylprostaglandin C_2

An important pathway for *in vivo* deactivation of prostaglandin A_2 is the rapid conversion in mammalian blood *via* prostaglandin C_2 to the more stable and biologically inactive prostaglandin B_2. 12-Methyl PGA_2 and 8-methyl PGC_2 were synthesized because they cannot be deactivated by this pathway.

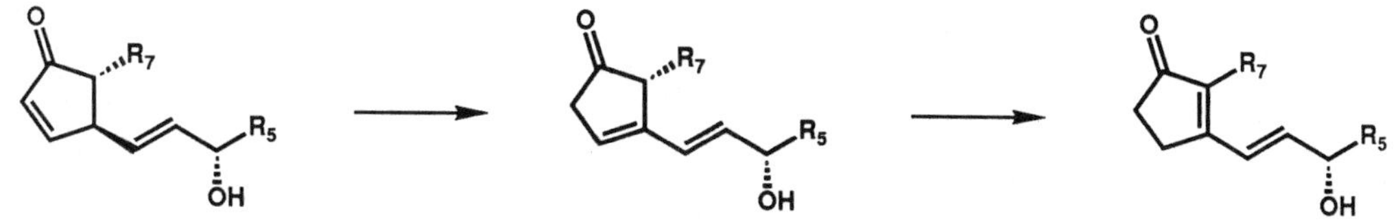

O
R7
R5
OH

PGA PGC PGB

1. Synthesis of 12-methylprostaglandin A_2 (Ref. 1):

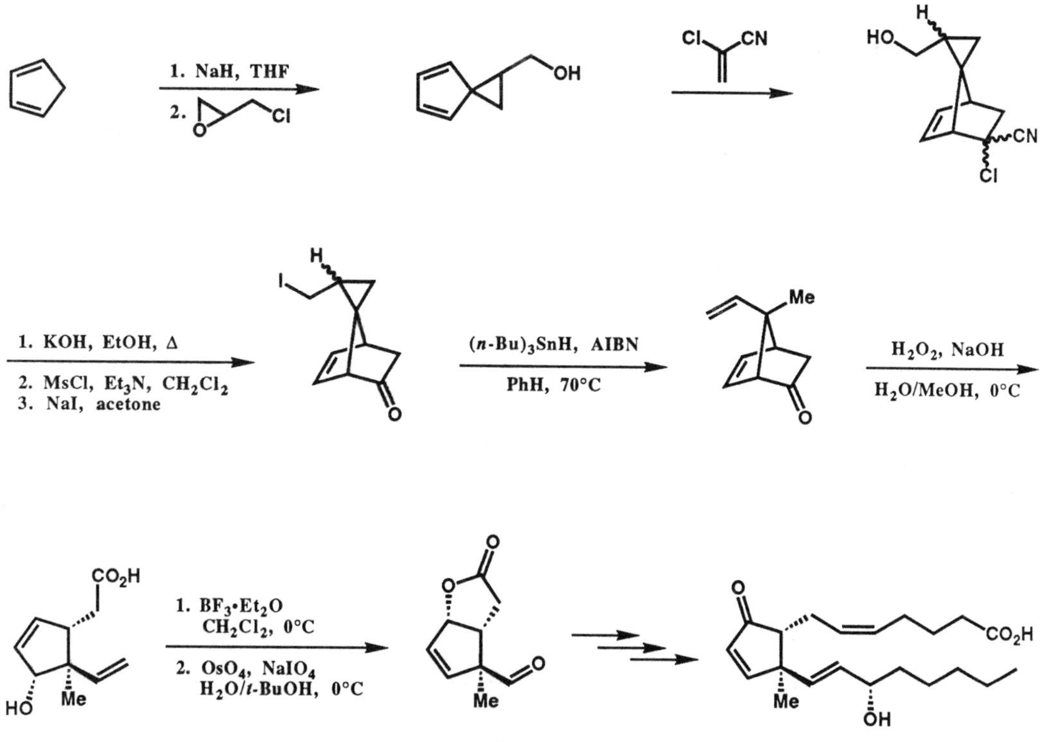

1. NaH, THF
2.
O
Cl
OH
Cl
CN
HO
H
CN
Cl
1. KOH, EtOH, Δ
2. MsCl, Et3N, CH2Cl2
3. NaI, acetone
I
(n-Bu)3SnH, AIBN
PhH, 70°C
Me
H2O2, NaOH
H2O/MeOH, 0°C
CO2H
HO
Me
1. BF3•Et2O
CH2Cl2, 0°C
2. OsO4, NaIO4
H2O/t-BuOH, 0°C
CO2H
OH

12-Methyl PGA_2

11.17

2. Synthesis of 8-methylprostaglandin C_2 (Ref. 2):

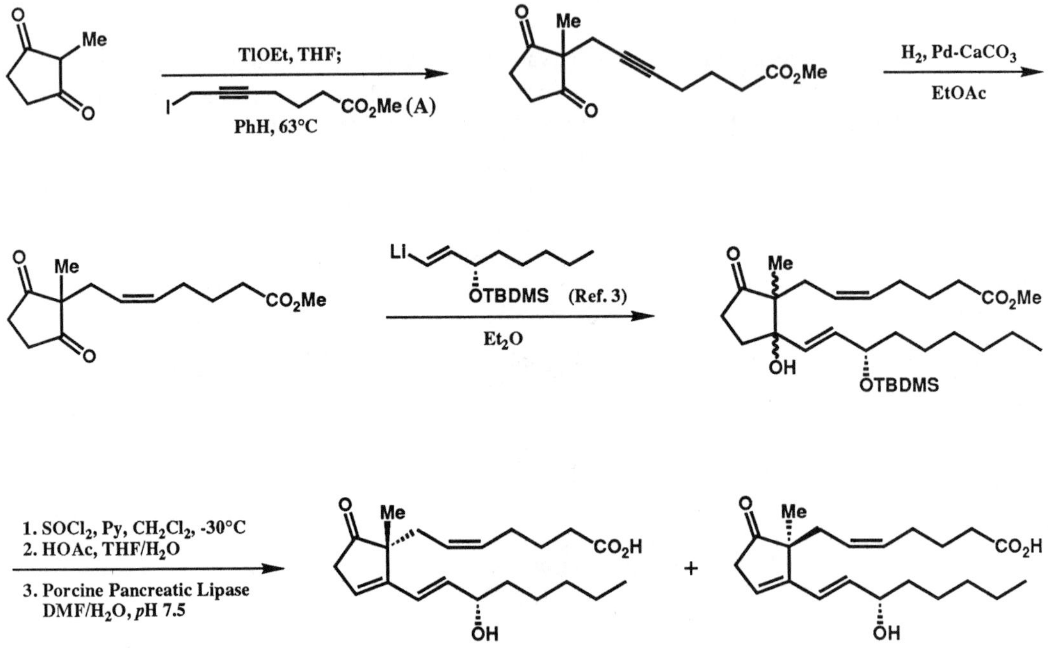

TlOEt, THF;
CO2Me (A)
PhH, 63°C
H2, Pd-CaCO3
EtOAc
OTBDMS (Ref. 3)
Et2O
1. SOCl2, Py, CH2Cl2, -30°C
2. HOAc, THF/H2O
3. Porcine Pancreatic Lipase
DMF/H2O, pH 7.5

$[\alpha]^{25}_D$ +2.45° (*c* 1.3, $CHCl_3$) $[\alpha]^{25}_D$ +13.6° (*c* 1.1, $CHCl_3$)

Preparation of the propargyl iodide **A**:

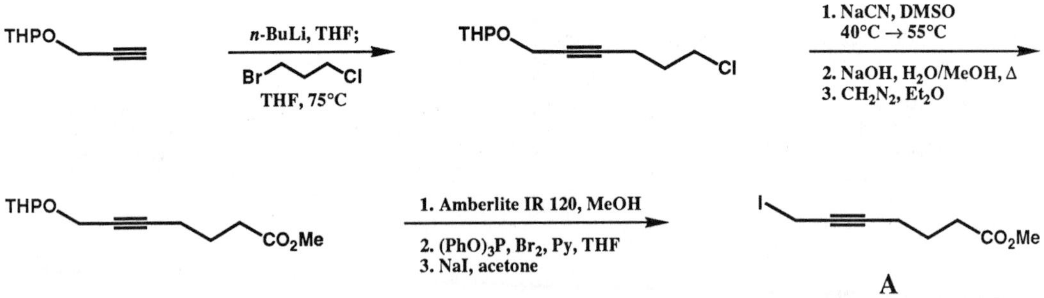

n-BuLi, THF;
THF, 75°C
1. NaCN, DMSO
40°C → 55°C
2. NaOH, H2O/MeOH, Δ
3. CH2N2, Et2O
1. Amberlite IR 120, MeOH
2. (PhO)3P, Br2, Py, THF
3. NaI, acetone

A

References:

1. E. J. Corey, C. S. Shiner, R. P. Volante, and C. R. Cyr, *Tetrahedron Lett.* **1975**, 1161.

2. E. J. Corey and H. S. Sachdev, *J. Am. Chem. Soc.* **1973**, *95*, 8483.

3. (a) E. J. Corey and D. J. Beames, *J. Am. Chem. Soc.* **1972**, *94*, 7210; (b) E. J. Corey and J. Mann, *ibid.* **1973**, *95*, 6832.

Synthesis of Two Stable Analogs of Thromboxane A_2

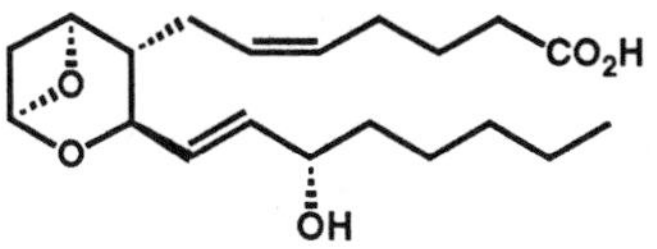
CO2H
OH

Thromboxane A_2

Thromboxane A_2 is a potent platelet aggregating agent and vasodilator which undergoes rapid hydrolysis under physiological conditions ($t_{1/2}$ 32 sec. at *p*H 7 and 37°C). The synthesis of stable analogs was of interest for biological studies of this potent but evanescent prostanoid.

1. Synthesis of the cyclobutyl analog **A** (Ref. 1):

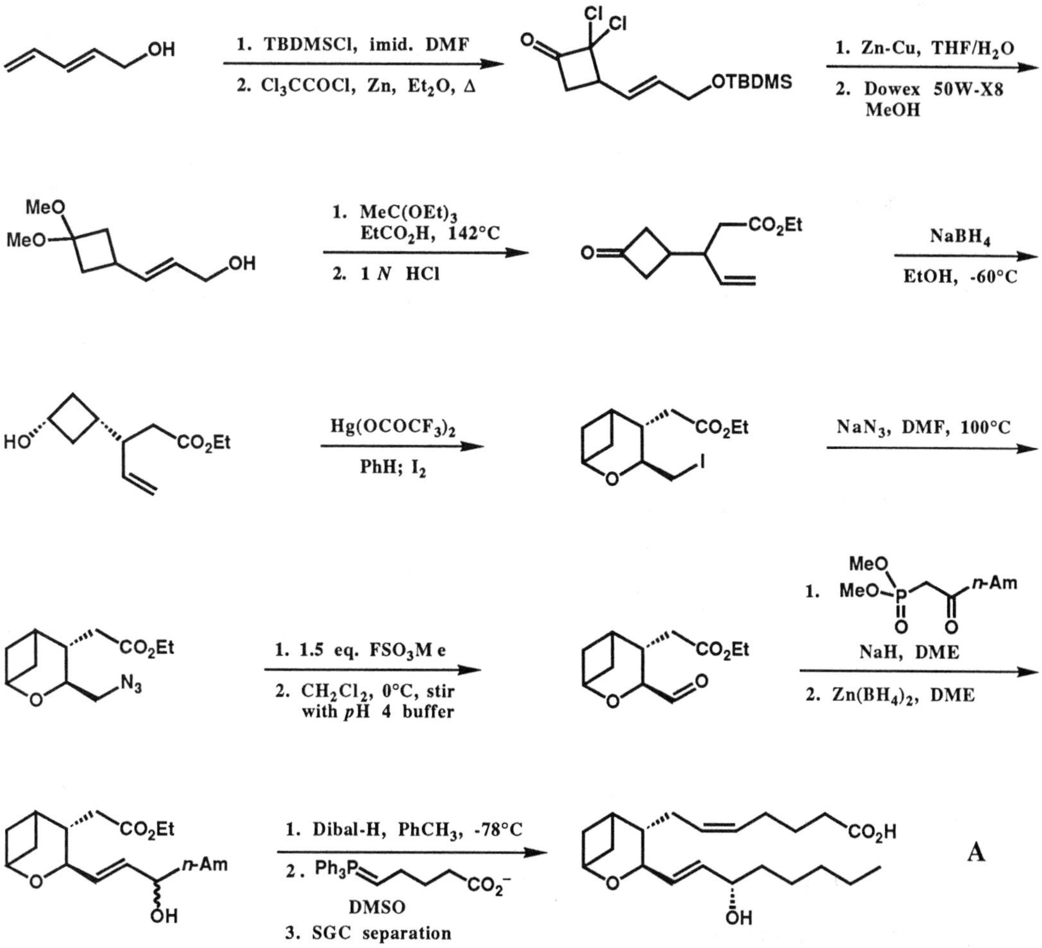
OH
1. TBDMSCl, imid. DMF
2. Cl3CCOCl, Zn, Et2O, Δ
Cl
Cl
OTBDMS
1. Zn-Cu, THF/H2O
2. Dowex 50W-X8
MeOH
MeO
MeO
OH
1. MeC(OEt)3
EtCO2H, 142°C
2. 1 N HCl
CO2Et
NaBH4
EtOH, -60°C
HO
CO2Et
Hg(OCOCF3)2
PhH; I2
CO2Et
I
NaN3, DMF, 100°C
CO2Et
N3
1. 1.5 eq. FSO3Me
2. CH2Cl2, 0°C, stir
with pH 4 buffer
CO2Et
1. MeO
MeO–P
n-Am
NaH, DME
2. Zn(BH4)2, DME
CO2Et
n-Am
OH
1. Dibal-H, PhCH3, -78°C
2. Ph3P
CO2–
DMSO
3. SGC separation
CO2H
OH
A

11.18

2. Synthesis of 9α,11α-epoxymethanothromboxane A_2 (Ref. 2):

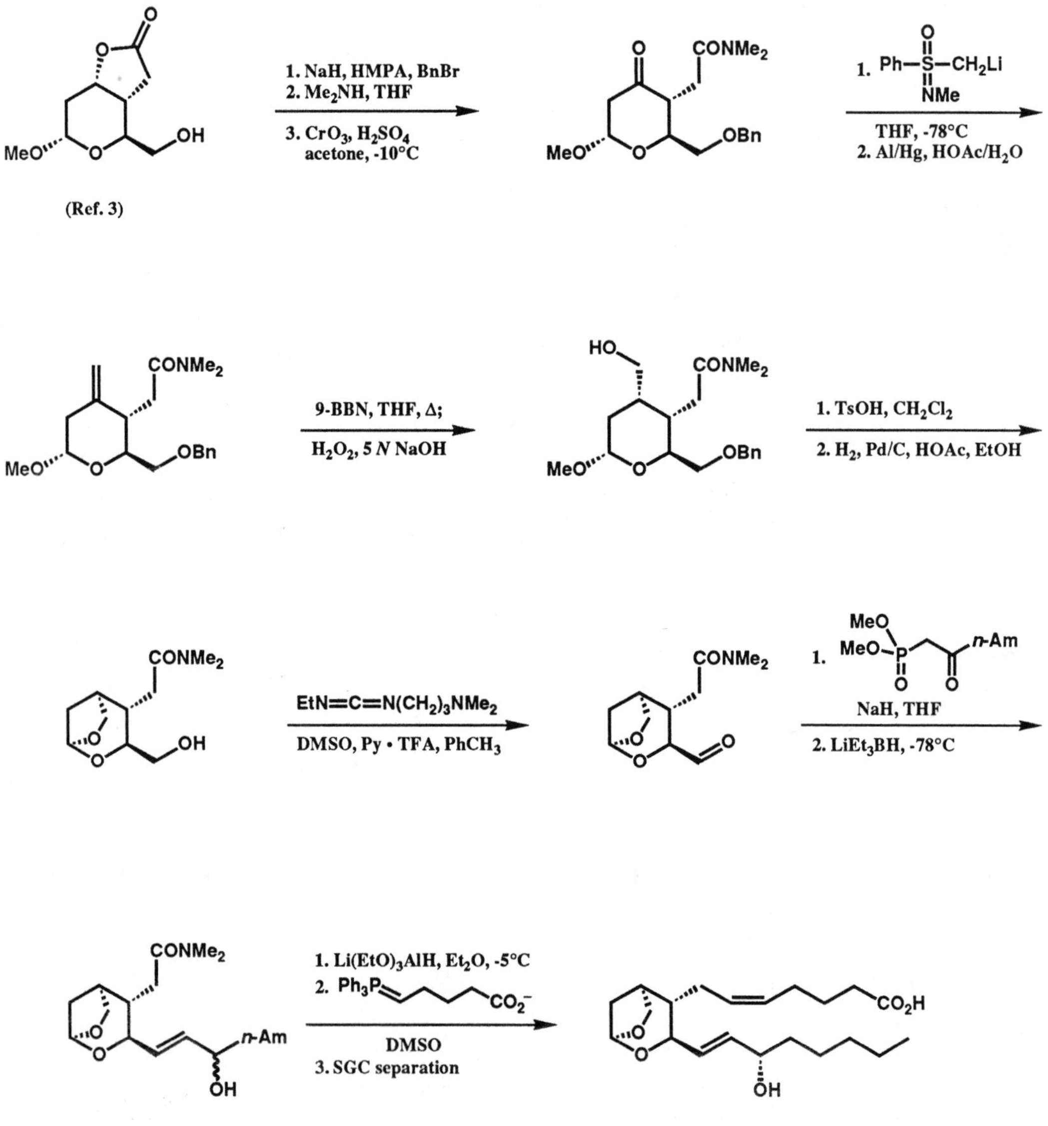
1. NaH, HMPA, BnBr
2. Me2NH, THF
3. CrO3, H2SO4
acetone, -10°C
(Ref. 3)
CONMe2
OBn
MeO
OH
1. Ph—S—CH2Li
NMe
THF, -78°C
2. Al/Hg, HOAc/H2O
9-BBN, THF, Δ;
H2O2, 5 N NaOH
HO
1. TsOH, CH2Cl2
2. H2, Pd/C, HOAc, EtOH
EtN=C=N(CH2)3NMe2
DMSO, Py • TFA, PhCH3
NaH, THF
2. LiEt3BH, -78°C
n-Am
1. Li(EtO)3AlH, Et2O, -5°C
2. Ph3P
DMSO
3. SGC separation
CO2H

Reference:

1. E. J. Corey, J. W. Ponder, and P. Ulrich, *Tetrahedron Lett.* **1980**, *21*, 137.
2. T. K. Schaaf, D. L. Bussolotti, M. J. Parry, and E. J. Corey, *J. Am. Chem. Soc.* **1981**, *103*, 6502.
3. E. J. Corey, M. Shibasaki, and J. Knolle, *Tetrahedron Lett.* **1977**, 1625.

Total Synthesis of (±)-Thromboxane B_2

The synthesis of thromboxane B_2, the hydrolytic deactivation product of thromboxane A_2, provided this material for studies of metabolism and bioactivity, and also for the development of a radioimmunassay. Two different synthetic routes were developed.

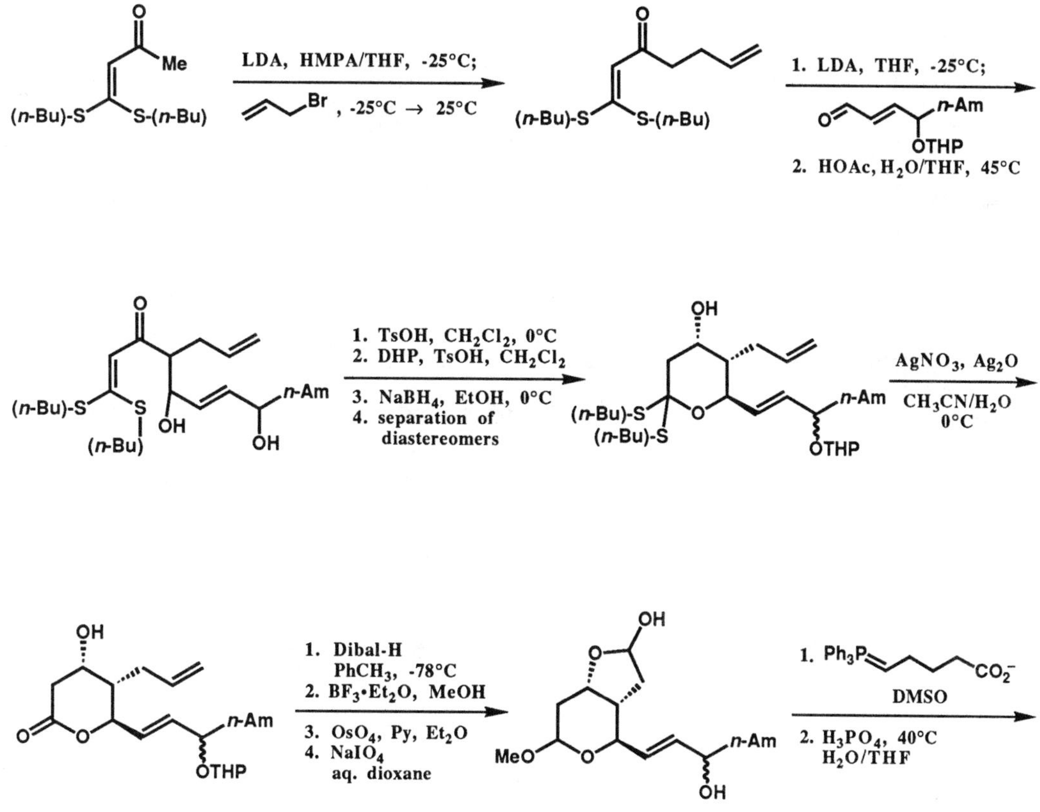
Me
(n-Bu)-S
S-(n-Bu)
LDA, HMPA/THF, -25°C;
Br , -25°C → 25°C
1. LDA, THF, -25°C;
n-Am
OTHP
2. HOAc, H2O/THF, 45°C
OH
1. TsOH, CH2Cl2, 0°C
2. DHP, TsOH, CH2Cl2
3. NaBH4, EtOH, 0°C
4. separation of diastereomers
AgNO3, Ag2O
CH3CN/H2O
0°C
1. Dibal-H
PhCH3, -78°C
2. BF3•Et2O, MeOH
3. OsO4, Py, Et2O
4. NaIO4
aq. dioxane
MeO
1. Ph3P
CO2−
DMSO
2. H3PO4, 40°C
H2O/THF

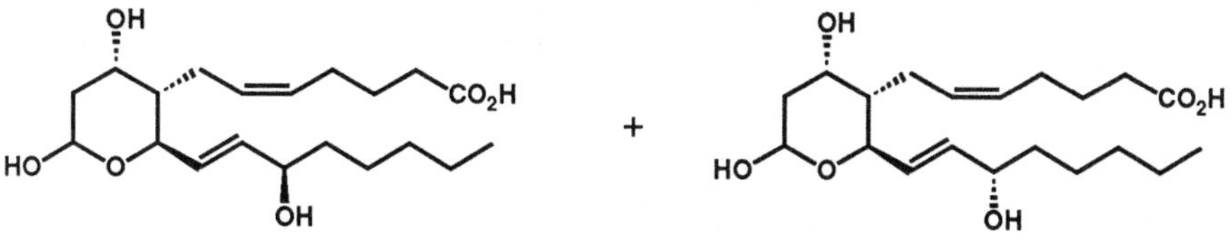
OH
CO2H
HO
+

15-*epi*-(±)-Thromboxane B_2 (±)-Thromboxane B_2

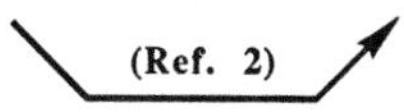
(Ref. 2)

11.19

A stereocontrolled route to the native form of thromboxane B2 was carried out starting with α-methyl-D-glucoside (Ref. 3).

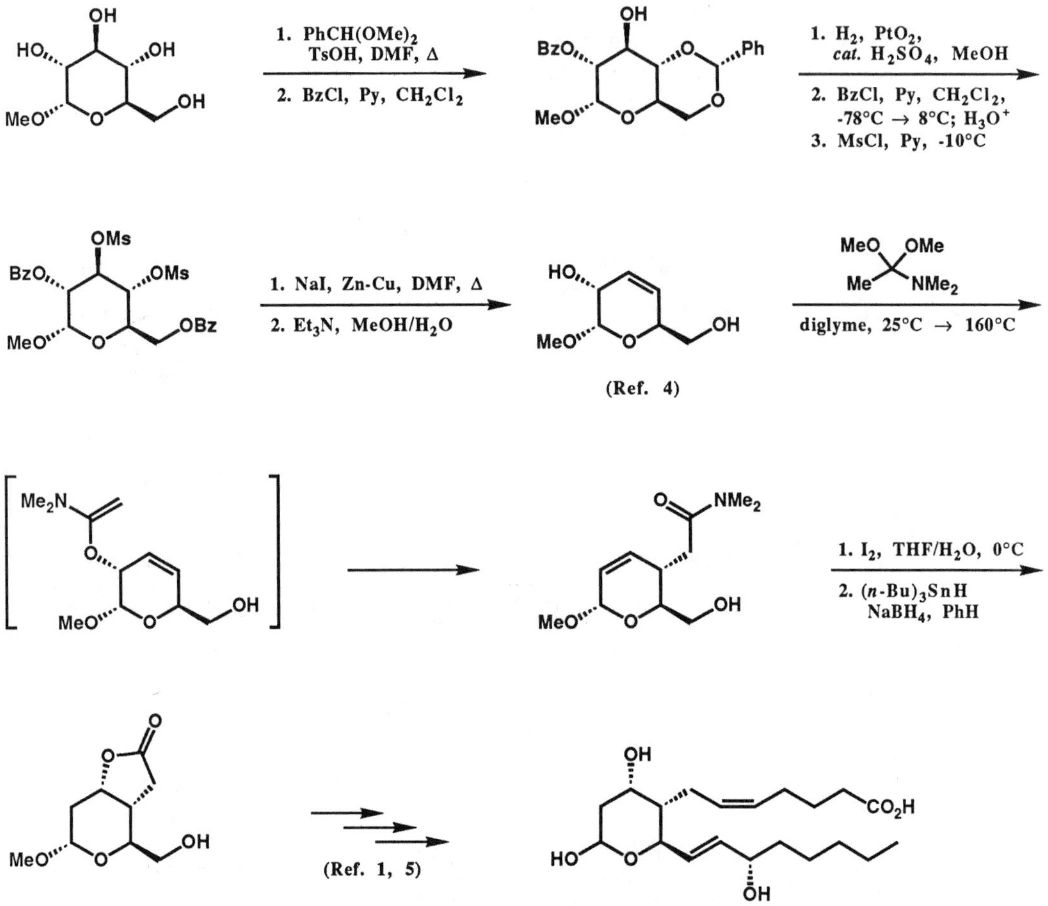
1. PhCH(OMe)2 TsOH, DMF, Δ
2. BzCl, Py, CH2Cl2
1. H2, PtO2, cat. H2SO4, MeOH
2. BzCl, Py, CH2Cl2, -78°C → 8°C; H3O+
3. MsCl, Py, -10°C
1. NaI, Zn-Cu, DMF, Δ
2. Et3N, MeOH/H2O
(Ref. 4)
diglyme, 25°C → 160°C
1. I2, THF/H2O, 0°C
2. (n-Bu)3SnH NaBH4, PhH
(Ref. 1, 5)

$[\alpha]^{23}_D$ +86.6° ($CHCl_3$)

Thromboxane B_2

References:

1. E. J. Corey, M. Shibasaki, J. Knolle, and T. Sugahara, *Tetrahedron Lett.* **1977**, 785.

2. (a) E. J. Corey, K. C. Nicolaou, and M. Shibasaki, *J. Chem. Soc. Chem. Commun.* **1975**, 658; (b) E. J. Corey, K. C. Nicolaou, M. Shibasaki, Y. Machida, and C. S. Shiner, *Tetrahedron Lett.* **1975**, 3183.

3. E. J. Corey, M. Shibasaki, and J. Knolle, *Tetrahedron Lett.* **1977**, 1625.

4. N. L. Holder and B. Fraser-Reid, *Can. J. Chem.* **1973**, *51*, 3357.

5. (a) N. A. Nelson and R. W. Jackson, *Tetrahedron Lett.* **1976**, 3275; (b) R. C. Kelly, I. Schletter, and S. J. Stein, *ibid*, **1976**, 3279; (c) W. P. Schneider and R. A. Morge, *ibid*, **1976**, 3283.

Synthesis of Prostaglandins *via* an Endoperoxide Intermediate
Stereochemical Divergence of
Enzymatic and Biomimetic Chemical Cyclization Reactions

Prostaglandin biosynthesis from C_{20} polyunsaturated fatty acids occurs by way of the endoperoxides PGG_2 and PGH_2.

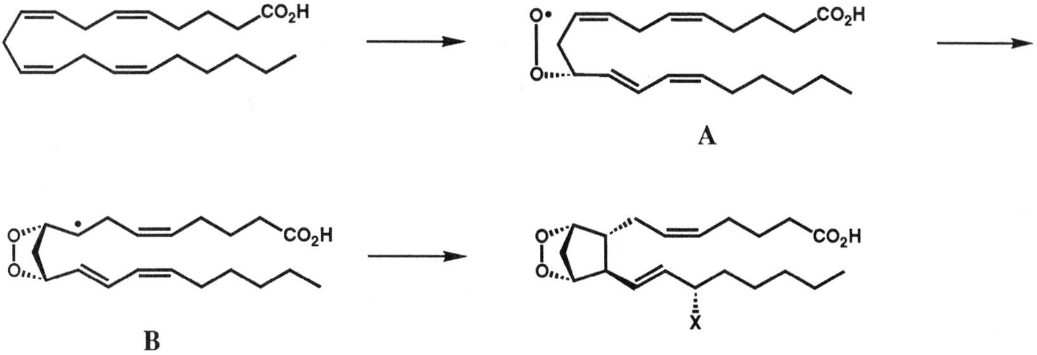
CO2H
A
B
X

PGG_2: X = OOH
PGH_2: X = OH

A chemical synthesis of prostaglandins by a free radical pathway through an endoperoxide intermediate showed a strong stereochemical preference for the formation of the endoperoxides having *cis* alpha and omega appendages.

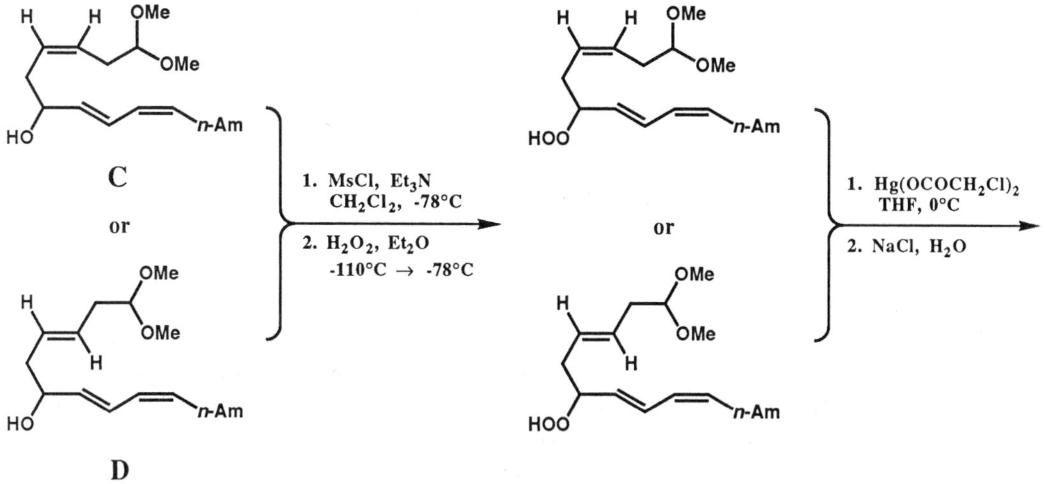
H
OMe
HO
HOO
n-Am
C
or
D
1. MsCl, Et3N CH2Cl2, -78°C
2. H2O2, Et2O -110°C → -78°C
1. Hg(OCOCH2Cl)2 THF, 0°C
2. NaCl, H2O

11.20

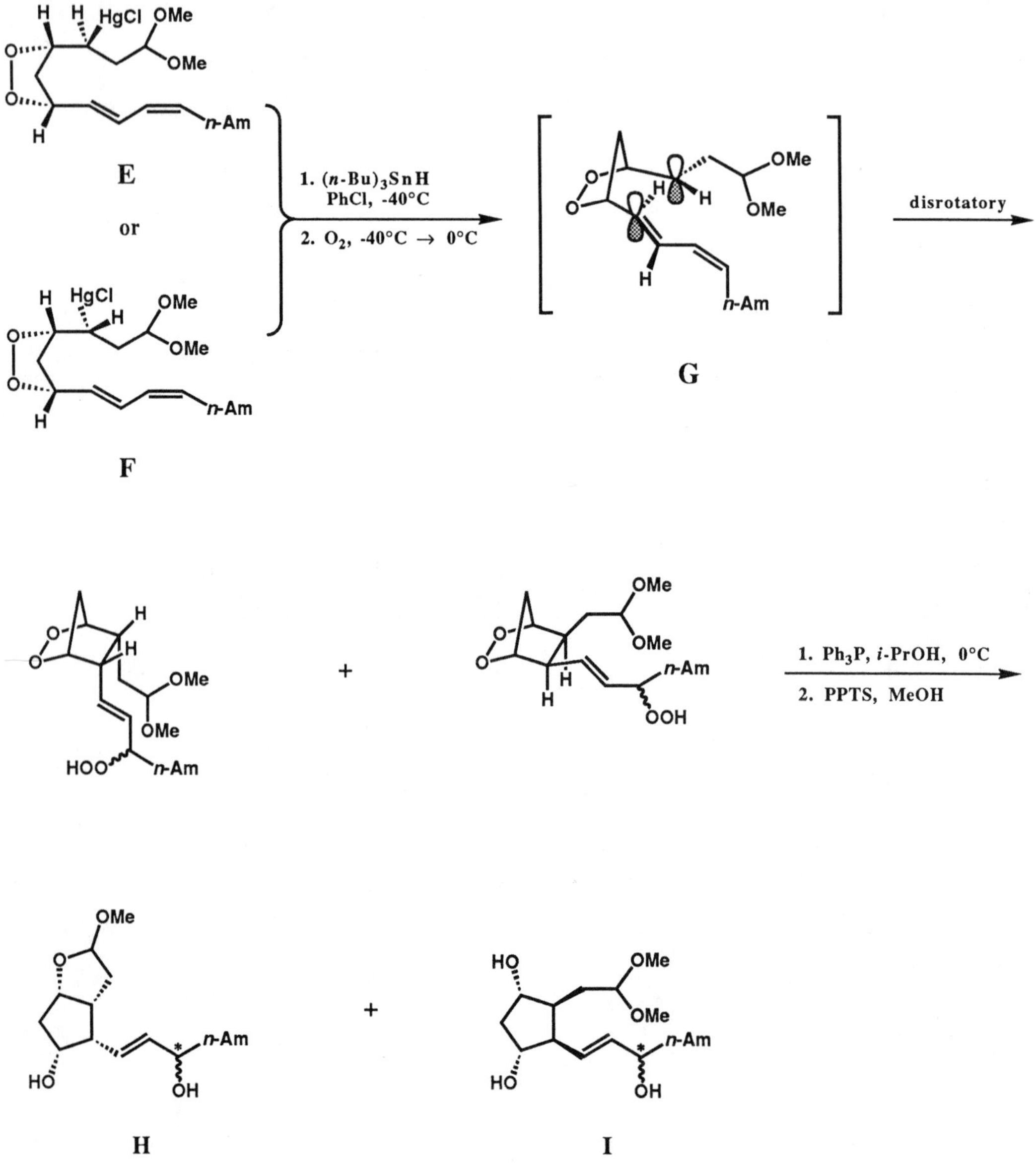
E
or
F
1. (n-Bu)3SnH PhCl, -40°C
2. O2, -40°C → 0°C
G
disrotatory
1. Ph3P, i-PrOH, 0°C
2. PPTS, MeOH
H
I

2 : 1

1 : 1 mixture at C for both* **H** *and* **I**.

The same mixture of **H** and **I** was obtained starting with either of the geometrically isomeric radical precursors **E** or **F**. A possible explanation is based on the assumption of a common radical conformer **G**, stabilized in the geometry shown by electron delocalization involving the radicaloid *p*-orbital, the β-peroxy oxygen and π* of the diene unit. The structure of the compounds **H** and **I** were determined by ^{1}H NMR spectra and the conversion of **H** to diol **J**, a known intermediate for the synthesis of prostaglandins.

11.20

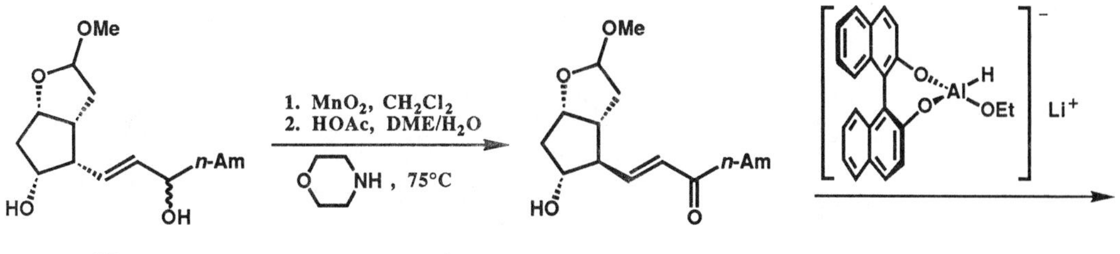
OMe
n-Am
HO
OH
1. MnO2, CH2Cl2
2. HOAc, DME/H2O
NH , 75°C
O
Li+
OEt

H

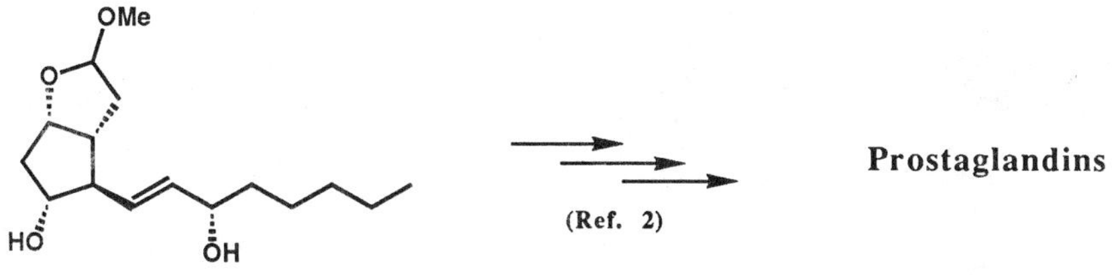
OMe
HO
OH
(Ref. 2)
Prostaglandins

J

diastereoselection >95%

A similar result was obtained with substrates having the full eicosanoid side chain and a 12,13(Z)-double bond regardless of whether the 14,15-double bond is *E* or *Z* (Ref. 3).

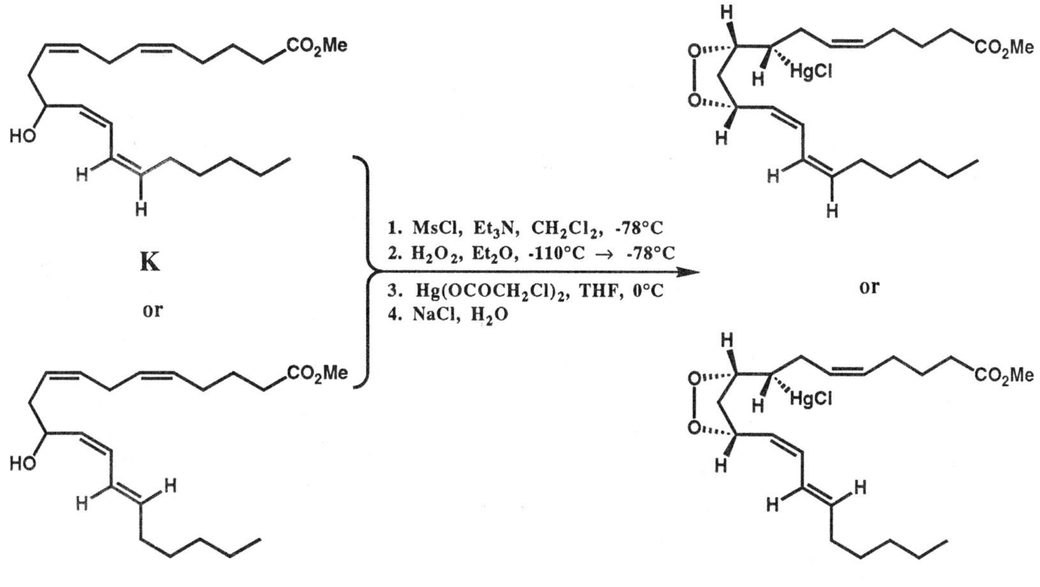
CO2Me
HO
K
or
L
1. MsCl, Et3N, CH2Cl2, -78°C
2. H2O2, Et2O, -110°C → -78°C
3. Hg(OCOCH2Cl)2, THF, 0°C
4. NaCl, H2O
HgCl
or

11.20

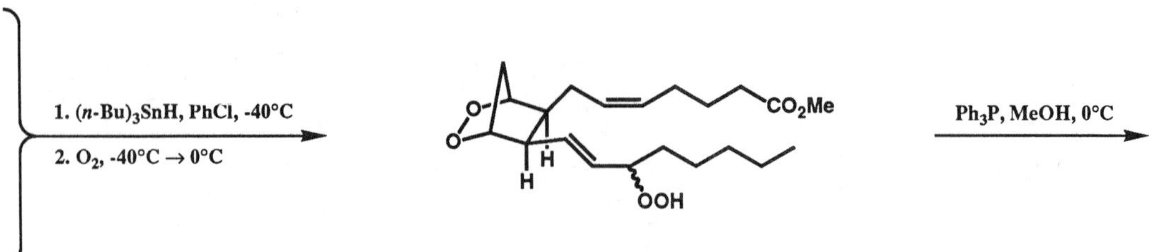

1. (n-Bu)3SnH, PhCl, -40°C
2. O2, -40°C → 0°C
CO2Me
H
H
OOH
Ph3P, MeOH, 0°C

both appendages exo

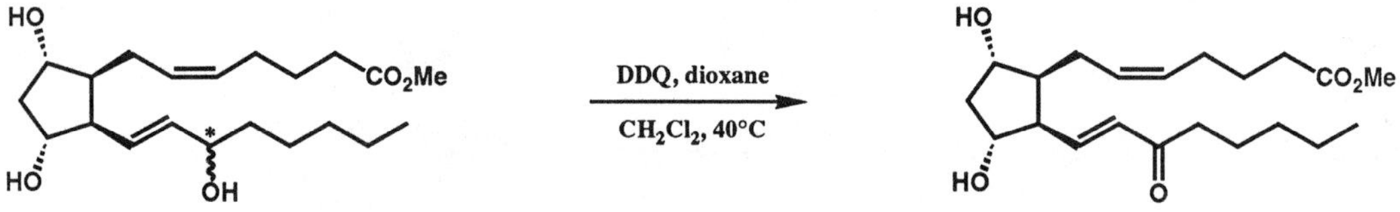

HO
HO
CO2Me
OH
DDQ, dioxane
CH2Cl2, 40°C
HO
HO
CO2Me
O

*1 : 1 mixture at C**

Preparation of the homoallylic alcohols **C** and **D**:

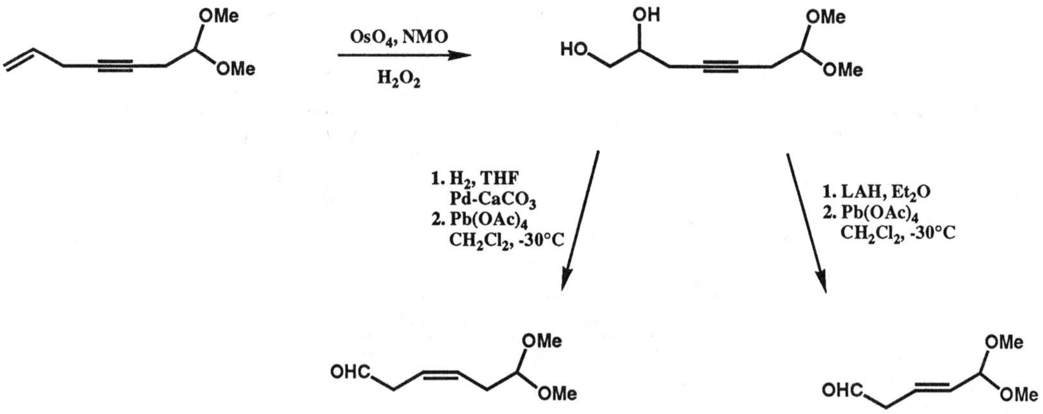

OMe
OMe
OsO4, NMO
H2O2
HO
OH
OMe
OMe
1. H2, THF
Pd-CaCO3
2. Pb(OAc)4
CH2Cl2, -30°C
1. LAH, Et2O
2. Pb(OAc)4
CH2Cl2, -30°C
OHC
OMe
OMe
OHC
OMe
OMe

Me
1. n-BuLi, Et2O, -40°C;
TIPSOTf
2. n-BuLi, TMEDA
Et2O, -15°C; TIPSOTf
TIPS
CH2TIPS
n-BuLi, THF, -20°C;
n-AmCHO, -78°C

11.20

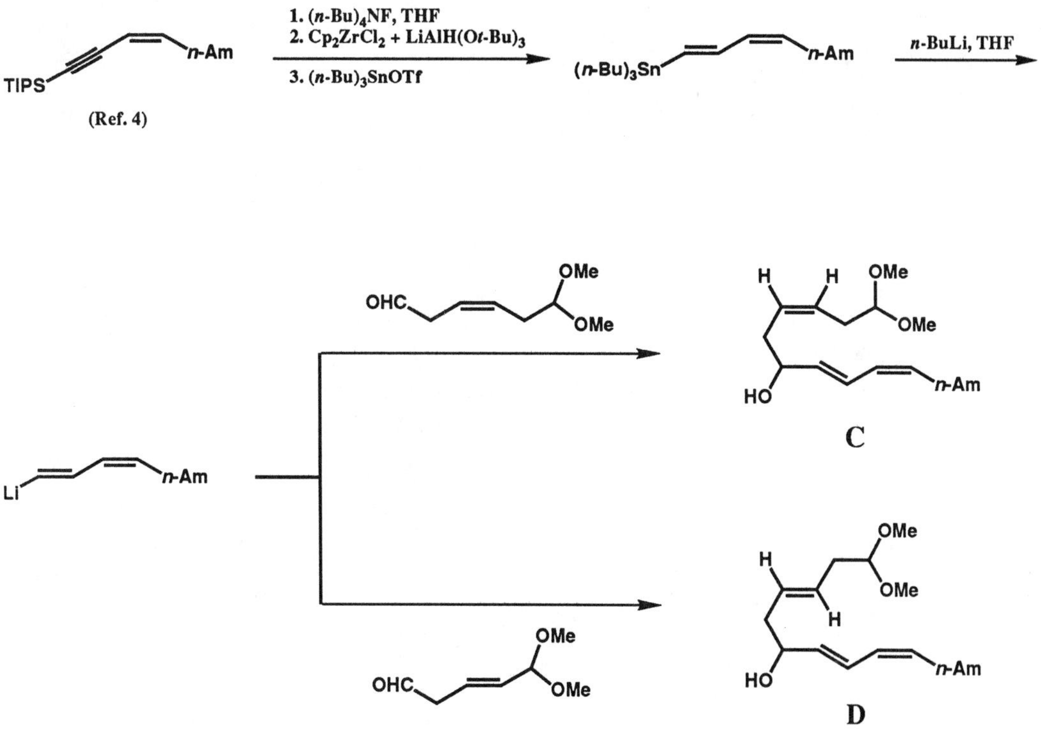
1. (n-Bu)4NF, THF
2. Cp2ZrCl2 + LiAlH(Ot-Bu)3
3. (n-Bu)3SnOTf
TIPS
n-Am
(Ref. 4)
(n-Bu)3Sn
n-BuLi, THF
Li
OHC
OMe
H
HO
C
D

Preparation of the homoallylic alcohols **K** and **L**:

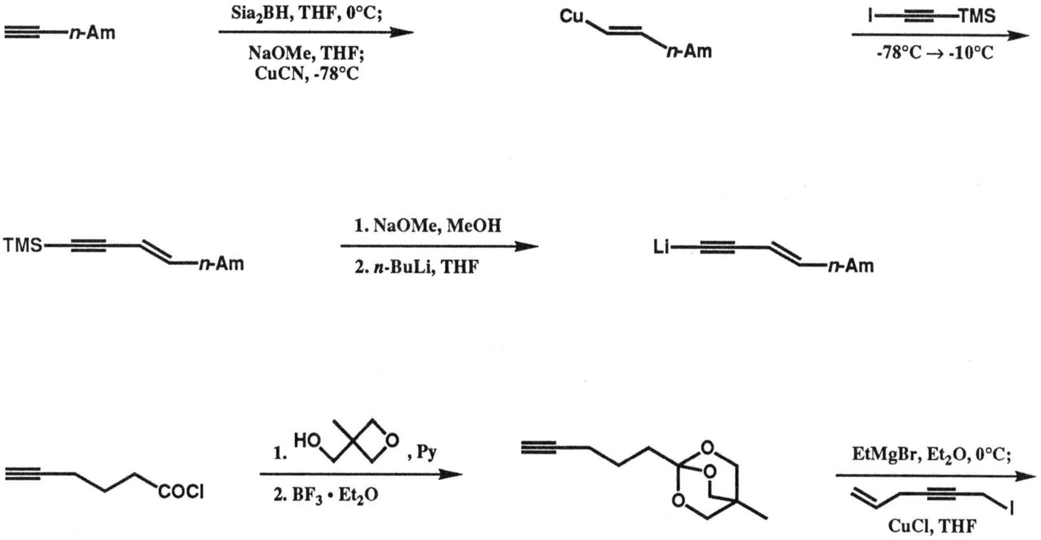
n-Am
Sia2BH, THF, 0°C;
NaOMe, THF;
CuCN, -78°C
Cu
I—≡—TMS
-78°C → -10°C
TMS
1. NaOMe, MeOH
2. n-BuLi, THF
Li
COCl
1. HO , Py
2. BF3 • Et2O
EtMgBr, Et2O, 0°C;
CuCl, THF
I

11.20

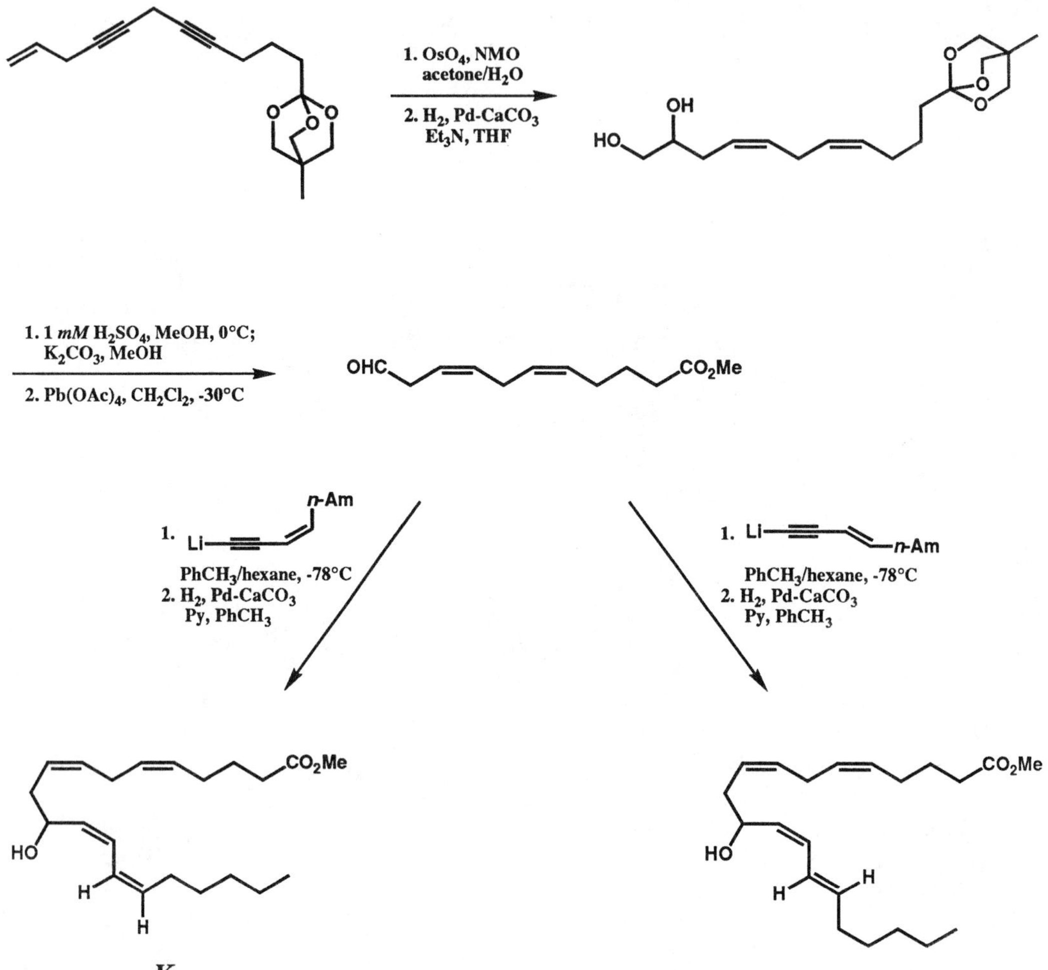
1. OsO_4, NMO
acetone/H_2O
2. H_2, Pd-$CaCO_3$
Et_3N, THF
OH
HO
1. 1 *mM* H_2SO_4, MeOH, 0°C;
K_2CO_3, MeOH
2. $Pb(OAc)_4$, CH_2Cl_2, -30°C
OHC
CO_2Me
1. Li
n-Am
$PhCH_3$/hexane, -78°C
2. H_2, Pd-$CaCO_3$
Py, $PhCH_3$
1. Li
n-Am
$PhCH_3$/hexane, -78°C
2. H_2, Pd-$CaCO_3$
Py, $PhCH_3$
CO_2Me
HO
H
H
CO_2Me
HO
H
H

K L

References:

1. E. J. Corey, K. Shimoji, and C. Shih, *J. Am. Chem. Soc.* **1984**, *106*, 6425.
2. E. J. Corey and R. Noyori, *Tetrahedron Lett.* **1970**, 311.
3. E. J. Corey, C. Shih, N.-Y. Shih, and K. Shimoji, *Tetrahedron Lett.* **1984**, *25*, 5013.
4. E. J. Corey and C. Rücker, *Tetrahedron Lett.* **1982**, *23*, 719.

(±)-Clavulone I (±)-Clavulone II

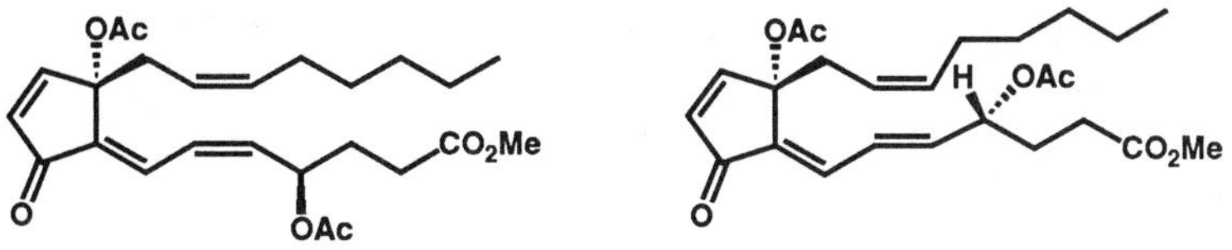
OAc
CO2Me
OAc
O
OAc
H
OAc
CO2Me
O

Clavulones I and II are members of an unusual family of marine prostanoids from the coral *Clavularia viridis* which are biosynthesized by a cationic (i.e. non-radical, non-endoperoxide) pathway. The total synthesis of clavulones I and II was accomplished from cyclopentadiene as SM goal.

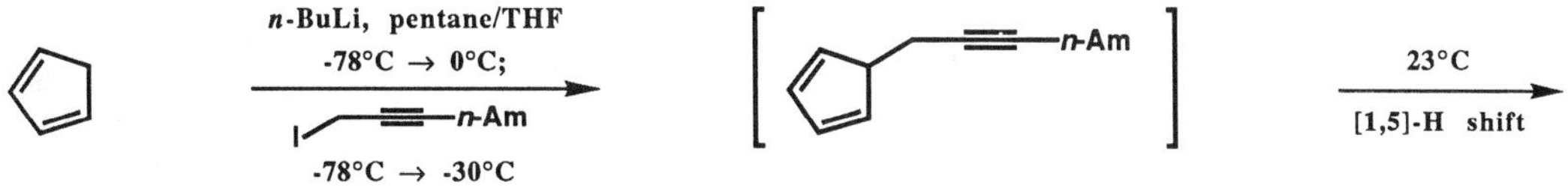
n-BuLi, pentane/THF
-78°C → 0°C;
I
n-Am
-78°C → -30°C
n-Am
23°C
[1,5]-H shift

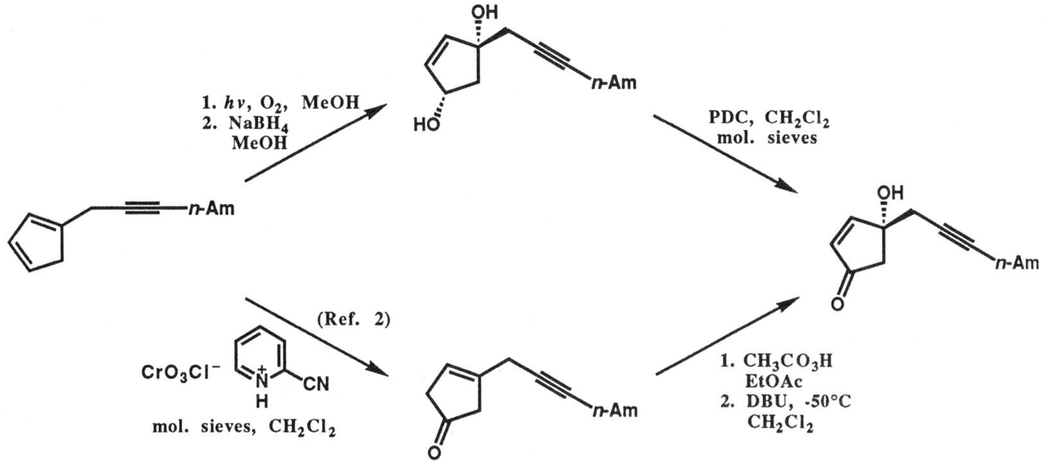
n-Am
1. hν, O2, MeOH
2. NaBH4
MeOH
OH
HO
n-Am
PDC, CH2Cl2
mol. sieves
OH
n-Am
O
(Ref. 2)
CrO3Cl⁻
N
H
CN
mol. sieves, CH2Cl2
n-Am
O
1. CH3CO3H
EtOAc
2. DBU, -50°C
CH2Cl2

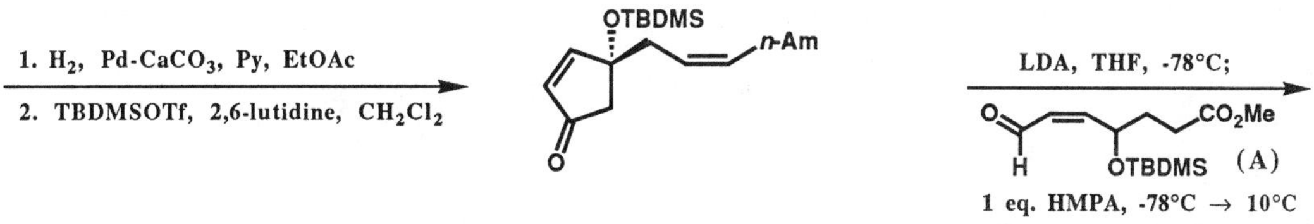
1. H2, Pd-CaCO3, Py, EtOAc
2. TBDMSOTf, 2,6-lutidine, CH2Cl2
OTBDMS
n-Am
O
LDA, THF, -78°C;
O
H
CO2Me
OTBDMS
(A)
1 eq. HMPA, -78°C → 10°C

11.21

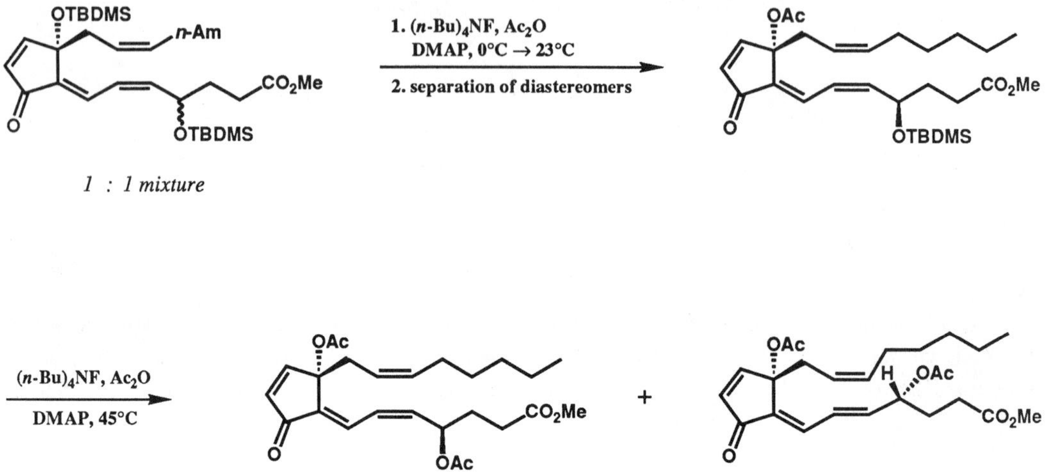
OTBDMS
n-Am
CO2Me
OTBDMS
1 : 1 mixture
1. (n-Bu)4NF, Ac2O
DMAP, 0°C → 23°C
2. separation of diastereomers
OAc
CO2Me
OTBDMS
(n-Bu)4NF, Ac2O
DMAP, 45°C
OAc
CO2Me
OAc
+
OAc
H
OAc
CO2Me

(±)-Clavulone I (±)-Clavulone II

60% *15%*

Preparation of the enal **A**:

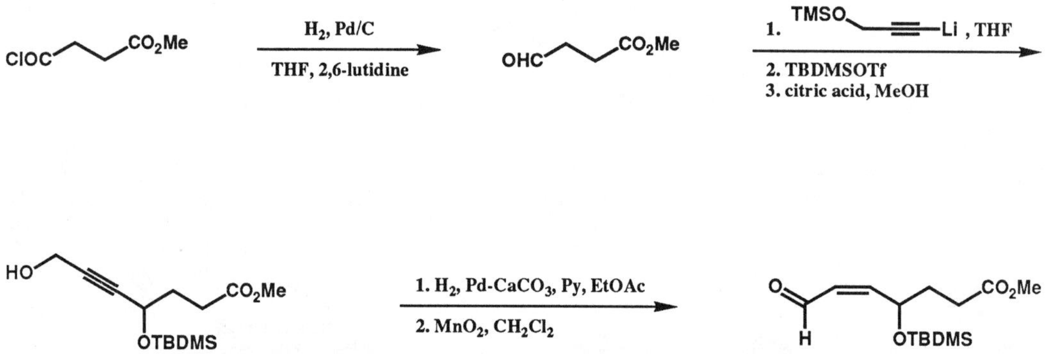
ClOC
CO2Me
H2, Pd/C
THF, 2,6-lutidine
OHC
CO2Me
1. TMSO —Li , THF
2. TBDMSOTf
3. citric acid, MeOH
HO
CO2Me
OTBDMS
1. H2, Pd-CaCO3, Py, EtOAc
2. MnO2, CH2Cl2
O
H
CO2Me
OTBDMS

A

References:

1. E. J. Corey and M. M. Mehrotra, *J. Am. Chem. Soc.* **1984**, *106*, 3384.

2. E. J. Corey and M. M. Mehrotra, *Tetrahedron Lett.* **1985**, *26*, 2411.

(-)-Preclavulone-A

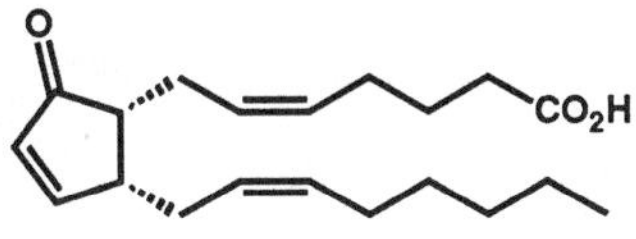
O
CO2H

Extracts from *Clavularia viridis* and also many other coral species convert arachidonic acid to the prostanoid preclavulone-A *via* 8-*(R)*-hydroperoxy-5,11,14*(Z)*, 9*(E)*-eicosatetraenoic acid. The carbocyclization is considered to occur from allene oxide and oxidopentadienyl cation intermediates. An enantioselective total synthesis of preclavulone-A was developed to assist the biosynthetic research.

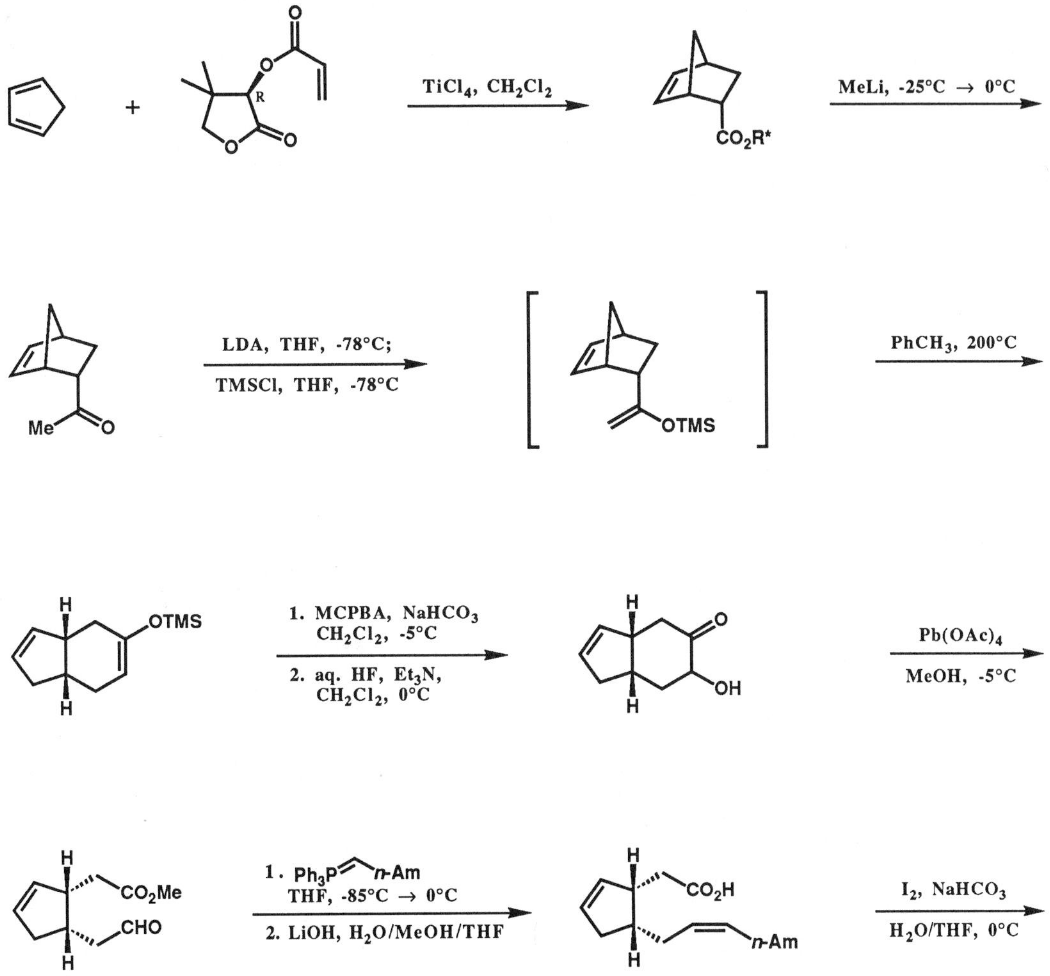
+
R
TiCl4, CH2Cl2
CO2R*
MeLi, -25°C → 0°C
Me
LDA, THF, -78°C;
TMSCl, THF, -78°C
OTMS
PhCH3, 200°C
H
OTMS
H
1. MCPBA, NaHCO3
CH2Cl2, -5°C
2. aq. HF, Et3N,
CH2Cl2, 0°C
OH
Pb(OAc)4
MeOH, -5°C
CO2Me
CHO
1. Ph3P n-Am
THF, -85°C → 0°C
2. LiOH, H2O/MeOH/THF
CO2H
n-Am
I2, NaHCO3
H2O/THF, 0°C

11.22

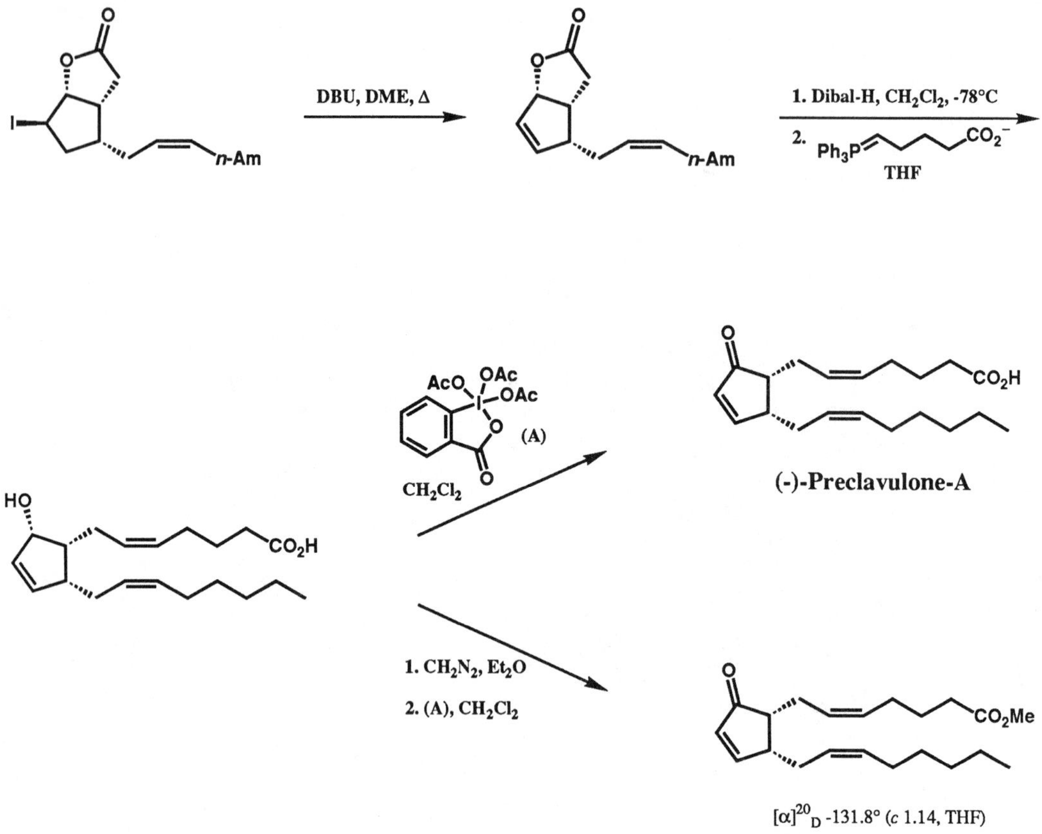

O
O
I
n-Am
DBU, DME, Δ
1. Dibal-H, CH2Cl2, -78°C
2. Ph3P CO2−
THF
HO
CO2H
AcO OAc
OAc
(A)
CH2Cl2
(-)-Preclavulone-A
1. CH2N2, Et2O
2. (A), CH2Cl2
CO2Me
[α]20D -131.8° (c 1.14, THF)

Reference:

E. J. Corey and Y. B. Xiang, *Tetrahedron Lett.* **1988**, *29*, 995.

Hybridalactone

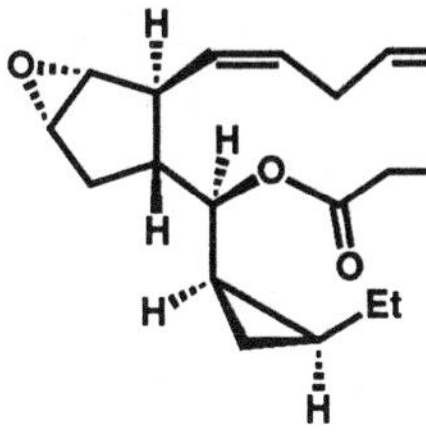
H
O
H
O
H
O
H
Et
H

Hybridalactone, a novel marine derived eicosanoid from *Laurencia hybrida,* appears to be biosynthesized by a unique pathway from eicosapentaenoic acid. The synthesis of hybridalactone was carried out enantiospecifically from (+)-bicyclo[3.2.0]hept-4-ene-1-one so as to provide proof of stereochemistry (Ref. 4).

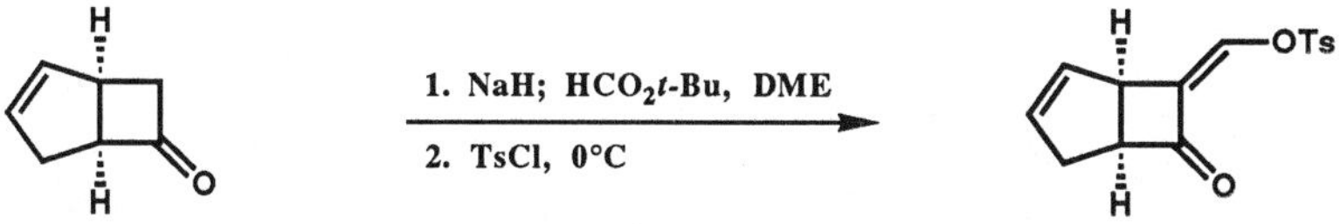
H
H
O
1. NaH; HCO2t-Bu, DME
2. TsCl, 0°C
H
OTs
H
O

$[\alpha]^{23}_D$ +61.4° (*c* 1.0, $CHCl_3$)

A

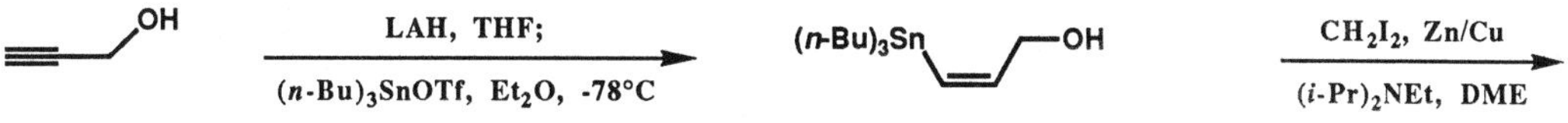
OH
LAH, THF;
(n-Bu)3SnOTf, Et2O, -78°C
(n-Bu)3Sn
OH
CH2I2, Zn/Cu
(i-Pr)2NEt, DME

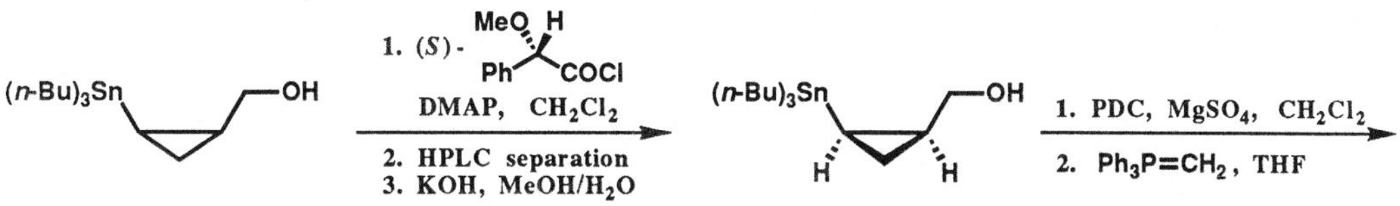
(n-Bu)3Sn
OH
MeO H
1. (S)-
Ph COCl
DMAP, CH2Cl2
2. HPLC separation
3. KOH, MeOH/H2O
(n-Bu)3Sn
OH
H
H
1. PDC, MgSO4, CH2Cl2
2. Ph3P=CH2, THF

$[\alpha]^{19}_D$ -9.46° ($CHCl_3$)

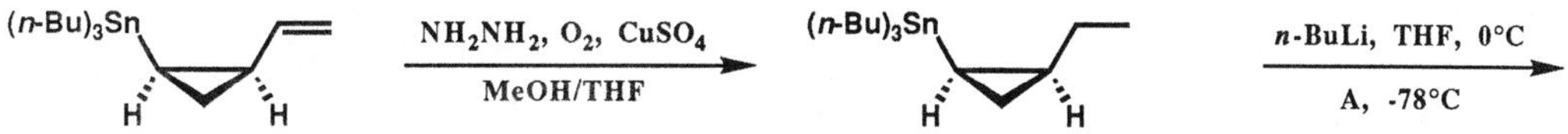
(n-Bu)3Sn
H
H
NH2NH2, O2, CuSO4
MeOH/THF
(n-Bu)3Sn
H
H
n-BuLi, THF, 0°C
A, -78°C

(Ref. 2)

11.23

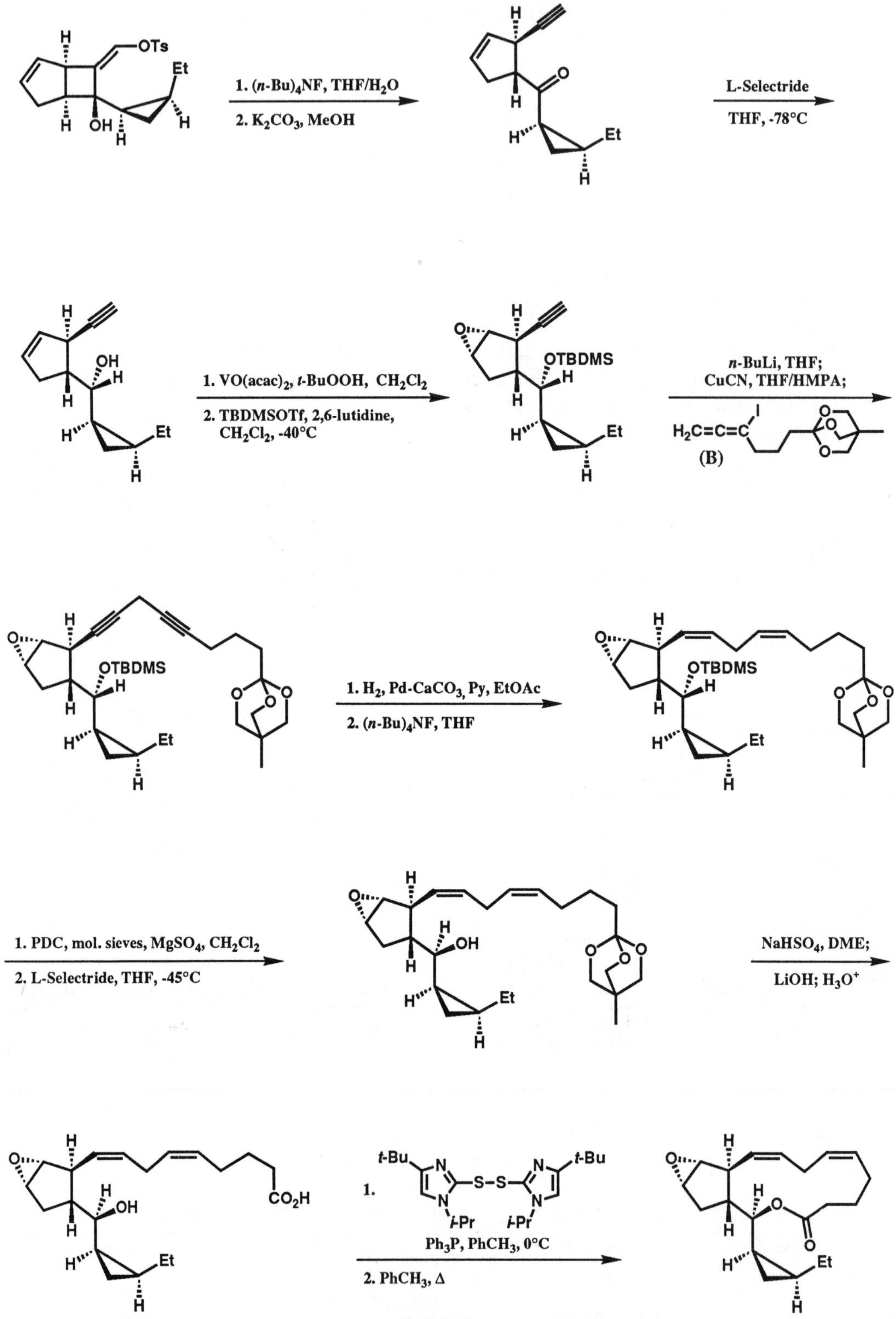

OTs
Et
OH
1. $(n\text{-Bu})_4NF$, THF/H_2O
2. K_2CO_3, MeOH
L-Selectride
THF, -78°C
1. $VO(acac)_2$, t-BuOOH, CH_2Cl_2
2. TBDMSOTf, 2,6-lutidine, CH_2Cl_2, -40°C
OTBDMS
n-BuLi, THF;
CuCN, THF/HMPA;
$H_2C{=}C{=}C$
(B)
1. H_2, Pd-$CaCO_3$, Py, EtOAc
2. $(n\text{-Bu})_4NF$, THF
1. PDC, mol. sieves, $MgSO_4$, CH_2Cl_2
2. L-Selectride, THF, -45°C
$NaHSO_4$, DME;
LiOH; H_3O^+
CO_2H
t-Bu
i-Pr
Ph_3P, $PhCH_3$, 0°C
2. $PhCH_3$, Δ

Preparation of the iodoallene **B** (Ref. 3):

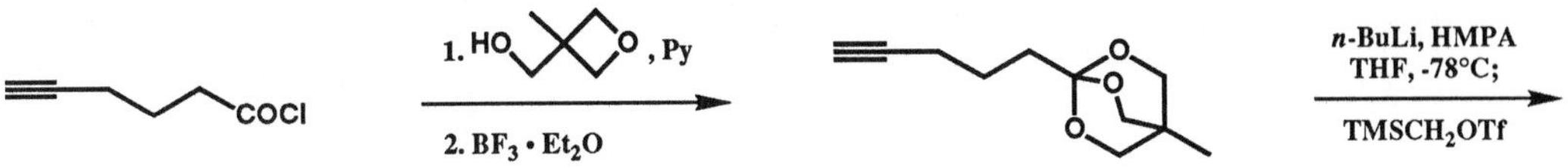
COCl
1. HO
O
, Py
2. BF3 • Et2O
O
O
O
n-BuLi, HMPA
THF, -78°C;
TMSCH2OTf

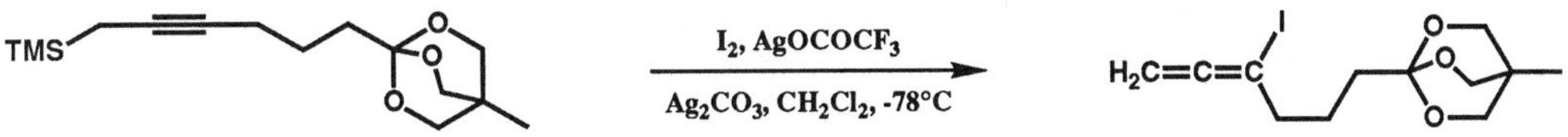
TMS
O
O
O
I2, AgOCOCF3
Ag2CO3, CH2Cl2, -78°C
I
H2C=C=C
O
O
O

B

References:

1. E. J. Corey and B. De, *J. Am. Chem. Soc.* **1984**, *106*, 2735.
2. E. J. Corey and T. M. Eckrich, *Tetrahedron Lett.* **1984**, *25*, 2415, 2419.
3. E. J. Corey and N. Raju, *Tetrahedron Lett.* **1983**, *24*, 5571.
4. E. J. Corey, B. De, J. W. Ponder, and J. M. Berg, *Tetrahedron Lett.* **1984**, *25*, 1015.

CHAPTER TWELVE

Leukotrienes and Other Bioactive Polyenes

12.1

Formation of Leukotrienes from Arachidonic Acid

Samuelsson's discovery of the bioconversion of arachidonic acid in neutrophils to the hydroperoxide 5-HPETE, the corresponding alcohol, and a 5*S*,12*R*-dihydroxy-6,8,10,14-tetraenoic acid opened a new chapter in the eicosanoid field.[1] On the basis of a surmise that the 5,12-diol might be formed from a 5,6-epoxide, a synthesis of the epoxide, now known as leukotriene A_4 (LTA_4), was developed.[2] It subsequently transpired that LTA_4 is the predecessor not only of the 5,12-diol (now called LTB_4) but also of the peptidic leukotrienes LTC_4, LTD_4 and LTE_4, which have important roles in cell and tissue biology.[1, 2]

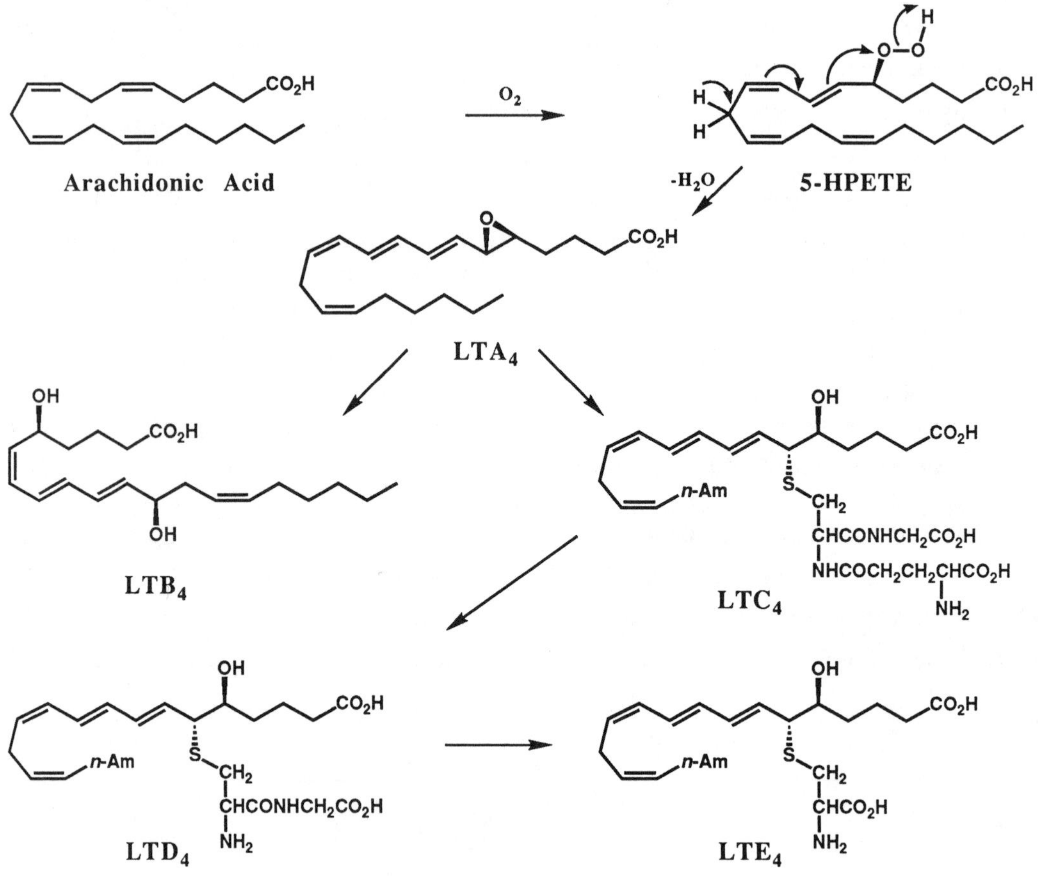

CO2H
O2
Arachidonic Acid
H
O–O
H
CO2H
H
5-HPETE
-H2O
O
CO2H
LTA4
OH
CO2H
OH
LTB4
OH
CO2H
n-Am
S
CH2
CHCONHCH2CO2H
NHCOCH2CH2CHCO2H
NH2
LTC4
OH
CO2H
n-Am
S
CH2
CHCONHCH2CO2H
NH2
LTD4
OH
CO2H
n-Am
S
CH2
CHCO2H
NH2
LTE4

References:

1. B. Samuelsson, *Science* **1983**, *220*, 568.

2. (a) E. J. Corey, *Experientia* **1982**, *38*, 1259; (b) E. J. Corey, In *Current Trends in Organic Synthesis*; H. Nozaki, Ed.; Pergamon Press: Oxford, 1984; (c) E. J. Corey, D. A. Clark, and A. Marfat, In *The Leukotrienes*; W. Chakrin and D. M. Bailey, Eds.; Academic Press: New York, 1984; p 13; (d) A. Marfat and E. J. Corey, In *Advances in Prostaglandin, Thromboxane, and Leukotriene Research*; J. E. Pike and D. R. Morton, Jr., Eds.; Raven Press: New York, 1985; Vol. 14, p 155.

Leukotriene A_4

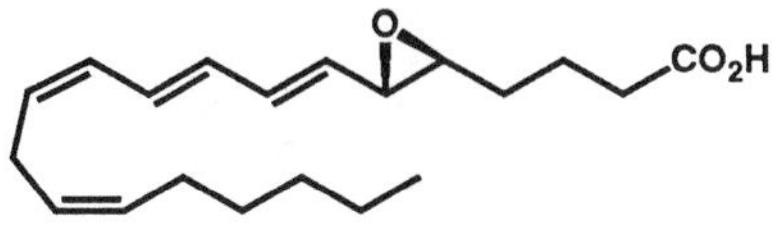
O
CO_2H

Leukotriene A_4 was first synthesized as the racemate and shortly thereafter in chiral form from D-(-)-ribose. The synthetic material was then used to confirm that this compound was indeed a link between 5-HPETE and LTB_4. In addition the constitution of the "slow reacting substance of anophylaxis" (SRSA) was proved by synthesis of the peptidic leukotrienes LTC_4, LTD_4 and LTE_4 from LTA_4 and comparison with naturally derived materials. Several other routes to LTA_4 were also developed, including a short synthesis from arachidonic acid.

1. The first synthesis of (±)-5,6-epoxy-7,9-(*E*)-11,14-(*Z*)-eicosatetraenoate, later identified as the primary leukotriene LTA_4 (Ref. 1).

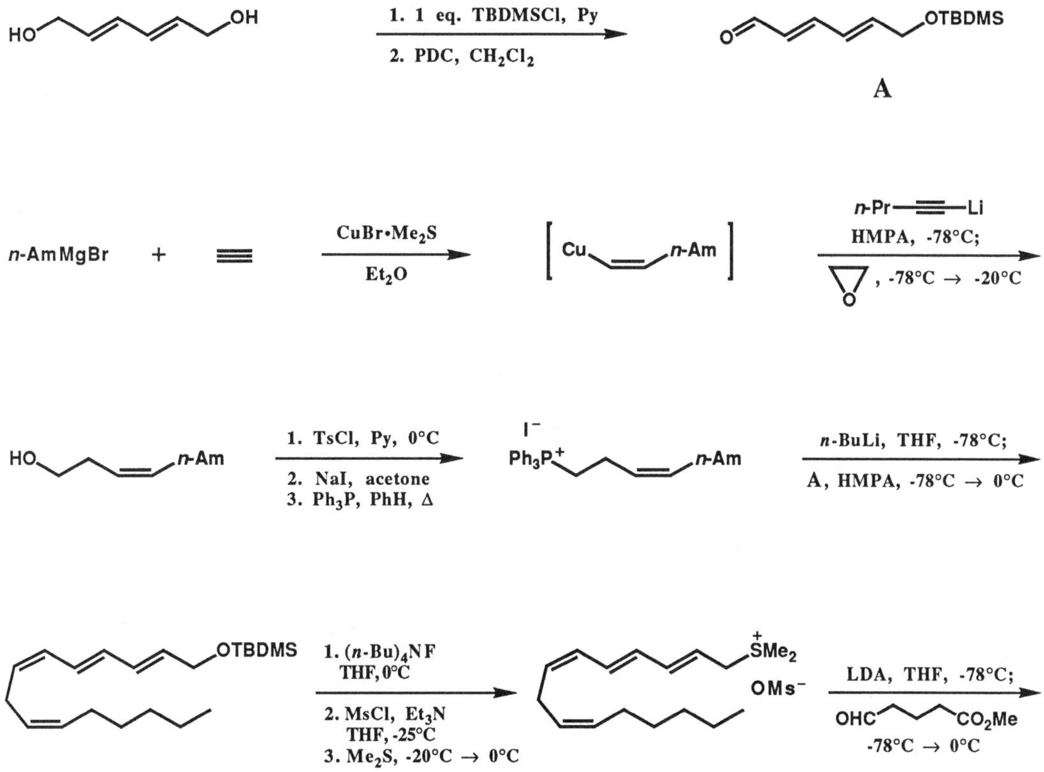
HO
OH
1. 1 eq. TBDMSCl, Py
2. PDC, CH_2Cl_2
O
OTBDMS
A
n-AmMgBr + ≡
$CuBr{\cdot}Me_2S$
Et_2O
Cu
n-Am
n-Pr—≡—Li
HMPA, -78°C;
, -78°C → -20°C
O
HO
n-Am
1. TsCl, Py, 0°C
2. NaI, acetone
3. Ph_3P, PhH, Δ
I^-
Ph_3P^+
n-Am
n-BuLi, THF, -78°C;
A, HMPA, -78°C → 0°C
OTBDMS
1. $(n\text{-Bu})_4NF$
THF, 0°C
2. MsCl, Et_3N
THF, -25°C
3. Me_2S, -20°C → 0°C
$\overset{+}{S}Me_2$
OMs^-
LDA, THF, -78°C;
OHC
CO_2Me
-78°C → 0°C

12.2

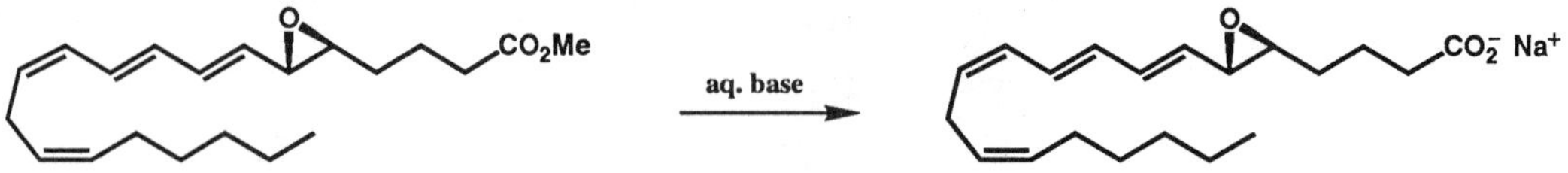
CO2Me
aq. base
CO2⁻ Na+

plus the 5,6-cis-epoxide (removed by HPLC)

λ_{max} 269, 278, 287 nm (MeOH)

(±)-LTA$_4$

2. Stereospecific synthesis (Ref. 2):

BzO
OH
BzO
OBz
1. Ph3P=CO2Et
PhCO2H, DME, Δ
2. Ac2O, H2SO4
OAc
OBz
BzO
CO2Et
OBz
Zn/Hg, HCl
Et2O

from D-(-)-ribose

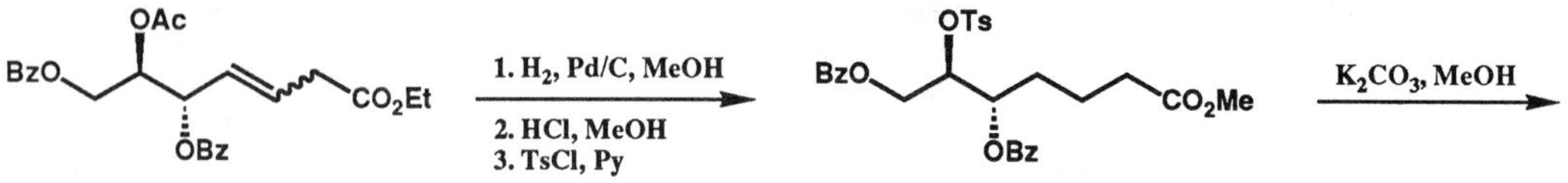
OAc
BzO
CO2Et
OBz
1. H2, Pd/C, MeOH
2. HCl, MeOH
3. TsCl, Py
OTs
BzO
CO2Me
OBz
K2CO3, MeOH

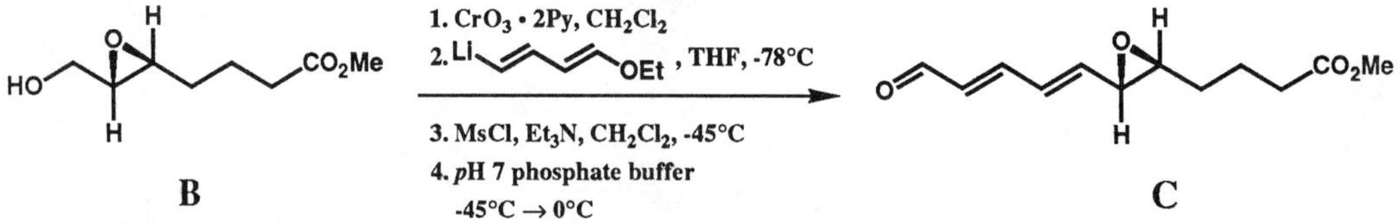
H
HO
CO2Me
H
1. CrO3 • 2Py, CH2Cl2
2. Li OEt, THF, -78°C
3. MsCl, Et3N, CH2Cl2, -45°C
4. pH 7 phosphate buffer
-45°C → 0°C
H
CO2Me
H

B $[\alpha]^{24}_D$ -37.4° (*c* 0.27, $CHCl_3$)

C $[\alpha]^{25}_D$ -27.3° (*c* 1.97, $CHCl_3$)

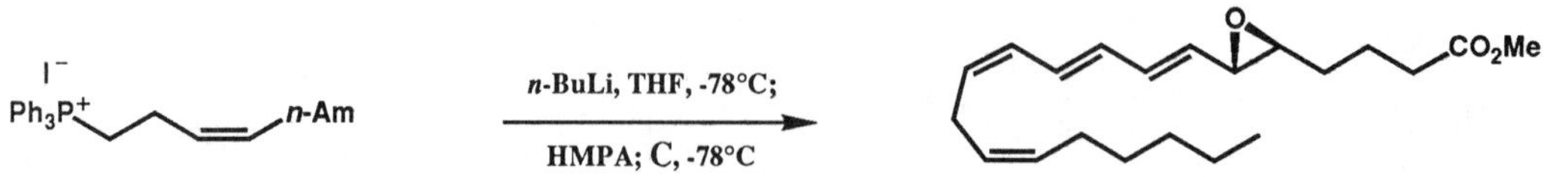
I⁻
Ph3P+
n-Am
n-BuLi, THF, -78°C;
HMPA; C, -78°C
CO2Me

(Ref. 1)

$[\alpha]^{25}_D$ -21.9° (*c* 0.32, cyclohexane)

λ_{max} 269, 278, 289 nm (MeOH)

12.2

A number of modifications were made for the synthesis of the compounds **B** and **C**, and other intermediates used for the synthesis of LTA4.

i. Synthesis of the hydroxy epoxide **B** (Ref. 3):

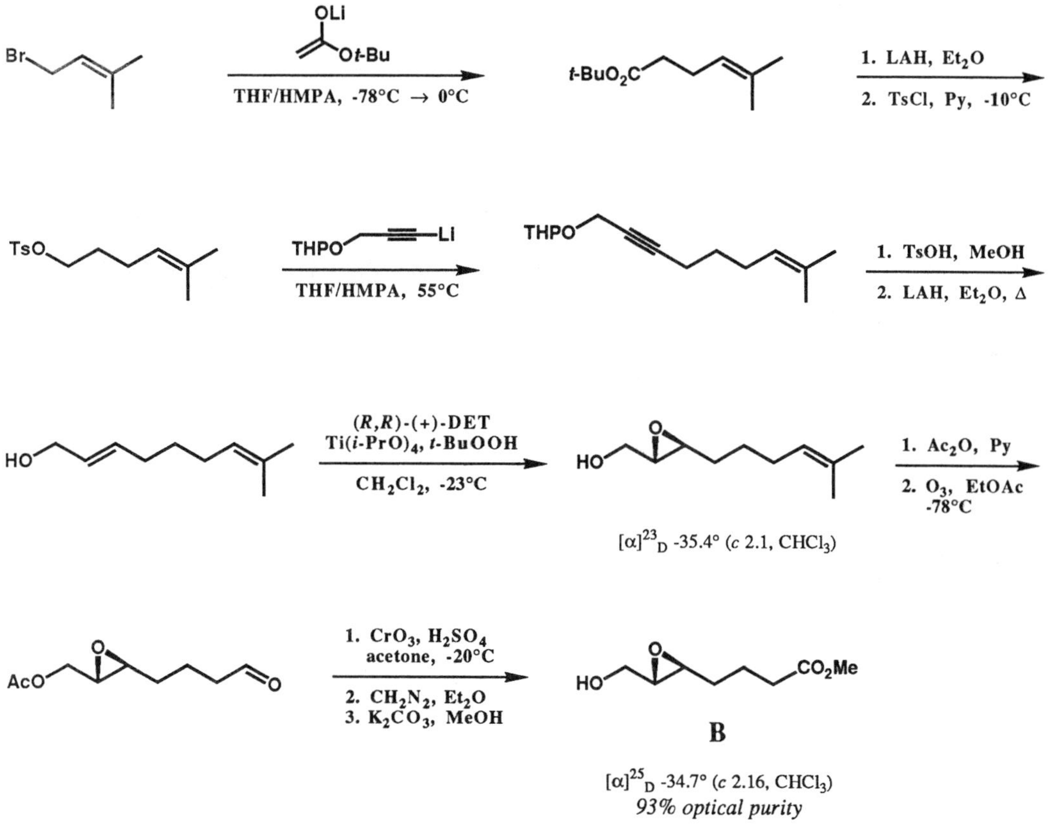

Br
OLi
Ot-Bu
THF/HMPA, -78°C → 0°C
t-BuO2C
1. LAH, Et2O
2. TsCl, Py, -10°C
TsO
THPO
Li
THF/HMPA, 55°C
THPO
1. TsOH, MeOH
2. LAH, Et2O, Δ
HO
(R,R)-(+)-DET
Ti(i-PrO)4, t-BuOOH
CH2Cl2, -23°C
HO
O
[α]23D -35.4° (c 2.1, CHCl3)
1. Ac2O, Py
2. O3, EtOAc
-78°C
AcO
O
O
1. CrO3, H2SO4
acetone, -20°C
2. CH2N2, Et2O
3. K2CO3, MeOH
HO
O
CO2Me
B
[α]25D -34.7° (c 2.16, CHCl3)
93% optical purity

ii. Synthesis of the dienal C (Ref. 4):

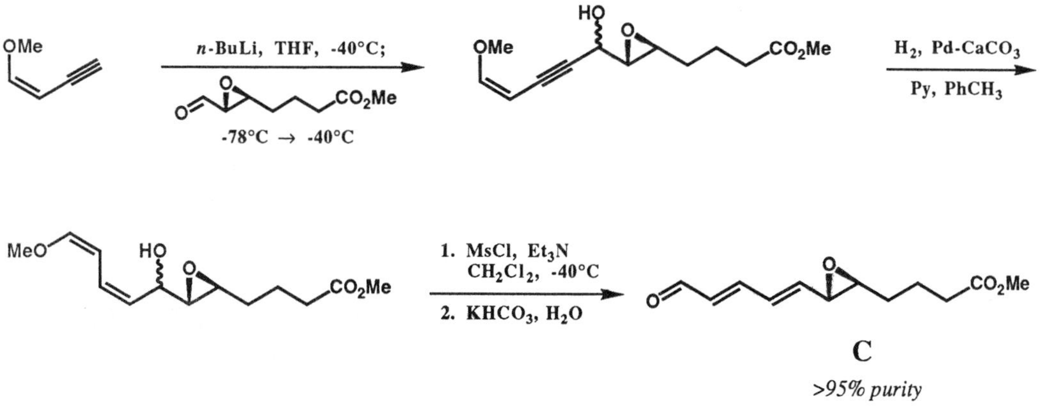

OMe
n-BuLi, THF, -40°C;
O
O
CO2Me
-78°C → -40°C
OMe
HO
O
CO2Me
H2, Pd-CaCO3
Py, PhCH3
MeO
HO
O
CO2Me
1. MsCl, Et3N
CH2Cl2, -40°C
2. KHCO3, H2O
O
O
CO2Me
C
>95% purity

3. Synthesis from 5-HPETE by hydroperoxide → oxiranyl carbinol rearrangement (Ref. 5):

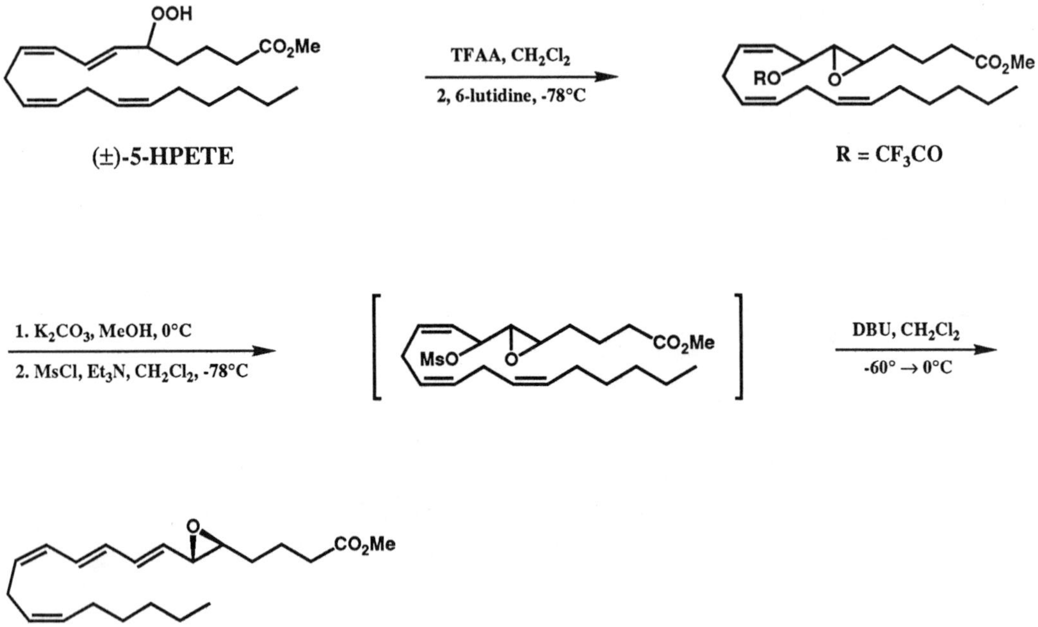
OOH
CO2Me
TFAA, CH2Cl2
2, 6-lutidine, -78°C
RO
O
(±)-5-HPETE
R = CF3CO
1. K2CO3, MeOH, 0°C
2. MsCl, Et3N, CH2Cl2, -78°C
MsO
DBU, CH2Cl2
-60° → 0°C

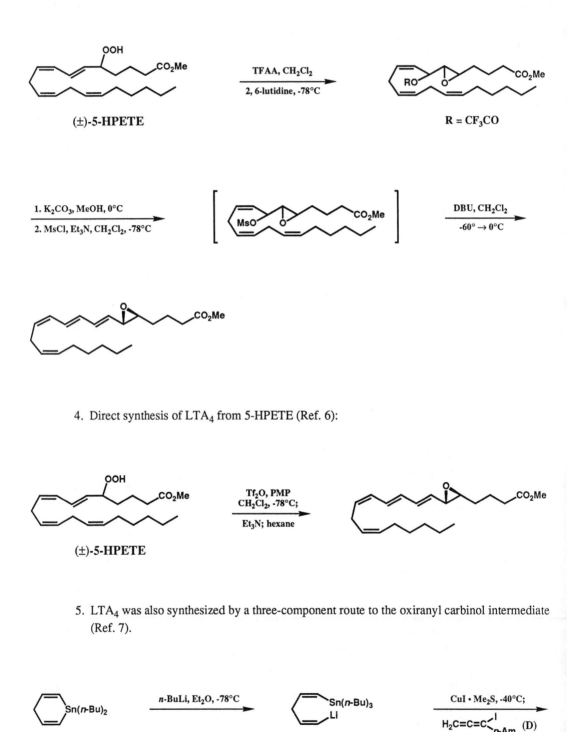

4. Direct synthesis of LTA_4 from 5-HPETE (Ref. 6):

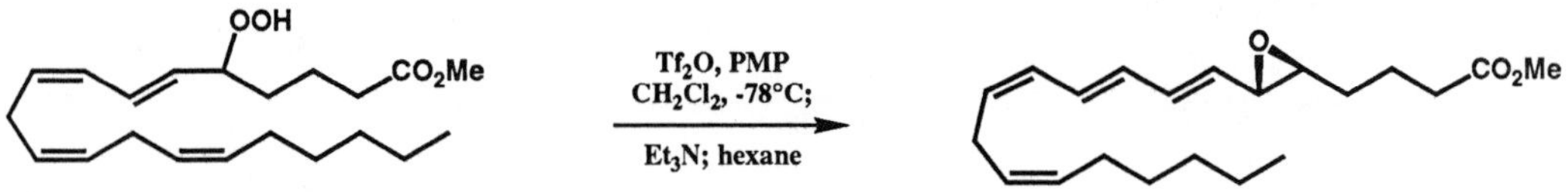
OOH
CO2Me
Tf2O, PMP
CH2Cl2, -78°C;
Et3N; hexane
O

(±)-5-HPETE

5. LTA_4 was also synthesized by a three-component route to the oxiranyl carbinol intermediate (Ref. 7).

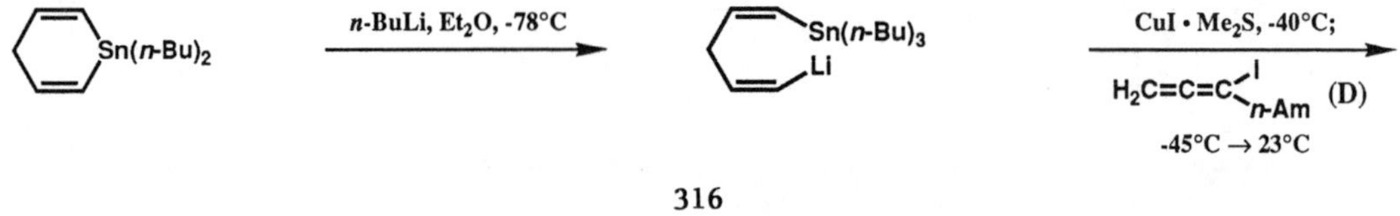
Sn(n-Bu)2
n-BuLi, Et2O, -78°C
Sn(n-Bu)3
Li
CuI • Me2S, -40°C;
H2C=C=C
I
n-Am
(D)
-45°C → 23°C

12.2

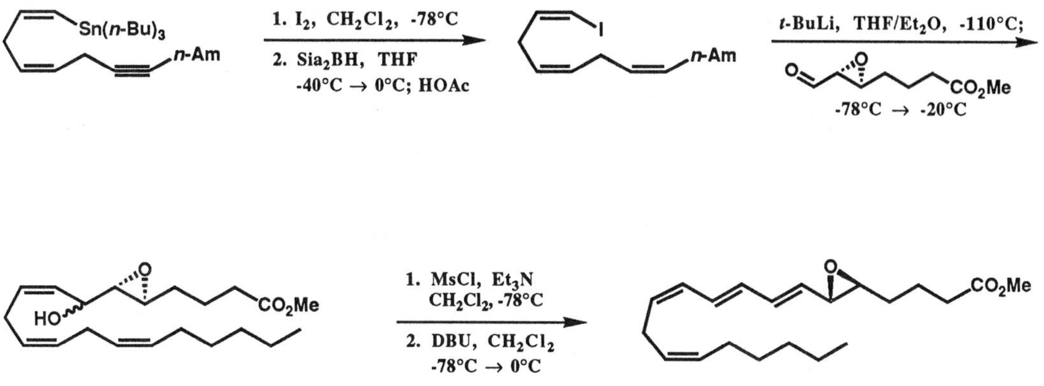

Sn(n-Bu)3
n-Am
1. I2, CH2Cl2, -78°C
2. Sia2BH, THF
-40°C → 0°C; HOAc
I
n-Am
t-BuLi, THF/Et2O, -110°C;
O
CO2Me
-78°C → -20°C
O
HO
CO2Me
1. MsCl, Et3N
CH2Cl2, -78°C
2. DBU, CH2Cl2
-78°C → 0°C
O
CO2Me

3 : 2 diastereomeric mixture

Preparation of the iodo allene **D**:

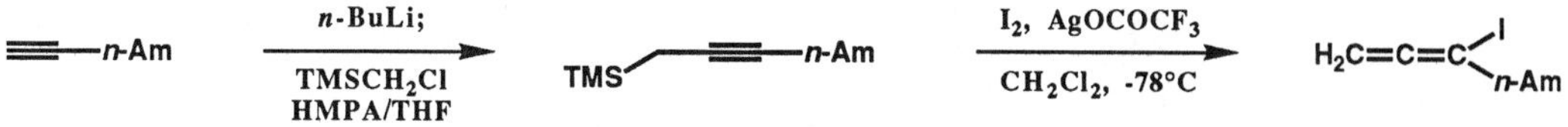

n-Am
n-BuLi;
TMSCH2Cl
HMPA/THF
TMS
n-Am
I2, AgOCOCF3
CH2Cl2, -78°C
H2C=C=C
I
n-Am

D

References:

1. E. J. Corey, Y. Arai, and C. Mioskowski, *J. Am. Chem. Soc.* **1979**, *101*, 6748.
2. E. J. Corey, D. A. Clark, G. Goto, A. Marfat, C. Mioskowski, B. Samuelsson, and S. Hammarström, *J. Am. Chem. Soc.* **1980**, *102*, 1436, 3663.
3. E. J. Corey, S.-i. Hashimoto, and A. E. Barton, *J. Am. Chem. Soc.* **1981**, *103*, 721.
4. E. J. Corey and J. O. Albright, *J. Org. Chem.* **1983**, *48*, 2114.
5. E. J. Corey, W.-g. Su, and M. M. Mehrotra, *Tetrahedron Lett.* **1984**, *25*, 5123.
6. E. J. Corey and A. E. Barton, *Tetrahedron Lett.* **1982**, *23*, 2351.
7. E. J. Corey, M. M. Mehrotra, and J. R. Cashman, *Tetrahedron Lett.* **1983**, *24*, 4917.

Leukotriene C_4 (Originally LTC-1) / Leukotriene D_4

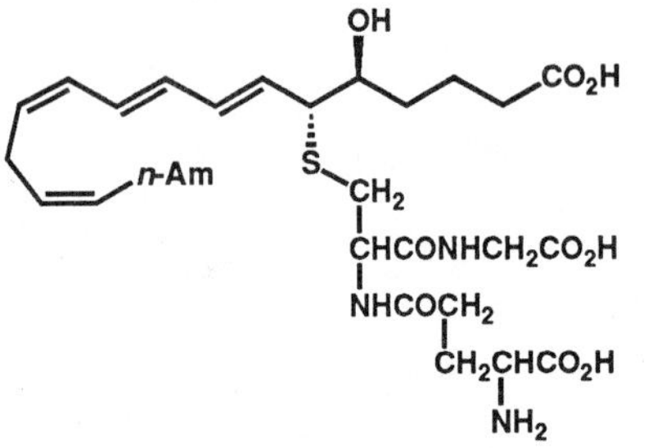
OH
CO_2H
n-Am
S
CH_2
$CHCONHCH_2CO_2H$
$NHCOCH_2$
CH_2CHCO_2H
NH_2

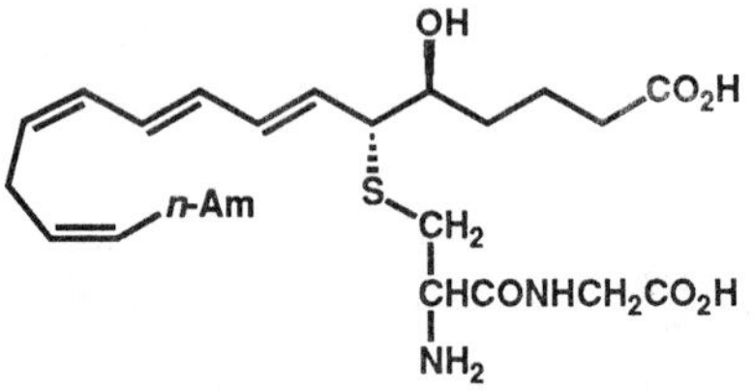
OH
CO_2H
n-Am
S
CH_2
$CHCONHCH_2CO_2H$
NH_2

The parent spasmogenic leukotriene, LTC_4, was synthesized from LTA_4 and glutathione thereby providing an abundant source of this rare and unstable compound for biological studies as well as a proof of structure and stereochemistry (Ref. 1). LTD_4 and LTE_4 were similarly prepared from LTA_4 and the *N*-trifluoroacetyl derivatives of cysteinylglycine methyl ester and cysteine methyl ester, respectively. LTC_4, LTD_4 and LTE_4 are unstable in solution because of facile radical-induced 11-*Z* → 11–*E* isomerization and also polymerization.

1. Synthesis of leukotriene C_4 (Ref. 1, 2):

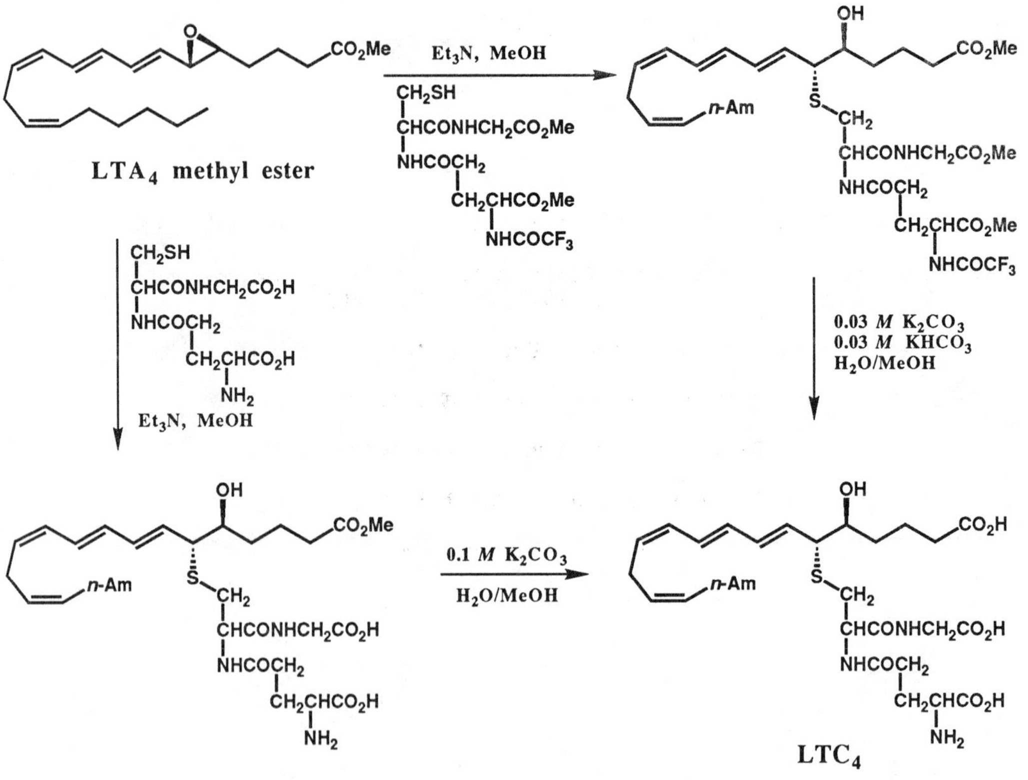
CO_2Me
LTA_4 methyl ester
Et_3N, MeOH
CH_2SH
$CHCONHCH_2CO_2Me$
$NHCOCH_2$
CH_2CHCO_2Me
$NHCOCF_3$
OH
n-Am
S
CH_2
$CHCONHCH_2CO_2H$
CH_2CHCO_2H
NH_2
0.03 *M* K_2CO_3
0.03 *M* $KHCO_3$
H_2O/MeOH
0.1 *M* K_2CO_3
CO_2H
LTC_4

λ_{max} 270, 280, 290 nm (MeOH)

2. Synthesis of leukotriene D_4 (Ref. 3):

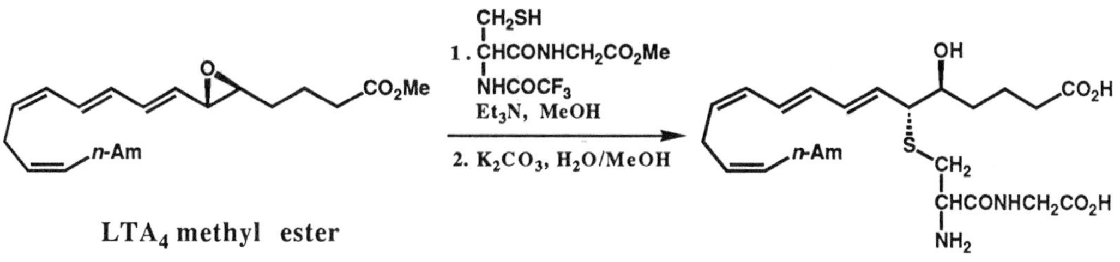
CO2Me
n-Am
1. CH2SH CHCONHCH2CO2Me NHCOCF3
Et3N, MeOH
2. K2CO3, H2O/MeOH
OH
CO2H
n-Am
S CH2 CHCONHCH2CO2H NH2

LTA$_4$ methyl ester

LTD$_4$

Preparation of the *N*-trifluoroacetylcysteinylglycine methyl ester:

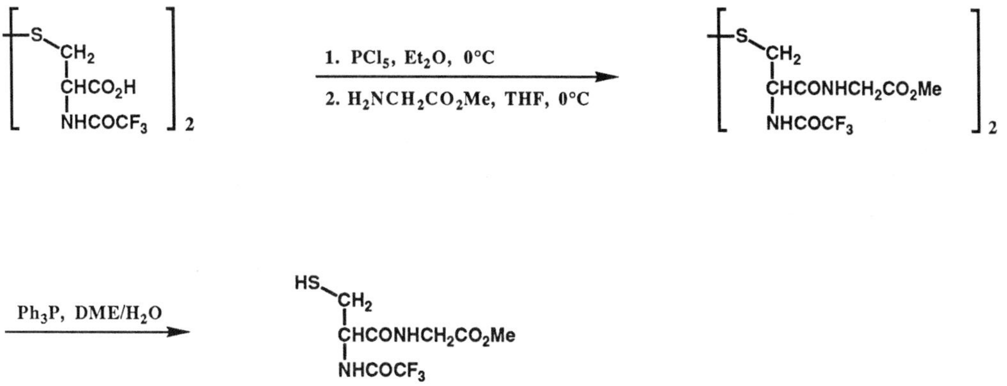
S CH2 CHCO2H NHCOCF3
2
1. PCl5, Et2O, 0°C
2. H2NCH2CO2Me, THF, 0°C
S CH2 CHCONHCH2CO2Me NHCOCF3
2
Ph3P, DME/H2O
HS CH2 CHCONHCH2CO2Me NHCOCF3

References:

1. E. J. Corey, D. A. Clark, G. Goto, A. Marfat, C. Mioskowski, B. Samuelsson, and S. Hammarström, *J. Am. Chem. Soc.* **1980**, *102*, 1436, 3663.

2. S. Hammarström, B. Samuelsson, D. A. Clark, G. Goto, A. Marfat, C. Mioskowski, and E. J. Corey, *Biochem. Biophys. Res. Commun.* **1980**, *92*, 946.

3. E. J. Corey, D. A. Clark, A. Marfat, and G. Goto, *Tetrahedron Lett.* **1980**, *21*, 3143.

12.4

Leukotriene B_4

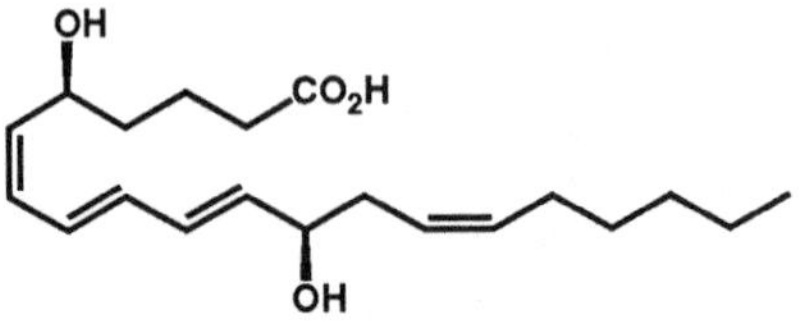
OH
CO2H
OH

Leukotriene B4, formed by enzymic hydrolysis of LTA4, is chemotactic for macrophages and neutrophils at concentrations as low as 1 ng/ml. The stereochemistry of the conjugated triene subunit was established by synthesis which also made LTB4 available in quantity for biological research.

First Synthesis (Ref. 1):

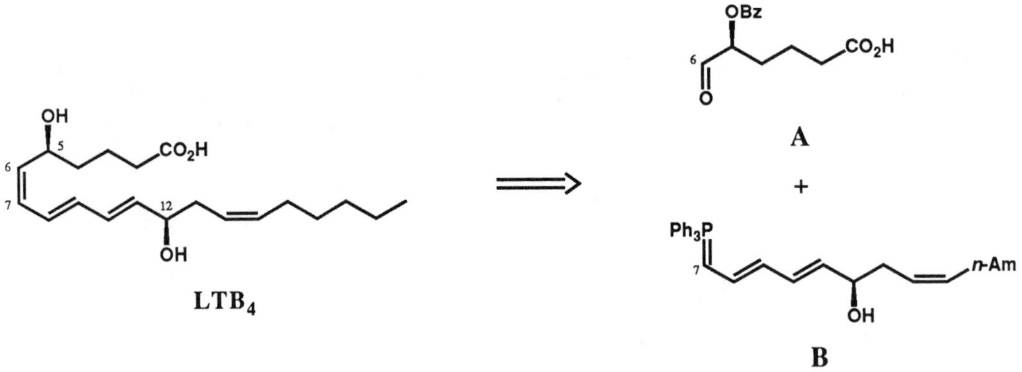
OH
5
CO2H
6
7
12
OH
LTB4
OBz
6
CO2H
O
A
+
Ph3P
7
n-Am
OH
B

1. Synthesis of the fragment **A**:

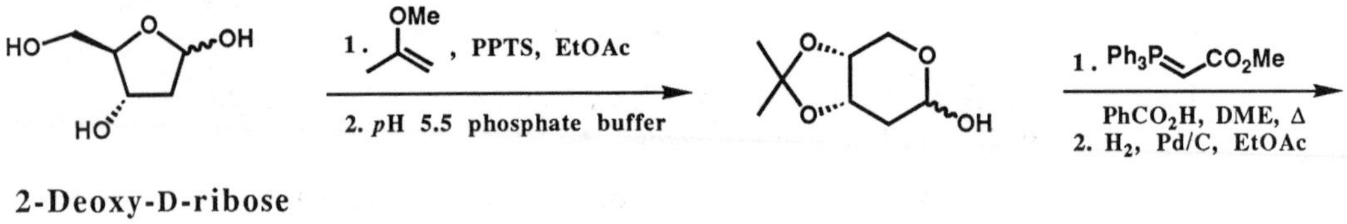
HO
O
OH
HO
2-Deoxy-D-ribose
OMe
1. , PPTS, EtOAc
2. pH 5.5 phosphate buffer
O
O
O
OH
1. Ph3P CO2Me
PhCO2H, DME, Δ
2. H2, Pd/C, EtOAc

O
O
OH
CO2Me
1. TsCl, Py
2. HCl, MeOH
HO
HO
OTs
CO2Me
K2CO3, MeOH

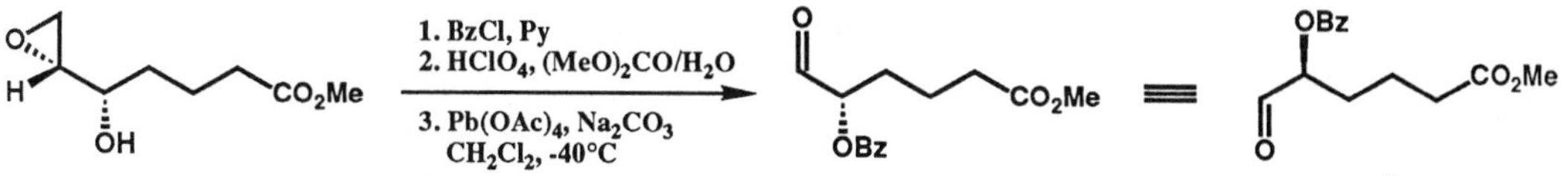

A

$[\alpha]^{25}_D$ -33.3° (*c* 2.5, $CHCl_3$)

2. Synthesis of the fragment **B**:

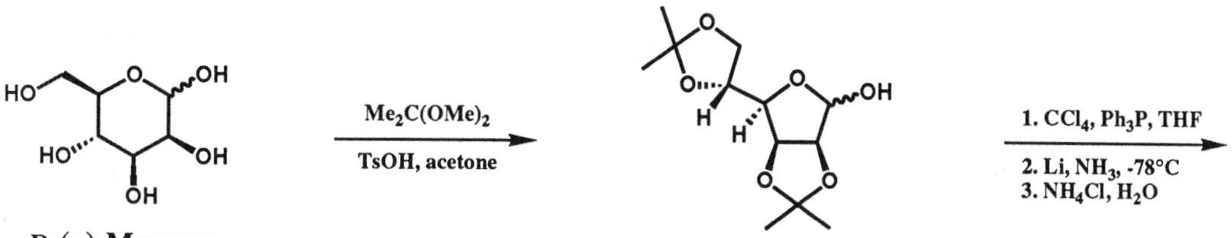

D-(+)-Mannose

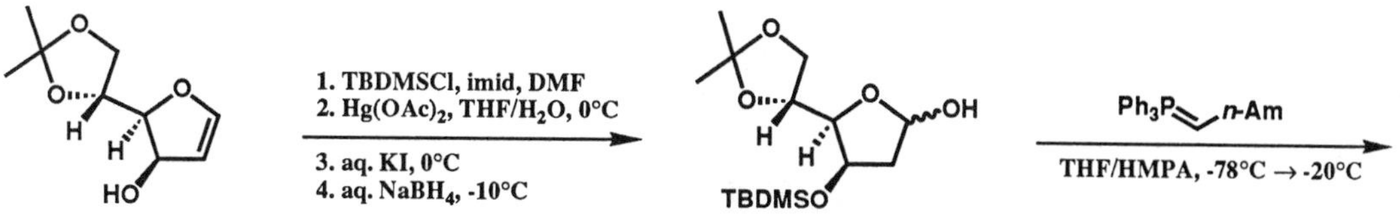

(Ref. 2)

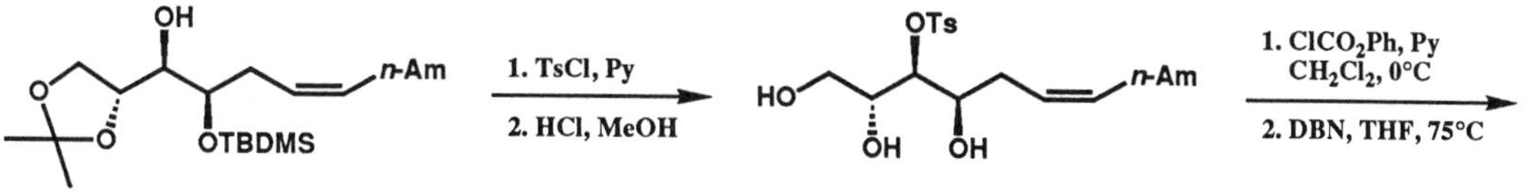

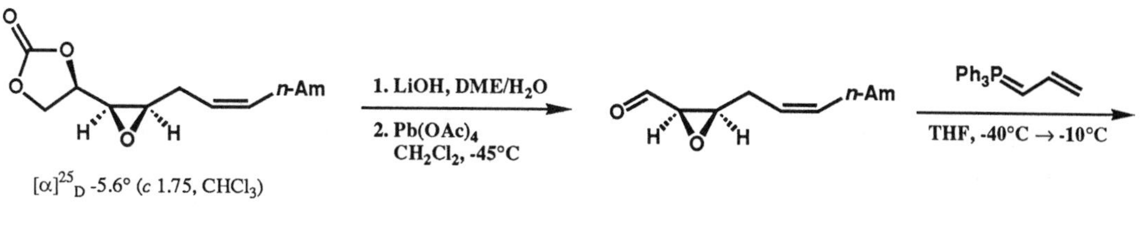

$[\alpha]^{25}_D$ -5.6° (*c* 1.75, $CHCl_3$)

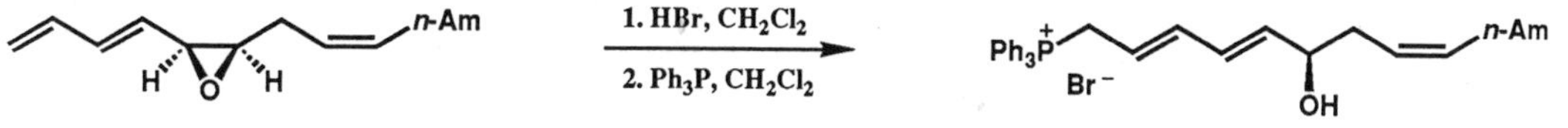

B

3. Coupling of the fragments **A** and **B** and the completion of the synthesis:

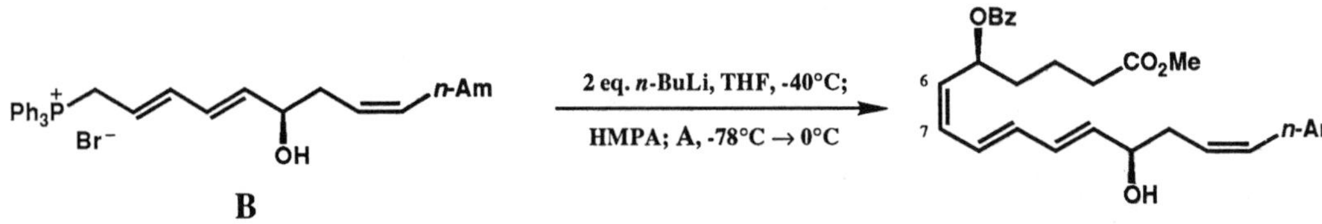
Ph3P+
Br−
n-Am
OH
2 eq. n-BuLi, THF, -40°C;
HMPA; A, -78°C → 0°C
OBz
CO2Me
6
7
n-Am
OH

B

$[\alpha]^{25}_{D}$ +164.4° (*c* 1.4, $CHCl_3$)

plus 15% 6,7-(E) isomer

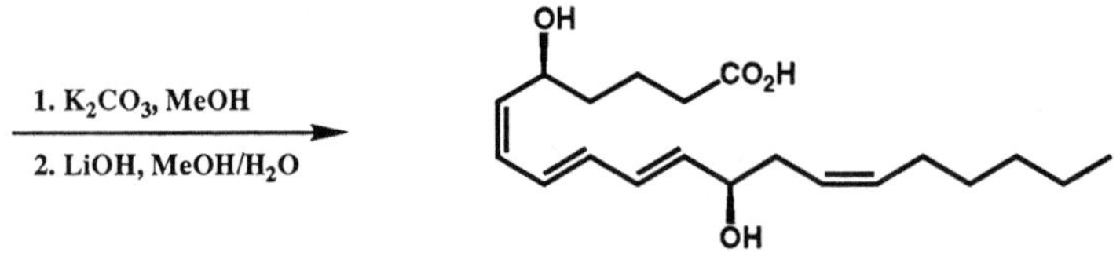
1. K2CO3, MeOH
2. LiOH, MeOH/H2O
OH
CO2H
OH

LTB$_4$

λ_{max} 260, 270.5, 281 nm (MeOH)

Second Synthesis (Ref. 3):

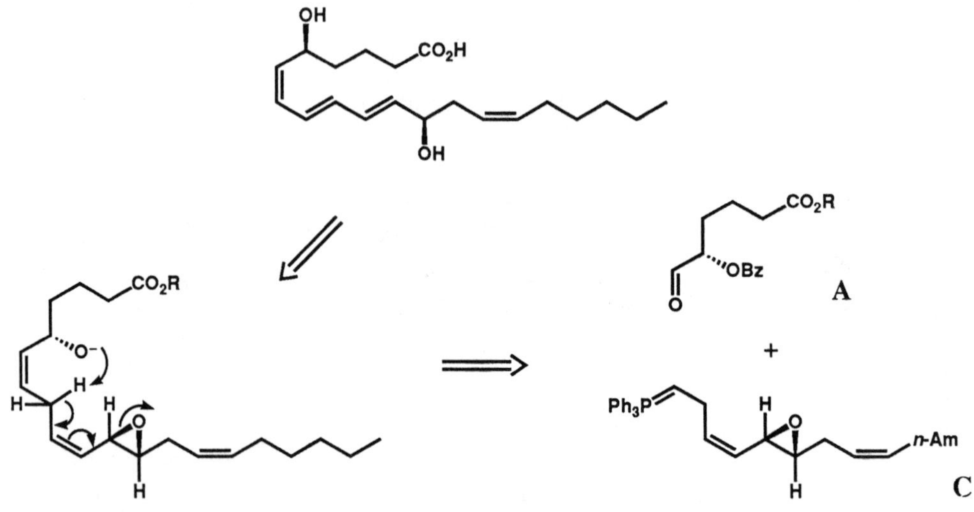
OH
CO2H
OH
CO2R
O−
H
H
H
O
H
CO2R
OBz
O
A
+
Ph3P
H
O
n-Am
H
C

Fragment **A** was synthesized from 2-deoxy-D-ribose by a modification of the approach previously described (Ref. 1):

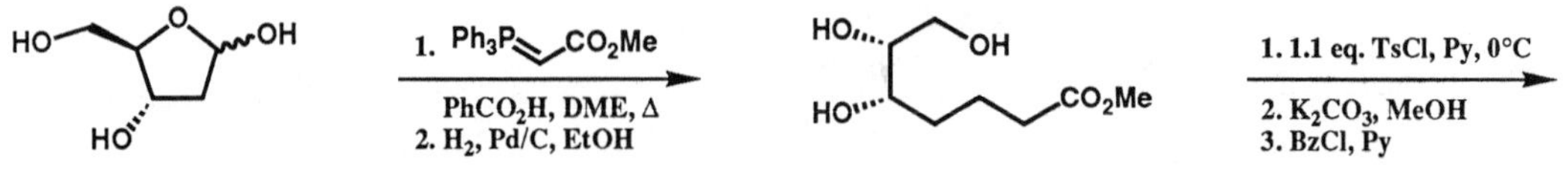
HO
OH
HO
1. Ph3P CO2Me
PhCO2H, DME, Δ
2. H2, Pd/C, EtOH
HO
OH
HO
CO2Me
1. 1.1 eq. TsCl, Py, 0°C
2. K2CO3, MeOH
3. BzCl, Py

2-Deoxy-D-ribose

12.4

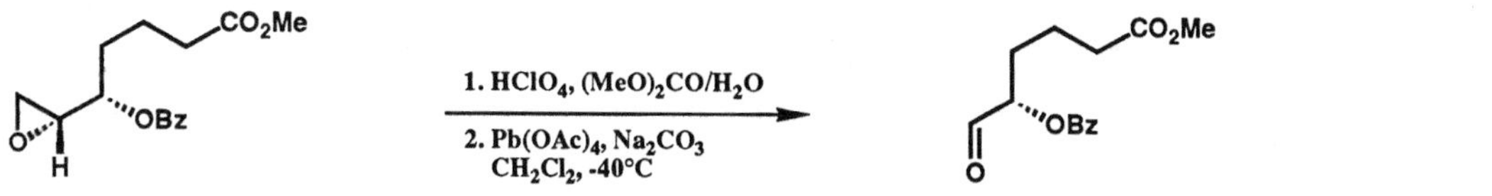
CO2Me
OBz
H
1. HClO4, (MeO)2CO/H2O
2. Pb(OAc)4, Na2CO3
CH2Cl2, -40°C
CO2Me
OBz
O

A

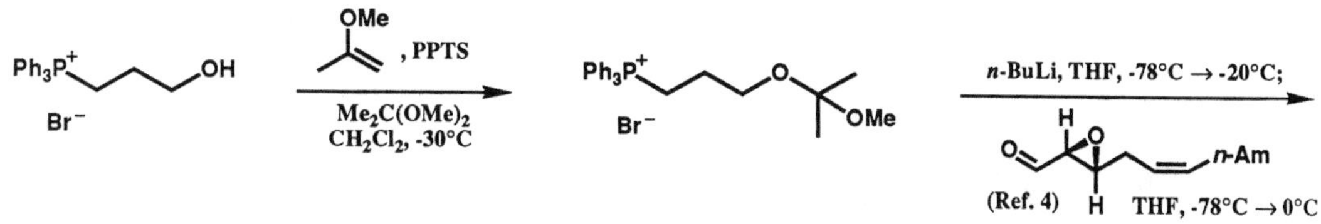
Ph3P+
OH
Br−
OMe
, PPTS
Me2C(OMe)2
CH2Cl2, -30°C
Ph3P+
O
OMe
Br−
n-BuLi, THF, -78°C → -20°C;
H
O
n-Am
(Ref. 4)
H
THF, -78°C → 0°C

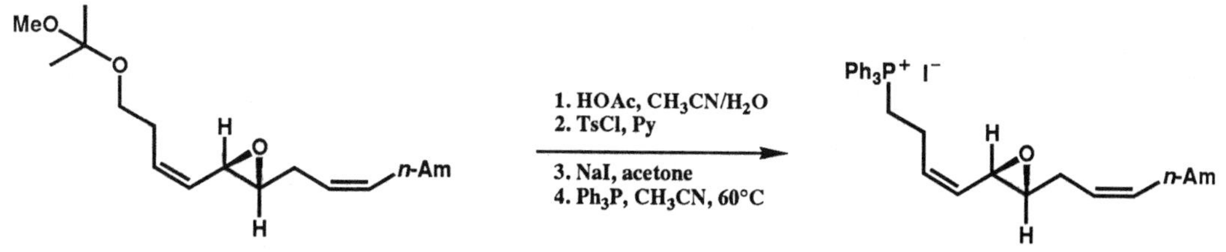
MeO
O
H
O
n-Am
H
1. HOAc, CH3CN/H2O
2. TsCl, Py
3. NaI, acetone
4. Ph3P, CH3CN, 60°C
Ph3P+ I−
H
O
n-Am
H

C

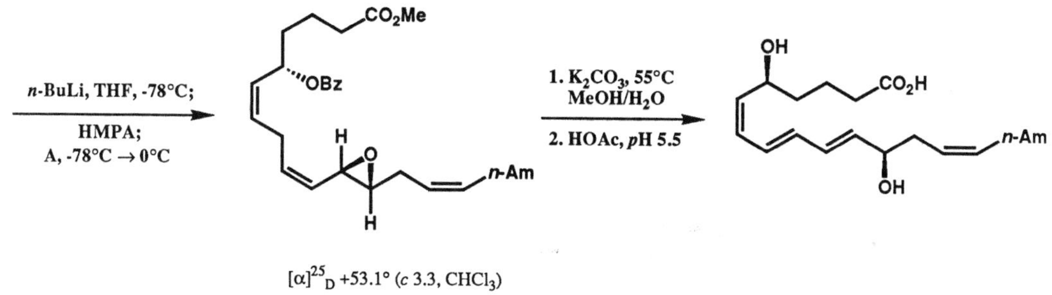
n-BuLi, THF, -78°C;
HMPA;
A, -78°C → 0°C
CO2Me
OBz
H
O
n-Am
H
1. K2CO3, 55°C
MeOH/H2O
2. HOAc, pH 5.5
OH
CO2H
n-Am
OH

$[\alpha]^{25}_{D}$ +53.1° (*c* 3.3, $CHCl_3$)

References:

1. E. J. Corey, A. Marfat, G. Goto, and F. Brion, *J. Am. Chem. Soc.* **1980**, *102*, 7984.
2. E. J. Corey and G. Goto, *Tetrahedron Lett.* **1980**, *21*, 3463.
3. E. J. Corey, A. Marfat, J. E. Munroe, K. S. Kim, P. B. Hopkins, and F. Brion, *Tetrahedron Lett.* **1981**, *22*, 1077.
4. E. J. Corey, A. Marfat, and G. Goto, *J. Am. Chem. Soc.* **1980**, *102*, 6607.

12.5

Synthesis of Stereoisomers of Leukotriene B_4 (for Assignment of Stereochemistry)

Since the stereochemistry of the triene system of LTB_4 had not been determined prior to synthesis, a number of stereoisomers of LTB_4 were prepared for purposes of definitive comparison of physical properties and bioactivity with biologically produced LTB_4. The various stereoisomers of LTB_4 were much less active biologically than LTB_4 itself.

1. 6-(*E*),10-(*Z*) Isomer of LTB_4 (Ref. 1):

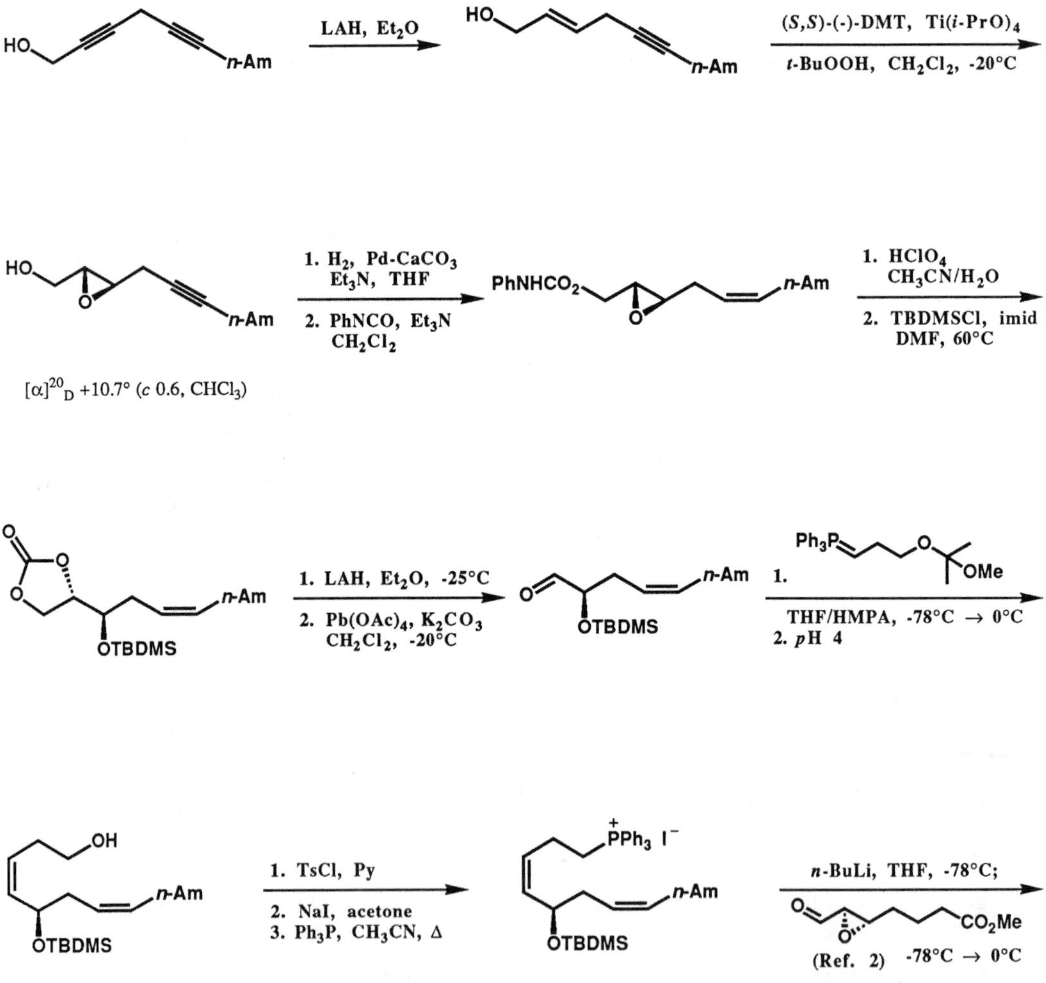
HO
n-Am
LAH, Et2O
(S,S)-(-)-DMT, Ti(i-PrO)4
t-BuOOH, CH2Cl2, -20°C
1. H2, Pd-CaCO3 Et3N, THF
2. PhNCO, Et3N CH2Cl2
PhNHCO2
1. HClO4 CH3CN/H2O
2. TBDMSCl, imid DMF, 60°C
[α]20D +10.7° (c 0.6, CHCl3)
OTBDMS
1. LAH, Et2O, -25°C
2. Pb(OAc)4, K2CO3 CH2Cl2, -20°C
Ph3P
OMe
1.
THF/HMPA, -78°C → 0°C
2. pH 4
OH
1. TsCl, Py
2. NaI, acetone
3. Ph3P, CH3CN, Δ
PPh3 I−
n-BuLi, THF, -78°C;
CO2Me
(Ref. 2) -78°C → 0°C

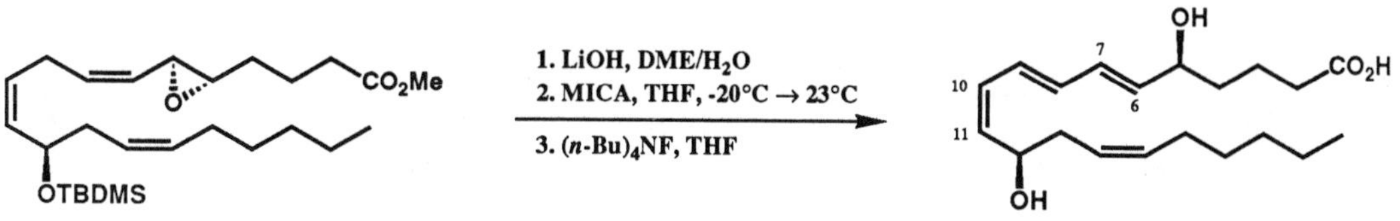

λ_{max} 260, 269.5, 280.5 nm (MeOH)

2. (±)-6-(*E*),8-(*Z*) Isomer of LTB_4 (Ref. 1):

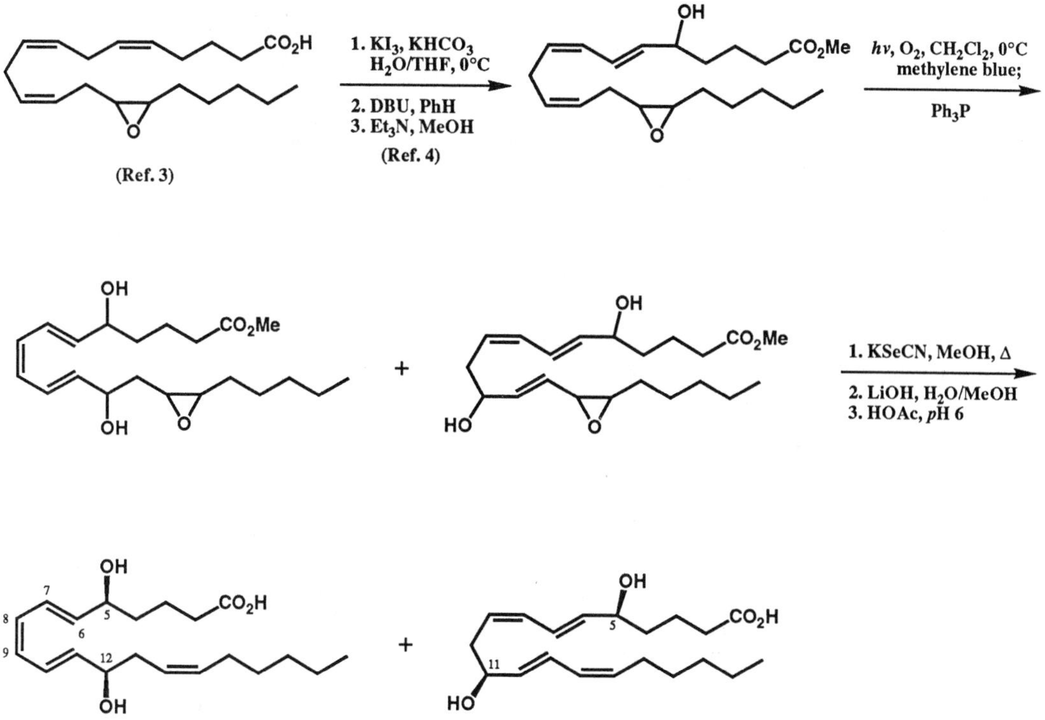

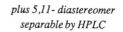

plus 5,12- diastereomer separable by HPLC

λ_{max} 258, 268, 278 nm (MeOH)

plus 5,11- diastereomer separable by HPLC

λ_{max} 233 nm (MeOH)

3. 12-(*S*)-Form of 6-(*E*) LTB_4 (Ref. 5):

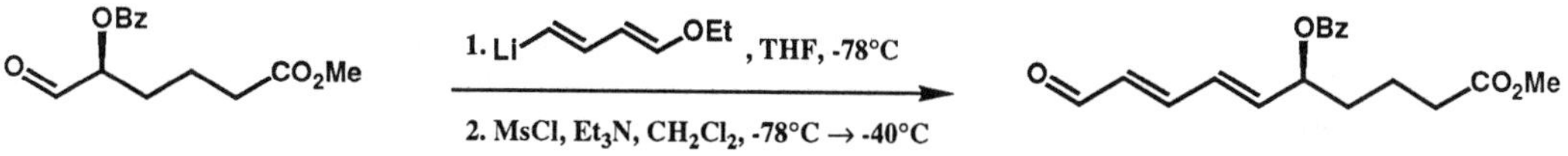

(Ref. 6)

A (Ref. 2)

12.5

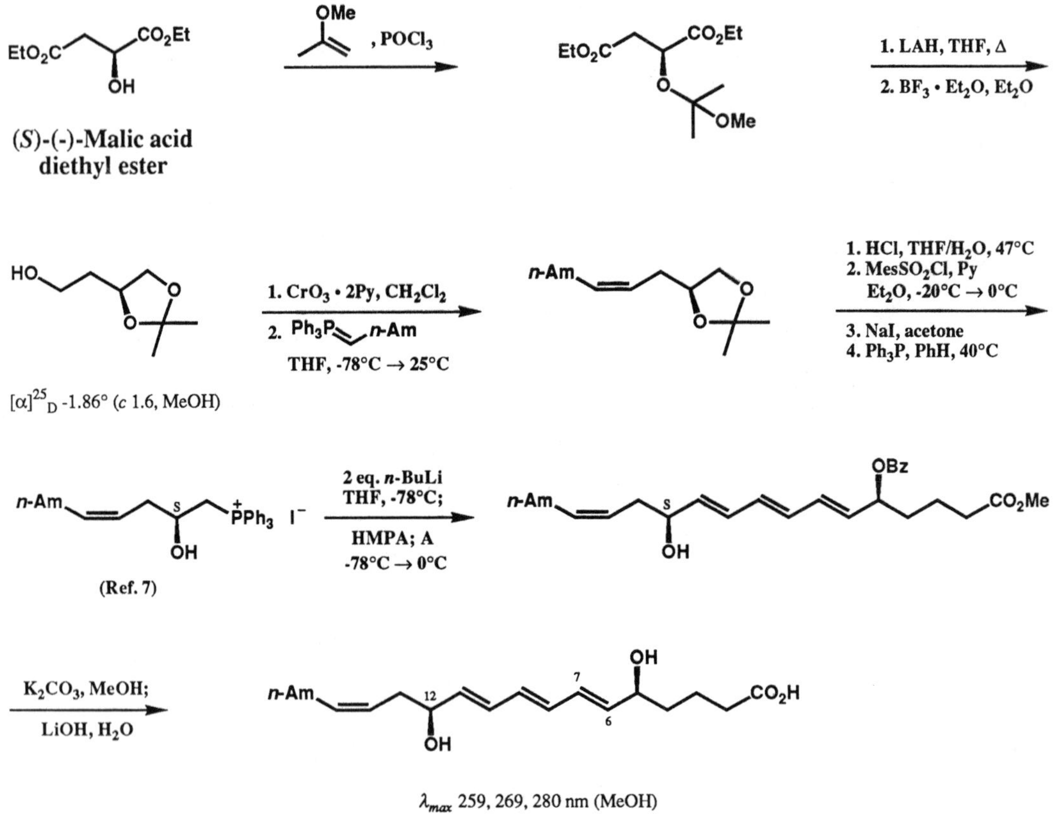
OMe
, POCl3
1. LAH, THF, Δ
2. BF3 • Et2O, Et2O
(S)-(-)-Malic acid diethyl ester
[α]25D -1.86° (c 1.6, MeOH)
1. CrO3 • 2Py, CH2Cl2
2. Ph3P n-Am
THF, -78°C → 25°C
1. HCl, THF/H2O, 47°C
2. MesSO2Cl, Py
Et2O, -20°C → 0°C
3. NaI, acetone
4. Ph3P, PhH, 40°C
(Ref. 7)
2 eq. n-BuLi
THF, -78°C;
HMPA; A
-78°C → 0°C
K2CO3, MeOH;
LiOH, H2O
λmax 259, 269, 280 nm (MeOH)

4. 12-(R)-Form of 6-(E) LTB_4 (Ref. 5):

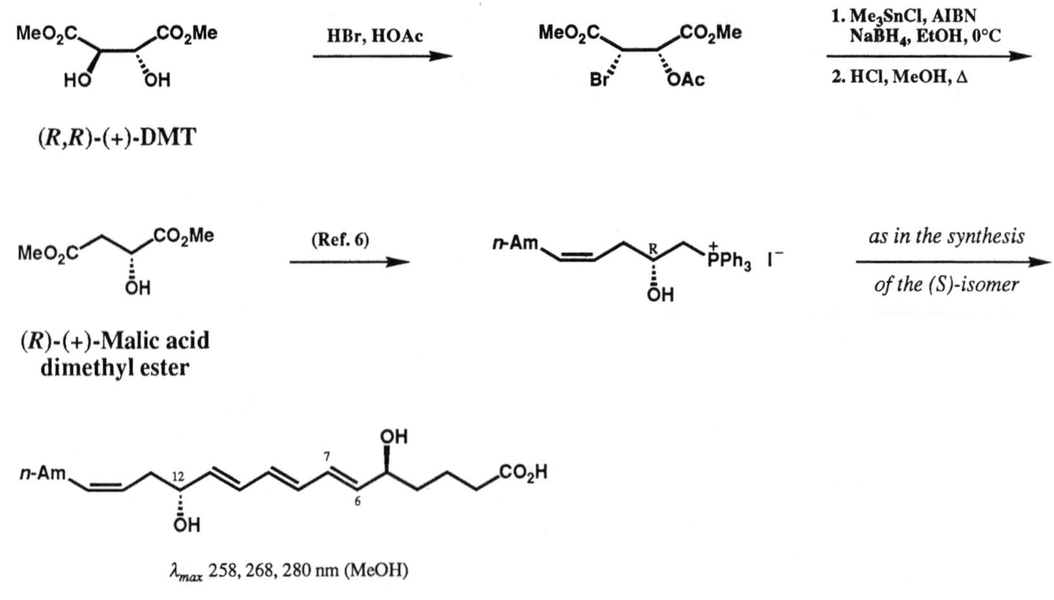
HBr, HOAc
1. Me3SnCl, AIBN
NaBH4, EtOH, 0°C
2. HCl, MeOH, Δ
(R,R)-(+)-DMT
(Ref. 6)
as in the synthesis
of the (S)-isomer
(R)-(+)-Malic acid dimethyl ester
λmax 258, 268, 280 nm (MeOH)

5. 5-(S),12-(S)-Dihydroxy-6-(E),8-(Z) isomer of LTB_4 (Ref. 8):

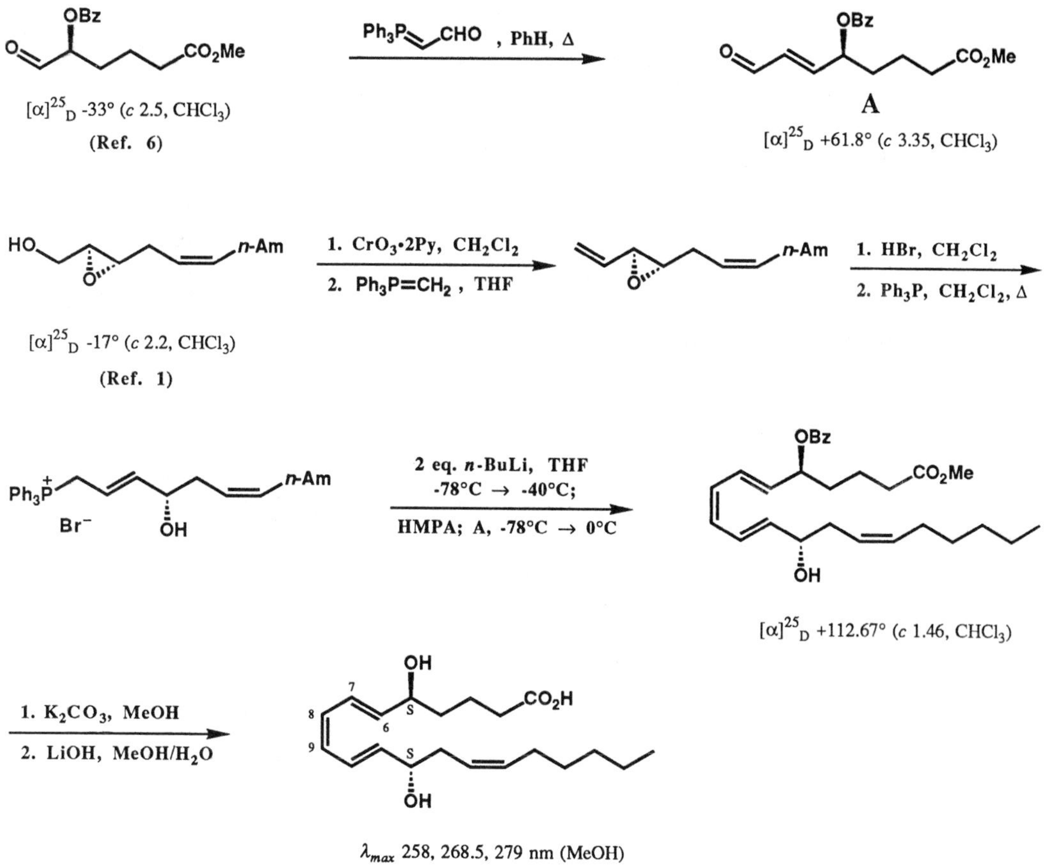
OBz
CO2Me
Ph3P=CHCHO , PhH, Δ
OBz
CO2Me
A
[α]25D -33° (c 2.5, CHCl3)
(Ref. 6)
[α]25D +61.8° (c 3.35, CHCl3)
HO
n-Am
1. CrO3•2Py, CH2Cl2
2. Ph3P=CH2 , THF
n-Am
1. HBr, CH2Cl2
2. Ph3P, CH2Cl2, Δ
[α]25D -17° (c 2.2, CHCl3)
(Ref. 1)
Ph3P+
Br−
OH
n-Am
2 eq. n-BuLi, THF
-78°C → -40°C;
HMPA; A, -78°C → 0°C
OBz
CO2Me
OH
[α]25D +112.67° (c 1.46, CHCl3)
1. K2CO3, MeOH
2. LiOH, MeOH/H2O
OH
CO2H
OH
λmax 258, 268.5, 279 nm (MeOH)

References:

1. E. J. Corey, P. B. Hopkins, J. E. Munroe, A. Marfat, and S.-i. Hashimoto, *J. Am. Chem. Soc.* **1980**, *102*, 7986.

2. E. J. Corey, D. A. Clark, G. Goto, A. Marfat, C. Mioskowski, B. Samuelsson, and S. Hammarström, *J. Am. Chem. Soc.* **1980**, *102*, 1436, 3663.

3. E. J. Corey, H. Niwa, and J. R. Falck, *J. Am. Chem. Soc.* **1979**, *101*, 1586.

4. E. J. Corey, J. O. Albright, A. E. Barton, and S.-i. Hashimoto, *J. Am. Chem. Soc.* **1980**, *102*, 1435.

5. E. J. Corey, A. Marfat, and D. J. Hoover, *Tetrahedron Lett.* **1981**, *22*, 1587.

6. E. J. Corey, A. Marfat, G. Goto, and F. Brion, *J. Am. Chem. Soc.* **1980**, *102*, 7984.

7. E. J. Corey, H. Niwa, and J. Knolle, *J. Am. Chem. Soc.* **1978**, *100*, 1942.

8. E. J. Corey, A. Marfat, and B. C. Laguzza, *Tetrahedron Lett.* **1981**, *22*, 3339.

12.6

Leukotriene B_5

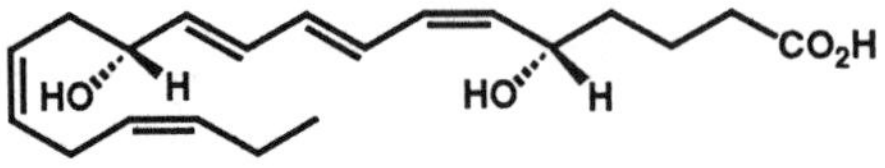

HO
H
HO
H
CO2H

Leukotriene B_5 can be biosynthesized in the body from eicosapentaenoic acid, which is ingested in the form of dietary fish lipid. Synthetic LTB_5 was synthesized as outlined below and found to have only 20% of the neutrophil chemotactic activity of LTB_4, a fact which may be relevant to the antiinflammatory effect of dietary marine lipid.

AcO
AcO
OAc
MCPBA, BF3•Et2O
CH2Cl2, -10°C
AcO
AcO
Zn/Hg, HCl
Et2O, 0°C

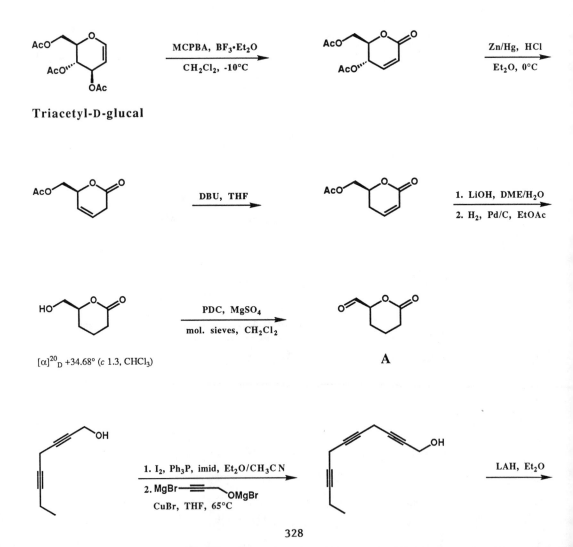

Triacetyl-D-glucal

AcO
DBU, THF
AcO
1. LiOH, DME/H2O
2. H2, Pd/C, EtOAc

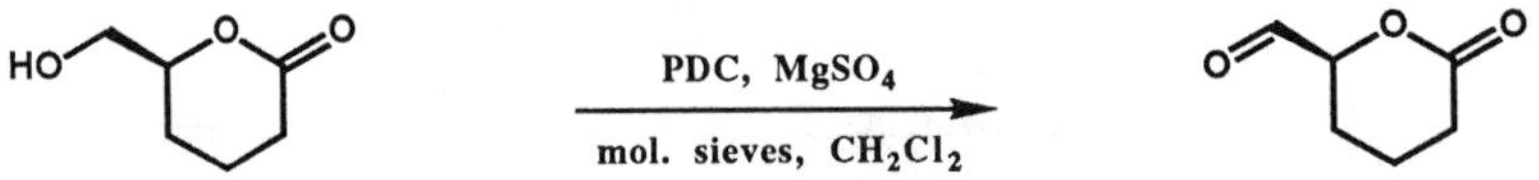

HO
PDC, MgSO4
mol. sieves, CH2Cl2

$[\alpha]^{20}_D$ +34.68° (*c* 1.3, $CHCl_3$)

A

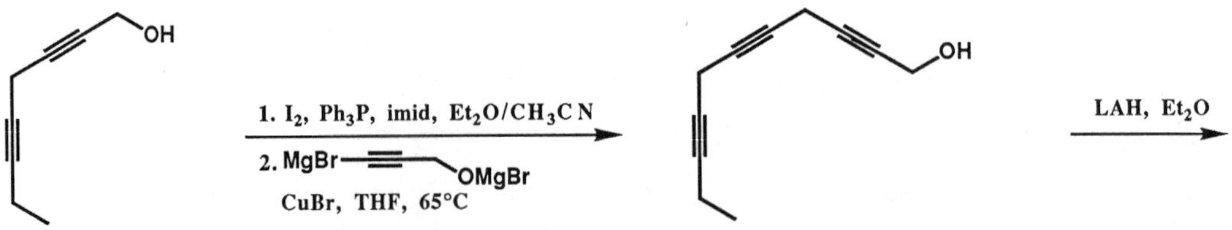

OH
1. I2, Ph3P, imid, Et2O/CH3CN
2. MgBr OMgBr
CuBr, THF, 65°C
OH
LAH, Et2O

12.6

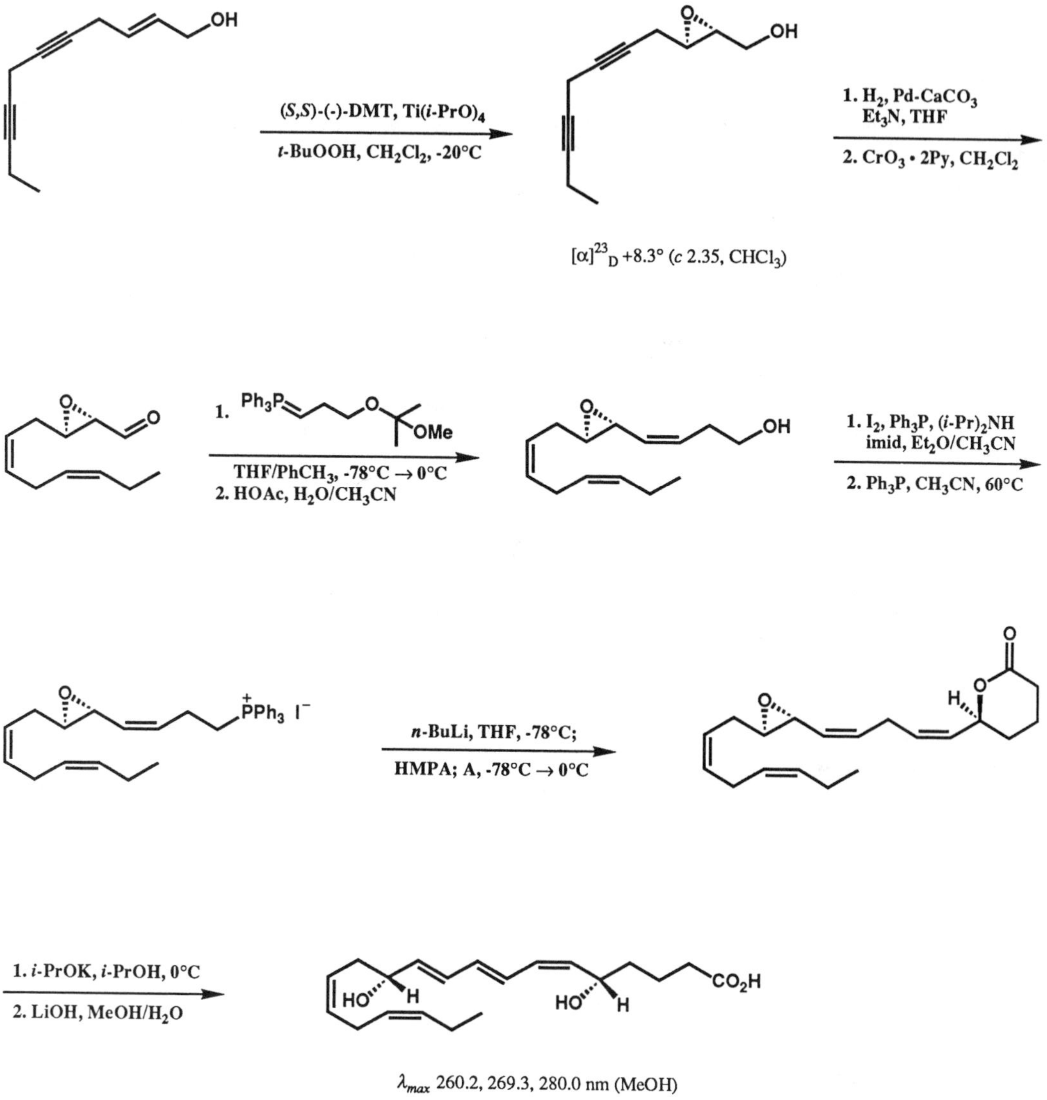
OH
(S,S)-(-)-DMT, Ti(i-PrO)4
t-BuOOH, CH2Cl2, -20°C
OH
1. H2, Pd-CaCO3
Et3N, THF
2. CrO3 • 2Py, CH2Cl2
[α]23D +8.3° (c 2.35, CHCl3)
1. Ph3P
OMe
THF/PhCH3, -78°C → 0°C
2. HOAc, H2O/CH3CN
OH
1. I2, Ph3P, (i-Pr)2NH
imid, Et2O/CH3CN
2. Ph3P, CH3CN, 60°C
PPh3 I⁻
n-BuLi, THF, -78°C;
HMPA; A, -78°C → 0°C
1. i-PrOK, i-PrOH, 0°C
2. LiOH, MeOH/H2O
HO H
HO H
CO2H
λmax 260.2, 269.3, 280.0 nm (MeOH)

Reference:

E. J. Corey, S. G. Pyne, and W.-g. Su, *Tetrahedron Lett.* **1983**, *24*, 4883.

12.7

5-Desoxyleukotriene D_4

5-Desoxyleukotriene D_4 was synthesized to determine whether the 5-hydroxyl group is necessary for biological activity. It is, since the bioactivity of 5-desoxyleukotriene D_4 is less than 1% that of LTD_4 itself. An interesting synthetic equivalent of the 4-formyl-*E,E*-1,3-butadienyl anion was utilized in the synthesis.

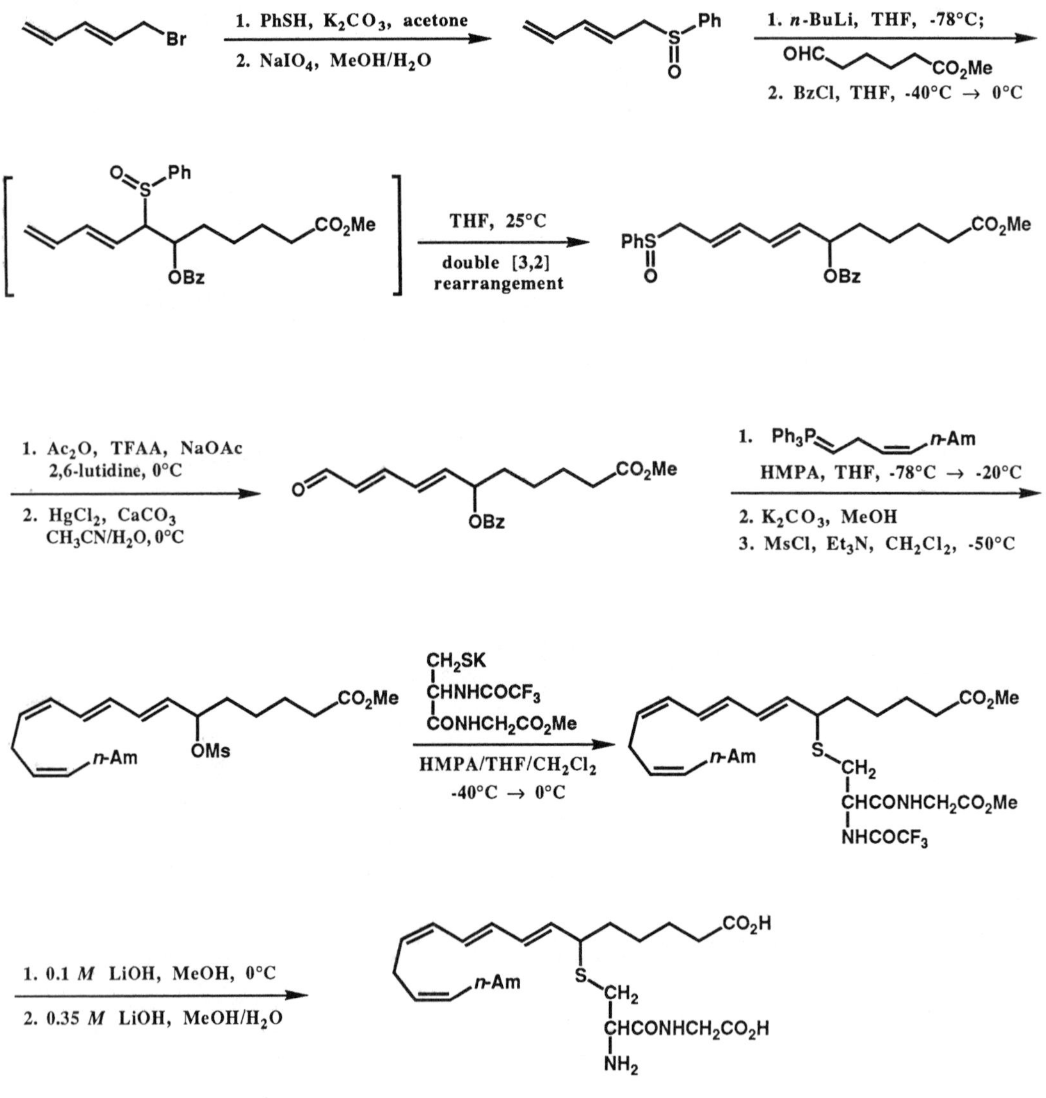

Br
1. PhSH, K2CO3, acetone
2. NaIO4, MeOH/H2O
S
Ph
O
1. n-BuLi, THF, -78°C;
OHC
CO2Me
2. BzCl, THF, -40°C → 0°C
O
S
Ph
CO2Me
OBz
THF, 25°C
double [3,2] rearrangement
PhS
O
CO2Me
OBz
1. Ac2O, TFAA, NaOAc 2,6-lutidine, 0°C
2. HgCl2, CaCO3 CH3CN/H2O, 0°C
O
CO2Me
OBz
1. Ph3P
n-Am
HMPA, THF, -78°C → -20°C
2. K2CO3, MeOH
3. MsCl, Et3N, CH2Cl2, -50°C
CO2Me
n-Am
OMs
CH2SK
CHNHCOCF3
CONHCH2CO2Me
HMPA/THF/CH2Cl2
-40°C → 0°C
CO2Me
n-Am
S
CH2
CHCONHCH2CO2Me
NHCOCF3
1. 0.1 M LiOH, MeOH, 0°C
2. 0.35 M LiOH, MeOH/H2O
CO2H
n-Am
S
CH2
CHCONHCH2CO2H
NH2

Reference:

E. J. Corey and D. J. Hoover, *Tetrahedron Lett.* **1982**, *23*, 3463.

Synthesis of the 11,12-Oxido and 14,15-Oxido Analogs of Leukotriene A_4 and the Corresponding Analogs of Leukotrienes C_4 and D_4

The 11,12-oxido and 14,15-oxido analogs of leukotriene A_4 were synthesized to help answer the question of whether these compounds might be biosynthesized from arachidonate by the 12- and 15-lipoxygenation pathways and serve as physiologic regulators. Hydrolysis products of the 14,15-oxide were later found to be formed in biological systems.

1. 11,12-Oxido analog of LTA_4 and the corresponding analogs of LTC_4 and LTD_4 (Ref. 1):

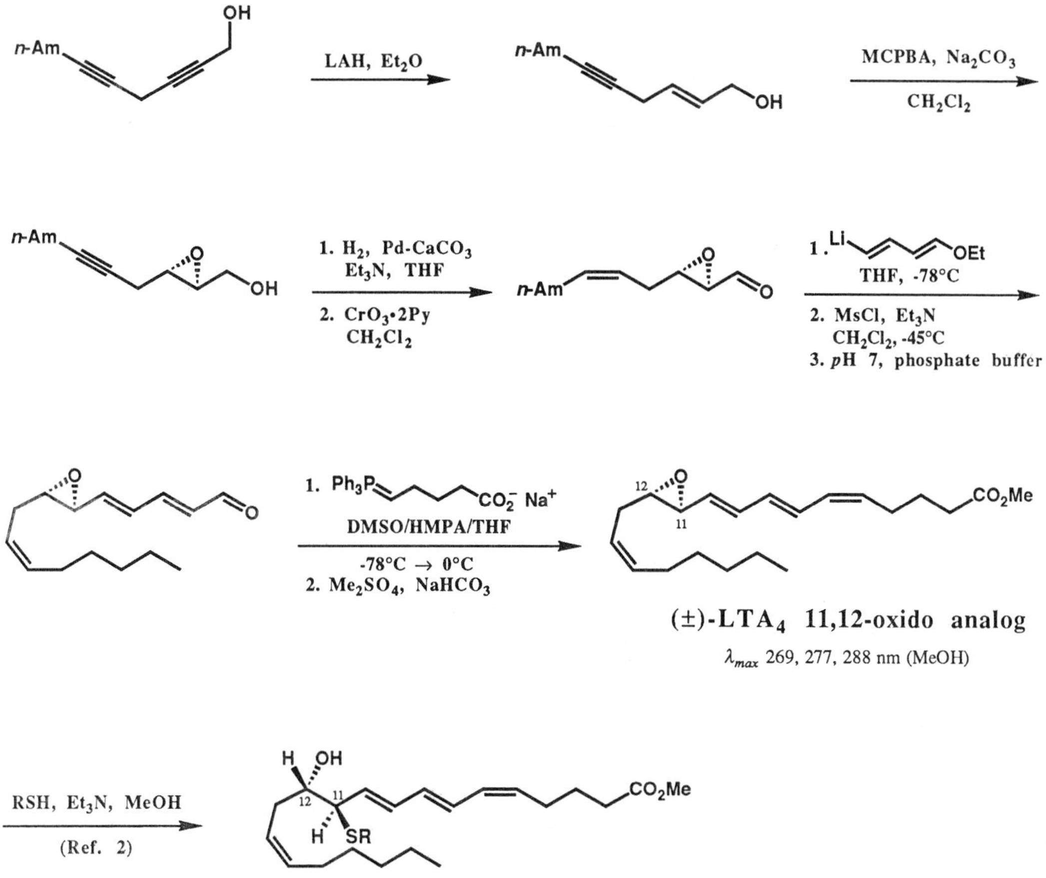
n-Am
OH
LAH, Et2O
MCPBA, Na2CO3
CH2Cl2
1. H2, Pd-CaCO3
Et3N, THF
2. CrO3•2Py
CH2Cl2
1. Li OEt
THF, -78°C
2. MsCl, Et3N
CH2Cl2, -45°C
3. pH 7, phosphate buffer
1. Ph3P CO2- Na+
DMSO/HMPA/THF
-78°C → 0°C
2. Me2SO4, NaHCO3
12
11
CO2Me
(±)-LTA4 11,12-oxido analog
λmax 269, 277, 288 nm (MeOH)
RSH, Et3N, MeOH
(Ref. 2)
H OH
H SR

SR = *S*-glutathionyl (LTC_4 analog)

or

SR = *S*-cysteinylglycyl (LTD_4 analog)

12.8

2. 14,15-Oxido analog of LTA_4 and the corresponding analogs of LTC_4 and LTD_4:

Method 1 (Ref. 1):

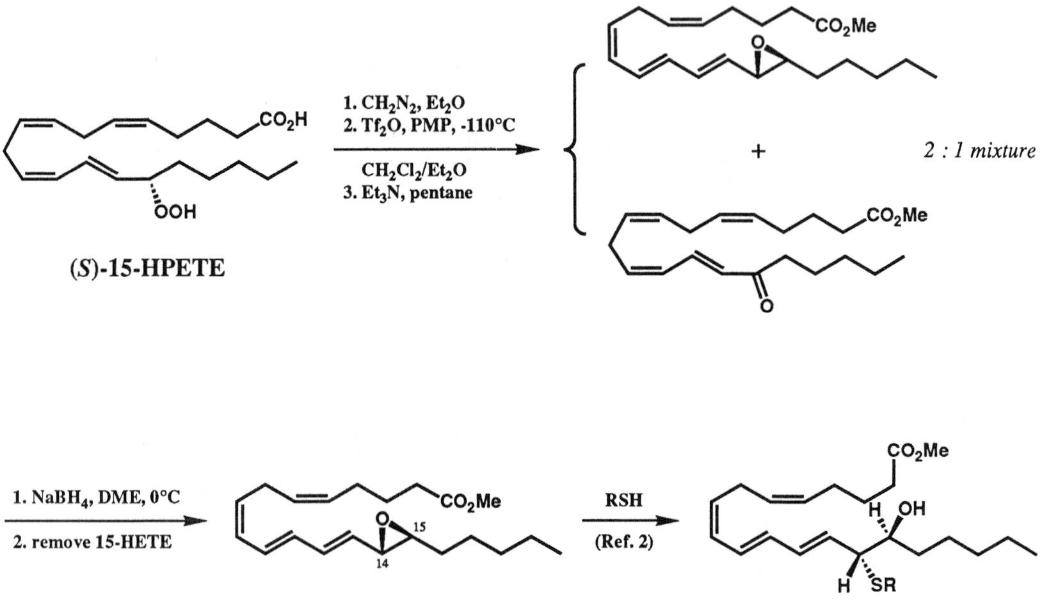

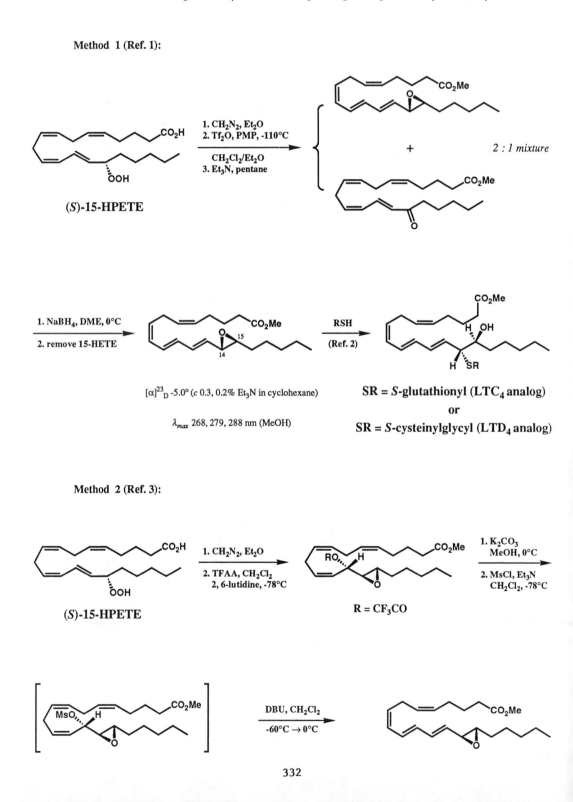

CO2H
OOH
(S)-15-HPETE
1. CH2N2, Et2O
2. Tf2O, PMP, -110°C
CH2Cl2/Et2O
3. Et3N, pentane
CO2Me
O
+
2 : 1 mixture
CO2Me
O
1. NaBH4, DME, 0°C
2. remove 15-HETE
CO2Me
O
15
14
RSH
(Ref. 2)
CO2Me
H OH
H SR

$[\alpha]^{23}_D$ -5.0° (*c* 0.3, 0.2% Et_3N in cyclohexane)

λ_{max} 268, 279, 288 nm (MeOH)

SR = *S*-glutathionyl (LTC_4 analog)

or

SR = *S*-cysteinylglycyl (LTD_4 analog)

Method 2 (Ref. 3):

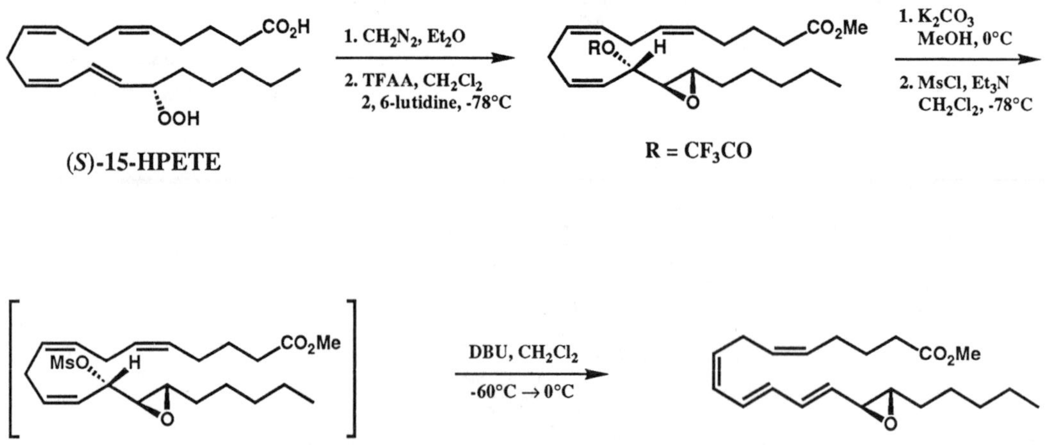

CO2H
OOH
(S)-15-HPETE
1. CH2N2, Et2O
2. TFAA, CH2Cl2
2, 6-lutidine, -78°C
RO
H
O
CO2Me
R = CF3CO
1. K2CO3
MeOH, 0°C
2. MsCl, Et3N
CH2Cl2, -78°C
MsO
H
O
CO2Me
DBU, CH2Cl2
-60°C → 0°C
CO2Me
O

Method 3 (Ref. 4):

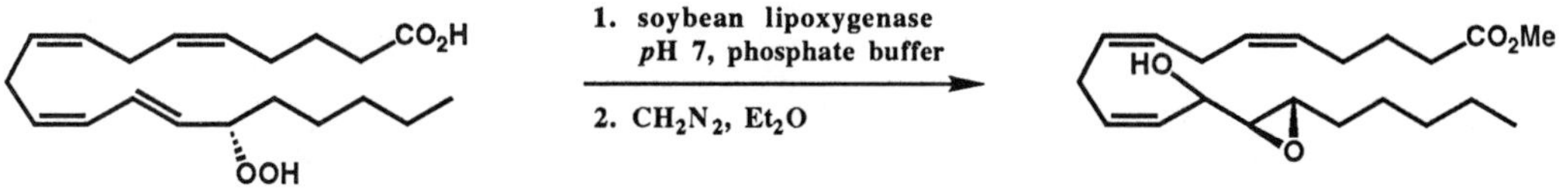
CO2H
OOH
1. soybean lipoxygenase
pH 7, phosphate buffer
2. CH2N2, Et2O
HO
CO2Me
O

(*S*)-15-HPETE

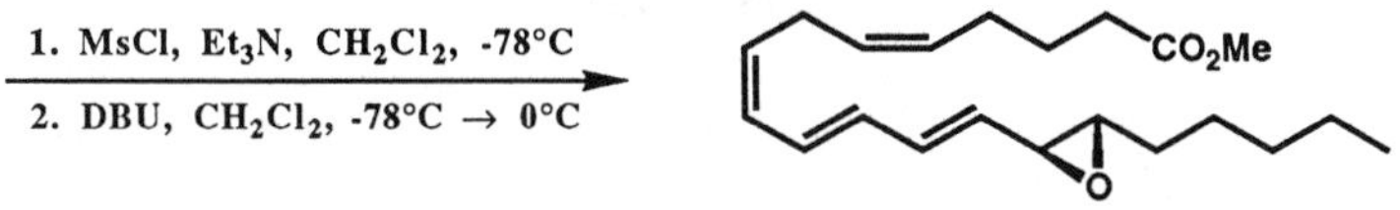
1. MsCl, Et3N, CH2Cl2, -78°C
2. DBU, CH2Cl2, -78°C → 0°C
CO2Me
O

References:

1. E. J. Corey, A. Marfat, and G. Goto, *J. Am. Chem. Soc.* **1980**, *102*, 6607.
2. E. J. Corey, D. A. Clark, G. Goto, A. Marfat, C. Mioskowski, B. Samuelsson, and S. Hammarström, *J. Am. Chem. Soc.* **1980**, *102*, 1436, 3663.
3. E. J. Corey, W.-g. Su, and M. M. Mehrotra, *Tetrahedron Lett.* **1984**, *25*, 5123.
4. E. J. Corey, M. M. Mehrotra, and J. R. Cashman, *Tetrahedron Lett.* **1983**, *24*, 4917.

12.9

12-Hydroxy-5,8,14-(*Z*)-10-(*E*)-eicosatetraenoic Acid (12-HETE)

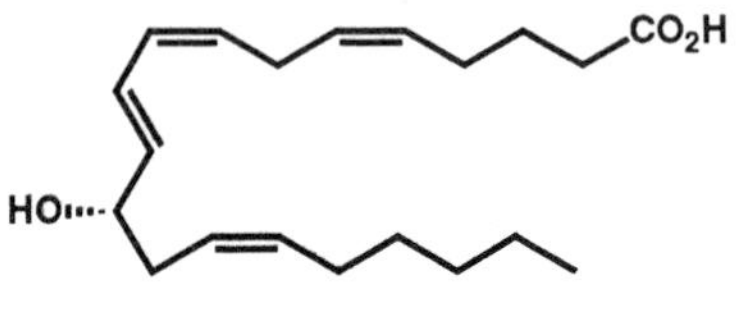
CO2H
HO

12-Lipoxygenation is the major pathway of dioxygenation of arachidonic acid in blood platelets and leads to the 12-*S*-hydroperoxy acid 12-HPETE and the corresponding 12-hydroxy acid 12-HETE. Several pathways for the synthesis of 12-HETE have been developed. However, despite the availability of this substance, its biological role remains undetermined.

1. Stereospecific synthesis (Ref. 1):

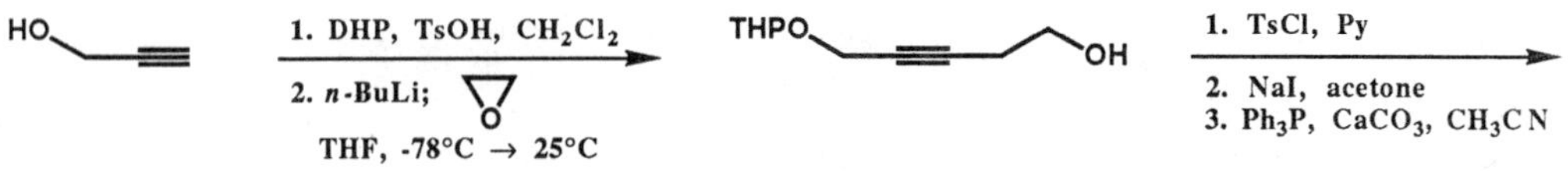
HO
1. DHP, TsOH, CH2Cl2
2. n-BuLi;
THF, -78°C → 25°C
THPO
OH
1. TsCl, Py
2. NaI, acetone
3. Ph3P, CaCO3, CH3CN

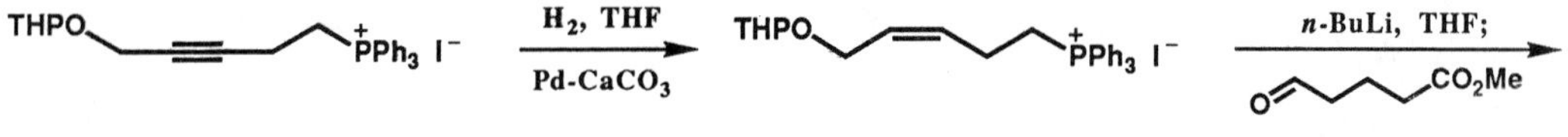
THPO
PPh3 I⁻
H2, THF
Pd-CaCO3
THPO
PPh3 I⁻
n-BuLi, THF;
CO2Me

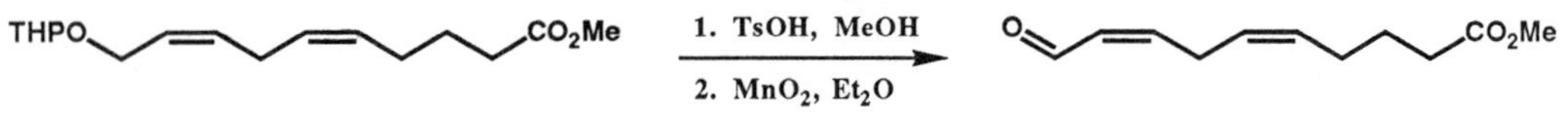
THPO
CO2Me
1. TsOH, MeOH
2. MnO2, Et2O
O
CO2Me

A

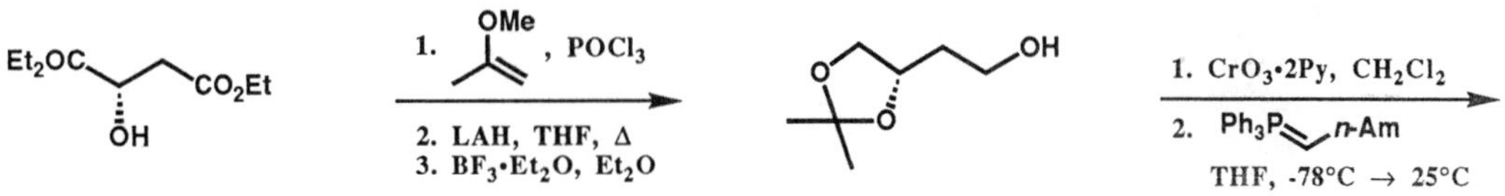
Et2OC
CO2Et
OH
OMe
1. , POCl3
2. LAH, THF, Δ
3. BF3•Et2O, Et2O
O
O
OH
1. CrO3•2Py, CH2Cl2
2. Ph3P n-Am
THF, -78°C → 25°C

(*S*)-(-)-Malic acid diethyl ester

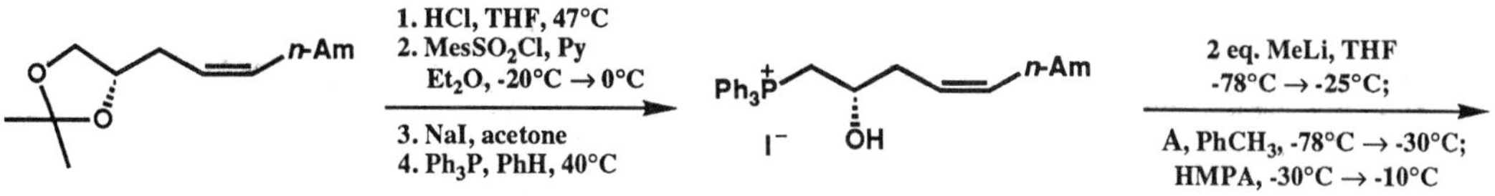
n-Am
1. HCl, THF, 47°C
2. MesSO2Cl, Py
Et2O, -20°C → 0°C
3. NaI, acetone
4. Ph3P, PhH, 40°C
Ph3P+
I−
OH
n-Am
2 eq. MeLi, THF
-78°C → -25°C;
A, PhCH3, -78°C → -30°C;
HMPA, -30°C → -10°C

CO2Me
HO
LiOH, EtOH
CO2H
HO
12

λ_{max} 237 nm (MeOH)

$[\alpha]^{25}_{D}$ +1.50° (*c* 0.2, $CHCl_3$)

(*S*)-12-HETE

2. Synthesis of (±)-12-HETE from the corresponding ketone (Ref. 2):

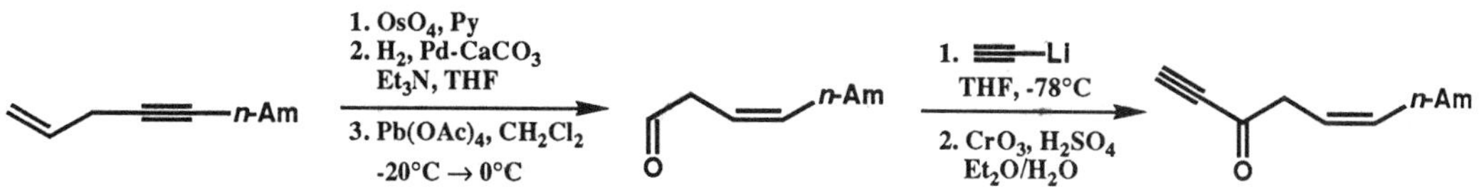
n-Am
1. OsO4, Py
2. H2, Pd-CaCO3
Et3N, THF
3. Pb(OAc)4, CH2Cl2
-20°C → 0°C
O
n-Am
1. ≡—Li
THF, -78°C
2. CrO3, H2SO4
Et2O/H2O
O
n-Am

B

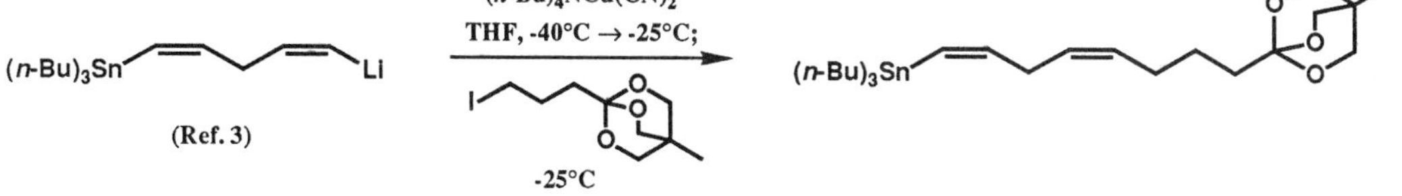
(n-Bu)3Sn
Li
(Ref. 3)
(n-Bu)4NCu(CN)2
THF, -40°C → -25°C;
I
-25°C
(n-Bu)3Sn

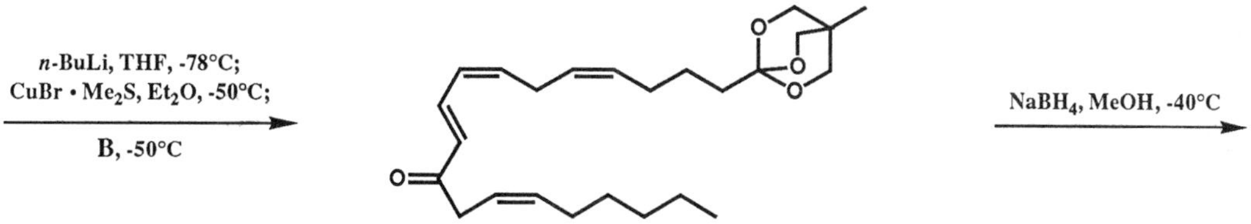
n-BuLi, THF, -78°C;
CuBr • Me2S, Et2O, -50°C;
B, -50°C
O
NaBH4, MeOH, -40°C

12.9

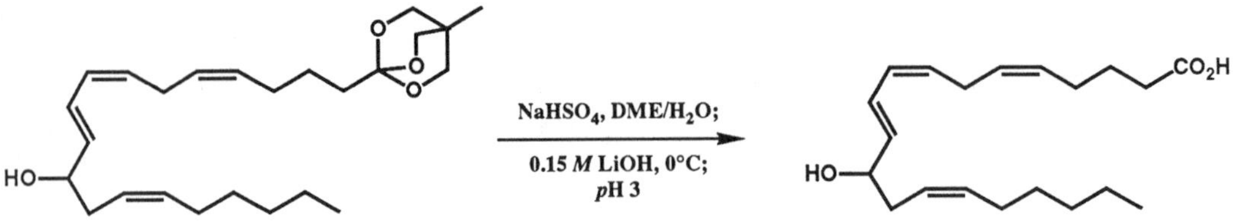
HO
NaHSO4, DME/H2O;
0.15 M LiOH, 0°C;
pH 3
HO
CO2H

(±)-12-HETE

3. Synthesis of (±)-12-HETE from 14,15-epoxyarachidonic acid (Ref. 4):

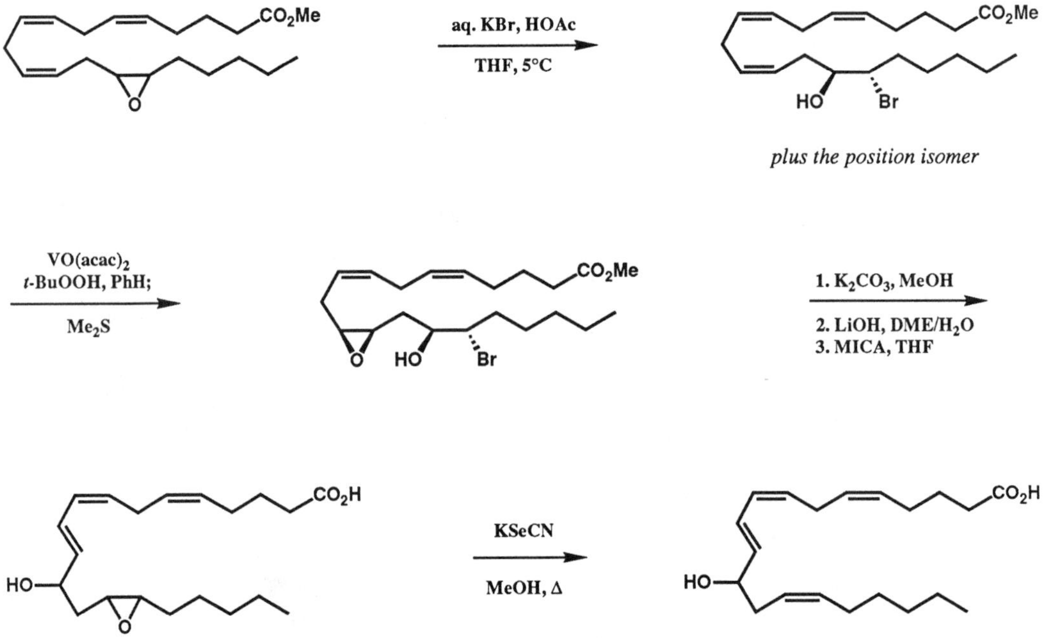
CO2Me
O
aq. KBr, HOAc
THF, 5°C
CO2Me
HO
Br
plus the position isomer
VO(acac)2
t-BuOOH, PhH;
Me2S
CO2Me
O
HO
Br
1. K2CO3, MeOH
2. LiOH, DME/H2O
3. MICA, THF
CO2H
HO
O
KSeCN
MeOH, Δ
CO2H
HO

(±)-12-HETE

(±)-12-HETE was also obtained together with (±)-11-HETE *via* the epoxide opening reaction of 11,12-epoxyarachidonic acid (Ref. 4).

References:

1. E. J. Corey, H. Niwa, and J. Knolle, *J. Am. Chem. Soc.* **1978**, *100*, 1942.
2. E. J. Corey, K. Kyler, and N. Raju, *Tetrahedron Lett.* **1984**, *25*, 5115.
3. E. J. Corey and J. Kang, *Tetrahedron Lett.* **1982**, *23*, 1651.
4. E. J. Corey, A. Marfat, J. R. Falck, and J. O. Albright, *J. Am. Chem. Soc.* **1980**, *102*, 1433.

Total Synthesis of Hepoxylins and Related Metabolites of Arachidonic Acid

Hepoxylins are metabolites of arachidonic acid which arise from 12-HPETE in tissues such as pancreatic islet cells (where they stimulate glucose-dependent insulin release) and brain (where they appear to have a neuromodulatory role). The structure of the hepoxylins was confirmed by synthesis which also has provided this scarce material for biological investigation.

1. Synthesis of 8-hydroxy-11,12(*S*,*S*)-epoxyeicosa-5,14-(*Z*)-9-(*E*)-trienoic acids (hepoxylins) (Ref. 1):

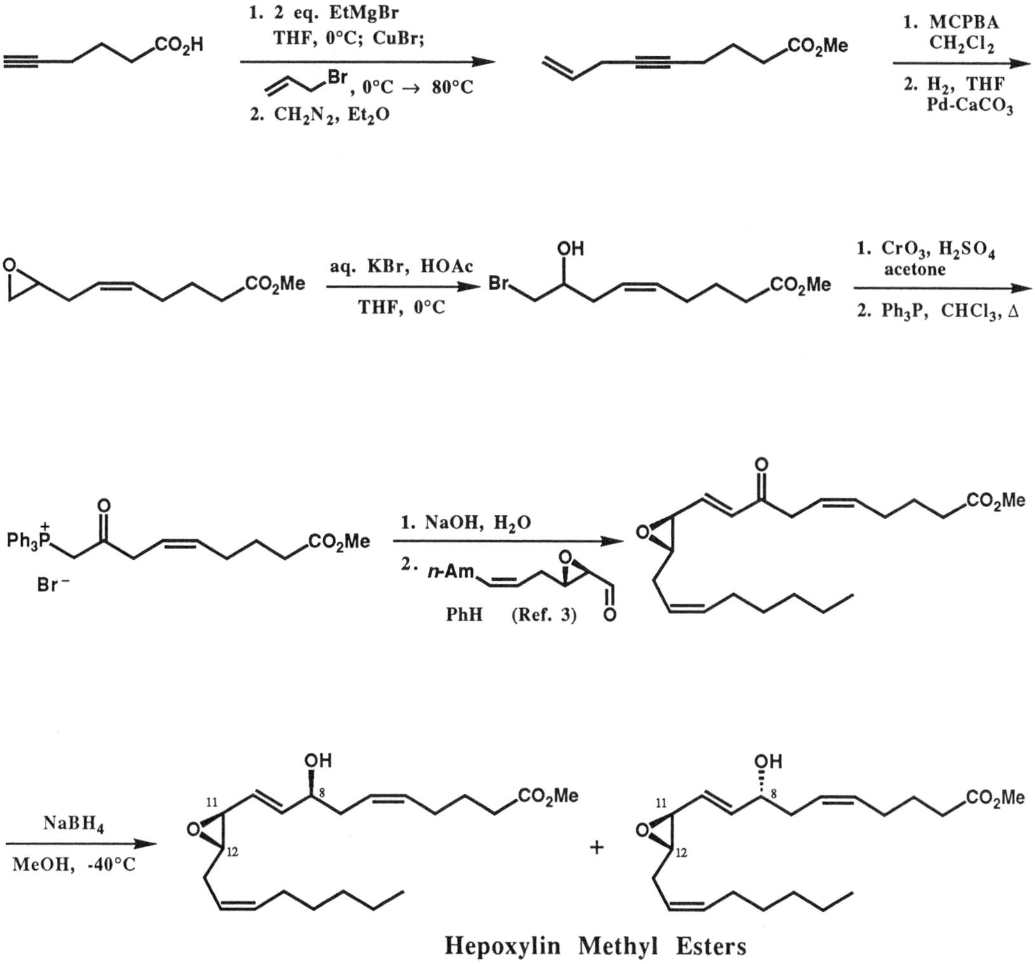

CO2H
1. 2 eq. EtMgBr THF, 0°C; CuBr;
Br, 0°C → 80°C
2. CH2N2, Et2O
CO2Me
1. MCPBA CH2Cl2
2. H2, THF Pd-CaCO3
O
CO2Me
aq. KBr, HOAc
THF, 0°C
Br
OH
CO2Me
1. CrO3, H2SO4 acetone
2. Ph3P, CHCl3, Δ
Ph3P
Br⁻
O
CO2Me
1. NaOH, H2O
2. n-Am
PhH (Ref. 3)
O
CO2Me
NaBH4
MeOH, -40°C
OH
11
8
12
CO2Me
+

Hepoxylin Methyl Esters

(*separable by SGC*)

12.10

2. Synthesis of 12-(*S*)-10-hydroxy-*trans*-11,12-epoxyeicosa-5,9,14-(*Z*)-trienoic acids, position isomers of hepoxylins isolated from blood platelets (Ref. 2):

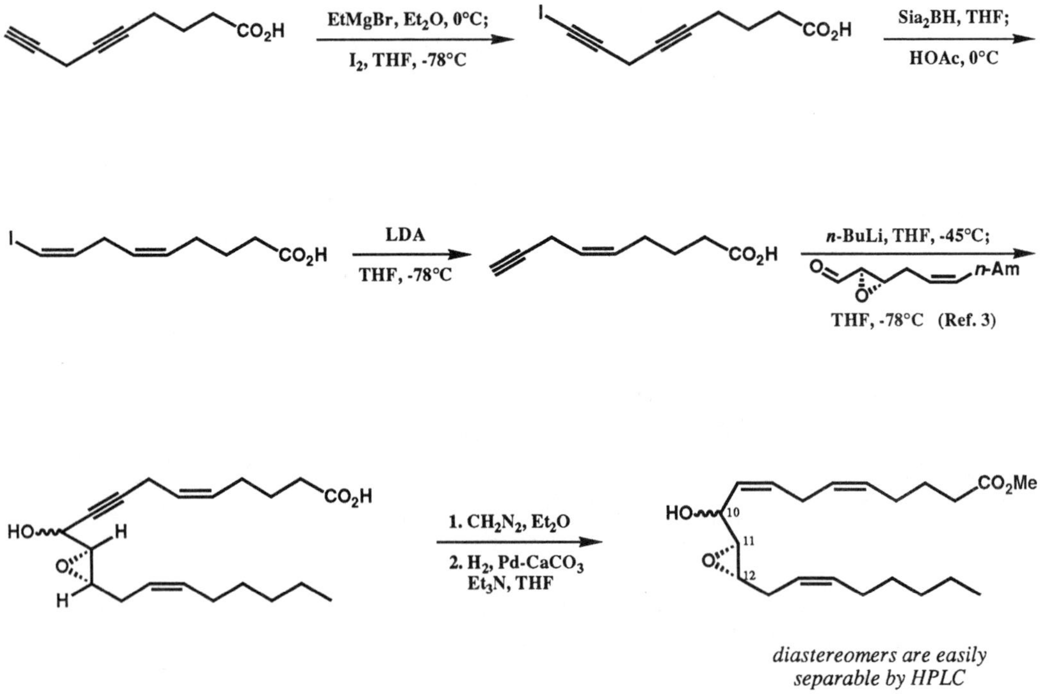
CO2H
EtMgBr, Et2O, 0°C;
I2, THF, -78°C
I
CO2H
Sia2BH, THF;
HOAc, 0°C
I
CO2H
LDA
THF, -78°C
CO2H
n-BuLi, THF, -45°C;
n-Am
THF, -78°C (Ref. 3)
CO2H
HO
H
H
1. CH2N2, Et2O
2. H2, Pd-CaCO3
Et3N, THF
CO2Me
HO
10
11
12
diastereomers are easily
separable by HPLC

References:

1. E. J. Corey and W.-g. Su, *Tetrahedron Lett.* **1984**, *25*, 5119.
2. E. J. Corey, J. Kang, and B. C. Laguzza, *Tetrahedron Lett.* **1983**, *24*, 4913.
3. E. J. Corey, A. Marfat, and B. C. Laguzza, *Tetrahedron Lett.* **1981**, *22*, 3339.

Synthesis of 5-, 11-, and 15-HETE's.
Conversion of HETE's into the Corresponding HPETE's

The first step in the biosynthesis of eicosanoids from arachidonic acid is generally a lipoxygenation reaction. The resulting hydroperoxides (HPETE's) can undergo reduction to the corresponding alcohols (HETE's). Preparative routes to the 5-, 11-, and 15-HETE's and HPETE's have been developed as outline below.

1. 5-HETE and 5-HPETE (Ref. 1):

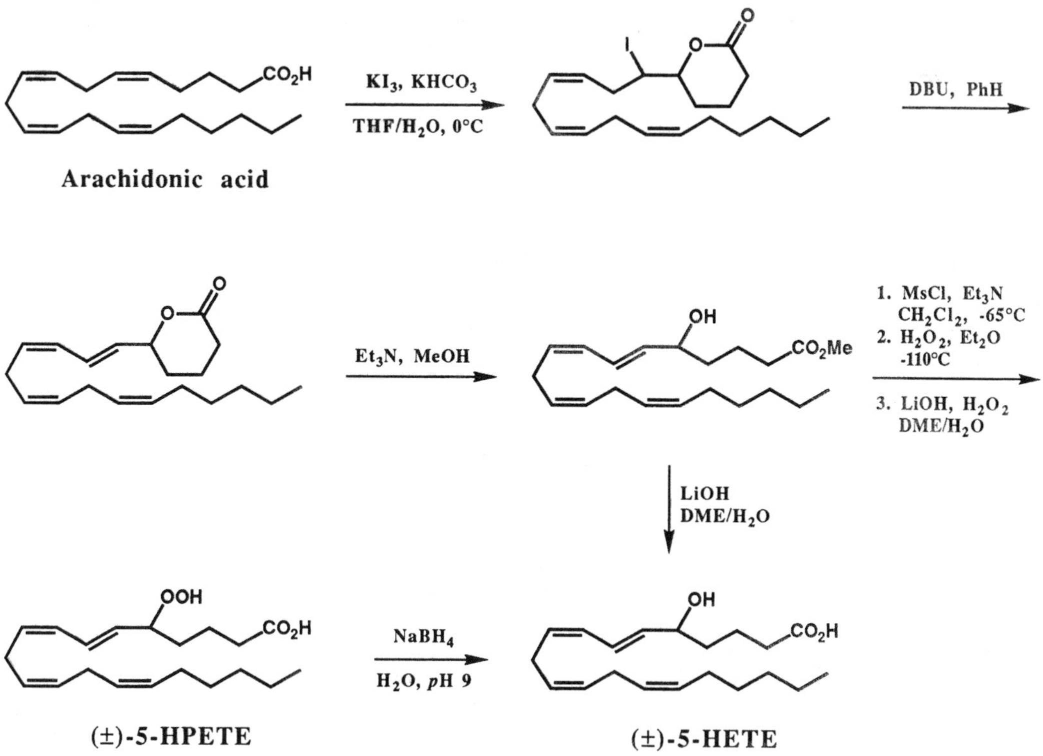
CO2H
Arachidonic acid
KI3, KHCO3
THF/H2O, 0°C
I
O
DBU, PhH
Et3N, MeOH
OH
CO2Me
1. MsCl, Et3N
CH2Cl2, -65°C
2. H2O2, Et2O
-110°C
3. LiOH, H2O2
DME/H2O
LiOH
DME/H2O
OOH
CO2H
NaBH4
H2O, pH 9
OH
CO2H
(±)-5-HPETE
(±)-5-HETE

Enzymic synthesis of (*S*)-5-HPETE:

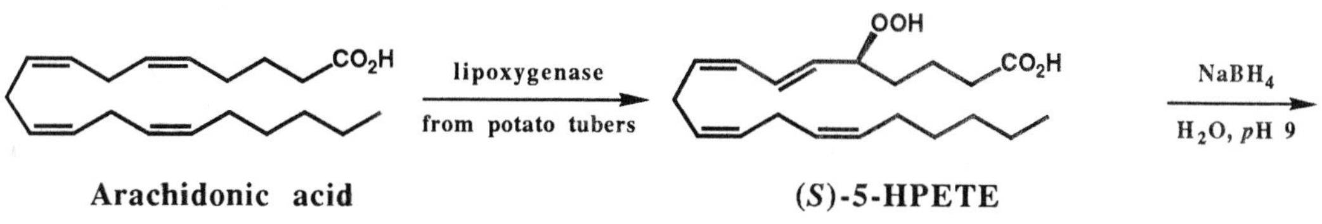
CO2H
lipoxygenase
from potato tubers
OOH
CO2H
NaBH4
H2O, pH 9
Arachidonic acid
(S)-5-HPETE

12.11

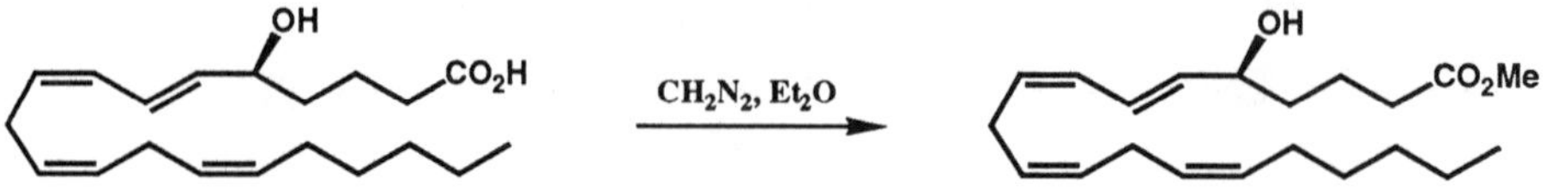
OH
CO2H
CH2N2, Et2O
OH
CO2Me

(*S*)-5-HETE

$[\alpha]^{23}_{436}$ +12.42° (*c* 0.99,EtOH)

$[\alpha]^{23}_{D}$ +4.73° (*c* 0.99, EtOH)

Optically pure 5-HETE can be made in quantity by resolution of racemic 5-HETE (Ref. 2).

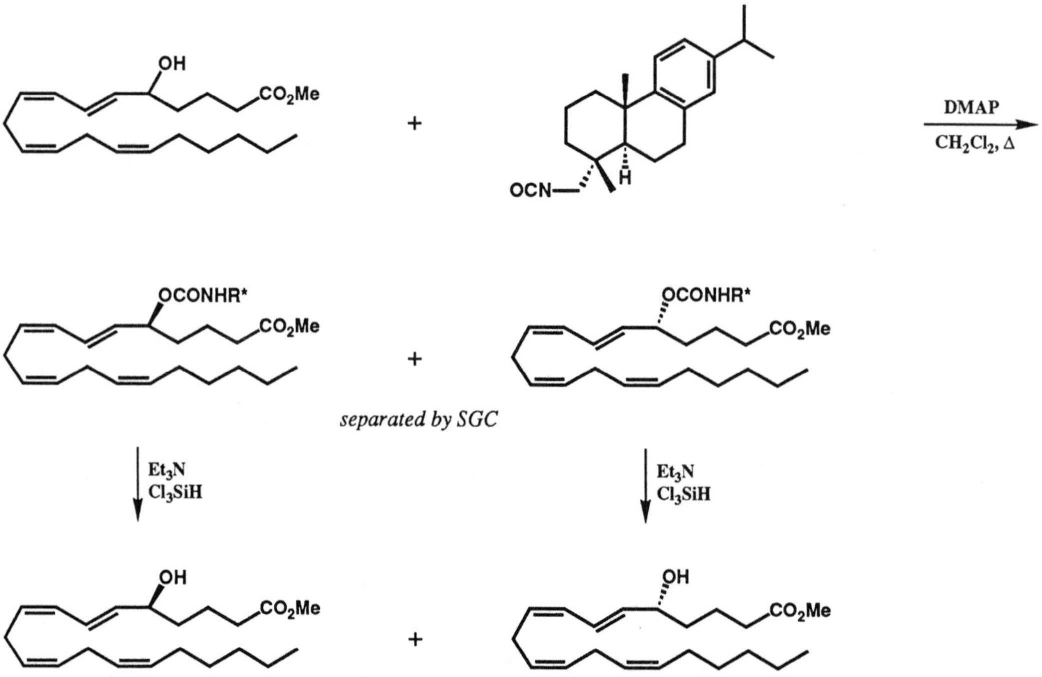
OH
CO2Me
+
OCN
H
DMAP
CH2Cl2, Δ
OCONHR*
CO2Me
+
OCONHR*
CO2Me
separated by SGC
Et3N
Cl3SiH
Et3N
Cl3SiH
OH
CO2Me
+
OH
CO2Me

(*S*)-(+)-5-HETE methyl ester

$[\alpha]^{23}_{D}$ +14.0° (*c* 2.0, PhH)

(*R*)-(-)-5-HETE methyl ester

$[\alpha]^{23}_{D}$ -13.5° (*c* 2.0, PhH)

2. (*R*)-11-HETE (Ref. 3):

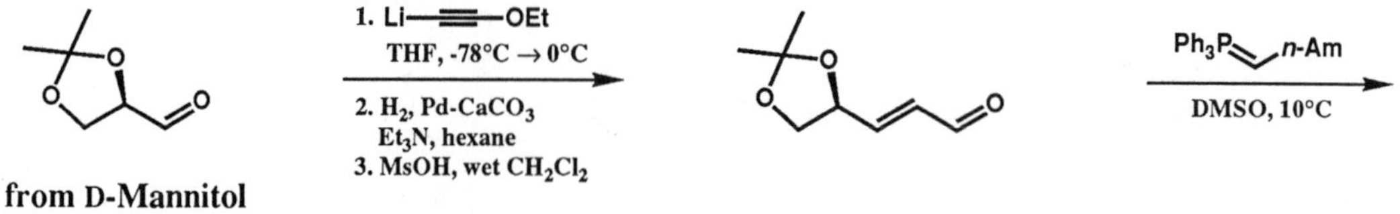
1. Li—≡—OEt
THF, -78°C → 0°C
2. H2, Pd-CaCO3
Et3N, hexane
3. MsOH, wet CH2Cl2
Ph3P n-Am
DMSO, 10°C

from D-Mannitol

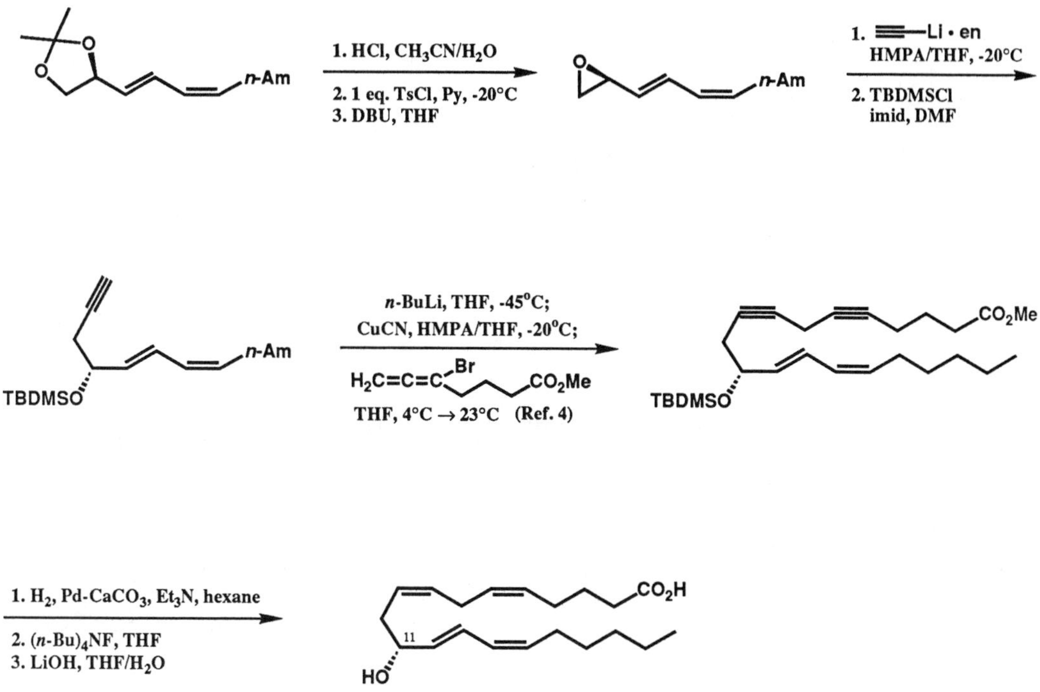
n-Am
1. HCl, CH3CN/H2O
2. 1 eq. TsCl, Py, -20°C
3. DBU, THF
n-Am
1. ≡—Li • en
HMPA/THF, -20°C
2. TBDMSCl
imid, DMF
TBDMSO
n-Am
n-BuLi, THF, -45°C;
CuCN, HMPA/THF, -20°C;
H2C=C=C Br CO2Me
THF, 4°C → 23°C (Ref. 4)
CO2Me
TBDMSO
1. H2, Pd-CaCO3, Et3N, hexane
2. (n-Bu)4NF, THF
3. LiOH, THF/H2O
CO2H
11
HO

(*R*)-11-HETE

(±)-11, and 12-HETE were obtained concurrently from 11,12-oxido-arachidonic acid (Ref. 5):

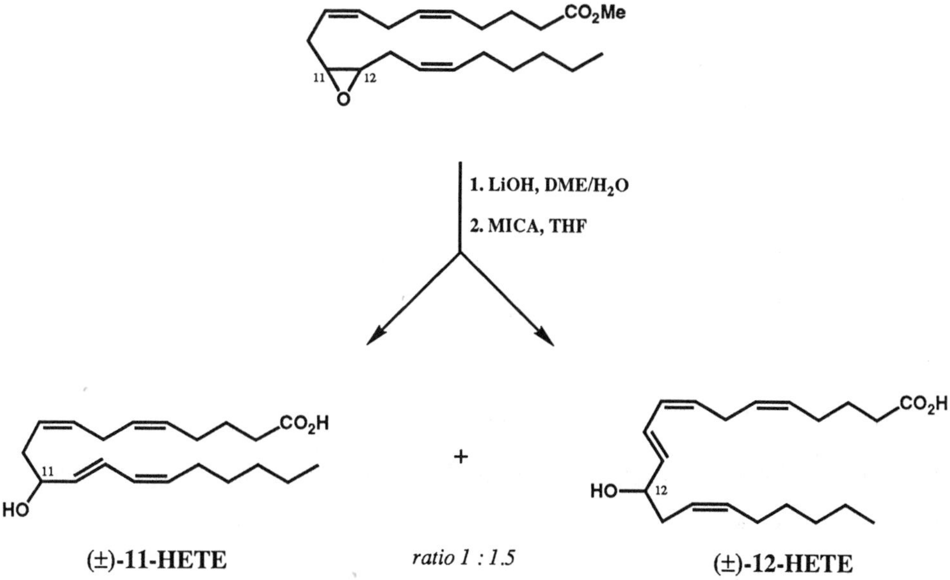
CO2Me
11
12
O
1. LiOH, DME/H2O
2. MICA, THF
CO2H
11
HO
+
CO2H
HO
12

(±)-11-HETE *ratio 1 : 1.5* **(±)-12-HETE**

12.11

3. 15-HETE (Ref. 5):

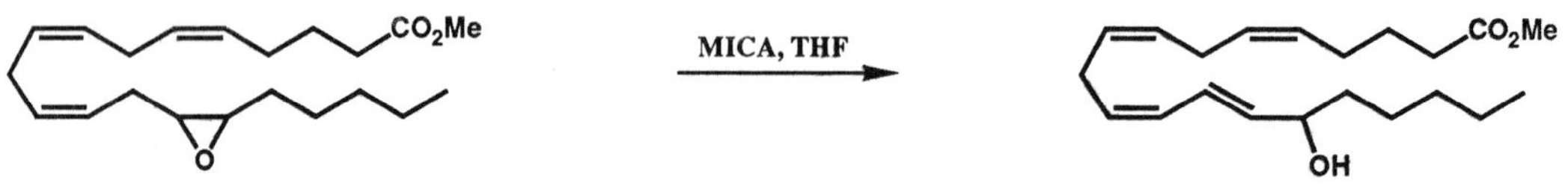
CO2Me
O
MICA, THF
CO2Me
OH

(Ref. 6) (±)-15-HETE methyl ester

Each HETE can be converted into the corresponding HPETE as shown in the following example.

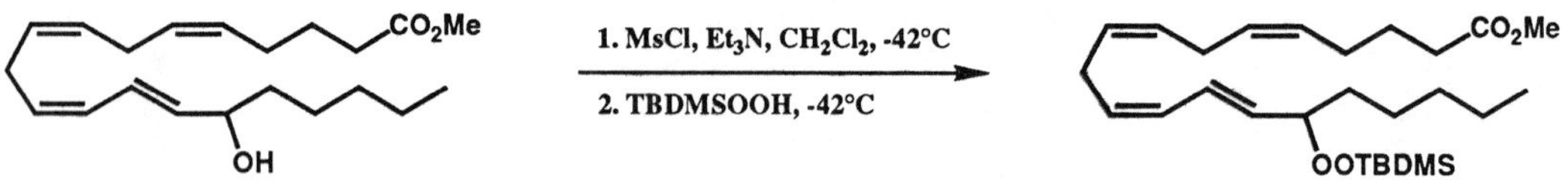
CO2Me
OH
1. MsCl, Et3N, CH2Cl2, -42°C
2. TBDMSOOH, -42°C
CO2Me
OOTBDMS

15-HETE methyl ester

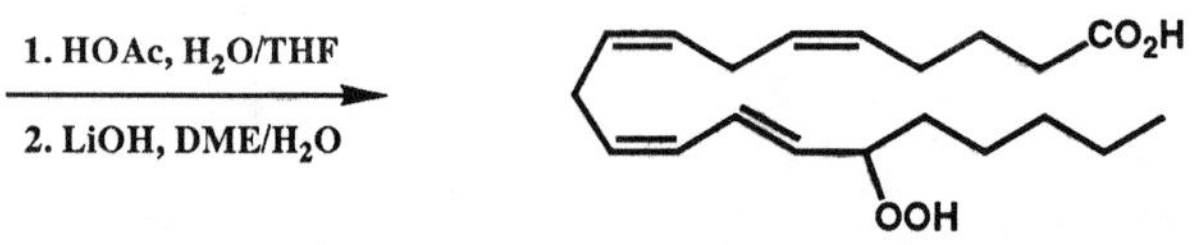
1. HOAc, H2O/THF
2. LiOH, DME/H2O
CO2H
OOH

15-HPETE

References:

1. E. J. Corey, J. O. Albright, A. E. Barton, and S.-i. Hashimoto, *J. Am. Chem. Soc.* **1980**, *102*, 1435.
2. E. J. Corey and S.-i. Hashimoto, *Tetrahedron Lett.* **1981**, *22*, 299.
3. E. J. Corey and J. Kang, *J. Am. Chem. Soc.* **1981**, *103*, 4618.
4. E. J. Corey and J. Kang, *Tetrahedron Lett.* **1982**, *23*, 1651.
5. E. J. Corey, A. Marfat, J. R. Falck, and J. O. Albright, *J. Am. Chem. Soc.* **1980**, *102*, 1433.
6. E. J. Corey, H. Niwa, and J. R. Falck, *J. Am. Chem. Soc.* **1979**, *101*, 1586.

Selective Epoxidation of Arachidonic Acid

1. 5,6-Oxido-arachidonic acid (Ref. 1):

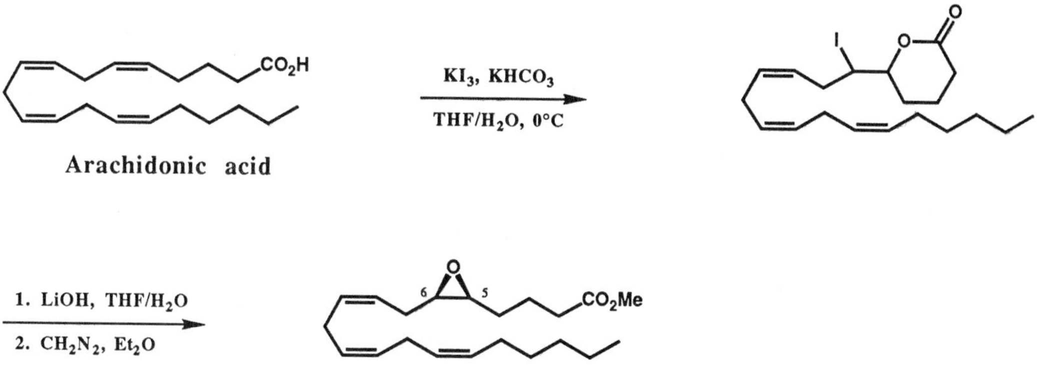

CO2H
Arachidonic acid
KI3, KHCO3
THF/H2O, 0°C
I
O
O
1. LiOH, THF/H2O
2. CH2N2, Et2O
O
6
5
CO2Me

2. 14,15-Oxido-arachidonic acid (Ref. 1):

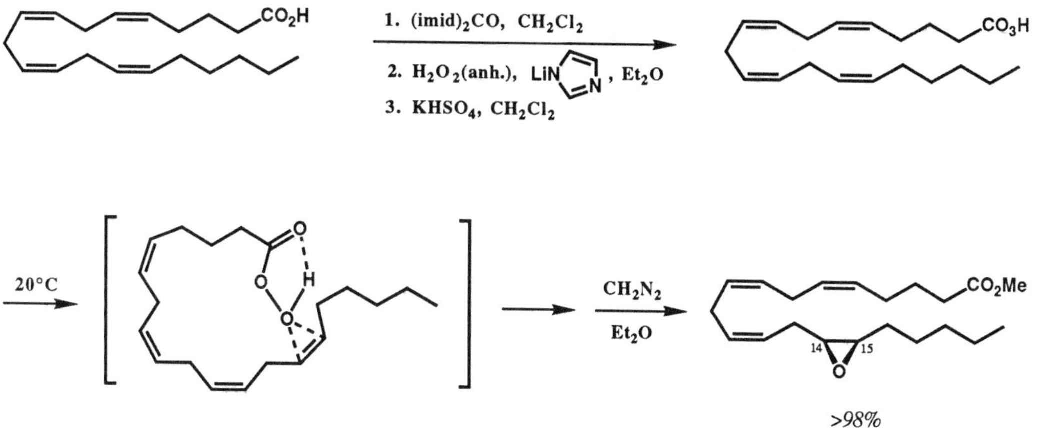

CO2H
1. (imid)2CO, CH2Cl2
2. H2O2(anh.), LiN, Et2O
N
3. KHSO4, CH2Cl2
CO3H
20°C
O
O
H
O
CH2N2
Et2O
CO2Me
14
15
O
>98%

This remarkably selective internal epoxidation of peroxyarachidonic acid to form 14,15-oxido-arachidonic acid occurs as shown because of unusually favorable stereoelectronics. The corresponding reaction sequence with eicosa-(*E*)-8,11,14-trienoic acid affords the $\Delta^{14,15}$-epoxide in 94% isolated yield and >95% purity.

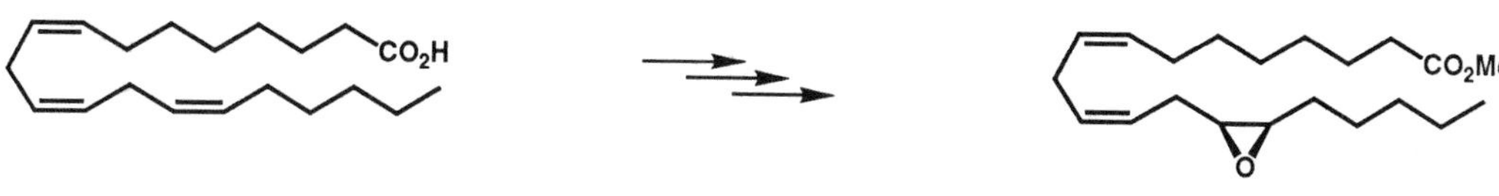

CO2H
CO2Me
O

3. 11,12-Oxido-arachidonic acid (Ref. 2):

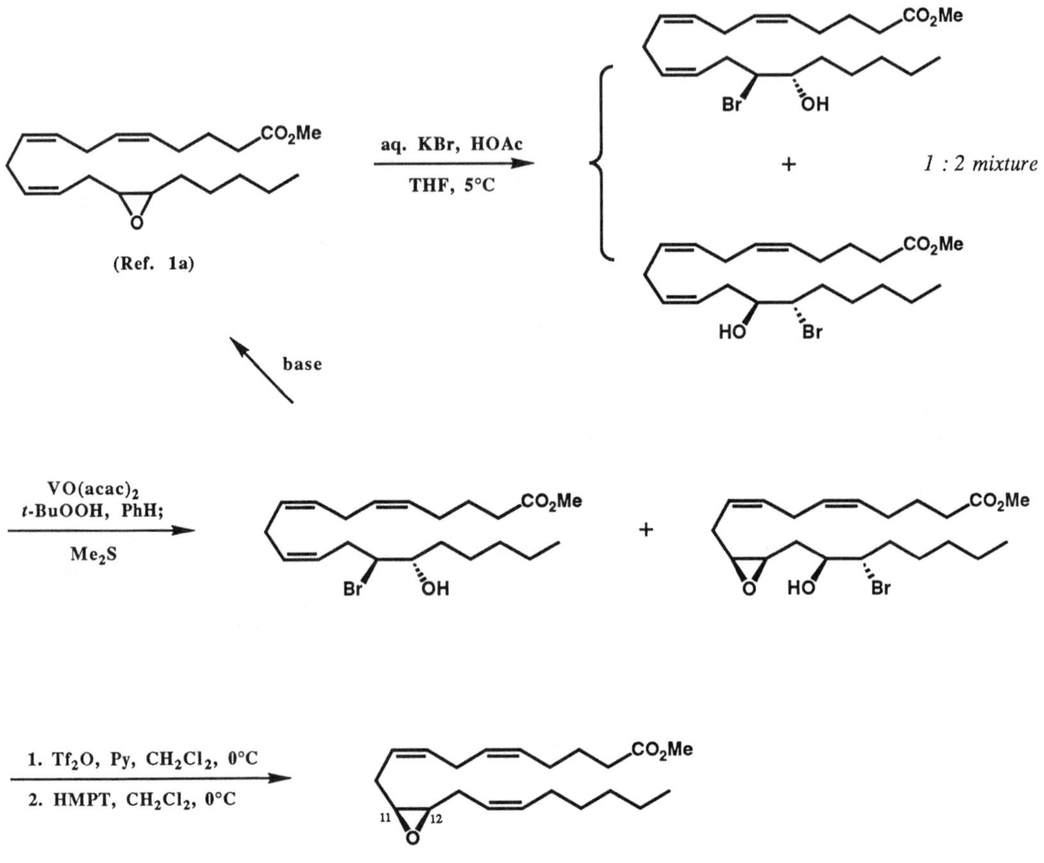
CO2Me
(Ref. 1a)
aq. KBr, HOAc
THF, 5°C
Br
OH
+
1 : 2 mixture
HO
Br
base
VO(acac)2
t-BuOOH, PhH;
Me2S
1. Tf2O, Py, CH2Cl2, 0°C
2. HMPT, CH2Cl2, 0°C
11
12
O

References:

1. (a) E. J. Corey, H. Niwa, and J. R. Falck, *J. Am. Chem. Soc.* **1979**, *101*, 1586; (b) E. J. Corey, H. Niwa, J. R. Falck, C. Mioskowski, Y. Arai, and A. Marfat, In *Advances in Prostaglandin, Thromboxane, and Leukotriene Research*; B. Samuelsson, P. W. Ramwell, and R. Paoletti, Eds; Raven Press: New York, 1980; Vol. 6, p 19.

2. E. J. Corey, A. Marfat, J. R. Falck, and J. O. Albright, *J. Am. Chem. Soc.* **1980**, *102*, 1433.

Synthesis of Irreversible Inhibitors of Eicosanoid Biosynthesis, 5,6-, 8,9-, and 11,12-Dehydroarachidonic Acid

Dehydroarachidonic acid analogs in which one *Z*-olefinic unit is replaced by a triple bond are irreversible inhibitors of the lipoxygenases which normally deliver dioxygen to the corresponding site of arachidonic acid. The inactivation appears to be a consequence of dioxygenation at the acetylinic unit to form a vinyl hydroperoxide which undergoes rapid O-O homolysis. Synthetic routes to these interesting enzyme inhibitors are outlined below.

1. Synthesis of 5,6-DHA:

Method 1 (Ref. 1):

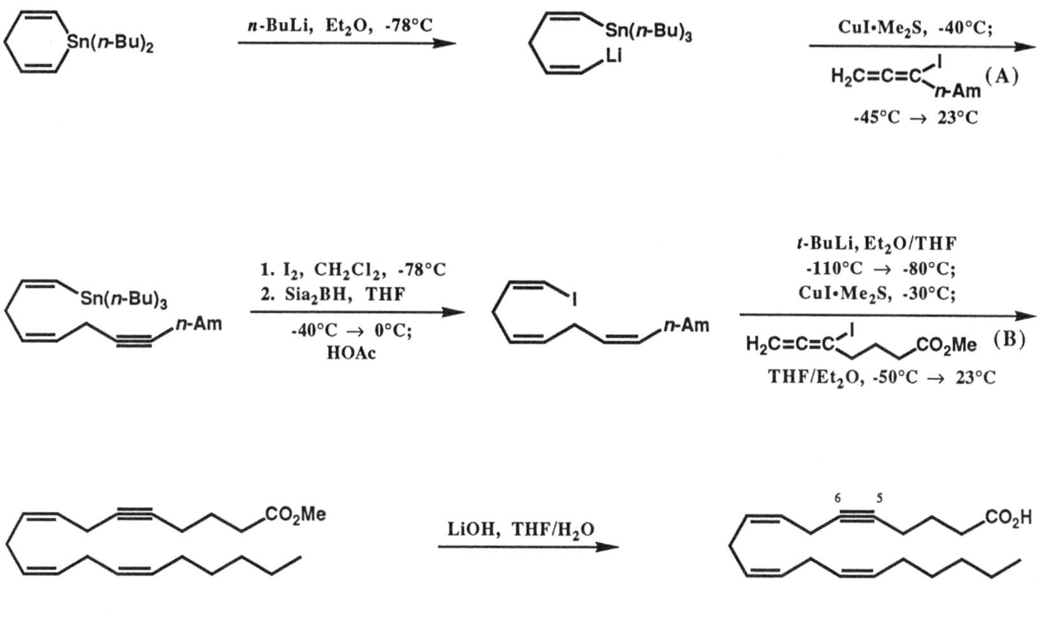
Sn(n-Bu)2
n-BuLi, Et2O, -78°C
Sn(n-Bu)3
Li
CuI•Me2S, -40°C;
H2C=C=C I n-Am (A)
-45°C → 23°C
Sn(n-Bu)3
n-Am
1. I2, CH2Cl2, -78°C
2. Sia2BH, THF
-40°C → 0°C;
HOAc
I
n-Am
t-BuLi, Et2O/THF
-110°C → -80°C;
CuI•Me2S, -30°C;
H2C=C=C I CO2Me (B)
THF/Et2O, -50°C → 23°C
CO2Me
LiOH, THF/H2O
6 5
CO2H

Method 2 (Ref. 2):

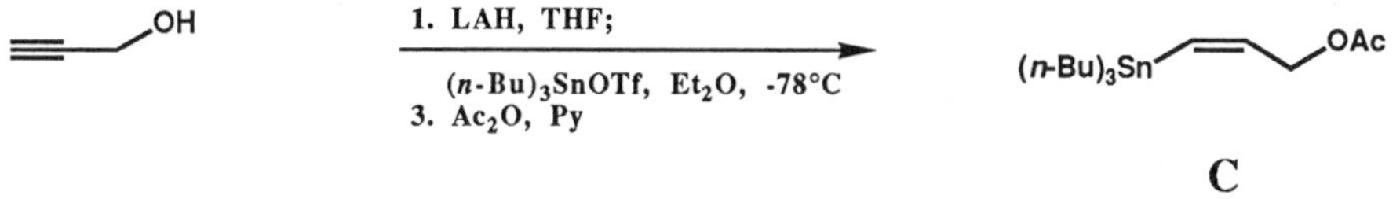
OH
1. LAH, THF;
(n-Bu)3SnOTf, Et2O, -78°C
3. Ac2O, Py
(n-Bu)3Sn
OAc

C

12.13

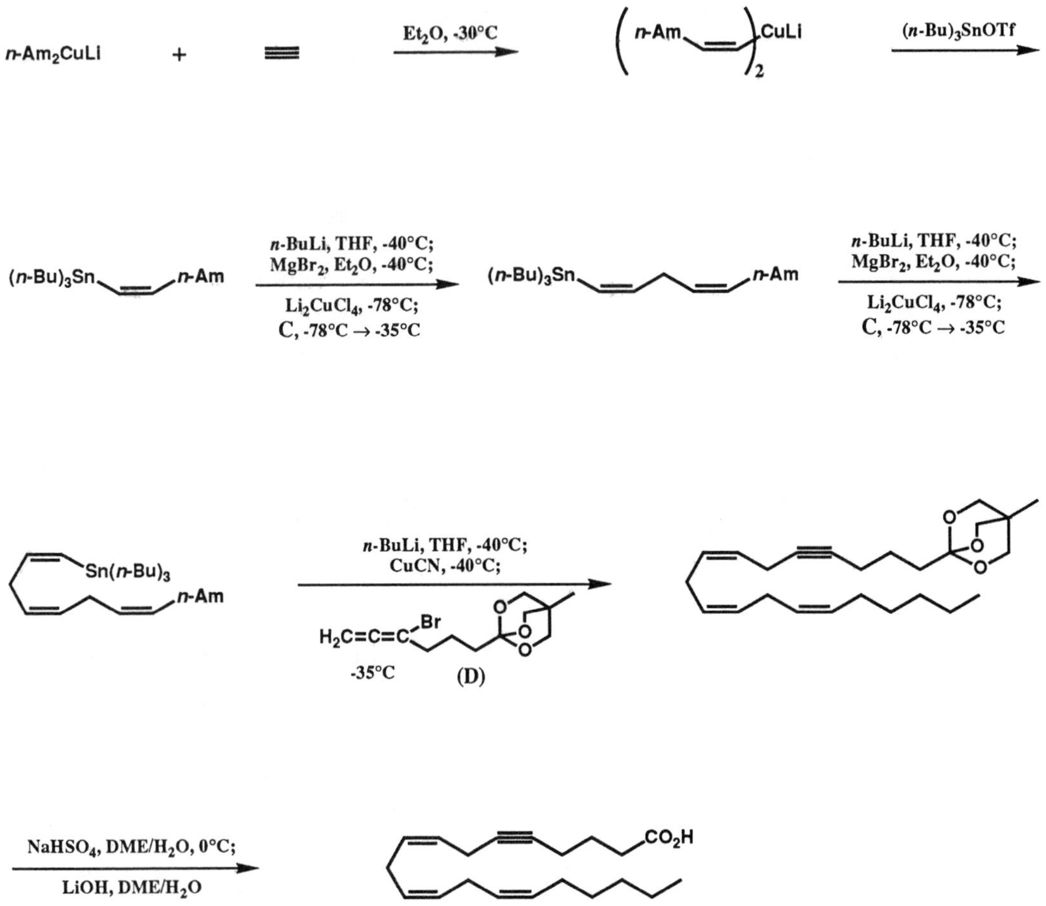

n-Am2CuLi
+
Et2O, -30°C
n-Am
CuLi
2
(n-Bu)3SnOTf
(n-Bu)3Sn
n-Am
n-BuLi, THF, -40°C;
MgBr2, Et2O, -40°C;
Li2CuCl4, -78°C;
C, -78°C → -35°C
(n-Bu)3Sn
n-Am
n-BuLi, THF, -40°C;
MgBr2, Et2O, -40°C;
Li2CuCl4, -78°C;
C, -78°C → -35°C
Sn(n-Bu)3
n-Am
n-BuLi, THF, -40°C;
CuCN, -40°C;
H2C=C=C
Br
O
O
O
-35°C
(D)
O
O
O
NaHSO4, DME/H2O, 0°C;
LiOH, DME/H2O
CO2H

Method 3 (Ref. 3):

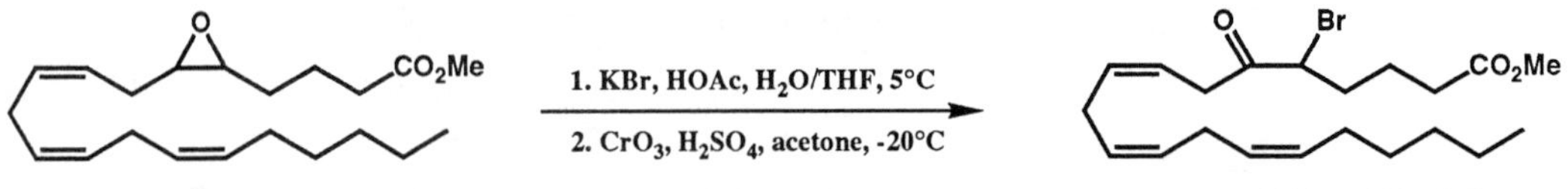

O
CO2Me
1. KBr, HOAc, H2O/THF, 5°C
2. CrO3, H2SO4, acetone, -20°C
O
Br
CO2Me

(Ref. 4)

plus the position isomer

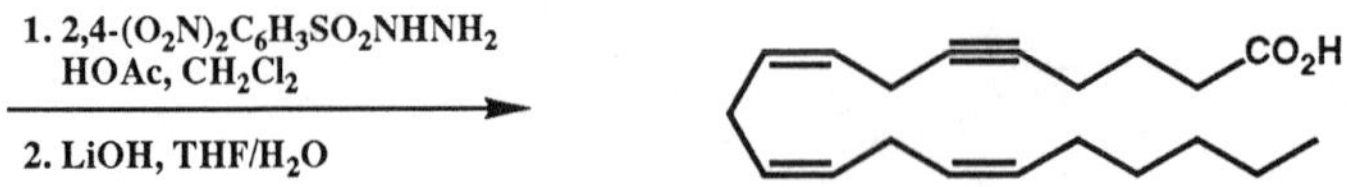

1. 2,4-(O2N)2C6H3SO2NHNH2
HOAc, CH2Cl2
2. LiOH, THF/H2O
CO2H

The (Z)-8-eicosen-5-ynoic acid **F**, another rationally devised irreversible inhibitor of the biosynthesis of leukotrienes and SRS-A's was synthesized as shown below (Ref. 3):

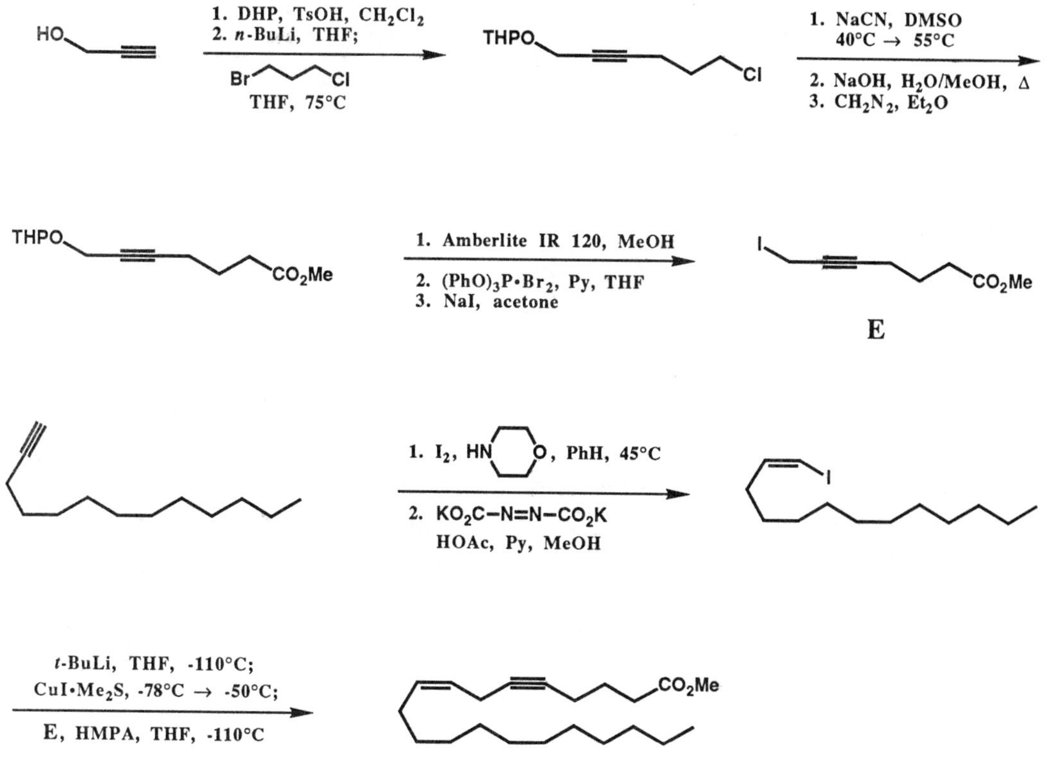
HO
1. DHP, TsOH, CH_2Cl_2
2. n-BuLi, THF;
Br
Cl
THF, 75°C
THPO
Cl
1. NaCN, DMSO
40°C → 55°C
2. NaOH, H_2O/MeOH, Δ
3. CH_2N_2, Et_2O
THPO
CO_2Me
1. Amberlite IR 120, MeOH
2. $(PhO)_3P\cdot Br_2$, Py, THF
3. NaI, acetone
I
CO_2Me
E
1. I_2, HN O, PhH, 45°C
2. $KO_2C-N=N-CO_2K$
HOAc, Py, MeOH
I
t-BuLi, THF, -110°C;
CuI•Me_2S, -78°C → -50°C;
E, HMPA, THF, -110°C
CO_2Me

F

2. Synthesis of 8,9-DHA:

Method 1 (Ref. 1):

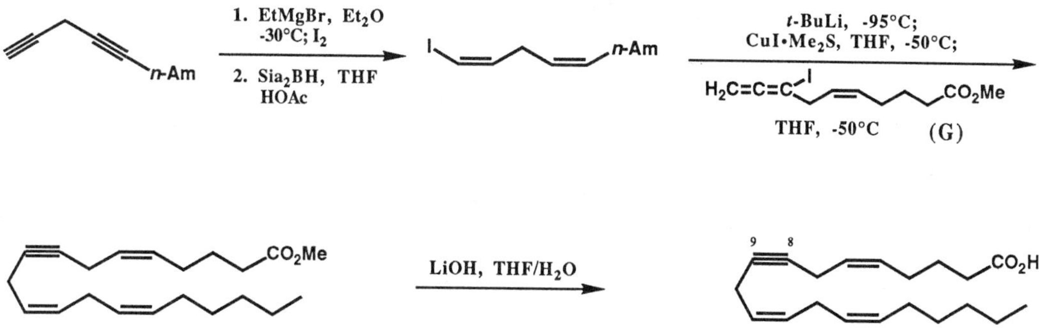
n-Am
1. EtMgBr, Et_2O
-30°C; I_2
2. Sia_2BH, THF
HOAc
I
n-Am
t-BuLi, -95°C;
CuI•Me_2S, THF, -50°C;
$H_2C=C=C$
I
CO_2Me
THF, -50°C
(G)
CO_2Me
LiOH, THF/H_2O
9
8
CO_2H

12.13

Method 2 (Ref. 5):

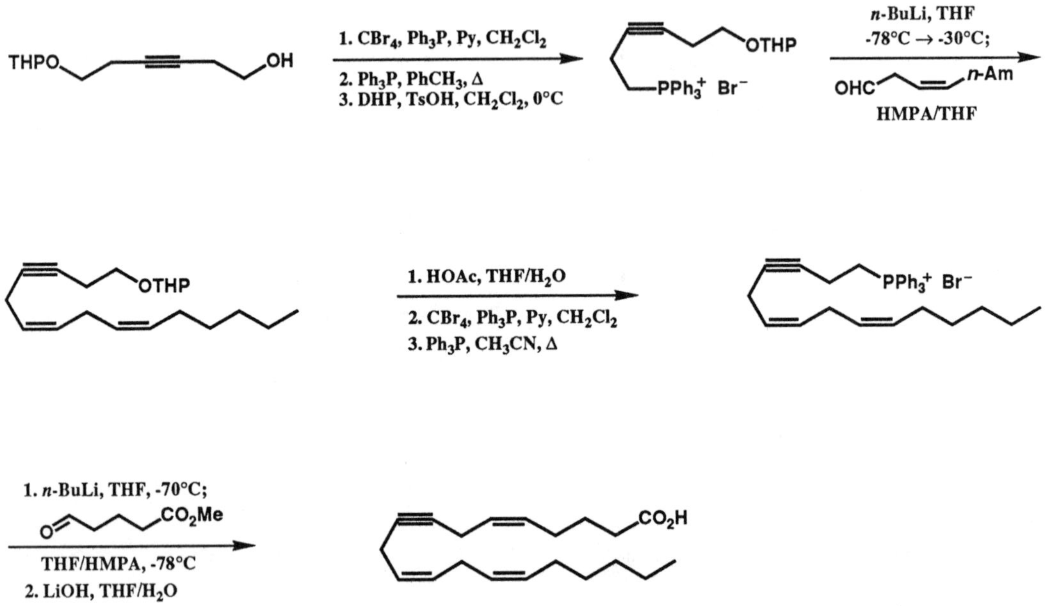
THPO
OH
1. CBr4, Ph3P, Py, CH2Cl2
2. Ph3P, PhCH3, Δ
3. DHP, TsOH, CH2Cl2, 0°C
OTHP
PPh3+ Br−
n-BuLi, THF
-78°C → -30°C;
OHC
n-Am
HMPA/THF
OTHP
1. HOAc, THF/H2O
2. CBr4, Ph3P, Py, CH2Cl2
3. Ph3P, CH3CN, Δ
PPh3+ Br−
1. n-BuLi, THF, -70°C;
O
CO2Me
THF/HMPA, -78°C
2. LiOH, THF/H2O
CO2H

3. Synthesis of 11,12-DHA:

Method 1 (Ref. 1):

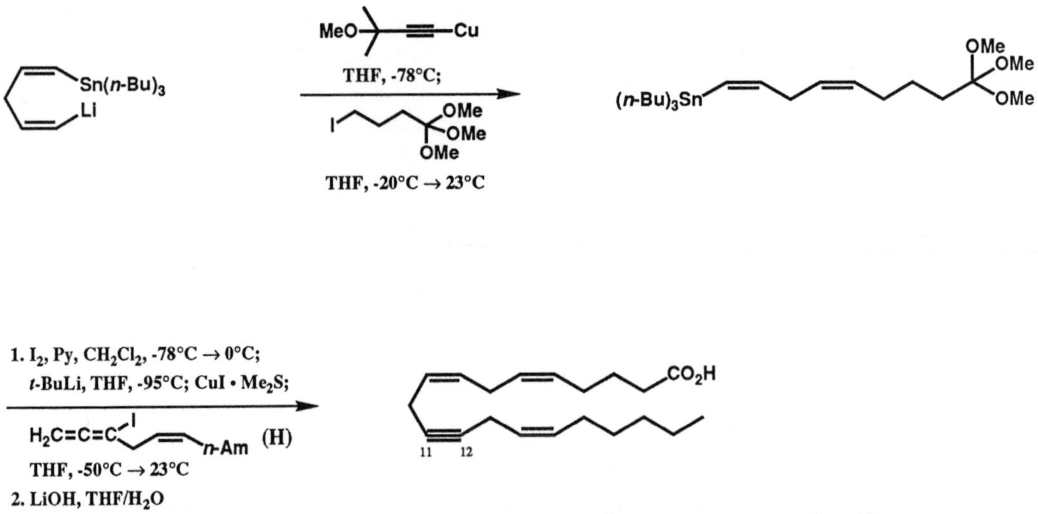
Sn(n-Bu)3
Li
MeO
Cu
THF, -78°C;
I
OMe
OMe
OMe
THF, -20°C → 23°C
(n-Bu)3Sn
OMe
OMe
OMe
1. I2, Py, CH2Cl2, -78°C → 0°C;
t-BuLi, THF, -95°C; CuI • Me2S;
H2C=C=C
I
n-Am
(H)
THF, -50°C → 23°C
2. LiOH, THF/H2O
CO2H
11
12

Method 2 (Ref. 5):

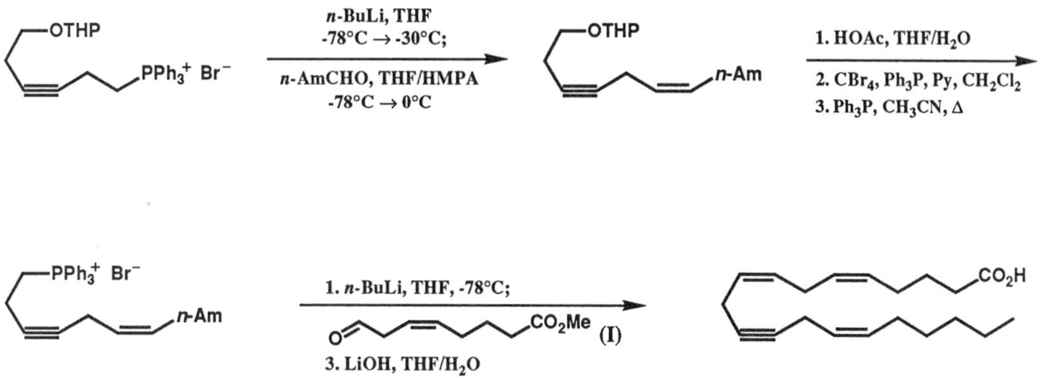
OTHP
PPh3+ Br−
n-BuLi, THF
-78°C → -30°C;
n-AmCHO, THF/HMPA
-78°C → 0°C
OTHP
n-Am
1. HOAc, THF/H2O
2. CBr4, Ph3P, Py, CH2Cl2
3. Ph3P, CH3CN, Δ
PPh3+ Br−
n-Am
1. n-BuLi, THF, -78°C;
CO2Me
(I)
3. LiOH, THF/H2O
CO2H

4. Synthesis of 14,15-DHA (Ref.6):

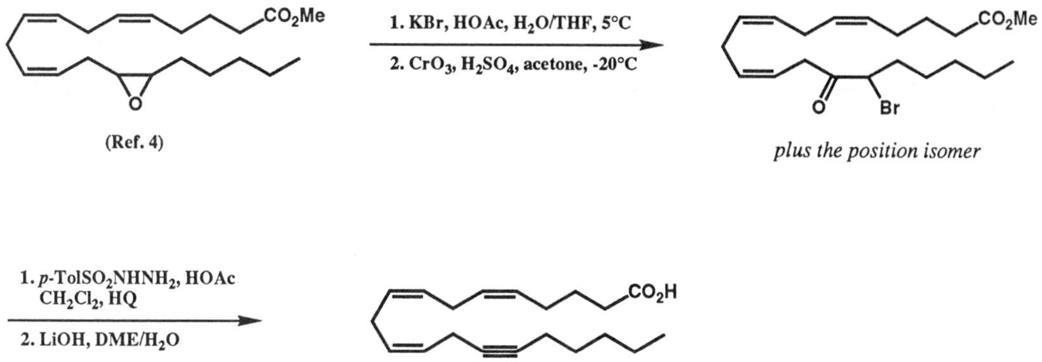
CO2Me
1. KBr, HOAc, H2O/THF, 5°C
2. CrO3, H2SO4, acetone, -20°C
CO2Me
O
Br
(Ref. 4)
plus the position isomer
1. p-TolSO2NHNH2, HOAc
CH2Cl2, HQ
2. LiOH, DME/H2O
CO2H

Preparation of the halo allenes **A**, **B** and **D**:

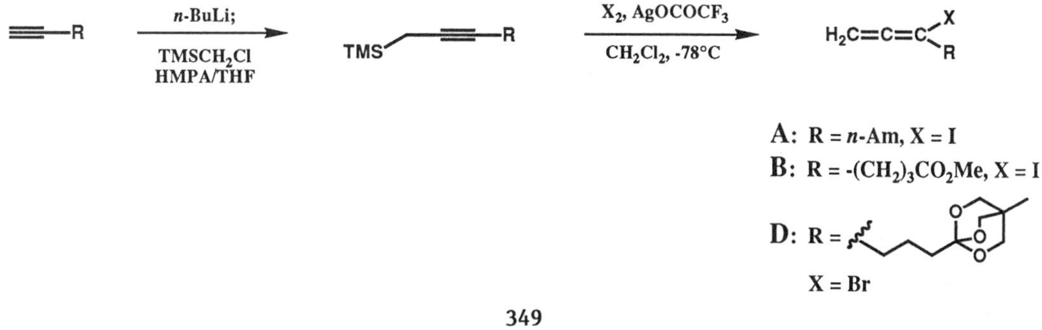
R
n-BuLi;
TMSCH2Cl
HMPA/THF
TMS
R
X2, AgOCOCF3
CH2Cl2, -78°C
H2C=C=C
X
R
A: R = n-Am, X = I
B: R = -(CH2)3CO2Me, X = I
D: R =
X = Br

12.13

Preparation of the iodo allenes **G** and **H**:

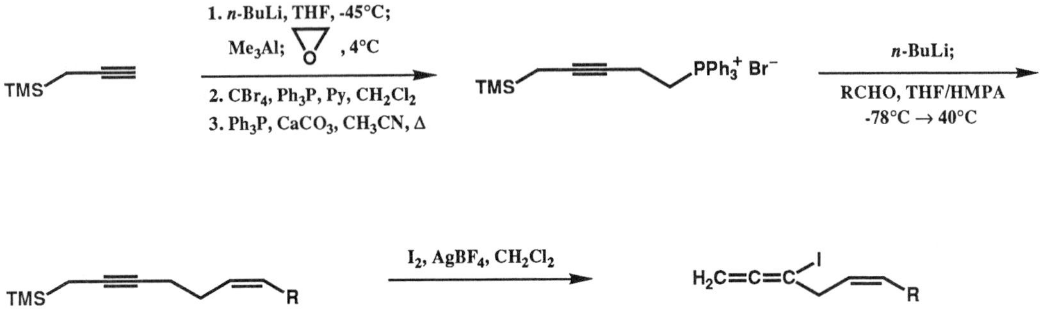

1. n-BuLi, THF, -45°C;
Me3Al; , 4°C
2. CBr4, Ph3P, Py, CH2Cl2
3. Ph3P, CaCO3, CH3CN, Δ
TMS
PPh3+ Br−
n-BuLi;
RCHO, THF/HMPA
-78°C → 40°C
I2, AgBF4, CH2Cl2
H2C=C=C
I
R

G: R = $-(CH_2)_3CO_2Me$

H: R = *n*-Am

Preparation of the aldehyde **I**:

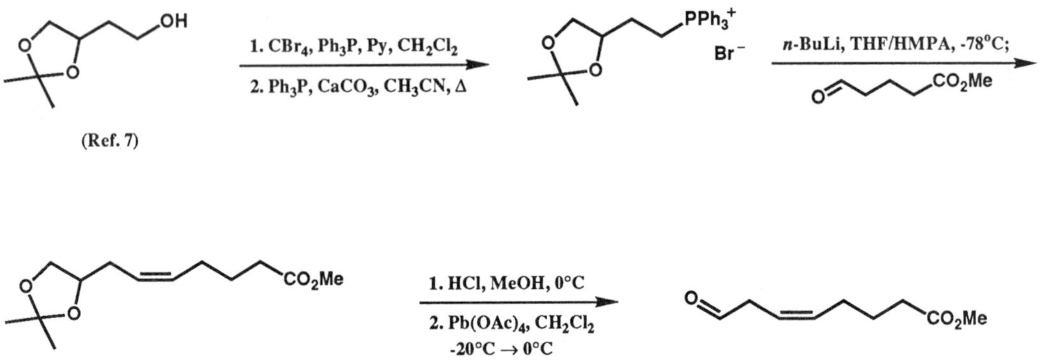

OH
(Ref. 7)
1. CBr4, Ph3P, Py, CH2Cl2
2. Ph3P, CaCO3, CH3CN, Δ
PPh3+
Br−
n-BuLi, THF/HMPA, -78°C;
CO2Me
1. HCl, MeOH, 0°C
2. Pb(OAc)4, CH2Cl2
-20°C → 0°C

I

References:

1. E. J. Corey and J. Kang, *Tetrahedron Lett.* **1982**, *23*, 1651.
2. E. J. Corey and T. M. Eckrich, *Tetrahedron Lett.* **1984**, *25*, 2419.
3. E. J. Corey, H. Park, A. Barton, and Y. Nii, *Tetrahedron Lett.* **1980**, *21*, 4243.
4. E. J. Corey, H. Niwa, and J. R. Falck, *J. Am. Chem. Soc.* **1979**, *101*, 1586.
5. E. J. Corey and J. E. Munroe, *J. Am. Chem. Soc.* **1982**, *104*, 1752.
6. E. J. Corey and H. Park, *J. Am. Chem. Soc.* **1982**, *104*, 1750.
7. E. J. Corey, H. Niwa, and J. Knolle, *J. Am. Chem. Soc.* **1978**, *100*, 1942.

Synthesis of a Class of Sulfur-Containing Lipoxygenase Inhibitors

1. Synthesis of 7-thiaarachidonic acid (Ref. 1):

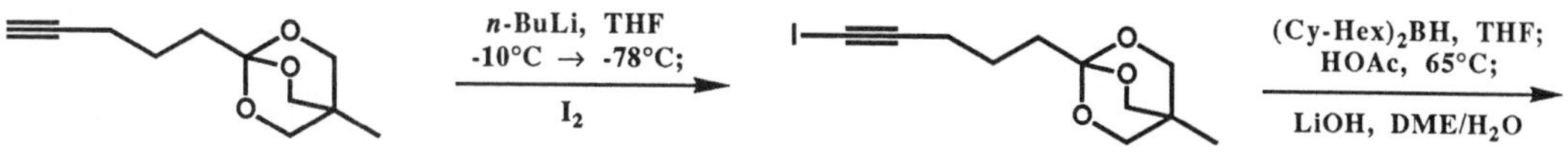

n-BuLi, THF
-10°C → -78°C;
I2
(Cy-Hex)2BH, THF;
HOAc, 65°C;
LiOH, DME/H2O

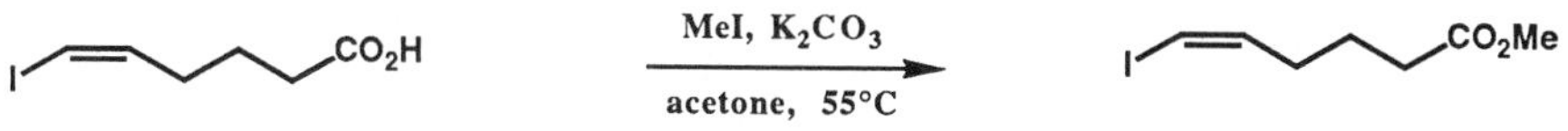

CO2H
MeI, K2CO3
acetone, 55°C
CO2Me

A

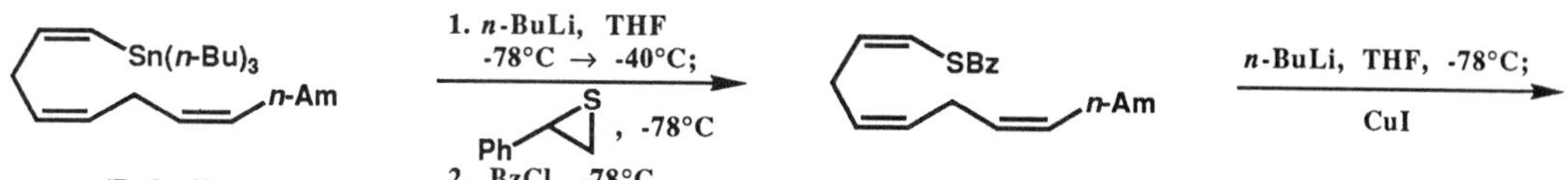

Sn(n-Bu)3
n-Am
1. n-BuLi, THF
-78°C → -40°C;
Ph
S
, -78°C
2. BzCl, -78°C
SBz
n-Am
n-BuLi, THF, -78°C;
CuI

(Ref. 2)

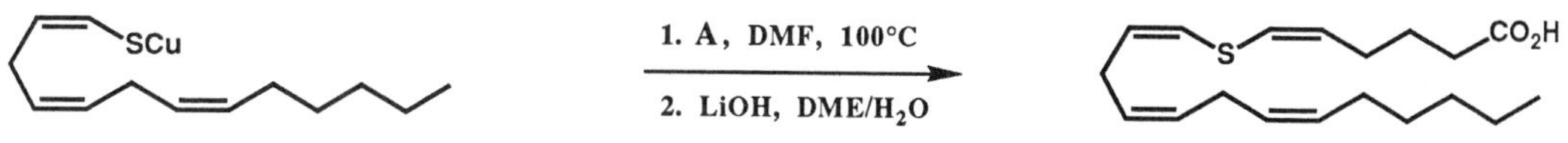

SCu
1. A, DMF, 100°C
2. LiOH, DME/H2O
S
CO2H

2. Synthesis of 10-thiaarachidonic acid (Ref. 3):

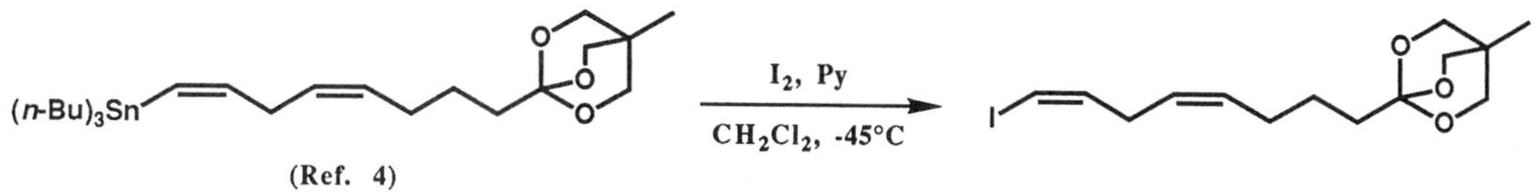

(n-Bu)3Sn
I2, Py
CH2Cl2, -45°C

(Ref. 4)

B

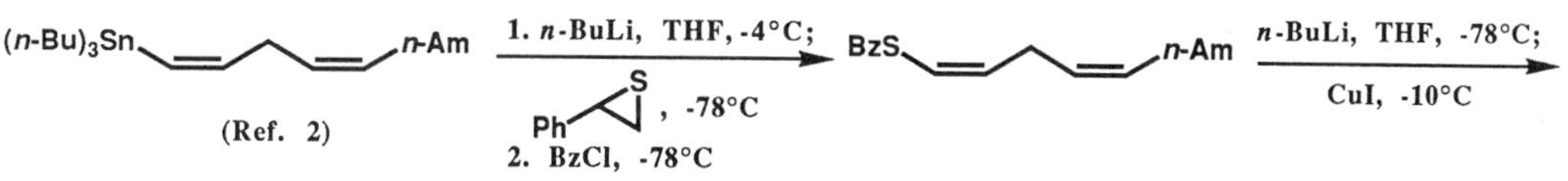

(n-Bu)3Sn
n-Am
1. n-BuLi, THF, -4°C;
Ph
S
, -78°C
2. BzCl, -78°C
BzS
n-Am
n-BuLi, THF, -78°C;
CuI, -10°C

(Ref. 2)

12.14

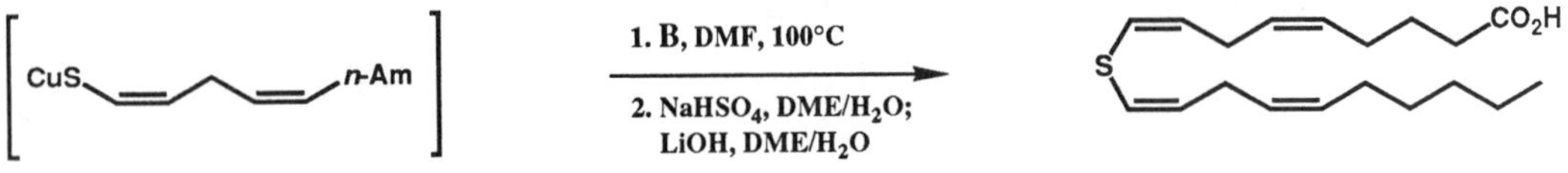
CuS
n-Am
1. B, DMF, 100°C
2. NaHSO4, DME/H2O;
LiOH, DME/H2O
S
CO2H

3. Synthesis of 13-thiaarachidonic acid (Ref. 3):

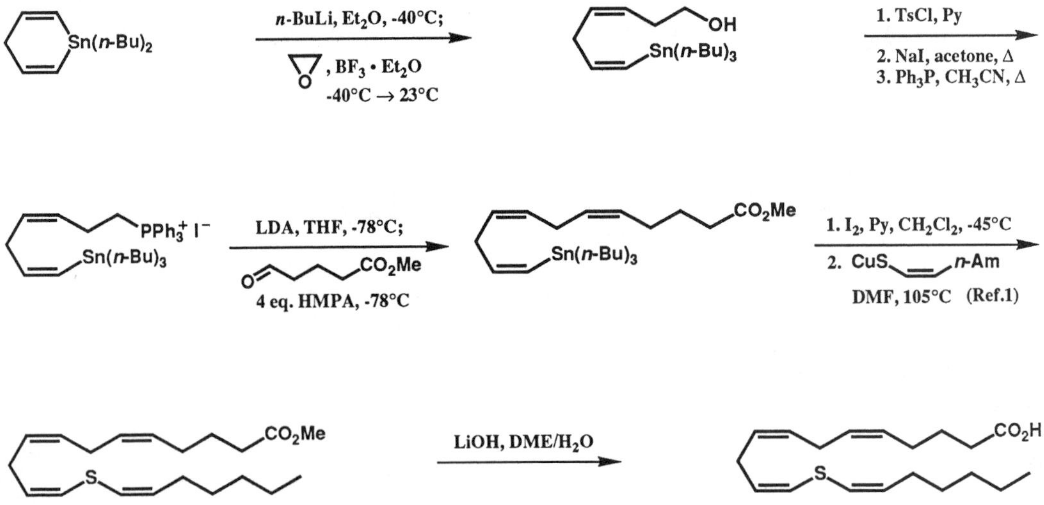
Sn(n-Bu)2
n-BuLi, Et2O, -40°C;
, BF3 • Et2O
O
-40°C → 23°C
OH
Sn(n-Bu)3
1. TsCl, Py
2. NaI, acetone, Δ
3. Ph3P, CH3CN, Δ
PPh3+ I−
Sn(n-Bu)3
LDA, THF, -78°C;
O
CO2Me
4 eq. HMPA, -78°C
CO2Me
Sn(n-Bu)3
1. I2, Py, CH2Cl2, -45°C
2. CuS
n-Am
DMF, 105°C (Ref.1)
CO2Me
S
LiOH, DME/H2O
CO2H
S

Two simpler acids were also synthesized for biological studies (Ref. 1):

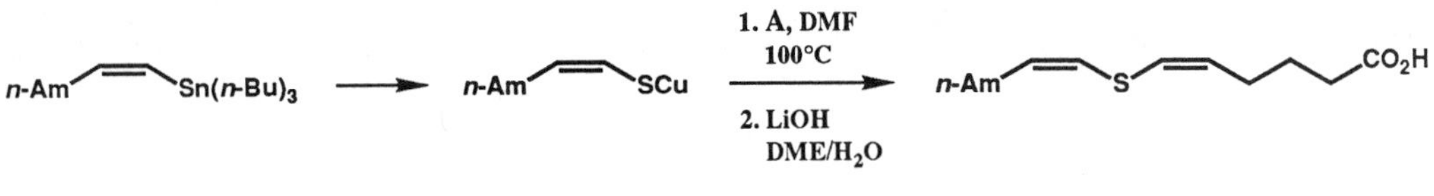
n-Am
Sn(n-Bu)3
n-Am
SCu
1. A, DMF
100°C
2. LiOH
DME/H2O
n-Am
S
CO2H

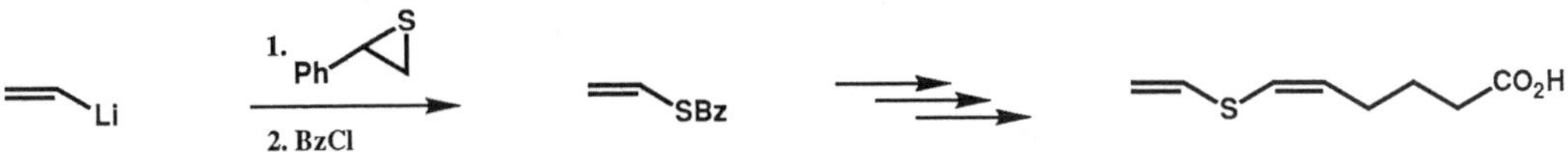
Li
1.
Ph
S
2. BzCl
SBz
S
CO2H

References:

1. E. J. Corey, J. R. Cashman, T. M. Eckrich, and D. R. Corey, *J. Am. Chem. Soc.* **1985**, *107*, 713.
2. E. J. Corey and T. M. Eckrich, *Tetrahedron Lett.* **1984**, *25*, 2419.
3. E. J. Corey, M. d'Alarcao, and K. S. Kyler, *Tetrahedron Lett.* **1985**, *26*, 3919.
4. E. J. Corey, K. S. Kyler, and N. Raju, *Tetrahedron Lett.* **1984**, *25*, 5115.

Total Synthesis of a Putative Precursor of the Lipoxins

A class of 5,6,15- and 5,14,15-trioxygenated metabolites of arachidonate has been described by Samuelsson *et al.* Two of these compounds, termed lipoxins A and B, have recently been assigned the structures shown below.[1] Outlined below is a synthesis of a putative biosynthetic precursor of these compounds.[2] Syntheses of the structures assigned to lipoxins A and B have also been accomplished.[3,4]

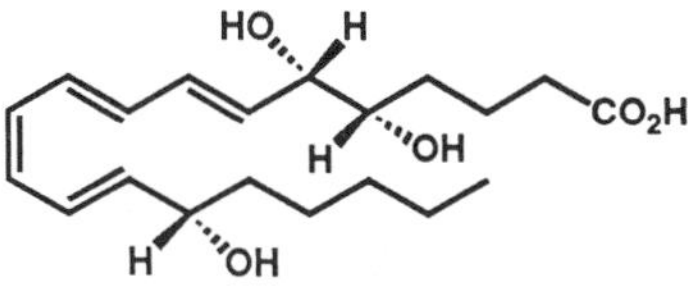

HO
H
H
OH
CO2H
H
OH

Lipoxin A

(Ref. 1)

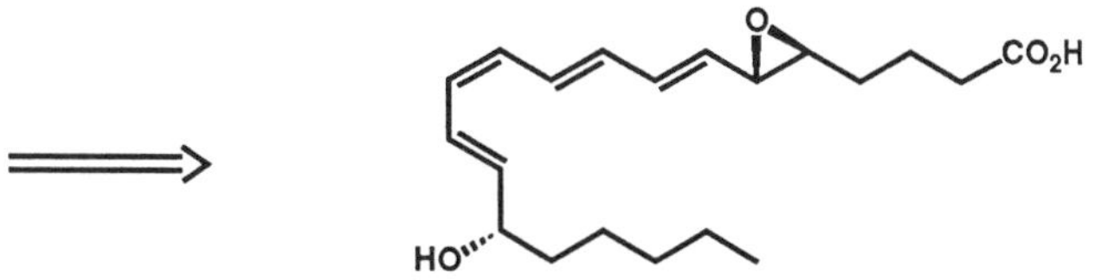

O
CO2H
HO

(Ref. 2)

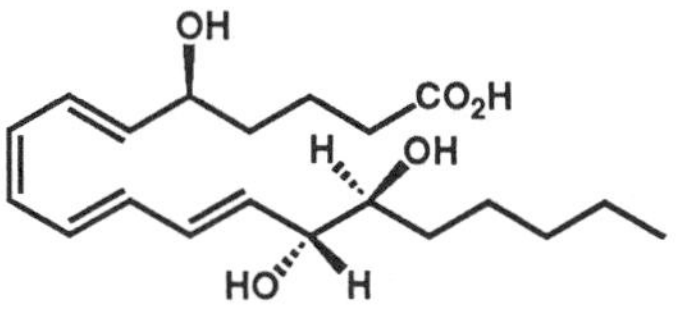

OH
CO2H
H
OH
HO
H

Lipoxin B

(Ref. 1)

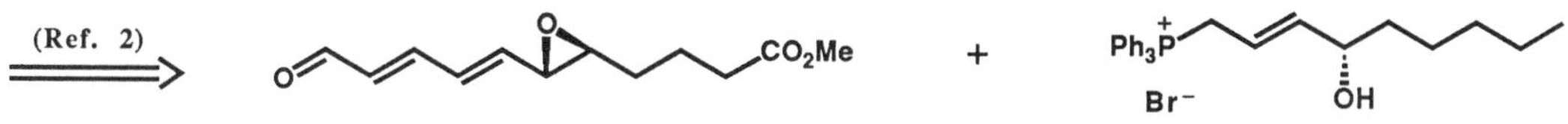

(Ref. 2)
O
O
CO2Me
+
Ph3P+
Br−
OH

C (Ref. 5) D

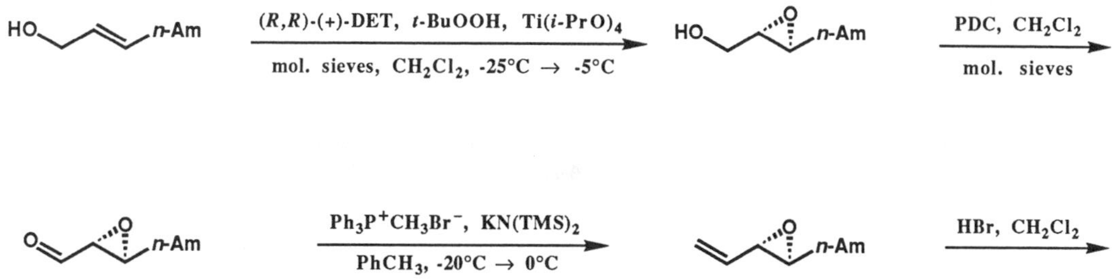

HO
n-Am
(R,R)-(+)-DET, t-BuOOH, Ti(i-PrO)4
mol. sieves, CH2Cl2, -25°C → -5°C
HO
O
n-Am
PDC, CH2Cl2
mol. sieves
O
O
n-Am
Ph3P+CH3Br−, KN(TMS)2
PhCH3, -20°C → 0°C
O
n-Am
HBr, CH2Cl2

12.15

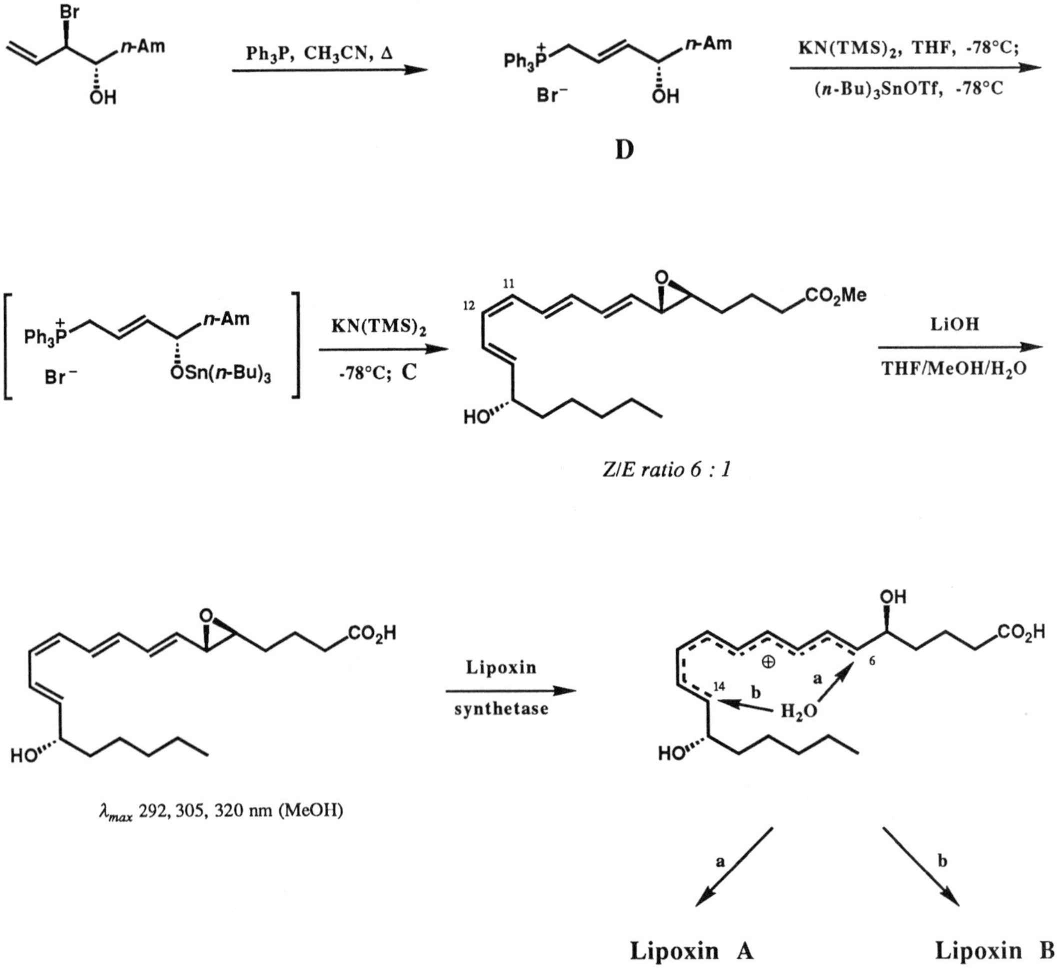

Br
n-Am
OH
Ph3P, CH3CN, Δ
Ph3P
Br−
n-Am
OH
D
KN(TMS)2, THF, -78°C;
(n-Bu)3SnOTf, -78°C
Ph3P
Br−
n-Am
OSn(n-Bu)3
KN(TMS)2
-78°C; C
11
12
CO2Me
HO
Z/E ratio 6 : 1
LiOH
THF/MeOH/H2O
CO2H
HO
λmax 292, 305, 320 nm (MeOH)
Lipoxin
synthetase
OH
CO2H
6
14
a
b
H2O
HO
a
b
Lipoxin A
Lipoxin B

References:

1. C. N. Serhan, M. Hamberg, B. Samuelsson, J. Morris, and D. G. Wishka, *Proc. Natl. Acad. Sci. USA*, **1986**, *83*, 1983.

2. E. J. Corey and M. M. Mehrotra, *Tetrahedron Lett.* **1986**, *27*, 5173.

3. E. J. Corey and W.-g. Su, *Tetrahedron Lett.* **1985**, *26*, 281.

4. E. J. Corey, M. M. Mehrotra, and W.-g. Su, *Tetrahedron Lett.* **1985**, *26*, 1919.

5. E. J. Corey, D. A. Clark, G. Goto, A. Marfat, C. Mioskowski, B. Samuelsson, and S. Hammarström, *J. Am. Chem. Soc.* **1980**, *102*, 1436, 3663.

Bongkrekic Acid

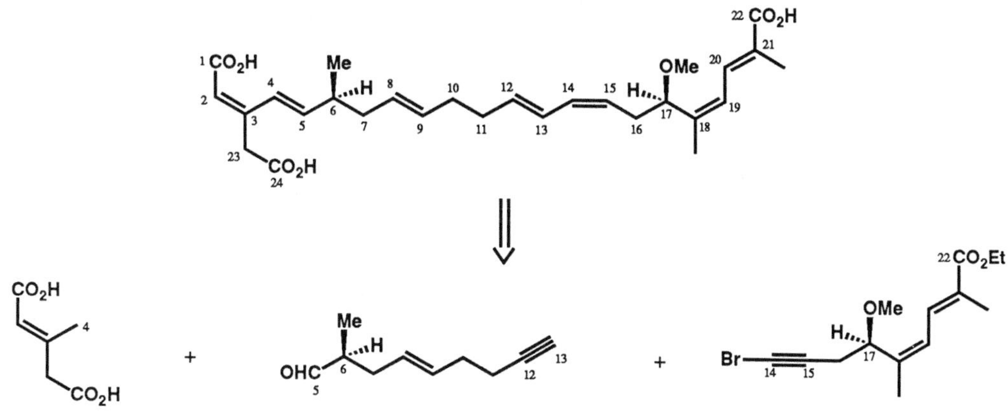

A B

The final synthetic target in this section is bongkrekic acid, a toxic polyenoic triacid which is produced by the microorganism *Pseudomonas cocovenenans*. Bongkrekic acid is a potent inhibitor of ATP export from mitochrondria as a result of its high affinity for the ATP translocator site. Because the original producing microorganism is no longer available, the synthesis outlined below was developed to provide research quantities of this biochemically useful agent.

1. Synthesis of the fragment **A**:

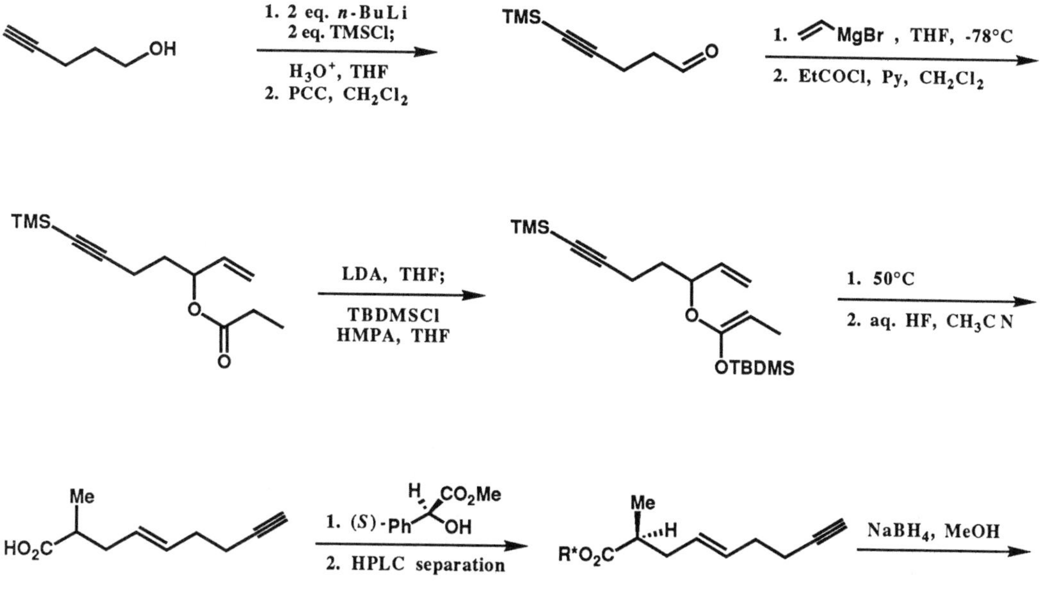

12.16

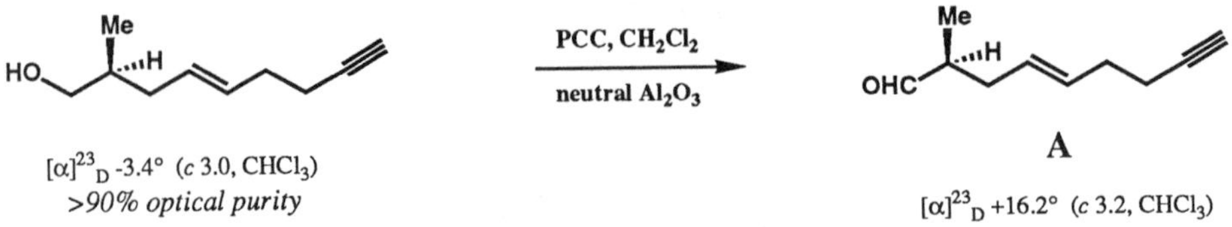
Me
HO
H
PCC, CH_2Cl_2
neutral Al_2O_3
Me
OHC
H
A
$[\alpha]^{23}_D$ -3.4° (*c* 3.0, $CHCl_3$)
>90% optical purity
$[\alpha]^{23}_D$ +16.2° (*c* 3.2, $CHCl_3$)

2. Synthesis of the fragment **B**:

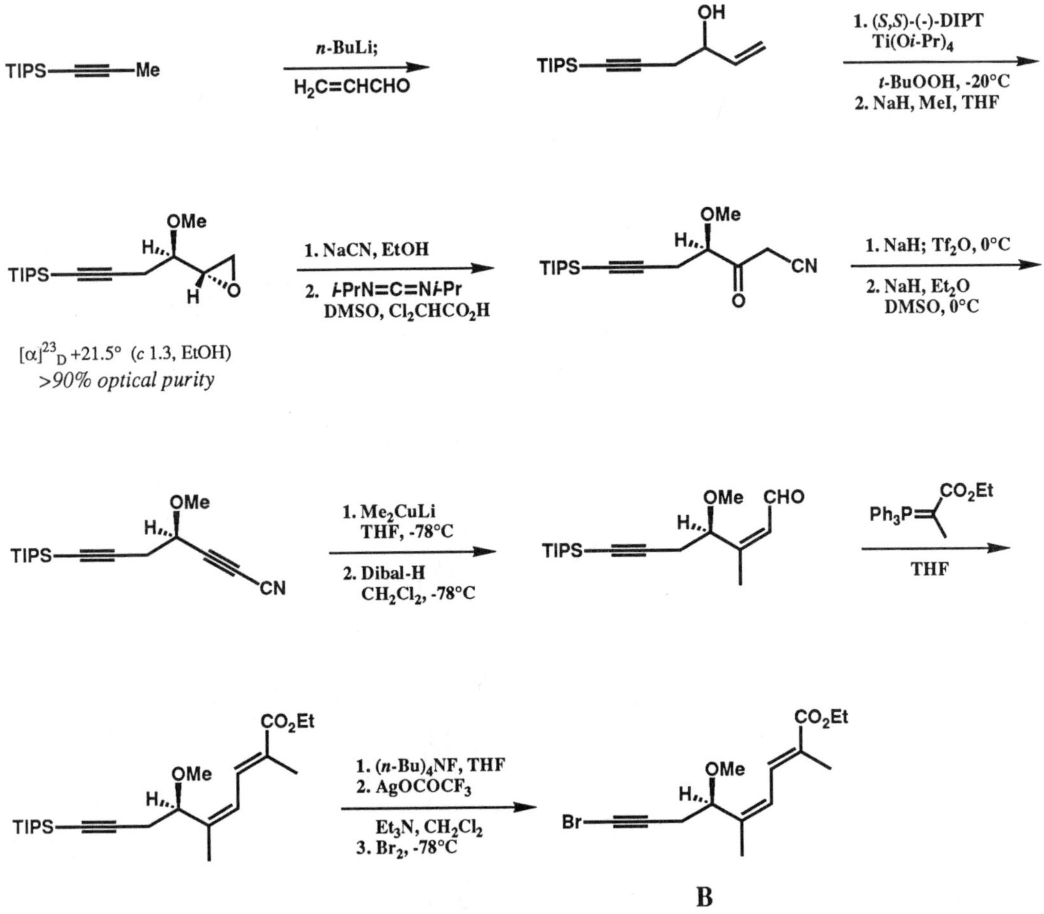
TIPS—≡—Me
n-BuLi;
$H_2C{=}CHCHO$
OH
TIPS
1. (*S,S*)-(-)-DIPT
$Ti(O\text{-}i\text{-}Pr)_4$
t-BuOOH, -20°C
2. NaH, MeI, THF
OMe
H
TIPS
H
O
$[\alpha]^{23}_D$ +21.5° (*c* 1.3, EtOH)
>90% optical purity
1. NaCN, EtOH
2. *i*-PrN=C=N*i*-Pr
DMSO, Cl_2CHCO_2H
OMe
H
TIPS
CN
O
1. NaH; Tf_2O, 0°C
2. NaH, Et_2O
DMSO, 0°C
OMe
H
TIPS
CN
1. Me_2CuLi
THF, -78°C
2. Dibal-H
CH_2Cl_2, -78°C
OMe
CHO
H
TIPS
CO_2Et
Ph_3P
THF
CO_2Et
OMe
H
TIPS
1. $(n\text{-}Bu)_4NF$, THF
2. $AgOCOCF_3$
Et_3N, CH_2Cl_2
3. Br_2, -78°C
CO_2Et
OMe
H
Br
B

3. Condensation of the fragment **A** with dimethyl *β*-methyl glutaconate, further condensation with the fragment **B**, and the completion of the synthesis:

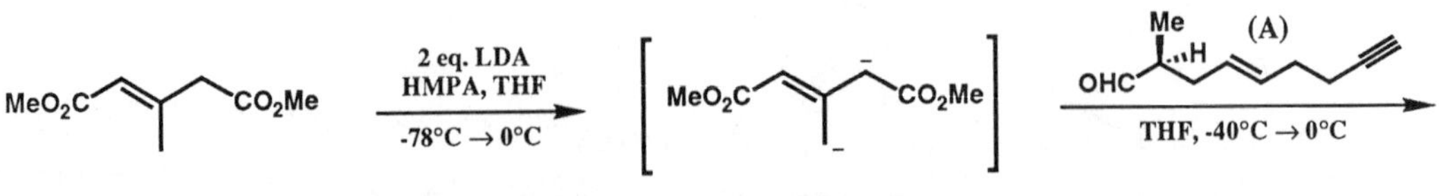
MeO_2C
CO_2Me
2 eq. LDA
HMPA, THF
-78°C → 0°C
MeO_2C
CO_2Me
Me
(A)
OHC
H
THF, -40°C → 0°C

12.16

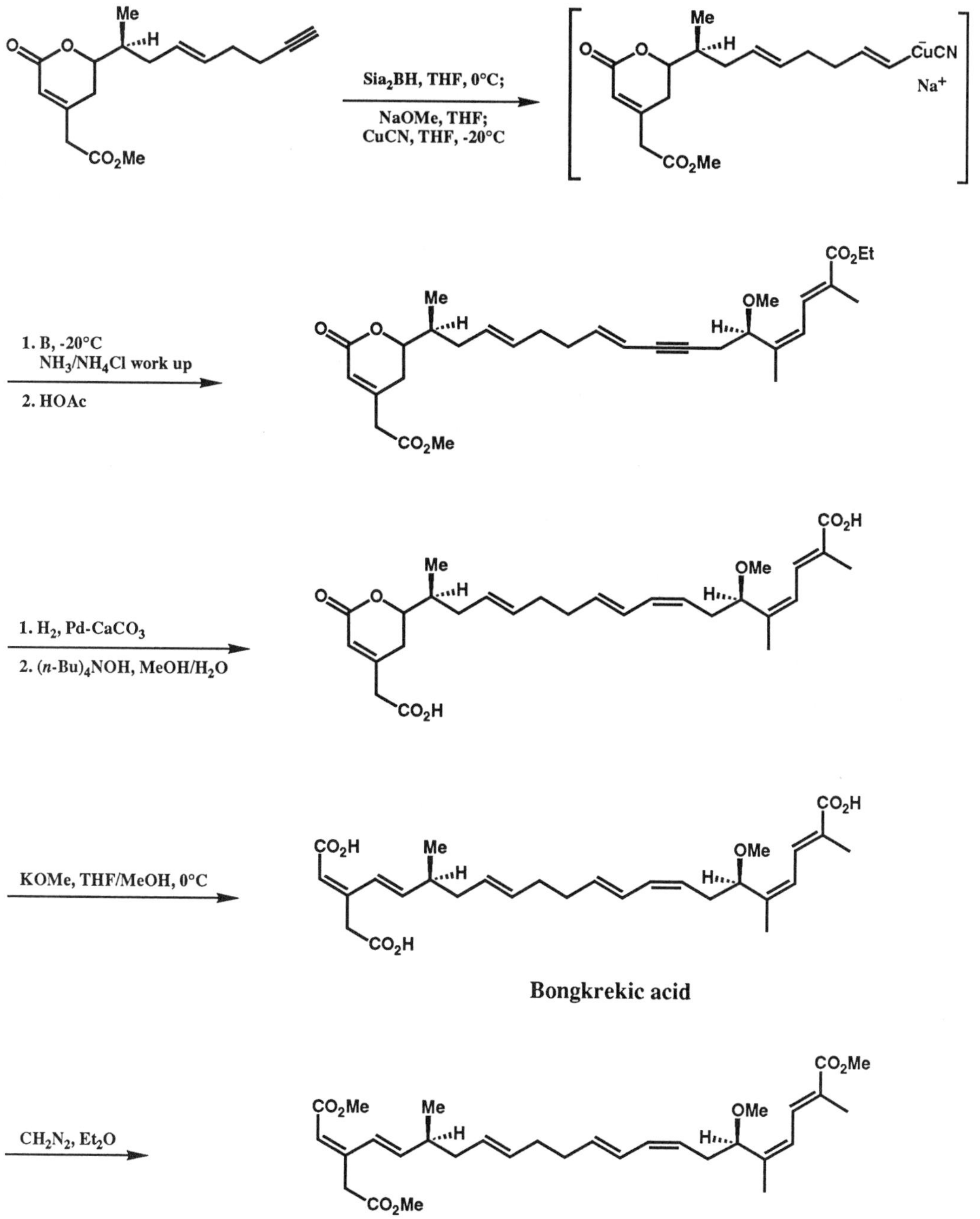
Me
H
CO_2Me
Sia_2BH, THF, 0°C;
NaOMe, THF;
CuCN, THF, -20°C
$\bar{C}uCN$
Na^+
1. B, -20°C
NH_3/NH_4Cl work up
2. HOAc
CO_2Et
OMe
1. H_2, Pd-$CaCO_3$
2. $(n\text{-Bu})_4NOH$, $MeOH/H_2O$
CO_2H
KOMe, THF/MeOH, 0°C
Bongkrekic acid
CH_2N_2, Et_2O

synthetic sample: $[\alpha]^{23}_D$ +80 ± 2°

derivative from the natural product: $[\alpha]^{23}_D$ +85 ± 2°

Reference:

E. J. Corey and A. Tramontano, *J. Am. Chem. Soc.* **1984**, *106*, 462.

PART THREE

Guide to the Original Literature of Multistep Synthesis

CHAPTER THIRTEEN

13.1 Introduction

Some six hundred structures of naturally occurring carbogenic molecules appear on the pages which follow, together with the name of each compound and references to the original literature of successful chemical synthesis. Thus, Part Three of this book is effectively a key to the literature of chemical synthesis as applied to the complex molecules of nature. The survey does not include oligomeric or polymeric structures, such as peptides, proteins, carbohydrates and polynucleotides, which fall outside the scope of this book because they can be assembled by repetitive procedures.

A quick scan of the collection of structures in Part Three will confirm the high level of activity and accomplishment which has been associated with the field of chemical synthesis over the past three decades. Upon closer inspection, it becomes apparent that the rate of achievement of successful syntheses has been increasing steadily with time. As of this writing (1988) there are many laboratories in which outstanding work on multistep synthesis of complex molecules is being carried out. One consequence of this intense effort is that the problem of keeping track of the literature of synthesis has become a difficult one for investigators and students. The compilation of literature which is contained in Part Three addresses this issue along a broad front and both extends and complements Parts One and Two.

The ordering of the target structures which appear on the pages which follow has been made on the basis of certain key molecular features, especially topology, ring types, and overall molecular complexity. Compounds with similar ring systems have been grouped together, and within each class there is a progression from simpler to more complex structures. The first section (13.3) presents acyclic or monocarbocyclic molecules. This unit is followed by Section 13.4 dealing with fused bi- and tricarbocyclic structures (mainly terpenoids) and Section 13.5 which contains bridged carbocyclic molecules (mainly tricarbocyclic terpenoids). The progression continues with Section 13.6 on the higher isoprenoids, steroids and steroidal alkaloids. There follows a block of eight pages of structures (13.7) in the N-heterocyclic, non-bridged, non-indole category. The next two sections, 13.8 and 13.9, contain fused- and bridged-ring indole alkaloids, respectively. Appearing immediately thereafter is a segment (13.10) on non-indole bridged-ring alkaloids and porphrins. Section 13.11 consists of polycyclic benzenoid structures. Collections of oxygen heterocycles (13.12, five pages), macrocyclic lactones (13.13, six pages), and macrocyclic lactams (13.14, three pages) appear next. Finally, Part Three is brought to a conclusion by Section 13.15 which contains structures in the polyether category (two pages).

There are a large number of naturally occurring molecules which have not yet been obtained by chemical synthesis. A convenient source of information on such compounds is *The Dictionary of Organic Compounds*, Fifth Addition (1982) and Supplements 1-5, published as a multivolume series by Chapman and Hall, New York and London; J. Buckingham, Executive Editor. This compendium contains references to syntheses which are not included in this collection, especially those involving simpler target structures.

In general, Part Three does not cover partial syntheses or syntheses which depend on simple interconversions. In the case of multiple syntheses of a compound by a particular author, reference is given to the first synthesis (both preliminary and full publications, if available) and also to the most recent synthesis. For syntheses of chiral targets, the sign of optical rotation of the product, (+) or (-), or the designations (*nat*) or (*unnat*) is given depending on the reported data. When such indications of chirality are lacking, it can be assumed that the racemic form of the product was synthesized.

No less astounding than the productivity and creativity of the synthetic chemists who devised and executed the syntheses referred to in Part Three is the capacity of nature to assemble such a staggering range of highly complicated molecules. In this connection it must be remembered that there are many more complex naturally occurring molecules which have not yet been synthesized. The amount of information used by nature to program the biosynthesis of so many complex carbogens is awesome to contemplate. Yet the comprehension of these naturally produced syntheses must surely be part of the unfinished business of synthetic and bioorganic chemists. An extensive cataloging of the chemical transformations and starting materials available for biosynthesis should make it possible for a chemist to derive reasonable biosynthetic pathways and to specify the nature of the catalytic events required for control of each step. Eventually, it should even be possible to explain the details of molecular processing once the organizational and manufacturing principles of multistep biosynthesis are understood for each class of naturally produced molecules.

We cannot now foresee the ultimate achievements of humankind in the fields of complex molecular synthesis. However, anyone familiar with the progress made in the field over the past century and one half would be foolish indeed to expect anything less than revolutionary advances in sophistication, efficiency and power.

ABBREVIATIONS OF JOURNAL NAMES

ACIE	*Angew. Chem. Int. Ed.*
ACR	*Acct. of Chem. Res.*
ACS, Ser. B	*Acta Chem. Scandinavica, Series B*
Ag. BC	*Agri. Biol. Chem.*
AJC	*Aust. J. Chem.*
Ann	*Annalen Chemie*
BCS, Jpn	*Bull. Chem. Soc., Japan*
Ber	*Chem. Ber.*
Bioorg. Chem.	*Bioorganic Chemistry*
BSC, Belg	*Bull. Soc. Chim., Belg.*
Chem. Ind.	*Chem. and Industry*
CJC	*Canadian J. Chem.*
CL	*Chem. Lett.*
CPB	*Chem. Pharm. Bull., Japan*
CSR	*Chem. Soc. Review*
HCA	*Helv. Chim. Acta*
Het	*Heterocycles*
J. Antibiot	*J. Antibiotics*
JACS	*J. Am. Chem. Soc.*
JCS	*J. Chem. Soc., London*
JCS, CC	*J. Chem. Soc., Chem. Commun.*
JCS, PI	*J. Chem. Soc., Perkin Trans. I*
JHC	*J. Heterocyclic Chem.*
JMC	*J. Med. Chem.*
JOC	*J. Org. Chem.*
P&AC	*Pure and Appl. Chem.*
PCS	*Proc. Chem. Soc., London*
PJPS	*Proc. Japan Pharm. Soc.*
PNAS	*Proc. Natl. Acad. Sci., USA*
Synth	*Synthesis*
Synth. Comm.	*Synthetic Commun.*
Tet	*Tetrahedron*
TL	*Tetrahedron Lett.*

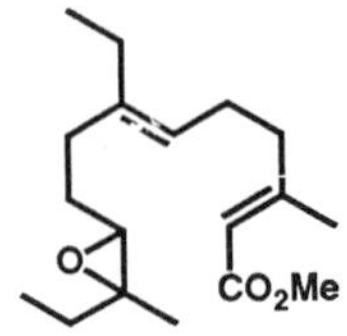
O
CO2Me

Juvenile Hormone (JH-1)
Trost, *JACS*, **1967**, *89*, 5292.
Siddall, *JACS*, **1968**, *90*, 6224.
Johnson, *JACS*, **1968**, *90*, 6225.
Mori, *Tet*, **1969**, *25*, 1667.
van Tamelen, *JACS*, **1970**, *92*, 737.
Henrick, *JACS*, **1972**, *94*, 5374, 5379.
Hanson, *JCS, PI*, **1972**, 361.
Stotter, *JACS*, **1973**, *95*, 4444.
Still, *TL*, **1979**, 593.

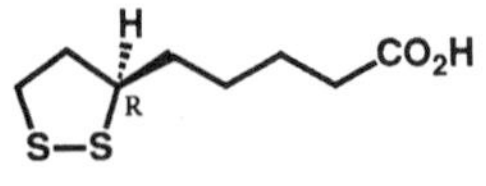
H
CO2H
R
S–S

α-(+)-Lipoic Acid (*nat*)
Walton, *JACS*, **1955**, *77*, 5144. (±)
Acker, *JACS*, **1957**, *79*, 6483. (±)
Tsuji, *JOC*, **1978**, *43*, 3606. (±)
Golding, *JCS, CC*, **1983**, 1051; (-)
JCS, PI, **1988**, 9. (+), (-)
Elliott, *TL*, **1985**, *26*, 2535. (+)
Sutherland, *JCS, CC*, **1986**, 1408. (+)
Ravindranathan, *TL*, **1987**, *28*, 5313. (+), (-)

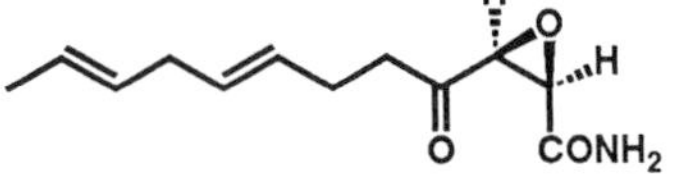
H
O
H
O
CONH2

(-)-Cerulenin (*nat*)
Boeckman, *JACS*, **1977**, *99*, 2805; *ibid.*, **1979**, *101*, 987. (±)
Ohrui, *TL*, **1979**, 2039. (+)
Sinay, *TL*, **1979**, 4741. (-)
Tishler, *JOC*, **1982**, *47*, 1221. (±)
Nozoe, *Het*, **1986**, *24*, 1137. (±)

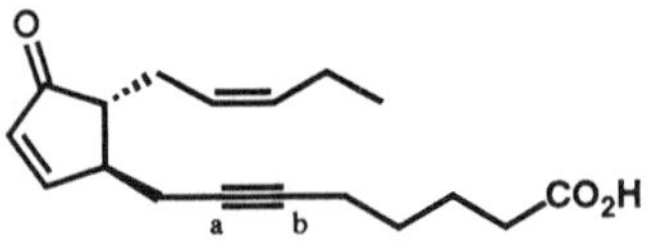
O
CO2H
a
b

Dicranenone A
Sakai, *TL*, **1985**, *26*, 2089.
Salaün, *JCS, CC*, **1985**, 1269.
(a,b)Tetrahydro-dicranenone A
Krüger, *TL*, **1985**, *26*, 6027.

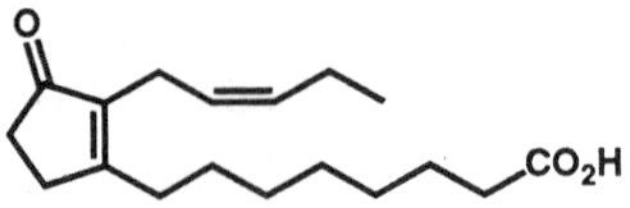
O
CO2H

Tetrahydrodicranenone B
Moody, *JCS, CC*, **1986**, 1292.

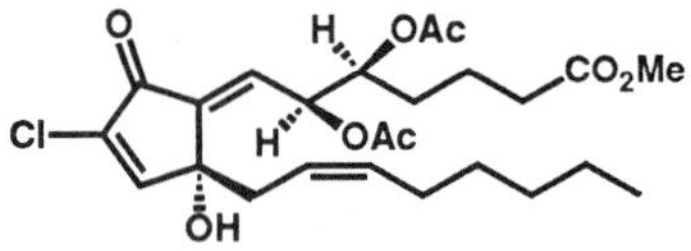
O
H
OAc
CO2Me
Cl
H
OAc
OH

Punaglandin 4
Yamada, *JACS*, **1986**, *108*, 5019.
Noyori, *JACS*, **1986**, *108*, 5021.

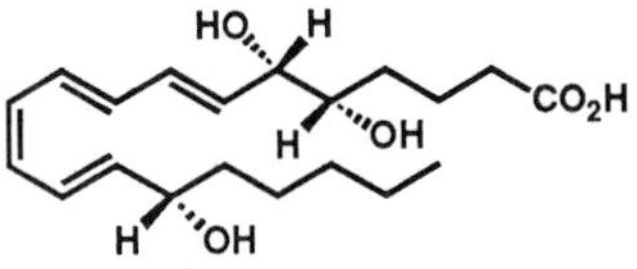
HO
H
CO2H
H
OH
H
OH

Lipoxin A
Adams, *JACS*, **1985**, *107*, 464.
Nicolaou, *JACS*, **1985**, *107*, 7515.

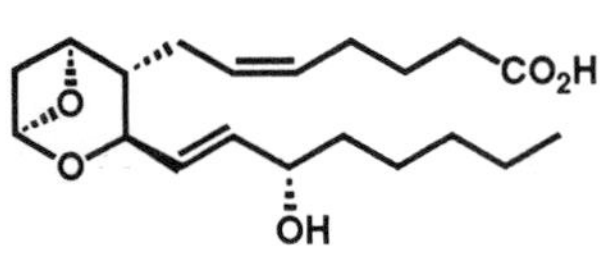
O
CO2H
O
OH

Thromboxane A_2
Still, *JACS*, **1985**, *107*, 6372.

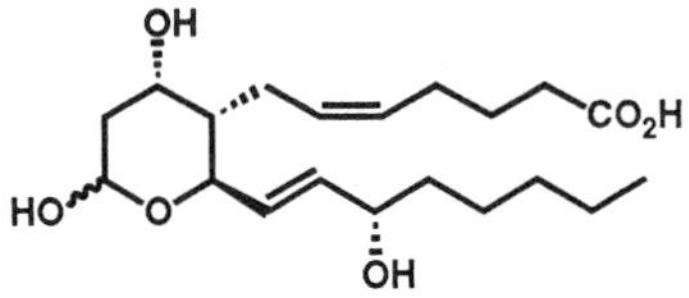
OH
CO2H
HO
O
OH

Thromboxane B_2
Hanessian, *CJC*, **1977**, *55*, 562; *ibid.*, **1981**, *59*, 870.

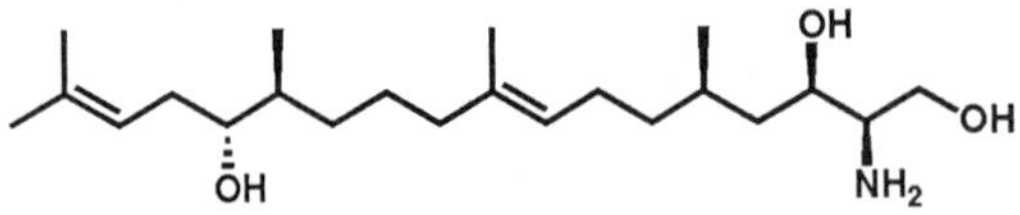
OH
OH
OH
NH2

(+)-Aplidiasphingosine
Mori, *TL*, **1982**, *23*, 3391.

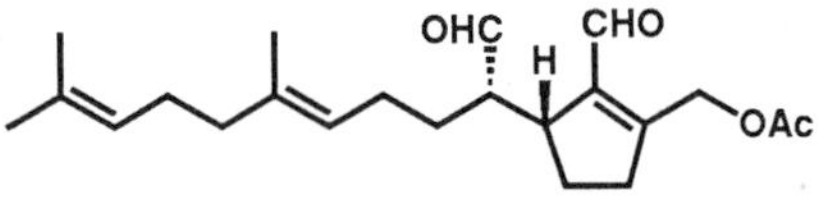
OHC
H
CHO
OAc

Petiodial
Trost, *JACS*, **1988**, *110*, 5233.

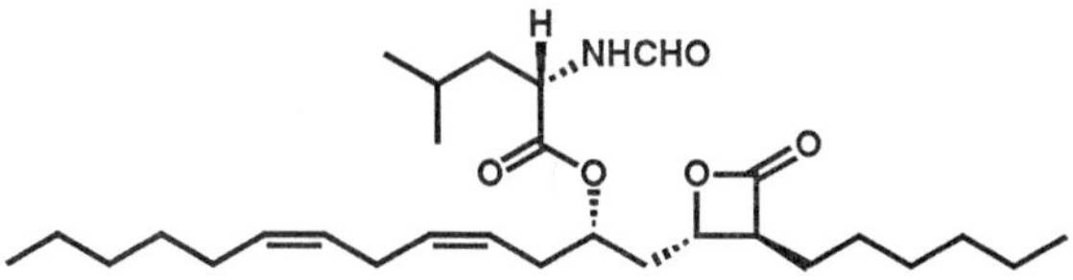

Lipstatin
Schneider, *HCA*, **1987**, *70*, 196.

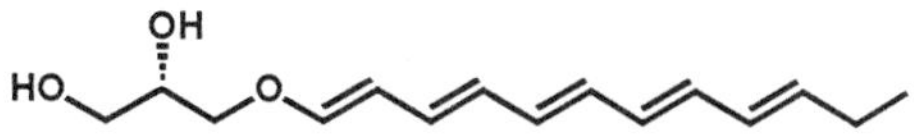

(*S*)-3-(Dodeca-1,3,5,7,9-pentaenyloxy)-propane-1,2-diol
Nicolaou, *JCS, CC*, **1984**, 349.

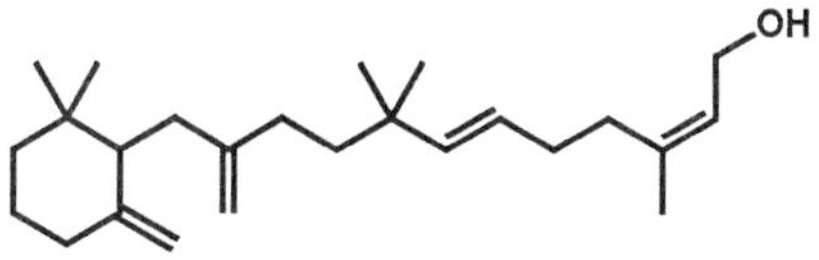

Diumycinol
Grieco, *JOC*, **1975**, *40*, 2261.
Kocienski, *JCS, CC*, **1982**, 1078.

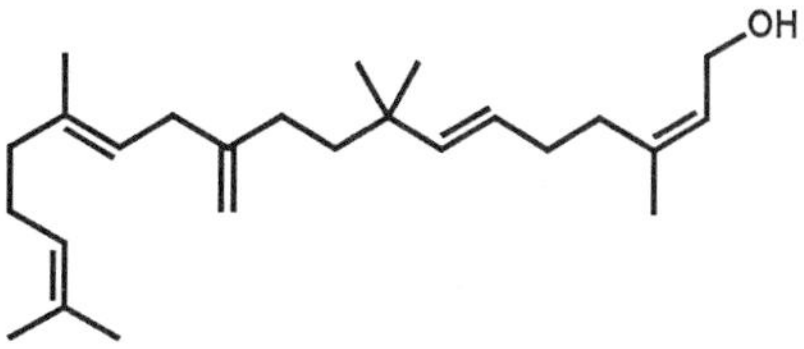

Moenocinol
Tschesche, *Ann*, **1974**, 853.
Grieco, *JACS*, **1975**, *97*, 1597.
Kocienski, *JOC*, **1980**, *45*, 2037.
Coates, *JOC*, **1980**, *45*, 2685.
Welzel, *TL*, **1983**, *24*, 5201.

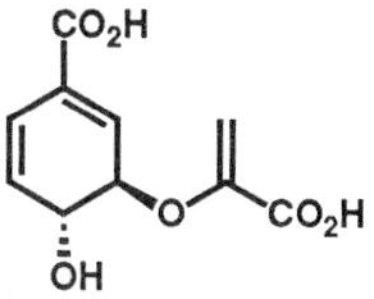

Chorismic Acid
Ganem, *JACS*, **1982**, *104*, 6787.

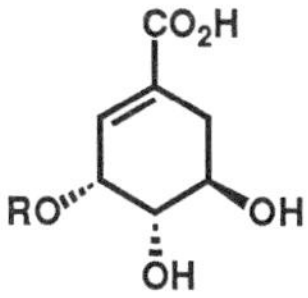

R = H: Shikimic Acid
Smissman, *JACS*, **1959**, *81*, 2909.
Raphael, *JCS*, **1960**, 1560.
Grewe, *Ber*, **1967**, *100*, 2546.
Bestmann, *ACIE*, **1971**, *10*, 336. (-)
Géro, *ACIE*, **1971**, *10*, 652. (-)
Rapoport, *JACS*, **1973**, *95*, 7821. (-)
Kitagawa, *Het*, **1982**, *17*, 209. (-)
Koreeda, *JACS*, **1982**, *104*, 2308.
Ganem, *JOC*, **1982**, *47*, 5041.
Fleet, *JCS, PI*, **1984**, 905. (-)
Campbell, *Tet*, **1984**, *40*, 2461.
Rodrigo, *CJC*, **1984**, *62*, 826.
Berchtold, *JOC*, **1987**, *52*, 1765. (-)
Birch, *JOC*, **1988**, *53*, 278. (+), (-)
R = H_2O_3P-:
Bartlett, *JACS*, **1984**, *106*, 7854.

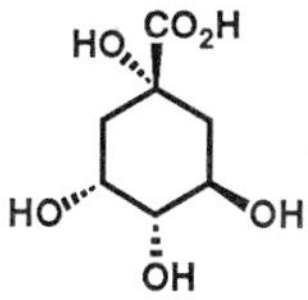

Quinic Acid
Grewe, *Ber*, **1954**, *87*, 793.
Smissman, *JACS*, **1963**, *85*, 2184.
Wolinsky, *JOC*, **1964**, *29*, 3596.
Bestmann, *ACIE*, **1971**, *10*, 336. (-)

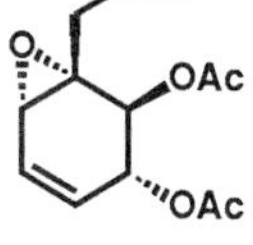

Senepoxide
Ichihara, *TL*, **1974**, 4235;
Tet, **1980**, *36*, 183.
Ganem, *JACS*, **1978**, *100*, 352.
Schlessinger, *JOC*, **1981**, *46*, 5252.
β-Senepoxide
Ogawa, *JOC*, **1985**, *50*, 2356. (+)

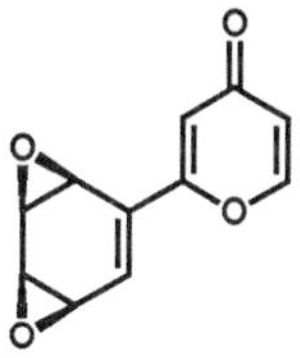

LL-Z1120
Vogel,
ACIE, **1984**, *23*, 966.

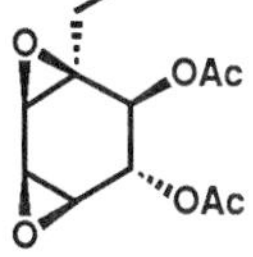

Crotepoxide
Ichihara, *TL*, **1975**, 3187;
Tet, **1980**, *36*, 183.
White, *JACS*, **1976**, *98*, 634.
Matsumoto, *TL*, **1977**, 3361.
Schlessinger, *JOC*, **1981**, *46*, 5252.
Ogawa, *BCS, Jpn*, **1987**, *60*, 800. (+)

13.3

(+)-Dictyopterene A

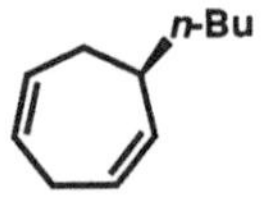

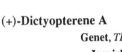

(-)-Dictyopterene C' (*nat*)

Genet, *TL*, **1985**, *26*, 2779. (+)
Jaenicke, *Ann*, **1979**, 986;
TL, **1986**, *27*, 2349. (+), (-)

1*R*,3*R*-Chrysanthemic Acid

Krief, *TL*, **1978**, 1847; (±)
JACS, **1982**, *104*, 4282;
TL, **1983**, *24*, 103;
ibid., **1988**, *29*, 1079.
Mulzer, *ACIE*, **1983**, *22*, 63.

Dictyopterene B

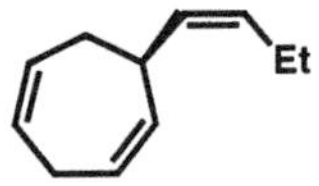

Dictyopterene D'

Weinstein, *JCS, CC*, **1971**, 940.
Jaenicke, *Ann*, **1979**, 986.
Schneider, *JACS*, **1980**, *102*, 6114.

Chrysanthemum-dicarboxylic Acid

Fraser-Reid,
JACS, **1979**, *101*, 6123. (+), (-)

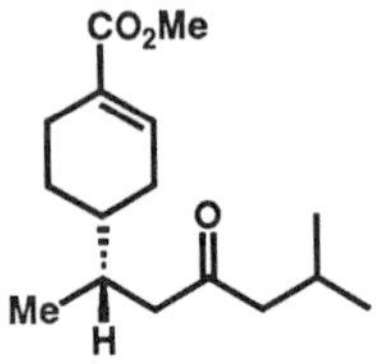

***dl-erythro*-Juvabione**

Ficini, *JACS*, **1974**, *96*, 1213.
Evans, *JACS*, **1980**, *102*, 774.
Schultz, *JOC*, **1984**, *49*, 2615.

***dl-threo*-Juvabione**

Morgans, *JACS*, **1983**, *105*, 5477.

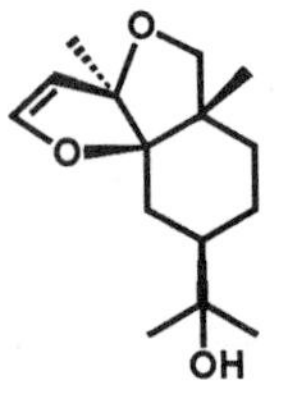

Phytuberin

Masamune, *JACS*, **1978**, *100*, 7751;
BCS, Jpn, **1982**, *55*, 1202.
Caine, *JACS*, **1980**, *102*, 7568. (-)
Findlay, *CJC*, **1980**, *58*, 2827. (-)
Yoshikoshi, *JOC*, **1986**, *51*, 1478.

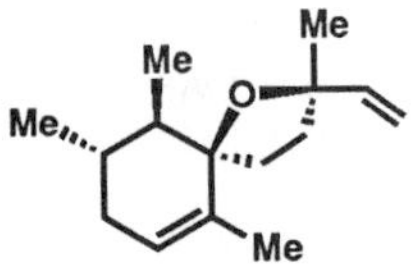

Dactyloxene B

Maurer,
HCA, **1980**, *63*, 293, 2503.

Karahana Ether

Coates, *JOC*, **1970**, *35*, 865.
Yamada, *TL*, **1979**, 1323.
Mukaiyama, *CL*, **1979**, 1175.
Weiler, *CJC*, **1986**, *64*, 584.

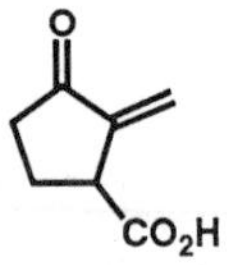

Sarkomycin

Toki, *BCS, Jpn*, **1958**, *31*, 333. (*R*), (*S*)
Marx, *TL*, **1979**, 4175.
Boeckman, *JOC*, **1980**, *45*, 752. (*R*)
Tsuji, *TL*, **1981**, *22*, 4295.
Kozikowski, *JACS*, **1982**, *104*, 4023.

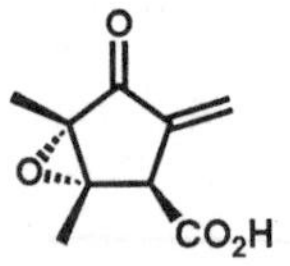

Methylenomycin A

Smith, *JACS*, **1977**, *99*, 7085;
ibid., **1980**, *102*, 3904.
Jernow, *JOC*, **1979**, *44*, 4210. (+)
Uda, *JCS, CC*, **1982**, 496.

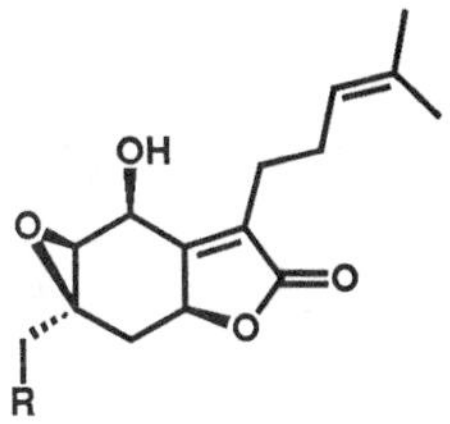

R = H: Paniculide A
R = OH: Paniculide B
Smith, *JOC*, **1981**, *46*, 4814;
JACS, **1983**, *105* , 575, (A&B).
Baker, *JCS, PI*, **1985**, 1509, (B).
Yoshikoshi, *Tet*, **1987**, *43*, 5467, (A).
Jacobi, *Tet*, **1987**, *43*, 5475, (A).

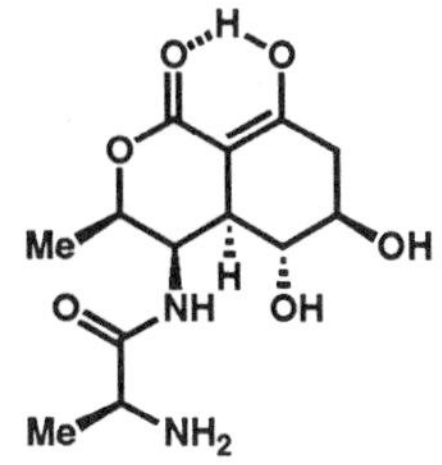

(+)-Actinobolin
Ohno, *JACS*, **1984**, *106*, 1133.
Weinreb, *JACS*, **1985**, *107*, 7790.
Kozikowski, *JACS*, **1987**, *109*, 5167.
***N*-Acetylactinobolamine**
Fraser-Reid, *JACS*, **1985**, *107*, 5576.
Danishefsky, *JOC*, **1985**, *50*, 5005. (±)

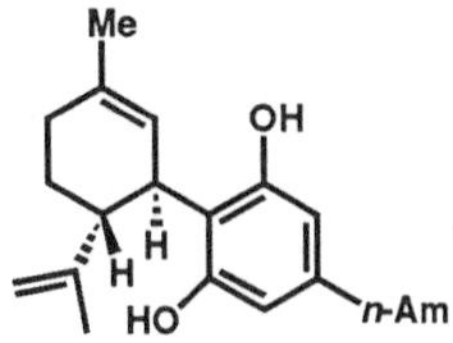

(-)-Cannabidiol
Eschenmoser,
HCA, **1967**, *50*, 719.

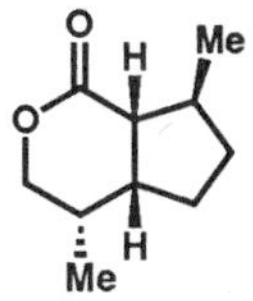

Dihydronepetalactone
Fleming, *TL*, **1984**, *25*, 5103.

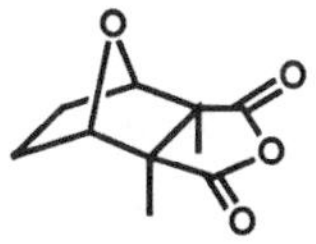

Cantharidin
Ziegler, *Ann*, **1942**, *551*, 1.
Stork, *JACS*, **1953**, *75*, 384.
Dauben, *JACS*, **1980**, *102*, 6893.

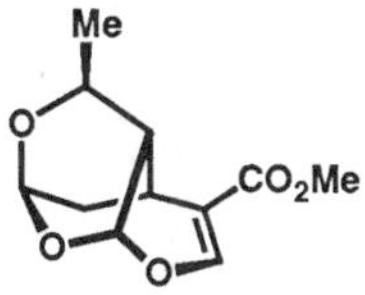

(-)-Sarracenin (*nat*)
Whitesell, *JOC*, **1978**, *43*, 784;
JACS, **1981**, *103*, 3468. (±)
Tietze, *ACIE*, **1982**, *21*, 70. (+)
Baldwin, *JACS*, **1982**, *104*, 1132. (-)
Takano, *TL*, **1983**, *24*, 401. (-)

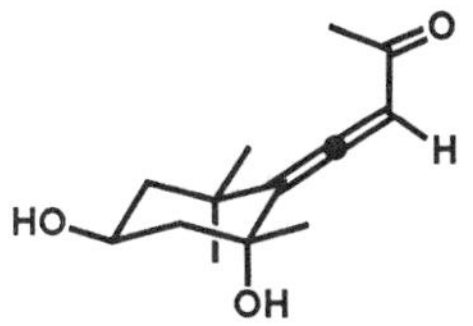

Grasshopper Ketone
Meinwald, *TL*, **1969**, 1657.
Weedon, *JCS, CC*, **1969**, 85.
Isoe, *TL*, **1971**, 1089.
Mori, *TL*, **1973**, 723.

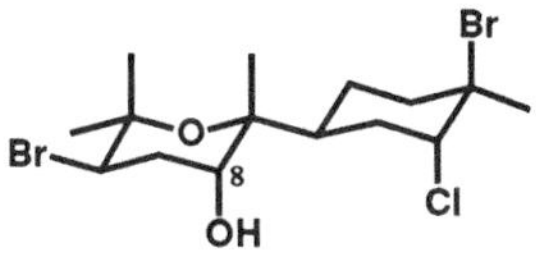

Isocaespitol
González, *TL*, **1976**, 2279.
8-Desoxyisocaespitol
González, *TL*, **1980**, *21*, 187.

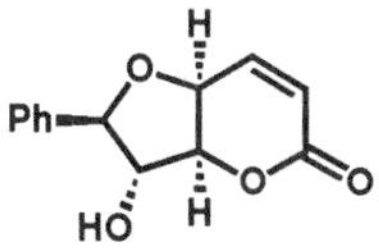

(+)-Altholactone
(Goniothalenol)
Gesson, *TL*, **1987**, *28*, 3949.
Tadano & Ogawa, *CL*, **1988**, 111.

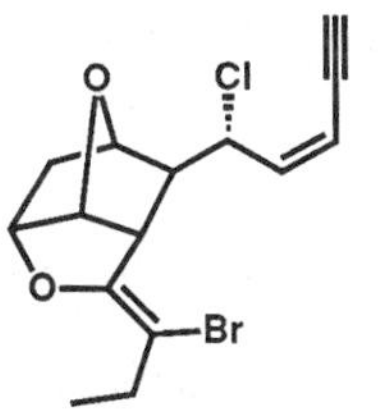

***cis*-Maneonene-A**
Holmes, *JCS, CC*, **1983**, 415.

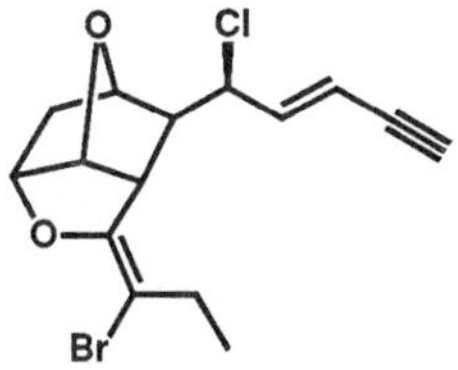

***trans*-Maneonene-B**
Holmes, *JCS, CC*, **1984**, 1594.

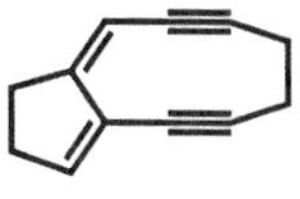

The Parent Carbocyclic Subunit of Neocarzinostatin Chromophore A
Wender, *TL*, **1988**, *29*, 909.

13.4

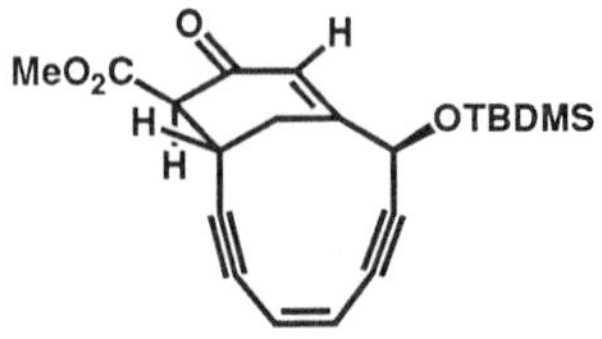

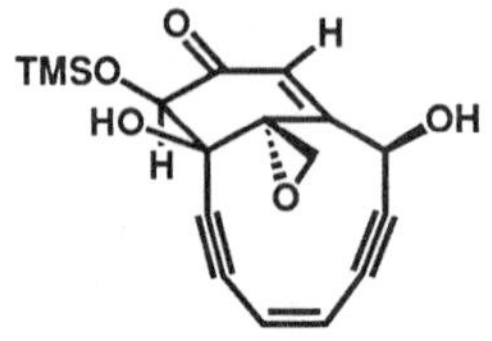

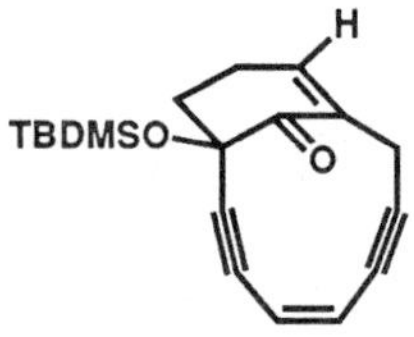

Bicyclic Core of the Esperamicin/Calichemicin Class of Antitumor Agents

Schreiber,
JACS, **1988**, *110*, 631.

Danishefsky,
JACS, **1988**, *110*, 6890.

Magnus,
JACS, **1988**, *110*, 6921.

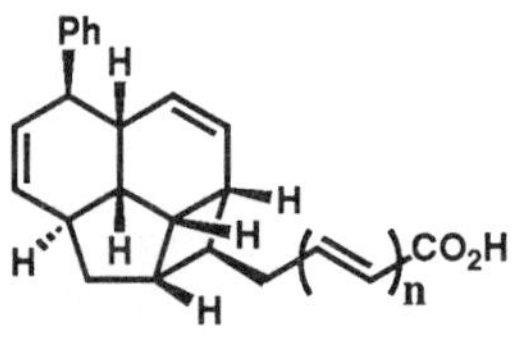

n = 0: Endiandric Acid A
n = 1: Endiandric Acid B
Nicolaou,
JACS, **1982**, *104*, 5555, 5558, 5560.

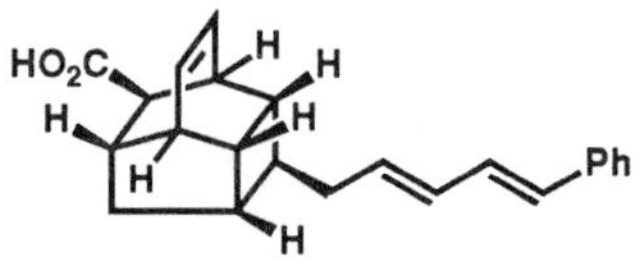

Endiandric Acid C
Nicolaou,
JACS, **1982**, *104*, 5557, 5558, 5560.

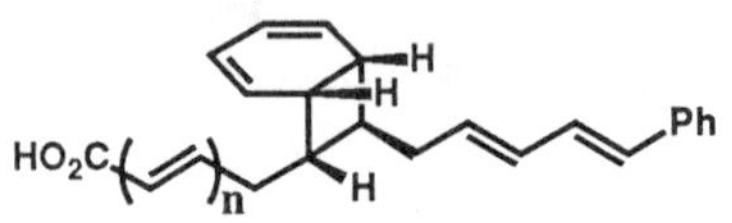

n = 0: Endiandric Acid D
n = 1: Endiandric Acid G
Nicolaou,
JACS, **1982**, *104*, 5557, 5558, 5560.

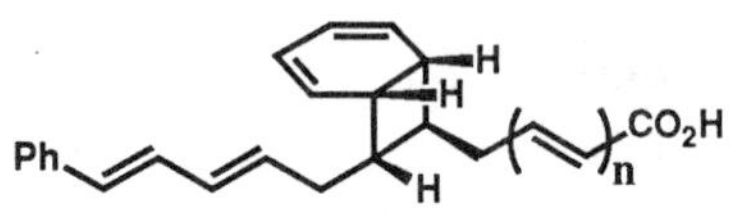

n = 0: Endiandric Acid E
n = 1: Endiandric Acid F
Nicolaou,
JACS, **1982**, *104*, 5557, 5558, 5560.

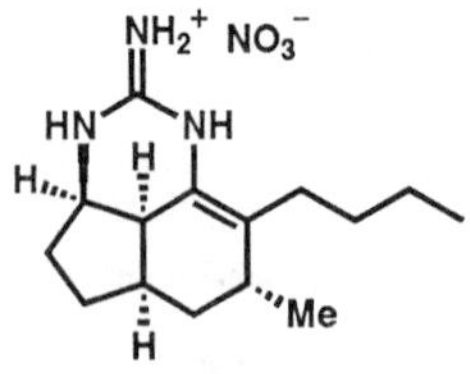

(-)-Ptilocaulin (*unnat*)
Snider, *TL*, **1983**, *24*, 861; (±)
JACS, **1984**, *106*, 1443. (±), (-)
Roush, *Tet*, **1985**, *41*, 3463. (-)
Hassner, *TL*, **1986**, *27*, 1407. (±)
Uyehara, *JCS, CC*, **1986**, 539. (±)

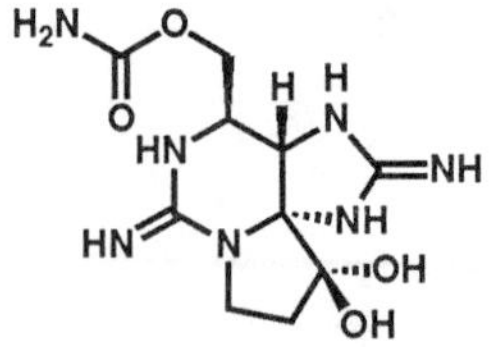

Saxitoxin
Kishi, *JACS*, **1977**, *99*, 2818;
Het, **1980**, *14*, 1477;
JOC, **1983**, *48*, 3833.
Jacobi, *JACS*, **1984**, *106*, 5594.

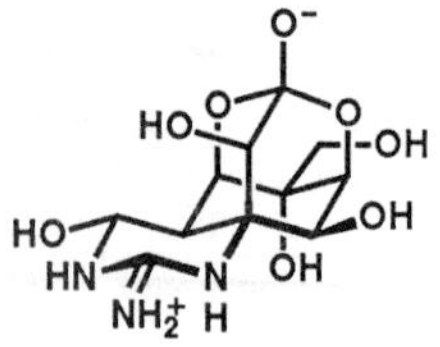

Tetrodotoxin
Kishi,
JACS, **1972**, *94*, 9219.

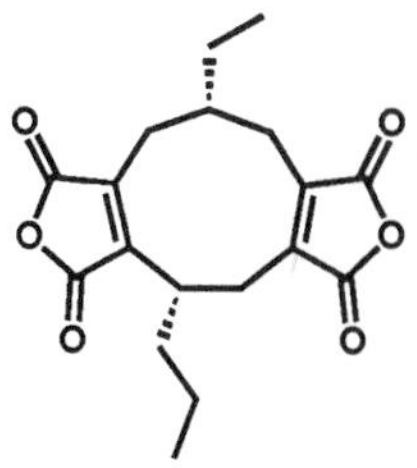

Byssochlamic Acid

Stork, *JACS*, **1972**, *94*, 4735.

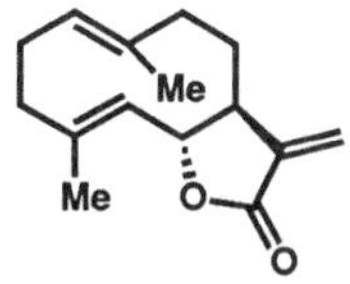

Costunolide

Grieco, *JOC*, **1977**, *42*, 1717. (+)
Takahashi, *JOC*, **1986**, *51*, 4315;
Tet, **1987**, *43*, 5499.
Kitagawa, *CL*, **1986**, 85.

Dihydrocostunolide

Fujimoto, *TL*, **1976**, 2041.
Raucher, *JOC*, **1986**, *51*, 5503. (+)

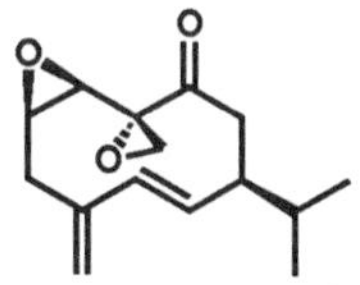

Periplanone-B

Still, *JACS*, **1979**, *101*, 2493.
Schreiber, *JACS*, **1984**, *106*, 4038.
Takahashi, *JOC*, **1986**, *51*, 3393.
Hauptmann, *TL*, **1986**, *27*, 1315.
Kitahara, *Tet*, **1987**, *43*, 2689. (-)

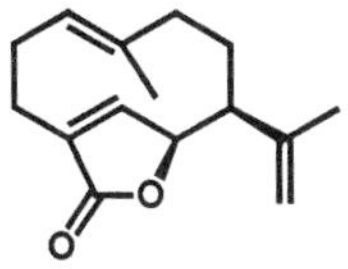

Aristolactone

Marshall,
TL, **1987**, *28*, 723; (±)
ibid., **1987**, *28*, 3323. (+)

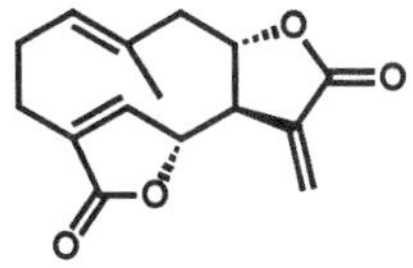

Isabelin

Wender,
JACS, **1980**, *102*, 6340.

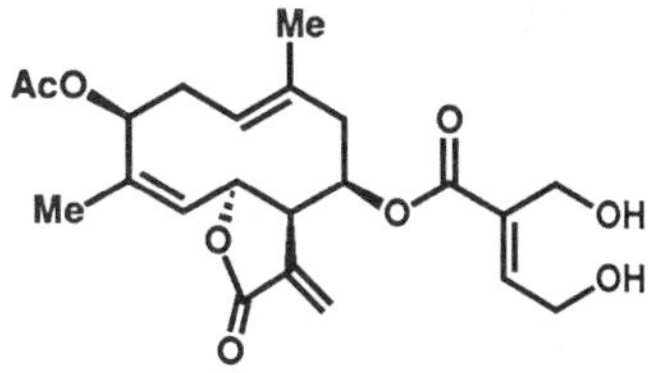

Eucannabinolide

Still, *JACS*, **1983**, *105*, 625. (*nat*)

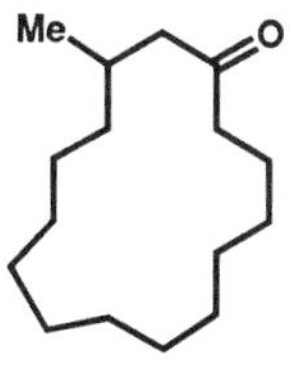

Muscone

Eschenmoser, *HCA*, **1971**, *54*, 2896.
Mookherjee, *JOC*, **1971**, *36*, 4124.
Baker & Cookson, *JCS, CC*, **1974**, 515.
Stork, *JACS*, **1975**, *97*, 1264.
Baumann, *TL*, **1976**, 3585.
Branca, *HCA*, **1977**, *60*, 925. (*R*), (*S*)
Taguchi, *BCS, Jpn*, **1977**, *50*, 1592.
Saegusa, *JOC*, **1977**, *42*, 2326.
Utimoto, *TL*, **1978**, 2301. (*R*)

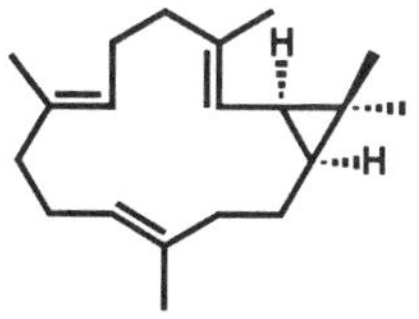

(-)-Casbene

Crombie & Pattenden,
JCS, CC, **1976**, 66;
JCS, PI, **1980**, 1711. (-)
Takahashi,
CL, **1982**, 863. (±)
McMurry,
JOC, **1987**, *52*, 4885. (+)

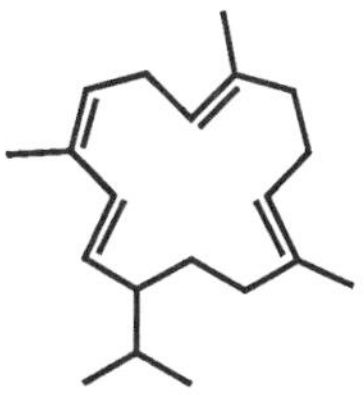

Cembrene

Dauben, *JACS*, **1974**, *96*, 4724.
Kato, *Synth. Comm*, **1976**, *6*, 365.

Cembrene-A

Ito, *TL*, **1975**, 3065.
Schwabe, *HCA*, **1988**, *71*, 292. (-)

13.4

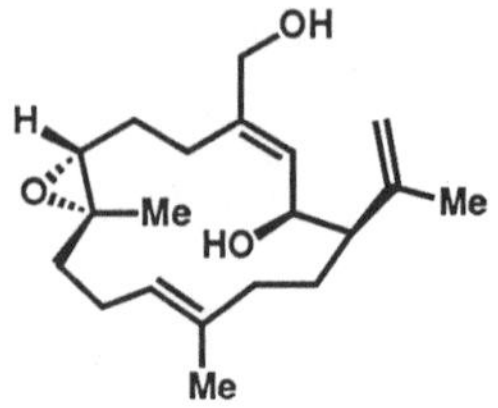

Asperdiol

Kato, *TL*, **1983**, *24*, 2267; *JOC*, **1987**, *52*, 1803.
Still, *JOC*, **1983**, *48*, 4785.
Tius, *JACS*, **1986**, *108*, 6389. (-)
Marshall, *JOC*, **1986**, *51*, 858.

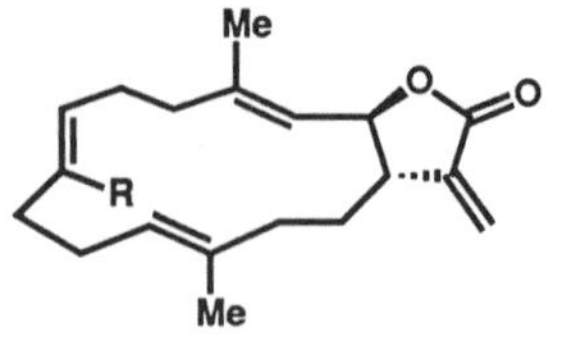

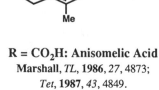

R = CO_2H: Anisomelic Acid

Marshall, *TL*, **1986**, *27*, 4873; *Tet*, **1987**, *43*, 4849.

R = Me:

Ito, *TL*, **1982**, *23*, 5175.
Marshall, *JOC*, **1988**, *53*, 1616.

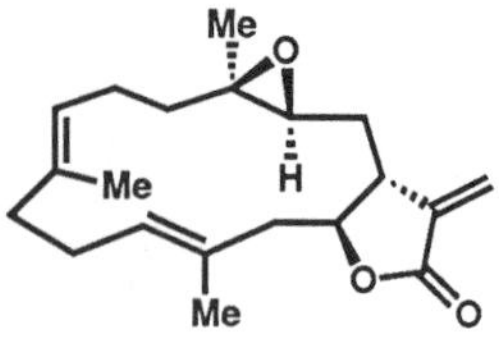

Isolobophytolide

Marshall, *TL*, **1986**, *27*, 5197; *JOC*, **1987**, *52*, 2378.

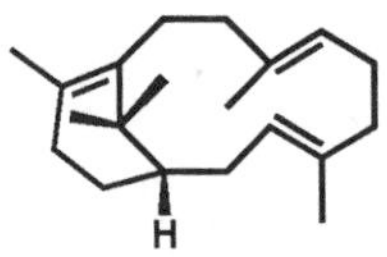

Verticillene

Pattenden, *TL*, **1985**, *26*, 3393.

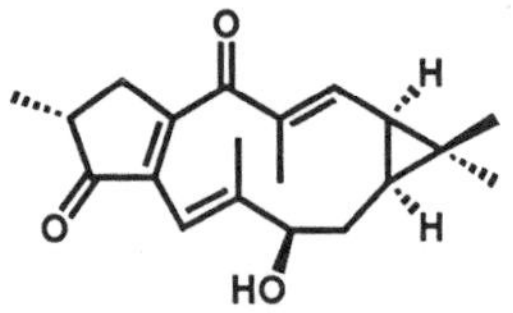

(-)-Bertyadionol

Smith, *JACS*, **1986**, *108*, 3110.

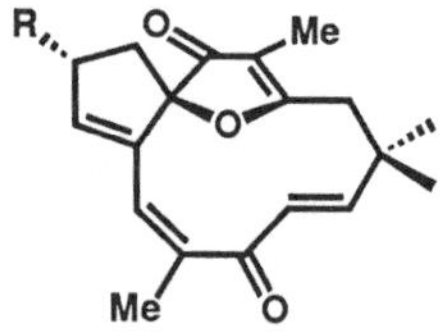

R = H: Normethyljatrophone

Smith, *JACS*, **1981**, *103*, 219.

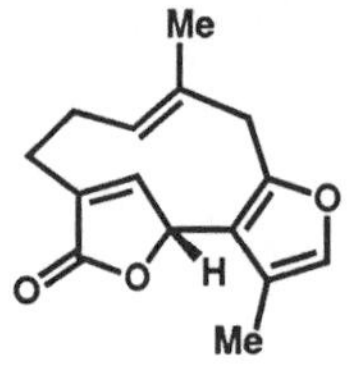

Linderalactone

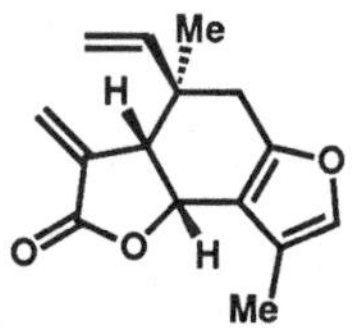

Isolinderalactone

Magnus, *JACS*, **1980**, *102*, 1756; *JOC*, **1984**, *49*, 2317.

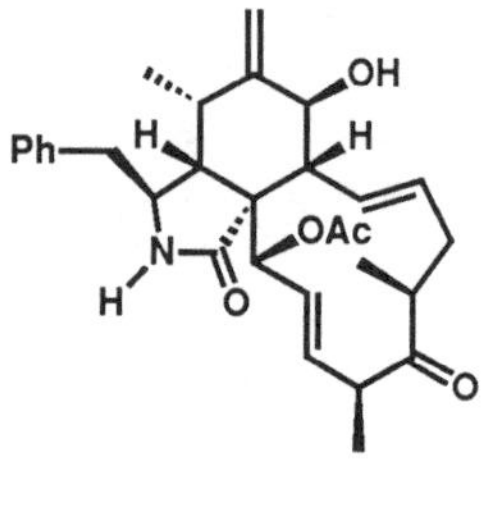

Zygosporin E

Vedejs, *JACS*, **1988**, *110*, 4822.

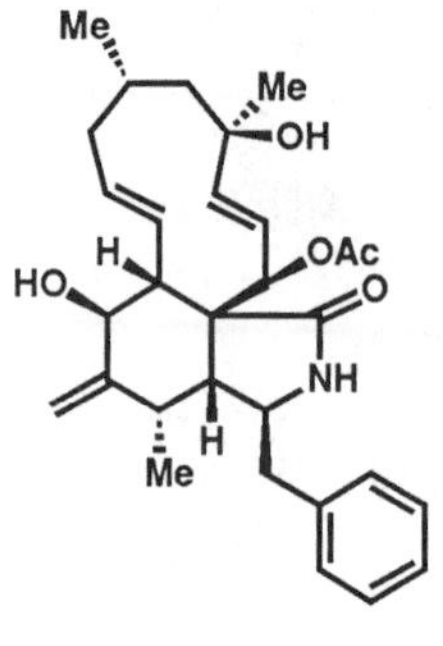

(-)-Cytochalasin H

Thomas, *JCS, CC*, **1986**, 727, 1449.

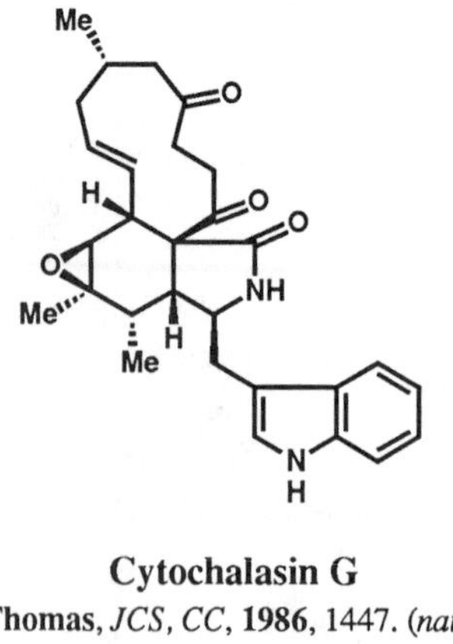

Cytochalasin G

Thomas, *JCS, CC*, **1986**, 1447. (*nat*)

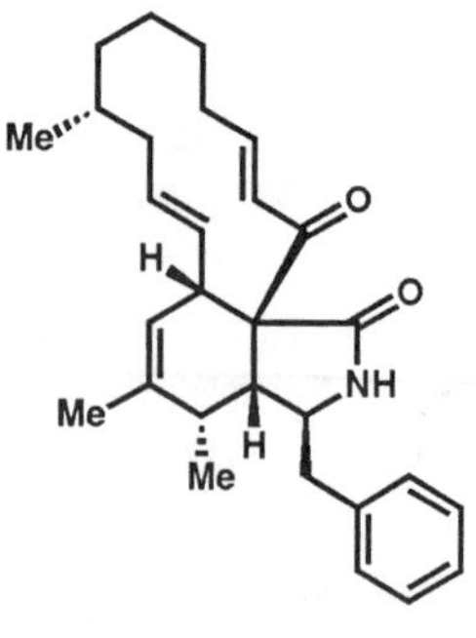

(-)-Proxiphomin

Thomas, *JCS, CC*, **1985**, 143.

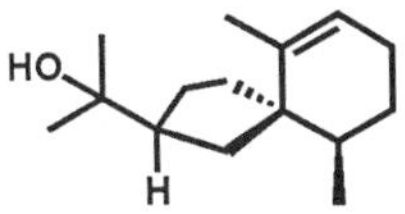
HO
H

Hinesol (H)

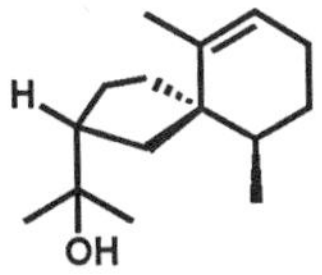
H
OH

Agarospirol (Ag)

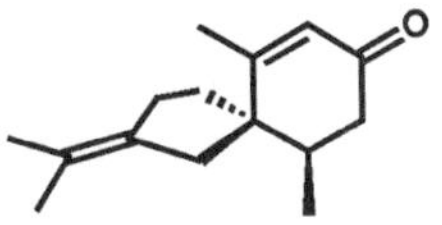
O

β-Vetivone (V)

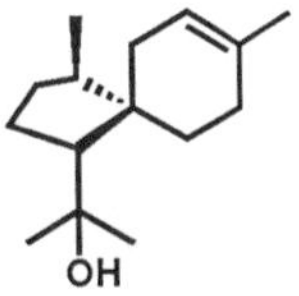
OH

β-Acorenol (Ac)

Marshall, *TL*, **1969**, 1387, (H); *JOC*, **1970**, *35*, 192, (V); *ibid*, **1970**, *35*, 4068, (V).
Deslongchamps, *CJC*, **1970**, *48*, 3273, (Ag). **Yamada**, *TL*, **1973**, 4963, (V); *ibid*, **1973**, 4967, (H).
Stork, *JACS*, **1973**, *95*, 3414, (V). **McCurry**, *TL*, **1973**, 3325, (V). **Oppolzer**, *HCA*, **1973**, *56*, 1812, (Ac).
Ramage, *JCS, CC*, **1975**, 662, (-)-(Ag&V). **Dauben**, *JACS*, **1975**, *97*, 1622; *ibid.*, **1977**, *99*, 7307, (H&V).
Büchi, *JOC*, **1976**, *41*, 3208, (H&V). **Wenkert**, *JACS*, **1978**, *100*, 1267, (V).
Magnus, *JOC*, **1978**, *43*, 1750, (+)-(H). **Inubushi**, *CJC*, **1979**, *57*, 1579, (H&Ag&V); *TL*, **1979**, 159, (H&Ag).

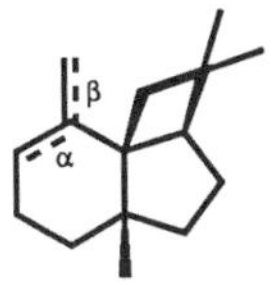
β
α

α- and β-Panasinsene
McMurry,
TL, **1980**, *21*, 2477.

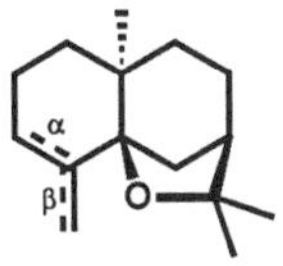
α
β
O

α- and β-Agarofuran
Büchi, *JACS*, **1967**, *89*, 5665;
JOC, **1979**, *44*, 546.
Marshall, *JOC*, **1968**, *33*, 435.
Deslongchamps, *CJC*, **1968**, *46*, 2817.

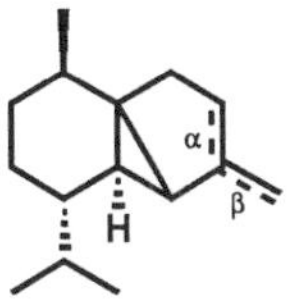
α
β
H

α- and β-Cubebene
Piers, *TL*, **1969**, 1251.
Yoshikoshi, *JCS, CC*, **1969**, 308.

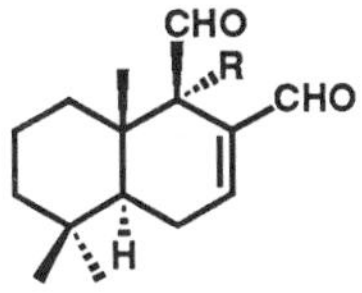
CHO
R
CHO
H

R = OH: Warburganal
Ohsuka, *CL*, **1979**, 635.
Nakanishi, *JACS*, **1979**, *101*, 4398.
Oishi, *JACS*, **1979**, *101*, 4400.
Kende, *TL*, **1980**, *21*, 3119.
Goldsmith, *TL*, **1980**, *21*, 3543.
Ley, *TL*, **1981**, *22*, 3909.
Ohno, *TL*, **1982**, *23*, 1087. (-)
Wender, *TL*, **1982**, *23*, 1871.
de Groot, *JOC*, **1988**, *53*, 855.
R = H: Polygodial
Kitahara, *TL*, **1971**, 1961.
Ley, *JCS, PI*, **1983**, 1579.
Lallemand, *Tet*, **1983**, *39*, 749.
Mori, *Tet*, **1986**, *42*, 273. (+), (-)

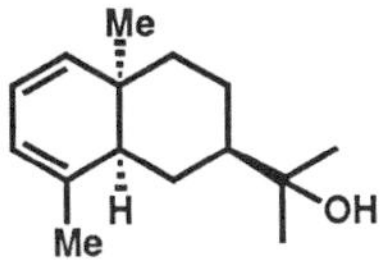
Me
H
Me
OH

(+)-Occidentalol
Ando, *TL*, **1970**, 3891. (-)
Deslongchamps, *CJC*, **1972**, *50*, 336. (+)
Heathcock, *CJC*, **1972**, *50*, 340. (+)
Hortmann, *JOC*, **1973**, *38*, 728. (+)
Marshall, *JOC*, **1977**, *42*, 1794. (±)

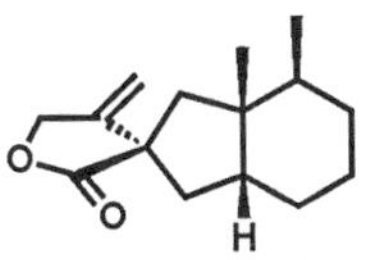
O
O
H

Bakkenolide-A
Evans, *TL*, **1973**, 4691;
JACS, **1977**, *99*, 5453.
Hayashi, *CPB*, **1973**, *21*, 2806.
Greene, *JOC*, **1985**, *50*, 3943.

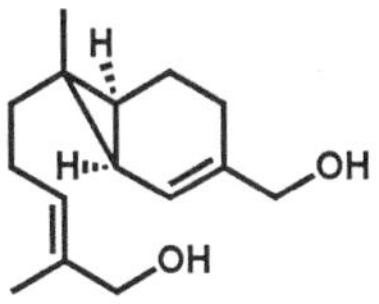
H
H
OH
OH

Sirenin
Rapoport, *JACS*, **1969**, *91*, 4933;
ibid., **1970**, *92*, 3429;
ibid., **1971**, *93*, 1758. (+), (-)
Grieco, *JACS*, **1969**, *91*, 5660.
Mori, *Tet*, **1970**, *26*, 2801.
Garbers, *TL*, **1975**, 3753.
Hiyama, *JACS*, **1976**, *98*, 2362.

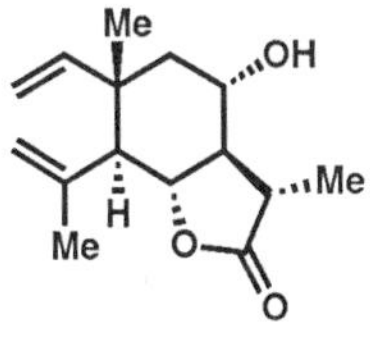
Me
OH
Me
H
Me
O
O

Temisin
Grieco, *JCS, CC*, **1978**, 76.

13.4

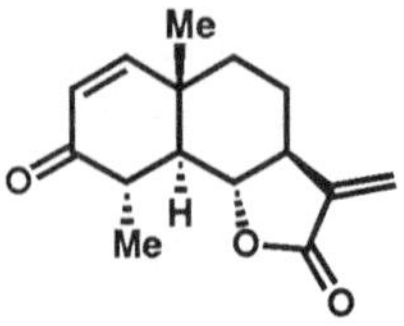
Me
O
H
Me
O
O

Tuberiferine
Grieco, *JCS, CC*, **1976**, 582.

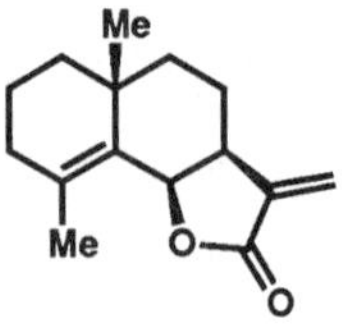
Me
Me
O
O

Frullanolide
Still, *JACS*, **1977**, *99*, 948.
Yoshikoshi, *JACS*, **1979**, *101*, 6420.
Semmelhack, *JACS*, **1981**, *103*, 3945.

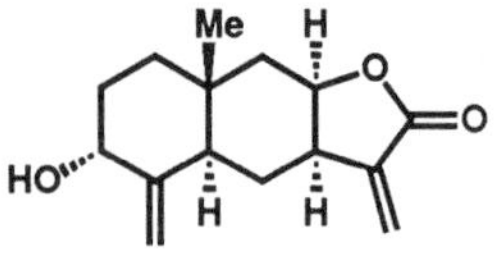
Me
H
O
O
HO
H
H

Isotelekin
Miller, *JACS*, **1974**, *96*, 8102.

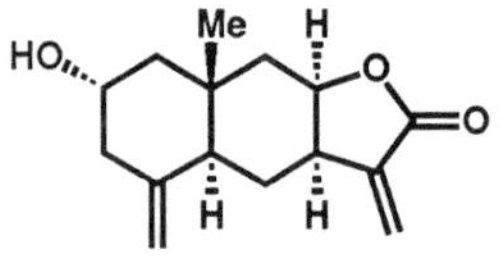
HO
Me
H
O
O
H
H

(+)-Ivalin
Koga, *TL*, **1984**, *25*, 333.

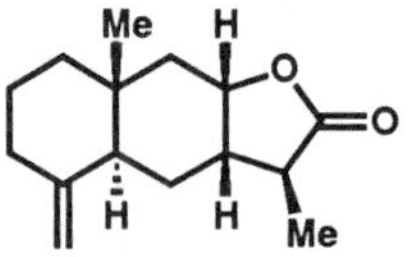
Me
H
O
O
H
H
Me

Dihydrocallitrisin
Schultz, *TL*, **1979**, 3241;
JACS, **1980**, *102*, 2414.

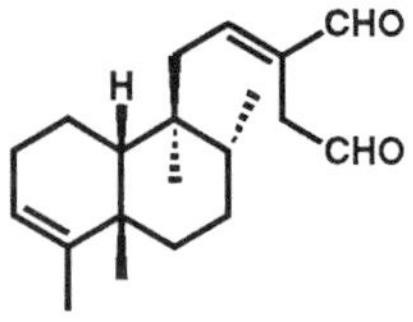
CHO
H
CHO

Linaridial
Tokoroyama,
TL, **1987**, *28*, 6645.

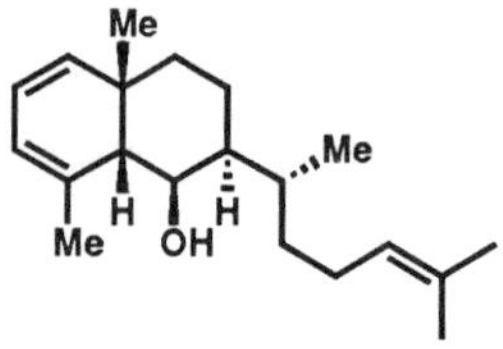
Me
Me
H
Me
OH
H

Dictyolene
Marshall, *JACS*, **1978**, *100*, 1627.

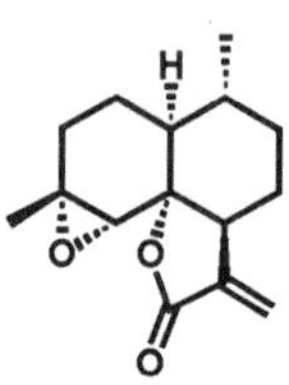
H
O
O
O

Arteannuin B
Lansbury,
TL, **1986**, *27*, 3967.

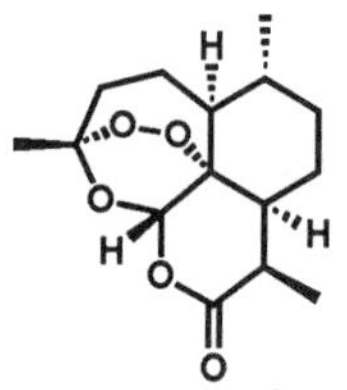
H
O–O
O
H
H
O
O

(+)-Arteannuin (Qinghaosu)
Hofheinz, *JACS*, **1983**, *105*, 624.
Zhou, *Tet*, **1986**, *42*, 819;
P&AC, **1986**, *58*, 817.
Avery, *TL*, **1987**, *28*, 4629.

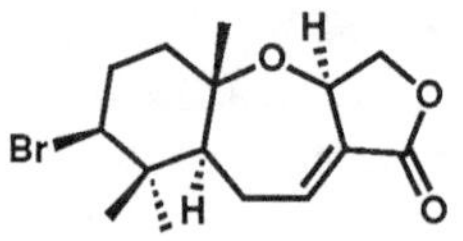
H
O
O
Br
H
O

Aplysistatin
Hoye, *JACS*, **1979**, *101*, 5065;
ibid., **1982**, *104*, 6704.
White, *JACS*, **1982**, *104*, 3923.
Prestwich, *TL*, **1982**, *23*, 4643. (-)
Gosselin & Rouessac,
TL, **1983**, *24*, 5515.
Kraus, *JOC*, **1983**, *48*, 5356.
Tanaka, *Ag. BC*, **1984**, *48*, 2535. (-)

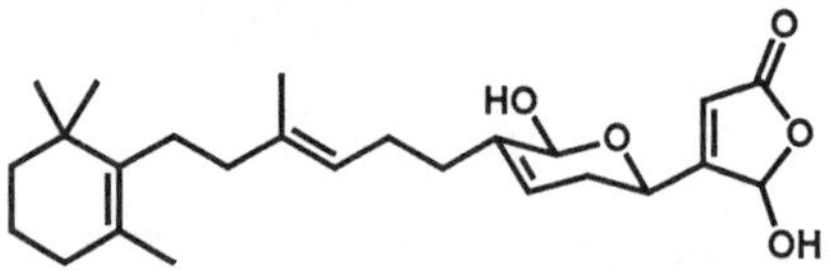
HO
O
O
O
OH

Manoalide
Katsumura & Isoe, *TL*, **1985**, *26*, 5827;
ibid., **1988**, *29*, 1173.
Garst, *TL*, **1986**, *27*, 4533.
(*S*)-Manoalide Diol
Weigele, *TL*, **1988**, *29*, 2401.

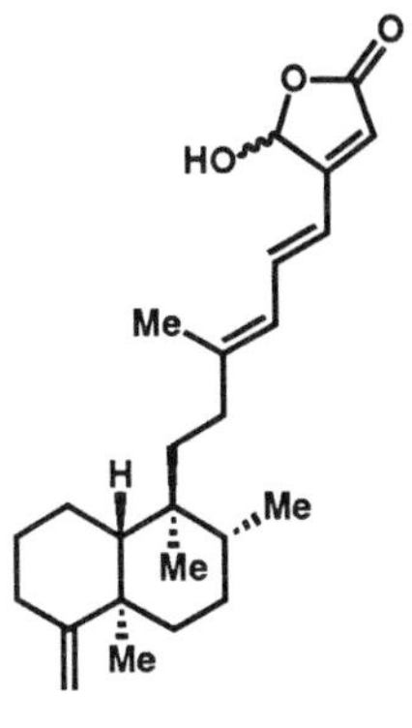

Palauolide
Piers, *JCS, CC*, **1987**, 1342.

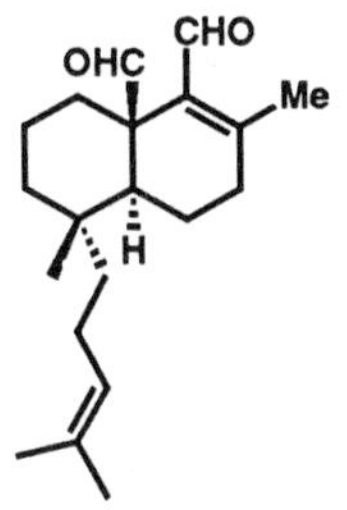

(+)-Perrottetianal A
Hagiwara,
JCS, CC, **1987**, 1351.

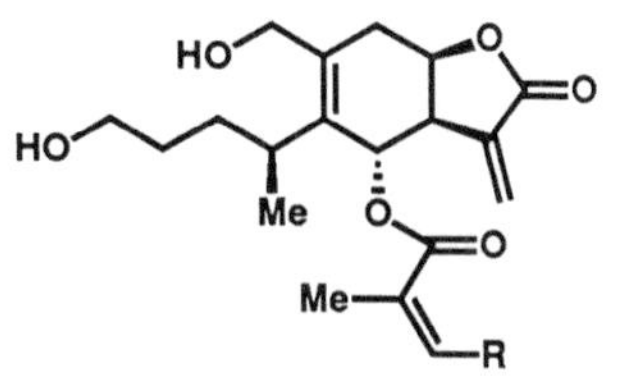

R = Me: Eriolangin
R = H: Eriolanin
Grieco, *JACS*, **1978**, *100*, 1616;
ibid., **1980**, *102*, 5886.
Schlessinger, *JACS*, **1981**, *103*, 724.

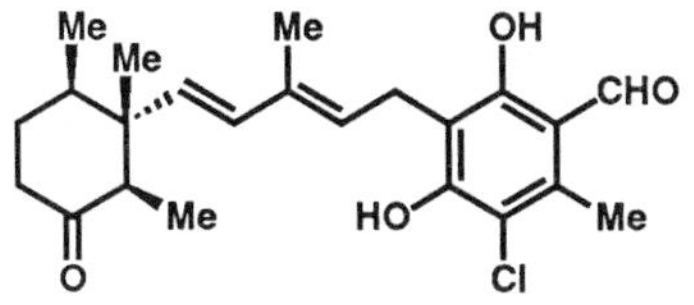

Ascochlorin
Mori, *TL*, **1982**, *23*, 5443.

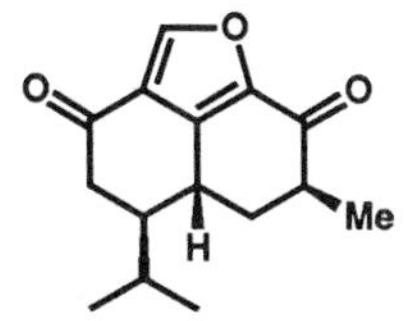

Hibiscone C
Smith,
JACS, **1982**, *104*, 5568;
ibid., **1984**, *106*, 2115.

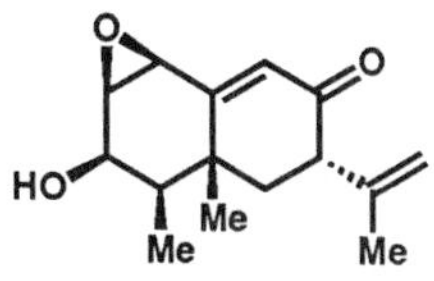

Eremofortin B
Yamakawa, *CL*, **1981**, 929.

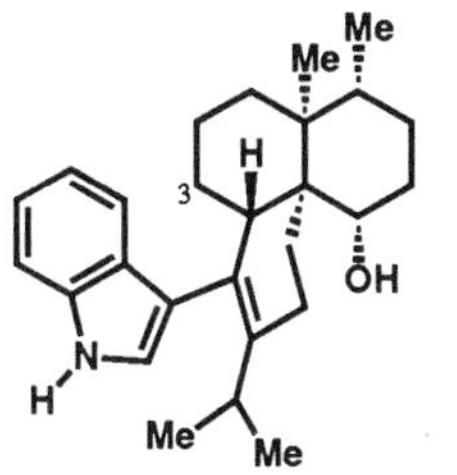

3-Desmethylaflavinine
Danishefsky, *JACS*, **1985**, *107*, 2474.

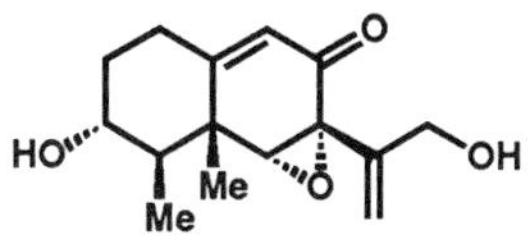

Phomenone
Yamakawa, *TL*, **1979**, 3871;
CPB, **1980**, *28*, 3265.

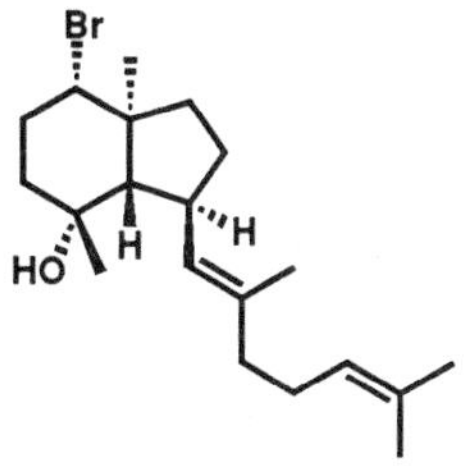

Prepinnaterpene
Masamune, *TL*, **1987**, *28*, 4303.

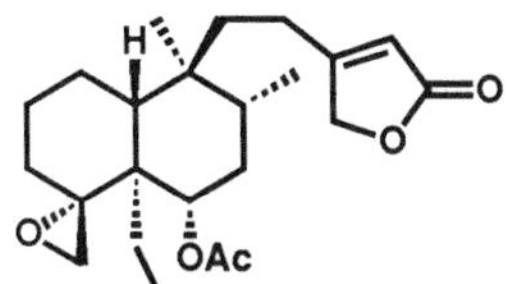

Ajugarin I
Ley, *JCS, CC*, **1983**, 503;
Tet, **1986**, *42*, 6519.

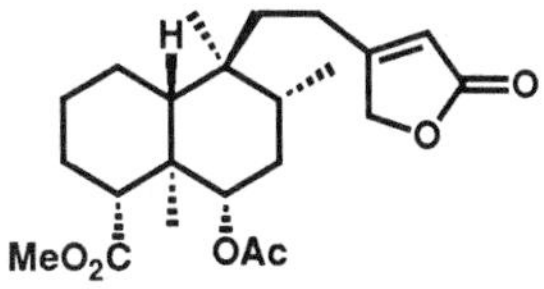

Ajugarin IV
Kende, *TL*, **1982**, *23*, 1751.

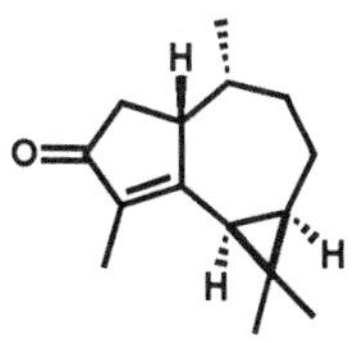

Cyclocolorenone
Caine, *JOC*, **1972**, *37*, 3751. (-)
Nicholas, *TL*, **1986**, *27*, 915.

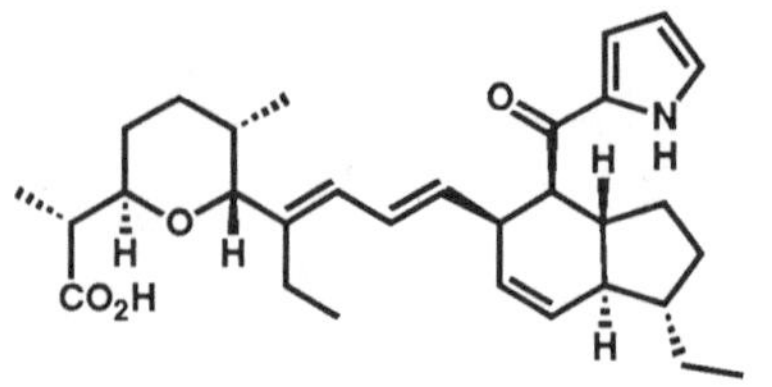

(-)-Antibiotic X-14547A (Indanomycin)

Nicolaou, *JACS*, **1981**, *103*, 6967, 6969; *JOC*, **1985**, *50*, 1440.
Roush, *JOC*, **1984**, *49*, 3429.
Ley, *JOC*, **1984**, *49*, 3503.
Boeckman, *JOC*, **1986**, *51*, 4743.

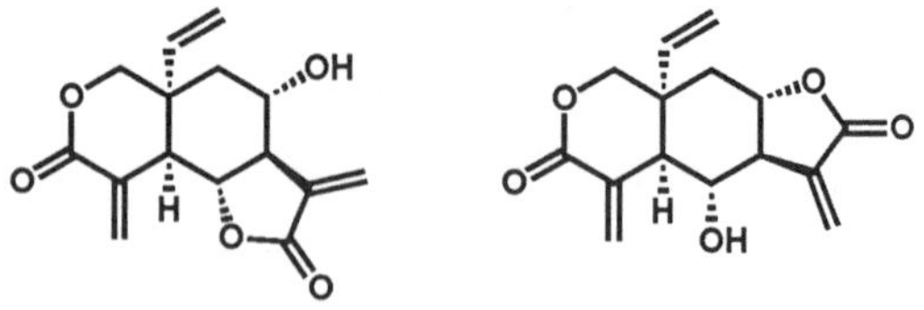

Vernolepin **Vernomenin**

Grieco, *JACS*, **1976**, *98*, 1612; *ibid.*, **1977**, *99*, 5773.
Danishefsky, *JACS*, **1976**, *98*, 3028; *ibid.*, **1977**, *99*, 6066.
Schlessinger, *JACS*, **1978**, *100*, 1938; *ibid.*, **1980**, *102*, 782.
Isobe, *JACS*, **1978**, *100*, 1940; *ibid.*, **1979**, *101*, 6076.
Vandewalle, *Tet*, **1979**, *35*, 2389.
Wakamatsu, *JOC*, **1985**, *50*, 108.

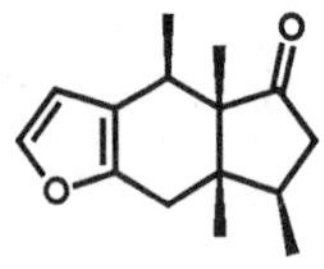

Pinguisone

Jommi, *JCS, PI*, **1981**, 2394.
Uyehara, *TL*, **1985**, *26*, 2343.

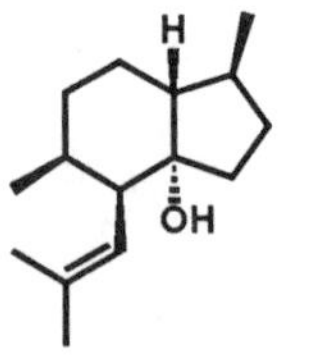

Pacifigorgiol

Clardy, *P&AC*, **1982**, *54*, 1915.

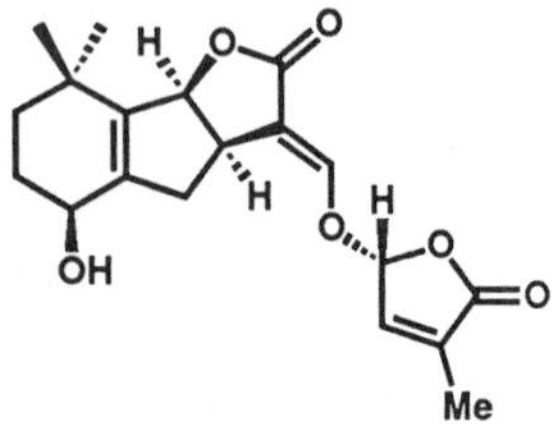

Strigol

Sih, *JACS*, **1974**, *96*, 1976; (±) *ibid.*, **1976**, *98*, 3661. (+)
Raphael, *JCS, CC*, **1974**, 834; *JCS, PI*, **1976**, 410.
Brooks, *JOC*, **1985**, *50*, 628.
Dailey, *JOC*, **1987**, *52*, 1984.
Welzel, *TL*, **1987**, *28*, 3091; (±) *ibid.*, **1987**, *28*, 3095. (+)

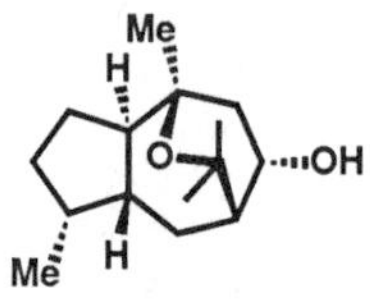

Kessanol

Andersen, *TL*, **1977**, 3783.

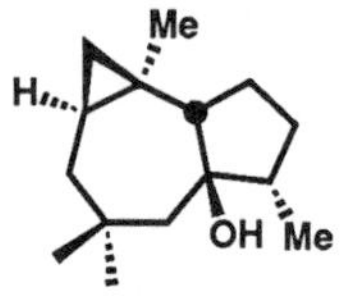

Africanol

Paquette, *TL*, **1986**, *27*, 2341; *JACS*, **1987**, *109*, 3025.

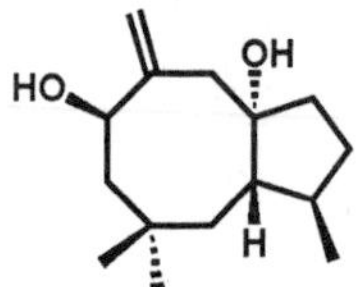

Poitediol

Gadwood, *JACS*, **1984**, *106*, 3869.

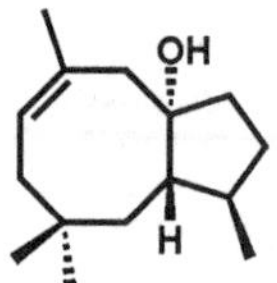

Dactylol

Gadwood, *JCS, CC*, **1985**, 123.
Hayasaka, *TL*, **1985**, *26*, 873.
Paquette, *TL*, **1985**, *26*, 4983.

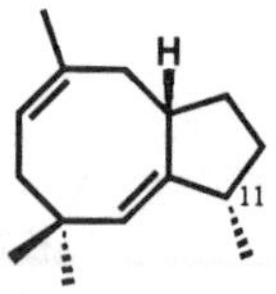

Precapnelladiene

Paquette, *JACS*, **1984**, *106*, 6868; *ibid.*, **1985**, *107*, 7352.
Mehta, *JCS, CC*, **1984**, 1058.

11-*epi*-Precapnelladiene

Pattenden, *JCS, PI*, **1983**, 1913.

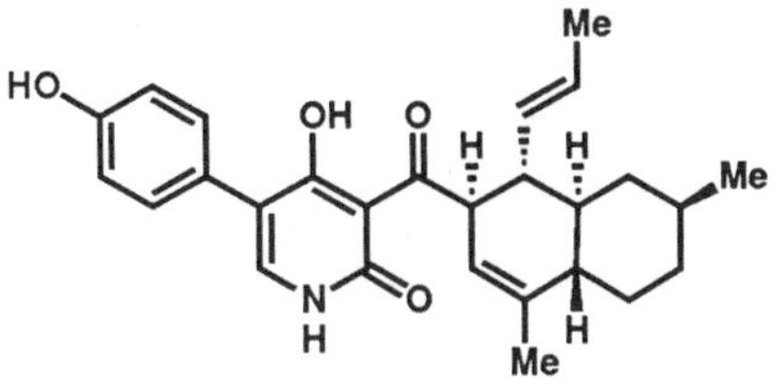

Ilicicolin H
Williams, *JOC*, **1985**, *50*, 2807.

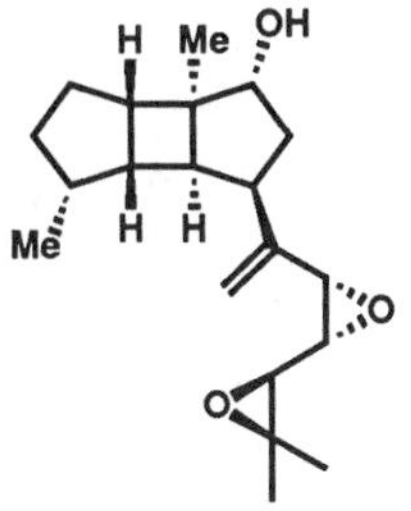

(+)-Spatol
Koga, *TL*, **1985**, *26*, 6109.

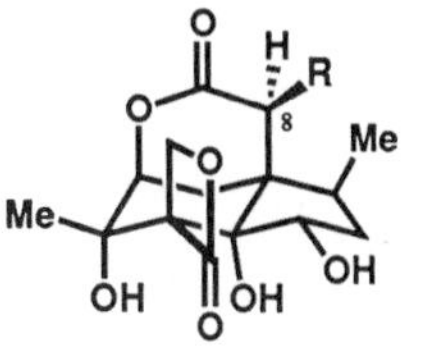

R = H: 8-Deoxyanisatin
Kende,
JACS, **1985**, *107*, 7184.

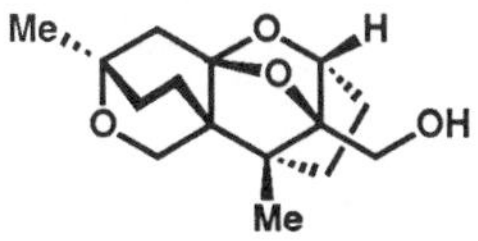

Neosporol
Ziegler, *TL*, **1988**, *29*, 1669.

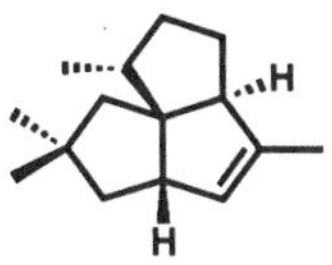

Pentalenene
Ohfune, *TL*, **1976**, 2869;
ibid., **1979**, 31.
Paquette, *JACS*, **1982**, *104*, 4504;
ibid., **1983**, *105*, 7358.
Piers, *JCS, CC*, **1984**, 959.
Crimmins, *JACS*, **1986**, *108*, 800.
Iwata, *CPB*, **1986**, *34*, 2268.
Hua, *JACS*, **1986**, *108*, 3835. (+)
Mehta, *JACS*, **1986**, *108*, 8015.
Hudlicky, *JOC*, **1987**, *52*, 4641.
Pattenden, *Tet*, **1987**, *43*, 5637.

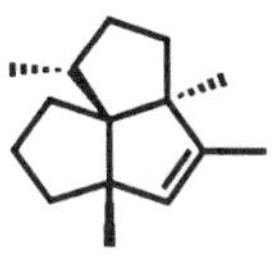

Isocomene
Oppolzer, *HCA*, **1979**, *62*, 1493.
Pirrung, *JACS*, **1979**, *101*, 7130;
ibid., **1981**, *103*, 82.
Paquette, *JOC*, **1979**, *44*, 4014;
JACS, **1981**, *103*, 1835.
Dauben, *JOC*, **1981**, *46*, 1103.
Wender, *Tet*, **1981**, *37*, 4445.
Wenkert, *JACS*, **1983**, *105*, 2030.
Hudlicky, *TL*, **1984**, *25*, 2447.
Tobe, *JCS, CC*, **1985**, 898.
Dreiding, *HCA*, **1986**, *69*, 659.

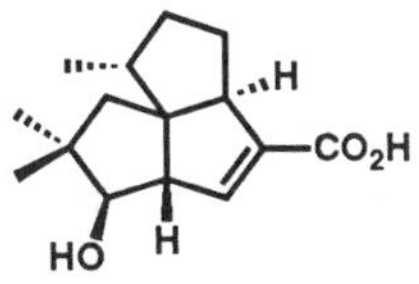

Pentalenic Acid
Matsumoto, *CL*, **1981**, 355.
Crimmins, *JACS*, **1986**, *108*, 800.
Fukumoto, *JCS, CC*, **1987**, 721.

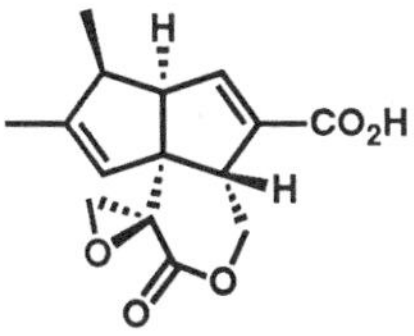

Pentalenolactone

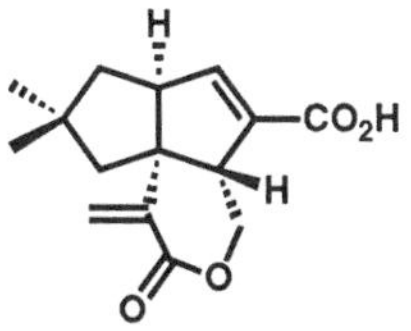

Pentalenolactone E

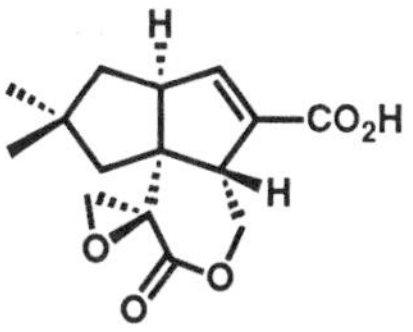

Pentalenolactone F

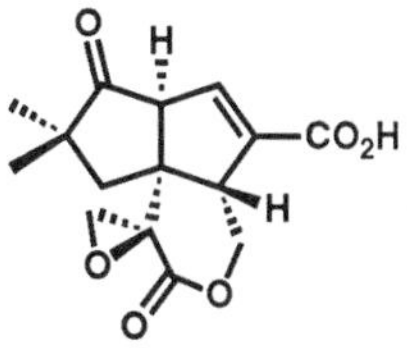

Pentalenolactone G

Danishefsky, *JACS*, **1978**, *100*, 6536; *ibid.*, **1979**, *101*, 7020. Schlessinger, *JACS*, **1980**, *102*, 889.
Paquette, *JACS*, **1981**, *103*, 6526, (E); *ibid.*, **1982**, *104*, 6646, (E). Matsumoto, *TL*, **1983**, *24*, 3851, (E&F).
Cane, *JACS*, **1984**, *106*, 5295, (E&F). Taber, *JACS*, **1985**, *107*, 5289, (E); *Tet*, **1987**, *43*, 5677, (E).
Hua, *TL*, **1987**, *28*, 5465, (E). Marino, *JOC*, **1987**, *52*, 4139, (E). Pirrung, *JOC*, **1988**, *53*, 227, (G).
Mori, *Tet*, **1988**, *44*, 2835, (-)-(E).

13.4

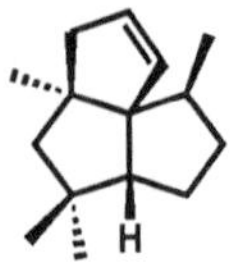

Silphinene

Paquette, *JOC*, **1982**, *47*, 4173;
JACS, **1983**, *105*, 7352.
Ito, *TL*, **1983**, *24*, 83.
Sternbach, *JACS*, **1985**, *107*, 2149.
Wender, *TL*, **1985**, *26*, 2625.
Crimmins, *JACS*, **1986**, *108*, 3435.

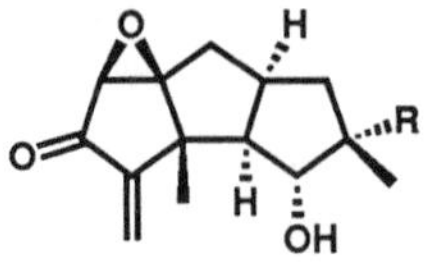

R = Me: Hypnophilin

Little, *JOC*, **1987**, *52*, 4647.

R = CO_2H: Complicatic Acid

Schuda, *JOC*, **1986**, *51*, 2742.

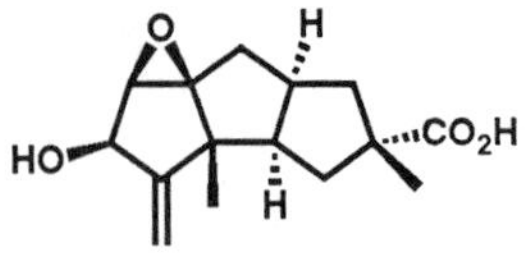

Hirsutic Acid C

Matsumoto, *TL*, **1974**, 3745.
Trost, *JACS*, **1979**, *101*, 1284.
Ikegami, *TL*, **1982**, *23*, 5311. (+)
Greene, *JACS*, **1983**, *105*, 2435;
JOC, **1985**, *50*, 3957. (+)
Schuda, *JOC*, **1986**, *51*, 2742.

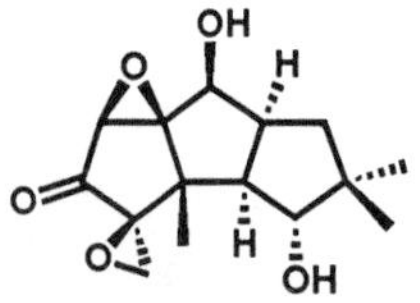

Coriolin

Danishfsky, *JACS*, **1980**, *102*, 2097;
ibid., **1981**, *103*, 3460.
Ikegami, *TL*, **1980**, *21*, 3587.
Tatsuta, *J. Antibiot*, **1980**, *33*, 100.
Trost, *JACS*, **1981**, *103*, 7380.
Magnus, *JACS*, **1983**, *105*, 2477.
Koreeda, *JACS*, **1983**, *105*, 7203.
Wender, *TL*, **1983**, *24*, 5325;
JACS, **1987**, *109*, 2523.
Matsumoto, *Tet*, **1984**, *40*, 241.
Schuda, *Tet*, **1984**, *40*, 2365.
Funk, *Tet*, **1985**, *41*, 3479.
Mehta, *JACS*, **1986**, *108*, 3443.
Demuth, *JACS*, **1986**, *108*, 4149, (-)
Little, *JOC*, **1987**, *52*, 4647.

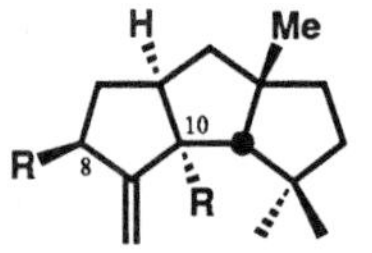

R = H: $\Delta^{9(12)}$-Capnellene

Paquette, *TL*, **1981**, *22*, 4393.
Oppolzer, *TL*, **1982**, *23*, 4669.
Dreiding, *HCA*, **1982**, *65*, 2413.
Mehta, *JCS, CC*, **1983**, 824.
Little, *JACS*, **1983**, *105*, 928.
Piers, *CJC*, **1984**, *62*, 629.
Stille, *JACS*, **1984**, *106*, 7500.
Curran, *TL*, **1985**, *26*, 4991.

R = OH: $\Delta^{9(12)}$-Capnellene-8β,10α-diol

Pattenden, *JCS, PI*, **1988**, 1077.

Modhephene

Dreiding, *TL*, **1980**, *21*, 4569.
Smith, *JACS*, **1981**, *103*, 194.
Paquette, *JACS*, **1981**, *103*, 722.
Oppolzer, *HCA*, **1981**, *64*, 1575, 2489.
Wender, *JACS*, **1982**, *104*, 5805.
Cook & Bertz, *JOC*, **1983**, *48*, 139.
Mundy, *JOC*, **1985**, *50*, 5727.
Mash, *TL*, **1988**, *29*, 2147. (+)

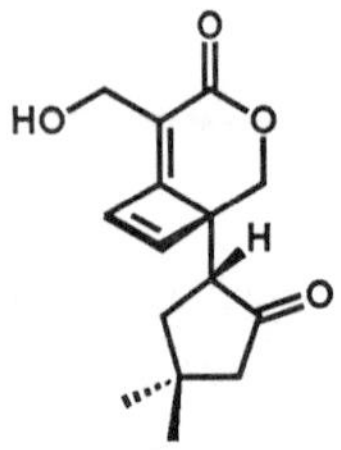

Fomannosin

Semmelhack,
JACS, **1981**, *103*, 2427.

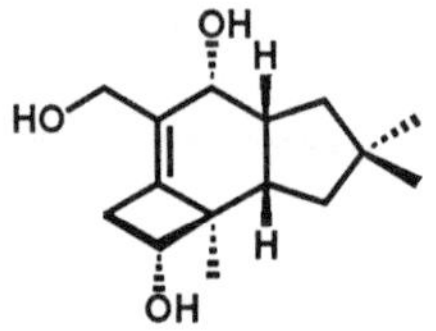

Illudol

Matsumoto, *TL*, **1971**, 3521.
Semmelhack, *JACS*, **1980**, *102*, 7567.

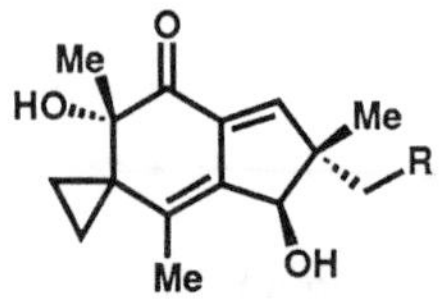

R = H: Illudin-M
R = OH: Illudin-S

Matsumoto,
JACS, **1968**, *90*, 3280, (M);
TL, **1971**, 2049, (S).

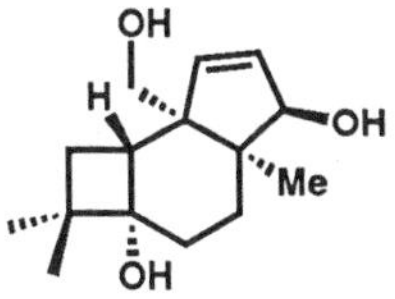

(-)-Punctatin A
(Punctaporonin A)
Paquette, *JACS*, **1986**, *108*, 3841;
ibid., **1987**, *109*, 3017.

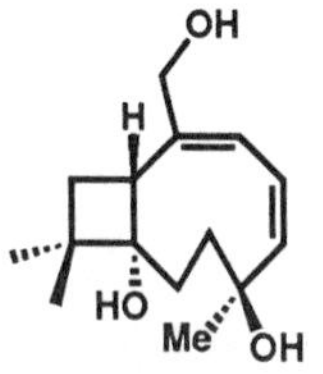

Punctaporonin B
Kende, *JACS*, **1988**, *110*, 6265.

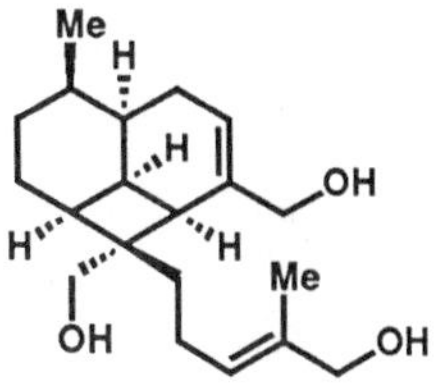

Trihydroxydecipiadiene
Greenlee, *JACS*, **1981**, *103*, 2425.
Dauben, *JOC*, **1984**, *49* , 4252.

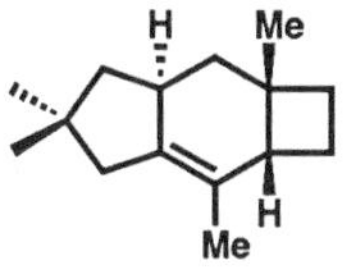

Sterpurene
Matsumoto, *TL*, **1981**, *22*, 4313.
Little, *JOC*, **1986**, *51*, 4497.
Okamura, *JACS*, **1988**, *110*, 4062. (+)

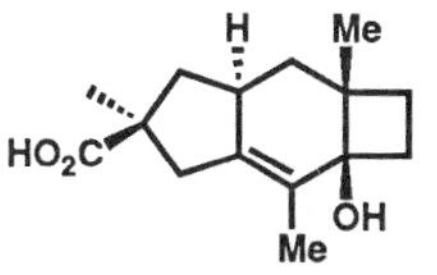

Sterpuric Acid
Paquette,
TL, **1987**, *28*, 5017.

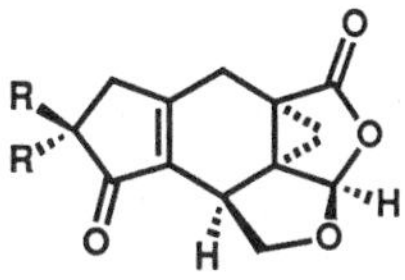

R = Me: Sterepolide
Trost, *JACS*, **1985**, *107*, 4586.
R = H: Nor-sterepolide
Arai, *CL*, **1985**, 1531.

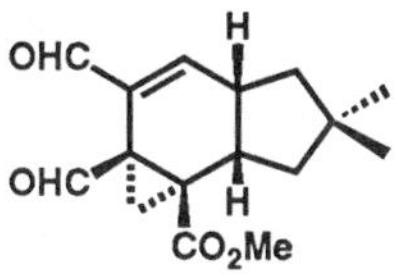

Methyl Isomarasmate
de Mayo, *TL*, **1970**, 349.
Woodward, *Tet*, **1980**, *36*, 3361.

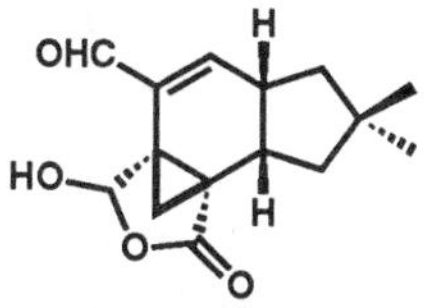

Marasmic Acid
Woodward, *JACS*, **1976**, *98*, 6075;
Tet, **1980**, *36*, 3367.
Boeckman, *JACS*, **1980**, *102*, 7146.

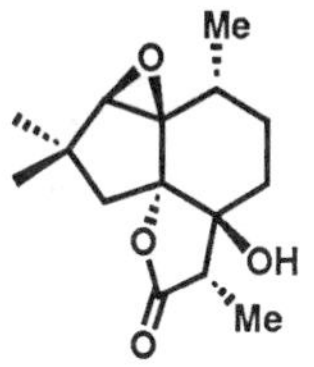

Alliacolide
Pattenden,
TL, **1985**, *26*, 4413;
JCS, PI, **1988**, 1107.

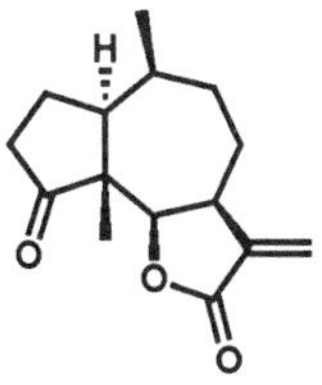

Damsin
Kretchmer, *JACS*, **1976**, *98*, 3379.
Vandewalle, *JOC*, **1977**, *42*, 3447.
Grieco, *JACS*, **1977**, *99*, 7393;
ibid., **1982**, *104*, 4226.
Schlessinger, *JACS*, **1979**, *101*, 7627.

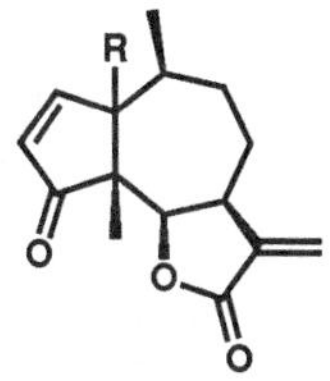

R = α-OH: Parthenin
R = β-OH: Hymenin
Vandewalle, *BSC, Belg*, **1978**, *87*, 615.
Heathcock, *JACS*, **1982**, *104*, 6081, (**P**).

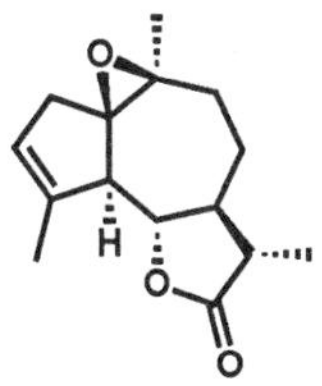

Arborescin
Ando, *CL*, **1978**, 727.

13.4

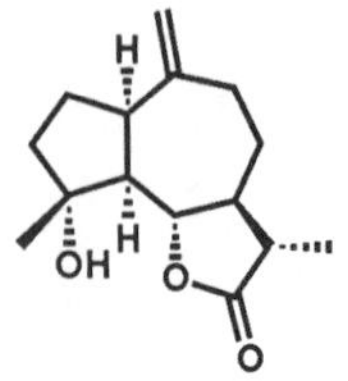

Compressanolide
Vandewalle, *TL*, **1980**, *21*, 4767.

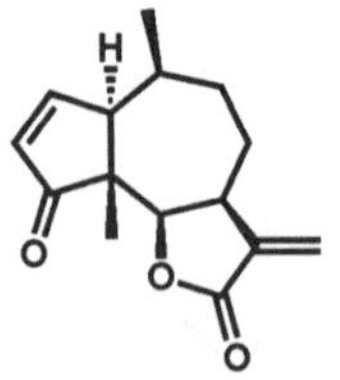

Ambrosin
Grieco, *JACS*, **1977**, *99*, 7393;
ibid., **1982**, *104*, 4226.

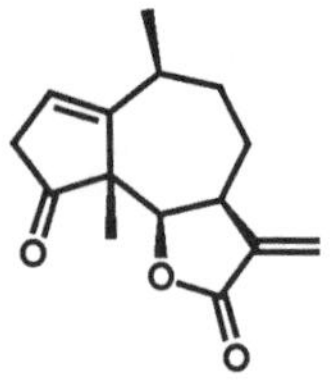

Neoambrosin
Vandewalle,
BSC, Belg, **1978**, *87*, 615.

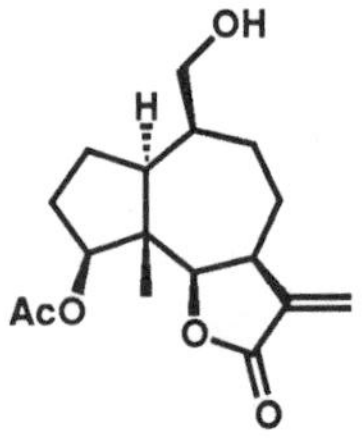

Hysterin
Vandewalle, *JOC*, **1979**, *44*, 4863.

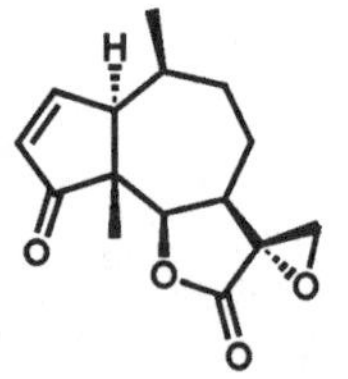

Stramonin B
Grieco, *JOC*, **1978**, *43*, 4552.

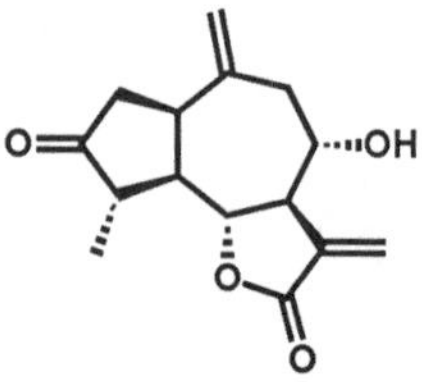

Grosshemin
Rigby, *JACS*, **1987**, *109*, 3147.

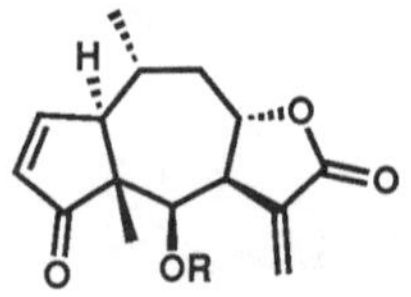

R = H: Mexicanin I
R = Ac: Linifolin A
Grieco, *TL*, **1979**, 3265.

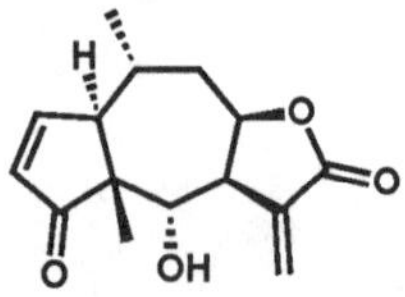

Helenalin
Grieco, *JACS*, **1978**, *100*, 5946;
ibid., **1982**, *104*, 4233.
Schlessinger, *JACS*, **1979**, *101*, 7626.
Bryson, *TL*, **1988**, *29*, 521.

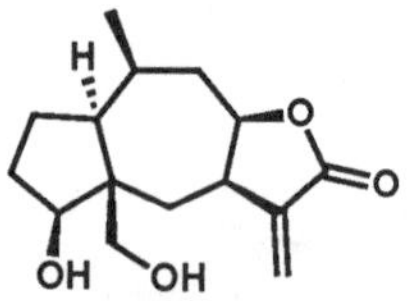

Rudmollin
Wender, *TL*, **1986**, *27*, 1857.

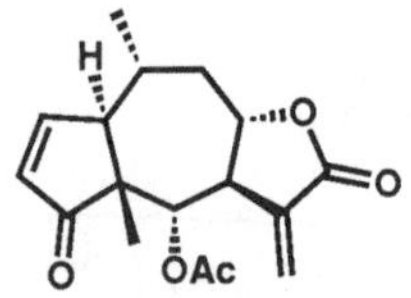

Bigelovin
Grieco, *JOC*, **1979**, *44*, 3092.

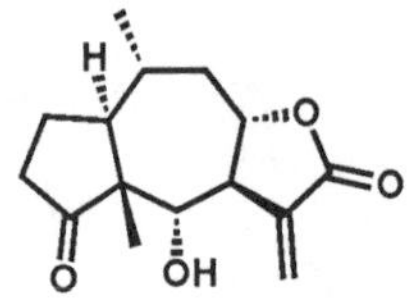

Carpesiolin
Vandewalle, *JOC*, **1979**, *44*, 4553.

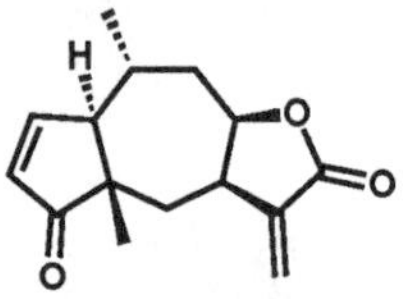

Aromatin
Lansbury, *JACS*, **1980**, *102*, 3964;
Tet, **1982**, *38*, 2797.
Ziegler, *JACS*, **1982**, *104*, 7174.
Schultz, *JACS*, **1986**, *108*, 1056.

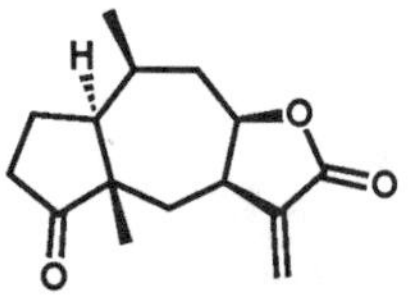

Confertin
Marshall, *JACS*, **1976**, *98*, 4312.
Semmelhack, *JACS*, **1978**, *100*, 5565.
Wender, *JACS*, **1979**, *101*, 2196.
Schlessinger, *JACS*, **1979**, *101*, 7627.
Heathcock, *JACS*, **1982**, *104*, 1907.
Schultz, *JACS*, **1982**, *104*, 5800;
ibid., **1986**, *108*, 1056.
Ziegler, *JACS*, **1982**, *104*, 7174.
Quinkert, *ACIE*, **1987**, *26*, 61. (+)
Bryson, *TL*, **1988**, *29*, 521.

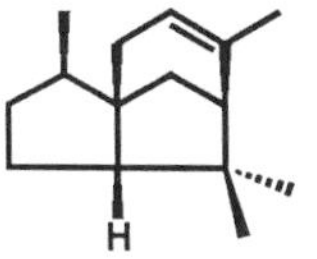
H
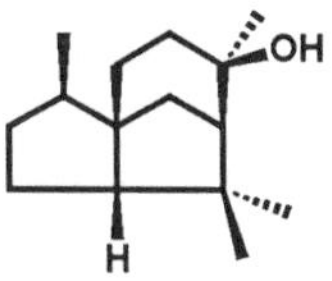
OH
H

Cedrene **Cedrol**

Stork, *JACS*, **1955**, *77*, 1072;
ibid., **1961**, *83*, 3114.
Crandall, *JACS*, **1969**, *91*, 2127.
Demole, *HCA*, **1971**, *54*, 1845.
Naegeli, *TL*, **1972**, 2013.
Lansbury, *JACS*, **1974**, *96*, 896.
Büchi, *JOC*, **1977**, *42*, 3323.
Fallis, *JOC*, **1978**, *43*, 1964.
Wender, *JACS*, **1981**, *103*, 688.

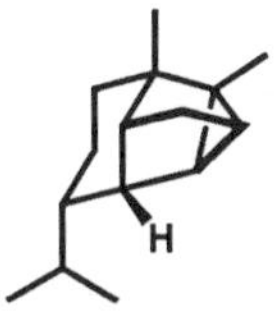
H

Cyclosativene

Baldwin, *TL*, **1975**, 1055.
Piers, *CJC*, **1975**, *53*, 2849. (+)

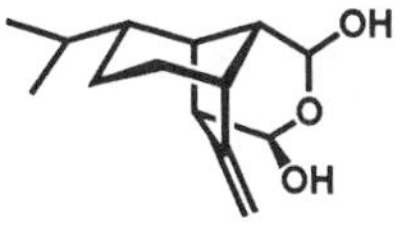
OH
O
OH

Prehelminthosporal

Piers, *CJC*, **1977**, *55*, 1039.

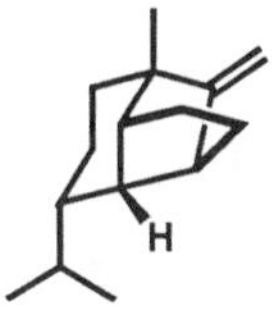
H

Sativene

McMurry, *JACS*, **1968**, *90*, 6821;
TL, **1969**, 55.
Money, *JCS, PI*, **1974**, 1938.
Piers, *CJC*, **1975**, *53*, 2849. (+)
Yoshikoshi, *BCS, Jpn*, **1975**, *48*, 3723. (+)
Bakuzis, *JOC*, **1976**, *41*, 3261.
Matsumoto, *TL*, **1979**, 1761.
Oppolzer, *HCA*, **1984**, *67*, 1154. (+)
Dreiding, *JCS, CC*, **1986**, 944.
Snowden, *Tet*, **1986**, *42*, 3277.

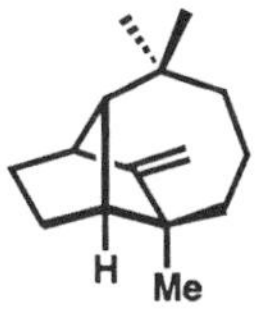
H
Me

Longifolene

McMurry, *JACS*, **1972**, *94*, 7132.
Johnson, *JACS*, **1975**, *97*, 4777.
Oppolzer, *JACS*, **1978**, *100*, 2583;
HCA, **1984**, *67*, 1154. (+)
Schultz, *JOC*, **1985**, *50*, 915. (±), (-)

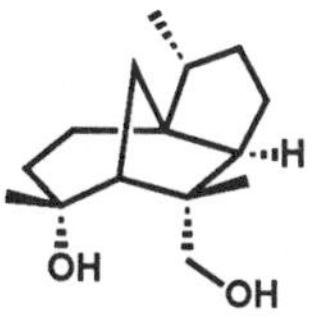
H
OH
OH

8*S*,14-Cedranediol

Landry,
Tet, **1983**, *39*, 2761.

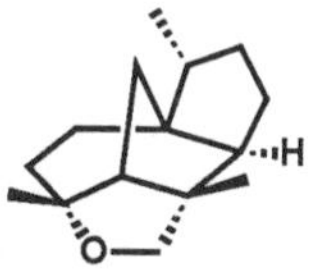
H
O

8*S*,14-Cedranoxide

Yamamura,
TL, **1987**, *28*, 6661.

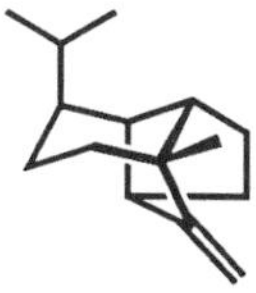

Copacamphene

McMurry, *JOC*, **1971**, *36*, 2826.
Money, *JCS, PI*, **1974**, 1938.
Piers, *CJC*, **1975**, *53*, 2849. (-)
Bakuzis, *JOC*, **1976**, *41*, 3261.
Matsumoto, *TL*, **1979**, 1761.

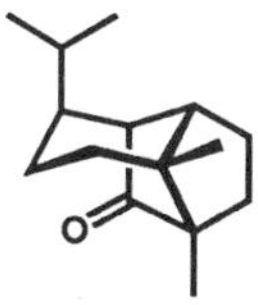
O
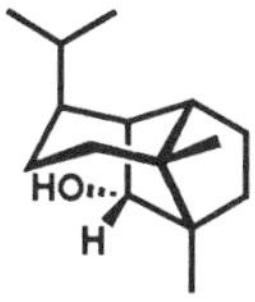
HO
H
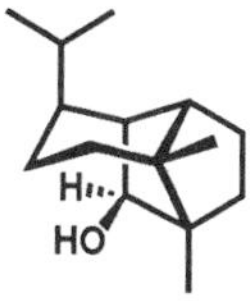
H
HO

Copacamphor **Copaborneol** **Copaisoborneol**

and the corresponding ylango series

Piers, *CJC*, **1971**, *49*, 2620; *ibid.*, **1975**, *53*, 2827. (+)
ibid., **1971**, *49*, 2623; *ibid.*, **1975**, *53*, 2838. (-)

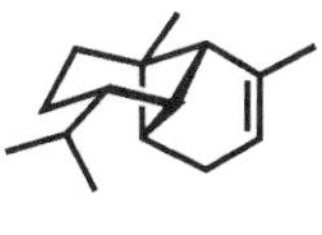

α-Copaene **α-Ylangene**

Heathcock,
JACS, **1966**, *88*, 4110; *ibid.*, **1967**, *89*, 4133.

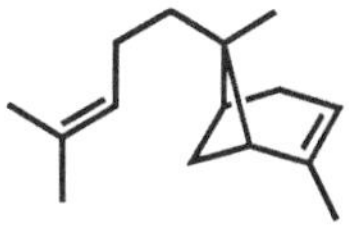

α-*trans*-Bergamotene

Monti
JACS, **1977**, *99*, 8015.

13.5

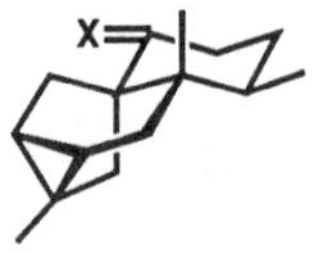

X = H_2: Ishwarane

Kelly, *JCS, CC*, **1971**, 479;
CJC, **1972**, *50*, 3455.
Cory, *JCS, CC*, **1977**, 587.
Hagiwara, *JCS, PI*, **1980**, 963.

X = O: Ishwarone

Piers, *JCS, CC*, **1977**, 880.
Cory, *TL*, **1979**, 4133.

Clovene

Raphael, *PCS*, **1963**, 239;
JCS, **1965**, 1344.
Becker, *JCS*, **1965**, 1338.
Schultz, *JOC*, **1983**, *48*, 2318.
Dreiding, *HCA*, **1984**, *67*, 1963.
Funk, *TL*, **1988**, *29*, 1493.

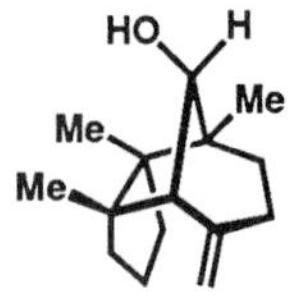
HO
H
Me
Me
Me

Gymnomitrol

Coates, *JACS*, **1979**, *101*, 6765;
ibid., **1982**, *104*, 2198.
Büchi, *JACS*, **1979**, *101*, 6767.
Welch, *JACS*, **1979**, *101*, 6768.
Kodama, *CJC*, **1979**, *57*, 3343.
Paquette, *JOC*, **1979**, *44*, 3731.

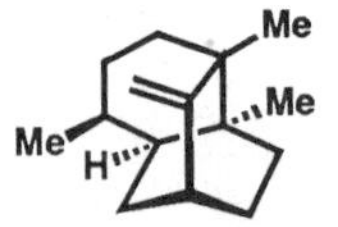
Me
Me
Me
H

Seychellene

Piers, *JCS, CC*, **1969**, 1069;
JACS, **1971**, *93*, 5113.
Mirrington, *JOC*, **1972**, *37*, 2877.
Yoshikoshi, *JCS, PI*, **1973**, 1843.
Fráter, *HCA*, **1974**, *57*, 172;
ibid., **1979**, *62*, 1893.
Yamada, *Tet*, **1979**, *35*, 293.
Jung, *JACS*, **1981**, *103*, 6677.
Welch, *JOC*, **1985**, *50*, 2668, 2676.
Hagiwara, *JCS, CC*, **1985**, 1047.

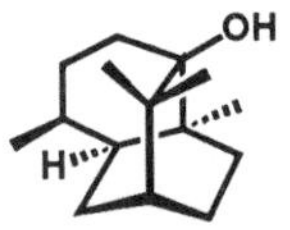
OH
H

Patchouli Alcohol

Büchi, *JACS*, **1964**, *86*, 4438.
Danishefsky, *JCS, CC*, **1968**, 1287.
Mirrington, *JOC*, **1972**, *37*, 2871.
Näf, *HCA*, **1974**, *57*, 1868.
Yamada, *Tet*, **1979**, *35*, 293.

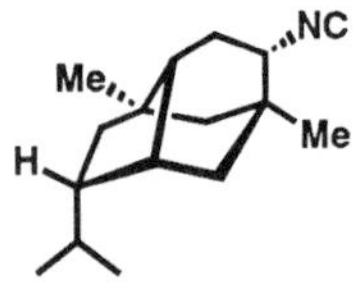
NC
Me
Me
H

9-Isocyanopupukeanane

Yamamoto,
JACS, **1979**, *101*, 1609.

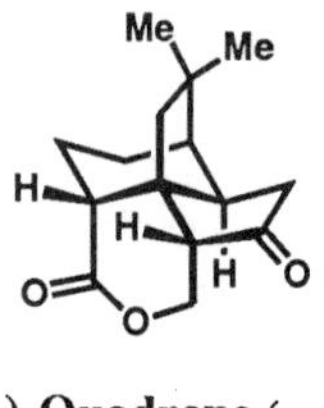
Me
Me
H
H
H
O
O
O

(-)-Quadrone (*nat*)

Danishefsky, *JACS*, **1980**, *102*, 4262;
ibid., **1981**, *103*, 4136. (±)
Helquist, *JACS*, **1981**, *103*, 4647. (±)
Kende, *JACS*, **1982**, *104*, 5808. (±)
Schlessinger, *JOC*, **1983**, *48*, 1146. (±)
Vandewalle, *Tet*, **1983**, *39*, 3235. (±)
Yoshii, *JACS*, **1983**, *105*, 563. (±)
Burke, *JACS*, **1984**, *106*, 4558. (±)
Smith, *JOC*, **1984**, *49*, 4094. (+)
Isoe, *TL*, **1984**, *25*, 3739. (+)
Wender, *JOC*, **1985**, *50*, 4418. (±)
Piers, *TL*, **1985**, *26*, 2735. (±)
Funk, *JOC*, **1986**, *51*, 3247. (±)
Iwata, *JCS, CC*, **1987**, 1802. (±)
Magnus, *JOC*, **1987**, *52*, 1483. (±)
Liu, *CJC*, **1988**, *66*, 528. (-)

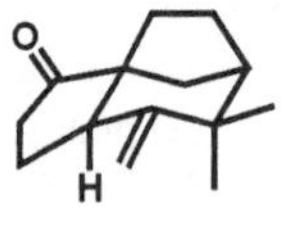
O
H

Khusimone

Büchi, *JOC*, **1977**, *42*, 3323.
Liu, *CJC*, **1979**, *57*, 708. (-)

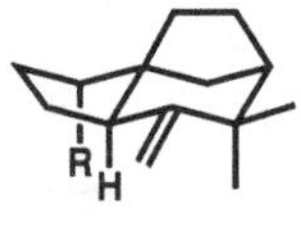
R
H

R = CO_2H: (+)-Zizanoic Acid

Yoshikoshi, *JCS, CC*, **1969**, 1335;
JCS, PI, **1972**, 1755.
MacSweeney, *Tet*, **1971**, *27*, 1481.

R = Me: Zizaene

Wiesner, *CJC*, **1972**, *50*, 726.
Coates, *JACS*, **1972**, *94*, 5386.
Piers, *JCS, CC*, **1979**, 1138.
Pattenden, *JCS, PI*, **1983**, 1901.

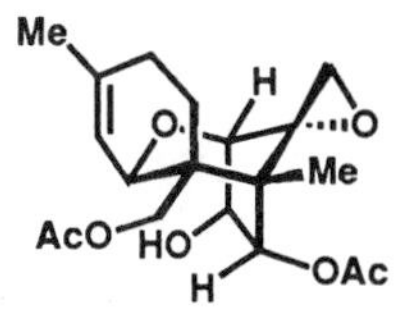

Anguidine
Brooks,
JACS, **1983**, *105*, 4472.

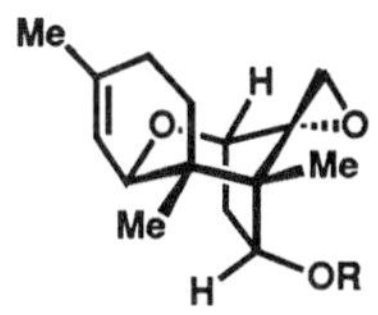

R = Ac: Trichodermin
Raphael, *JCS, CC*, **1971**, 858;
JCS, PI, **1973**, 1989.
R = H: Trichodermol
Still, *JACS*, **1980**, *102*, 3654.

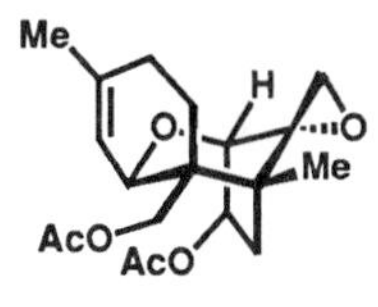

Calonectrin
Kraus, *JACS*, **1982**, *104*, 1114.
Koga, *CPB*, **1987**, *35*, 906. (+), (-)

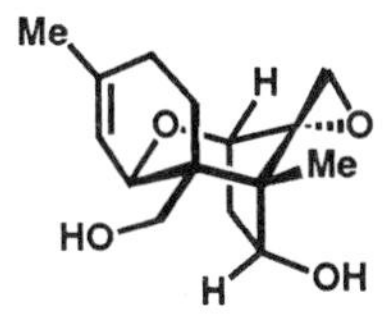

Verrucarol
Schlessinger, *JACS*, **1982**, *104*, 1116.
Trost, *JACS*, **1982**, *104*, 6110;
ibid., **1984**, *106*, 383.
Roush, *JACS*, **1983**, *105*, 1058.
Koreeda, *JOC*, **1988**, *53*, 5586.

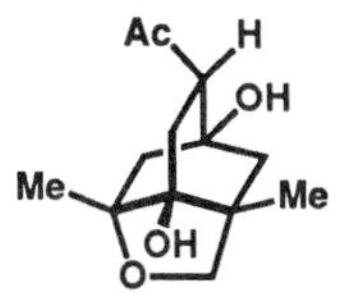

Drummondone A
Takahashi,
Ag. BC, **1988**, *52*, 297.

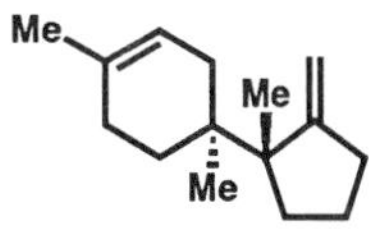

Trichodiene
Welch, *JOC*, **1980**, *45*, 4077.
Suda, *TL*, **1982**, *23*, 427.
Schlessinger, *JOC*, **1983**, *48*, 407.
Snowden, *JOC*, **1984**, *49*, 1464.
Harding, *JOC*, **1984**, *49*, 3870.
Kraus, *JOC*, **1986**, *51*, 503.
Gilbert, *JOC*, **1986**, *51*, 4485.
VanMiddlesworth, *JOC*, **1986**, *51*, 5019.
Pearson, *JCS, CC*, **1987**, 1445.

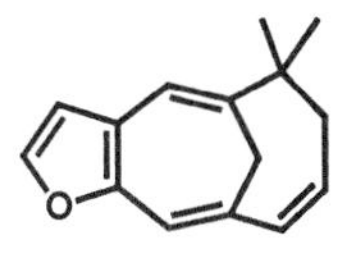

Spiniferin-1
Marshall,
JACS, **1980**, *102*, 4274;
ibid., **1983**, *105*, 5679.

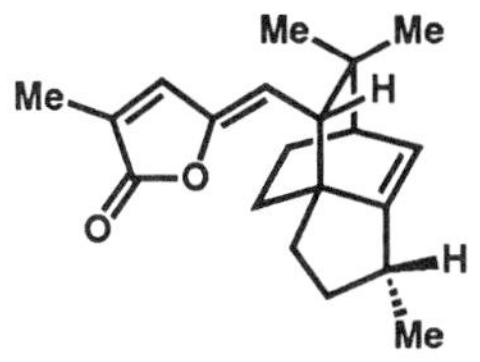

Eremolactone
Takei, *TL*, **1983**, *24*, 5127.

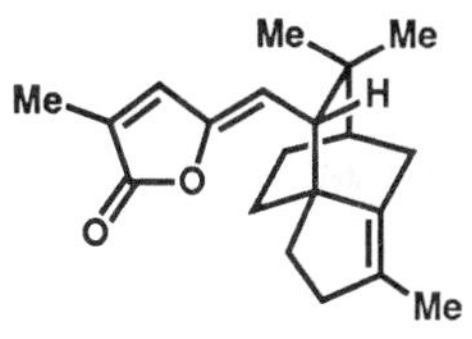

(+)-Isoeremolactone
Ramage, *TL*, **1983**, *24*, 4487.
Spitzner, *JCS, PI*, **1988**, 373.

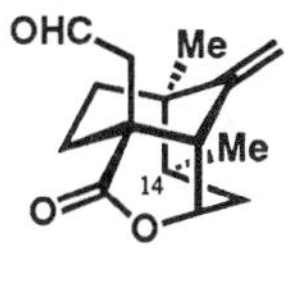

(-)-Upial
Taschner, *JACS*, **1985**, *107*, 5570.
14-*epi*-Upial
Paquette, *Tet*, **1987**, *43*, 5567.

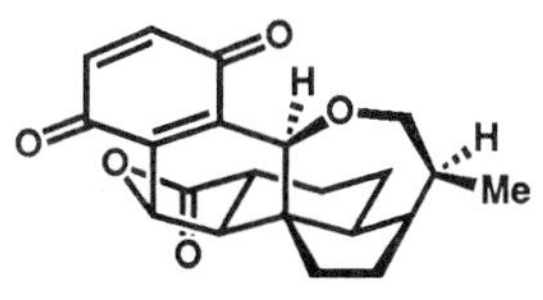

Pleurotin
Hart, *JACS*, **1988**, *110*, 1634.

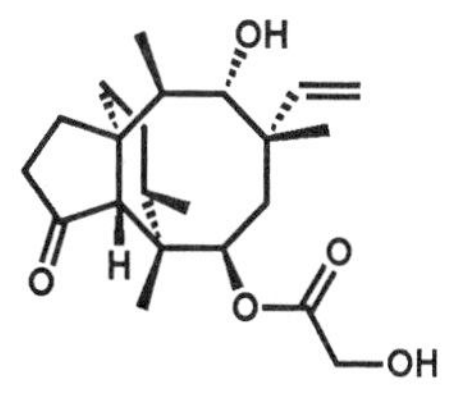

Pleuromutilin
Gibbons,
JACS, **1982**, *104*, 1767.

13.5

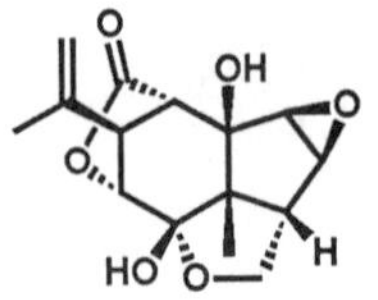

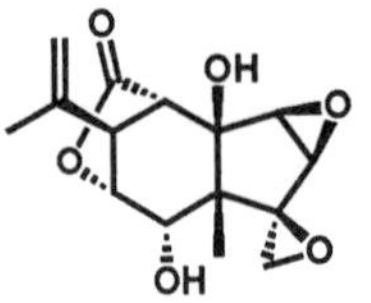

(+)-Asteromurin A (+)-Tutin

Yamada, *TL*, **1984**, *25*, 3873, (**T**);
CL, **1984**, 1763, (**A**);
Tet, **1986**, *42*, 5551, (**A & T**).

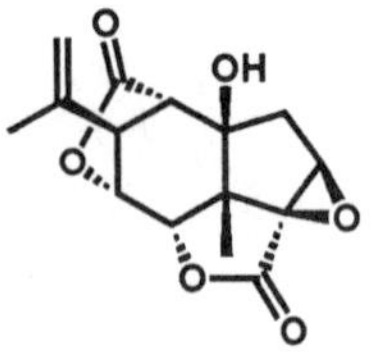

(-)-Picrotoxinin
Yamada,
JACS, **1984**, *106*, 4547.

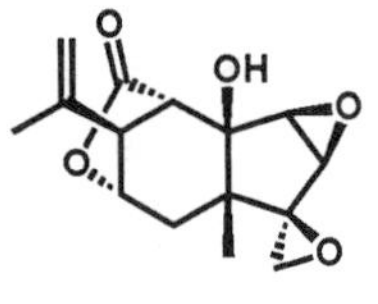

Coriamyrtin
Inubushi,
JACS, **1982**, *104*, 4965;
CPB, **1983**, *31*, 1972.
Yamada,
JACS, **1984**, *106*, 4547. (+)

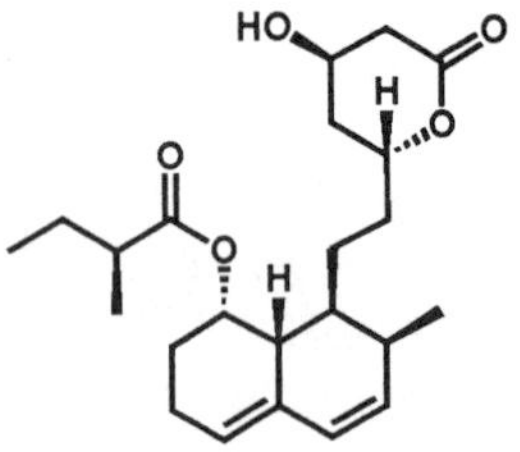

(+)-Compactin
Sih, *JACS*, **1981**, *103*, 6538;
ibid., **1983**, *105*, 593.
Hirama, *JACS*, **1982**, *104*, 4251.
Girotra, *TL*, **1983**, *24*, 3687.
Heathcock, *JACS*, **1985**, *107*, 3731.
Grieco, *JACS*, **1986**, *108*, 5908.
Keck, *JOC*, **1986**, *51*, 2487.
Kozikowski, *JOC*, **1987**, *52*, 3541.
Clive, *JACS*, **1988**, *110*, 6914.

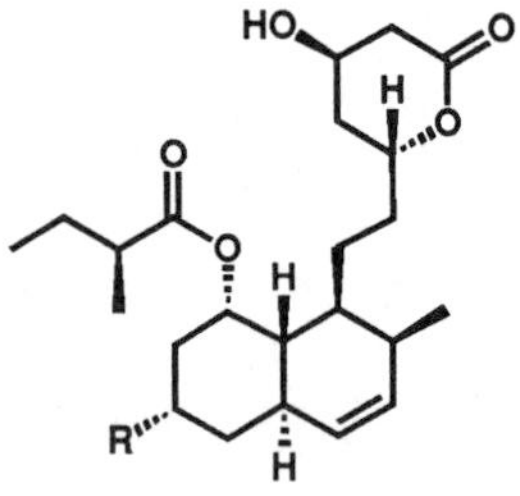

R = Me: (+)-Dihydromevinolin
Falck, *TL*, **1984**, *25*, 3563.
Heathcock, *JACS*, **1986**, *108*, 4586.
Davidson, *JCS, CC*, **1987**, 1786.
R = H: (+)-Dihydrocompactin
Falck, *JACS*, **1984**, *106*, 3811.

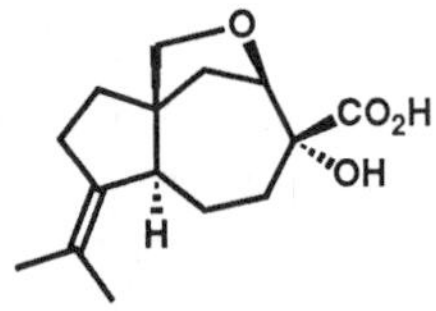

Aspterric Acid
Harayama,
TL, **1983**, *24*, 5241;
CPB, **1987**, *35*, 1434.

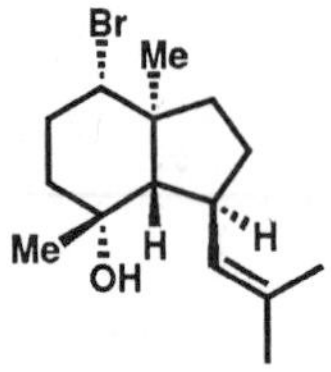

Oppositol
Yamamura, *TL*, **1986**, *27*, 57.
Masamune, *TL*, **1987**, *28*, 4303.

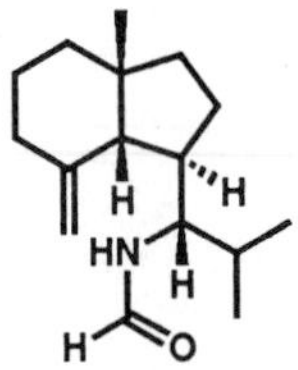

Axamide-1
Piers, *CJC*, **1986**, *64*, 2475.

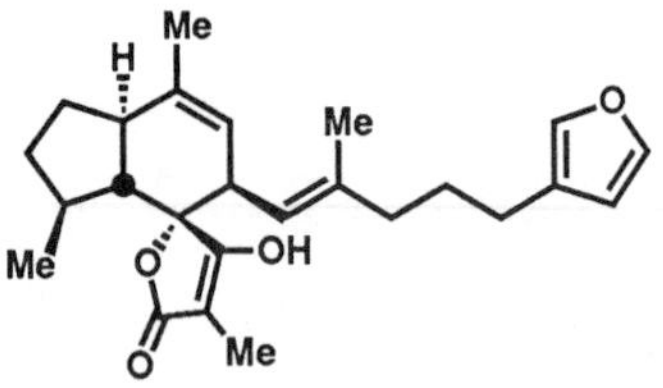

Ircinianin
Yoshii, *TL*, **1986**, *27*, 3903.

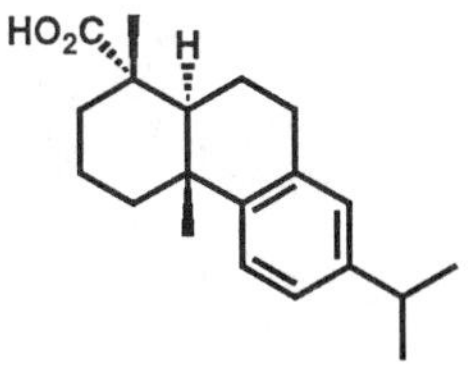
HO$_2$C
H

Dehydroabietic Acid
Stork, *JACS*, **1956**, *78*, 250;
ibid., **1962**, *84*, 284.
Ireland, *JOC*, **1966**, *31*, 2543.
Meyer, *JOC*, **1977**, *42*, 2769.
Abietic Acid
Burgstahler, *JACS*, **1961**, *83*, 2587.

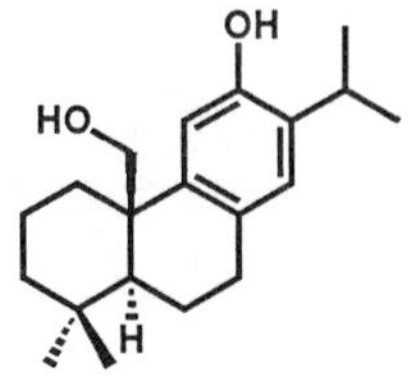
OH
HO
H

(+)-Pisiferol
Matsumoto, *BCS, Jpn*, **1982**, *55*, 1599;
ibid., **1983**, *56*, 2018.
Uda, *JCS, PI*, **1986**, 1311.

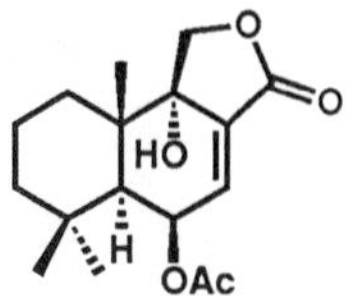
O
O
HO
H
OAc

Cinnamosmolide
Naito, *CL*, **1980**, 445.

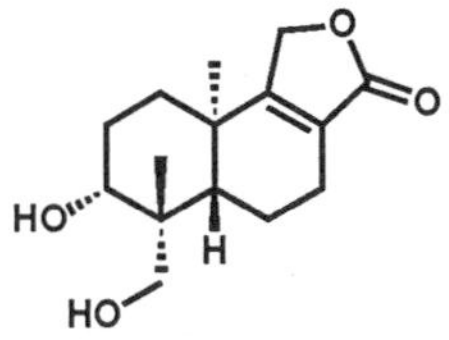
O
O
HO
H
HO

Isoiresin
Pelletier,
JACS, **1968**, *90*, 5318.

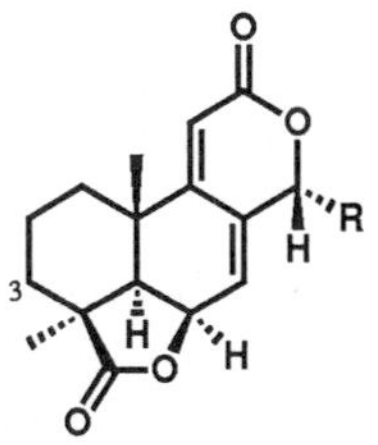
O
O
R
H
3
H
O
H
O

R = *i*-Pr: Nagilactone F
Hayashi, *JOC*, **1982**, *47*, 3428.
3β-Hydroxynagilactone F
de Groot, *JOC*, **1986**, *51*, 4594.
R = OMe: LL-Z1271α
Welch, *JACS*, **1977**, *99*, 549.

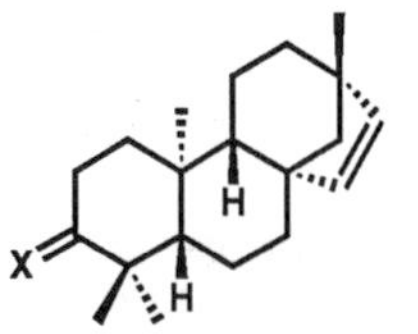
H
X
H

X = O: Stachenone
Monti, *JOC*, **1979**, *44*, 897.
X = H_2: Hibaene
Ireland, *TL*, **1965**, 2627.

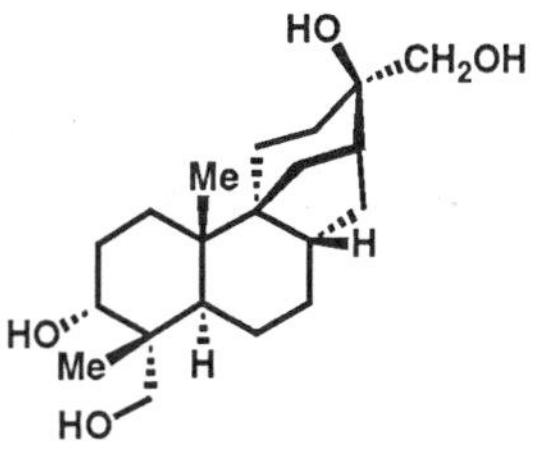
HO
CH$_2$OH
Me
H
HO
Me
H
HO

Aphidicolin
Trost, *JACS*, **1979**, *101*, 1328.
McMurry, *JACS*, **1979**, *101*, 1330.
Ireland, *JACS*, **1981**, *103*, 2446.
van Tamelen, *JACS*, **1983**, *105*, 142.
Bettolo & Lupi, *HCA*, **1983**, *66*, 1922.
Holton, *JACS*, **1987**, *109*, 1597. (+)

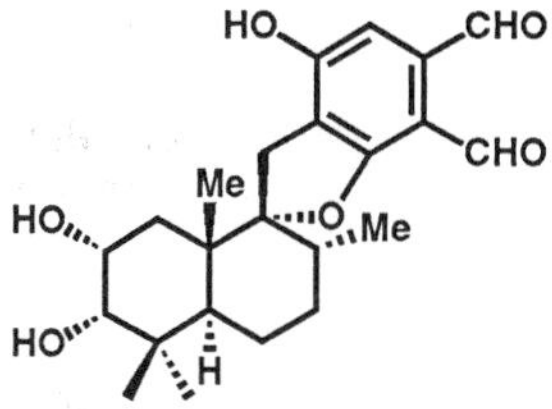
HO
CHO
CHO
Me
HO
O
Me
HO
H

K-76
McMurry, *JACS*, **1985**, *107*, 2712.
Mori, *Ann*, **1988**, 107. (-)

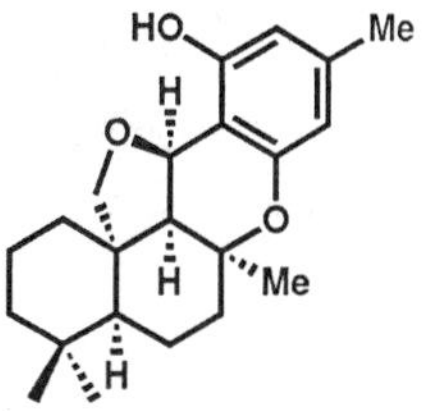
HO
Me
H
O
O
H
Me
H

Siccanin
Yoshikoshi,
JACS, **1981**, *103*, 2434;
Tet, **1987**, *43*, 711.

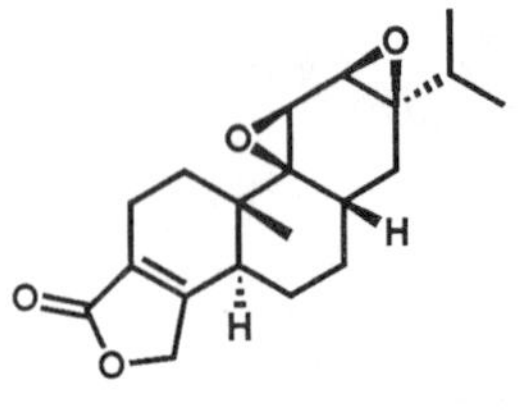

(-)-Stemolide
van Tamelen,
JACS, **1980**, *102*, 1202.

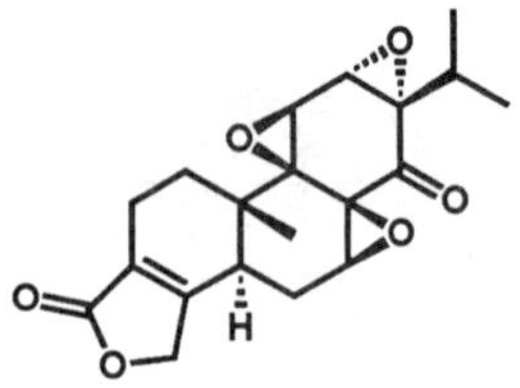

Triptonide

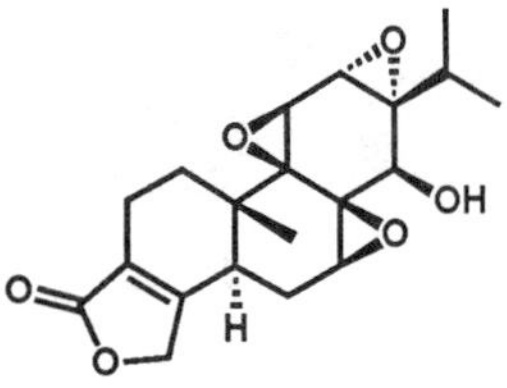

Triptolide

Berchtold,
JACS, **1980**, *102*, 1200; *JOC*, **1982**, *47*, 2364.
van Tamelen,
JACS, **1980**, *102*, 5424, (-); *ibid.*, **1982**, *104*, 867, 1785.

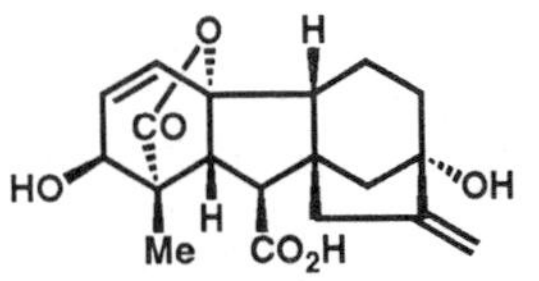

Gibberellic Acid
Mander,
JACS, **1980**, *102*, 6626, 6628;
JOC, **1984**, *49*, 3250.

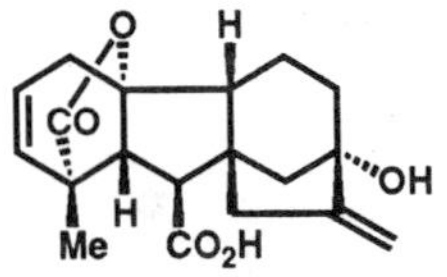

Gibberellin A_5
De Clercq,
TL, **1986**, *27*, 1731.

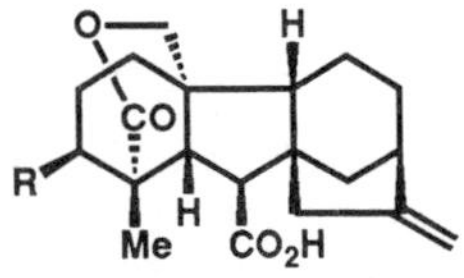

R = H: Gibberellin A_{15}
Nagata, *JACS*, **1970**, *92*, 3202;
ibid., **1971**, *93*, 5740.
R = OH: Gibberellin A_{37}
Fujita, *JCS, CC*, **1975**, 898;
JCS, PI, **1977**, 611.

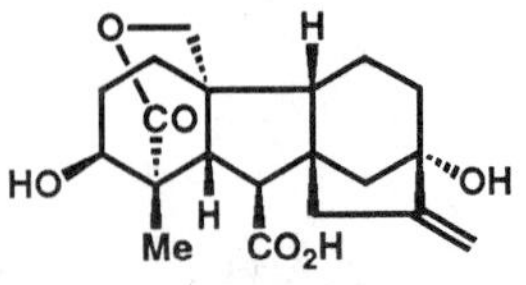

Gibberellin A_{38}
Mander, *JOC*, **1983**, *48*, 2298.

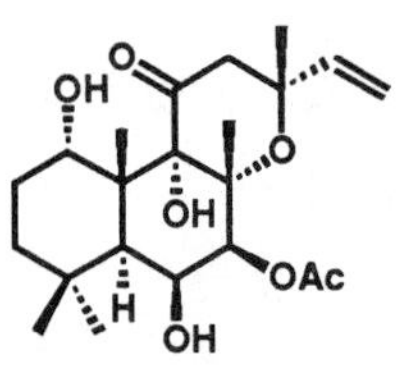

Forskolin
Ziegler, *JACS*, **1987**, *109*, 8115.

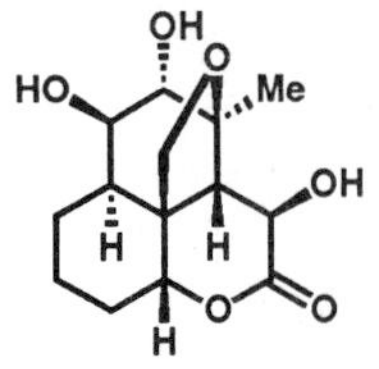

De-A-quassimarin
Grieco, *JOC*, **1987**, *52*, 3346.

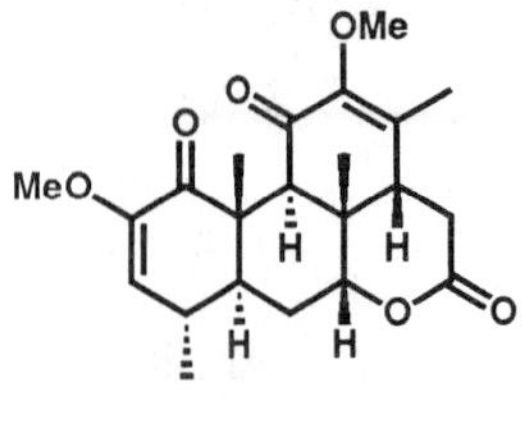

Quassin
Grieco, *JACS*, **1980**, *102*, 7586;
ibid., **1984**, *106*, 3539.

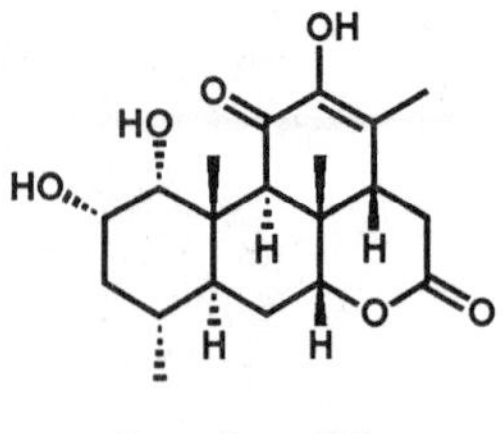

Castelanolide
Grieco, *JOC*, **1982**, *47*, 601;
ibid., **1984**, *49*, 2342.

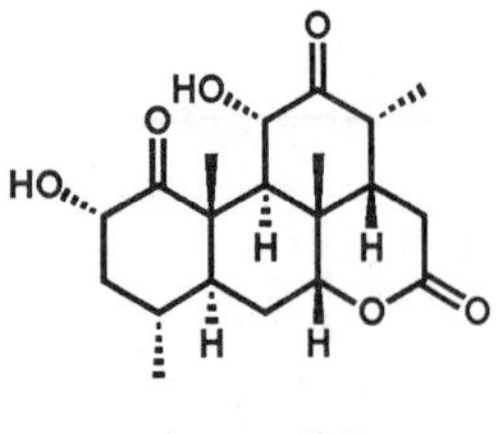

Amarolide
Hirota & Takahashi,
TL, **1987**, *28*, 435.

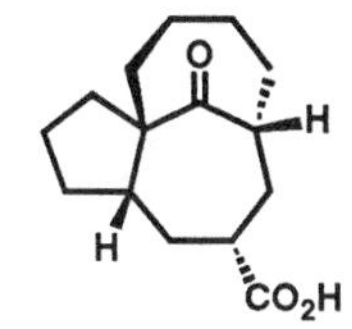

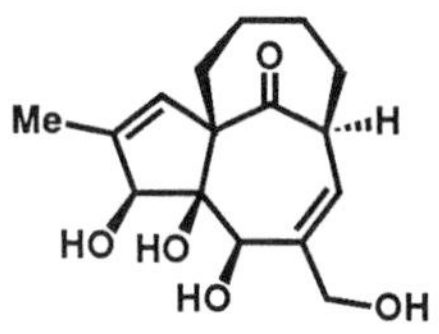

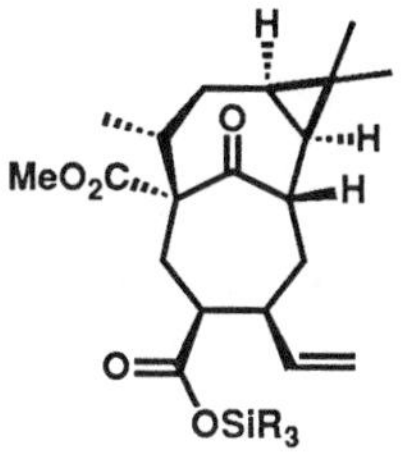

Ingenane Synthetic Studies

Winkler,
JACS, **1987**, *109*, 2850.

Paquette,
JACS, **1984**, *106*, 1446;
JOC, **1987**, *52*, 5497.

Funk,
JACS, **1988**, *110*, 3298.

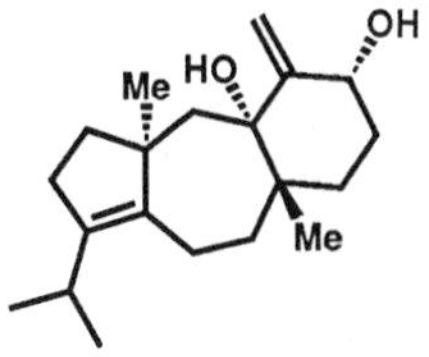

Isoamijiol
Pattenden,
TL, **1986**, *27*, 399.

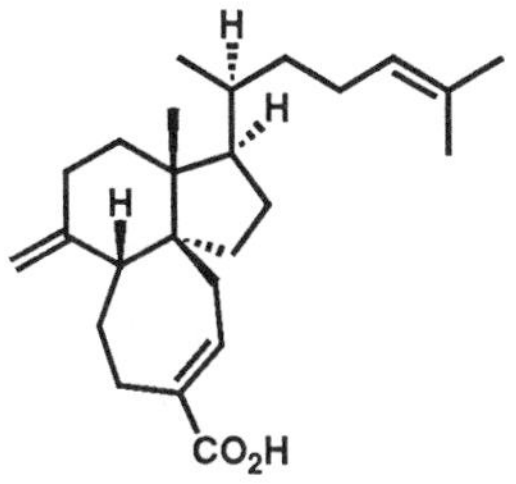

Gascardic Acid
Boeckman,
JACS, **1979**, *101*, 5060.

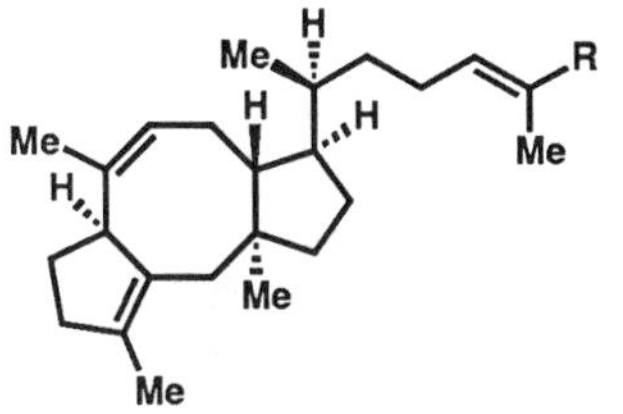

R = CO_2H: Albolic Acid
R = CH_2OH: Ceroplastol II
Kato & Takeshita,
JCS, CC, **1988**, 354. (*nat*)

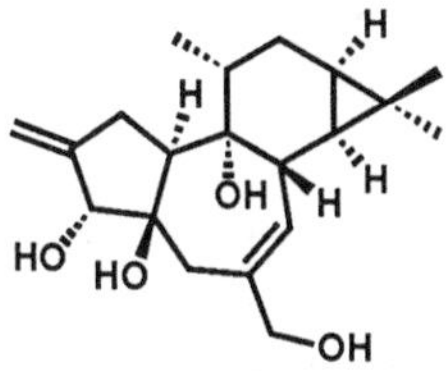

Phorbol Skeleton
Wender,
JACS, **1987**, *109*, 4390.

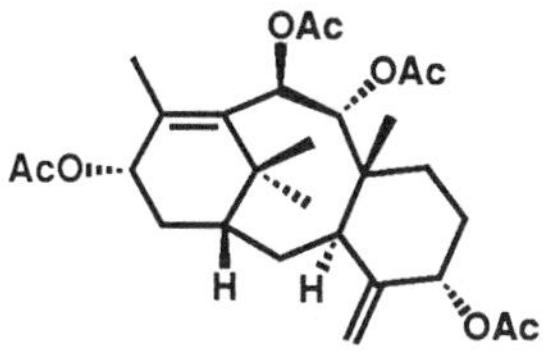

(+)-Taxusin (*nat*)
Holton,
JACS, **1988**, *110*, 6558. (-)

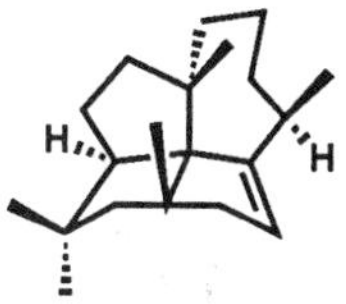

Laurenene
Crimmins, *JACS*, **1987**, *109*, 6199.
Tsunoda, *TL*, **1987**, *28*, 2537.
Paquette, *JOC*, **1988**, *53*, 477.
Wender, *JACS*, **1988**, *110*, 4858.

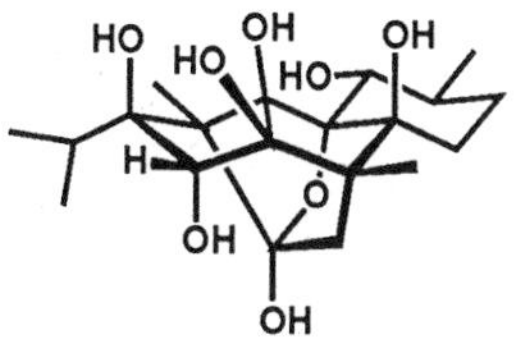

(+)-Ryanodol
Deslongchamps,
CJC, **1979**, *57*, 3348.

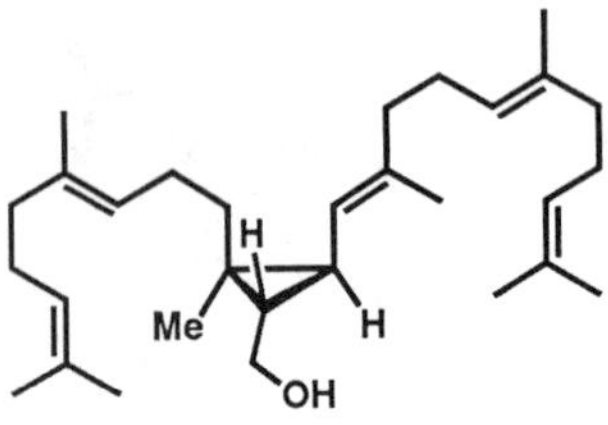
H
Me
H
OH

Presqualene Alcohol
Altman, *JACS*, **1971**, *93*, 1782.
Coates, *JACS*, **1971**, *93*, 1785.
Crombie, *JCS, CC*, **1971**, 218; *JCS, PI*, **1975**, 897.

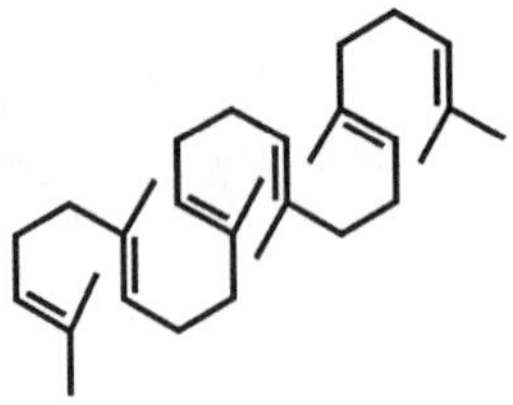

Squalene
Cornforth, *JCS*, **1959**, 2539.
Johnson, *JACS*, **1970**, *92*, 741; *PNAS*, **1970**, *67*, 1465, 1810.
van Tamelen, *JACS*, **1970**, *92*, 2139.
Grieco, *JOC*, **1974**, *39*, 2135.

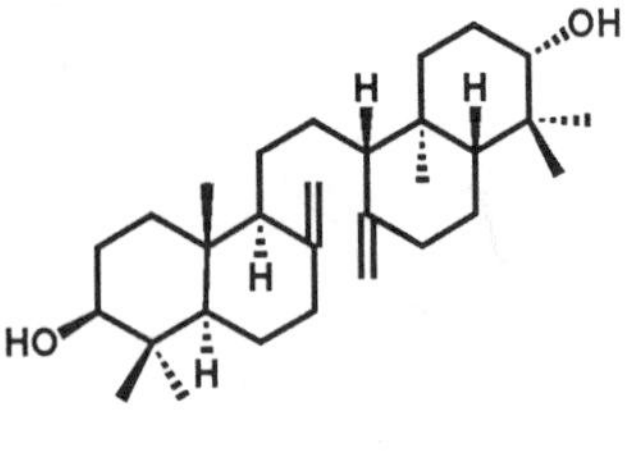
OH
HO

(+)-α-Onocerin
Stork, *JACS*, **1959**, *81*, 5516; *ibid.*, **1963**, *85*, 3419.

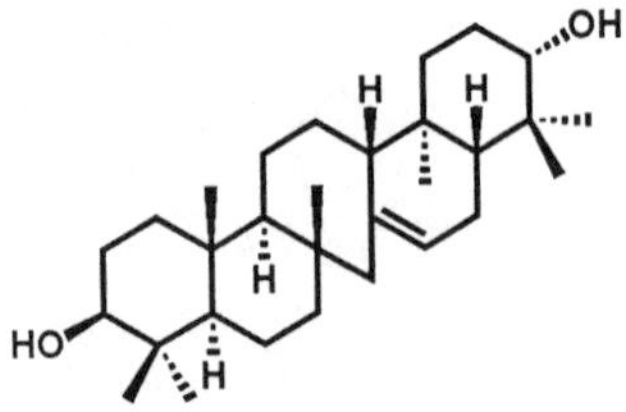
OH
HO

Serratenediol
Johnson, *JACS*, **1974**, *96*, 7103.

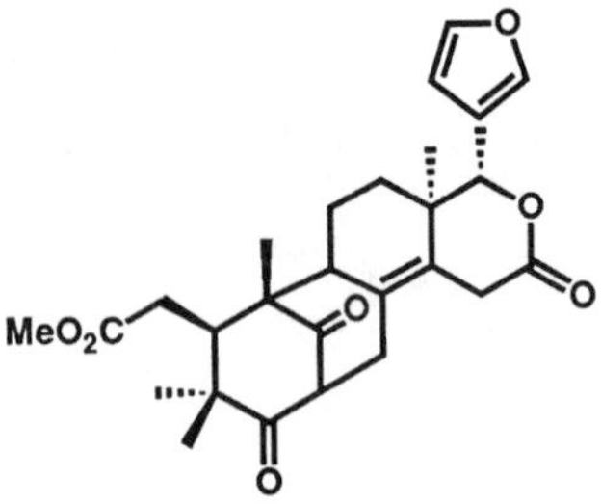
MeO2C
O

Mexicanolide
Connolly, *JCS, CC*, **1971**, 17; *JCS, PI*, **1973**, 2407.
Ekong, *JCS, PI*, **1972**, 1943.

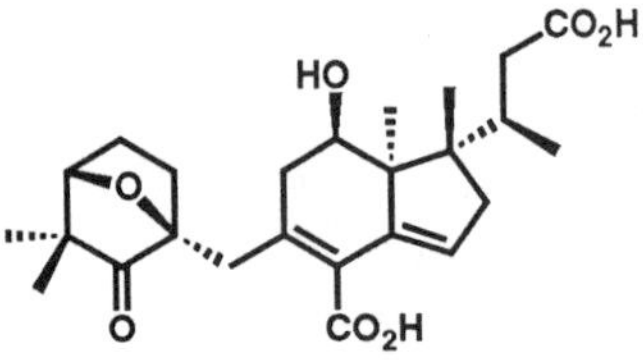
CO2H
HO
O
CO2H

Glycinoeclepin A
Murai, *JACS*, **1988**, *110*, 1985.

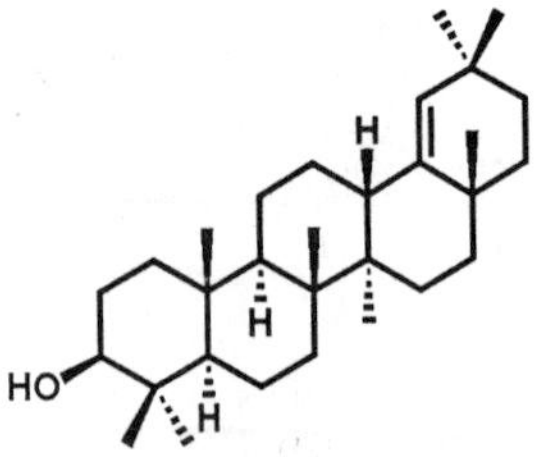
HO

Germanicol
Ireland, *JACS*, **1970**, *92*, 5743.

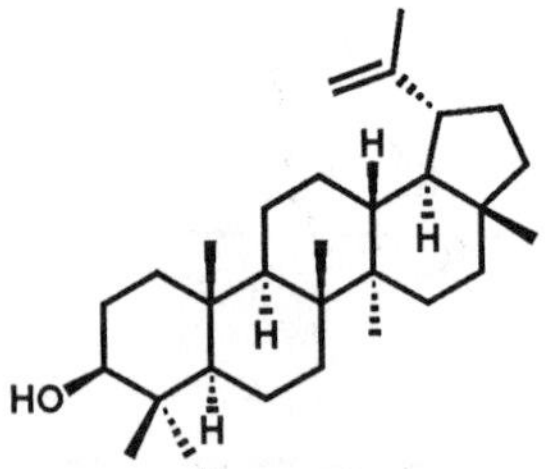
HO

Lupeol
Stork, *JACS*, **1971**, *93*, 4945.

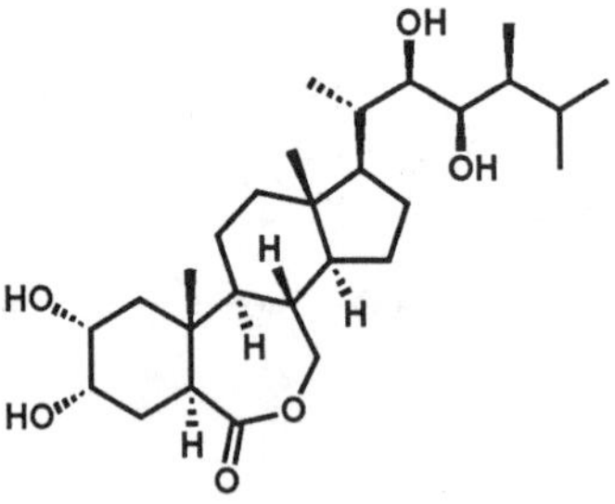
OH
OH
HO
HO
O
O

Brassinolide
Siddall, *JACS*, **1980**, *102*, 6580.
Ikekawa, *JCS, PI*, **1984**, 139.
Mori, *Tet*, **1984**, *40*, 1767.
McMorris, *JOC*, **1984**, *49*, 2833.
Fiecchi, *JOC*, **1984**, *49*, 4297.
Kametani, *JACS*, **1986**, *108*, 7055.

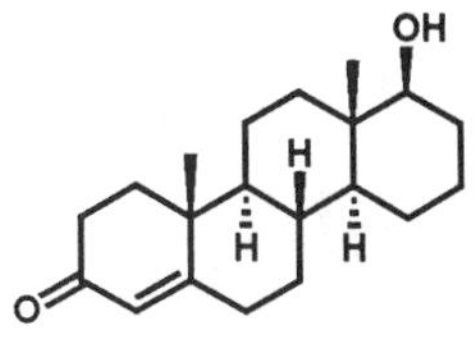

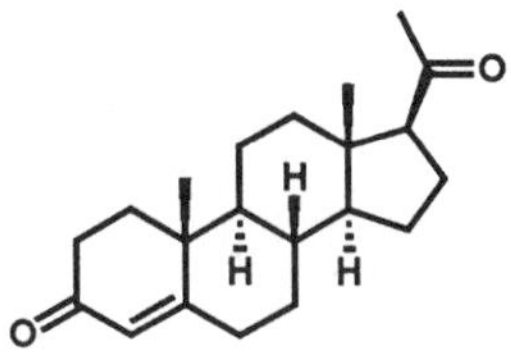

D-Homotestosterone **Progesterone**

Stork, *JACS*, **1967**, *89*, 5464.

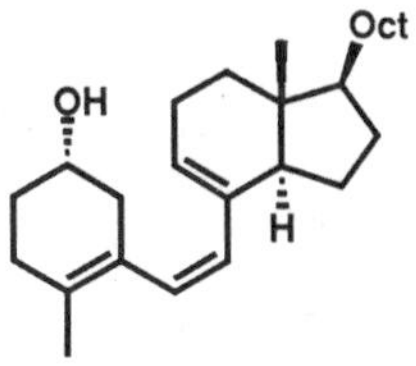

Precalciferol$_3$

Lythgoe, *JCS (C)*, **1971**, 2960.

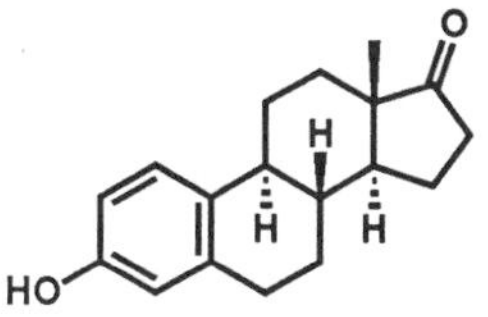

Estrone

Torgov, *TL*, **1963**, 1553.
Smith, *JCS*, **1963**, 5072.
Johnson, *JACS*, **1973**, *95*, 7501.
Cohen, *JOC*, **1975**, *40*, 681. (+)
Danishefsky, *JACS*, **1976**, *98*, 4975.
Kametani, *JACS*, **1977**, *99*, 3461.
Oppolzer, *HCA*, **1980**, *63*, 1703. (+)
Vollhardt, *JACS*, **1980**, *102*, 5253.
Grieco, *JOC*, **1980**, *45*, 2247.
Quinkert, *ACIE*, **1980**, *19*, 1027.
Bryson, *TL*, **1980**, *21*, 2381.
Saegusa, *JACS*, **1981**, *103*, 476.
Ziegler, *JOC*, **1982**, *47*, 5229.
Money, *TL*, **1985**, *26*, 1819. (-)
Posner, *JACS*, **1986**, *108*, 1239. (+)

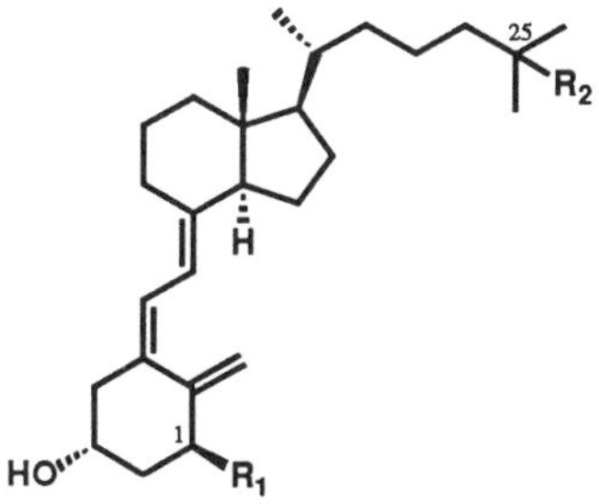

$R_1 = R_2 = H$: Vitamin D$_3$
R_1 or/and R_2 = OH:
Hydroxyvitamin D$_3$'s

DeLuca, *TL*, **1972**, 4147.
Hesse, *JCS, CC*, **1974**, 203.
Mazur, *JOC*, **1976**, *41*, 2651.
Sato, *CPB*, **1978**, *26*, 2933.
Vandewalle, *TL*, **1982**, 995;
Tet, **1985**, *41*, 141.
Fukumoto & Kametani,
TL, **1984**, *25*, 3095; *CL*, **1985**, 1131.

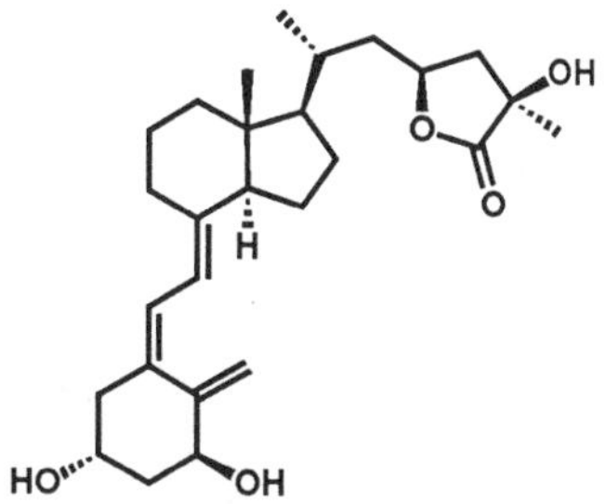

Calcitriol Lactone

Wovkulich, *JOC*, **1983**, *48*, 4433.

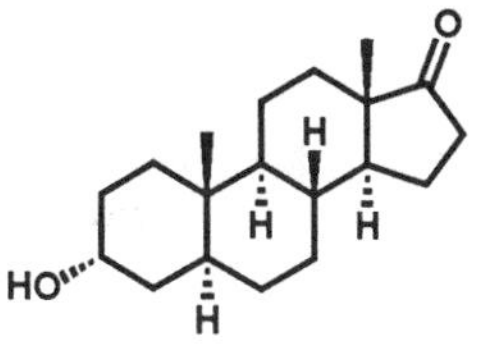

Androsterone

Fukumoto, *JOC*, **1985**, *50*, 144;
JCS, PI, **1986**, 117.

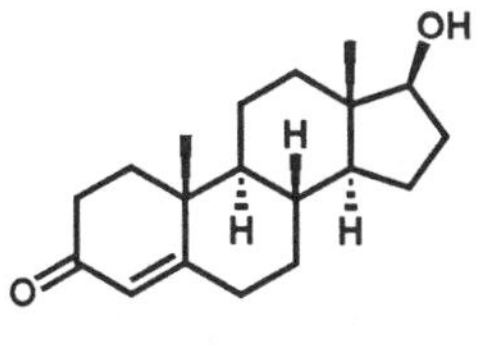

Testosterone

Johnson, *JACS*, **1956**, *78*, 6354;
ibid., **1960**, *82*, 3409; (+), (-)
ibid., **1973**, *95*, 4419.
Fukumoto, *JCS, PI*, **1986**, 117.

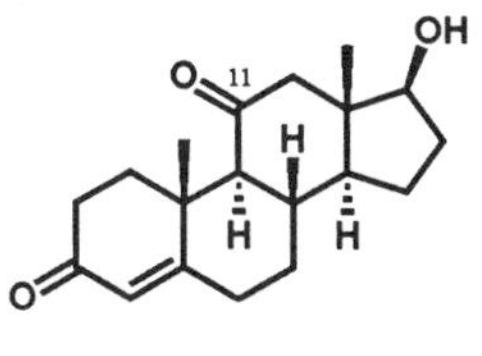

11-Ketotestosterone

Stork, *JACS*, **1981**, *103*, 4948.
for other 11-oxygenated steroids, see:
Stork, *JACS*, **1956**, *78*, 501;
ibid., **1980**, *102*, 1219.
Magnus, *JACS*, **1980**, *102*, 6885.

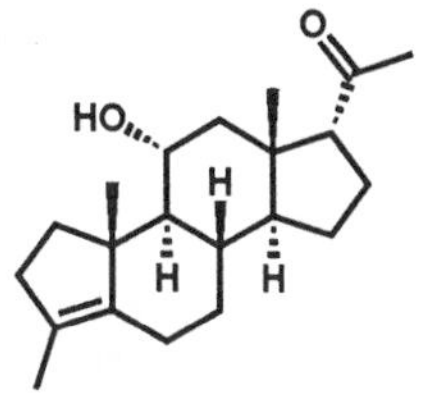

Corticoids

Johnson, *JACS*, **1977**, *99*, 8341;
JOC, **1981**, *46*, 1512;
ibid., **1982**, *47*, 161.

13.6

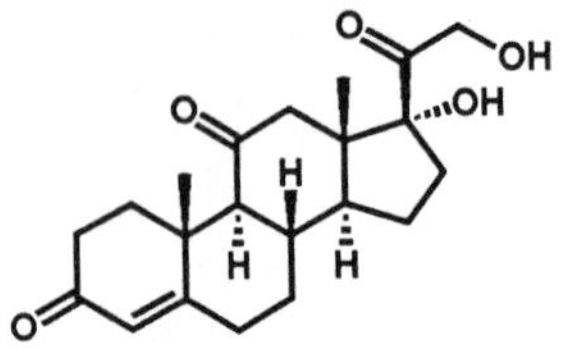

OH
OH
H
H
H

Cortisone

Woodward, *JACS*, **1951**, *73*, 4057.
Sarett, *JACS*, **1952**, *74*, 4974.
Kuwajima, *JOC*, **1986**, *51*, 4323.

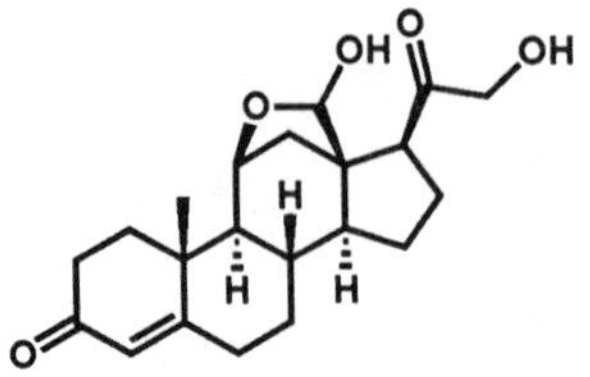

OH
OH
H
H
H

Aldosterone

Johnson, *JACS*, **1958**, *80*, 2585.
Heusler, *HCA*, **1959**, *42*, 1586.
Miyano, *JOC*, **1981**, *46*, 1846.

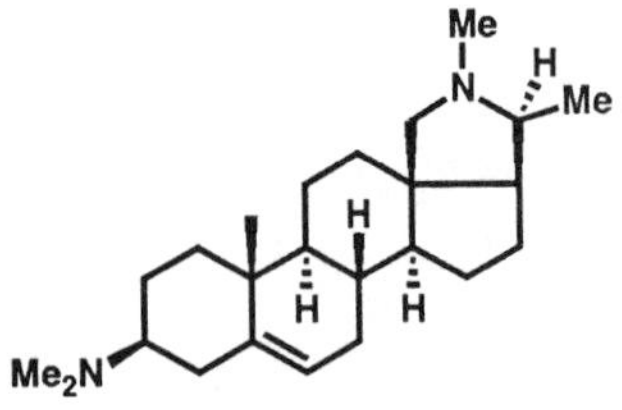

Me
N
H
Me
H
H
H
Me2N

Conessine

Barton, *PCS*, **1961**, 206.
Johnson, *JACS*, **1962**, *84*, 1485.
Stork, *JACS*, **1962**, *84*, 2018.

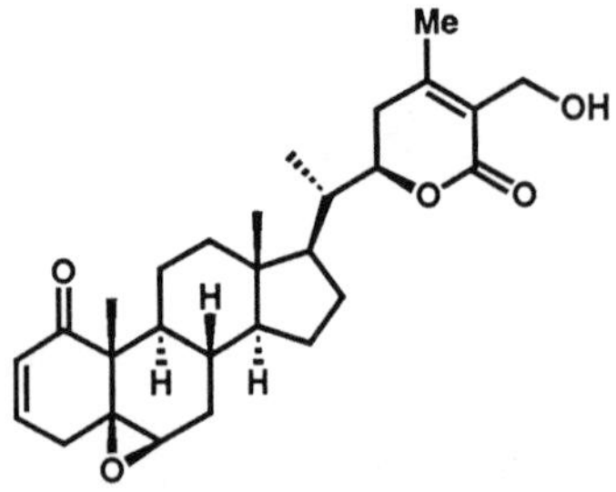

Me
OH
O
O
O
H
H
H
O

Jaborosalactone A

Ikekawa, *JACS*, **1982**, *104*, 3735.

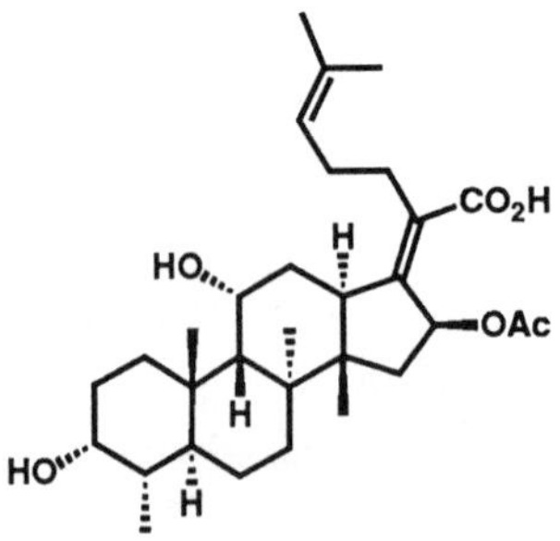

CO2H
H
HO
OAc
H
HO
H

Fusidic Acid

Dauben, *JACS*, **1982**, *104*, 303.

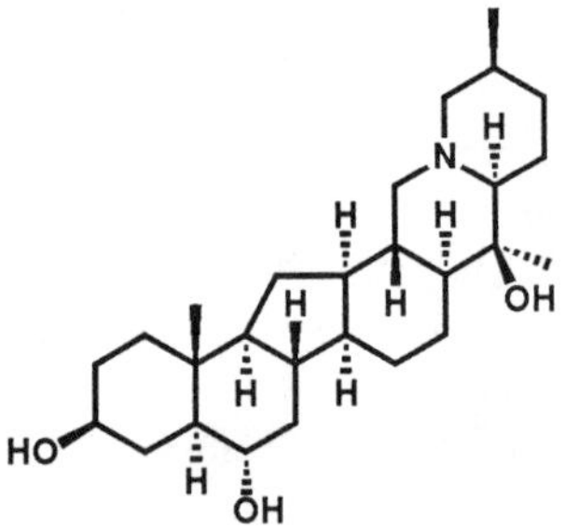

H
N
H
H
OH
H
H
H
H
HO
H
OH

Verticine

Kutney, *JACS*, **1977**, *99*, 964.

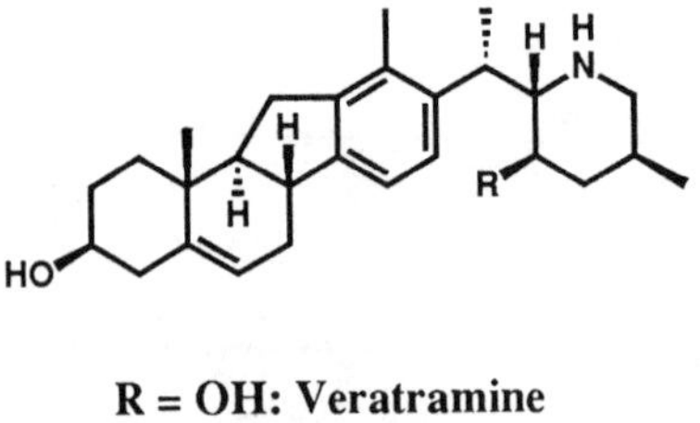

H
H
N
H
H
R
HO

R = OH: Veratramine
R = H: Verarine

Masamune, *JACS*, **1967**, *89*, 4521.
Johnson, *JACS*, **1967**, *89*, 4523.
Kutney, *JACS*, **1968**, *90*, 5332;
CJC, **1975**, *53*, 1796.

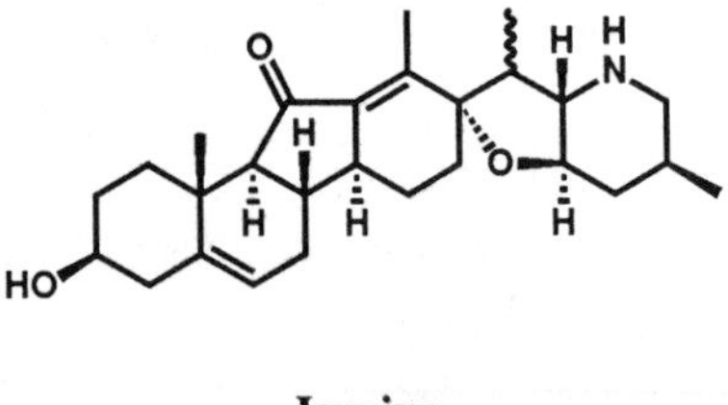

O
H
H
N
H
O
H
H
H
HO

Jervine

Masamune, *JACS*, **1967**, *89*, 4521.
Kutney, *CJC*, **1975**, *53*, 1796.

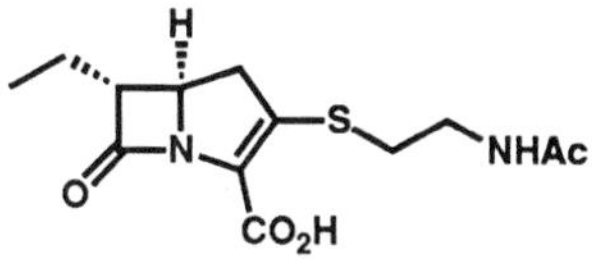

PS-5

Southgate, *JCS, CC*, **1980**, 1084.
Kametani, *JCS, PI*, **1981**, 2228.
Favara, *TL*, **1982**, *23*, 3105. (+)
Ohno, *TL*, **1983**, *24*, 217. (+)
Hatanaka, *TL*, **1984**, *25*, 2387.
Wasserman, *TL*, **1984**, *25*, 3747.
Evans, *TL*, **1986**, *27*, 3119. (+)
Hart, *JACS*, **1986**, *108*, 6054. (+)
Georg, *JOC*, **1988**, *53*, 692. (+)
Tanner, *Tet*, **1988**, *44*, 619. (+)

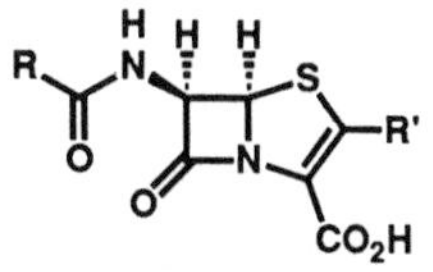

Penems

Woodward, *JACS*, **1978**, *100*, 8214;
ibid., **1980**, *102*, 2039.
Afonso, *JACS*, **1982**, *104*, 6138.
Girijavallabhan, *JCS, CC*, **1983**, 908.
Hanessian, *JACS*, **1985**, *107*, 1438.

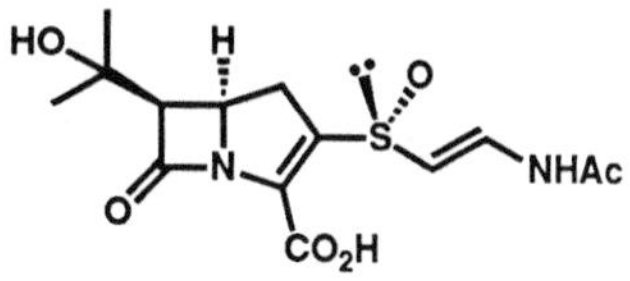

Carpetimycin A

Ohno, *JACS*, **1983**, *105*, 1659. (-)
Shibasaki, *TL*, **1985**, *26*, 2217. (-)
Buynak, *JOC*, **1986**, *51*, 1571.

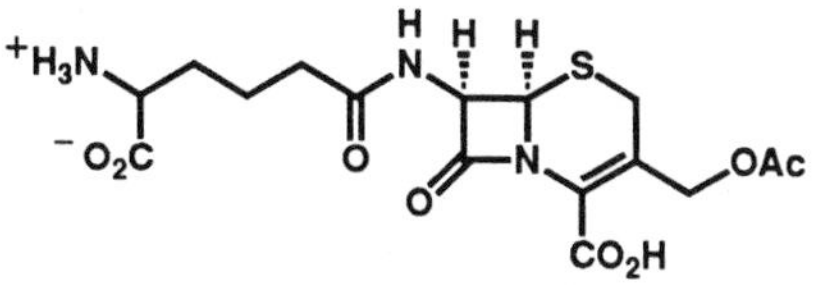

Cephalosporin C

Woodward, *JACS*, **1966**, *88*, 852. (+)

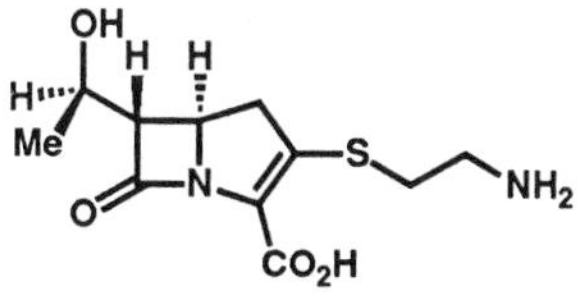

Thienamycin

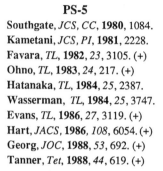

Christensen, *JACS*, **1978**, *100*, 313;
ibid., **1980**, *102*, 6161. (+)
Kametani, *JACS*, **1980**, *102*, 2060.
Shiozaki, *TL*, **1980**, *21*, 4473.
Hanessian, *CJC*, **1982**, *60*, 2292. (+)
Ikegami, *TL*, **1982**, *23*, 2875.
Shinkai, *TL*, **1982**, *23*, 4899.
Yoshikoshi, *JCS, CC*, **1982**, 1354. (+)
Koga, *CPB*, **1982**, *30*, 1929. (+)
Grieco, *JACS*, **1984**, *106*, 6414.
Ley, *Tet*, **1985**, *41*, 5871. (+)
Shibasaki, *TL*, **1985**, *26*, 1523. (+)
Hart, *TL*, **1985**, *26*, 5493. (+)
Hiraoka, *JOC*, **1986**, *51*, 399.
Buynak, *JCS, CC*, **1986**, 941.
Evans, *TL*, **1986**, *27*, 4961. (+)
Fleming, *JCS, CC*, **1986**, 1198.
Georg, *JACS*, **1987**, *109*, 1129. (+)
Hatanaka, *TL*, **1987**, *28*, 69. (+)
Ohno, *TL*, **1988**, *29*, 1057. (+)

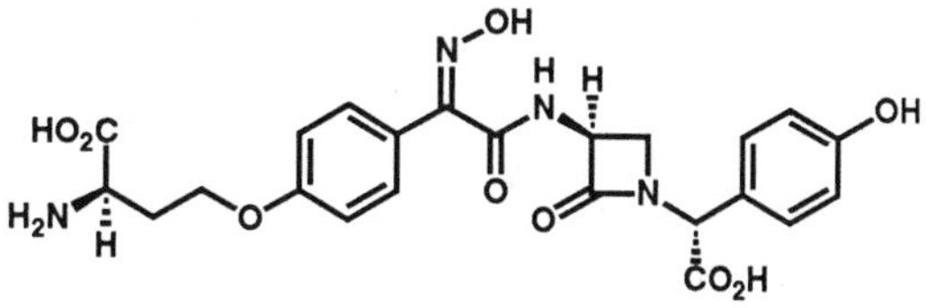

(-)-Nocardicin A

Koppel, *JACS*, **1978**, *100*, 3933.
Kamiya, *Tet*, **1979**, *35*, 323.
Hofheinz, *Tet*, **1983**, *39*, 2591.
Townsend, *TL*, **1986**, *27*, 3819.

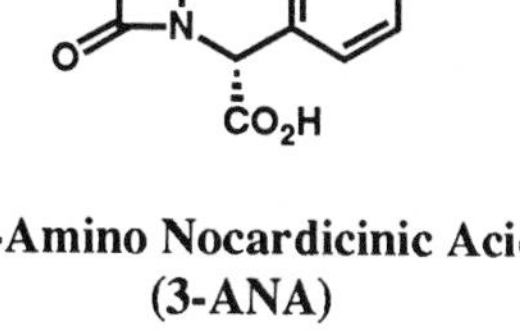

3-Amino Nocardicinic Acid (3-ANA)

Wasserman, *JACS*, **1978**, *100*, 6780;
JOC, **1981**, *46*, 2999.
Townsend, *JACS*, **1981**, *103*, 4582. (-)
Hatanaka, *BCS, Jpn*, **1982**, *55*, 1234. (-)
Hanessian, *CJC*, **1985**, *63*, 3613. (-)

13.7

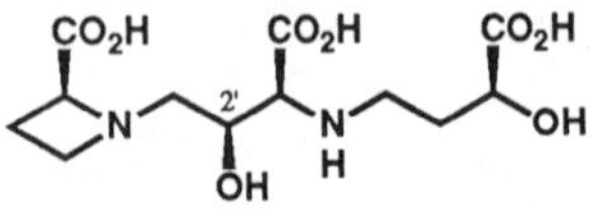

Mugineic Acid
Shioiri, *JOC*, **1986**, *51*, 5489.
(-)-2'-Deoxymugineic Acid
Ohfune, *JACS*, **1981**, *103*, 2409.
Nozoe, *CL*, **1981**, 909.

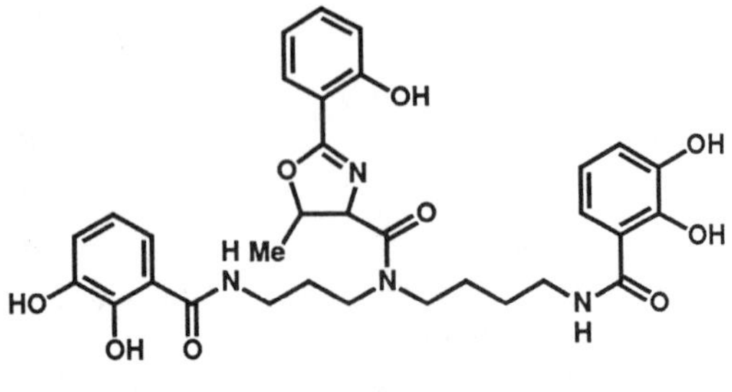

(+)-Parabactin
Bergeron, *JACS*, **1982**, *104*, 4489.

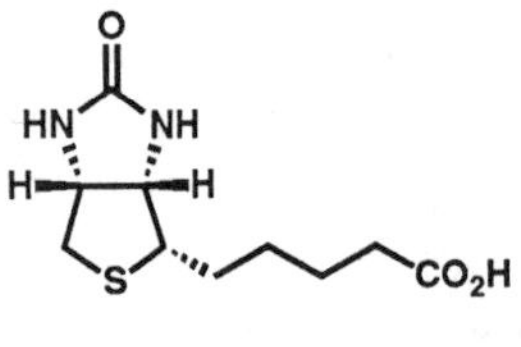

Biotin
Confalone, *JACS*, **1975**, *97*, 5936; (+)
ibid., **1980**, *102*, 1954. (+)
Ohrui, *TL*, **1975**, 2765. (+)
Marx, *JACS*, **1977**, *99*, 6754.
Marquet, *JACS*, **1978**, *100*, 1558.
Field, *JACS*, **1978**, *100*, 7423.
Hohenlohe-Oehringen,
Ber, **1980**, *113*, 607.
Schmidt, *Synth*, **1982**, 747. (+)
Baggiolini, *JACS*, **1982**, *104*, 6460. (+)
Volkmann, *JACS*, **1983**, *105*, 5946. (+)
Whitney, *CJC*, **1983**, *61*, 1158.

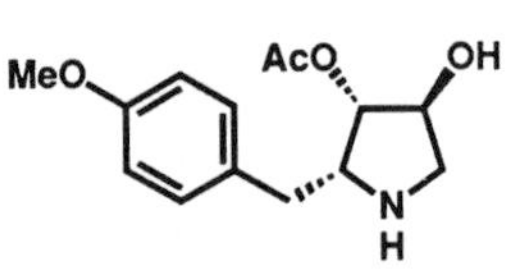

Anisomycin
Wong, *CJC*, **1968**, *46*, 3091;
ibid., **1969**, *47*, 2421. (±), (+), (-)
Oida, *CPB*, **1969**, *17*, 1405.
Felner, *HCA*, **1970**, *53*, 754. (-)
Verheyden, *P&AC*, **1978**, *50*, 1363. (-)
Hall, *JACS*, **1982**, *104*, 6076.
Buchanan, *JCS, CC*, **1983**, 486. (-)
Shono, *CL*, **1987**, 697.(-)

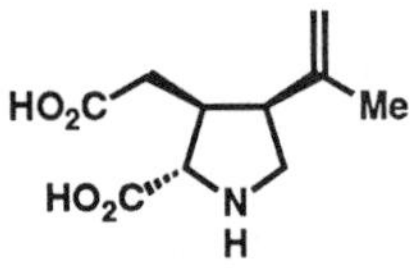

(-)-α-Kainic Acid
Oppolzer, *HCA*, **1979**, *62*, 2282;
JACS, **1982**, *104*, 4978.
Baldwin, *JCS, CC*, **1987**, 166.
Knight, *JCS, CC*, **1987**, 1220.

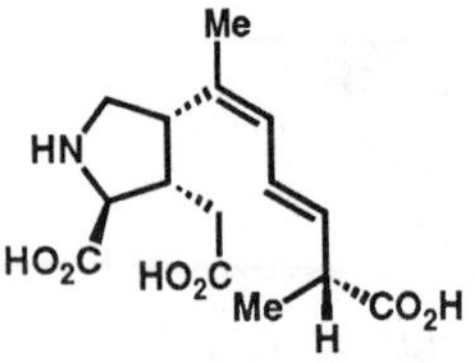

(-)-Domoic Acid
Ohfune,
JACS, **1982**, *104*, 3511.

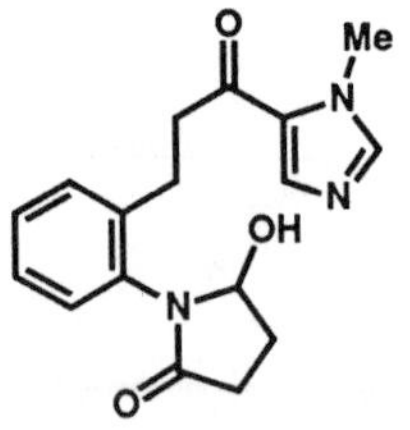

Isolongistrobine
Woodward,
JACS, **1973**, *95*, 5098.

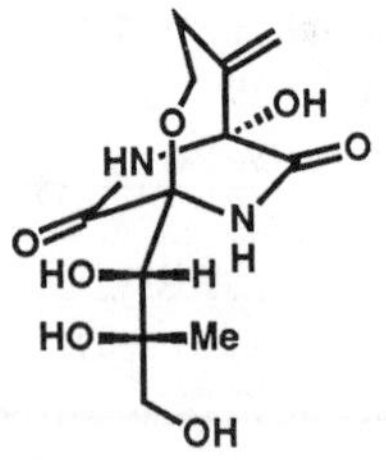

Bicyclomycin
Nakatsuka, *TL*, **1983**, *24*, 5627;
Het, **1984**, *21*, 61.
Yoshimura, *CL*, **1984**, 2157. (+)
Williams, *JACS*, **1984**, *106*, 5748;
ibid., **1985**, *107*, 3253. (±), (+)

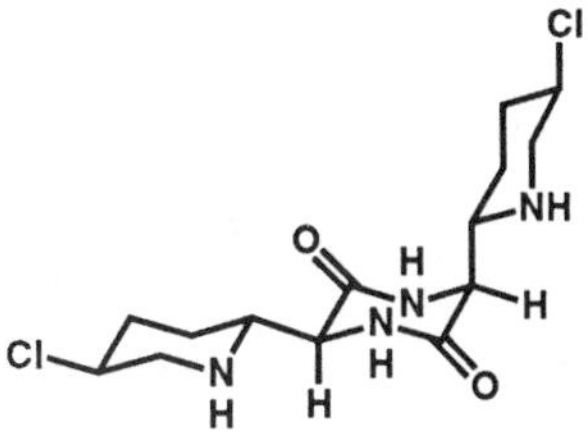

Antibiotic 593A
Fukuyama, *JACS*, **1980**, *102*, 2122.

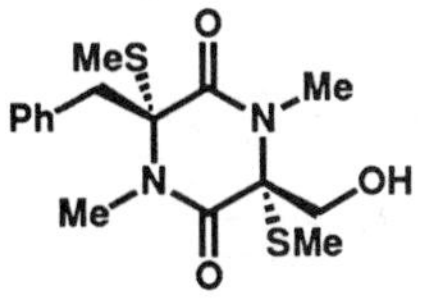

Gliovictin
Rastetter, *TL*, **1979**, 1187;
JOC, **1980**, *45*, 2625.

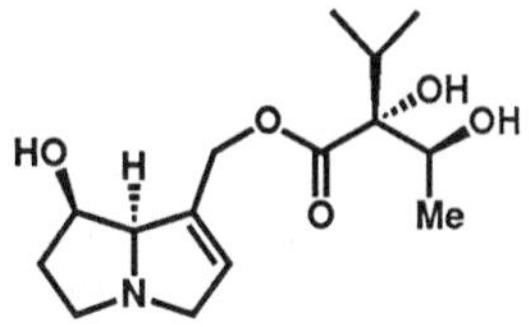

Indicine
Vedejs, *JOC*, **1985**, *50*, 2170.

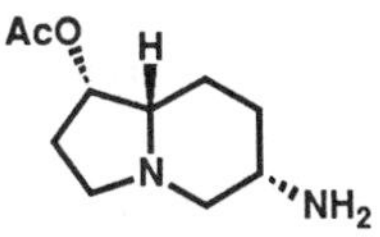

Slaframine
Rinehart, *JACS*, **1970**, *92*, 7615.
Gensler, *JOC*, **1973**, *38*, 3848.
Weinreb, *JACS*, **1982**, *104*, 7065.

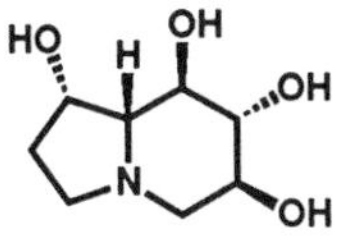

(+)-Castanospermine
Ganem, *TL*, **1984**, *25*, 165;
JOC, **1987**, *52*, 5492.
Hashimoto, *TL*, **1985**, *26*, 4617.

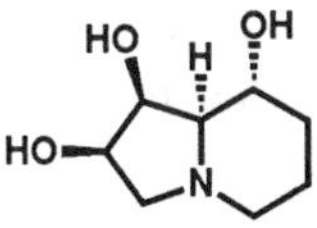

(-)-Swainsonine
Fleet, *TL*, **1984**, *25*, 1853.
Suami, *CL*, **1984**, 513.
Takaya, *CL*, **1984**, 1201.
Richardson, *JCS, CC*, **1984**, 447.
Sharpless, *JOC*, **1985**, *50*, 420.

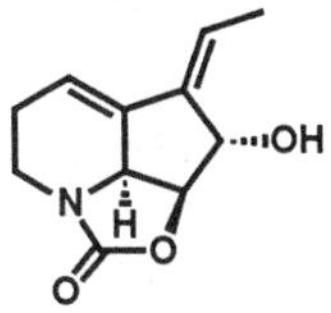

Streptazolin
Kozikowski, *JACS*, **1985**, *107*, 1763.
Overman, *JACS*, **1987**, *109*, 6115. (+)

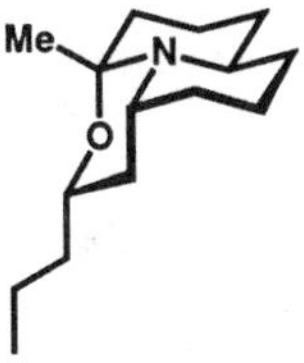

Porantheridin
Gössinger,
TL, **1980**, *21*, 2229.

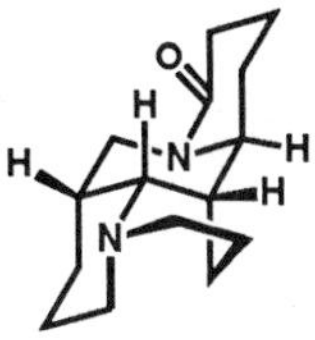

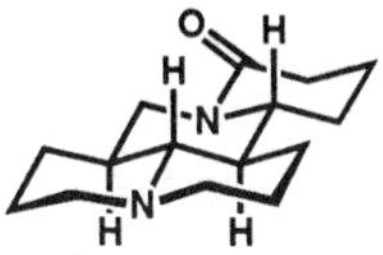

Matrine **Leontine**
Mandell, *JACS*, **1963**, *85*, 2682;
ibid., **1965**, *87*, 5234.

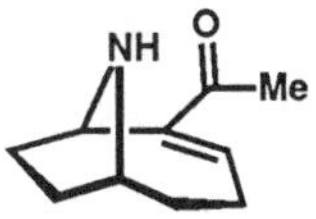

Anatoxin-a
Campbell, *CJC*, **1977**, *55*, 1372.
Rapoport, *JACS*, **1979**, *101*, 1259; (±)
ibid., **1984**, *106*, 4539. (+), (-)
Tufariello, *Tet*, **1985**, *41*, 3447.
Danheiser, *JACS*, **1985**, *107*, 8066.
Wiseman, *JOC*, **1986**, *51*, 2485.
Hiemstra & Speckamp,
Het, **1987**, *26*, 75.
Gallagher, *JCS, CC*, **1987**, 245.
Shono, *CL*, **1987**, 919.

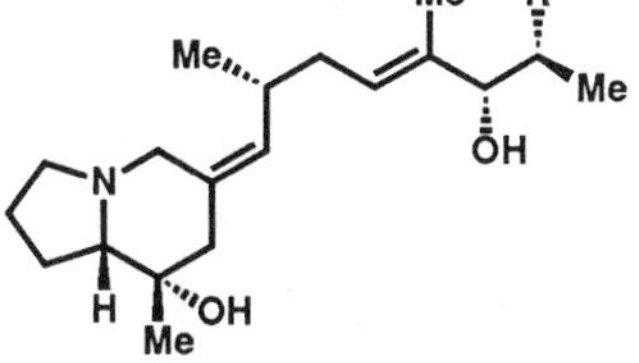

R = OH: (+)-Pumiliotoxin B
Overman, *JACS*, **1984**, *106*, 4192.
R = H: (+)-Pumiliotoxin A
Overman, *JOC*, **1985**, *50*, 3669;
TL, **1988**, *29*, 901.
also, see: **(+)-Pumiliotoxin 251D**
Overman, *JACS*, **1981**, *103*, 1851.

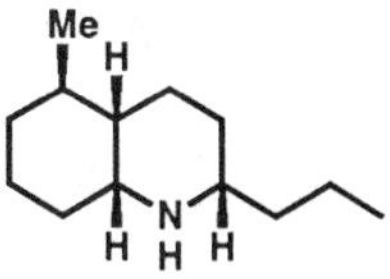

(+)-Pumiliotoxin C
Schultz,
JACS, **1987**, *109*, 6493.

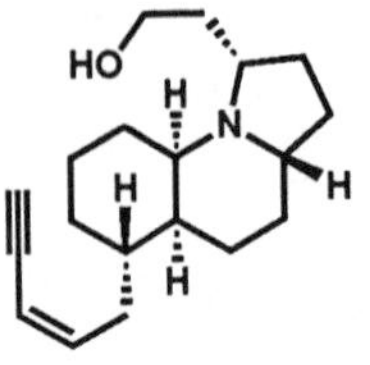

Gephyrotoxin

Kishi, *JACS*, **1980**, *102*, 7154.
Saegusa, *TL*, **1983**, *24*, 2881.
Hart, *JACS*, **1983**, *105*, 1255.
Overman, *JACS*, **1983**, *105*, 5373.

Perhydrogephyrotoxin

Overman, *JACS*, **1980**, *102*, 1454.

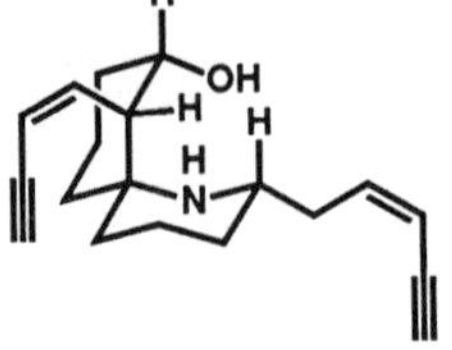

Histrionicotoxin

Kishi,
TL, **1985**, *26*, 5887.

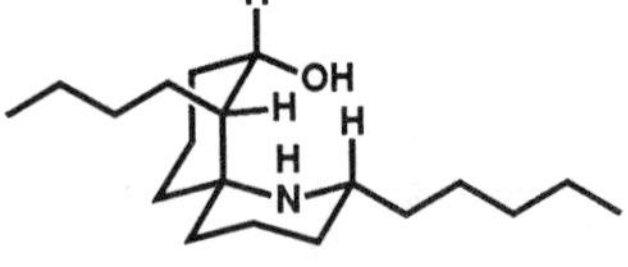

Perhydrohistrionicotoxin

Kishi, *JOC*, **1975**, *40*, 2009.
Speckamp, *TL*, **1978**, 4841.
Evans, *TL*, **1979**, 411.
Inubushi, *Het*, **1982**, *17*, 507.
Keck, *JOC*, **1982**, *47*, 3590.
Pearson, *JCS, PI*, **1983**, 1421.
Godleski, *JOC*, **1983**, *48*, 2101.
Holmes, *TL*, **1984**, *25*, 4163.
Tanner, *Tet*, **1986**, *42*, 5657.
Tanis, *TL*, **1987**, *28*, 2495.

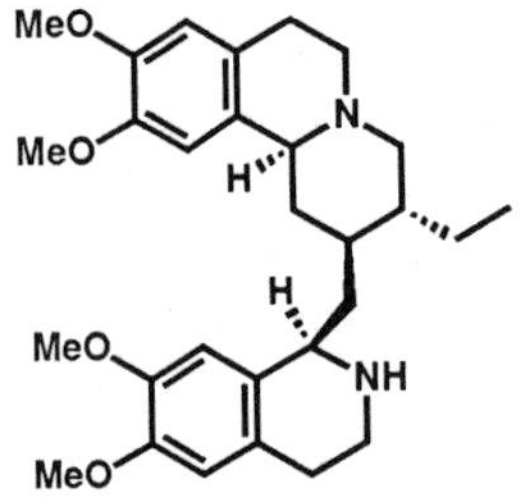

Emetine

Barash, *Chem. Ind.*, **1958**, 490.
Battersby, *JCS*, **1960**, 717. (-)
van Tamelen, *JACS*, **1969**, *91*, 7359.
Takano, *JOC*, **1978**, *43*, 4169.
Fujii, *CPB*, **1979**, *27*, 1486.
Kametani, *JCS, PI*, **1981**, 920.
Yamazaki, *Het*, **1986**, *24*, 571.
Ninomiya, *CPB*, **1986**, *34*, 3530.

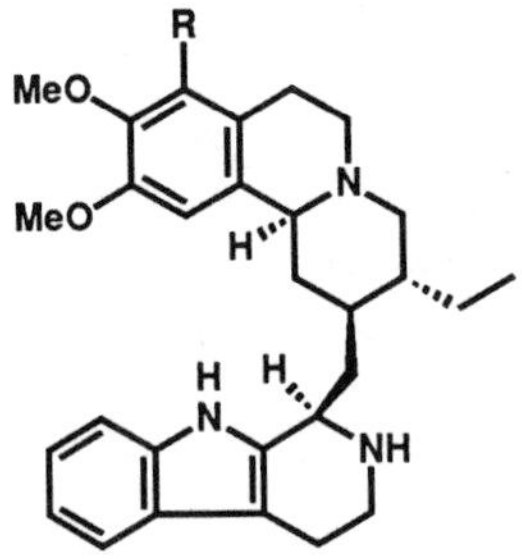

R = OH: Alangimarckine

Fujii, *TL*, **1977**, 3477;
ibid., **1978**, 3111. (-)

R = H: Deoxytubulosine

Kametani, *Het*, **1978**, *11*, 415;
CJC, **1979**, *57*, 1679.
Brown, *TL*, **1984**, *25*, 3127.

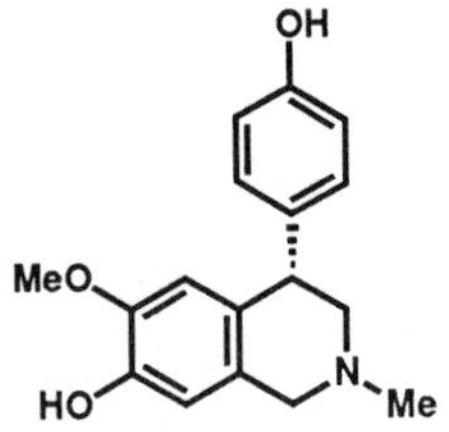

Cherylline

Brossi, *JOC*, **1970**, *35*, 3559. (-)
Schwartz, *JOC*, **1971**, *36*, 1827.
Evans, *JACS*, **1978**, *100*, 1548.
Irie, *CL*, **1980**, 875.
Kametani, *JCS, PI*, **1982**, 2935.
Umezawa, *CPB*, **1985**, *33*, 3107.

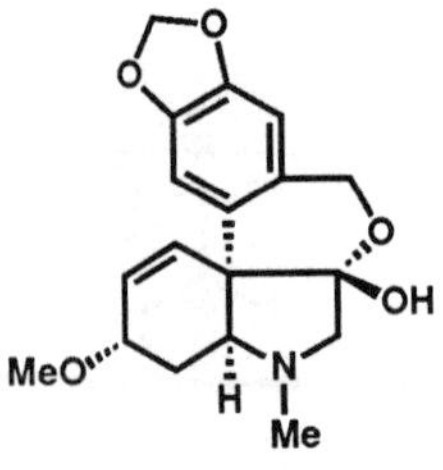

Tazettine

Hendrickson, *JACS*, **1970**, *92*, 5538.
Tsuda, *TL*, **1972**, 3153;
Het, **1978**, *10*, 555.
Danishefsky, *JACS*, **1980**, *102*, 2838.

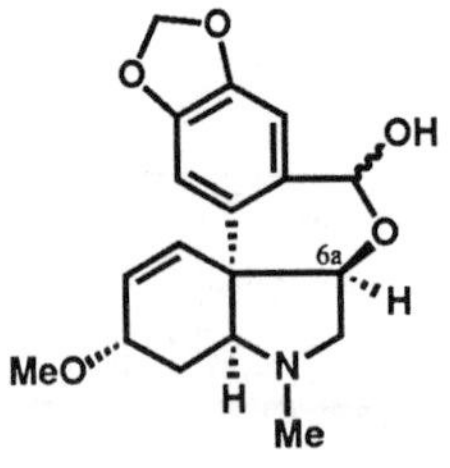

Pretazettine

Tsuda, *TL*, **1972**, 3153;
Het, **1978**, *10*, 555.
Danishefsky, *JACS*, **1982**, *104*, 7591.
Martin, *JOC*, **1987**, *52*, 1962.

6a-*epi*-Pretazettine

White, *JOC*, **1983**, *48*, 2300.

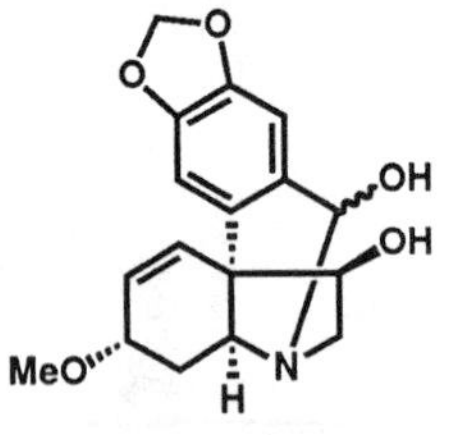

Haemanthidine

Hendrickson, *JCS, CC*, **1965**, 165;
JACS, **1970**, *92*, 5538;
ibid., **1974**, *96*, 7781.
Tsuda, *TL*, **1972**, 3153;
Het, **1978**, *10*, 555.
Martin, *JOC*, **1987**, *52*, 1962.

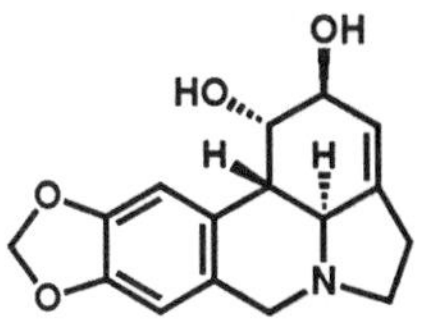

Lycorine
Tsuda, *JCS, CC*, **1975**, 933;
JCS, PI, **1979**, 1358.
Torssell, *ACS, Ser. B*, **1978**, *32*, 98.
Umezawa, *Het*, **1979**, *12*, 1475.
Martin, *JOC*, **1982**, *47*, 3634,

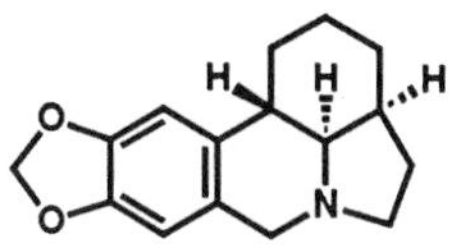

α-Lycorane
Wang,
TL, **1984**, *25*, 4613.

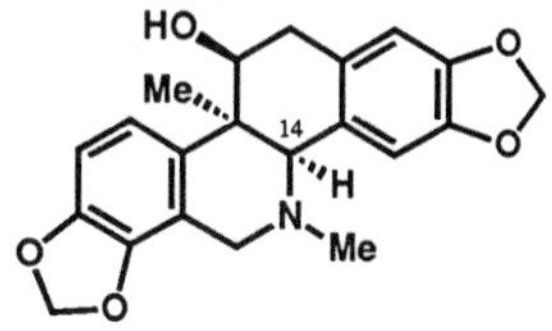

Corynoline
Ninomiya, *JCS, PI*, **1980**, 212.
Cushman, *JACS*, **1983**, *105*, 2873.
14-*epi*-Corynoline
Falck, *JACS*, **1983**, *105*, 631.

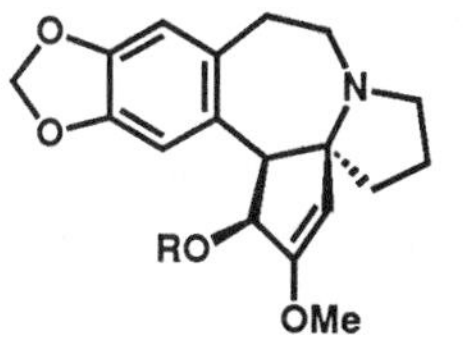

R = H: Cephalotaxine
Weinreb, *JACS*, **1972**, *94*, 7172;
ibid., **1975**, *97*, 2503.
Semmelhack, *JACS*, **1972**, *94*, 8629;
ibid., **1975**, *97*, 2507.
Weinreb & Semmelhack,
ACR, **1975**, *8*, 158.
Hanaoka, *TL*, **1986**, *27*, 2023.
Fuchs, *JACS*, **1988**, *110*, 2341.

R = 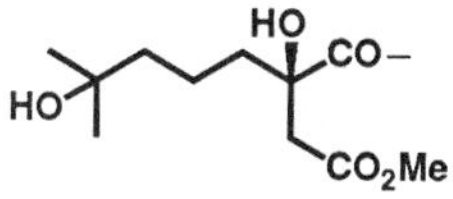

Homoharringtonine
Hudlicky, *JOC*, **1983**, *48*, 5321.

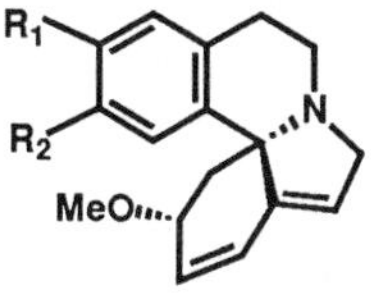

R_1, R_2 =  **: Erythraline**
Sano, *CPB*, **1987**, *35*, 479.
R_1 = R_2 = OMe: Erysotine
Mondon, *Ber*, **1979**, *112*, 1329.
also see: Haruna, *JCS, CC*, **1978**, 733.

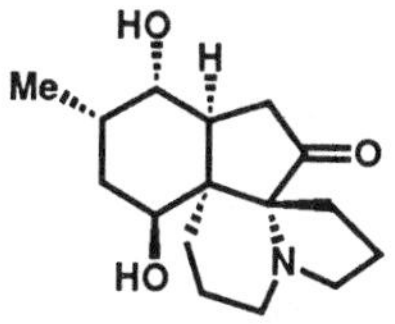

Serratinine
Inubushi,
CPB, **1975**, *23*, 1511.

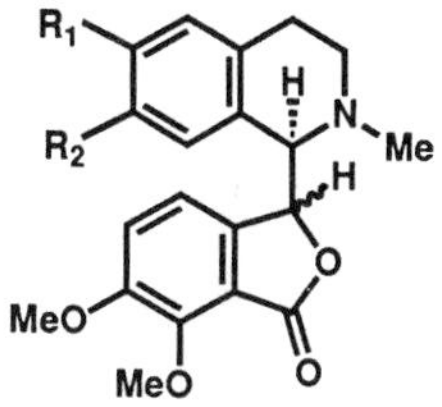

R_1, R_2 =  **:**
(-)-α- or β-Hydrastine
Shamma, *JOC*, **1979**, *44*, 4337.
Hanaoka, *CPB*, **1979**, *27*, 1947.
R_1 = R_2 = OMe:
Cordrastine, I, II
MacLean, *CJC*, **1973**, *51*, 3287;
ibid., **1977**, *55*, 922.
Kametani, *Het*, **1975**, *3*, 1091.
Snieckus, *TL*, **1978**, 5107.
Shono, *TL*, **1980**, *21*, 1351.

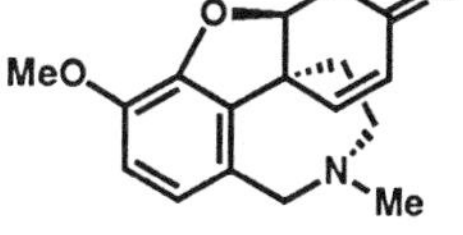

X = O: Narwedine
Holton, *JACS*, **1988**, *110*, 314.

X = 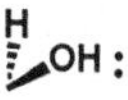 **: Galanthamine**
Barton, *JCS*, **1962**, 806. (-)
Kametani, *JCS, CC*, **1969**, 425;
JCS (C), **1969**, 2602;
JOC, **1971**, *36*, 1295.
Koga, *Het*, **1977**, *8*, 277; (+), (-)
CPB, **1978**, *26*, 3765. (+), (-)

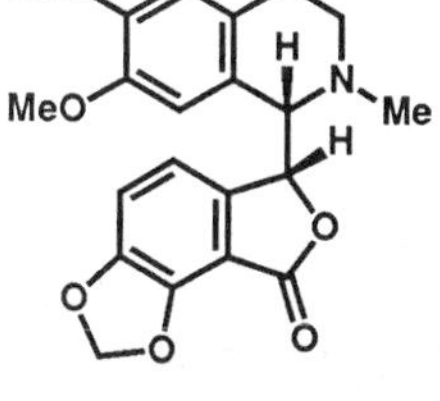

(+)-Corlumine
Seebach,
HCA, **1987**, *70*, 1944.

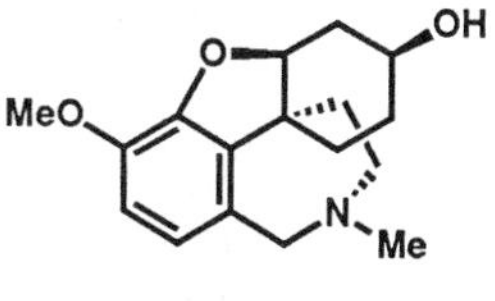

Lycoramine
Barton, *JCS*, **1962**, 806. (-)
Uyeo, *JCS (C)*, **1968**, 2947, 2954.
Schultz, *JACS*, **1977**, *99*, 8065.
Martin, *JOC*, **1982**, *47*, 1513.

13.7

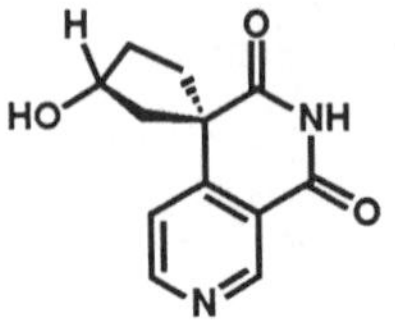

Sesbanine
Kende, *TL*, **1980**, *21*, 715.
Koga, *TL*, **1980**, *21*, 2321.
Berchtold, *JOC*, **1980**, *45*, 1176.
Pandit, *Tet*, **1982**, *38*, 2741.
Akiba, *TL*, **1985**, *26*, 3267.

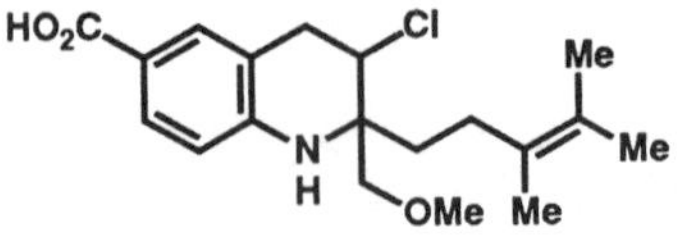

Virantmycin
Raphael,
TL, **1986**, *27*, 1293.

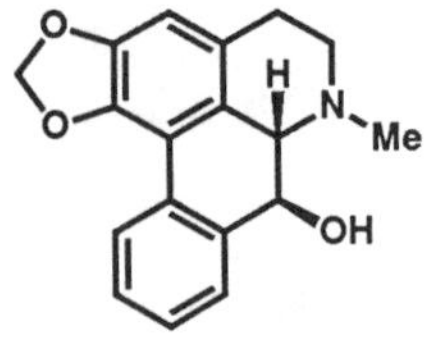

Oliveroline
Kessar, *TL*, **1980**, *21*, 3307.
Seebach, *HCA*, **1987**, *70*, 1357.

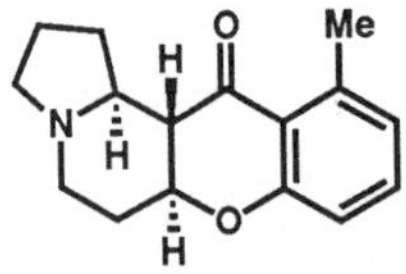

Elaeocarpine
Tanaka, *TL*, **1970**, 3963;
Tet, **1973**, *29*, 1285.
Onaka, *TL*, **1971**, 4395.
Tufariello, *JACS*, **1979**, *101*, 7114.

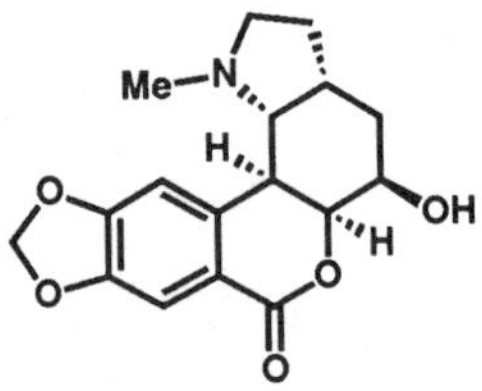

Clividine
Irie, *JCS, CC*, **1973**, 302;
JCS, PI, **1979**, 535.

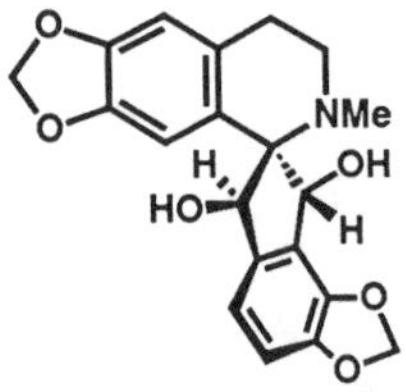

Ochrobirine
Kametani, *JCS, PI*, **1972**, 391.
Nalliah, *CJC*, **1972**, *50*, 1819.
MacLean, *CJC*, **1972**, *50*, 3028.

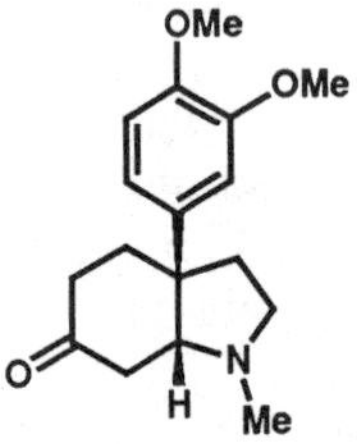

(-)-Mesembrine (*nat*)
Shamma, *Tet*, **1968**, *24*, 6583. (±)
Curphey, *TL*, **1968**, 1441. (±)
Stevens, *JACS*, **1968**, *90*, 5580. (±)
Keely, *JACS*, **1968**, *90*, 5584. (±)
Martin, *JOC*, **1979**, *44*, 3391. (±)
Wiechers, *TL*, **1979**, 4495. (-)
Keck, *JOC*, **1982**, *47*, 1302. (±)
Jeffs, *JOC*, **1983**, *48*, 3861. (±)
Sánchez, *TL*, **1983**, *24*, 551. (±)
Pinnick, *TL*, **1983**, *24*, 4785. (±)
Meyers, *JACS*, **1985**, *107*, 7776. (+)
Gramain, *TL*, **1985**, *26*, 4083. (±)
Livinghouse, *JOC*, **1986**, *51*, 1629. (±)
Winkler, *JACS*, **1988**, *110*, 4831. (±)

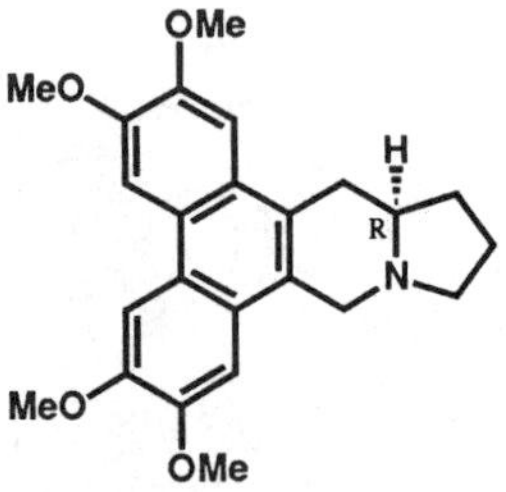

(-)-Tylophorine (*nat*)
Govindachari, *Tet*, **1961**, *14*, 284. (±)
Herbert, *JCS, CC*, **1970**, 121. (±)
Chauncy, *AJC*, **1970**, *23*, 2503. (±)
Liepa, *JCS, CC*, **1977**, 826. (±)
Weinreb, *JACS*, **1979**, *101*, 5073. (±)
Rapoport, *JOC*, **1983**, *48*, 4222. (+)
Kibayashi, *JOC*, **1984**, *49*, 2412. (±)
Njoroge, *JOC*, **1987**, *52*, 1627. (+)
Fukumoto, *TL*, **1988**, *29*, 4135. (-)

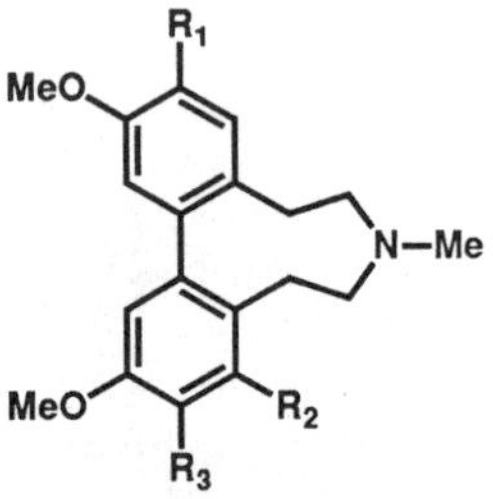

R_1 = OH, R_2 =H, R_3 = OMe: Erybidine
Kupchan, *JACS*, **1975**, *97*, 5623.
R_1 = OMe, R_2 = R_3 = H: Laurifonine
Kupchan, *Het*, **1976**, *4*, 235.
Ito, *Het*, **1978**, *9*, 485.
R_1 = R_2 = OMe, R_3 = H: Protostephanine
Pecherer, *JOC*, **1967**, *32*, 1053.
Battersby, *JCS, PI*, **1981**, 2002.

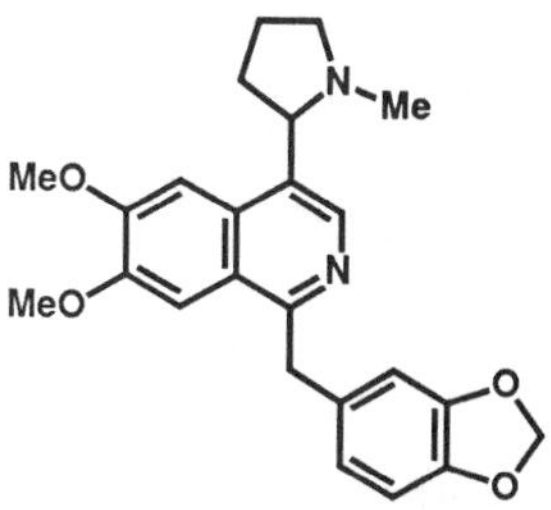

Macrostomine
Seebach, *TL*, **1980**, *21*, 1927.

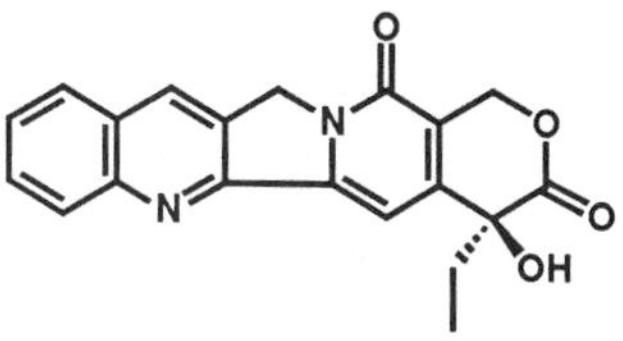

Camptothecin
Stork, *JACS*, **1971**, *93*, 4074.
Danishefsky, *JACS*, **1971**, *93*, 5576.
Wani, *JACS*, **1972**, *94*, 3631;
JMC, **1987**, *30*, 2317. (*R*), (*S*)
Kende, *TL*, **1973**, 1307.
Meyers, *JOC*, **1973**, *38*, 1974.
Rapoport, *JACS*, **1975**, *97*, 159.
Kametani, *JCS, PI*, **1981**, 1563.
Vollhardt, *JOC*, **1984**, *49*, 4786.

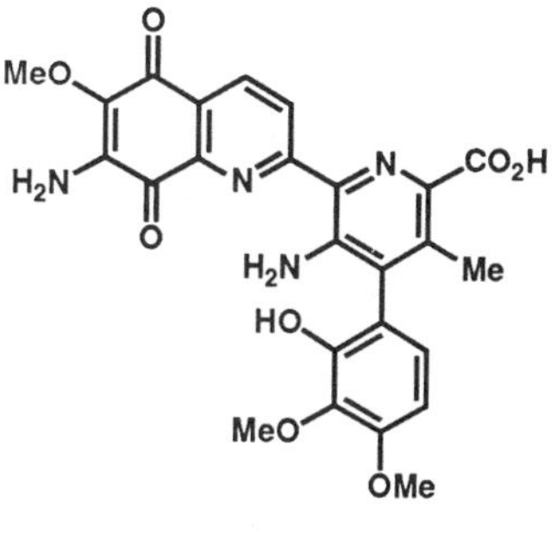

Streptonigrin
Weinreb, *JACS*, **1980**, *102*, 3962;
ibid., **1982**, *104*, 536.
Kende, *JACS*, **1981**, *103*, 1271.
Boger, *JACS*, **1985**, *107*, 5745.

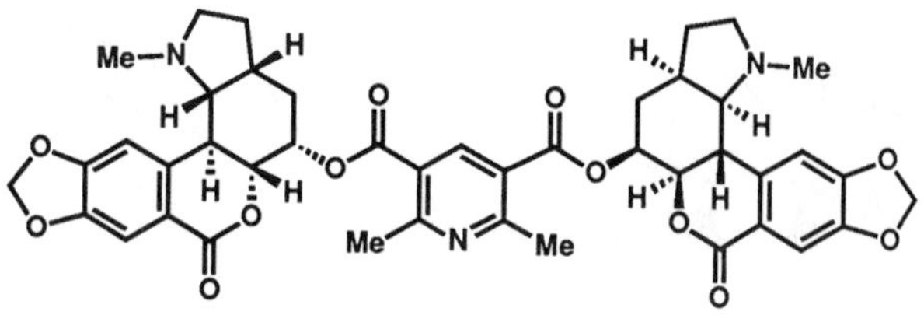

Clivimine
Kobayashi, *Het*, **1980**, *14*, 751.

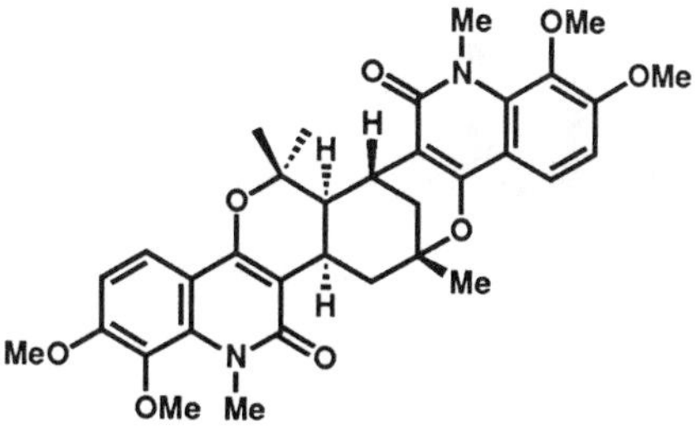

Vepridimerine A
Ayafor, *TL*, **1985**, *26*, 4529.

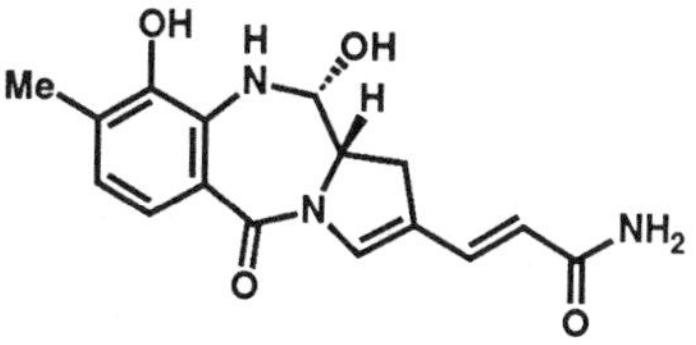

(+)-Anthramycin
Leimgruber, *JACS*, **1968**, *90*, 5641.

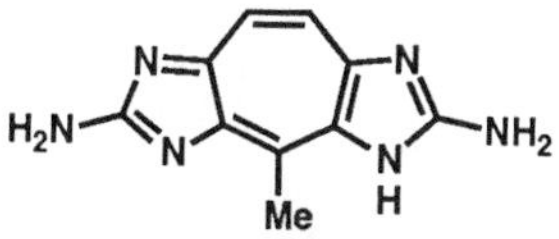

Parazoanthoxanthin A
Büchi, *JACS*, **1978**, *100*, 4208.

13.7

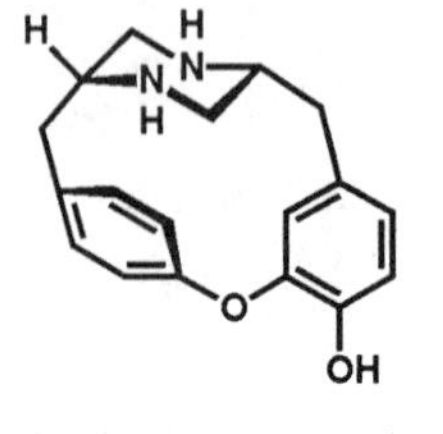

(+)-Piperazinomycin
Yamamura,
TL, **1986**, *27*, 4481.

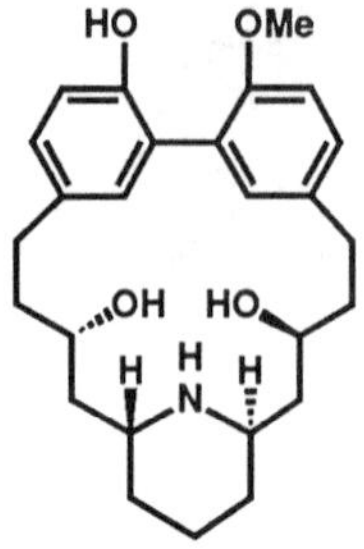

Lythranidine
Fujita, *TL*, **1979**, 361.

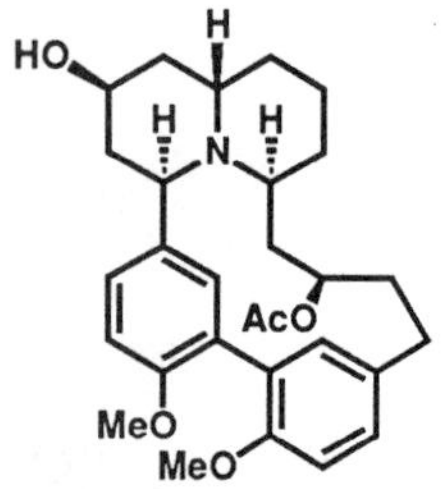

Lythrancepine II
Hart, *JOC*, **1985**, *50*, 3670;
ibid., **1987**, *52*, 4665.

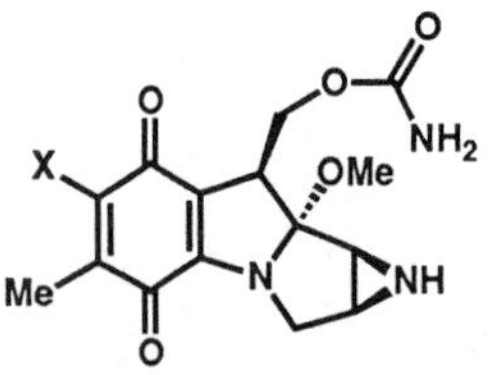

X = OMe: Mitomycin A
X = NH_2: Mitomycin C
Kishi, *TL*, **1977**, 4295.
Fukuyama,
JACS, **1987**, *109*, 7881.

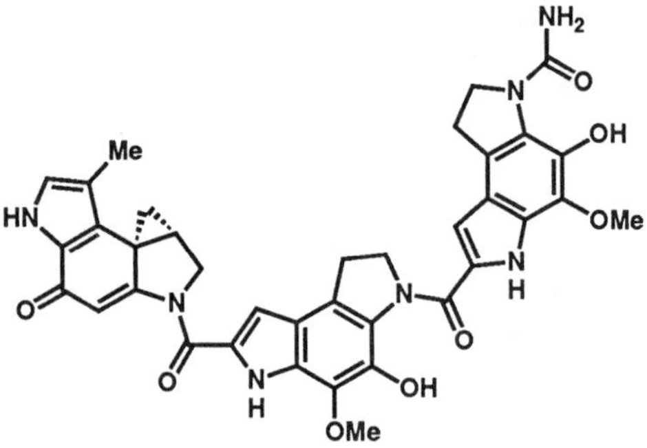

(+)-CC-1065
Kelly, *JACS*, **1987**, *109*, 6837.
Boger, *JACS*, **1988**, *110*, 1321.

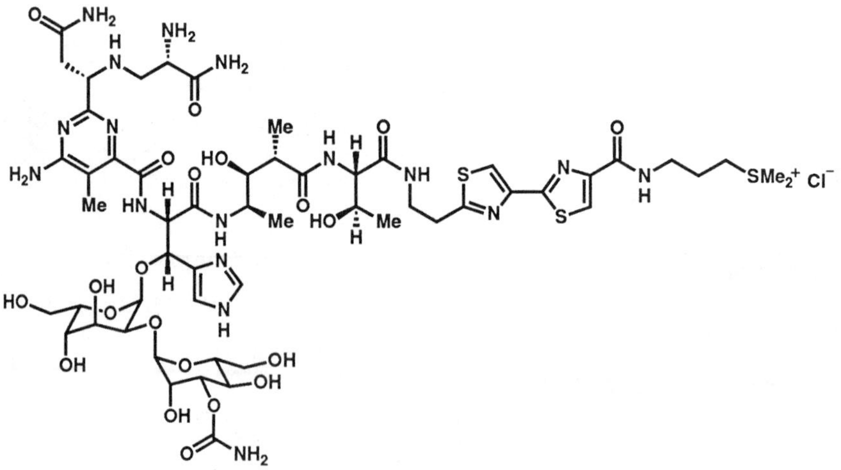

Bleomycin A_2
Ohno, *TL*, **1982**, *23*, 521. (*nat*)
Hecht, *JACS*, **1982**, *104*, 5537. (*nat*)
(-)-Deglyco-Bleomycin A_2
Takita, *TL*, **1981**, *22*, 671.

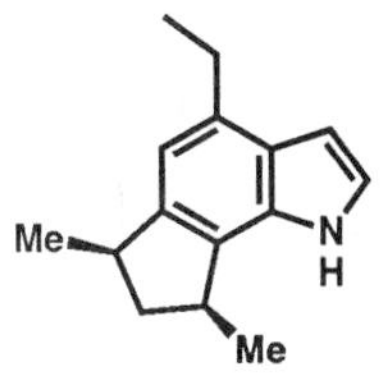

cis-Trikentrin A

MacLeod, *TL*, **1988**, *29*, 391.

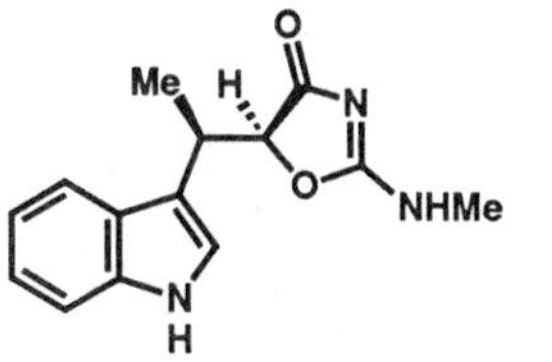

Indolmycin

Preobrazhenskaya, *Tet*, **1968**, *24*, 6131.

Mukaiyama, *CL*, **1980**, 163. (-)

Dirlam, *JOC*, **1986**, *51*, 4920.

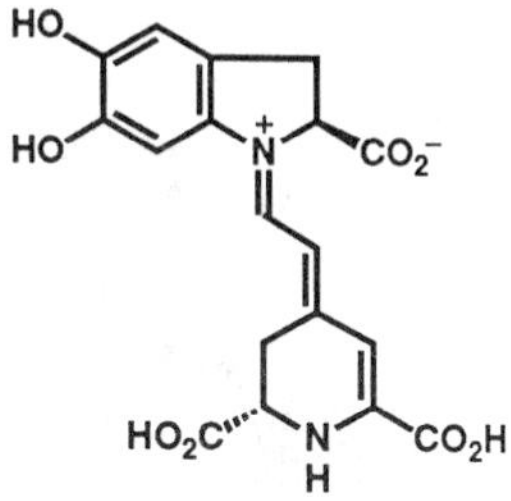

Betanidin

Dreiding, *HCA*, **1975**, *58*, 1805.

Büchi, *JOC*, **1978**, *43*, 4765.

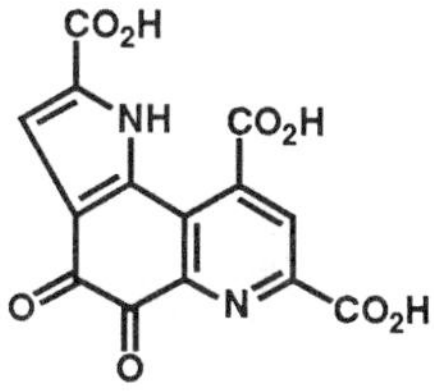

Methoxatin

Weinreb, *JOC*, **1981**, *46*, 4317; *ibid.*, **1982**, *47*, 2833.

Hendrickson, *JOC*, **1982**, *47*, 1148.

Büchi, *JACS*, **1985**, *107*, 5555.

Moody, *Tet*, **1986**, *42*, 3259.

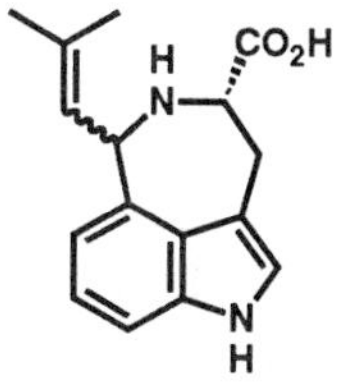

Clavicipitic Acid

Natsume, *Het*, **1983**, *20*, 1963.

Kozikowski, *JOC*, **1984**, *49*, 2310.

Hegedus, *JACS*, **1987**, *109*, 4335.

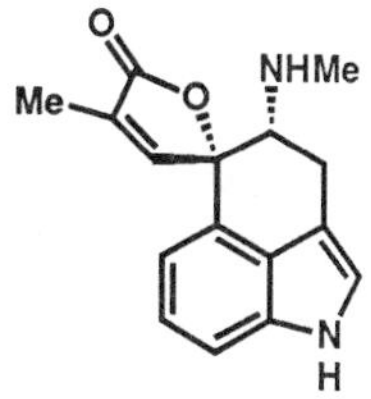

Rugulovasine

Rebek, *JACS*, **1980**, *102*, 5426.

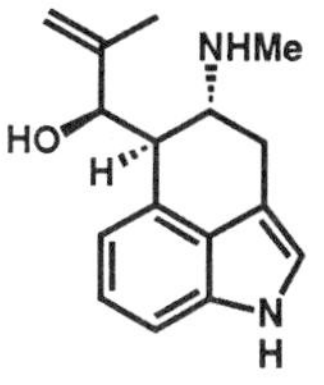

(+)-Paliclavine

Kozikowski, *JOC*, **1981**, *46*, 5248; *Tet*, **1984**, *40*, 2345. (+)

Oppolzer, *Tet*, **1983**, *39*, 3695.

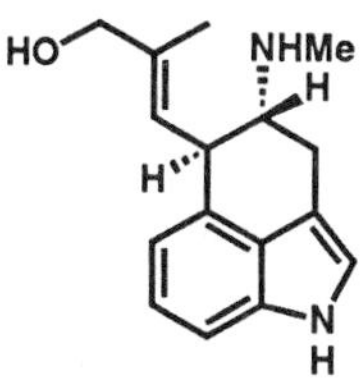

Chanoclavine I

Plieninger, *Ber*, **1976**, *109* , 2140.

Kozikowski, *JACS*, **1980**, *102*, 4265.

Natsume, *Het*, **1981**, *16*, 375.

Oppolzer, *Tet*, **1983**, *39*, 3695.

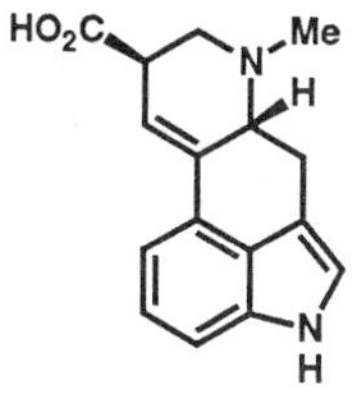

Lysergic Acid

Woodward, *JACS*, **1954**, *76*, 5256; *ibid.*, **1956**, *78*, 3087.

Julia, *TL*, **1969**, 1569.

Ramage, *TL*, **1976**, 4311; *Tet*, **1981**, *37*, *Suppl*, 157.

Oppolzer, *HCA*, **1981**, *64*, 478.

Rebek, *JACS*, **1984**, *106*, 1813.

Ninomyia, *JCS, PI*, **1985**, 941.

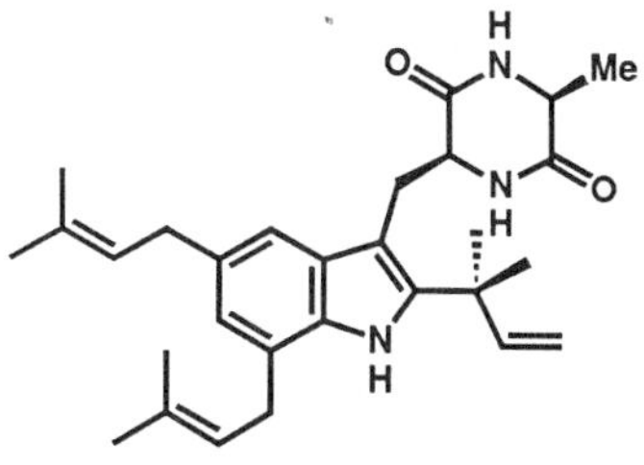

Echinulin

Kishi, *Yakugaku Zasshi*, **1977**, *97*, 558.

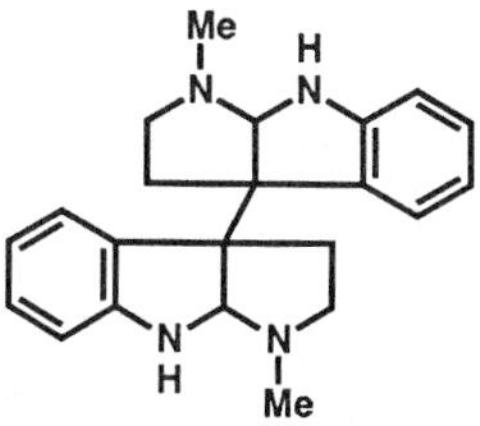

Chimonanthine

Hendrickson, *PCS*, **1962**, 383; *Tet*, **1964**, *20*, 565.

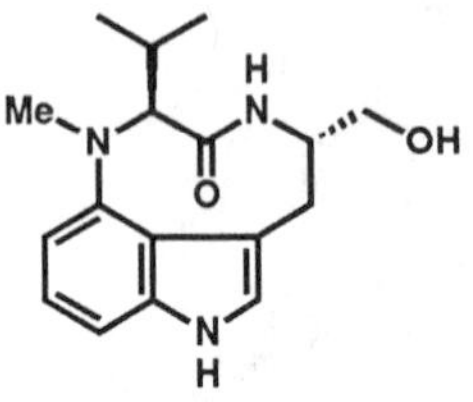

Indolactam V

Okamoto, *CPB*, **1982**, *30*, 3457.
Shudo, *CPB*, **1984**, *32*, 358;
Tet, **1986**, *42*, 5905. (+), (-)
Ley, *JCS, CC*, **1986**, 344.

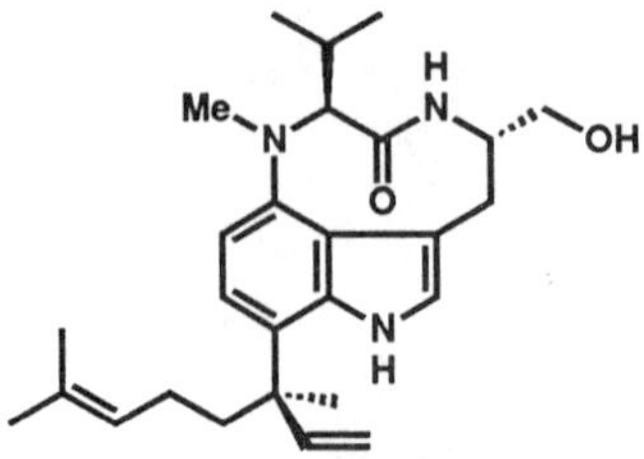

Lyngbyatoxin A

Natsume, *TL*, **1987**, *28*, 2265.

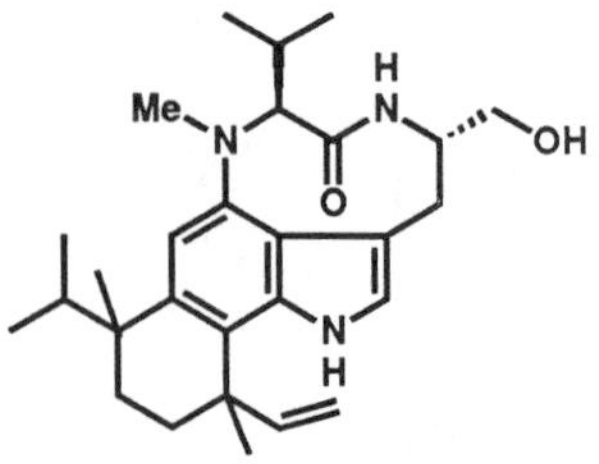

Teleocidin B's

Nakatsuka, *TL*, **1987**, *28*, 3671.

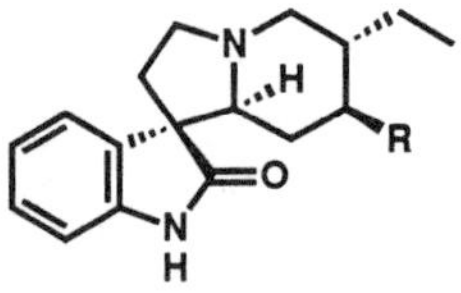

R = CH_2CH_2OH: Rhynchophyllol

van Tamelen, *JACS*, **1969**, *91*, 7333.

R = 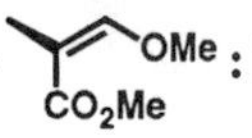 **: Rhynchophylline**

Ban, *TL*, **1972**, 2113.

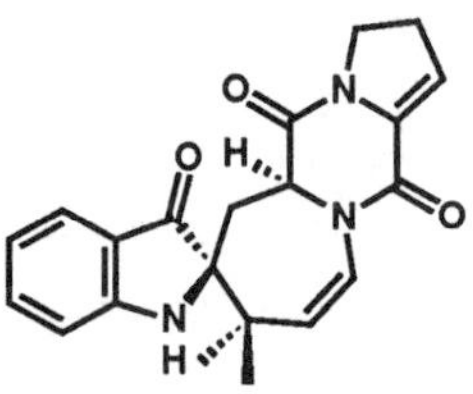

Austamide

Kishi,
JACS, **1979**, *101*, 6786.

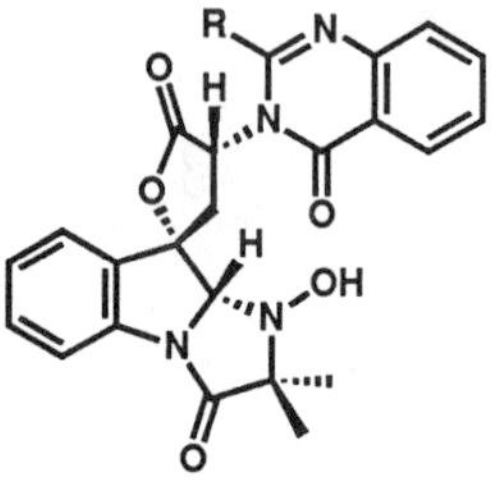

R = 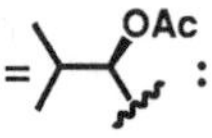 **: Tryptoquivaline**

Nakagawa, *TL*, **1984**, *25*, 3865.

R = H: Tryptoquivaline G

Büchi, *JACS*, **1979**, *101*, 5084.
Ban, *TL*, **1981**, *22*, 4969.
Nakagawa, *JACS*, **1983**, *105*, 3709;
TL, **1984**, *25*, 3865.

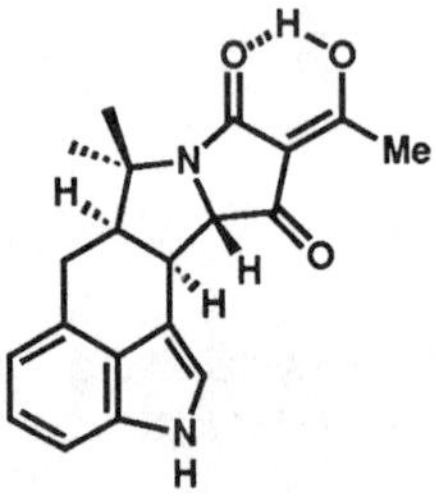

α-Cyclopiazonic Acid

Kozikowski, *JACS*, **1984**, *106*, 6873.
Natsume, *Het*, **1985**, *23*, 1111.

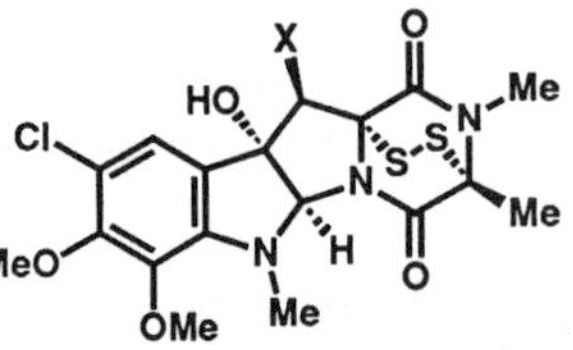

X = OH: Sporidesmin A

Kishi, *JACS*, **1973**, *95*, 6493.

X = H: Sporidesmin B

Kishi, *TL*, **1974**, 1549.

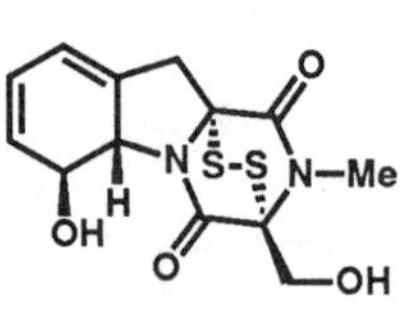

Gliotoxin

Kishi, *JACS*, **1976**, *98*, 6723.

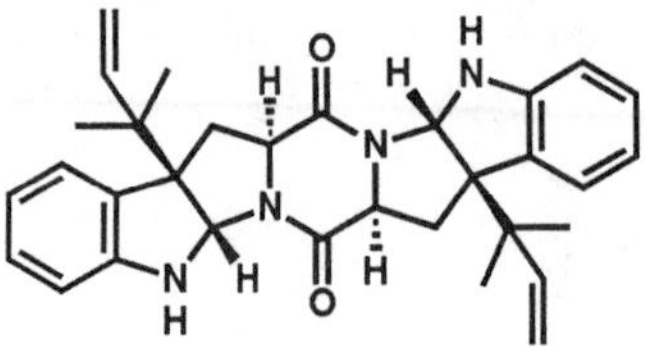

Amauromine

Takase, *TL*, **1985**, *26*, 847;
Tet, **1986**, *42*, 5887.

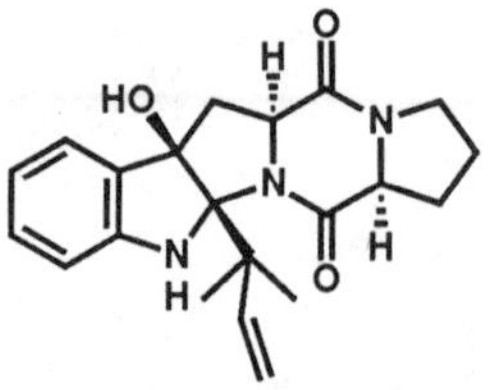

(-)-Brevianamide E

Kametani, *JACS*, **1980**, *102*, 3974;
JCS, PI, **1981**, 959.

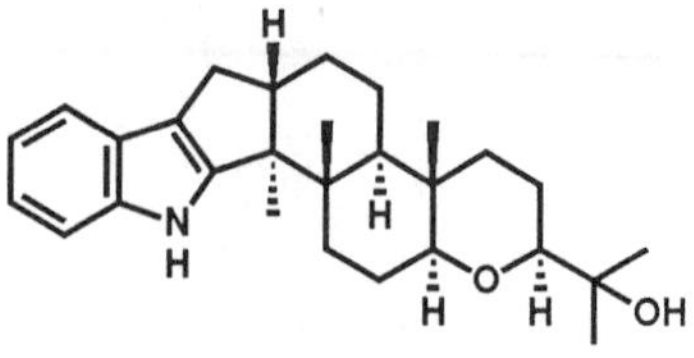

(-)-Paspaline

Smith, *JACS*, **1985**, *107*, 1769.

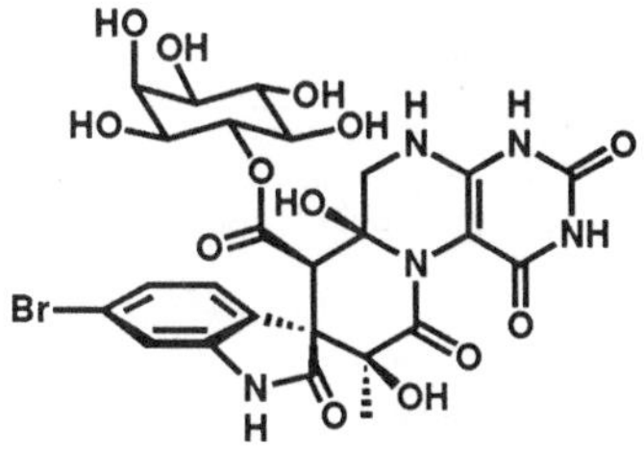

Surugatoxin
Inoue, *TL*, **1984**, *25*, 4407.

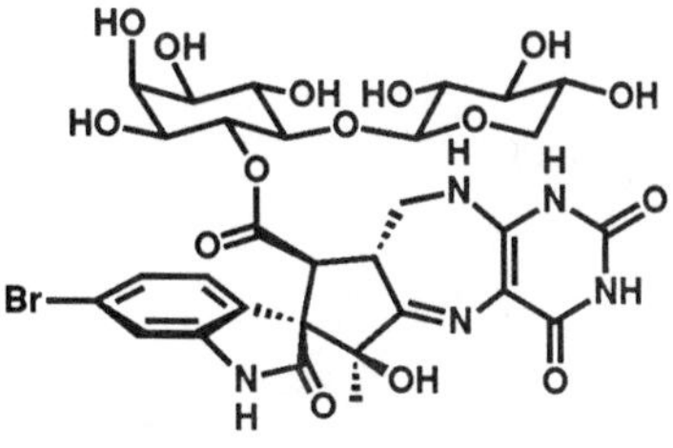

Neosurugatoxin
Inoue, *TL*, **1986**, *27*, 5225. (*nat*)

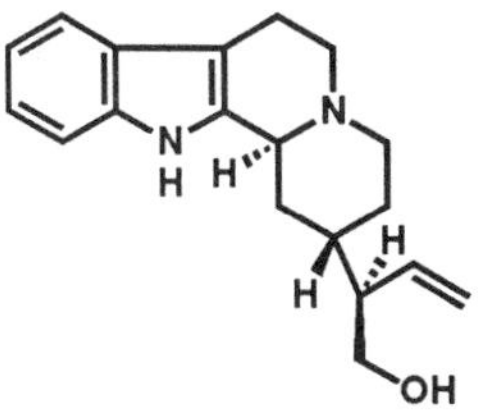

Antirhine
Takano,
JACS, **1980**, *102*, 4282.

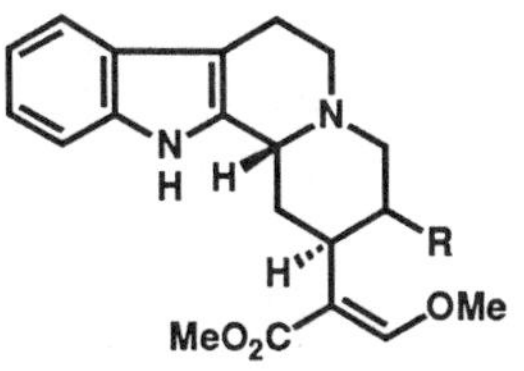

R = : Hirsutine
Wenkert, *JACS*, **1980**, *102*, 7971.
Brown, *JCS, CC*, **1984**, 847.
R = : Hirsuteine
Ninomiya, *Het*, **1987**, *26*, 1739.

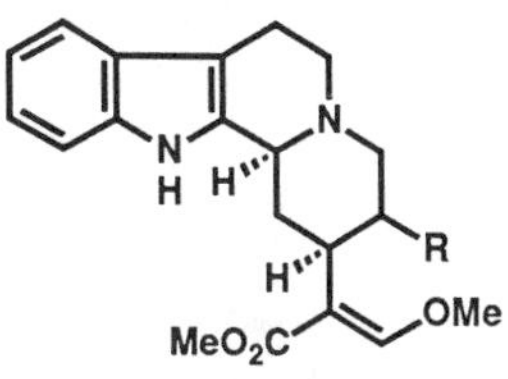

R = : Corynantheidine
Weisbach, *TL*, **1965**, 3457.
Wenkert, *JACS*, **1967**, *89*, 6741;
ACR, **1968**, *1*, 78.
Szántay, *TL*, **1968**, 1405. (-)
Sakai, *CPB*, **1978**, *26*, 2596.
R = : Corynantheine
van Tamelen, *TL*, **1964**, 295.
Autrey, *JACS*, **1968**, *90*, 4917. (+)
Kametani, *Het*, **1981**, *16*, 925.
Takano, *Het*, **1981**, *16*, 1125.

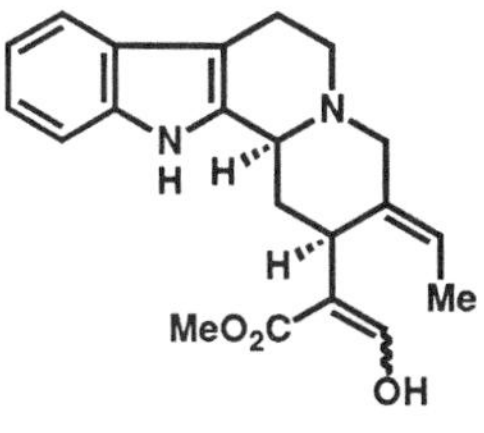

Geissoschizine
Yamada, *JCS, CC*, **1974**, 908.
Winterfeldt, *ACIE*, **1979**, *18*, 862;
Ann, **1985**, 1752. (+)
Wenkert, *JACS*, **1980**, *102*, 7971.
Harley-Mason, *TL*, **1981**, *22*, 1631.
Martin, *JACS*, **1988**, *110*, 5925.

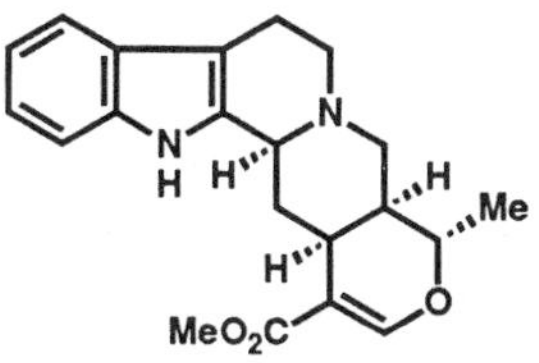

Tetrahydroalstonine
Winterfeldt, *Ber*, **1968**, *101*, 3172.
Brown, *JCS, CC*, **1977**, 636.
Uskokovic, *JACS*, **1979**, *101*, 6742.
Takano, *JCS, CC*, **1988**, 59. (-)
Martin, *JACS*, **1988**, *110*, 5925.

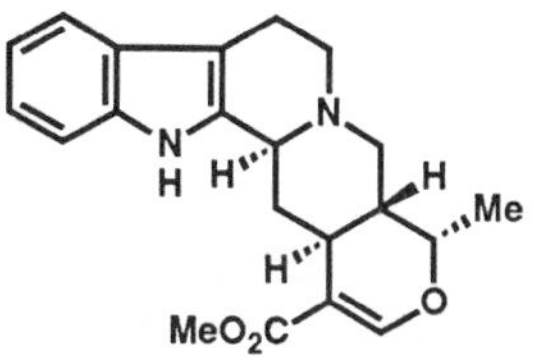

Ajmalicine
van Tamelen, *JACS*, **1961**, *83*, 2594;
ibid., **1969**, *91*, 7359;
ibid., **1973**, *95*, 7155. (-)
Goutarel, *Tet*, **1975**, *31*, 2695. (-)
Kametani, *Het*, **1981**, *16*, 925.
Massiot, *JCS, CC*, **1984**, 715. (-)
Takano, *TL*, **1985**, *26*, 865. (-)
Ninomiya, *Het*, **1986**, *24*, 2117.

13.8

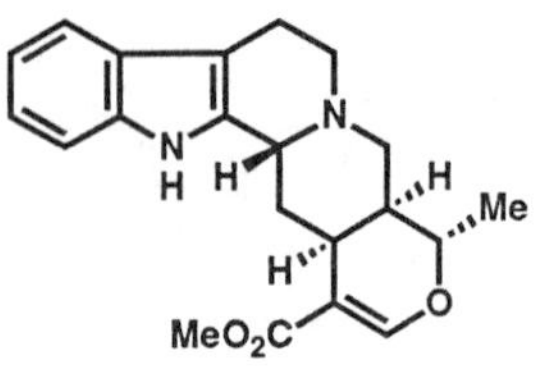

Akuammigine

Sakai, *CPB*, **1978**, *26*, 2596.

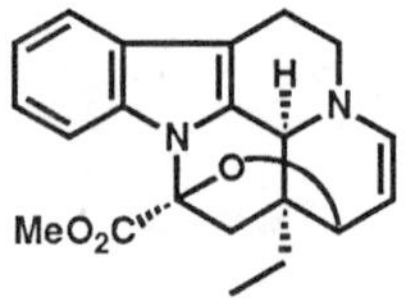

Criocerine

Le Men, *Tet*, **1980**, *36*, 511.

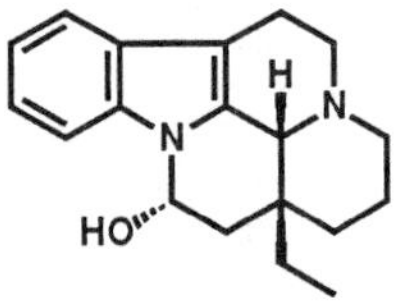

Eburnamine

Harley-Mason, *JCS, CC*, **1965**, 298.
Saxton, *JCS, CC*, **1969**, 799.
Winterfeldt, *Ber*, **1979**, *112*, 1902.
Takano, *JCS, PI*, **1985**, 305. (+)
Fuji, *JACS*, **1987**, *109*, 7901. (+)

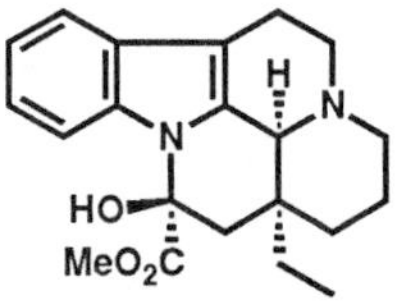

Vincamine

Kuehne, *JACS*, **1964**, *86*, 2946.
Oppolzer, *HCA*, **1977**, *60*, 1801.
Schlessinger, *JACS*, **1979**, *101*, 1540.
Ban, *Het*, **1982**, *18*, 255.
Szántay, *Tet*, **1983**, *39*, 3737. (+)
Winterfeldt, *Tet*, **1987**, *43*, 2035. (+), (-)

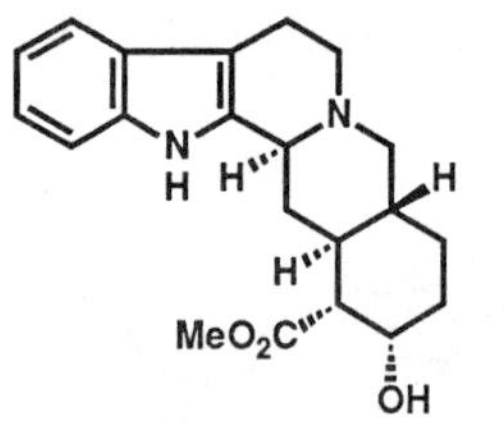

Yohimbine

van Tamelen, *JACS*, **1958**, *80*, 5006;
ibid., **1969**, *91*, 7315.
Szántay, *TL*, **1965**, 1665;
Ann, **1986**, 655. (+), (-)
Stork, *JACS*, **1972**, *94*, 5109.
Kametani, *Het*, **1975**, *3*, 179;
ibid., **1976**, *4*, 29.
Brown, *JCS, CC*, **1980**, 165.
Wenkert, *JACS*, **1982**, *104*, 2244.
Ninomiya, *JCS, CC*, **1983**, 1231.
Martin, *JACS*, **1987**, *109*, 6124.

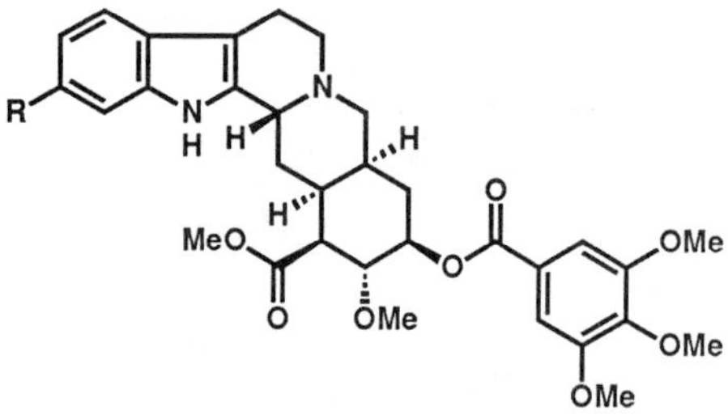

R = OMe: Reserpine

Woodward, *JACS*, **1956**, *78*, 2023, 2657;
Tet, **1958**, *2*, 1.
Pearlman, *JACS*, **1979**, *101*, 6398, 6404.
Wender, *JACS*, **1980**, *102*, 6157;
Het, **1987**, *25*, 263.
Martin, *JACS*, **1987**, *109*, 6124.

R = H: Deserpidine

Szántay, *Ann*, **1983**, 1292.
Ninomiya, *Het*, **1984**, 22, 1041.

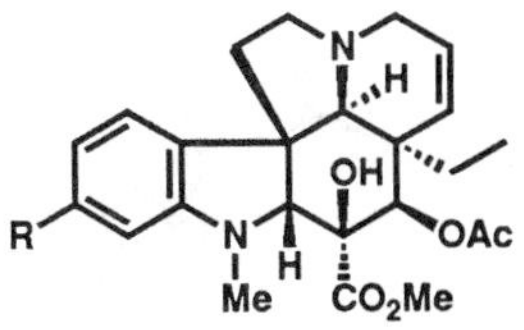

R = OMe: Vindoline

Büchi, *JACS*, **1975**, *97*, 6880.
Kutney, *JACS*, **1978**, *100*, 4220.
Ban, *TL*, **1978**, 151.
Danieli, *JCS, CC*, **1984**, 909.
Langlois, *JOC*, **1985**, *50*, 961.
Rapoport, *JOC*, **1986**, *51*, 3882;
JACS, **1987**, *109*, 1603. (-)
Kuehne, *JOC*, **1987**, *52*, 347. (+), (-)

R = H: Vindorosine

Büchi, *JACS*, **1971**, *93*, 3299.
Danieli, *JCS, CC*, **1984**, 909.
Natsume, *CPB*, **1984**, *32*, 2477.
Langlois, *JOC*, **1985**, *50*, 961.
Kuehne, *JOC*, **1987**, *52*, 347. (-)

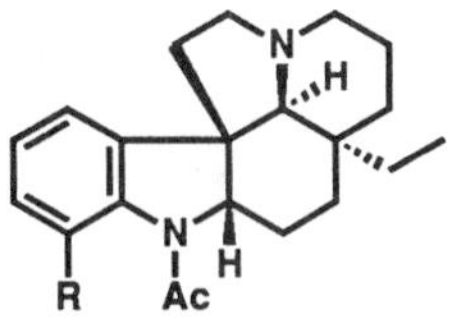

R = OMe: Aspidospermine

Stork, *JACS*, **1963**, *85*, 2872.
Ban, *TL*, **1965**, 2261;
ibid., **1975**, 723.
Stevens, *JCS, CC*, **1971**, 857.
Martin, *JACS*, **1980**, *102*, 3294.

R = H: Aspidospermidine

Harley-Mason, *JCS, CC*, **1967**, 915.
Kutney, *JACS*, **1970**, *92*, 1727.
Magnus, *JACS*, **1983**, *105*, 4750.
Fuji, *JACS*, **1987**, *109*, 7901. (-)

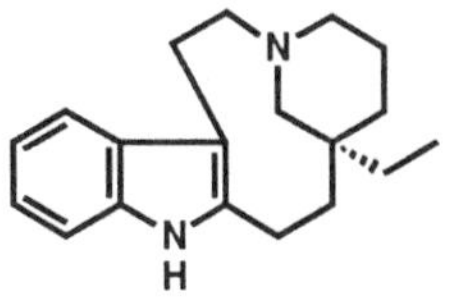

Quebrachamine

Stork, *JACS*, **1963**, *85*, 2872.
Ziegler, *JACS*, **1969**, *91*, 2342.
Kutney, *JACS*, **1970**, *92*, 1727.
Takano, *JACS*, **1979**, *101*, 6414;
JCS, CC, **1981**, 1153. (+), (-)
Pakrashi, *JHC*, **1980**, *17*, 1133.
Wenkert, *Tet*, **1981**, *37*, 4017.
Ban, *Tet*, **1983**, *39*, 3657.

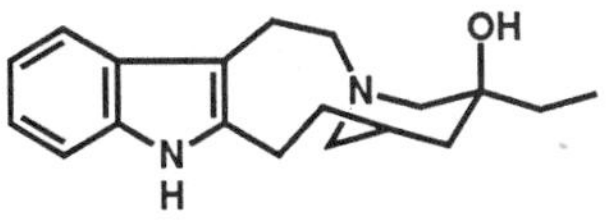

Velbanamine

Büchi, *JACS*, **1968**, *90*, 2448;
ibid., **1970**, *92*, 999.
Kutney, *JACS*, **1970**, *92*, 6090.
Nagata, *TL*, **1971**, 3681.

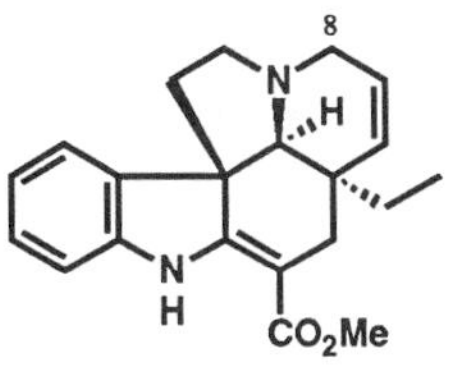

Tabersonine

Ziegler, *JACS*, **1971**, *93*, 5930.
Lévy, *TL*, **1978**, 1579. (-)
Takano, *JACS*, **1979**, *101*, 6414.
Hanaoka, *TL*, **1980**, *21*, 3285.
Kuehne, *JOC*, **1986**, *51*, 2913.

8-Oxotabersonine

Magnus, *JCS, CC*, **1986**, 1756.

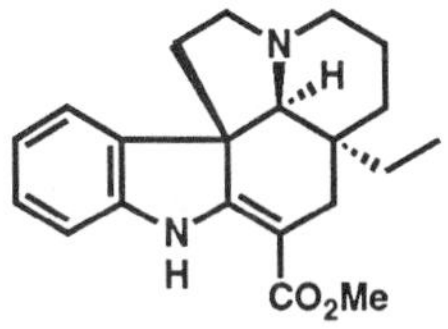

Vincadifformine

Kutney, *JACS*, **1968**, *90*, 3891.
Le Men, *TL*, **1974**, 491.
Kuehne, *JOC*, **1978**, *43*, 3705.
Ban, *Tet*, **1985**, *41*, 5495.

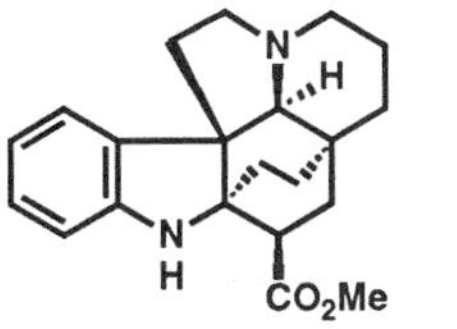

Kopsinine

Kuehne, *JOC*, **1985**, *50*, 4790.
Magnus, *JCS, CC*, **1985**, 184. (-)
Natsume, *TL*, **1987**, *28*, 3985.

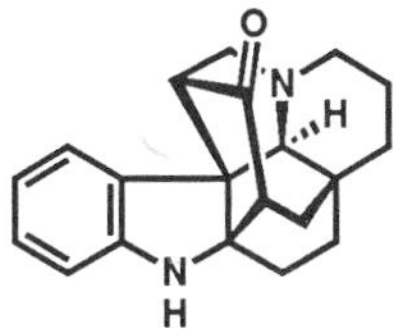

Kopsanone

Magnus, *JACS*, **1983**, *105*, 2086;
ibid., **1984**, *106*, 2105.
Kuehne, *JOC*, **1985**, *50*, 4790.

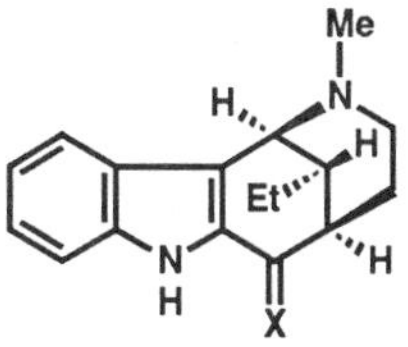

X = O: Dasycarpidone

Dolby, *JACS*, **1968**, *90*, 2699.

X = CH_2: Uleine

Joule, *JCS, CC*, **1968**, 584;
JCS (C), **1969**, 2738.
Büchi, *JACS*, **1971**, *93*, 2492.

13.9

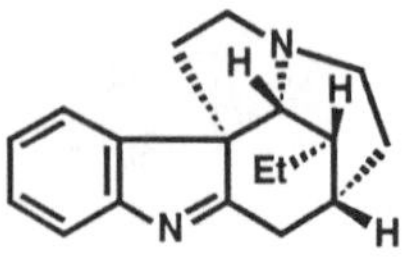

Condyfoline
Harley-Mason, *JCS, CC*, **1968**, 1233.
Ban, *JACS*, **1981**, *103*, 6990; *Tet*, **1983**, *39*, 3657.

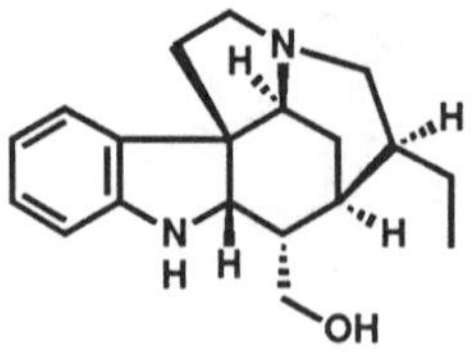

Geissoschizoline
Harley-Mason, *JCS, CC*, **1969**, 665.

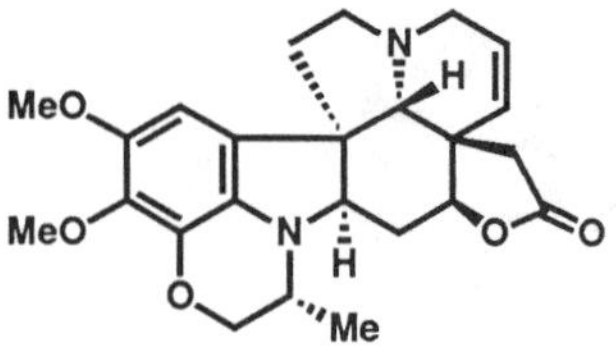

Obscurinervidine
Saxton, *Tet*, **1987**, *43*, 191.

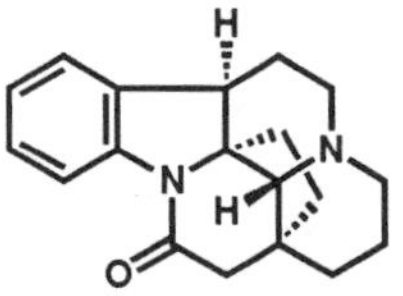

(-)-Strempeliopine
Hájícek & Trojánek, *TL*, **1981**, *22*, 2927; *ibid.*, **1982**, *23*, 365.

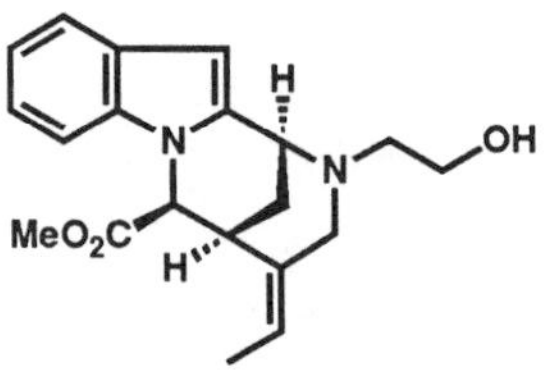

Vinoxine
Bosch, *TL*, **1984**, *25*, 3119.

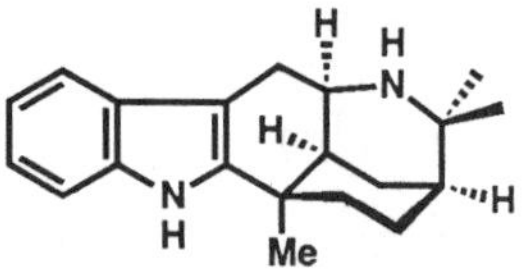

(+)-Aristoteline
Lévy, *JOC*, **1982**, *47*, 4169. (±)
Stevens, *JCS, CC*, **1983**, 384.
Borschberg, *HCA*, **1984**, *67*, 1040.
Gribble, *JOC*, **1985**, *50*, 5900.

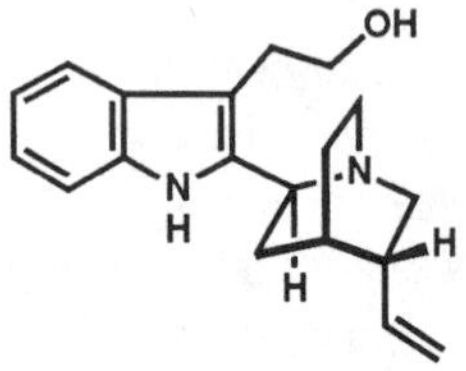

(+)-Cinchonamine
Uskokovic, *HCA*, **1976**, *59*, 2268.
Smith, *Tet*, **1986**, *42*, 2957.

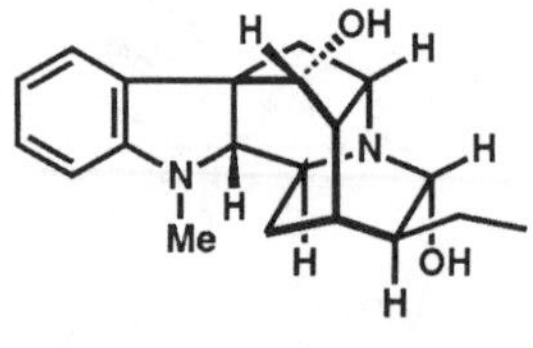

Ajmaline
Masamune, *JACS*, **1967**, *89*, 2506.
Mashimo, *CPB*, **1970**, *18*, 353.
van Tamelen, *JACS*, **1970**, *92*, 2136; *Bioorg. Chem*, **1976**, *5*, 309.

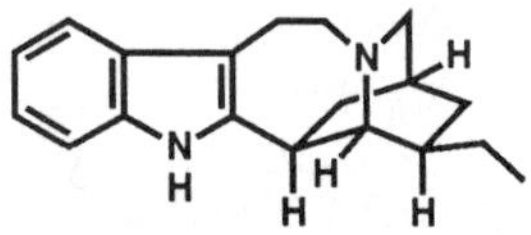

Ibogamine
Büchi, *JACS*, **1965**, *87*, 2073; *ibid.*, **1966**, *88*, 3099.
Sallay, *JACS*, **1967**, *89*, 6762.
Nagata, *JACS*, **1968**, *90*, 1651.
Ban, *JCS, CC*, **1969**, 88.
Rosenmund, *Ber*, **1975**, *108*, 1871.
Trost, *JACS*, **1978**, *100*, 3930. (+), (-)
Atta-ur-Rahman, *Tet*, **1980**, *36*, 1063.
Hanaoka, *CPB*, **1985**, *33*, 4202.
Huffman, *JOC*, **1985**, *50*, 1460.
Kuehne, *JOC*, **1985**, *50*, 1464.

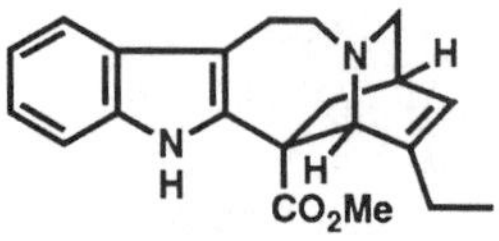

Catharanthine
Büchi, *JACS*, **1970**, *92*, 999.
Kutney, *JACS*, **1970**, *92*, 6090.
Trost, *JOC*, **1979**, *44*, 2052.
Atta-ur-Rahman, *Tet*, **1980**, *36*, 1063.
Hanaoka, *TL*, **1980**, *21*, 3285; *CPB*, **1982**, *30*, 4052.
Das, *JCS, CC*, **1981**, 389.
Raucher, *JOC*, **1985**, *50*, 3236.
Kuehne, *JOC*, **1986**, *51*, 2913.

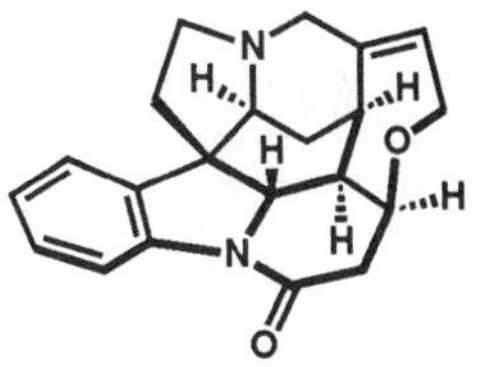

(-)-Strychnine
Woodward,
JACS, **1954**, *76*, 4749;
Tet, **1963**, *19*, 247.

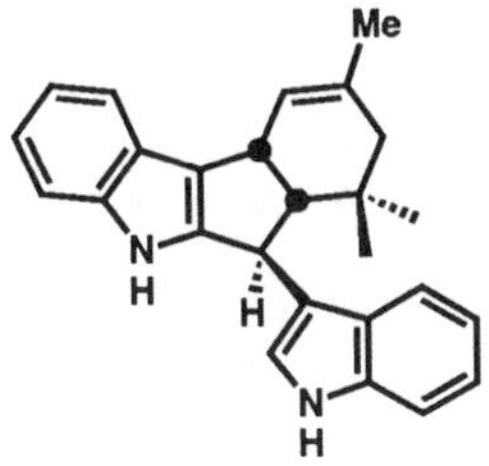

Yuehchukene
Cheng, *JCS, CC*, **1985**, 48.
Bergman, *TL*, **1988**, *29*, 2993.

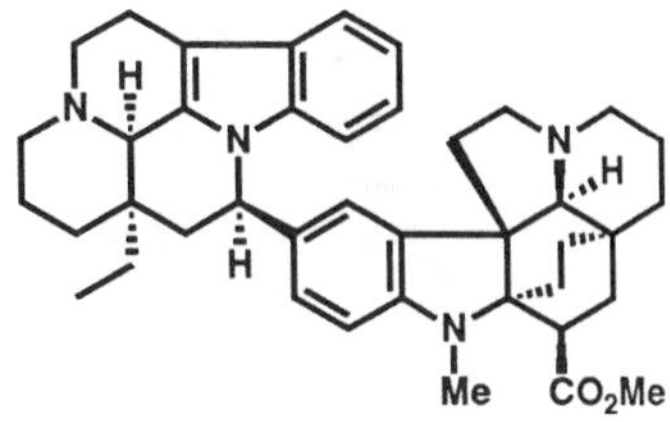

(-)-Pleiomutine
Magnus, *JCS, CC*, **1985**, 184.

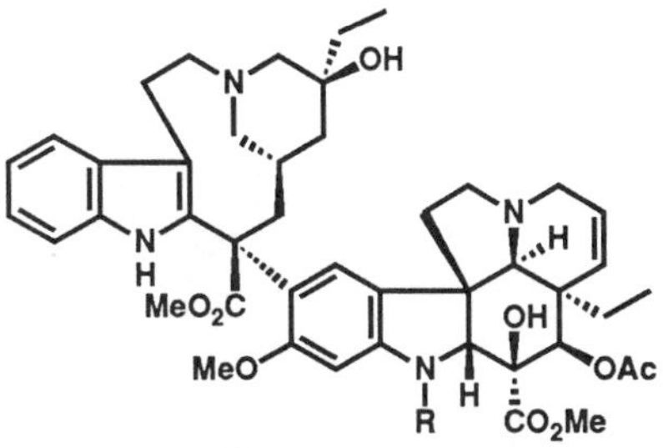

R = Me: Vinblastine
R = CHO: Vincristine
Potier, *JACS*, **1976**, *98*, 7017;
ibid., **1979**, *101*, 2243.

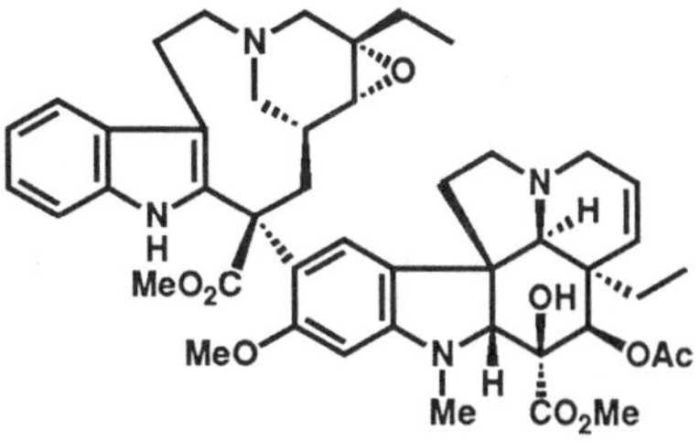

Leurosine
Kutney, *Het*, **1976**, *4*, 997;
CJC, **1977**, *55*, 3235;
ibid., **1978**, *56*, 62.

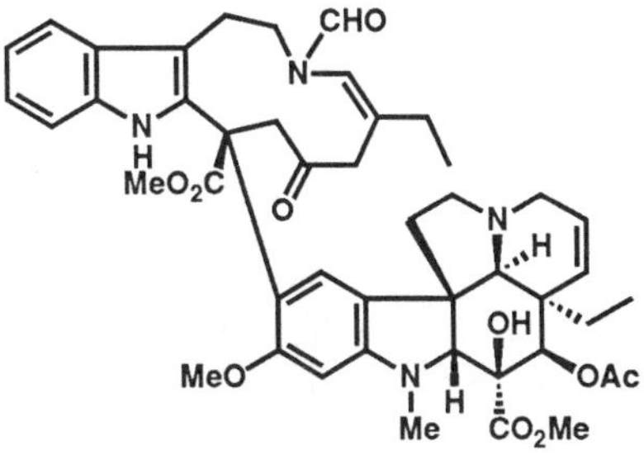

Catharine
Kutney, *CJC*, **1979**, *57*, 1682.

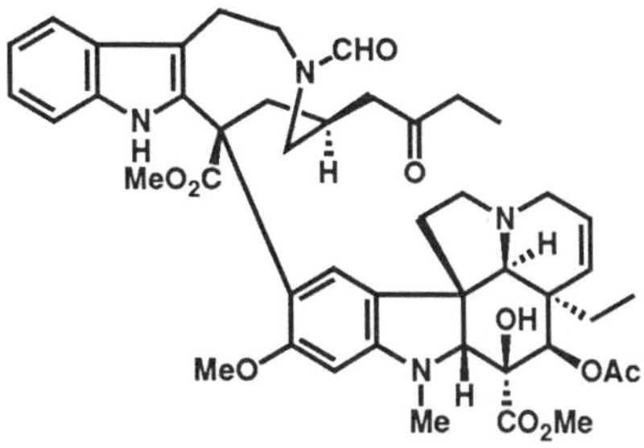

Catharinine (Vinamidine)
Kutney, *CJC*, **1979**, *57*, 1682.

13.9

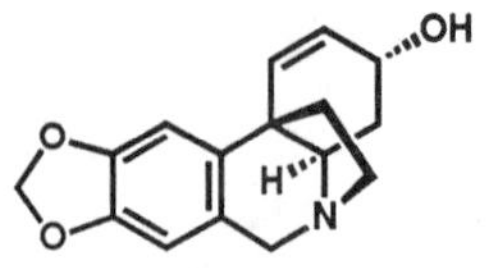

Crinine

Muxfeldt, *JACS*, **1966**, *88*, 3670.
Smith, *JACS*, **1967**, *89*, 3600.
Overman, *JACS*, **1981**, *103*, 5579;
ibid., **1983**, *105*, 6629;
HCA, **1985**, *68*, 745. (-)
Martin, *JOC*, **1988**, *53*, 3184.

***epi*-Crinine**

Kametani, *CPB*, **1972**, *20*, 1488.

Dihydrocrinine (Elwesine)

Irie, *JCS (C)*, **1968**, 1802;
Het, **1979**, *12*, 1311.
Stevens, *JOC*, **1972**, *37*, 977.

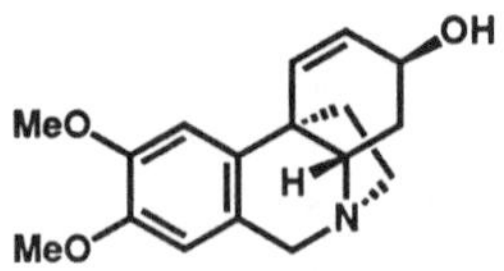

Maritidine

Schwartz, *JACS*, **1970**, *92*, 1090.
Kametani, *JCS, CC*, **1971**, 774;
Tet, **1971**, *27*, 5441.
Koga, *CPB*, **1977**, *25*, 2681. (+)

Dihydromaritidine

Speckamp, *Tet*, **1978**, *34*, 2579.

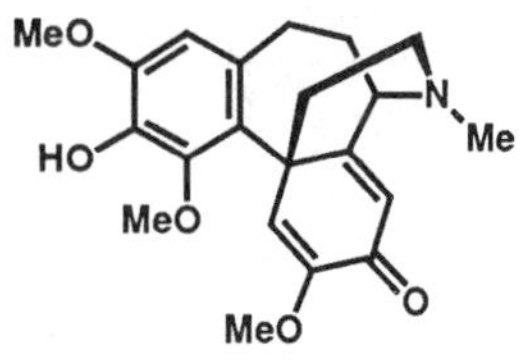

Androcymbine

Kametani, *JOC*, **1971**, *36*, 3729.
Umezawa, *JCS, PI*, **1979**, 2657.

***O*-Methylandrocymbine**

Schwartz, *JACS*, **1973**, *95*, 612.

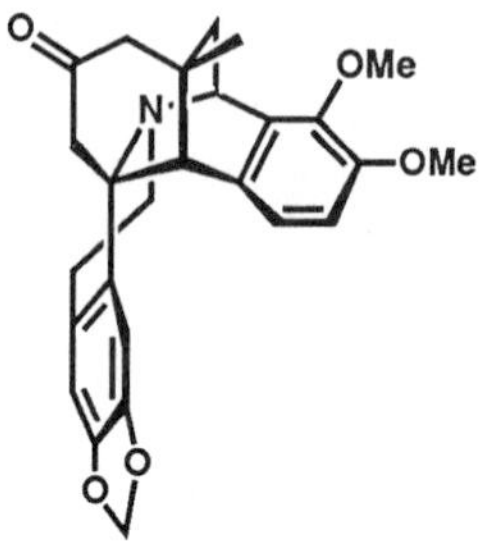

Karachine

Stevens, *JCS, CC*, **1983**, 1425.

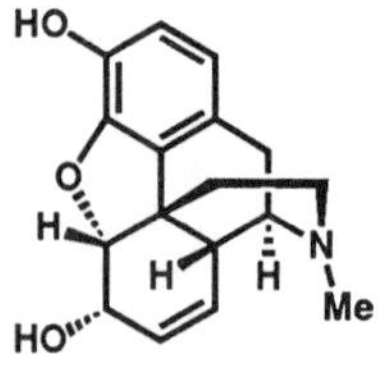

Morphine

Gates, *JACS*, **1952**, *74*, 1109;
ibid., **1956**, *78*, 1380. (-)
Ginsburg, *JACS*, **1954**, *76*, 312.
Morrison, *TL*, **1967**, 4055.
Kametani, *JCS (C)*, **1969**, 2030.
Schwartz, *JACS*, **1975**, *97*, 1239.
Rice, *JOC*, **1980**, *45*, 3135.
Evans, *TL*, **1982**, *23*, 285.
Fuchs, *JOC*, **1987**, *52*, 473.

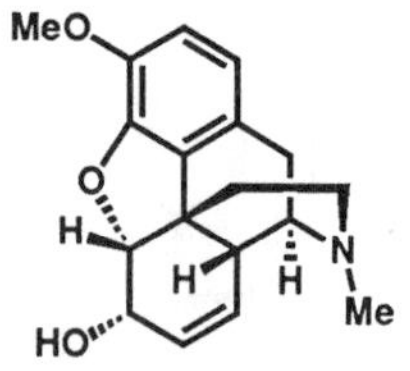

Codeine

Schwartz, *JACS*, **1975**, *97*, 1239.
Rice, *JOC*, **1980**, *45*, 3135.
Rapoport, *JOC*, **1983**, *48*, 227.
White, *Tet*, **1983**, *39*, 2393. (-)

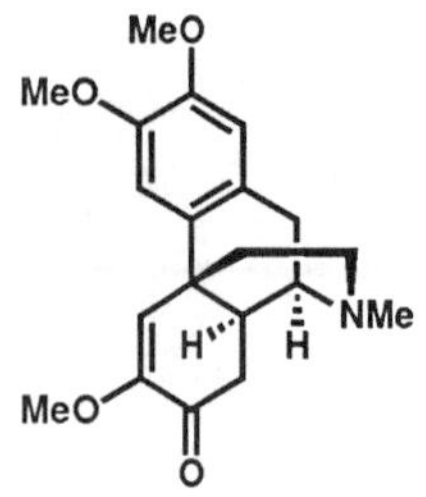

***O*-Methylpallidinine**

McMurry, *JOC*, **1984**, *49*, 3803.
Kano, *JACS*, **1986**, *108*, 6746.

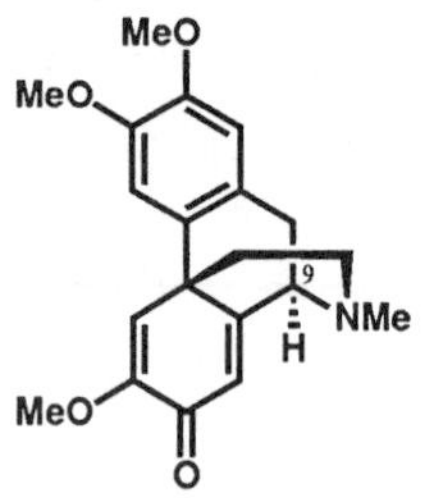

9-(*R*)-*O*-Methylflavinantine

Gawley, *TL*, **1988**, *29*, 301.

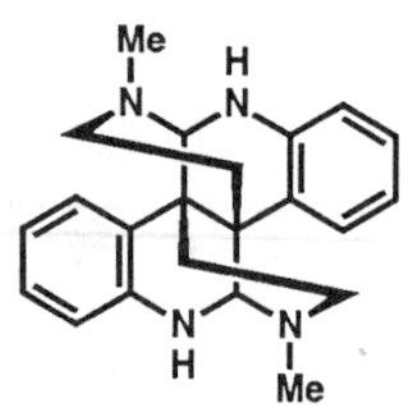

Calycanthine

Hendrickson, *Tet*, **1964**, *20*, 565.
Hall, *Tet*, **1967**, *23*, 4131.

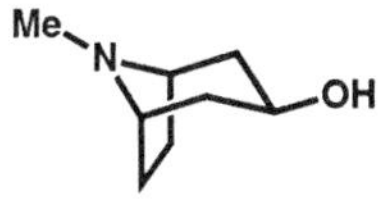

Pseudotropine
Tufariello,
JACS, **1979**, *101*, 2435.

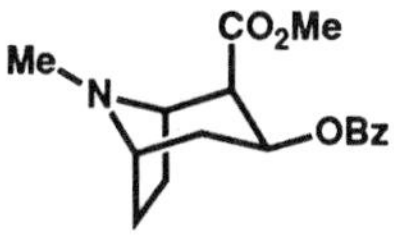

(-)-Cocaine (*nat*)
Willstatter, *Ann*, **1923**, *434*, 111. (+)
Tufariello, *JACS*, **1979**, *101*, 2435. (±)
Lewin & Carroll, *JHC*, **1987**, *24*, 19. (+)

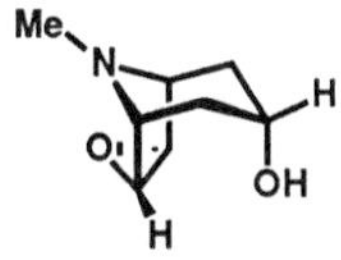

Scopine
Noyori,
JACS, **1974**, *96*, 3336.

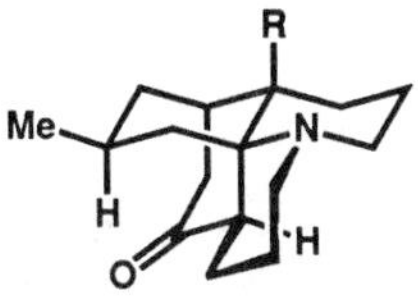

R = H: Lycopodine
Stork, *JACS*, **1968**, *90*, 1647.
Ayer, *JACS*, **1968**, *90*, 1648.
Kim, *TL*, **1978**, 2293.
Heathcock, *JACS*, **1978**, *100*, 8036; *ibid.*, **1982**, *104*, 1054.
Schumann, *Ann*, **1982**, 1700.
Wenkert, *JCS, CC*, **1984**, 714.
Kraus, *JACS*, **1985**, *107*, 4341.
R = OH: Lycodoline
Heathcock, *JACS*, **1981**, *103*, 222; *ibid.*, **1982**, *104*, 1054.

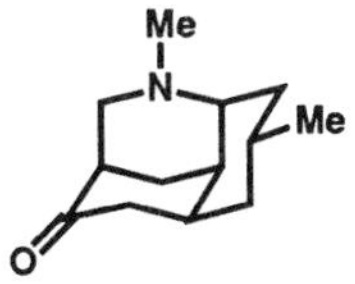

Luciduline
Evans, *JACS*, **1972**, *94*, 4779.
Oppolzer, *JACS*, **1976**, *98*, 6722.
MacLean, *CJC*, **1979**, *57*, 1631.
Schumann, *Ann*, **1984**, 1519.

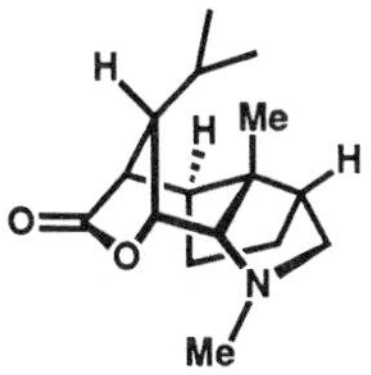

Dendrobine
Yamada, *JACS*, **1972**, *94*, 8278; *TL*, **1973**, 331.
Inubushi, *CPB*, **1974**, *22*, 349.
Kende, *JACS*, **1974**, *96*, 4332.
Roush, *JACS*, **1978**, *100*, 3599; *ibid.*, **1980**, *102*, 1390.

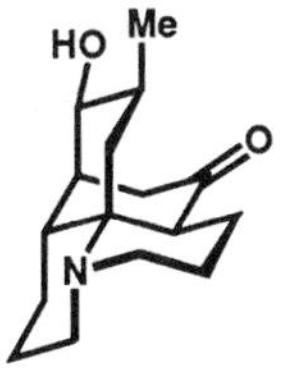

Clavolonine
Wenkert,
JCS, CC, **1984**, 714.

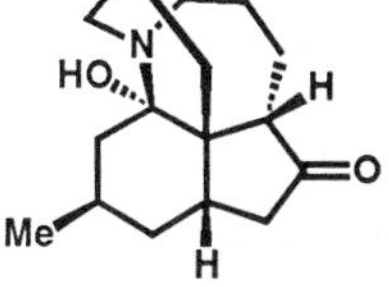

Fawcettimine
Harayama,
CPB, **1980**, *28*, 2394.
Heathcock,
JACS, **1986**, *108*, 5022.

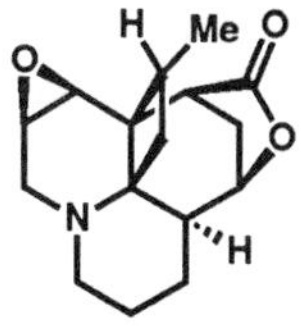

Annotinine
Wiesner, *TL*, **1967**, 4937; *CJC*, **1969**, *47*, 433. (*nat*)

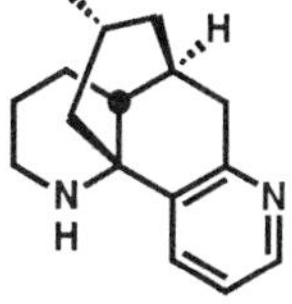

Lycodine
Heathcock,
TL, **1979**, 4125;
JACS, **1982**, *104*, 1054.

13.10

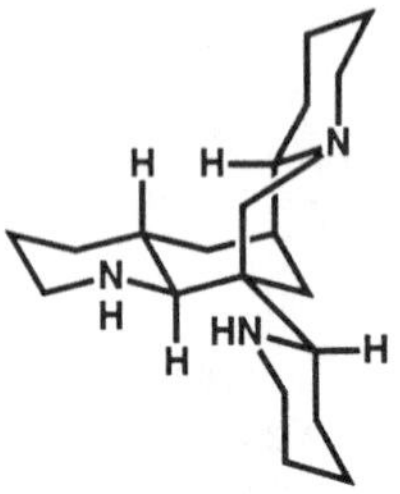
H
H
N
N
H
H
HN
H

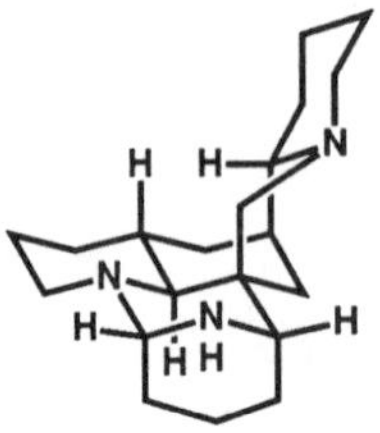
H
H
N
N
N
H
H H
H

Ormosanine **Panamine**

Liu, *CJC*, **1976**, *54*, 97.

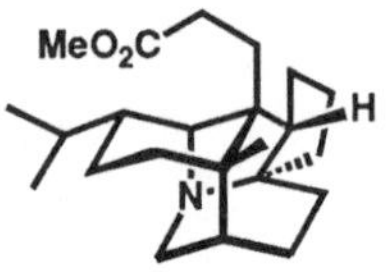
MeO2C
H
N

(+)-Methyl Homo-daphniphyllate

Heathcock,

JACS, **1986**, *108*, 5650.

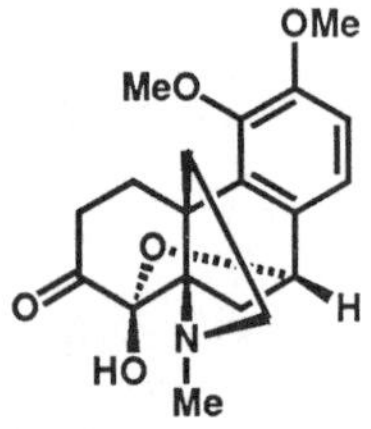
OMe
MeO
O
O
H
N
HO
Me

Metaphanine

Ibuka, *TL*, **1972**, 1393;

CPB, **1974**, *22*, 907.

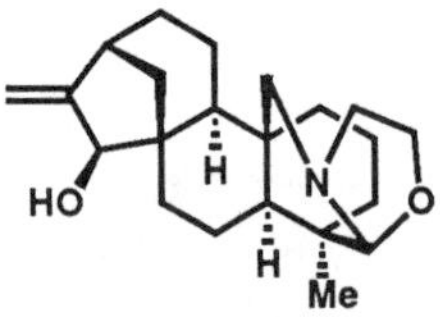
HO
H
N
O
H
Me

Garryine

Masamune, *JACS*, **1964**, *86*, 290.

Nagata, *JACS*, **1964**, *86*, 929;

ibid., **1967**, *89*, 1499.

Wiesner, *TL*, **1964**, 2437.

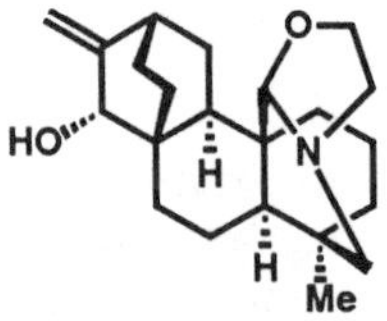
O
HO
H
N
H
Me

Atisine

Nagata, *JACS*, **1963**, *85*, 2342.

Masamune, *JACS*, **1964**, *86*, 291.

Tahara, *TL*, **1966**, 1453.

Wiesner, *TL*, **1966**, 4645.

Fukumoto, *JACS*, **1988**, *110*, 1963.

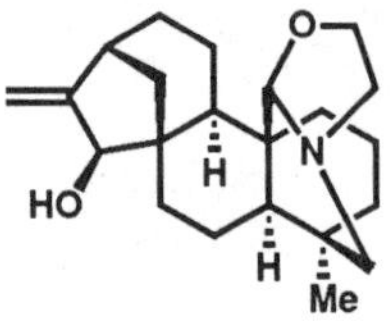
O
HO
H
N
H
Me

Veatchine

Nagata, *JACS*, **1964**, *86*, 929;

ibid., **1967**, *89*, 1499.

Wiesner, *TL*, **1964**, 2437;

ibid., **1968**, 6279.

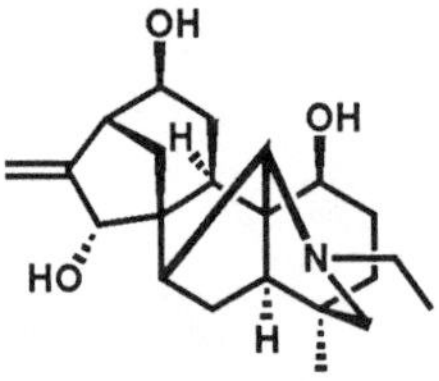
OH
H
OH
N
HO
H

Napelline

Wiesner,

CJC, **1974**, *52*, 2355;

ibid., **1980**, *58*, 1889.

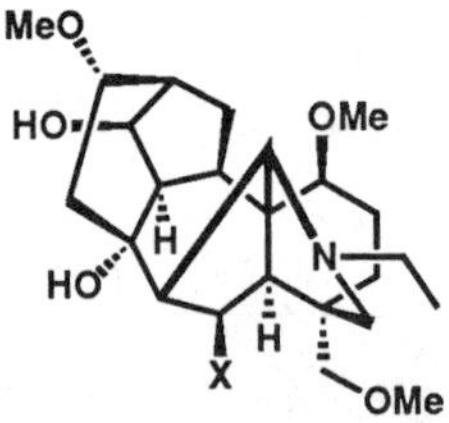
MeO
HO
OMe
H
N
HO
H
X
OMe

X = H: Talatisamine

Wiesner, *JACS*, **1974**, *96*, 4990;

P&AC, **1975**, *41*, 93.

X = OMe: Chasmanine

Wiesner, *Het*, **1977**, *7*, 217;

CJC, **1978**, *56*, 1451.

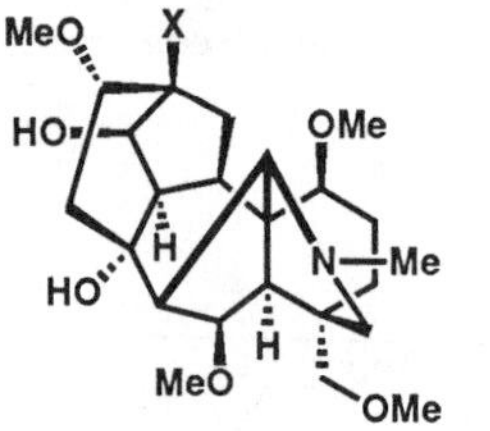
MeO
X
HO
OMe
H
N
Me
HO
H
MeO
OMe

X = H: 13-Desoxydelphinine

(X = OH: Delphinine)

Wiesner,

P&AC, **1979**, *51*, 689.

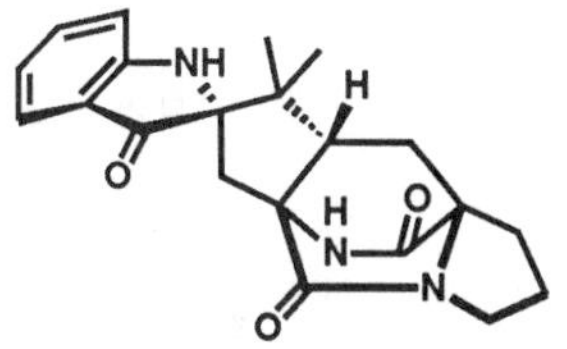

(-)-Brevianamide B
Williams, *JACS*, **1988**, *110*, 5927.

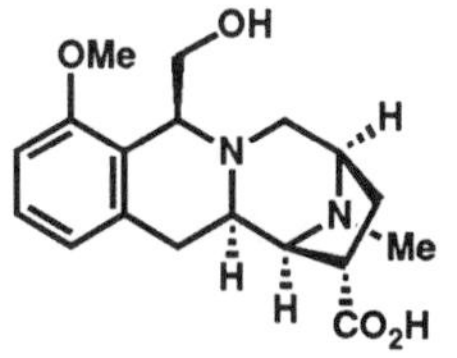

Quinocarcinol
Danishefsky,
JACS, **1985**, *107*, 1421.

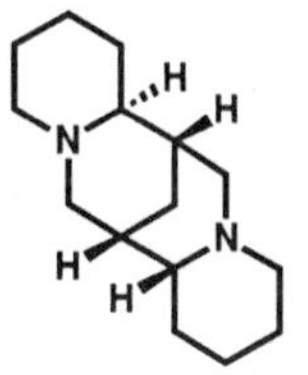

Sparteine
Leonard, *JACS*, **1950**, *72*, 1316.
van Tamelen, *JACS*, **1960**, *82*, 2400.

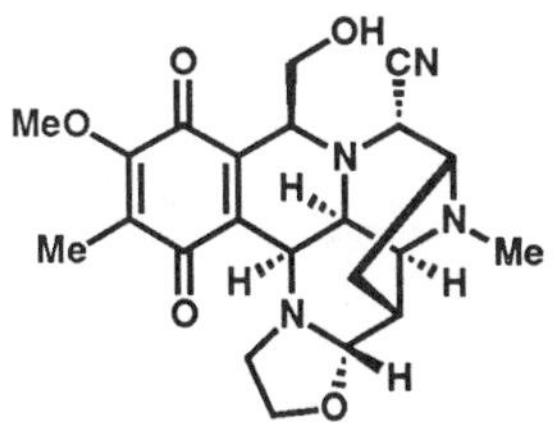

Cyanocycline A
Evans, *JACS*, **1986**, *108*, 2478.
Fukuyama, *JACS*, **1987**, *109*, 1587.

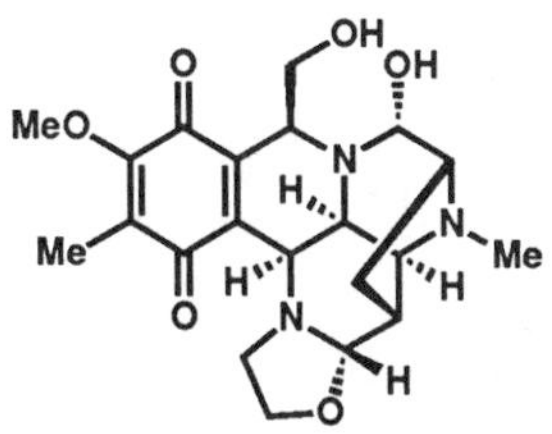

Naphthyridinomycin
Evans, *TL*, **1985**, *26*, 1907, 1911.

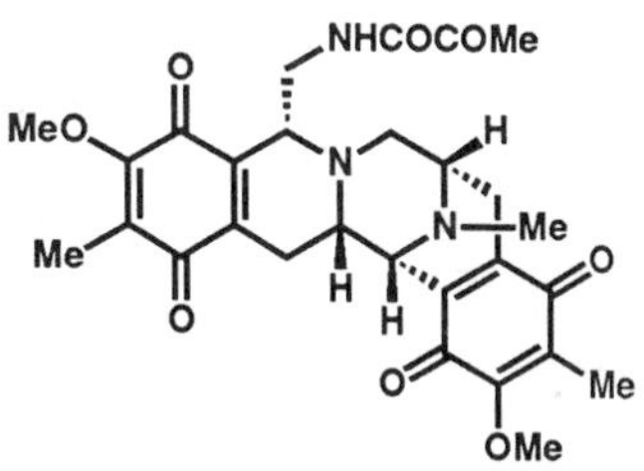

Saframycin B
Fukuyama, *JACS*, **1982**, *104*, 4957.
Kubo, *CPB*, **1987**, *35*, 2158;
JOC, **1988**, *53*, 4295.

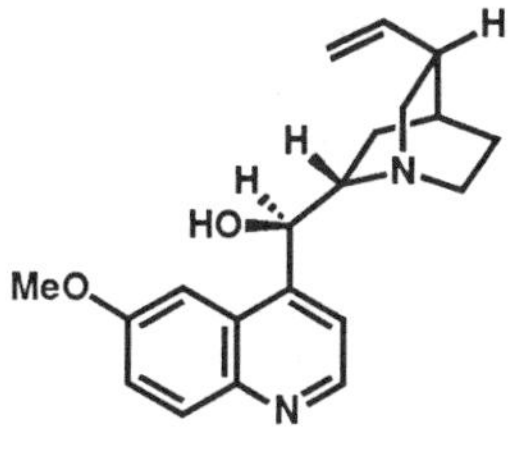

Quinine
Woodward, *JACS*, **1944**, *66*, 849; (-)
ibid., **1945**, *67*, 860. (-)
Gates, *JACS*, **1970**, *92*, 205. (-)
Uskokovic, *JACS*, **1970**, *92*, 203, 204;
ibid., **1971**, *93*, 5904;
ibid., **1978**, *100*, 576. (-)
Taylor, *JACS*, **1972**, *94*, 6218.
Hanaoka, *CPB*, **1982**, *30*, 1925.

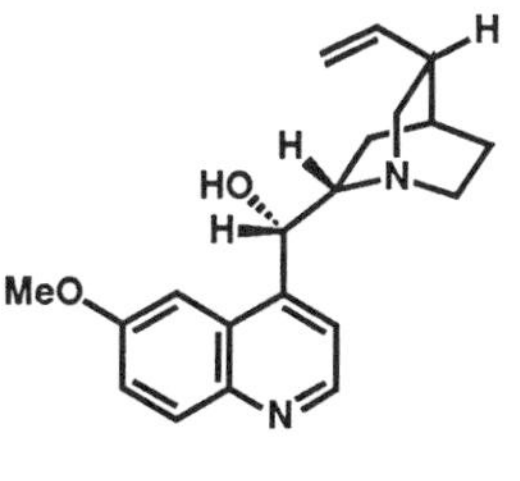

Quinidine
Uskokovic,
JACS, **1970**, *92*, 203, 204;
ibid., **1971**, *93*, 5904;
ibid., **1978**, *100*, 576. (+), (-)

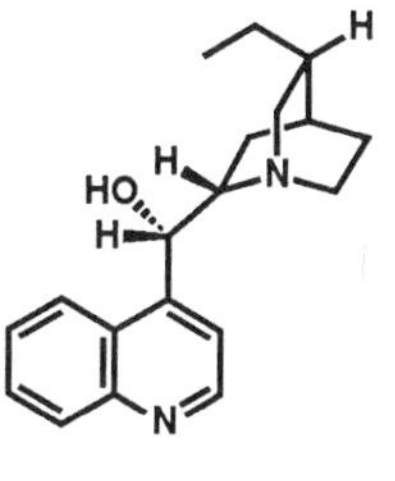

Hydrocinchonidine
Fukumoto,
JCS, CC, **1986**, 573.

13.10

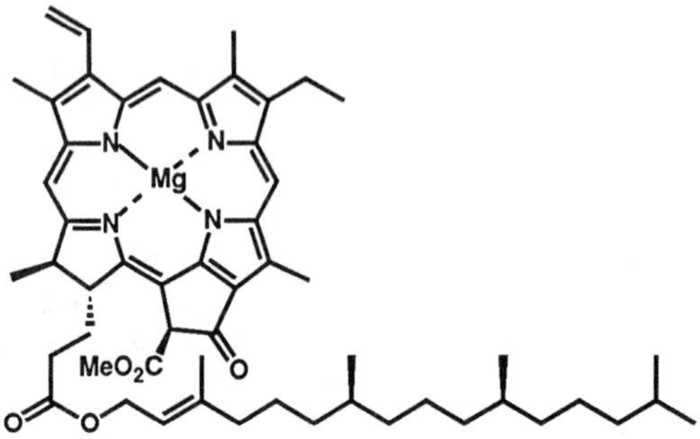

Chlorophyll a
Woodward, *JACS*, **1960**, *82*, 3800; *P&AC*, **1961**, *2*, 383.

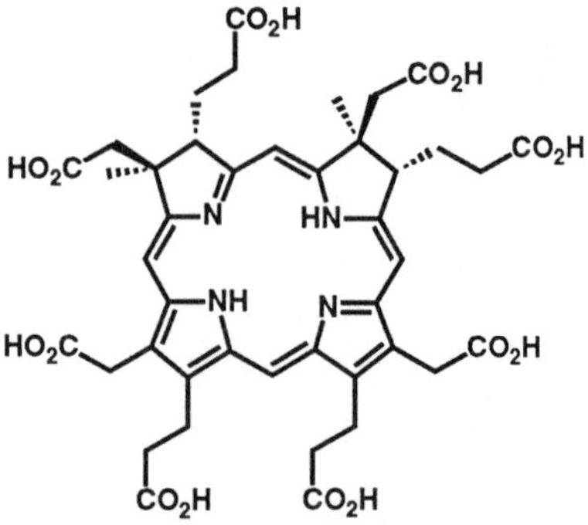

Sirohydrochlorin
Battersby, *JCS, CC*, **1985**, 1061. (*nat*)

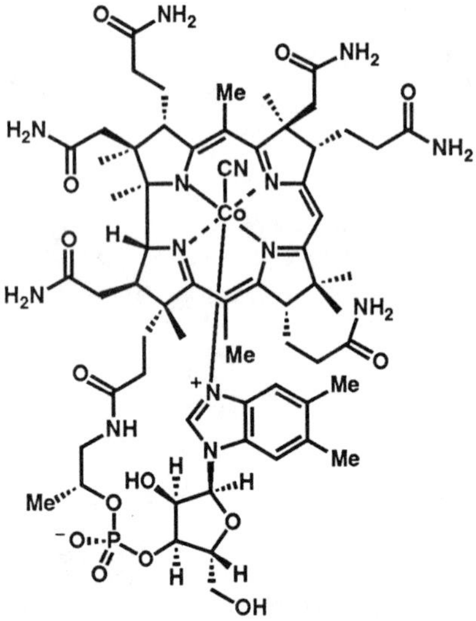

Vitamin B_{12}
Woodward, *P&AC*, **1973**, *33*, 145.
Eschenmoser, *Science*, **1977**, *196*, 1410; *ACIE*, **1988**, *27*, 5.

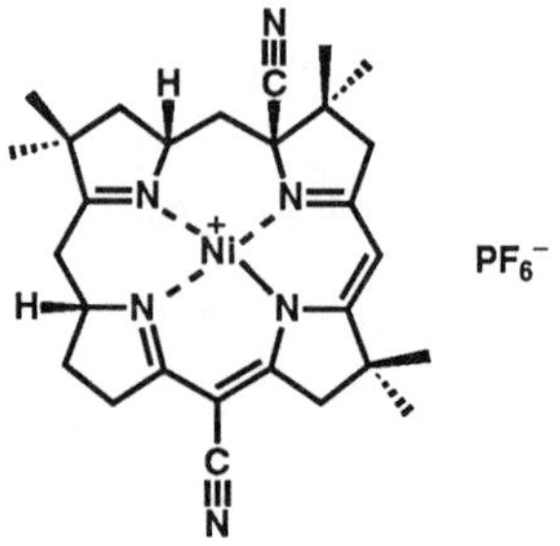

Chemical Model for Coenzyme F430
Eschenmoser, *JCS, CC*, **1984**, 1365; *ACIE*, **1988**, *27*, 5.

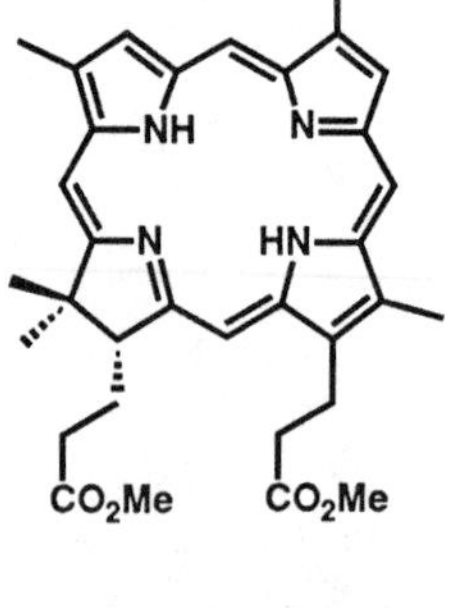

Bonellin Dimethyl Ester
Battersby, *JCS, CC*, **1983**, 1237.

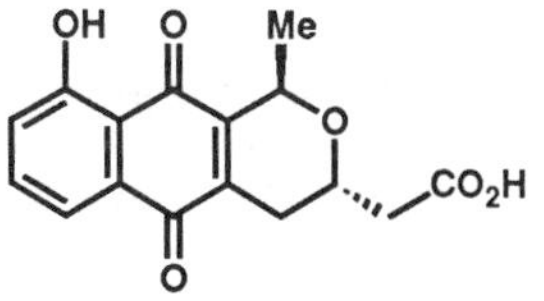

Nanaomycin A
Li, *JACS*, **1978**, *100*, 6263.
Ichihara, *TL*, **1980**, *21*, 4469.
Yoshii, *JCS, PI*, **1981**, 1197.
Maruyama, *CL*, **1982**, 609.
Semmelhack, *JACS*, **1982**, *104*, 5850.

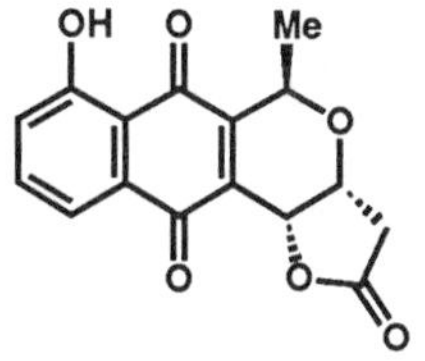

Kalafungin
Li, *JACS*, **1978**, *100*, 6263.
Deoxykalafungin
Kraus, *JOC*, **1978**, *43*, 4923.

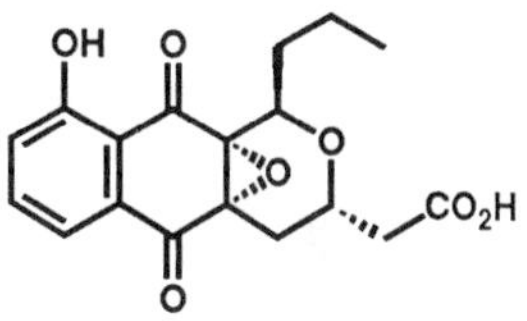

Frenolicin
Ichihara, *TL*, **1980**, *21*, 4469.

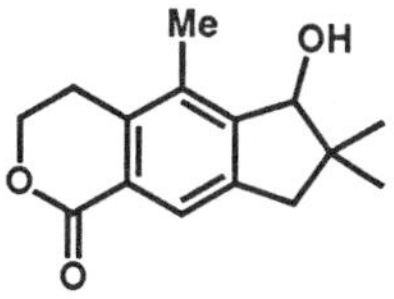

Calomelanolactone
Stevenson, *TL*, **1988**, *29*, 813.

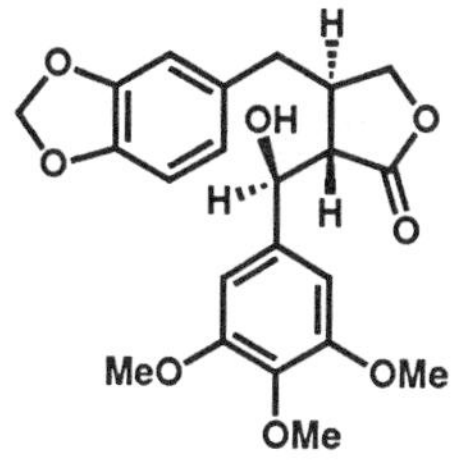

Podorhizol
Ziegler, *JOC*, **1978**, *43*, 985.

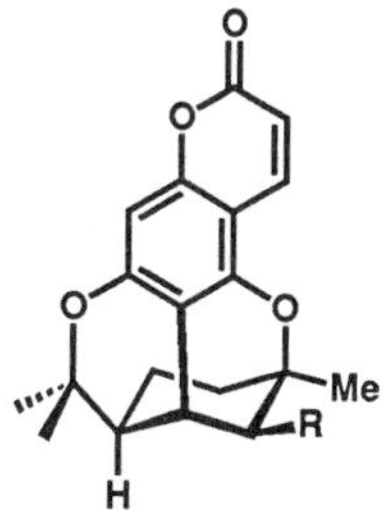

R = H: Deoxybruceol
Crombie, *JCS, CC*, **1968**, 368.

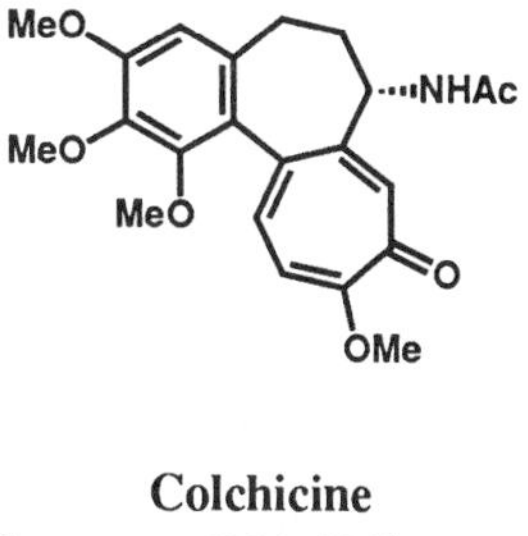

Colchicine
Eschenmoser, *HCA*, **1961**, *44*, 540.
van Tamelen, *Tet*, **1961**, *14*, 8.
Nakamura, *CPB*, **1962**, *10*, 299.
Scott, *Tet*, **1965**, *21*, 3605.
Woodward, *The Harvey Lectures*, p 31, **1965**, Academic Press: New York.
Martel, *JOC*, **1965**, *30*, 1752.
Kaneko, *Ag. BC*, **1968**, *32*, 995.
Tobinaga, *JCS, CC*, **1974**, 300.
Evans, *JACS*, **1981**, *103*, 5813.
Boger, *JACS*, **1986**, *108*, 6713.

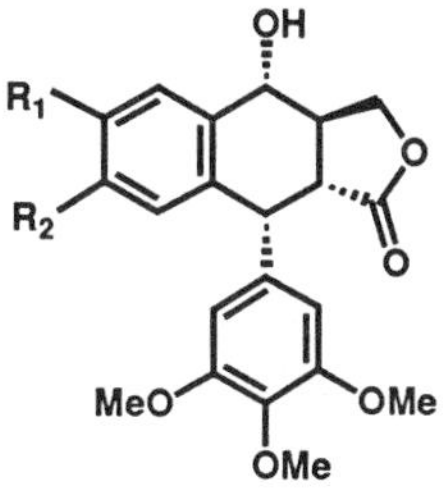

R_1, R_2 =  **:**
Podophyllotoxin
Murphy, *JCS, CC*, **1980**, 262.
Kende, *JOC*, **1981**, *46*, 2826.
Rodrigo, *JACS*, **1981**, *103*, 6208.
Vandewalle, *Tet*, **1986**, *42*, 4297.
Kaneko, *TL*, **1987**, *28*, 517.
R_1 = R_2 = OMe:
Sikkimotoxin
Takano, *Het*, **1987**, *25*, 69.

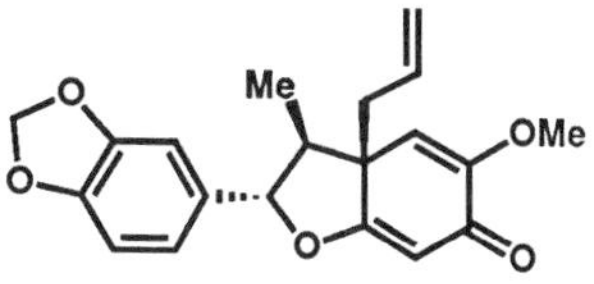

Burchellin
Büchi, *JACS*, **1977**, *99*, 8073.

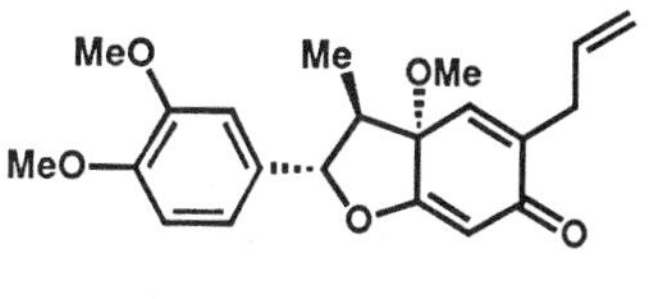

Kadsurenone
Ponpipom, *TL*, **1986**, *27*, 309.

13.11

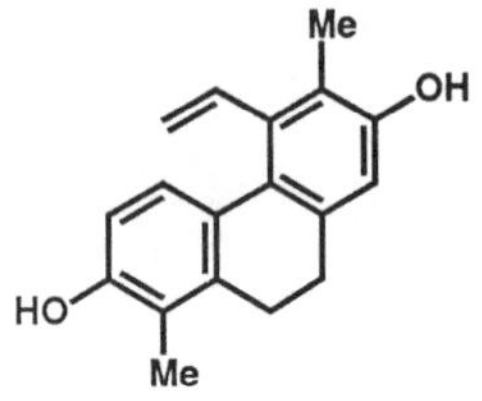

Juncusol

Kende, *TL*, **1978**, 3003;
JACS, **1979**, *101*, 1857.
McDonald, *TL*, **1978**, 4723.
Schultz, *TL*, **1981**, *22*, 1775.
Boger, *JOC*, **1984**, *49*, 4045.

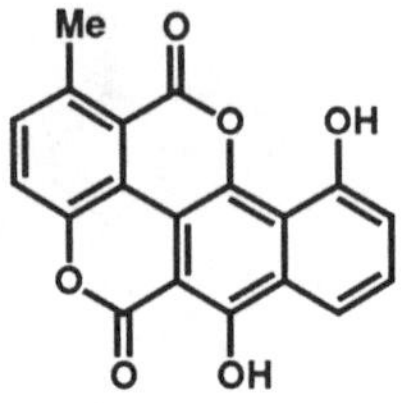

Chartreusin Aglycone

Kelly, *JACS*, **1980**, *102*, 798.
Hauser, *JOC*, **1980**, *45*, 4071.

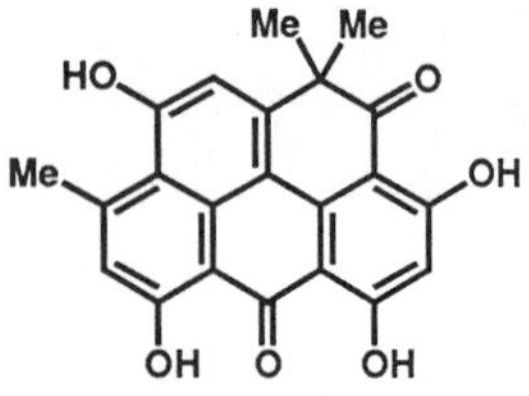

Resistomycin

Rodrigo, *JACS*, **1982**, *104*, 4725.
Kelly, *JACS*, **1985**, *107*, 3879.

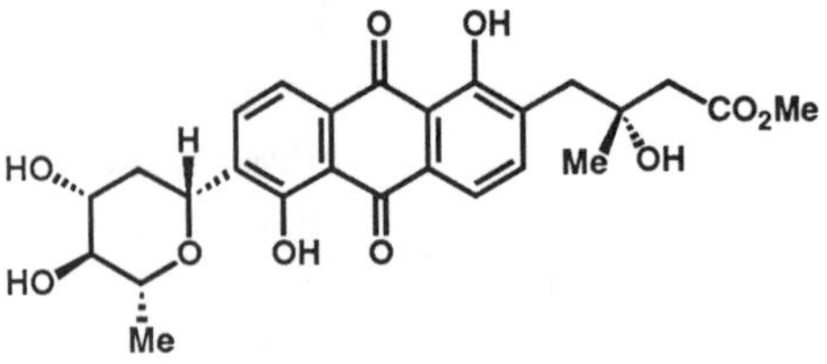

(+)-Vineomycinone B_2 Methyl Ester

Danishefsky, *JACS*, **1985**, *107*, 1285.

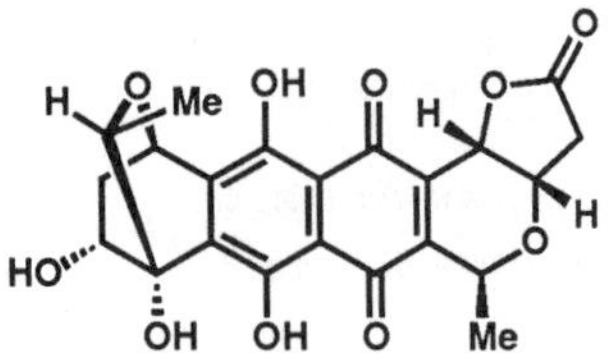

Granaticin

Yoshii, *JACS*, **1987**, *109*, 3402.

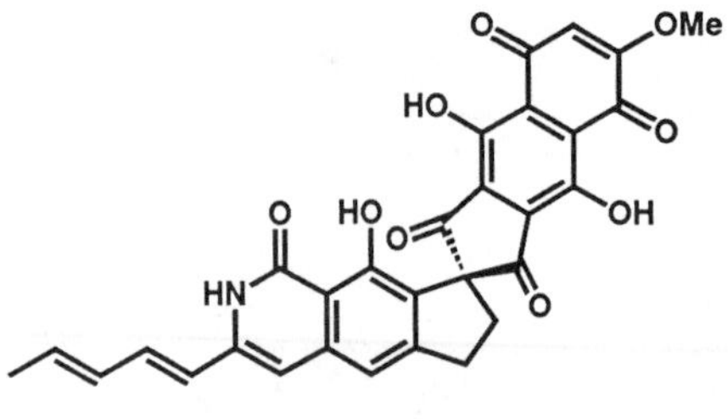

Fredericamycin A

Kelly, *JACS*, **1986**, *108*, 7100;
ibid., **1988**, *110*, 6471.

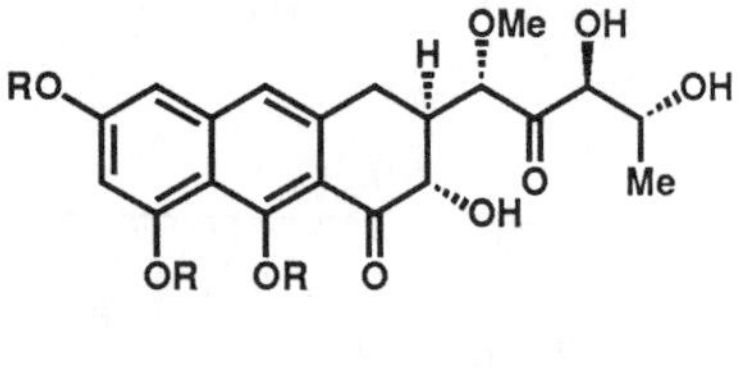

R = H: Olivin

Roush, *JACS*, **1987**, *109*, 7575. (+)

R = Me: Tri-*O*-methylolivin

Weinreb, *JACS*, **1984**, *106*, 1811.
Franck, *JACS*, **1986**, *108*, 2455. (+)

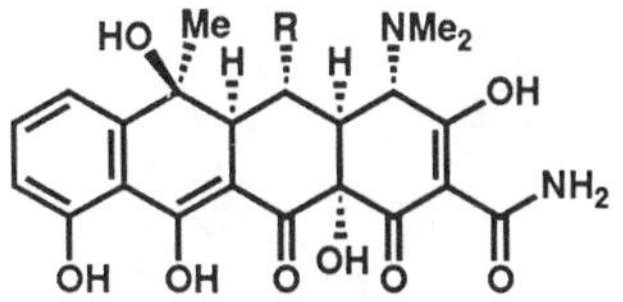

R = OH: Terramycin

Muxfeldt, *JACS*, **1968**, *90*, 6534;

ibid., **1979**, *101*, 689.

R = H: Tetracycline

Gurevich, *TL*, **1967**, 131.

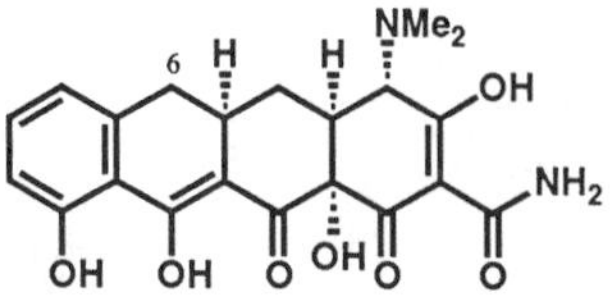

6-Demethyl-6-deoxytetracycline

Woodward, *JACS*, **1962**, *84*, 3222;

P&AC, **1963**, *6*, 561; *JACS*, **1968**, *90*, 439.

Muxfeldt, *JACS*, **1965**, *87*, 933.

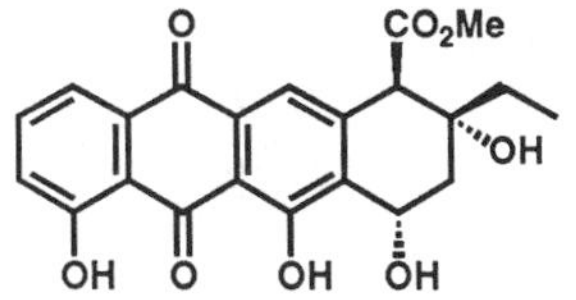

Aklavinone

Kende, *JACS*, **1981**, *103*, 4247;

Tet, **1984**, *40*, 4693. (+)

Kishi, *JACS*, **1981**, *103*, 4248;

ibid., **1982**, *104*, 7371. (+)

Confalone, *JACS*, **1981**, *103*, 4251.

Li, *JACS*, **1981**, *103*, 7007.

Boeckman, *JACS*, **1982**, *104*, 4604.

Krohn, *Ann*, **1983**, 2151.

Hauser, *JACS*, **1984**, *106*, 1098.

Maruyama, *Tet*, **1984**, *40*, 4725.

Rapoport, *JOC*, **1985**, *50*, 1569.

Kraus, *TL*, **1986**, *27*, 1873.

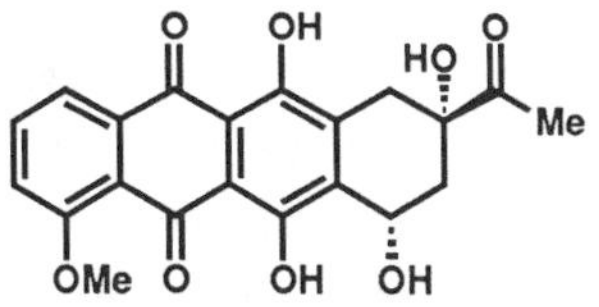

Daunomycinone

Wong, *CJC*, **1973**, *51*, 466.

Kende, *JACS*, **1976**, *98*, 1967;

TL, **1979**, 1201.

Swenton, *JACS*, **1978**, *100*, 6188.

Kelly, *JACS*, **1980**, *102*, 5983.

Hauser, *JACS*, **1981**, *103*, 6378.

Vogel, *Tet*, **1984**, *40*, 4549.

Rodrigo, *Tet*, **1984**, *40*, 4597.

also see: *Tet*, **1984**, *40*, #22,

Recent Aspects of

Anthracyclinone Chemistry

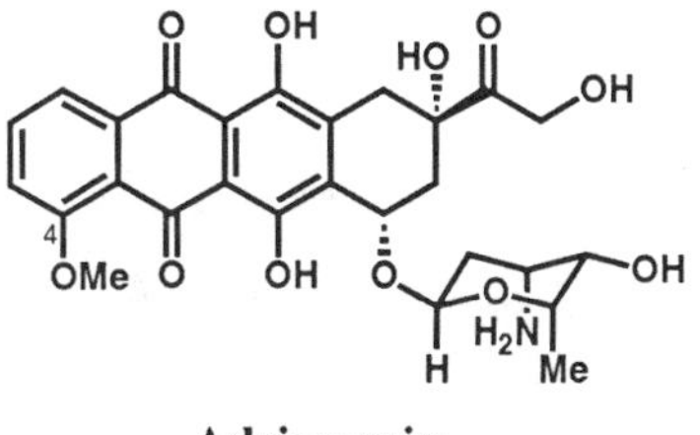

Adriamycin

Smith, *JOC*, **1977**, *42*, 3653.

4-Deoxyadriamycinone

Hauser, *JOC*, **1988**, *53*, 4515.

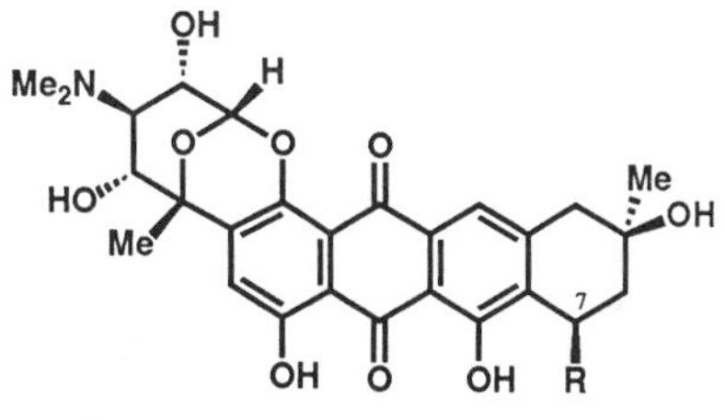

R = H: (+)-7-Deoxynogarol

R = OMe: (+)-7-Con-*O*-methylnogarol

Terashima, *TL*, **1988**, *29*, 791.

13.12

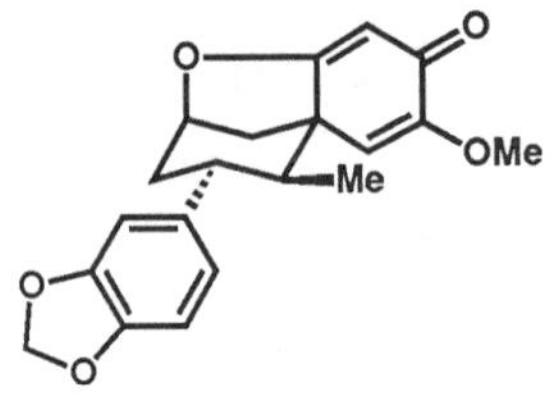

Futoenone
Yamamura,
TL, **1983**, *24*, 5011.

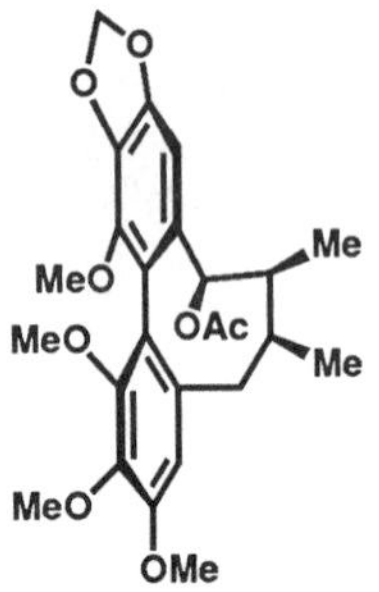

Kadsurin
Ghera, *JACS*, **1977**, *99*, 7673.

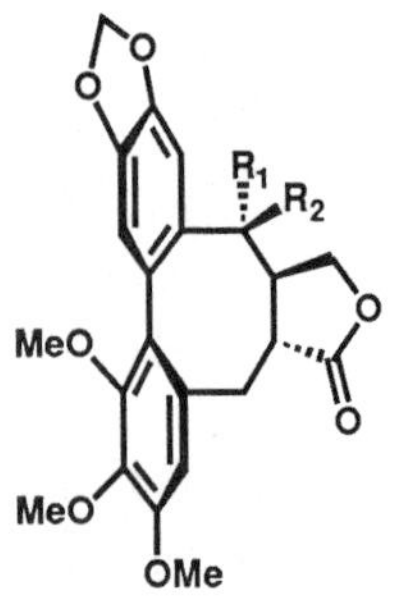

R_1, R_2 = O: (-)-Steganone (*nat*)
Raphael, *TL*, **1976**, 1543;
ibid., **1979**, 5041; *JCS, PI*, **1982**, 521. (-)
Krow, *JOC*, **1978**, *43*, 3950.
Robin, *TL*, **1980**, *21*, 2709. (-)
Ghera, *TL*, **1981**, *22*, 5091.
Magnus, *JACS*, **1985**, *107*, 4984.
Meyers, *JACS*, **1987**, *109*, 5446. (-)

R_1 = OAc, R_2 = H: Steganacin
Kende, *JACS*, **1976**, *98*, 267.
Raphael, *JCS, PI*, **1977**, 1674.
Ziegler, *JACS*, **1980**, *102*, 790.
Koga, *Tet*, **1984**, *40*, 1303. (+), (-)
Magnus, *JCS, CC*, **1984**, 1179.

R_1 = R_2 = H: Stegane
Koga, *TL*, **1979**, 1409.
Brown, *Tet*, **1983**, *39*, 2787.

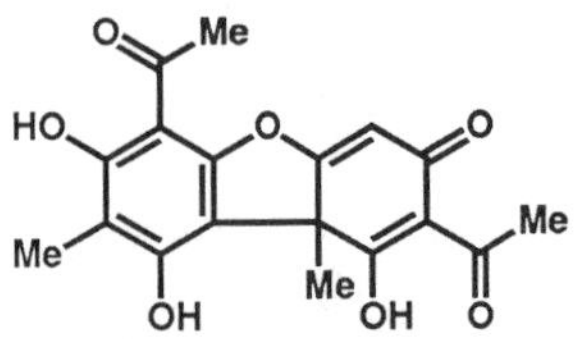

Usnic Acid
Barton, *Chem. Ind.*, **1955**, 1039;
JCS, **1956**, 530.

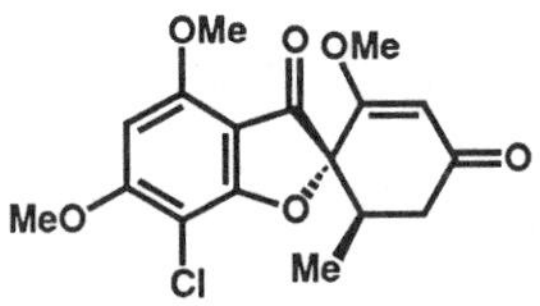

Griseofulvin
Brossi, *HCA*, **1960**, *43*, 2071.
Scott, *JCS*, **1961**, 4067.
Stork, *JACS*, **1962**, *84*, 310.
Taub, *Tet*, **1963**, *19*, 1. (+), (-)
Danishefsky,
JACS, **1979**, *101*, 7018.

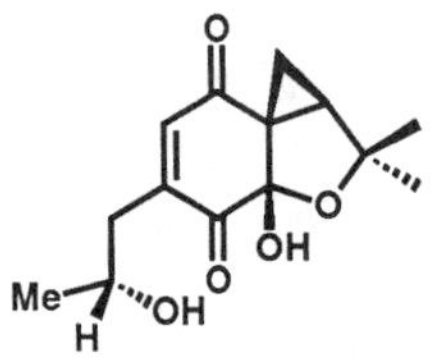

(-)-Gilmicolin
Smith, *JOC*, **1985**, *50*, 1342.

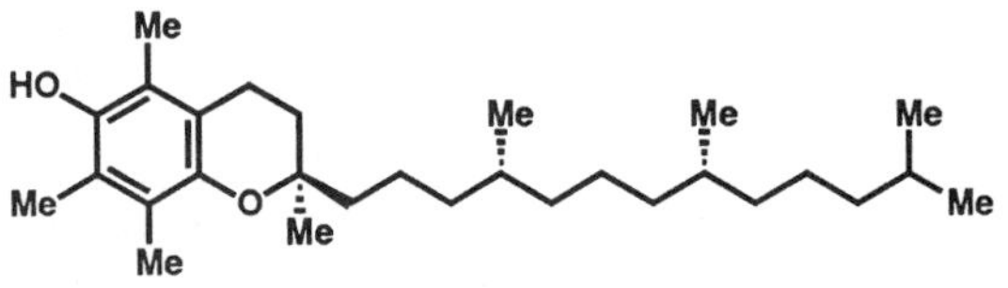

Vitamin E
Scott, *HCA*, **1976**, *59*, 290. (+)
Chan, *JOC*, **1978**, *43*, 3435. (+)
Cohen, *JACS*, **1979**, *101*, 6710. (+)
Heathcock, *TL*, **1982**, *23*, 2825.

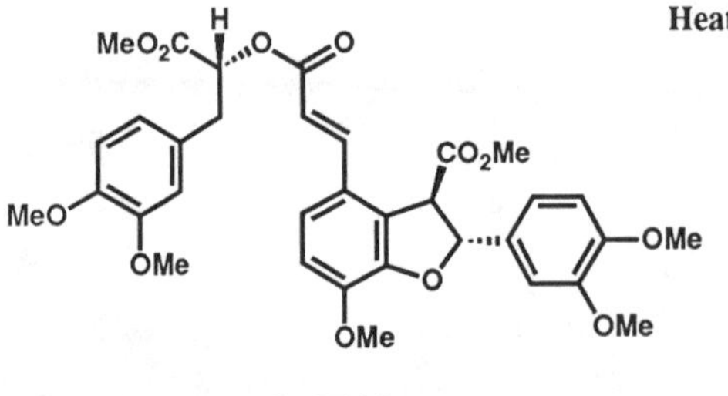

Heptamethyl Lithospermate
Jacobson, *JOC*, **1979**, *44*, 4013.

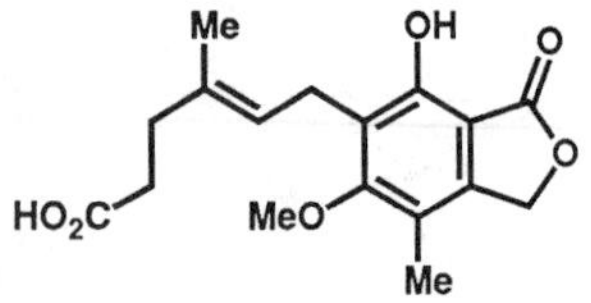

Mycophenolic Acid
Birch, *AJC*, **1969**, *22*, 2635.
Canonica, *Tet*, **1972**, *28*, 4395.
Danheiser, *JACS*, **1986**, *108*, 806.

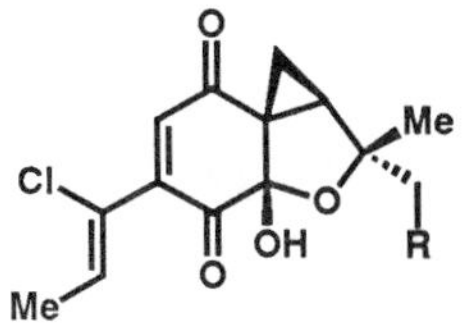

R = H: Mycorrhizin A
Smith, *JACS*, **1982**, *104*, 2659.
R = OH: (+)-Mikrolin
Smith, *TL*, **1987**, *28*, 3659.

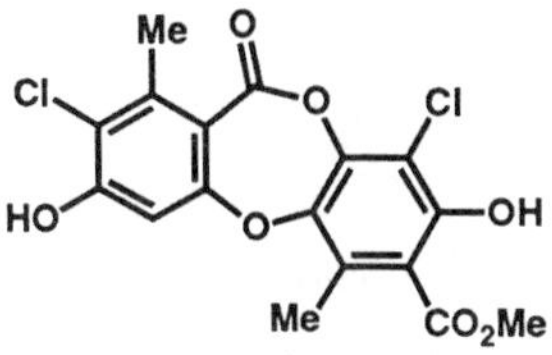

Lecideoidin
Sargent, *JCS, PI*, **1981**, 883.

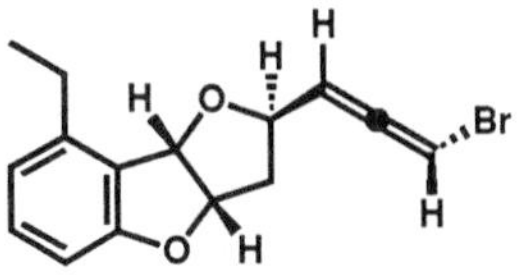

Panacene
Feldman, *JACS*, **1982**, *104*, 4011;
TL, **1982**, *23*, 3031.

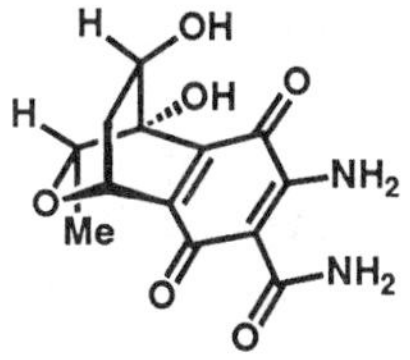

Sarubicin A
Yoshii, *JOC*, **1983**, *48*, 4151.
Semmelhack, *JACS*, **1985**, *107*, 4577.

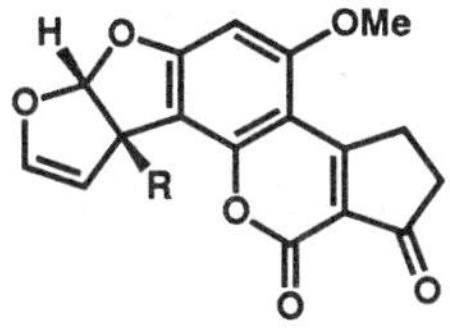

R = H: Aflatoxin B_1
Büchi, *JACS*, **1966**, *88*, 4534;
ibid., **1967**, *89*, 6745.
R = OH: Aflatoxin M_1
Büchi, *JACS*, **1969**, *91*, 5408;
ibid., **1971**, *93*, 746;
ibid., **1981**, *103*, 3497.

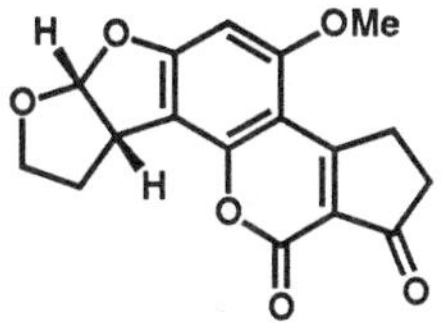

Aflatoxin B_2
Roberts, *JCS (C)*, **1968**, 22.
Rapoport, *JOC*, **1986**, *51*, 1006.
also, see: Schuda,
Top. Curr. Chem., **1980**, *91*, 75.

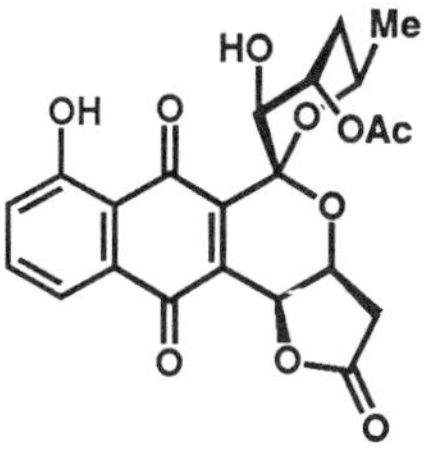

(+)-Griseusin A
Yoshii,
JOC, **1983**, *48*, 2311.

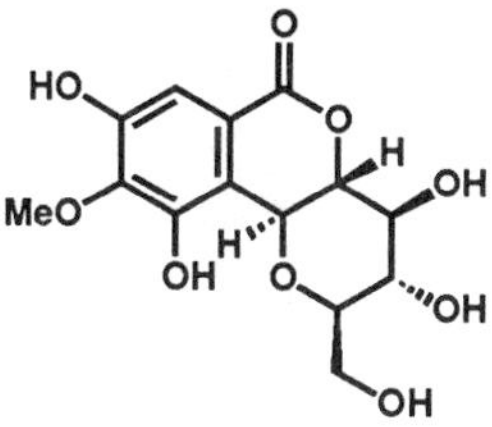

Bergenin
Hay, *JCS*, **1958**, 2231.

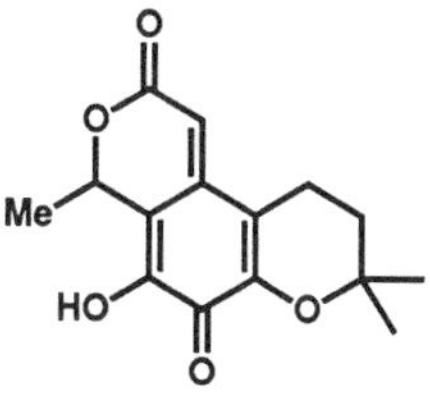

Fuscin
Barton, *JCS*, **1956**, 1028.
Scolastico, *Tet*, **1973**, *29*, 2849.

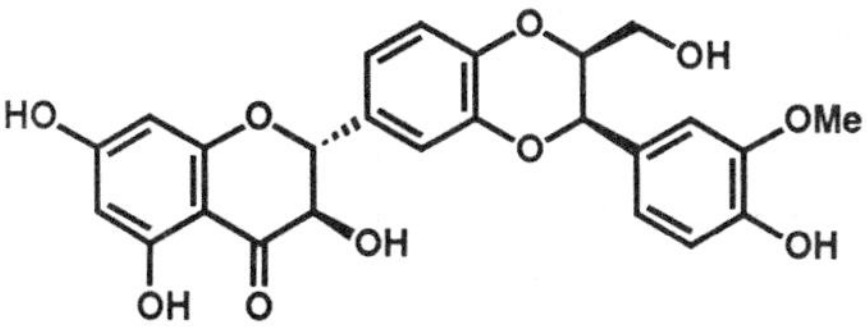

Silybin
Merlini, *JCS, CC*, **1979**, 695;
JCS, PI, **1980**, 775.
Ito, *CPB*, **1985**, *33*, 1419.

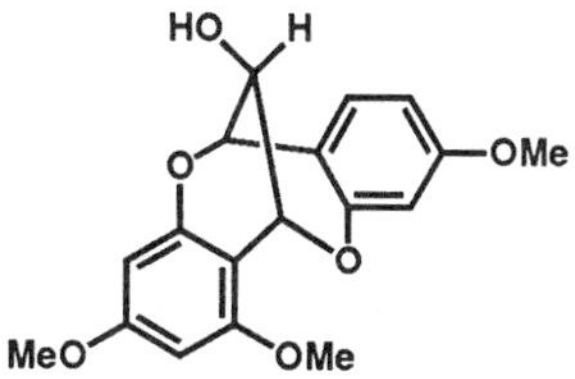

Trimethylcyanomaclurin
Bhatia, *Tet, Suppl*, **1966**, No. 8, 531.
Marathe, *Tet*, **1975**, *31*, 1011.

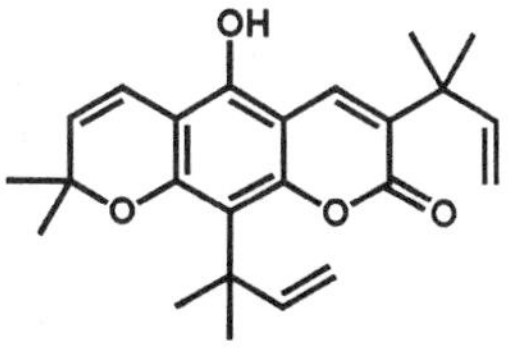

Clausarin
Murray,
TL, **1983**, *24*, 3773.

13.12

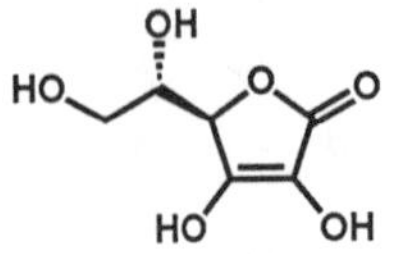

***L*-Ascorbic Acid (Vitamin C)**

Reichstein, *HCA*, **1934**, *17*, 311.
Helferich, *Ber*, **1937**, *70*, 465.
Theander, *JCS, CC*, **1971**, 175.
Ogawa, *Carbohyd. Res.*, **1976**, *51*, C1.
Crawford, *JCS, CC*, **1979**, 388.

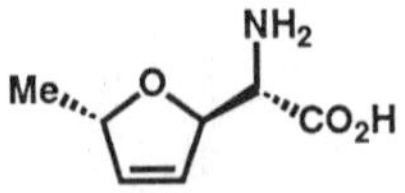

(+)-Furanomycin
Joullié,
JACS, **1980**, *102*, 887, 7505.

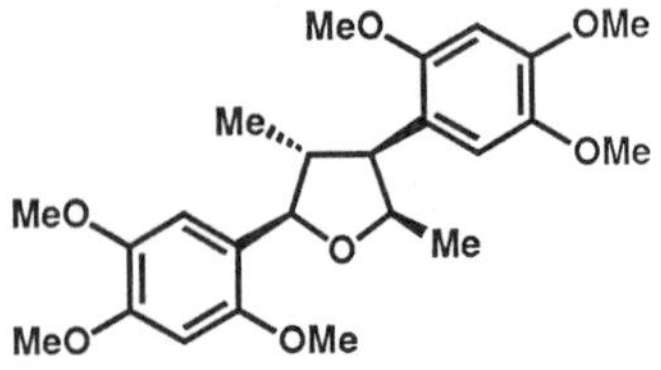

Magnosalicin
Mori, *Tet*, **1986**, *42*, 523.

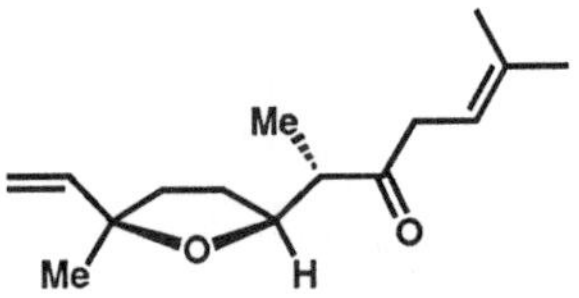

Davanone

Naegeli, *TL*, **1970**, 959.
Birch, *AJC*, **1970**, *23*, 1811.
Ohloff, *HCA*, **1970**, *53*, 841. (+)
Thomas, *HCA*, **1974**, *57*, 2062, 2066, 2076.
Bartlett, *TL*, **1983**, *24*, 1365.

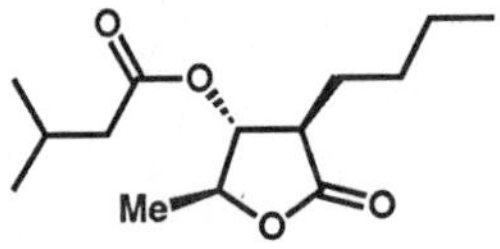

Blastmycinone

Kinoshita, *BCS, Jpn*, **1973**, *46*, 1279.
Heathcock, *JOC*, **1981**, *46*, 2290.
Nakata & Oishi, *TL*, **1983**, *24*, 2657.
Fujisawa, *TL*, **1984**, *25*, 5155. (+)
Kozikowski, *JOC*, **1984**, *49*, 2762.
Wasserman, *JACS*, **1985**, *107*, 1423. (+)
Sato, *CL*, **1985**, 467. (+)
Nakai, *CL*, **1985**, 1723. (+)

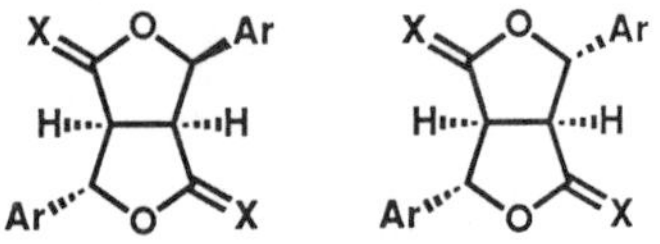

X = O and/or H_2:
Furofuran Lignans

Pelter, *TL*, **1979**, 2275; *JCS, PI*, **1985**, 587.
Chan, *TL*, **1980**, *21*, 3427.
Snieckus, *TL*, **1982**, *23*, 3975.
Whiting, *JCS, CC*, **1984**, 590; *TL*, **1986**, *27*, 4629.
Ishibashi, *CL*, **1986**, 1771.
Takano, *JCS, CC*, **1988**, 189.
also, see: **Ward**, *CSR*, **1982**, *11*, 75.

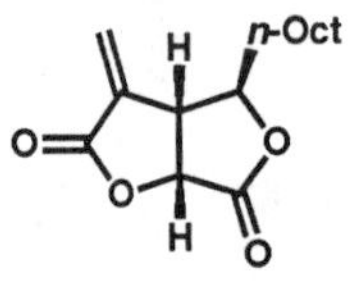

(-)-Avenaciolide (*nat*)

Johnson, *JACS*, **1969**, *91*, 7208; (±) *JOC*, **1973**, *38*, 2489. (±)
Ohrui, *TL*, **1975**, 3657. (+), (-)
Schlessinger, *JACS*, **1979**, *101*, 1544. (±)
Takeda, *JOC*, **1980**, *45*, 2039. (±)
Takei, *CL*, **1980**, 1311. (±)
Masamune, *JCS, CC*, **1981**, 221. (±)
Yoshikoshi, *CL*, **1983**, 881. (±)
Schreiber, *JACS*, **1984**, *106*, 7200. (±)
Kallmerten, *JOC*, **1985**, *50*, 1128. (±)
Fraser-Reid, *JOC*, **1985**. *50*, 4781. (-)
Kotsuki, *BCS, Jpn*, **1986**, *59*, 3881. (±)
Burke, *TL*, **1986**, *27*, 3345. (±)
Suzuki, *TL*, **1986**, *27*, 6237. (-)

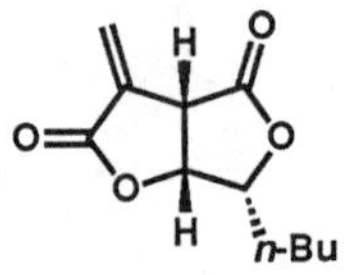

Canadensolide

Yoshikoshi, *JOC*, **1975**, *40*, 1932.
Carlson, *JOC*, **1976**, *41*, 4065.
Fraser-Reid, *JOC*, **1985**, *50*, 4786. (-)
Takeda, *JOC*, **1986**, *51*, 4944.

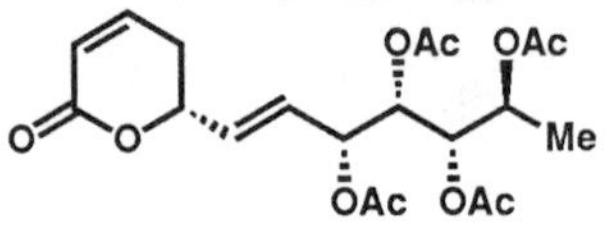

(+)-Anamarine
Lichtenthaler, *TL*, **1987**, *28*, 6437.

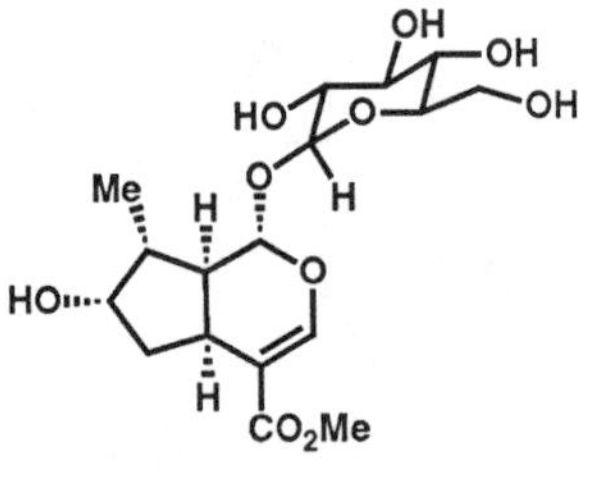

Loganin

Büchi, *JACS*, **1970**, *92*, 2165; *ibid.*, **1973**, *95*, 540. (-)
Partridge, *JACS*, **1973**, *95*, 532. (-)
Fleming, *JCS, CC*, **1977**, 81.
Hutchinson, *JOC*, **1980**, *45*, 4233.
Hiroi, *CL*, **1981**, 559.
Isoe, *HCA*, **1983**, *66*, 755.
Trost, *JACS*, **1985**, *107*, 1293.
Vandewalle & Oppolzer, *Tet*, **1986**, *42*, 4035. (-)

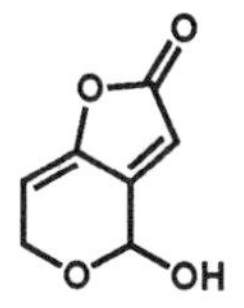

Patulin
Woodward,
JACS, **1950**, *72*, 1428.
Gill & Pattenden,
TL, **1988**, *29*, 2875.

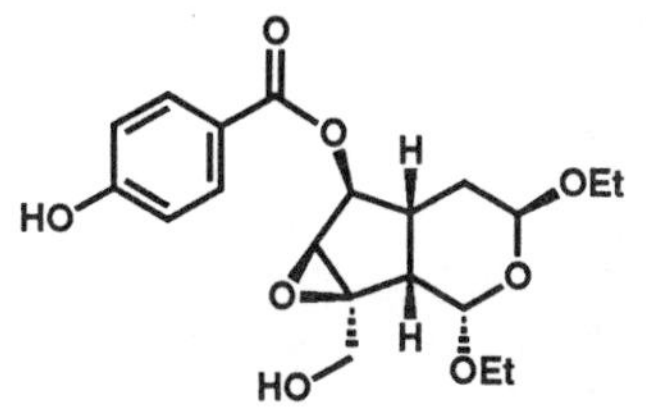

Specionin
Vandewalle, *JCS, CC*, **1985**, 1719;
Tet, **1986**, *42*, 5385; *TL*, **1987**, *28*, 3519. (-)
Leonard, *TL*, **1987**, *28*, 4871.
Curran, *JACS*, **1987**, *109*, 5280. (-)

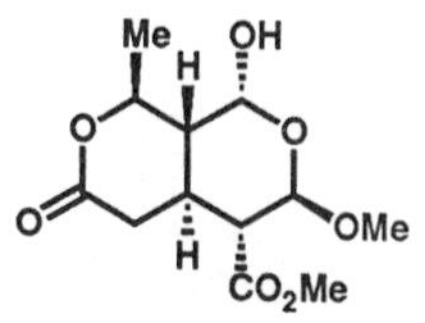

(-)-Xylomollin
Whitesell,
JACS, **1986**, *108*, 6802.

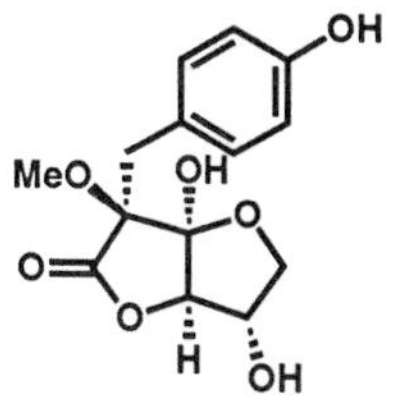

Delesserine
Seebach, *ACIE*, **1984**, *23*, 530. (+)
Poss, *TL*, **1987**, *28*, 2555.

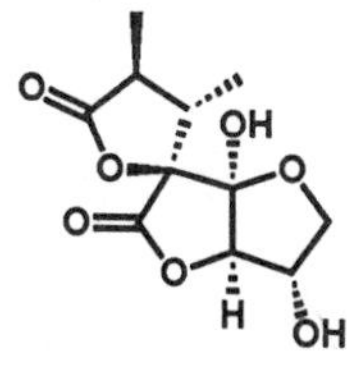

Piptosidin
Poss, *TL*, **1987**, *28*, 5469.

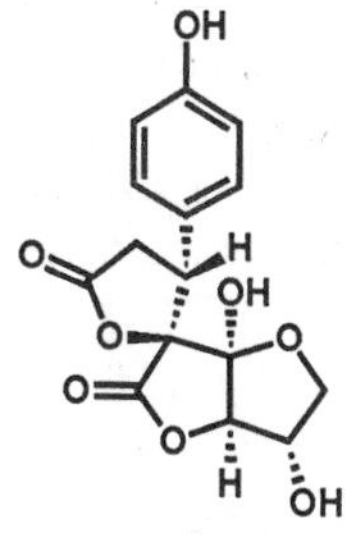

Dilaspirolactone
Poss, *TL*, **1987**, *28*, 2555.

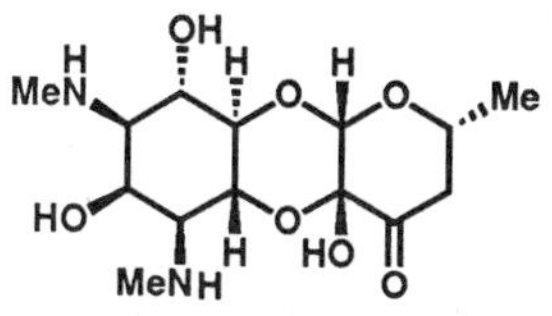

(+)-Spectinomycin
White, *TL*, **1979**, 2737.
Hanessian, *JACS*, **1979**, *101*, 5839;
CJC, **1985**, *63*, 163.

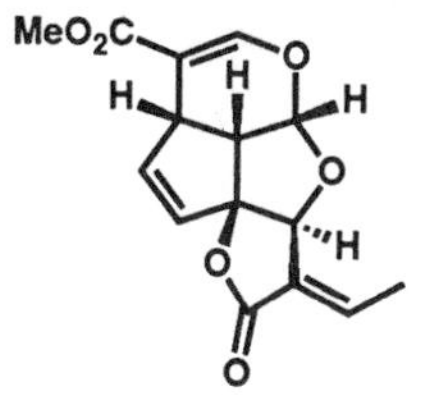

Plumericin
Trost,
JACS, **1983**, *105*, 6755;
ibid., **1986**, *108*, 4965, 4974.

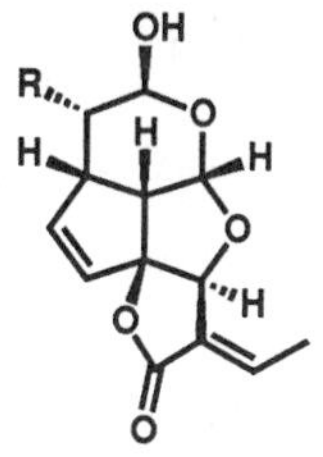

R = H: Allamcin
R = CO_2Me: Allamandin
Trost, *TL*, **1985**, *26*, 1807;
JACS, **1986**, *108*, 4974.
Pattenden, *TL*, **1986**, *27*, 1305;
JCS, PI, **1988**, 1119.

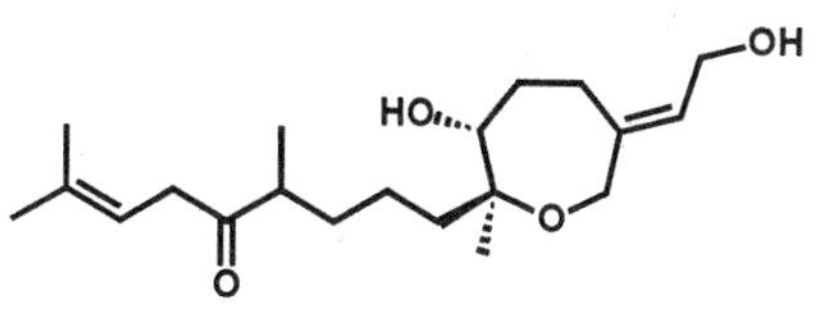

Zoapatanol
Chen, *JACS*, **1980**, *102*, 6609.
Nicolaou, *JACS*, **1980**, *102*, 6611.
Kane, *TL*, **1981**, *22*, 3027, 3031.
Cookson, *JCS, PI*, **1985**, 1589.

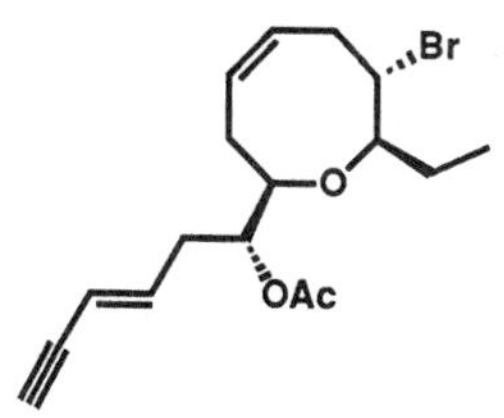

Laurencin
Masamune, *TL*, **1977**, 2507;
BCS, Jpn, **1979**, *52*, 127.

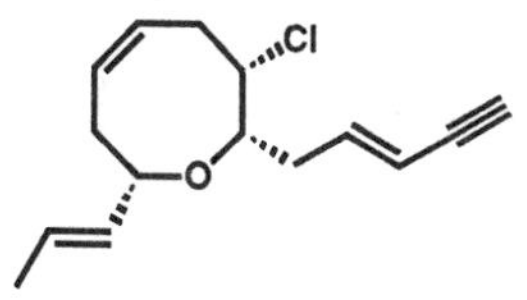

(-)-Laurenyne
Overman,
JACS, **1988**, *110*, 2248.

13.12

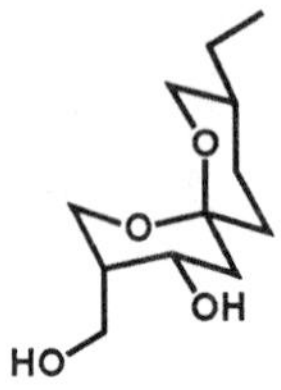

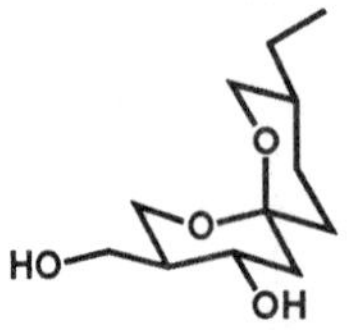

(-)-Talaromycin A **(-)-Talaromycin B**

Schreiber, *TL*, **1983**, *24*, 4781, (±)-(**B**);
ibid., **1985**, *26*, 17, (±)-(**A**&**B**).
Kozikowski, *JACS*, **1984**, *106*, 353, (±)-(**B**).
Smith, *JOC*, **1984**, *49*, 1469, (-)-(**A**&**B**).
Kay, *TL*, **1984**, *25*, 2035, (±)-(**B**).
Midland, *JOC*, **1985**, *50*, 1143, (-)-(**A**).
Kocienski, *JCS, PI*, **1985**, 1879, (±)-(**B**);
JCS, CC, **1987**, 906, (±)-(**A**&**B**).
Mori, *Tet*, **1987**, *43*, 45, (-)-(**A**&**B**).
Iwata, *TL*, **1987**, *28*, 3135, (+)-(**A**)&(-)-(**B**).

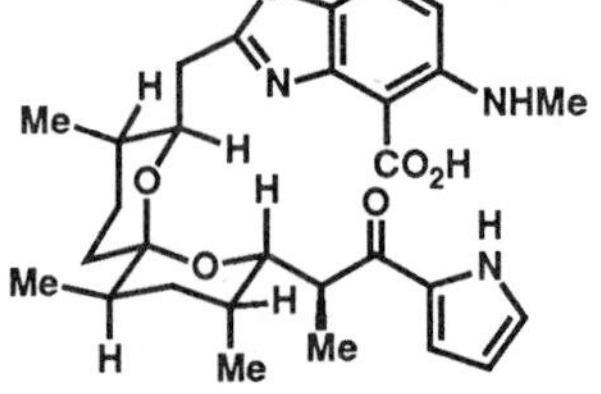

(-)-Antibiotic A-23187 (Calcimycin)

Evans, *JACS*, **1979**, *101*, 6789.
Grieco, *JACS*, **1982**, *104*, 1436.
Nakahara & Ogawa,
Tet, **1986**, *42*, 6465.
Kishi, *TL*, **1987**, *28*, 1063.
Boeckman, *JACS*, **1987**, *109*, 7553.

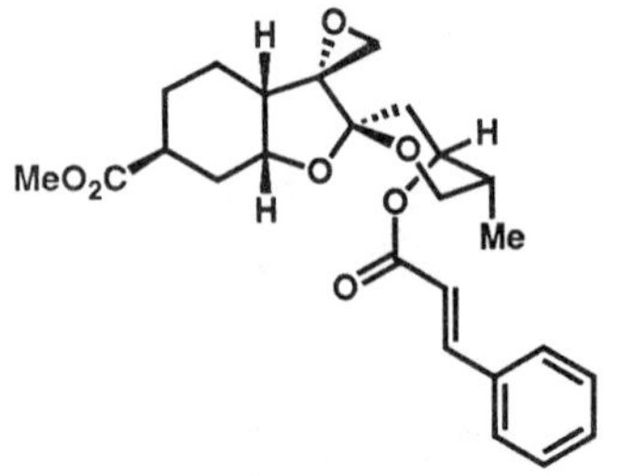

(+)-Phyllanthocin

Collum, *JACS*, **1982**, *104*, 4496.
Williams, *JACS*, **1984**, *106*, 2949.
Burke, *JOC*, **1985**, *50*, 3420;
TL, **1986**, *27*, 4237.
Martin, *JOC*, **1987**, *52*, 3706.
Smith, *JACS*, **1987**, *109*, 1269.

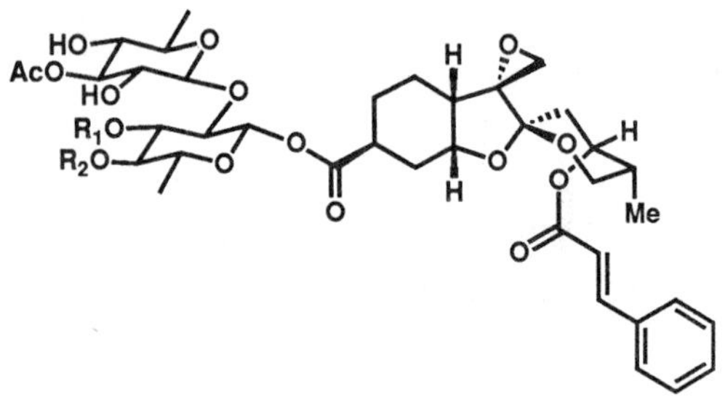

R_1 = Ac, R_2 = H: (+)-Phyllanthoside
Smith, *JACS*, **1987**, *109*, 1272.
R_1 = H, R_2 = Ac: (-)-Phyllanthostatin-1
Smith, *JCS, CC*, **1987**, 1026.

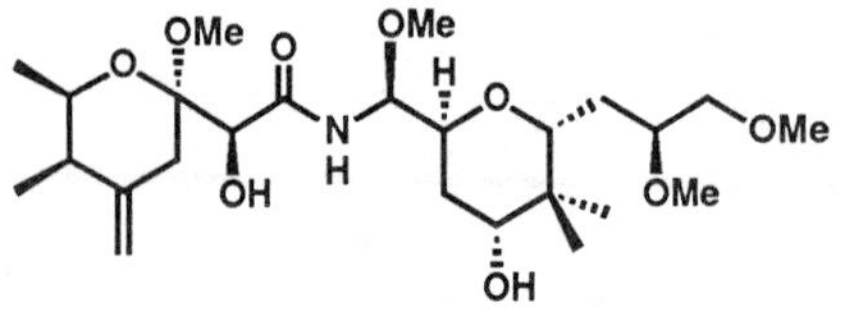

Pederin

Matsumoto, *TL*, **1982**, *23*, 4043;
ibid., **1983**, *24*, 1277. (+)
Nakata & Oishi, *TL*, **1985**, *26*, 6465. (+)
Kocienski, *JCS, CC*, **1987**, 106.

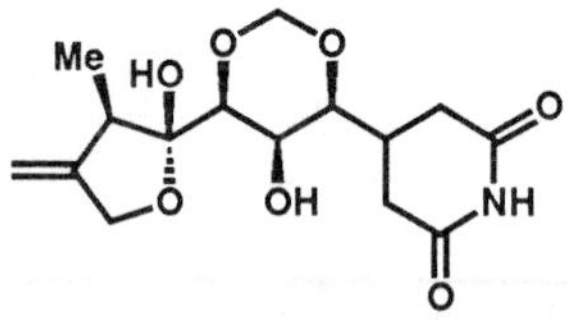

(+)-Sesbanimide A (*nat*)

Pandit, *JCS, CC*, **1986**, 396; (-)
Tet, **1987**, *43*, 2549. (+), (-)
Schlessinger, *JOC*, **1986**, *51*, 2621. (-)
Terashima, *TL*, **1986**, *27*, 3407. (+)
Koga, *TL*, **1988**, *29*, 3095. (+)

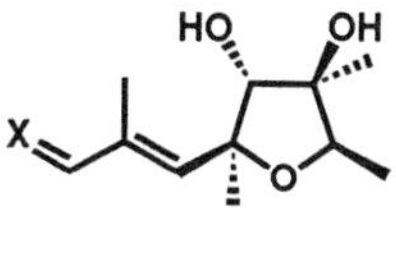

X = O: Citreoviral
Yamamura, *TL*, **1985**, *26*, 231. (+)
Williams, *TL*, **1985**, *26*, 2529.
Pattenden, *TL*, **1985**, *26*, 4793;
ibid., **1988**, *29*, 711.
Takano, *TL*, **1985**, *26*, 6485. (+)
Trost, *TL*, **1987**, *28*, 375. (+)
Wilcox, *JACS*, **1988**, *110*, 470. (+)

X =

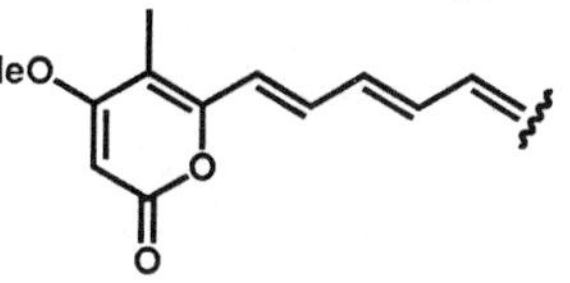

Citreoviridin
Yamamura, *TL*, **1985**, *26*, 231. (-)
Williams, *JOC*, **1987**, *52*, 5067.
Wilcox, *JACS*, **1988**, *110*, 470. (-)

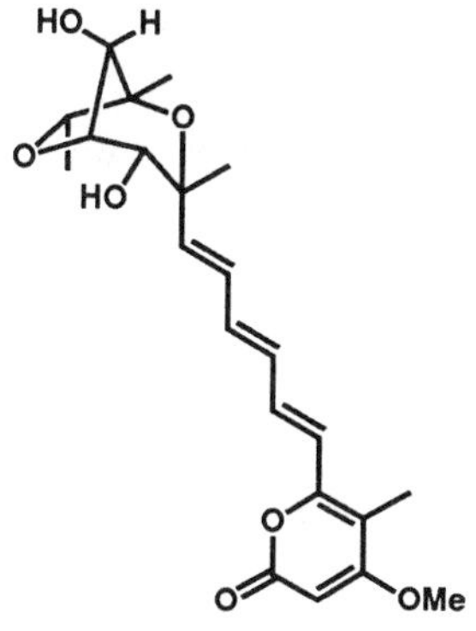

Citreoviridinol
Yamamura,
CL, **1986**, 1973.

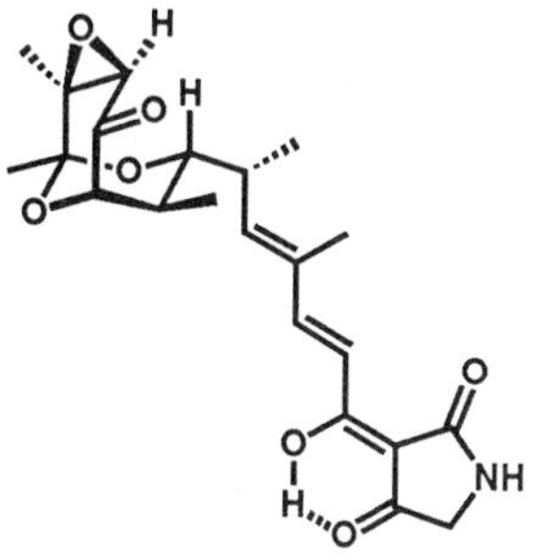

Tirandamycin A
Schlessinger, *JACS*, **1985**, *107*, 1777. (-)
DeShong, *JACS*, **1985**, *107*, 5219.
Boeckman, *JACS*, **1986**, *108*, 5549.
Bartlett, *JACS*, **1986**, *108*, 5559.

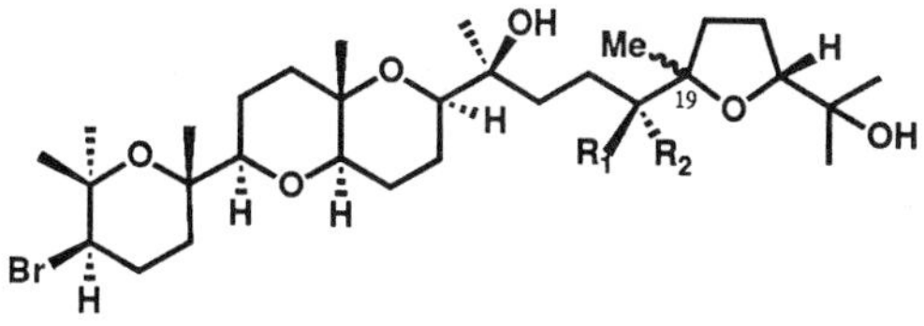

R_1 = OH, R_2 = H, α-Me at C(19): (+)-Thyrsiferol
R_1 = H, R_2 = OH, β-Me at C(19): (+)-Venustatriol
Shirahama, *TL*, **1988**, *29*, 1143.

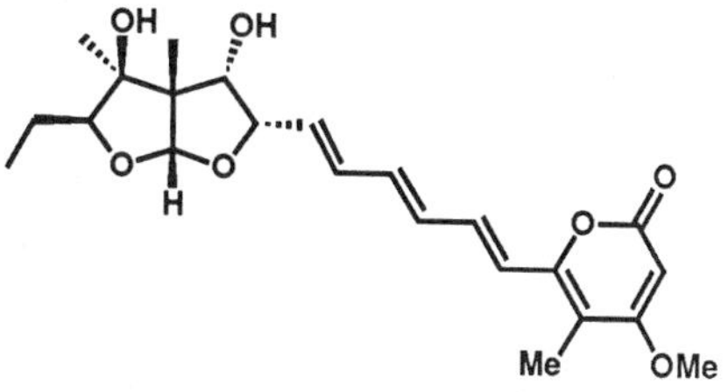

Asteltoxin
Schreiber, *JACS*, **1984**, *106*, 4186.

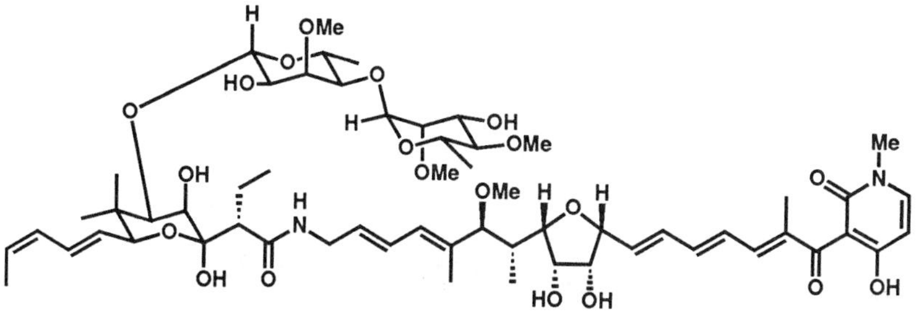

Efrotomycin
Nicolaou, *JACS*, **1985**, *107*, 1691, 1695. (*nat*)

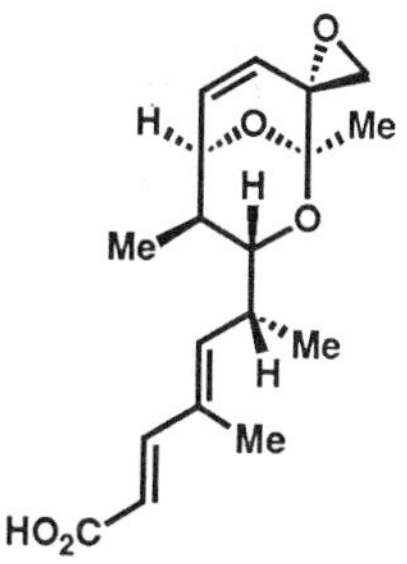

(+)-Streptolic Acid
Ireland,
JACS, **1988**, *110*, 854.

13.12

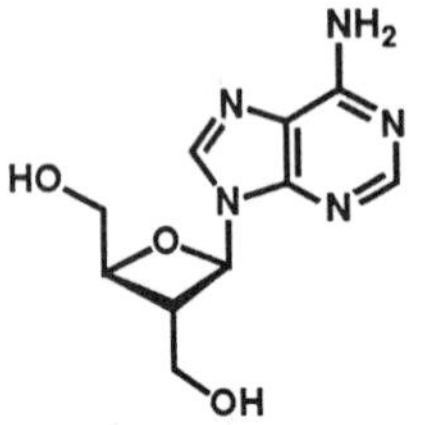

NH2
N
N
N
N
HO
O
OH

(-)-Oxetanocin
Niitsuma, *TL*, **1987**, *28*, 3967, 4713.
Norbeck, *JACS*, **1988**, *110*, 7217.

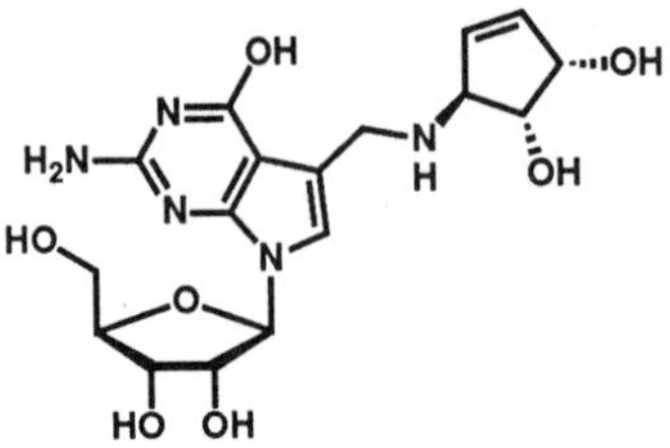

OH
N
H2N
N
N
HO
O
HO OH
N
H
OH
OH

(+)-Nucleoside Q
Goto, *JACS*, **1979**, *101*, 3629.

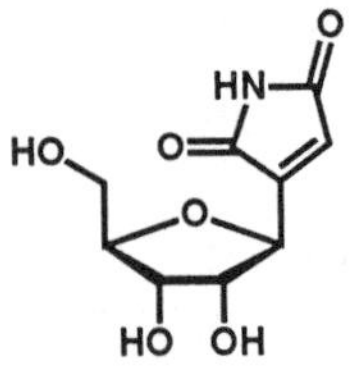

O
HN
O
HO
O
HO OH

(+)-Showdomycin
Kalvoda, *TL*, **1970**, 2297.
Moffatt, *JOC*, **1973**, *38*, 1841.
Noyori, *JACS*, **1978**, *100*, 2561;
BCS, Jpn, **1984**, *57*, 2515.
Buchanan, *JCS, PI*, **1979**, 225.
Just, *CJC*, **1980**, *58*, 2024. (±)
Kuwajima, *JCS, CC*, **1980**, 251.
Kozikowski, *JACS*, **1981**, *103*, 3923.
Ohno, *Tet*, **1984**, *40*, 145.
Katagiri, *Het*, **1984**, *22*, 2195.
Barrett, *JOC*, **1986**, *51*, 495.
Kametani, *JACS*, **1987**, *109*, 3010.

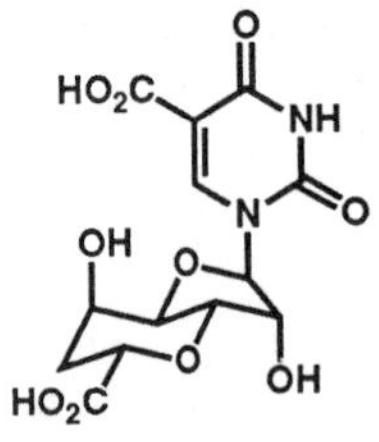

O
HO2C
NH
N
O
OH
O
O
HO2C
OH

Octosyl Acid A
Danishefsky,
JACS, **1986**, *108*, 2486.
Hanessian,
JACS, **1986**, *108*, 2758.

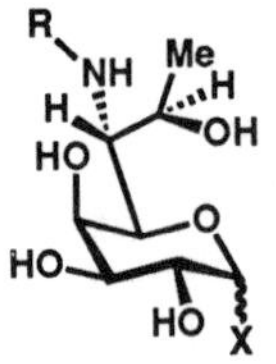

R
NH
Me
H
H
OH
HO
O
HO
HO
X

X = α-SMe, R =

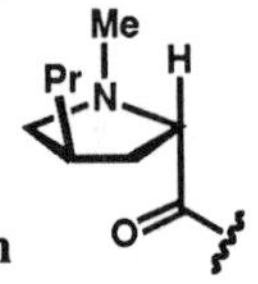

Me
Pr
N
H
O

Lincomycin
Magerlein, *TL*, **1970**, 33.
Szarek, *JCS (C)*, **1970**, 2218.

X = OH, R = H:
Lincosamine
Saeki, *CPB*, **1970**, *18*, 789.
Danishefsky, *JACS*, **1983**, *105*, 6715.

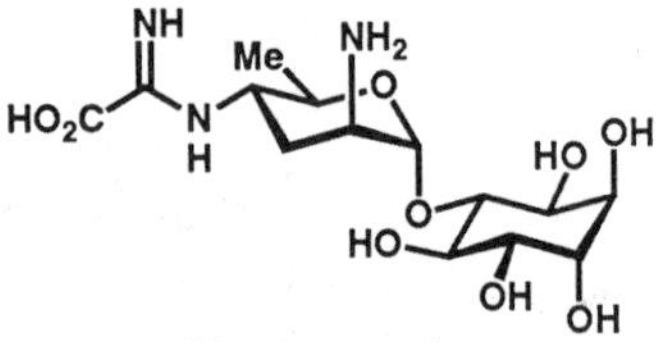

NH
Me
NH2
HO2C
N
H
O
O
HO OH
HO
OH
OH

Kasugamycin
Ohno, *JACS*, **1972**, *94*, 6501.

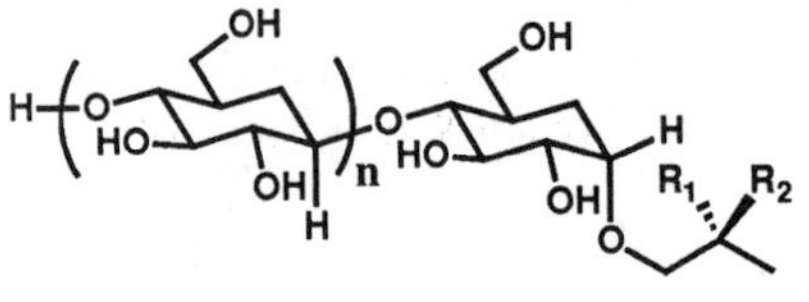

OH
OH
H
O
HO
OH
H
n
O
HO
H
OH
R1
R2
O

Rhynchosporosides
n = 0-4, R_1 = H, R_2 = OH (*S*)
R_1 = OH, R_2 = H (*R*)
Nicolaou, *JACS*, **1985**, *107*, 5556.

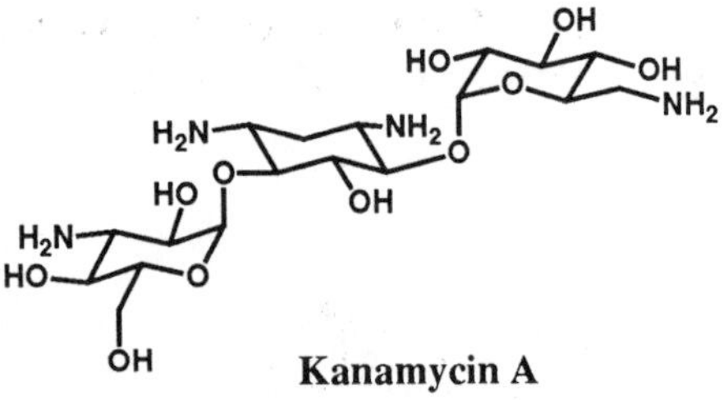

OH
HO
OH
O
NH2
H2N
NH2
O
HO
O
OH
H2N
HO
O
OH

Kanamycin A
Nakajima, *TL*, **1968**, 623.

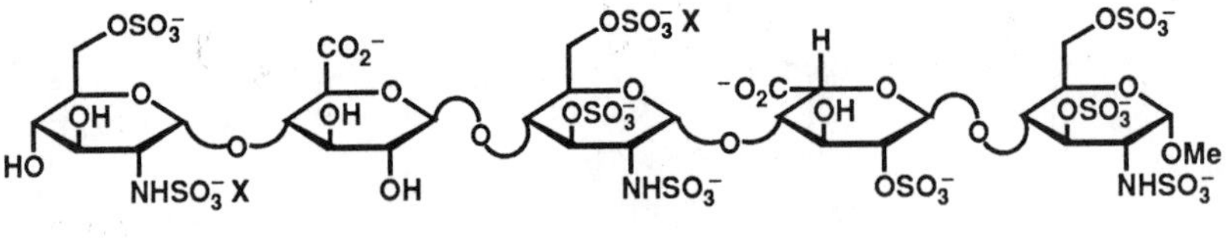

OSO3-
CO2-
OSO3- X
H
OSO3-
O
OH
HO
NHSO3- X
O
OH
OH
OSO3-
NHSO3-
-O2C
OH
OSO3-
OSO3-
OMe
NHSO3-

Heparin-like Pentasaccharide
van Boeckel, *TL*, **1988**, *29*, 803.

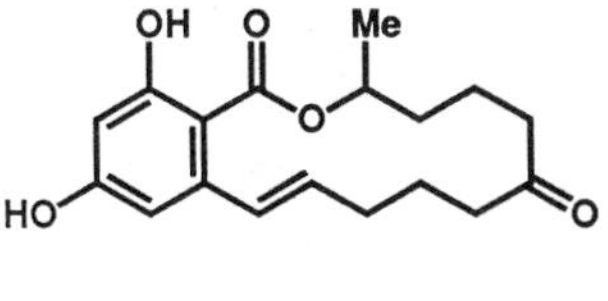

Zearalenone

Taub, *JCS, CC*, **1967**, 225;
Tet, **1968**, *24*, 2443.
Girotra, *Chem. Ind.*, **1967**, 1493.
Harrison, *JOC*, **1968**, *33*, 4176.
Tsuji, *JACS*, **1979**, *101*, 5072.

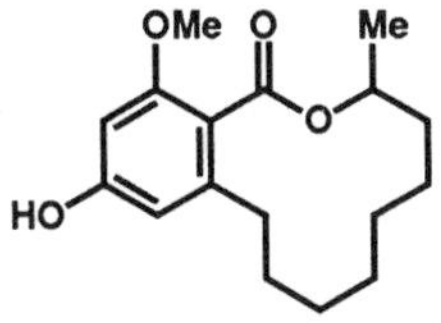

Lasiodiplodin

Gerlach, *HCA*, **1977**, *60*, 2866.
Tsuji, *TL*, **1978**, 4917.
Danishefsky, *JOC*, **1979**, *44*, 4716

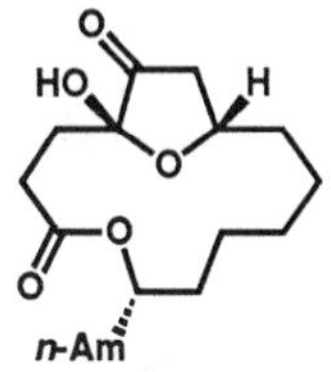

(+)-Gloeosporone

Seebach,
JACS, **1987**, *109*, 6176.

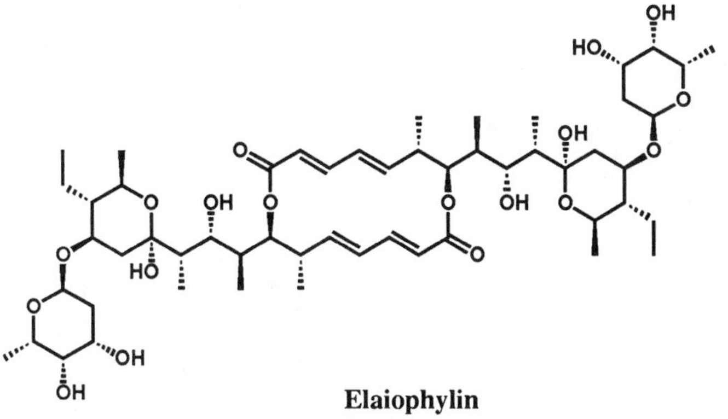

Elaiophylin

Kinoshita, *TL*, **1986**, *27*, 4741.

(+)-Elaiophylin Aglycone

Seebach, *JACS*, **1985**, *107*, 5292.

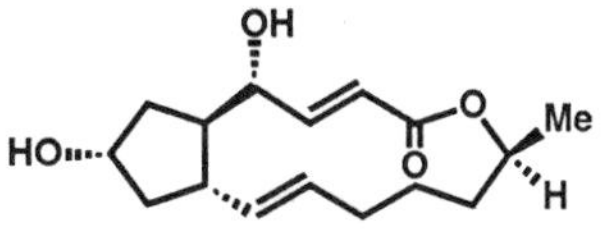

Brefeldin A

Greene, *TL*, **1977**, 2973;
JACS, **1982**, *104*, 5473. (±), (+)
Bartlett, *JACS*, **1978**, *100*, 4858.
Winterfeldt, *ACIE*, **1980**, *19*, 472. (+)
Honda, *TL*, **1981**, *22*, 2679.
Kitahara, *Tet*, **1984**, *40*, 2935. (+)
Gais, *ACIE*, **1984**, *23*, 145. (+)
Sakai, *CPB*, **1985**, *33*, 4021. (+)
Isoe, *TL*, **1985**, *26*, 2209.
Trost, *JACS*, **1986**, *108*, 284. (+)

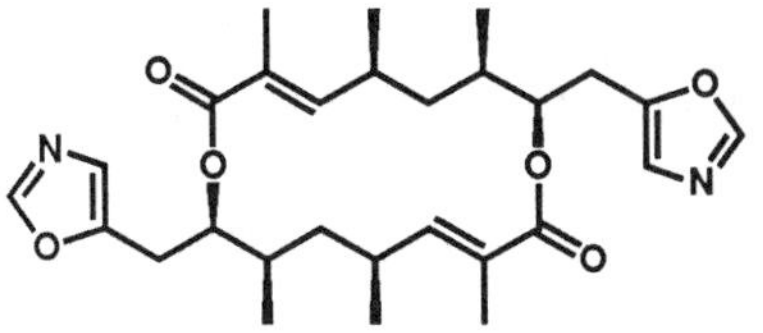

(+)-Conglobatin

Seebach, *TL*, **1984**, *25*, 5881.

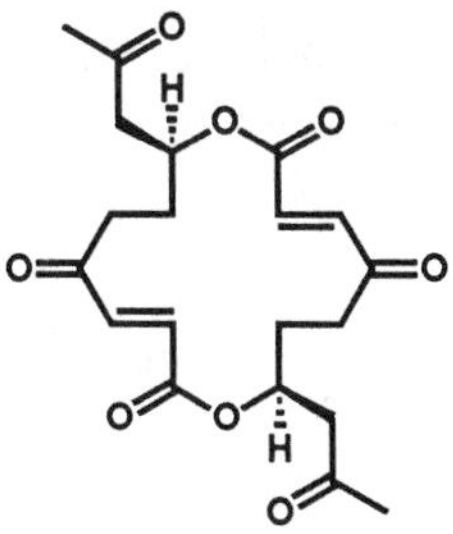

(-)-Vermiculine (*nat*)

White, *JACS*, **1977**, *99*, 646. (±)
Seebach, *ACIE*, **1977**, *16*, 264;
Ann, **1978**, 2044. (+)
Burri, *JACS*, **1978**, *100*, 7069. (-)
Hase, *JOC*, **1981**, *46*, 3137. (±)
Pollini, *JOC*, **1983**, *48*, 1297. (±)
Wakamatsu, *Het*, **1986**, *24*, 309. (±)

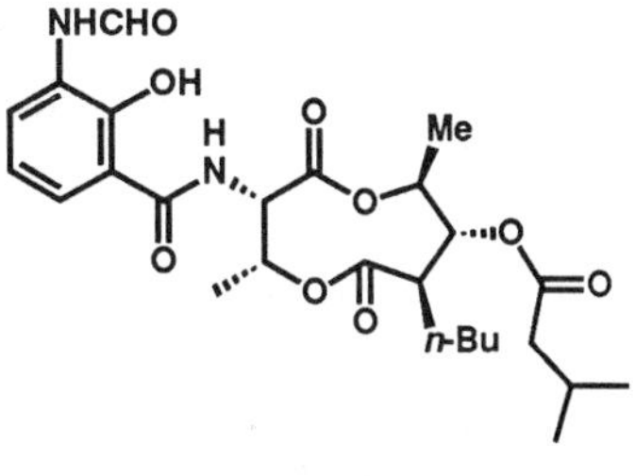

Antimycin A_3

Kinoshita, *BCS, Jpn*, **1979**, *52*, 198.
Nakata & Oishi, *TL*, **1983**, *24*, 2657.
Wasserman, *JACS*, **1985**, *107*, 1423. (+)

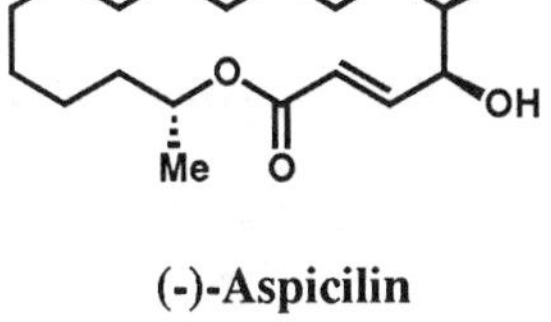

(-)-Aspicilin

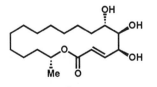

Zwanenburg, *TL*, **1987**, *28*, 2409.

13.13

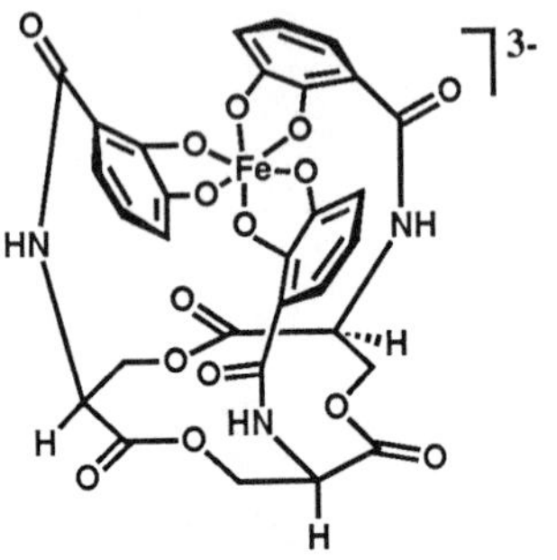

Enterobactin
Rastetter,
JOC, **1980**, *45*, 5011. (*nat*)

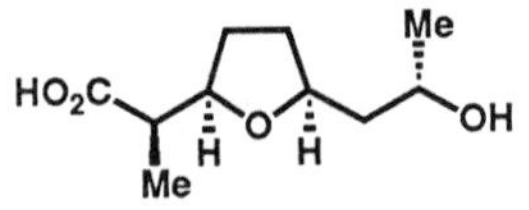

(-)-Nonactic Acid
Beck, *Ber*, **1971**, *104*, 21.
Schmidt, *ACIE*, **1975**, *14*, 432. (±), (+)
White, *JOC*, **1976**, *41*, 2075.
Bartlett, *TL*, **1980**, *21*, 1607;
JACS, **1984**, *106*, 5304. (+), (-)
Ireland, *CJC*, **1981**, *59*, 572. (+), (-)
Vogel, *TL*, **1986**, *27*, 5615.
Lygo, *TL*, **1987**, *28*, 3597.

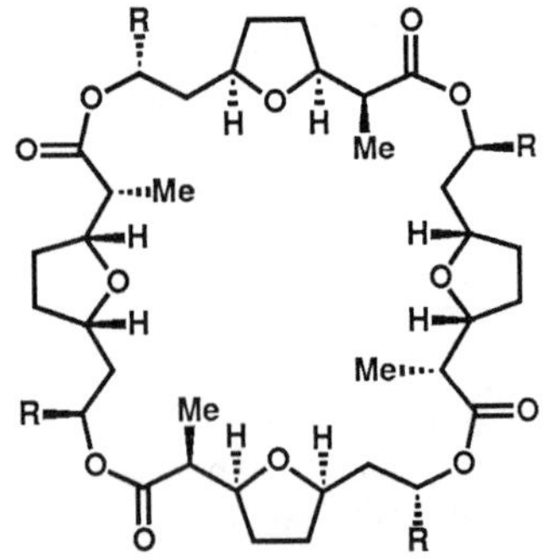

R = Me: Nonactin
Gerlach, *HCA*, **1975**, *58*, 2036.
Schmidt, *Ber*, **1976**, *109*, 2628. (*nat*)
Bartlett, *JACS*, **1984**, *106*, 5304. (*nat*)
R = Et: Tetranactin
Schmidt, *JCS, CC*, **1986**, 996.

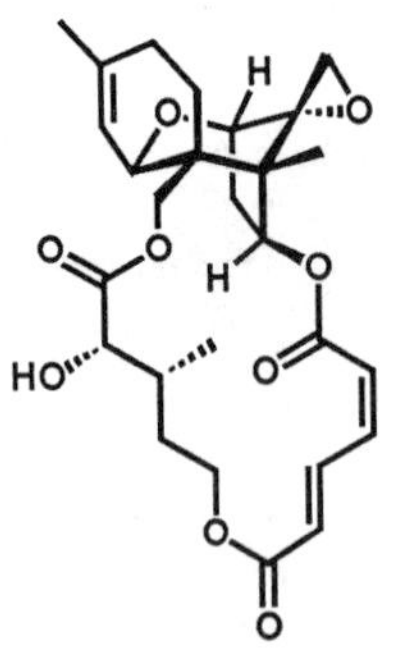

Verrucarin A
Still, *JOC*, **1981**, *46*, 5242.
Tamm, *HCA*, **1982**, *65*, 1412.

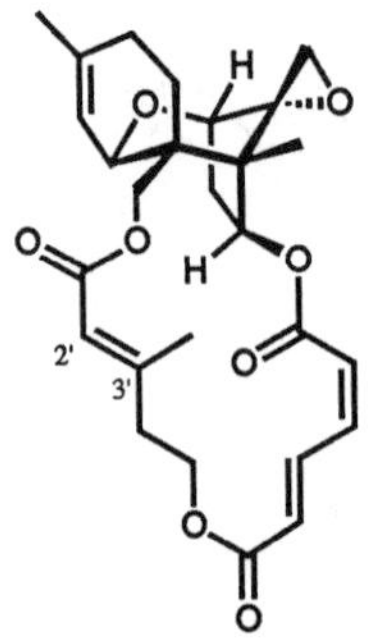

Verrucarin J
Fraser-Reid, *JOC*, **1982**, *47*, 3358.
Roush, *JOC*, **1983**, *48*, 758.
ibid., **1984**, *49*, 1772.
2',3'-α-epoxide: Verrucarin B
Roush, *JOC*, **1984**, *49*, 4332.

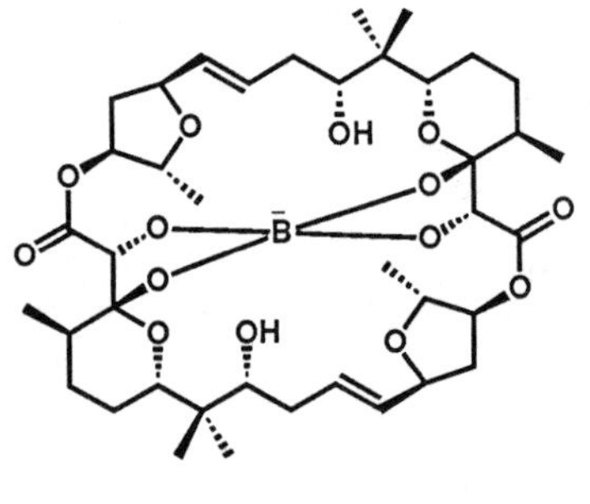

(+)-Aplasmomycin
White, *JACS*, **1986**, *108*, 8105.
Nakata & Oishi, *TL*, **1986**, *27*, 6341, 6345.

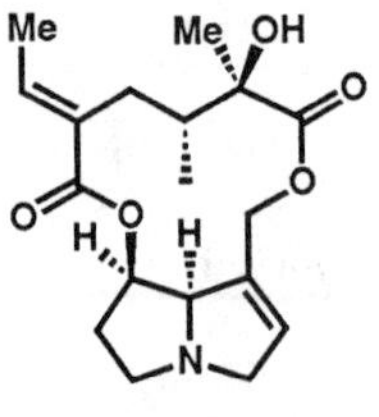

Integerrimine
Narasaka, *CL*, **1982**, 455;
JACS, **1984**, *106*, 2954.
Niwa & Yamada, *TL*, **1986**, *27*, 4609. (-)
White, *JOC*, **1986**, *51*, 5492. (-)

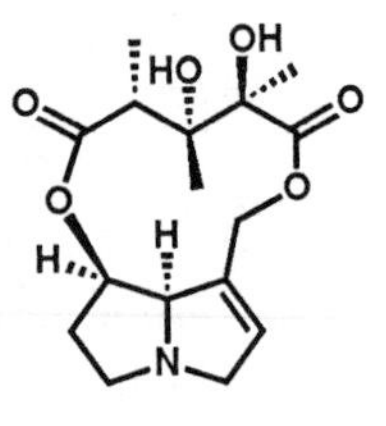

Monocrotaline
Vedejs,
JOC, **1987**, *52*, 3937.

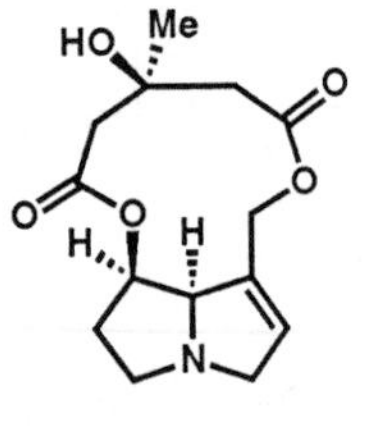

(+)-Dicrotaline
Robins,
JCS, PI, **1983**, 1819.

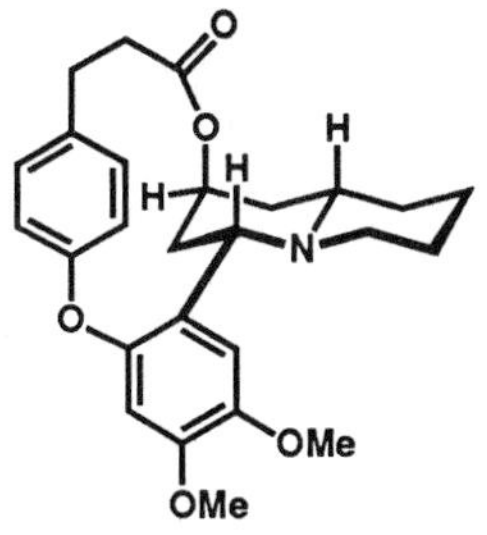
O
O
H
H
H
N
O
OMe
OMe

Decaline
Hanaoka, *TL*, **1973**, 2355.
Wróbel, *TL*, **1973**, 4293.
Fukumoto, *CPB*, **1985**, *33*, 532.

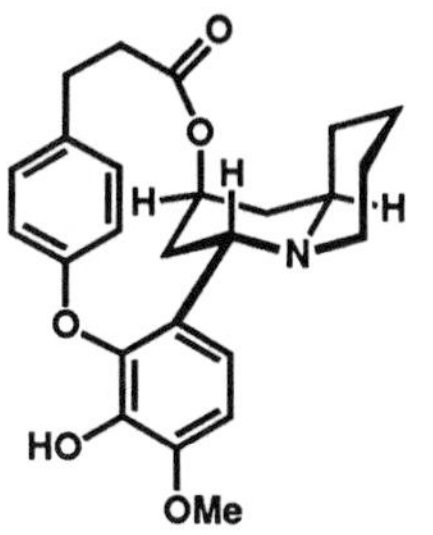
O
O
H
H
H
N
O
HO
OMe

Lagerine
Hanaoka, *CPB*, **1975**, *23*, 2191.

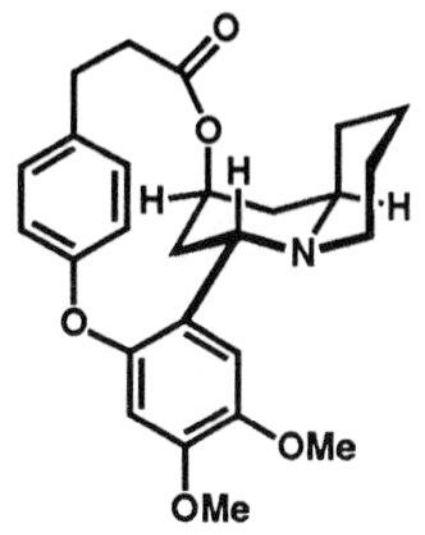
O
O
H
H
H
N
O
OMe
OMe

Vertaline
Hanaoka, *CPB*, **1974**, *22*, 973;
ibid., **1976**, *24*, 1045.
Hart, *JOC*, **1982**, *47*, 1555.
Fukumoto, *CPB*, **1985**, *33*, 532.

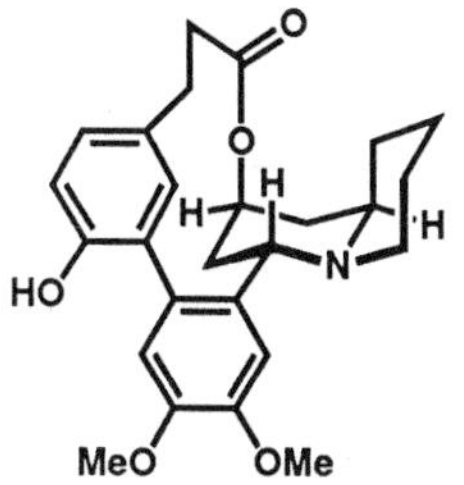
O
O
H
H
H
N
HO
MeO
OMe

Decamine
Lantos, *JOC*, **1977**, *42*, 228.

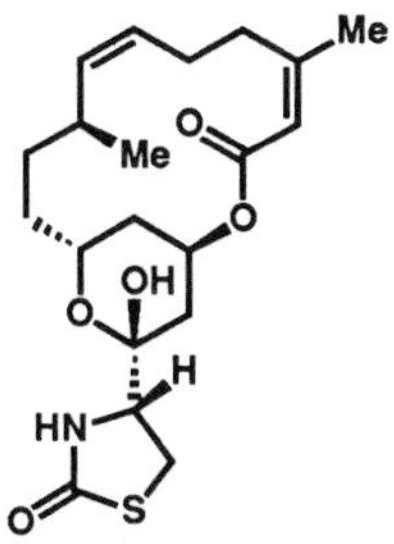
Me
Me
O
O
OH
O
H
HN
O
S

(+)-Latrunculin B
Smith, *JACS*, **1986**, *108*, 2451.

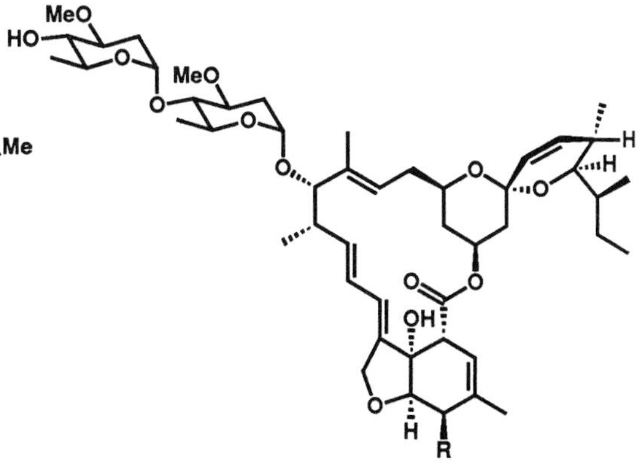
MeO
HO
O
MeO
O
O
O
O
H
H
O
O
O
O
OH
O
H
R

R = OH: (+)-Avermectin B_{1a}
Hanessian, *JACS*, **1986**, *108*, 2776;
P&AC, **1987**, *59*, 299.
R = OMe: Avermectin A_{1a}
Danishefsky, *JACS*, **1987**, *109*, 8117, 8119.

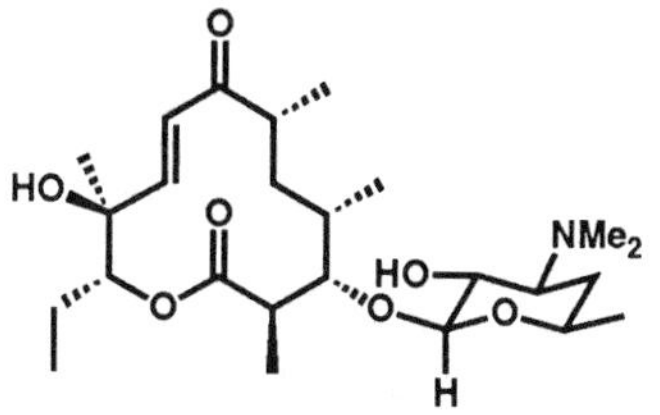
O
HO
O
O
HO
NMe2
O
O
H

Methymycin
Masamune,
JACS, **1975**, *97*, 3512, 3513.

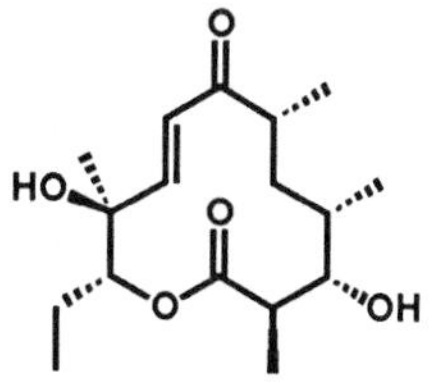
O
HO
O
O
OH

Methynolide
Yamaguchi, *CL*, **1979**, 1019, 1021. (+)
Grieco, *JACS*, **1979**, *101*, 4749.
Ireland, *JOC*, **1983**, *48*, 1312.
Yonemitsu, *TL*, **1986**, *27*, 3647;
CPB, **1987**, *35*, 2203. (+)
Vedejs, *JACS*, **1987**, *109*, 5878.

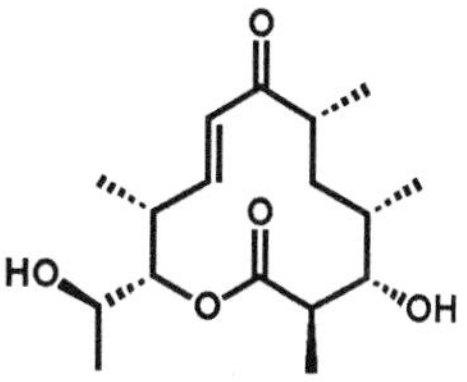
O
O
HO
O
OH

(+)-Neomethynolide
Yamaguchi,
CL, **1981**, 1415;
BCS, Jpn, **1986**, *59*, 1521.

13.13

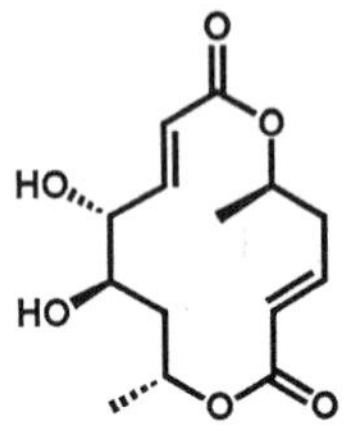

(+)-Colletodiol
Seebach,
TL, **1984**, *25*, 2209;
Ann, **1987**, 733.

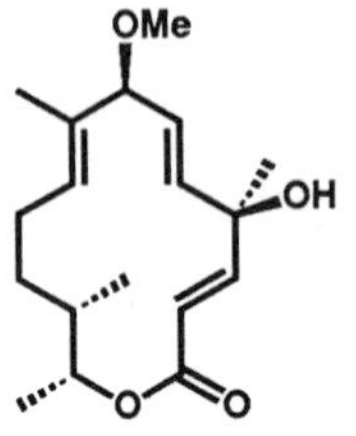

(-)-Ingramycin
Tanner,
Tet, **1987**, *43*, 4395.

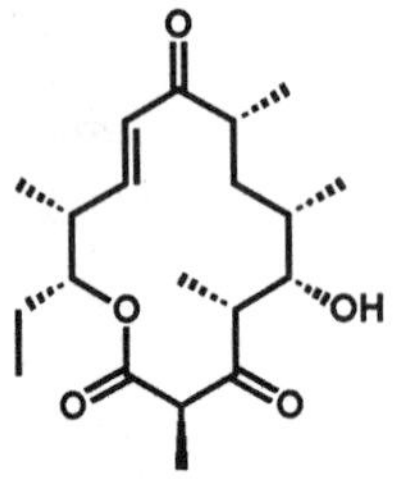

Narbonolide
Masamune,
JOC, **1982**, *47*, 1612.

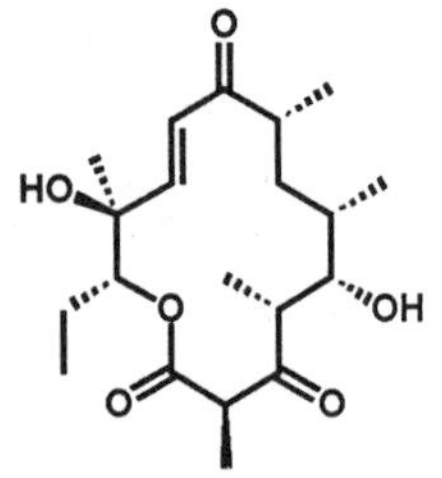

(+)-Pikronolide
Yonemitsu,
JACS, **1986**, *108*, 4645;
CPB, **1987**, *35*, 2228.

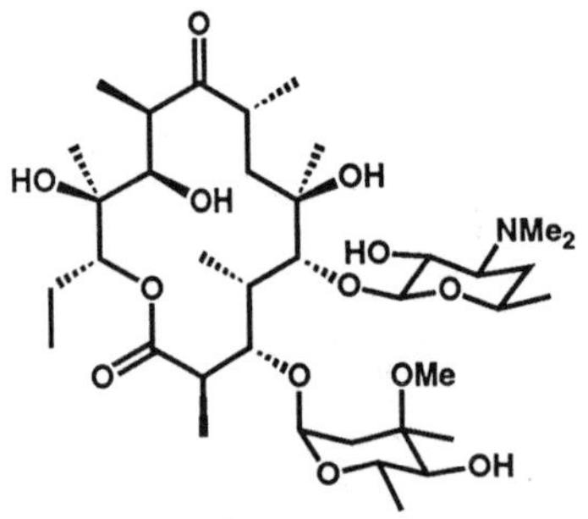

Erythromycin A
Woodward,
JACS, **1981**, *103*, 3210, 3213, 3215.

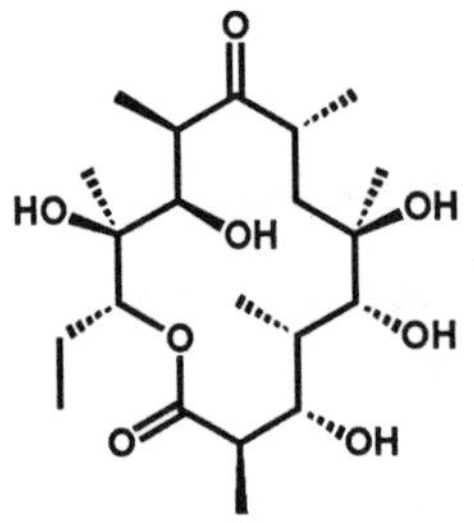

Erythronolide A
Deslongchamps,
CJC, **1985**, *63*, 2810, 2814, 2818.
Kinoshita, *TL*, **1986**, *27*, 1815.
Stork, *JACS*, **1987**, *109*, 1565;
P&AC, **1987**, *59*, 345.

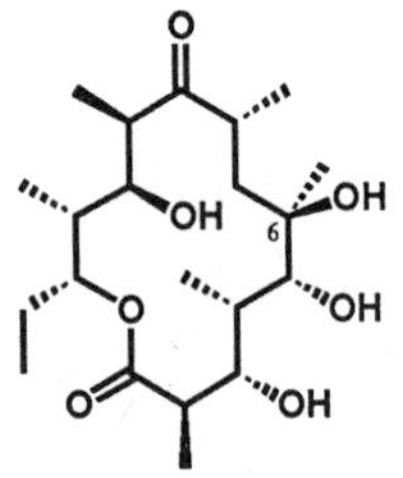

Erythronolide B
Kochetkov,
TL, **1987**, *28*, 3835, 3839.
6-Deoxyerythronolide B
Masamune,
JACS, **1981**, *103*, 1568.

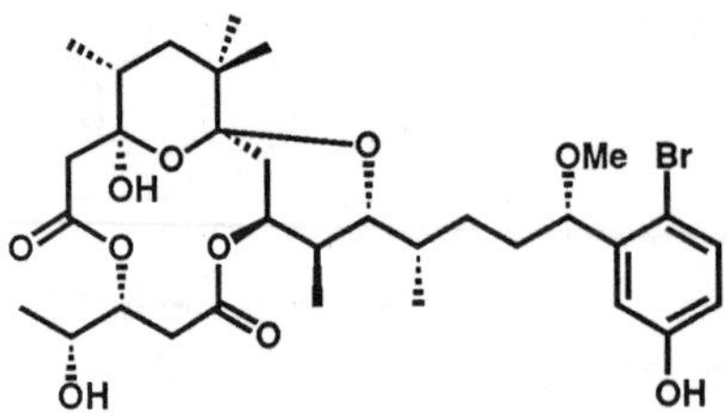

Aplysiatoxin
Kishi, *JACS*, **1987**, *109*, 6205.

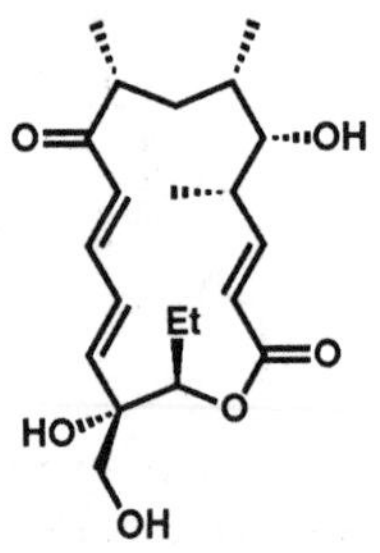

(+)-Mycinolide V
Hoffmann,
ACIE, **1986**, *25*, 1028.

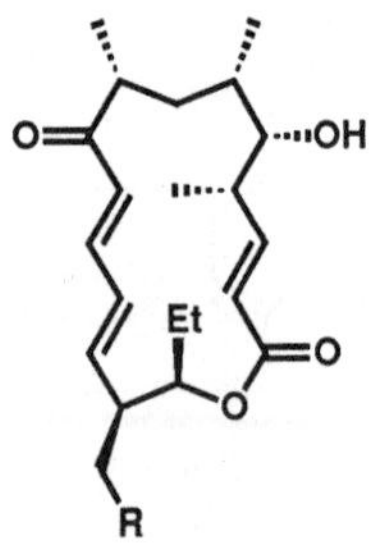

R = H: (+)-Protomycinolide IV
Yamaguchi, *TL*, **1984**, *25*, 3857.
Suzuki, *JACS*, **1986**, *108*, 5221.
R = OH: (+)-Mycinolide IV
Suzuki, *CL*, **1987**, 113.

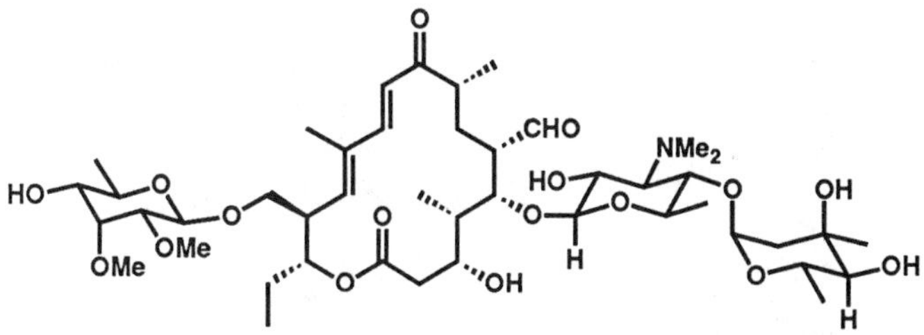

(-)-Tylosin

Tatsuta, *TL*, **1982**, *23*, 3375.

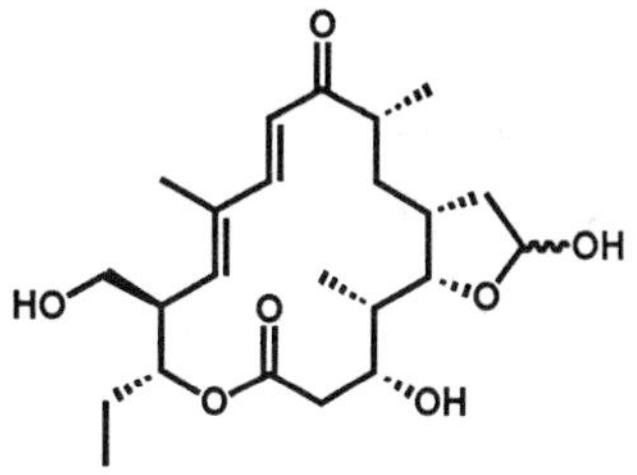

(+)-Tylonolide

Tatsuta, *TL*, **1981**, *22*, 3997.
Masamune, *JACS*, **1982**, *104*, 5523.
Grieco, *JACS*, **1982**, *104*, 5781.
Yonemitsu, *CPB*, **1987**, *35*, 2219.

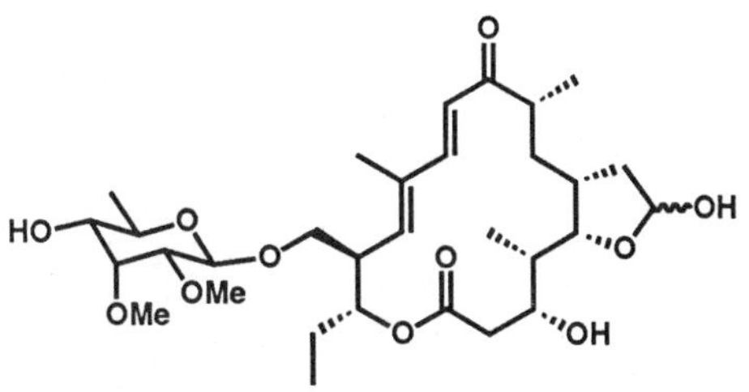

***O*-Mycinosyltylonolide**

Nicolaou, *JACS*, **1982**, *104*, 2027, 2030.

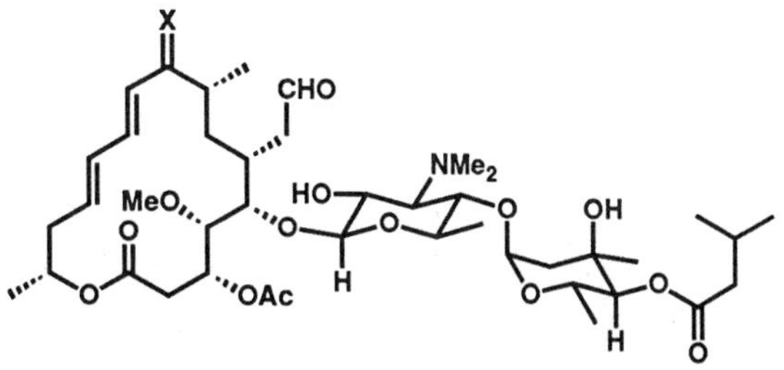

X = H, OH : Leucomycin A_3 (Josamycin)
X = O: Carbomycin B (Magnamycin B)

Nicolaou, *TL*, **1979**, 2327; *JOC*, **1979**, *44*, 4011;
JACS, **1981**, *103*, 1222, 1224. (*nat*)
Tatsuta, *TL*, **1980**, *21*, 2837. (*nat*)

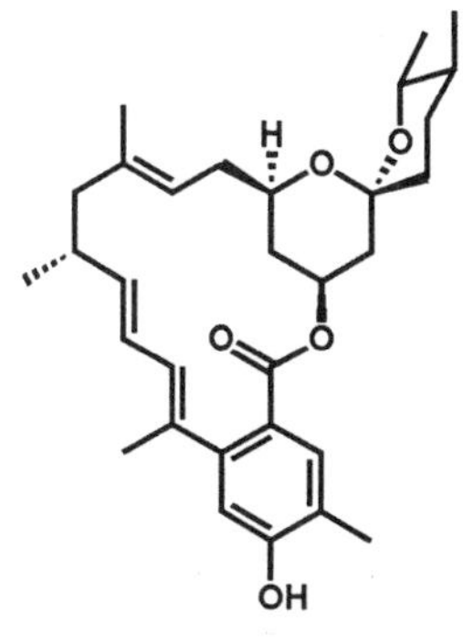

Milbemycin β_3

Smith, *JACS*, **1982**, *104*, 4015;
ibid., **1986**, *108*, 2662.
Williams, *JACS*, **1982**, *104*, 4708. (+)
Barrett, *JOC*, **1986**, *51*, 4840. (+)
Baker, *JCS, PI*, **1987**, 1623. (+)
Kocienski, *JCS, PI*, **1987**,
2171, 2183, 2189. (+)
Crimmins, *JOC*, **1988**, *53*, 652.

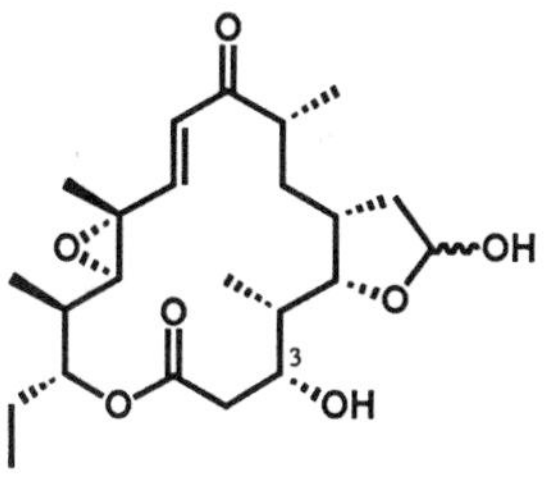

(+)-Rosaramicin Aglycone

Schlessinger, *JACS*, **1986**, *108*, 3112.

3-Deoxyrosaramicin Aglycone

Still, *JACS*, **1984**, *106*, 1148.

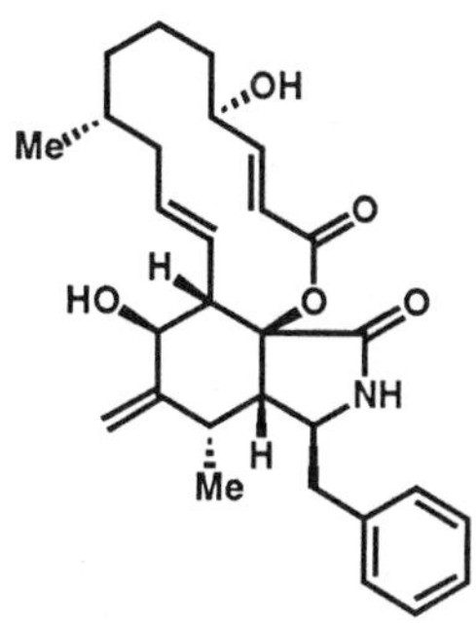

Cytochalasin B

Stork, *JACS*, **1978**, *100*, 7775;
ibid., **1983**, *105*, 5510. (*nat*)

13.14

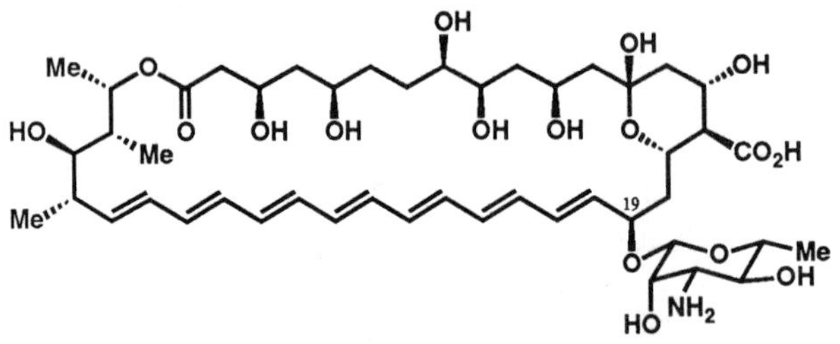

(+)-Amphotericin B
Nicolaou, *JACS*, **1987**, *109*, 2205, 2208, 2821;
ibid., **1988**, *110*, 4660, 4672, 4685, 4696.
19-Dehydroamphoteronolide B
Masamune, *TL*, **1988**, *29*, 451.

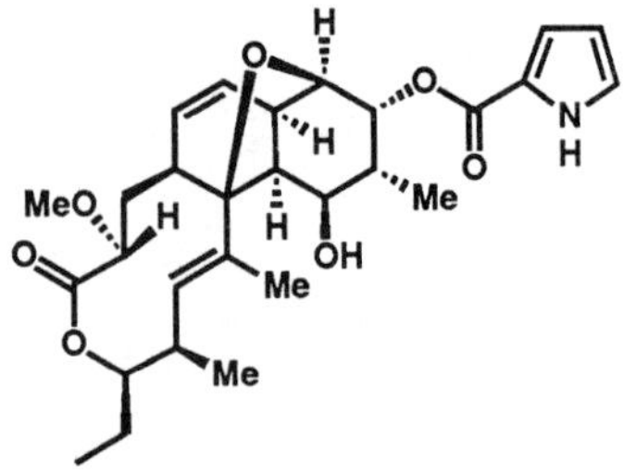

(+)-18-Deoxynargenicin A_1
Kallmerten, *JACS*, **1988**, *110*, 4041.

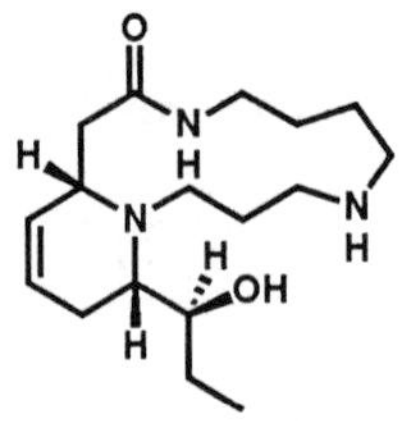

Palustrine
Natsume, *CPB*, **1984**, *32*, 812, 3789.
Dihydropalustrine
Eugster, *HCA*, **1978**, *61*, 928.
Wasserman, *TL*, **1984**, *25*, 2391.

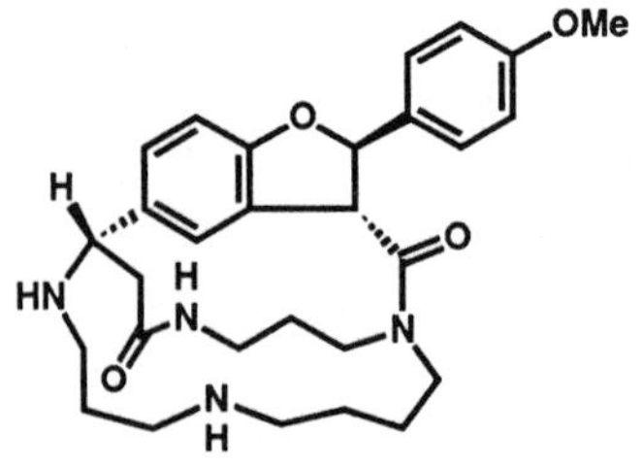

***O*-Methylorantine**
Wasserman,
JACS, **1985**, *107*, 519.

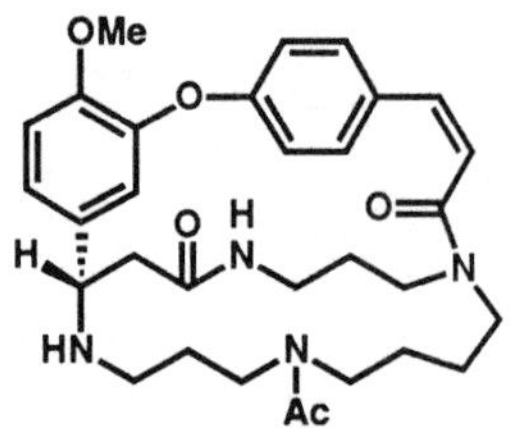

Chaenorhine
Wasserman,
JACS, **1983**, *105*, 1697.

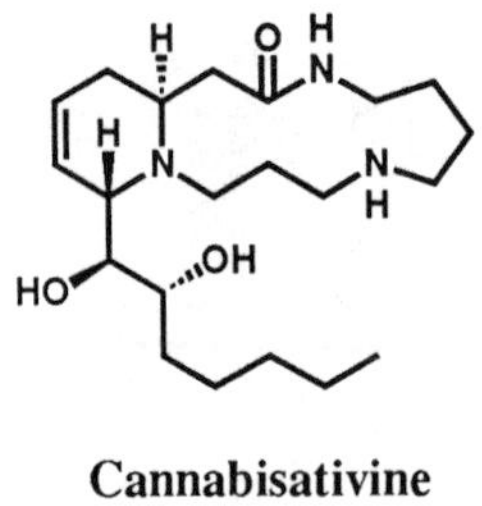

Cannabisativine
Natsume, *TL*, **1984**, *25*, 969.
Wasserman, *TL*, **1985**, *26*, 2241.
Anhydrocannabisativene
Weinreb, *JACS*, **1984**, *106*, 3240.
Wasserman, *TL*, **1985**, *26*, 2237.

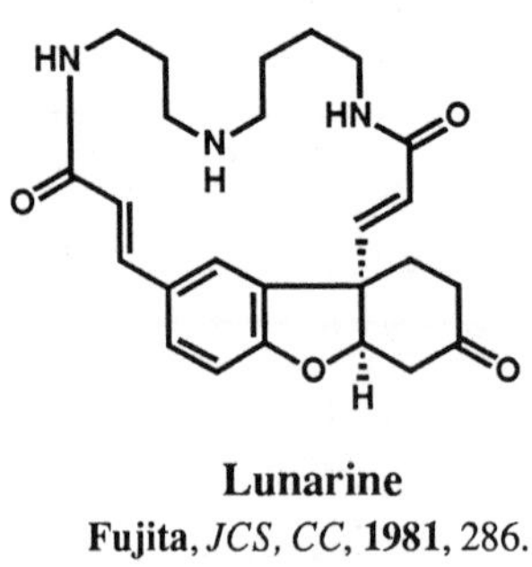

Lunarine
Fujita, *JCS, CC*, **1981**, 286.

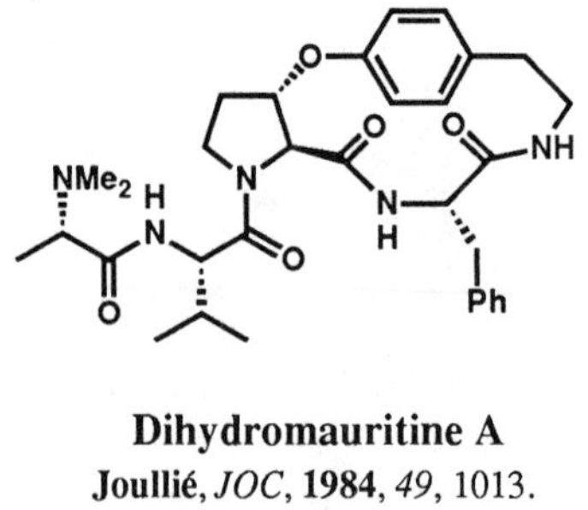

Dihydromauritine A
Joullié, *JOC*, **1984**, *49*, 1013.

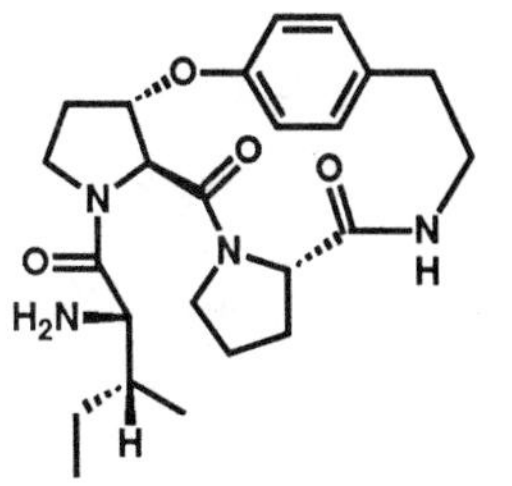

(-)-Dihydrozizyphin G
Schmidt, *ACIE*, **1981**, *20*, 281.

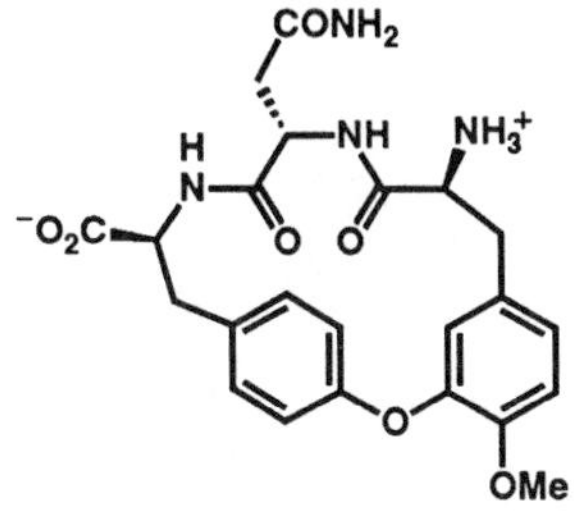

(-)-OF4949-III
Yamamura, *TL*, **1988**, *29*, 559.

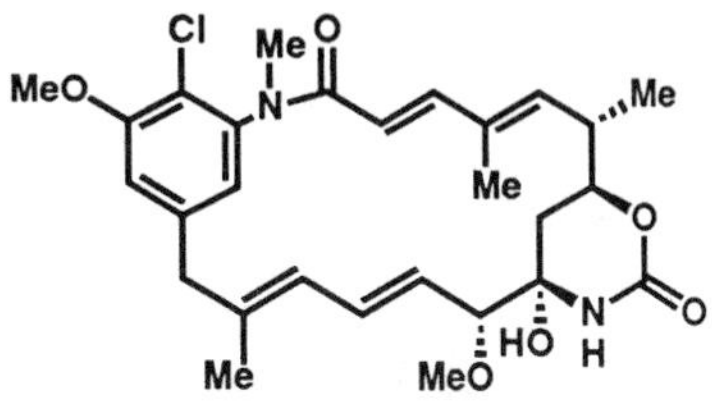

***N*-Methylmaysenine**
Meyers, *JACS*, **1979**, *101*, 4732.
Isobe, *JOC*, **1984**, *49*, 3517.

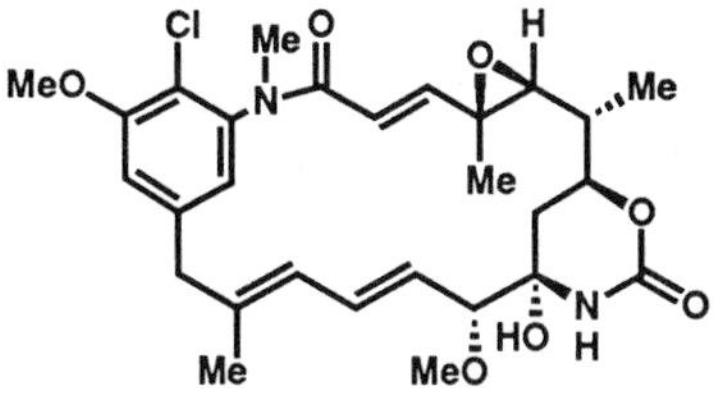

Maysine
Meyers, *JACS*, **1979**, *101*, 7104;
ibid., **1983**, *105*, 5015. (-)
Isobe, *JOC*, **1984**, *49*, 3517.

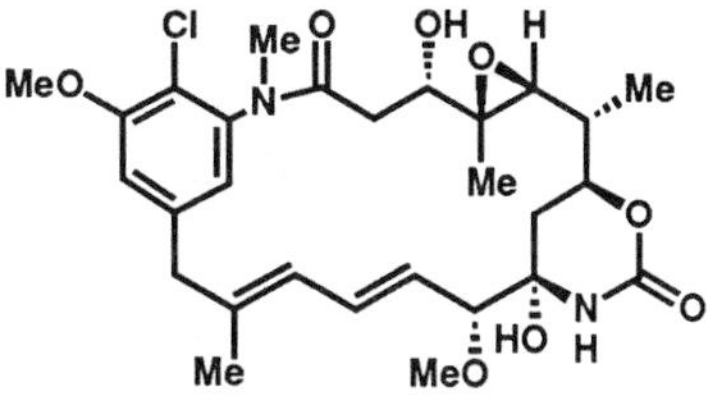

Maytansinol
Meyers, *JACS*, **1980**, *102*, 6597.
Isobe, *JACS*, **1982**, *104*, 4997;
ibid., **1984**, *106*, 3252. (-)

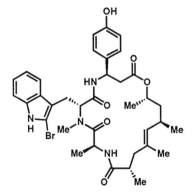

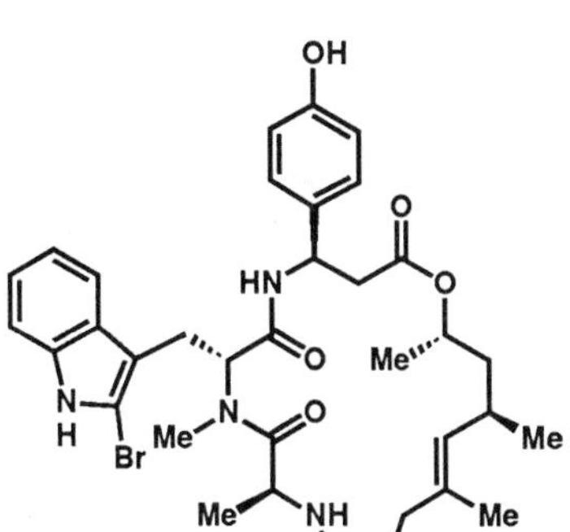

(+)-Jasplakinolide
Grieco, *JACS*, **1988**, *110*, 1630.

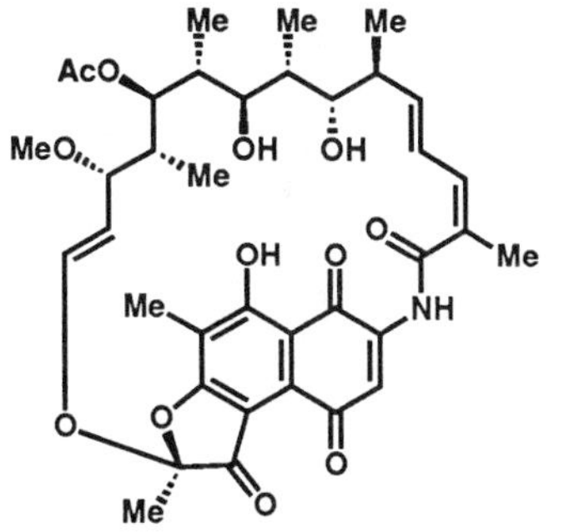

(+)-Rifamycin S
Kishi, *JACS*, **1980**, *102*, 7962, 7965;
Tet, **1981**, *37*, 3873;
P&AC, **1981**, *53*, 1163.
Hanessian, *JACS*, **1982**, *104*, 6164.

13.14

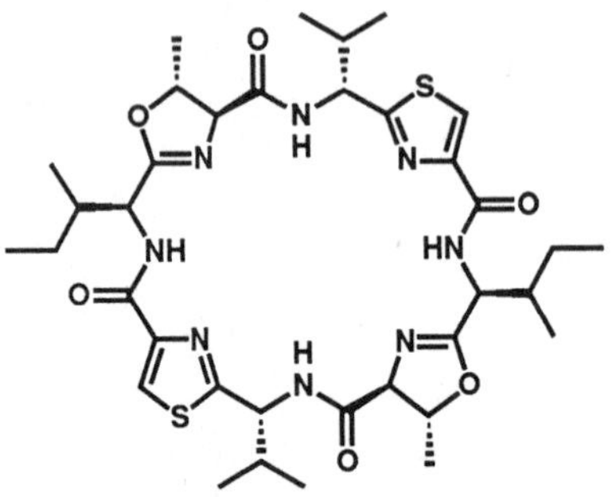

(+)-Ascidiacyclamide
Hamada & Shioiri,
TL, **1985**, *26*, 3223.

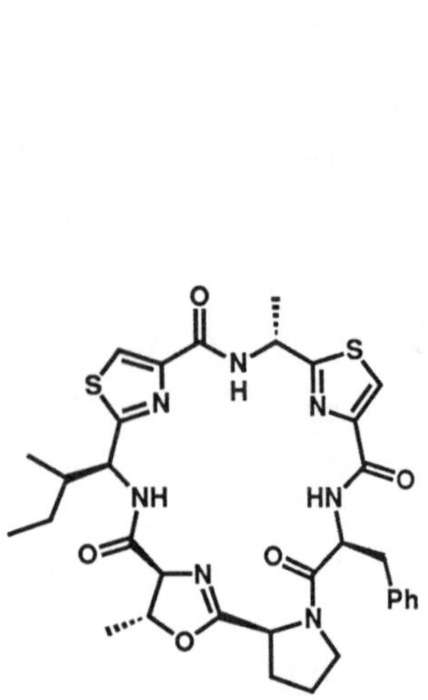

Ulicyclamide
Schmidt, *ACIE*, **1985**, *24*, 569.

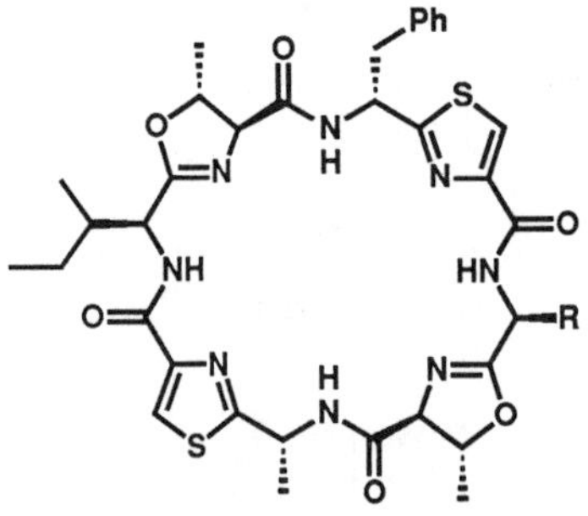

R = *i*-Bu: Patellamide B
R = *i*-Pr: Patellamide C
Hamada & Shioiri,
TL, **1985**, *26*, 5155, 5159, 6501.
Schmide, *TL*, **1986**, *27*, 163.

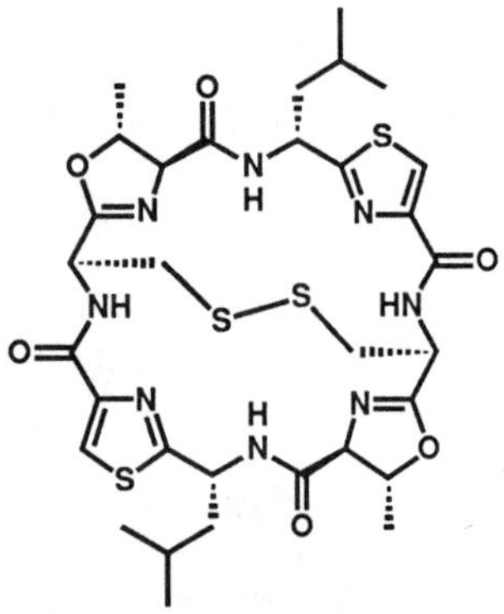

Ulithiacyclamide
Shioiri, *TL*, **1986**, *27*, 2653.
Schmidt, *TL*, **1986**, *27*, 3495.

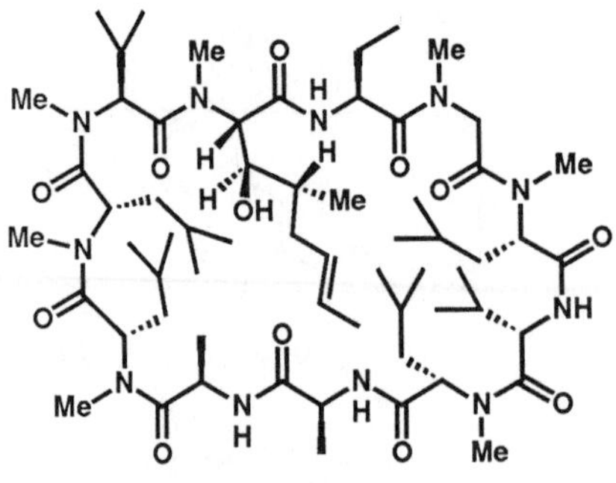

Cyclosporine
Wenger, *HCA*, **1983**, *66*, 2672;
ACIE, **1985**, *24*, 77.

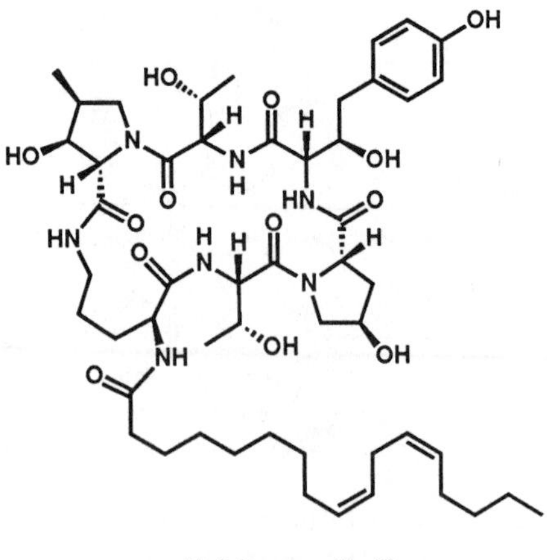

Echinocandin D
Ohfune, *JACS*, **1986**, *108*, 6041, 6043.
Evans, *JACS*, **1987**, *109*, 7151.

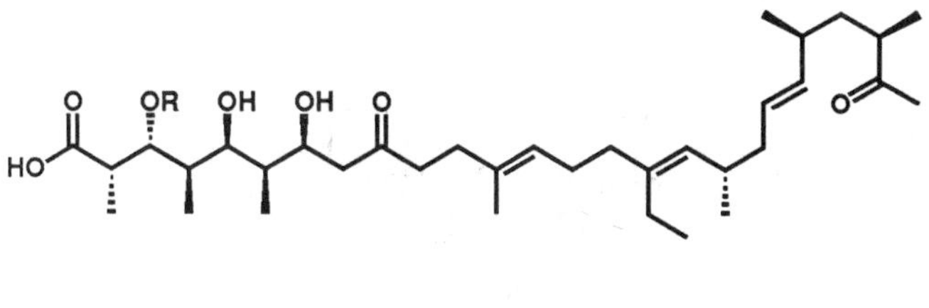
HO
O
OR
OH
OH
O
O

R = H: Premonensin
Evans, *JACS*, **1986**, *108*, 2476.
R = Me: Premonensin Methyl Ether
Sih, *JACS*, **1985**, *107*, 2996.

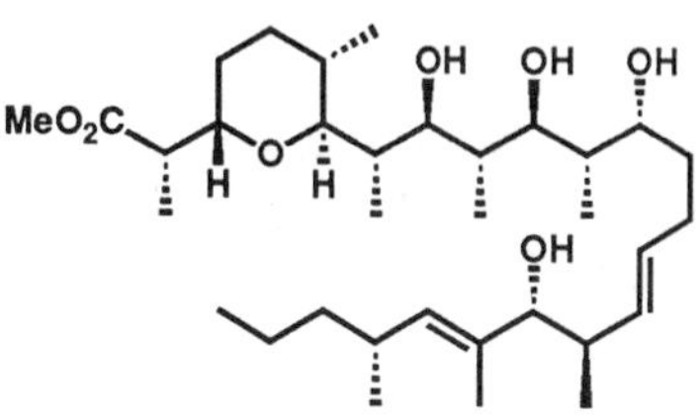
MeO2C
O
H
H
OH
OH
OH
OH

(+)-Zincophorin
Danishefsky, *JACS*, **1987**, *109*, 1572.

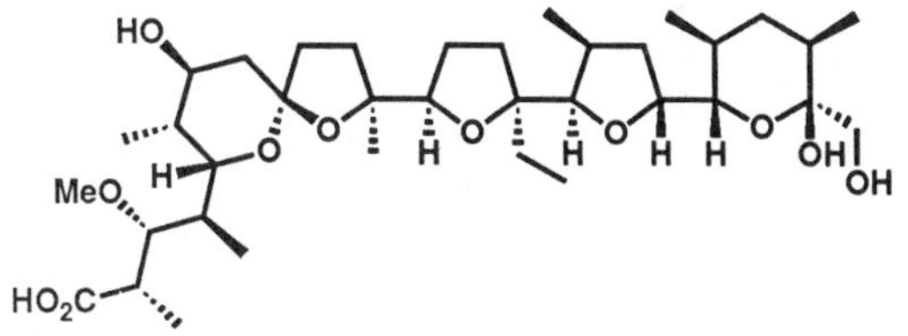
HO
MeO
H
HO2C
O
O
O
O
O
H
H
H
H
OH
OH

Monensin
Kishi, *JACS*, **1979**, *101*, 259, 260, 262.
Still, *JACS*, **1980**, *102*, 2117, 2118, 2120.

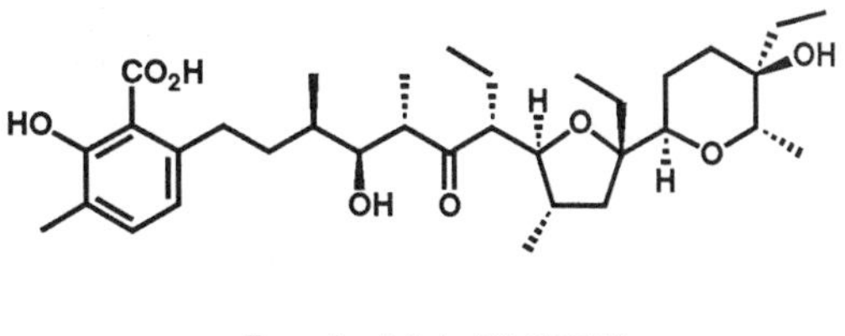
CO2H
HO
OH
O
H
O
H
O
H
OH

Lasalocid A (X-537A)
Kishi, *JACS*, **1978**, *100*, 2933.
Ireland, *JACS*, **1980**, *102*, 1155, 6178; *ibid.*, **1983**, *105*, 1988.

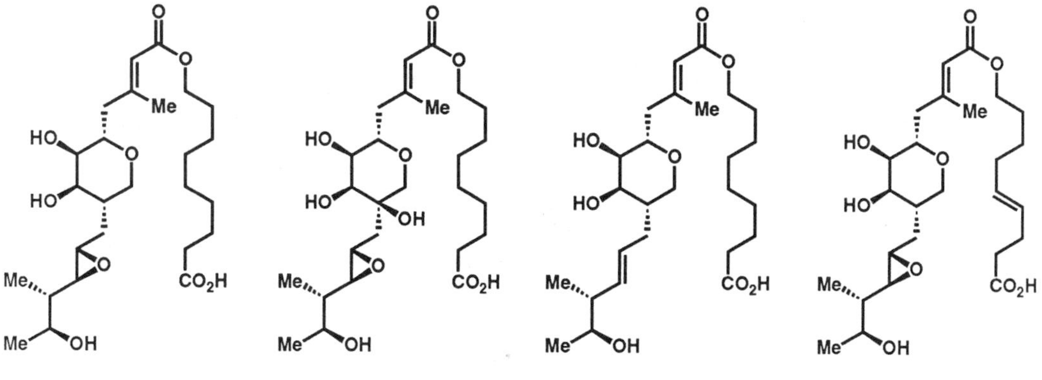
O
O
Me
HO
O
HO
O
Me
CO2H
Me
OH
OH

Pseudomonic Acid A **Pseudomonic Acid B** **Pseudomonic Acid C** **Pseudomonic Acid D**

Kozikowski, *JACS*, **1980**, *102*, 6577, (C); *TL*, **1981**, *22*, 2059, (A).
Snider, *JACS*, **1982**, *104*, 1113; *JOC*, **1983**, *48*, 3003, (A&C).
Schönenberger, *HCA*, **1982**, *65*, 2333, (Key intermediate to all pseudomonic acids).
Sinaÿ, *JACS*, **1983**, *105*, 621, (+)-(A&C).
Fleet, *TL*, **1983**, *24*, 3661, (+)-(A&C).
Keck, *JOC*, **1984**, *49*, 1462; *ibid.*, **1985**, *50*, 4317, (+)-(C).
Raphael, *JCS, PI*, **1984**, 2159, (A&C).
Willams, *JOC*, **1986**, *51*, 3916, (+)-(C).
Bates, *JOC*, **1986**, *51*, 2637, (C).
Nagarajan, *JOC*, **1988**, *53*, 1432, (+)-(C).

13.15

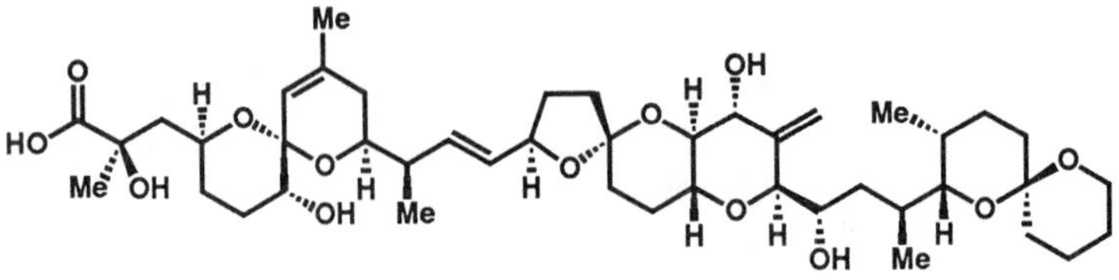

Me
O
HO
Me
OH
H
O
O
OH
H
Me
H
O
H
OH
O
H
O
H
H
OH
Me
Me
H
O
O

(+)-Okadaic Acid

Isobe, *TL*, **1986**, *27*, 963; *Tet*, **1987**, *43*, 4767.

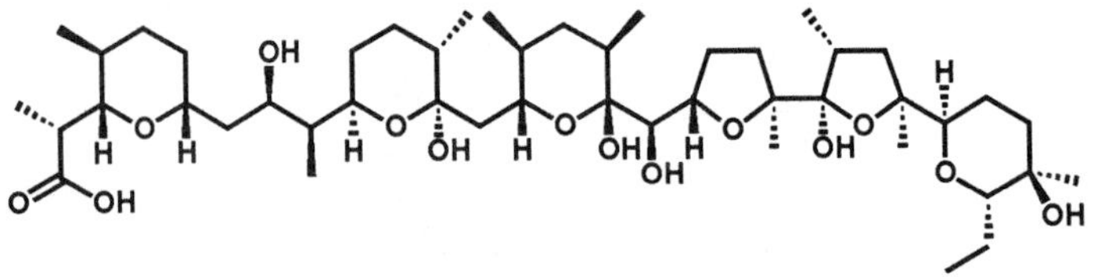

OH
O
H
H
O
OH
H
O
OH
H
O
OH
OH
H
O
OH
O
H
O
OH

Antibiotic X-206

Evans, *JACS*, **1988**, *110*, 2506.

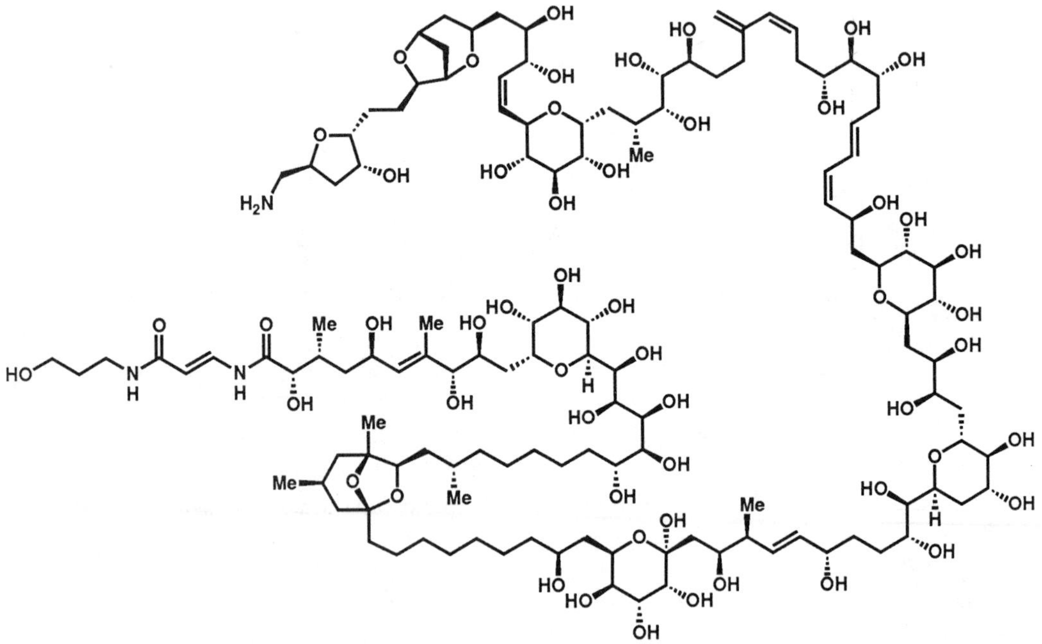

OH
O
O
OH
HO
HO
OH
OH
O
OH
O
OH
OH
Me
OH
H₂N
HO
OH
OH
OH
OH
OH
OH
OH
O
O
Me
OH
Me
HO
HO
OH
OH
HO
N
H
N
H
OH
OH
O
H
OH
HO
OH
HO
OH
HO
OH
Me
OH
Me
O
O
Me
OH
O
OH
HO
H
OH
OH
Me
HO
HO
OH
OH
OH
OH

Palytoxin

Kishi, *Chemica Scripta*, **1987**, *27*, 573.

Subject Index

Subject Index

Subject Index

Subject Index

Subject Index

Subject Index

Subject Index

Subject Index

Subject Index